Tutorials in Molecular Reaction Dynamics

Tutorials in Molecular Reaction Dynamics

Edited by

Mark Brouard and Claire Vallance
Department of Chemistry, University of Oxford, Oxford, UK

RSC Publishing

ISBN: 978-1-84973-530-8

A catalogue record for this book is available from the British Library

Published by The Royal Society of Chemistry,
Thomas Graham House, Science Park, Milton Road,
Cambridge CB4 0WF, UK

Registered Charity Number 207890

For further information see our web site at www.rsc.org

Preface

Molecular reaction dynamics is a hugely wide-ranging field of study, encompassing many branches of physical chemistry and atomic and molecular physics. At the heart of what motivates many researchers in this area is a desire to provide a description of chemistry on the fundamental scale of individual atoms and molecules. Such an approach might not be appropriate to all branches of chemistry, but it has been immensely successful in shedding light on gas-phase molecular collisions and chemical reactivity, and is now being extended to investigations of chemistry in the condensed phase. Femtochemistry, for example, a topic discussed in Chapter 11 of this book, has its roots in gas-phase studies of relatively simple molecular systems, but it is now being used to unravel the dynamics of much more complex systems in the condensed phase. The field of reaction dynamics has also yielded many new experimental and theoretical techniques. To give just a few illustrations, many readers will be familiar with concepts such as the potential energy surface or the transition state, and will have encountered examples in their studies of experiments using flash photolysis or laser pump-probe techniques.

Given the broad scope of the field, we decided when planning this book that rather than attempting to write our own account of Molecular Reaction Dynamics in its entirety, we would approach a range of experts in their fields of study to provide their own versions of specific aspects of the subject. Encouraged by the enthusiastic support we received from the writers we approached, as well as from the Royal Society of Chemistry, we proceeded to provide authors with a general plan of what we envisaged for the book. Other than that, authors were given complete freedom to choose the material to include in their chapters. Finalised chapter synopses were exchanged between the authors early in 2008, prior to commencement of writing. Authors have approached their topics in a variety of different ways, which we hope brings freshness to each chapter. We have not tried too hard to iron out such differences in approach or style between each chapter, and we believe that as a result we have all contributed to producing a unique textbook that could not have been written by any one individual.

As editors, we have viewed our main role as helping to bring the chapters together into a unified book, while at the same time retaining the individual character of each chapter. To help make this into a textbook, as opposed to a selection of research monographs, we have provided numerous links between the various chapters, and have produced study boxes and problems where appropriate. Authors have also provided similar material of their own, as well as worked examples in some cases. You will find that the Tutorials also have a single bibliography and index, and lists of

Tutorials in Molecular Reaction Dynamics
Edited by Mark Brouard and Claire Vallance

Published by the Royal Society of Chemistry, www.rsc.org

further reading are provided for each chapter. The editing of the book took us rather longer than it should have done, and in this regard we thank all the authors firstly for their patience in waiting for us to complete our task and secondly for their promptness at responding to our increasingly frequent requests as the final submission deadline approached.

This text is supplemented by further resources, particularly worked solutions to problems, which can be found by visiting the accompanying website at http://reaction-dynamics.chem.ox.ac.uk/.

Claire Vallance and Mark Brouard
Oxford

Contents

List of Study Boxes xiii

Acknowledgements xv

Biographies xvii

Chapter 1 Introduction 1
Mark Brouard and Claire Vallance

1.1 Introduction 1
1.2 Cross Sections and Rate Constants 3
1.2.1 The Impact Parameter and Orbital Angular Momentum 4
1.2.2 Opacity Functions and Cross Sections 5
1.2.3 Differential Cross Sections and Angular Distributions 7
1.2.4 State-to-state Cross Sections and Rate Constants 8
1.2.5 Thermal Rate Constants 9
1.3 Experimental Considerations 9
1.3.1 Single-collision Conditions 9
1.3.2 Crossed Molecular Beams 10
1.3.3 Motion in the Laboratory and Centre-of-mass Frames 11
1.3.4 Pump-probe Experiments 13
1.4 Guiding Principles in Dynamical Studies 15
1.4.1 The Born-Oppenheimer Approximation and Potential Energy Surfaces 18
1.4.2 Energy and Momentum Conservation and Kinematics 19
1.4.3 Effective Potentials and Centrifugal Barriers 21
1.4.4 Statistical *versus* Non-statistical Reactions 23
1.4.5 Transition States and Transition State Spectroscopy 24
1.5 Summary 26
1.6 Problems 26

Tutorials in Molecular Reaction Dynamics
Edited by Mark Brouard and Claire Vallance
© Royal Society of Chemistry 2010
Published by the Royal Society of Chemistry, www.rsc.org

Chapter 2 Potential Energy Surfaces: the Forces of Chemistry **28**
Matthew A. Addicoat and Michael A. Collins

2.1 Introduction 28
2.2 The Born-Oppenheimer Approximation 28
2.2.1 Dynamics 31
2.3 Coordinates 31
2.4 Methods to Calculate the Energy 33
2.5 Simple Examples 35
2.5.1 Diatomic Molecule 35
2.5.2 Constrained Triatomic 36
2.6 Polyatomic PESs 37
2.7 Methods for Constructing PESs 42
2.7.1 Functional Form Fitting 43
2.7.2 Interpolation 45
2.8 Outlook 47
2.9 Problems 47

Chapter 3 Scattering Theory: Predicting the Outcome of Chemical Events **49**
Anthony J. H. M. Meijer and Evelyn M. Goldfield

3.1 Introduction 49
3.2 Classical Mechanics 49
3.2.1 Newton's Laws and Conservation Laws 49
3.2.2 Lagrangian & Hamiltonian Mechanics 50
3.2.3 Example: Scattering in a Central Potential 51
3.3 Quantum Scattering 53
3.3.1 Preamble 53
3.3.2 Fundamental Theory 54
3.3.3 Overview of Methods 67
3.3.4 Time-independent Methods 68
3.3.5 Time-dependent Methods 73
3.3.6 Approximation Methods 82
3.3.7 Case study: $H+O_2$ 83
3.4 Outlook 85
3.5 Problems 86

Chapter 4 Processes Involving Multiple Potential Energy Surfaces **88**
Bertrand Retail and Andrew J. Orr-Ewing

4.1 Introduction 88
4.2 Breakdown of the Born-Oppenheimer Approximation 89
4.2.1 Non-adiabatic Couplings between Adiabatic PESs 90
4.2.2 Diabatic and Adiabatic Representations of the Coupled PESs 92
4.2.3 The Landau-Zener Model for Non-Adiabatic Dynamics 95
4.2.4 A Case Study in Non-Adiabatic Dynamics – The Ultraviolet Photodissociation of HCl 95
4.2.5 Multi-dimensional PESs and Conical Intersections 99
4.2.6 Vibronic, Coriolis and Spin-orbit Couplings between PESs 106

4.3 Experimental Probes of Non-Adiabatic Reactivity and Dynamics 109
4.4 Outlook 113
4.5 Problems 114

Chapter 5 Elastic and Inelastic Scattering: Energy Transfer in Collisions 116
David W. Chandler and Steven Stolte

5.1 Introduction 116
5.2 Modelling Energy Transfer Processes 117
5.2.1 Vibration to Translation Energy Transfer 117
5.2.2 Dynamics of a Single Hard Sphere Collision 120
5.2.3 Translation to Translation Energy Transfer 121
5.2.4 Classical Treatment of Translational to Rotational Energy Transfer 130
5.3 Experimental Studies of Energy Transfer 135
5.3.1 Early Measurements of DCSs and Rotational Rainbows 135
5.3.2 Molecular Beam Studies of the NO – Rare Gas System 139
5.3.3 Scalar Measurements 144
5.3.4 Vector Measurements of Rotational Energy Transfer 148
5.4 Outlook 163
5.5 Problems 165

Chapter 6 Reactive Scattering: Reactions in Three Dimensions 167
Piergiorgio Casavecchia, Kopin Liu and Xueming Yang

6.1 Introduction 167
6.1.1 Crossed Molecular Beams 167
6.1.2 Determining Cross Sections from CMB Experiments 168
6.2 CMB Experiments with Mass Spectrometric Detection 169
6.2.1 Angular Scattering and Reaction Mechanisms 174
6.2.2 Soft-Ionization Detection 179
6.2.3 Multichannel Polyatomic Reactions 180
6.3 CMB Experiments with Rydberg-Tagging Detection 186
6.3.1 Quantum Phenomena and Reaction Resonances 188
6.3.2 The $OH+D_2$ Reaction: Mode Specific Chemistry 200
6.4 CMB Experiments with REMPI Detection 201
6.4.1 Product Pair Correlations 202
6.4.2 Product Pair-Correlated DCSs 208
6.5 Outlook 210
6.6 Problems 212

Chapter 7 Reactive Scattering: Quantum State-Resolved Chemistry 214
F. Fleming Crim

7.1 Introduction 214
7.2 Motion over a Potential Energy Surface 215
7.3 State-to-state Bimolecular Reactions 218
7.3.1 Changing the Reaction Cross Section 218
7.3.2 Changing the Reaction Pathways: Populating Different States . . 220
7.3.3 Changing the Reaction Pathways: Breaking Different Bonds . . . 223

7.4 Vibrational Energy Flow 224
7.4.1 Chemical Activation: Rabinovitch's Bicycle 225
7.4.2 Laser Excitation 226
7.4.3 Intramolecular Vibrational Redistribution: Observing Energy Flow 228
7.5 Statistical Reactions 230
7.5.1 The Cumulative Reaction Probability 230
7.5.2 The Statistical Assumption: Microcanonical Transition State Theory 232
7.5.3 Quantum State Structure in Unimolecular Reactions 233
7.6 Outlook 235
7.7 Problems 237

Chapter 8 Photodissociation Dynamics: the Fragmentation of Molecules by Light 240
David H. Parker, André T. J. B. Eppink and Claire Vallance

8.1 Introduction 240
8.2 Photodissociation and Potential Energy Surfaces 240
8.3 An Introduction: the Photodissociation of Br_2 244
8.4 Femtosecond Probes of Photodissociation Dynamics 249
8.5 Femtosecond Dynamics of the Photodissociation of O_2 to O^+–O^- Ion Pairs 251
8.6 Photofragment Angular Distributions in Br_2 Photodissociation 254
8.6.1 Above-threshold Photodissociation of Br_2 at 266 nm 256
8.6.2 Near-threshold Photodissociation of Br_2 at 510 nm 257
8.7 Atomic Product Polarization 258
8.7.1 Atomic Polarization Effects in the Photodissociation of Br_2 and I_2 260
8.8 Theoretical Treatment of Photodissociation 263
8.8.1 Adiabatic Model of Photodissociation 265
8.8.2 Prediction of Atomic Polarization using the Adiabatic Model 266
8.8.3 Sudden Model for Photodissociation 267
8.8.4 Example: Photodissociation of the OH Radical 267
8.8.5 Predissociation of the $A^2\Sigma^+$ state of OH: Interference Effects 271
8.8.6 The 'Bad News': Fully Quantum Analysis of Models for Photodissociation of HX 274
8.9 Outlook 274
8.10 Problems 275

Chapter 9 Stereodynamics: Orientation and Alignment in Chemistry 278
F. Javier Aoiz and Marcelo P. de Miranda

9.1 Introduction 278
9.2 Concepts and Quantities 279
9.2.1 Polarization, Orientation, Alignment 279
9.2.2 Vector Correlations 280
9.2.3 Probability Density Functions 283
9.2.4 Polarization Moments of Probability Density Functions 285
9.2.5 Density Matrices 289
9.2.6 Polarization Moments of Density Matrices 290
9.2.7 Limiting Values of Real Polarization Moments 292

9.3 Experimental and Theoretical Stereodynamics 292
9.3.1 Theoretical Methods 293
9.3.2 Experimental Production of Polarized Reactants 299
9.4 A Detailed Example 310
9.4.1 Quantification of Known Reaction Mechanisms 311
9.4.2 Rationalization of the Complete Set of Stereodynamical Parameters 314
9.4.3 Incomplete Information 321
9.4.4 Manipulation of Collision (stereo)Dynamics 325
9.5 Conclusion and Outlook 328
9.6 Problems 329

Chapter 10 Surface Scattering: Molecular Collisions at Interfaces **333**
Andrew Hodgson and George Darling

10.1 Introduction 333
10.1.1 Overview of Different Gas-Surface Scattering Channels 334
10.1.2 Importance of Building an Atomic-Level Understanding of these Channels 335
10.1.3 A Brief Comparison with Gas Phase Scattering 336
10.2 The Molecule-Surface Potential Energy Surface 336
10.3 Scattering and Trapping 342
10.3.1 Elastic Scattering 342
10.3.2 Inelastic Scattering 344
10.4 Non-Activated Dissociation 348
10.5 Direct Dissociation 351
10.5.1 Early *versus* Late Barriers – Polanyi Rules 351
10.5.2 Incidence Angle Dependence and Surface Corrugation 352
10.5.3 Molecular Rotations 353
10.5.4 Mode/Bond Specific Surface Chemistry 354
10.5.5 Excitations of the Surface and Dissociation 354
10.5.6 Non-Adiabatic Effects in Molecular Dissociation Reactions 355
10.6 Recombinative Desorption 355
10.6.1 Detailed Balance 356
10.6.2 One Dimensional Models for Activated Adsorption-Desorption 359
10.6.3 Weakly Coupled Systems (Hydrogen Recombination) 359
10.6.4 Heavy Molecules 361
10.7 Outlook 362
10.8 Problems 362

Chapter 11 Femtochemistry and the Control of Chemical Reactivity **363**
Helen H. Fielding and Abigail D. G. Nunn

11.1 Introduction 363
11.2 Femtosecond lasers 365
11.2.1 Femtosecond Laser Oscillators 365
11.2.2 Femtosecond Amplifiers 370
11.2.3 Wavelength Tuning of Femtosecond Pulses 373
11.3 Probes for Ultrafast Chemical Dynamics 374
11.4 Ultrafast Photoinduced Processes 377

11.4.1 Wavepackets ... 378
11.4.2 Photodissociation ... 378
11.4.3 Femtochemistry of Larger Molecules ... 381
11.4.4 The Condensed Phase ... 382
11.5 Femtosecond Coherent Control ... 382
11.5.1 Pulse Trains ... 383
11.5.2 Chirped Pulses ... 384
11.5.3 Programmable Pulse Shaping ... 385
11.6 Outlook ... 389
11.7 Problems ... 390

Chapter 12 Cold and Ultracold Collisions **392**
Gerrit C. Groenenboom and Liesbeth M. C. Janssen

12.1 Introduction ... 392
12.2 Classical Capture Theory ... 393
12.2.1 Classical Central Force Problem ... 393
12.2.2 Cross Sections ... 401
12.2.3 Canonical Reaction Rates ... 401
12.2.4 Isotropic Interactions ... 403
12.2.5 Anisotropic Interactions ... 406
12.3 Quantum Capture Theory ... 407
12.3.1 Quantum Scattering Theory ... 408
12.3.2 Connection with Classical Capture Theory ... 410
12.3.3 Coupled Channels Capture Theory ... 411
12.3.4 Quantum Adiabatic Capture Theory ... 414
12.3.5 Thermal Capture Rates ... 414
12.3.6 Total Angular Momentum Representation ... 415
12.4 Wigner Threshold Laws ... 416
12.4.1 Bouncing Off a Cliff ... 417
12.4.2 *s*-Wave Elastic Scattering ... 418
12.4.3 Scattering Length ... 419
12.4.4 Inelastic Scattering at Low Energy ... 420
12.5 Ultracold Phenomena ... 424
12.5.1 Particle in a Box ... 426
12.5.2 Bose-Einstein Condensation ... 428
12.5.3 A Bose-Einstein Condensate in a Harmonic Trap ... 431
12.5.4 The Gross-Pitaevskii Equation ... 431
12.5.5 Thomas-Fermi Approximation ... 438
12.5.6 Bose-enhancement and Pauli-blocking ... 438
12.6 Outlook ... 440
12.7 Problems ... 440

Further Reading **442**

Glossary of Acronyms **447**

Bibliography **450**

Subject Index **474**

List of Study Boxes

- Chapter 1
  - 1.1 Changes of variables and Jacobian determinants 14
  - 1.2 Lasers in reaction dynamics 16
- Chapter 2
  - 2.1 Polanyi's rules 38
- Chapter 3
  - 3.1 Time-dependent expectation value of $\hat{A}$ 57
  - 3.2 Commutators 58
  - 3.3 Continuous *versus* discrete bases 62
  - 3.4 A qualitative introduction to wavepackets 64
  - 3.5 Derivation of local r-matrices 75
- Chapter 4
  - 4.1 The coupling of diabatic states and the derivation of adiabatic states 93
  - 4.2 The geometric phase at conical intersections 103
  - 4.3 Angular momentum coupling and Hund's coupling cases 108
- Chapter 5
  - 5.1 An introduction to rainbow and glory scattering 126
  - 5.2 LIF and REMPI detection schemes 136
  - 5.3 Velocity-map ion imaging 153
  - 5.4 Monte Carlo simulation techniques 156
- Chapter 6
  - 6.1 The classic CMB technique: the LAB to CM transformation 172
  - 6.2 The H-atom Rydberg tagging technique 189
  - 6.3 PHOTOLOC 191
  - 6.4 Quantum bottleneck states and scattering resonances 194
  - 6.5 Differential cross section measurements in the CM-frame 203
- Chapter 7
  - 7.1 Kinematics and skew angles 216
  - 7.2 Local modes 222
  - 7.3 Phase space 233
  - 7.4 Dynamics in the condensed phase 236
- Chapter 8
  - 8.1 Photodissociation of polyatomic molecules 242
  - 8.2 An introduction to electric dipole transitions 245
  - 8.3 The translational anisotropy 252
  - 8.4 Polarization parameters and molecular photodissociation 262

- Chapter 9
  - 9.1 The detection of angular momentum orientation and alignment 303
- Chapter 10
  - 10.1 Lattice vectors and Miller indices 340
- Chapter 11
  - 11.1 Phase and chirped pulses 367
  - 11.2 Non-linear optics 371
  - 11.3 Femtosecond laser optical layout 375
- Chapter 12
  - 12.1 Making molecules cold 394
  - 12.2 Long range interaction potentials 404
  - 12.3 An introduction to the theory of Bose-Einstein condensation 427
  - 12.4 A simulation of Bose-Einstein condensation 432
  - 12.5 Cold and ultracold chemistry 433

Acknowledgements

The editors, and authors of Chapter 1, would first and foremost like to thank all of their coauthors for their hard work in putting together these Tutorials. We would also particularly like to thank the U.K. Royal Society of Chemistry, both for their help in producing the book, and for their enthusiastic support for the project. We would like to thank John Freeman for his considerable help with converting the figures to a uniform and editable format. We acknowledge our numerous research collaborators, and the longstanding support of the U.K. EPSRC, the Royal Society, STFC and the EU, without which we would have been unable to do much of the research that has influenced this book. Finally, our two research groups deserve a special thanks, both for the hard work they do each day, and also for their patience while we have been too busy with the Tutorials to help them in our laboratories!

Piergiorgio Casavecchia thanks all the collaborators who have participated in the work discussed in Chapter 6, and whose names appear in the relevant references, and in particular his longstanding collaborator and colleague Nadia Balucani.

F. Fleming Crim thanks the École Polytechnique Fédérale de Lausanne (EPFL) for their kind hospitality during the preparation of Chapter 7. His research in chemical reaction dynamics at the University of Wisconsin is made possible by support from the National Science Foundation, the Department of Energy, and the Air Forces Office of Scientific Research.

Professor F. Javier Aoiz would like to acknowledge the longstanding collaboration on the subject of Chapter 9 with his friends and colleagues Professors Vicente Sáez Rábanos and Mark Brouard. The fruitful interaction with Jesús Aldegunde is also acknowledged.

The authors of Chapter 12 acknowledge the Council for Chemical Sciences of the Netherlands Organization for Scientific Research CW-NWO for financial support and Professor Ad van der Avoird for carefully reading the manuscript.

Biographies

CHAPTER 1

The authors of Chapter 1, and the editors of these Tutorials, Mark Brouard (left) and Claire Vallance (right).

Mark Brouard undertook his D.Phil. at the University of Oxford with Professor M. J. Pilling, before moving to a postdoctoral position with Professor J. P. Simons at Nottingham University. He was appointed to a lecturership there before moving back to Oxford in 1993, where he is now a Professor of Physical Chemistry and a Tutorial Fellow at Jesus College. His main interests lie in experimental studies of gas phase photodissociation and reaction dynamics, with a particular interest in angular momentum polarization effects.

Claire Vallance received B.Sc.(hons) and Ph.D. degrees from the University of Canterbury in Christchurch, New Zealand, before moving to the University of Oxford, where she currently holds the joint posts of University Lecturer in the Department of Chemistry and Tutorial Fellow in Physical Chemistry at Hertford College. Her current research interests include reaction dynamics, application of velocity and spatial-map imaging to mass spectrometry, and the development of laser spectroscopic techniques for microfluidics applications.

CHAPTER 2

Matthew Addicoat completed his Ph.D. at the University of Adelaide with Professor Greg Metha. He is currently a postdoctoral fellow at the Australian National University. His other research interests include transition metal clusters and finding efficient means to search potential energy surfaces.

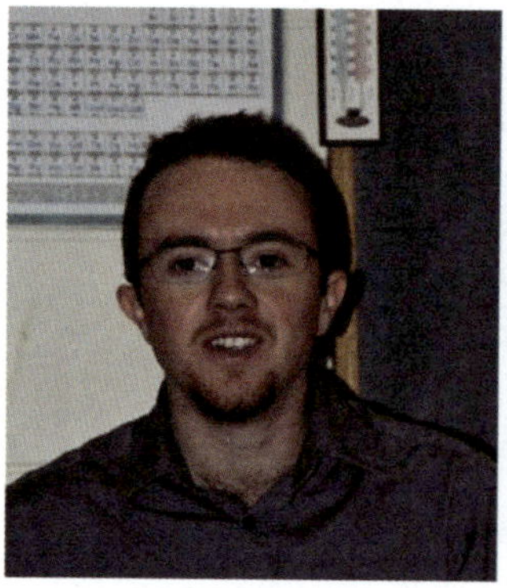

The authors of Chapter 2, Matthew Addicoat (left) and Michael Collins (right).

Michael Collins (PhD, University of Sydney) gained postdoctoral experience at MIT, ANU, the University of Chicago, and Cambridge, before taking up a permanent position at ANU. His research interests have included semi-classical mechanics and solitary wave phenomena in molecules and crystals, but are now concentrated on molecular potential energy surfaces and on the energetics of large molecules, crystals and surfaces.

CHAPTER 3

The authors of Chapter 3, Anthony J. H. M. Meijer (left) and Evelyn M. Goldfield (right).

Anthony J. H. M. Meijer (Ph.D., University of Nijmegen) is a senior lecturer in Theoretical Chemistry at the University of Sheffield. Before his appointment in Sheffield he was a postdoctoral researcher at UCL (UK) and WSU in Detroit (USA). His research interests lie in quantum reaction dynamics in general with an emphasis on using parallel computers to tackle gas-surface and gas-phase reactions. The reactions studied using these methods are in general of importance to our understanding of the interstellar medium (in the former case) and our understanding of atmospheric or combustion reactions (in the latter case).

Evelyn M. Goldfield is currently a program director in the Chemistry Division of the U.S. National Science Foundation. For many years she was an associate professor (research) at Wayne State University, and although she has retired, she maintains a research affiliation with the university. Her primary research interests are in quantum reaction dynamics, both methods development and applications to chemical systems, as well as in the design of large-scale parallel computing algorithms for quantum dynamics. Recently, she has become very interested in using the techniques of chemical dynamics, both quantum and classical, to study the effects of nanoscale confinement on chemical reactivity.

CHAPTER 4

The authors of Chapter 4, Bertrand Retail (left) and Andrew J. Orr-Ewing (right).

Bertrand Retail graduated from the University of Bordeaux with a degree in Chemistry, and moved to the University of Bristol to undertake postgraduate research on the dynamics of chemical reactions involving organic molecules. He employed velocity map imaging methods to study adiabatic and non-adiabatic scattering dynamics, and was awarded his Ph.D. in 2008. Bertrand subsequently accepted a research and development position with Hiden Analytical.

Andrew Orr-Ewing is a Professor of Physical Chemistry at the University of Bristol. He obtained his D.Phil. from the University of Oxford, and spent two years at Stanford University as a post-doctoral researcher before moving to Bristol. His research interests include chemical reaction dynamics in the gas phase and in liquid solutions, photodissociation dynamics, and atmospheric and organic photochemistry.

CHAPTER 5

The authors of Chapter 5, David W. Chandler (left) and Steven Stolte (right).

David W. Chandler received his Ph.D. from Indiana University in 1980, working with Professor George Ewing to study vibrational energy transfer in cryogenic liquids. After a postdoctoral fellowship in the laboratory of Professor R. N. Zare at Stanford he joined the technical staff at

Sandia National Laboratory, where he has studied gas phase chemical dynamics at the Combustion Research Facility. In 1987 he built the first ion imaging apparatus for the visualization of chemical reactions and in collaboration with Professor Paul L. Houston (Cornell) performed the first experiments using this technique on the photofragmentation of CH_3I. Ion imaging is a technique for the visualization and simultaneous determination of the velocity of a quantum-state-selectively ionized molecules or atoms and has become widely used in many areas of research. In addition to his work in photochemistry he has participated in research in non-linear optics, crossed molecular beam scattering, photoelectron spectroscopy and optical microscopy.

Steven Stolte obtained his Ph.D. from the Catholic University in Nijmegen under the supervision of Jörg Reuss. After a postdoctoral fellowship with Professor R. B. Bernstein, he returned to Nijmegen in 1974, where he developed his interest in the orientation and alignment dependence in state selected reactions and, together with Aart Kleyn, initiated the field of surface scattering with oriented and state selected molecules. In 1989 he was appointed to the chair of Physical Chemistry at the Free University of Amsterdam, where he studied the steric and state-to-state dependence of inelastic scattering and molecular photodissociation, both on the nanosecond and femtosecond timescale (the latter in collaboration with Maurice Janssen). In 2008 he was appointed as a professor in Molecular Physics at the Atomic and Molecular Physics Institute of the Jilin University in Changchun (China), and a Senior Chair of the Triangle du Physique at the Laboratoire Francis Perrin in CEA/Saclay, and was distinguished as the first recipient of the 'R. B. Bernstein award' for stereodynamics.

CHAPTER 6

The authors of Chapter 6, Piergiorgio Casavecchia (left), Kopin Liu (middle) and Xueming Yang(right).

Piergiorgio Casavecchia is a Professor of Physical Chemistry at the University of Perugia. He worked with Nobel Laureate Yuan T. Lee in Berkeley where he learned the subtleties of crossed molecular beam scattering with universal mass spectrometric detection. He has studied extensively elastic, inelastic and reactive collision processes using this technique. Over the past 20 years his main research interests have focused on the reaction dynamics of both simple triatomic and complex polyatomic elementary reactions. Recently, Professor Casavecchia has implemented 'soft' electron-ionization for product detection in crossed molecular beam experiments, which has proven invaluable for the investigation of gas-phase multichannel reactions. In 2008 he was awarded the Polanyi Medal for his contributions to gas-phase reaction dynamics and kinetics.

Kopin Liu is a Distinguished Research Fellow of the Institute of Atomic and Molecular Sciences, Academia Sinica, Taiwan. He received his Ph.D. in 1977 from Ohio State University, USA. His current research interests focus on product pair correlation in polyatomic reactions, mode and bond selective chemistry, and hydration dynamics.

Xueming Yang received his B.S. degree in Physics from Zhejiang Normal University in 1982 and his M.Sc. degree in Chemical Physics from Dalian Institute of Chemical Physics, CAS in 1985. He obtained his Ph.D. in chemistry from University of California at Santa Barbara in 1991. After postdoctoral works in Princeton University and University of California at Berkeley, he became an associated research fellow in the Institute of Atomic and Molecular Sciences in Taipei in 1995, and was promoted to a full research fellow with tenure in 2000. In 2001, he made a move to the Dalian Institute of Chemical Physics, Chinese Academy of Sciences. His main research interests are in the area of experimental chemical dynamics in the gas phase and at interfaces.

CHAPTER 7

The author of Chapter 7, F. Fleming Crim.

F. Fleming Crim received his Ph.D. from Cornell University, and after a brief period at the Western Electric Engineering Research Centre and a post-doctoral appointment at Los Alamos National Laboratory, he joined the faculty at the University of Wisconsin, where he is now the John E. Willard and Hilldale Professor of Chemistry. The unifying theme of his research is understanding the role of vibrational energy in chemical reactions. A pioneering aspect of his work is controlling reactions using vibrational excitation to break bonds selectively in chemical reactions. Professor Crim and his coworkers use both high resolution lasers and ultrafast lasers to probe the fundamental details of chemical reactivity in gases and in liquids.

CHAPTER 8

David H. Parker received his Ph.D. at the University of California Los Angeles under Professor M. A. El-Sayed and worked as a postdoctoral scholar at Columbia University with Professor R. B. Bernstein. He joined the University of California Santa Cruz as Assistant Professor and started his present position at the University of Nijmegen, the Netherlands as Professor of Physics in 1991. His work combines laser ionization spectroscopy and two-dimensional imaging to study the dynamics of molecular photodissociation, energy transfer, and crossed-beam reaction in systems of relevance to atmospheric chemistry and astrochemistry.

The authors of Chapter 8, David H. Parker (left), André Eppink (middle), and Claire Vallance (right).

Andre T. J. B. Eppink received his Ph.D. at the University of Nijmegen under professor D. H. Parker and worked as a postdoctoral scholar at the University of Bielefeld with Professor P. Andresen. After working in industry (LaVision in Gottingen and NXP/Philips in Nijmegen) and teaching at high school level he started his present position at the Department of Physics of the University of Nijmegen in 2009. Andre Eppink and David Parker co-discovered the velocity map imaging method, which is used in the study of molecular photodissociation as described in this chapter.

Claire Vallance – see author biographies for chapter 1.

CHAPTER 9

The authors of Chapter 9, F. Javier Aoiz (left) Marcelo P. de Miranda (right).

F. Javier Aoiz received his Ph.D. from Universidad Complutense de Madrid. He spent a two years period at Columbia, working with the late Professor R. B. Bernstein, before joining the faculty of the Physical Chemistry Department at Universidad Complutense de Madrid, where he now has a chair in Physical Chemistry as full professor. He has been visiting professor in the universities of Oxford (UK) and UCLA (Ca, USA). His research activities are related to experimental and theoretical chemical reaction dynamics and photodissociation processes. His present work is focussed on fundamental aspects of dynamics and stereodynamics of chemical reactions and inelastic processes.

Marcelo P. de Miranda was born in Rio de Janeiro, Brazil, where he also received his first degree – in Chemistry, at Universidade Federal (UFRJ). He completed his Ph.D. at Université de Paris-Sud (France) under the supervision of J. Alberto Beswick. Before taking a lectureship at

Leeds in 2001, he held positions at University College London (UK), Unicamp (Brazil) and New York University (USA). The first and third were postdoctoral positions (in the groups of David C. Clary and John Z. H. Zhang, respectively). As his students say, 'Marcelo loves the quantum'.

CHAPTER 10

The authors of Chapter 10, George Darling (left) and Andrew Hodgson (right).

George Darling undertook a Ph.D. at Imperial College. He then joined the Interdisciplinary Research Centre in Surface Science at Liverpool shortly after it was formed, working mainly in computational gas-surface dynamics. He now has interests ranging widely in computational materials chemistry and surface science.

Andrew Hodgson worked on laser spectroscopy and molecular photodissociation dynamics, before setting up a group to study gas-surface reaction dynamics at Liverpool. His current research interests centre around the structure and reactivity of ice surfaces, and particularly how water wets metal surfaces.

CHAPTER 11

The authors of Chapter 11, Helen H. Fielding (left) and Abigail D. G. Nunn (right).

Abigail D. G. Nunn carried out her Ph.D. at University College London, with Helen Fielding. She developed a pulse-shaper for creating arbitrarily shaped light pulses in the ultraviolet and employed femtosecond pump-probe methods to investigate the ultrafast dynamics of small organic molecules in the gas-phase.

Helen H. Fielding is a professor of physical chemistry at University College London. Her research is focussed on the spectroscopy, femtosecond dynamics and coherent control of atoms, molecules and biomolecules in the gas-phase and on surfaces.

CHAPTER 12

The authors of Chapter 12, Liesbeth M. C. Janssen (left) and Gerrit C. Groenenboom (right).

Gerrit C. Groenenboom obtained his Ph.D. at the Eindhoven University of Technology in 1991, after which he was a postdoctoral researcher with Professor W. H. Miller in Berkeley. In 1992 he obtained a position in the theoretical chemistry group of the University of Nijmegen, initially as a fellow of the Royal Dutch Academy of Sciences, later as assistant (1997) and associate Professor (2005). His first "cold molecule paper" in 2000 concerned a He-O_2 potential energy surface. He also worked on quantum scattering calculations of cold collisions in magnetic fields and in 2006 he was a coauthor on a Science publication on near threshold inelastic Xe-OH collisions. (Homepage: www.theochem.ru.nl/gerritg)

Liesbeth M. C. Janssen obtained her master's degree in chemistry *summa cum laude* from the Radboud University in Nijmegen. She has worked on photodissociation of small radicals with Professor D. H. Parker and Dr G. C. Groenenboom (Radboud University Nijmegen) and on quantum reaction dynamics in the group of Professor D. C. Clary (University of Oxford). Currently she is a Ph.D. student in the theoretical chemistry group in Nijmegen, working with Dr G. C. Groenenboom and Professor A. van der Avoird on cold collisions of NH radicals.

CHAPTER 1

Introduction

MARK BROUARD[a] AND CLAIRE VALLANCE[b]

[a] Department of Chemistry, Physical and Theoretical Chemistry Laboratory, University of Oxford, UK; [b] Department of Chemistry, Chemistry Research Laboratory, University of Oxford, UK

1.1 INTRODUCTION

The field of molecular reaction dynamics encompasses a broad range of research areas united by the common goal of gaining a truly fundamental understanding of chemical reactivity. A single chemical event may be thought of in terms of a collision in which the species involved change their chemical identity through the cleavage and formation of chemical bonds. When we look at many such collisions, we find that every chemical reaction bears its own unique fingerprint, embodied in the kinetic energy, angular distribution, and rotational and vibrational motion of the newly formed reaction products. These quantities reflect the forces acting during the reaction, particularly in the transition state region, and their measurement often provides unparalleled insight into the basic physics governing chemical reactivity.

The 'basic physics' we are interested in is in principle very simple. Any collision must satisfy certain physical laws, such as conservation of energy and momentum, with the consequence that the final result of a chemical reaction is determined completely by the forces and energetics involved in the reactive collision. However, despite this apparent simplicity, the intricacies of molecular structure and dynamics mean that gaining a detailed understanding of even a fairly simple reaction presents a challenging problem. The outcome of a reactive collision may be exquisitely sensitive not only to the chemical identity and structure of the reactants, but also to their relative orientation and velocity, their electronic, vibrational and rotational states, and so on. The many variables affecting reactive collisions give rise to a rich and varied spectrum of chemical reactivity. Gaining an appreciation of the various factors at play is the focus of this book.

As an example of the types of questions that a reaction dynamics study can answer, consider an S_N2 type reaction (*e.g.*, $F^- + CH_3Cl \rightarrow CH_3F + Cl^-$) and contemplate the type of scattering dynamics we might expect to see. If we could control the reactant orientation so that the F^- ion

Tutorials in Molecular Reaction Dynamics
Edited by Mark Brouard and Claire Vallance

Published by the Royal Society of Chemistry, www.rsc.org

strikes either the Cl or the CH_3 end of the methyl chloride molecule (which is not beyond current experimental technology) we might expect to find that in the former case we would see very little reaction, while in the latter case we would see Cl^- products scattered forwards with respect to the initial direction of the F^- ion. These products might be expected to have fairly high velocities since they are likely to receive most of the energy released in the reaction as the C-Cl bond breaks. We might find that exciting the C-Cl stretch (say with a pulse of laser light) leads to a much higher reaction cross section by making the bond easier to break, or that increasing the velocity of the attacking F^- increases the range of angles over which reaction can occur (often called the 'cone of acceptance'). This would occur if it is the velocity component along the breaking bond that is important for reaction, and would lead to a broader angular distribution for the scattered Cl^- product.

Molecular reaction dynamics is one area of chemistry in which the link between theory and experiment is particularly strong. The bridge between experiment and theory is the reaction *potential energy surface* (PES), a key concept in reaction dynamics, as we will see in Chapter 2. A PES describes the potential energy of the system under study as a function both of the nuclear coordinates of each atom involved in the reaction and of the electronic state of the system. The potential energy arises from the electrostatic interactions between the electrons and nuclei involved in the reaction in a given electronic state, and is generally a complicated multidimensional function. As explained in Chapter 3, from a theoretical point of view, once the PES is known, in principle the dynamics of the reaction may be understood completely, provided that the reaction does not involve more than one electronic state of the system. Reactions involving multiple electronic states, and therefore multiple PESs, are considered in Chapter 4.

A single point on the surface represents the electronic energy of the system at a set of fixed atomic positions. Some regions of the surface correspond to reactants, some to products, and others to the transition state structures linking the two (see Section 1.4.5). The gradient of the surface at a given point determines the forces acting on the atoms at that point, defining their subsequent motion. An analogy may be drawn between the dynamics of a reaction over a potential energy surface and the trajectory of a ball bearing or marble rolling over a curved surface (this analogy simply replaces the electrostatic potential relevant to chemistry with a gravitational potential). We explore the idea of nuclear motion over a potential energy 'landscape' using the simple example of a three-atom reaction proceeding over a reaction PES. This is shown in 'cartoon' form in Figure 1.1. Since each position on the PES represents the potential energy at a particular molecular geometry, then as

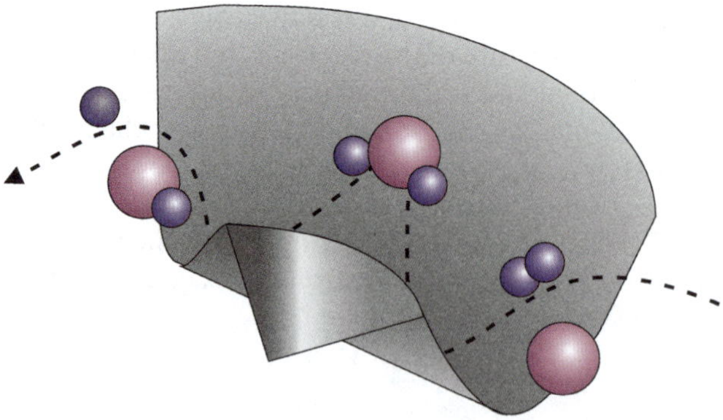

Figure 1.1 Simplified view of a three-atom reaction proceeding over a reaction potential energy surface, in which it is assumed that the atom and molecule approach each other in a colinear configuration. The dashed line maps out one example of a classical trajectory for the reactive system.

shown schematically in the figure, if we place a marble on the surface and track its path as it rolls, it traces out the changing geometry of the reacting molecules as they collide and scatter, effectively allowing us to construct a 'movie' of the sequence of events played out in the reaction. The initial position of the marble on the surface and the amount of energy imparted to it affects its trajectory in the same way as the initial geometry, relative positions and energies of the reacting molecules define the way in which they react.

Using state-of-the-art theoretical techniques, some of which are explored in Chapter 2 of this book, potential energy surfaces for reactions involving a few atoms may now be calculated from first principles to a high degree of accuracy. This involves solving the Schrödinger for the electronic motion at many geometries and interpolating the result to produce a smooth continuous surface. The dynamics of the reaction may then be predicted by carrying out either quasi-classical trajectory calculations (similar in spirit to the 'rolling the marble' approach described above) or full quantum mechanical scattering calculations over the surface, as discussed primarily in Chapters 3 to 7.

The ultimate test of our understanding of a chemical system, and the accuracy of a calculated potential energy surface, arises when the predictions of theory are compared with the results of experiment. There are a number of experimental probes of the potential energy surface, each of which provide information on different features of the reaction dynamics. Product quantum state distributions, velocity distributions, angular distributions, and even preferred axes of rotation are all amenable to experimental measurement and may be compared with the predictions of theory. The techniques developed to perform these measurements are the subjects of Chapters 4 to 7 of this book. The topic of stereochemistry, discussed in Chapter 9, introduces the notion of performing a complete experiment, in which the vectorial nature of linear and angular momenta are both recognized and exploited. This allows the effect of reactant bond-axis or angular momentum orientation on chemical reactivity, and the disposal of angular momentum after molecular collisions, to be fully quantified.

The above discussion has been focussed on gas phase bimolecular reactions. However, similar ideas involving the motion of particles over potential energy surfaces can be used to interpret other processes, such as collisional energy transfer (Chapter 5) and molecular photodissociation (Chapter 8), *i.e.* the break up of molecules subsequent to absorption of light. Chapter 10 illustrates how scattering of molecules at solid surfaces can also be tackled in an atomistic fashion, similar to that employed to interpret processes occurring in the gas phase. The final two chapters of these Tutorials provide a forward look to two important emerging areas in reaction dynamics; namely the study and control of chemical processes using femtosecond laser pump-probe techniques (Chapter 11), and the behaviour of molecules at extremely low temperatures, where only a few molecular quantum states are accessible, and classical models of reactivity are likely to be of little use (Chapter 12).

Having briefly introduced the topics to be covered over the course of the book, the remainder of this introductory chapter aims to provide an overview of some of the concepts and principles that form the basis of molecular reaction dynamics, setting the stage for the more detailed treatments of different aspects of the field provided in later chapters. We begin by introducing the idea of a reaction cross section, the 'single collision' version of a rate constant, and explore some of its properties. We then look at some of the experimental techniques used in reaction dynamics measurements, before considering a number of factors involved in interpreting the data from such experiments.

1.2 CROSS SECTIONS AND RATE CONSTANTS

The rate at which reactions occur is usually quantified by either a *cross section* or a *rate constant*, depending on the type of measurement being made.* Rate constants are most commonly used for

*The following assumes that the reader is familiar with undergraduate-level reaction kinetics, and has some knowledge of concepts such as rate equations and rate constants, simple collision theory, and transition state theory, as described in many standard text books.[1,2]

cases in which the reactants are in thermal equilibrium, and are then usually referred to as *thermal rate constants*. Rate constants will be employed particularly in the discussions of Chapters 7 and 12. Cross sections are more commonly encountered in situations in which the collision energy is well defined (see Section 1.3.3), or when the system is not at thermal equilibrium. Understanding cross sections, and their dependence on energy, quantum state, or the polarization of reactants or products, is a major goal of reaction dynamics studies. You should already be familiar with the concept of a rate constant, but for many readers the idea of a reaction cross section will require further explanation.

1.2.1 The Impact Parameter and Orbital Angular Momentum

To gain an understanding of what a cross section represents, we start with a simple classical picture of a binary collision between two reactants A and B, of mass m_1 and m_2, respectively, as illustrated in Figure 1.2. We define the impact parameter, b, as the distance of closest approach in the absence of an interaction between the two species. Although the impact parameter is a difficult quantity to control experimentally, we will see throughout this book that it plays an important role in molecular collisions.

Let us assume that the two species approach each other with a fixed relative velocity, $\boldsymbol{v}_{\rm rel}$, where $\boldsymbol{v}_{\rm rel} = \boldsymbol{v}_2 - \boldsymbol{v}_1$ and $\boldsymbol{v}_1$ and $\boldsymbol{v}_2$ are the velocities of the two reactants measured with respect to some reference point (*e.g.*, in the laboratory). Classically, the impact parameter defines the magnitude of the orbital angular momentum, ℓ, of the two reactants

$$|\ell| = |\boldsymbol{R} \times \boldsymbol{p}| = \mu v_{\rm rel} b, \tag{1.1}$$

where $\mu = m_1 m_2/(m_1 + m_2)$ is the reduced mass of the two particles, and $\boldsymbol{R}$ is the relative position of particle B with respect to particle A. Quantum mechanically, the square of the orbital angular momentum is quantized, and the magnitude of the orbital angular momentum is defined by

$$|\ell| = \hbar\sqrt{\ell(\ell + 1)}, \tag{1.2}$$

where ℓ is the orbital angular momentum quantum number. In the correspondence limit, in which ℓ takes integer values much greater than zero, Eqs. (1.1) and (1.2) provide a useful link between the classical and quantum descriptions of angular momentum. As we shall see in Chapters 3, 5, and 12, in the specific case of a collision between two structureless particles, the orbital angular momentum is conserved, but in general for more complex systems it is not (see Section 1.4.2).

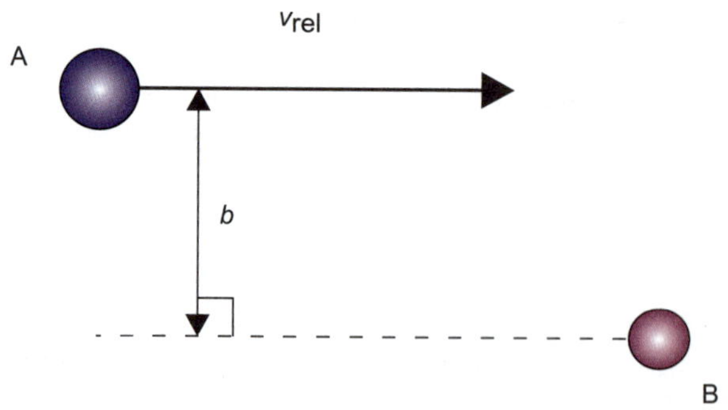

Figure 1.2 Definition of the impact parameter as the distance of closest approach in the absence of an interaction potential.

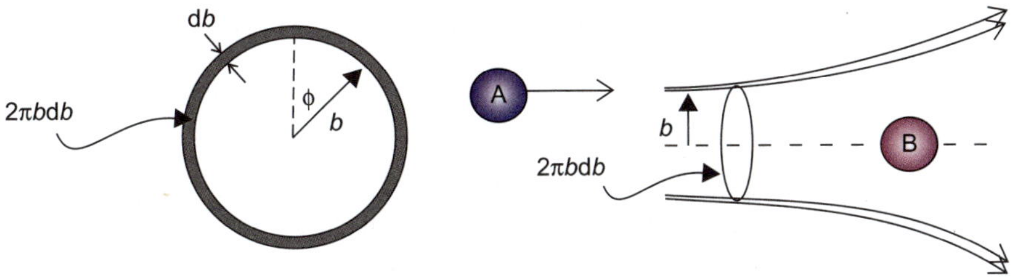

Figure 1.3 The cross section is an azimuthal (or dart-board) average of the reaction probability over impact parameter. The figure shows the approach of two spherical particles A and B at a well defined impact parameter b. As discussed in the text, the cross section is the integral of the reaction probability over all impact parameters, b and azimuthal angles, ϕ.

1.2.2 Opacity Functions and Cross Sections

The collision depicted in Figure 1.2 immediately raises two important general issues about scattering processes. It is clear that the probability of a scattering event, whether it be reaction or merely an energy transfer process, is likely to vary with impact parameter or orbital angular momentum. Collisions at high impact parameters, *i.e.* those which involve a 'glancing blow' between the two particles, are inherently less likely to lead to reaction or energy transfer than 'head on' collisions, which correspond to low-impact-parameter encounters. As we shall see in Section 1.4.3, the underlying reason for the difference between low and high impact parameter collisions is that only the kinetic energy associated with the relative motion of the particles along their line-of-centres is available for overcoming the reaction barrier. For high-impact-parameter collisions, the large amount of centrifugal energy associated with the orbiting motion of the two reactants cannot easily be used, for example, to promote reaction. For inelastic collisions (those in which kinetic energy is not conserved – see Section 1.4.2) and reactions, the variation in the probability of the process of interest with impact parameter is know as the *opacity function*, and is generally written $P(b)$. The equivalent quantum mechanical opacity function, $P(\ell)$, defines the variation in the probability with orbital angular momentum quantum number, ℓ.

The second important issue suggested by Figure 1.2 concerns the azimuthal angle, ϕ, associated with $\boldsymbol{v}_{\text{rel}}$ lying out of the plane of the page. It turns out that if the reactants are not oriented in space, or indeed are spherical particles, then the probability of the process in question becomes independent of ϕ, and we simply need to integrate over this angle when determining the reaction cross section. Figure 1.3 illustrates the averaging involved.

As discussed in more detail in Chapter 5, the cross section for the process of interest can therefore be written as a 'dart-board' average over the reaction probability*

$$\sigma = \int_0^{2\pi} \int_0^{b_{\max}} P(b) b \, \mathrm{d}b \, \mathrm{d}\phi = \int_0^{b_{\max}} P(b) 2\pi b \, \mathrm{d}b. \tag{1.3}$$

Note that the integral over the opacity function runs from $b=0$, for head on collisions, to some maximum value, $b_{\max}$. Beyond this value the collision becomes too glancing a blow for inelastic scattering or reaction to take place.† As a particularly simple example, if we set the reaction

*The quantity $P(b)2\pi b\,\mathrm{d}b$ can be interpreted as a partial cross section for a collision occurring between impact parameters b and $b+\mathrm{d}b$. This indicates that in reality direct head-on collisions virtually never take place.

†In the case of classical elastic scattering a problem arises, in that collisions that only change the direction of the velocity do not have a defined cut-off in impact parameter, because an elastic 'collision' still takes place no matter how small the deflection. This leads to a divergence in the classical elastic scattering cross section (see Study Box 5.1).

probability to unity below this cut-off impact parameter the cross section becomes

$$\sigma = \pi b_{max}^2. \tag{1.4}$$

In spite of the rather unfamiliar form of Eq. (1.3), it reassuringly yields a cross section with the correct dimensions of area, as expected on the basis of simple collision theory[1,2] (see Figure 1.4 and Figure 1.5). The cross section can therefore be thought of as an effective target area within which the colliding particles must approach for the particular process of interest to occur. Different types of collisional process have very different cross sections, reflecting the different effective target areas, resulting from the relevant probabilities for the various possible outcomes of a collision and the range of impact parameter for each process. It will not come as a particular surprise that reactions taking place on PESs with high barriers tend to have small reaction cross sections, since the probability of reaction is small unless reactants of sufficient energy approach in some well-defined direction. On the other hand, reactions on attractive surfaces, *e.g.*, those occurring between oppositely charged ions or between species possessing large dipole moments, tend to have large cross sections.

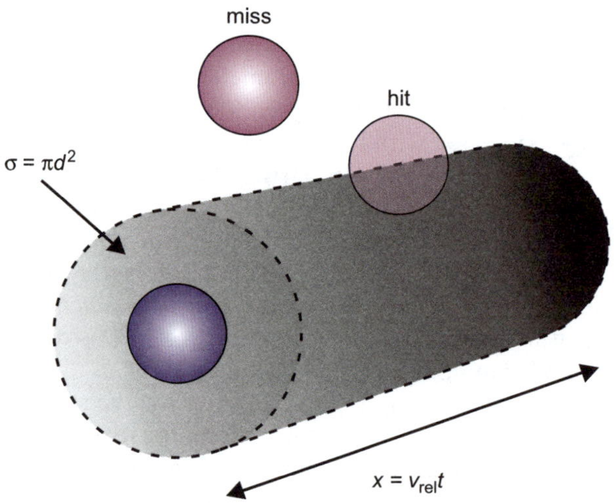

Figure 1.4 Simple collision theory: the cross section can be thought of as the effective target area of the reactants, while the rate constant is the effective collision volume swept out per unit time.

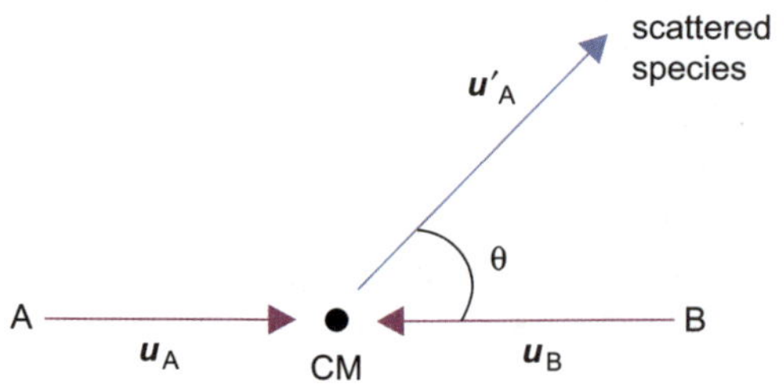

Figure 1.5 The CM scattering angle, θ, is defined as the angle between the relative velocity vectors of the reactants and products. In the figure $\boldsymbol{u}_A$ and $\boldsymbol{u}_B$ are the CM velocities of reactants, and $\boldsymbol{u}'_A$ is the CM velocity of the scattered species (in the case of an elastic collision shown, species A) after the collision. See section 1.3.3 for a discussion of the LAB to CM transformation.

By transforming the integral in Eq. (1.3) to a sum over ℓ, and using the fact that the momentum associated with the relative motion of the colliding particles is $p = \mu v_{\text{rel}} = k\hbar$, where k is the wavenumber, ($k = 2\pi/\lambda$, *i.e.* 2π times the number of waves per meter), it can be shown that the corresponding quantum expression for the cross section can be written (see Chapter 12)

$$\sigma = \frac{\pi}{k^2} \sum_{\ell} P(\ell)\,(2\ell + 1). \tag{1.5}$$

In quantum mechanics the cross section can be expressed as the weighted sum of reaction probabilities for each orbital angular momentum, with the $(2\ell + 1)$ factor arising from the degeneracy of the m_ℓ projections, reflecting the different orientations of the collision plane indicated in Figure 1.3.

1.2.3 Differential Cross Sections and Angular Distributions

The right hand panel of Figure 1.3 shows the products scattering at a particular angle relative to the initial approach direction of the atom. The angular distribution of the products turns out to be an important characteristic feature of a given collision process. In the simple process under consideration, namely the collision between two spherical particles, the figure indicates, and indeed it is the case, that there is a one-to-one correspondence between impact parameter and the angle at which the products depart. This angle is known as the scattering angle, or specifically in this case, the centre-of-mass (CM) scattering angle (we return to a discussion of the centre-of-mass in Section 1.3.3).

For more complex systems, there is no longer a direct one-to-one relationship between impact parameter and scattering angle, but nonetheless the angular distribution of the scattered products provides valuable clues about the mechanism of the collisional process under study. As will be evident from the discussion in the preceding subsection, provided that the reactants are initially randomly oriented, the angular distribution of the products is independent of azimuthal angle ϕ, *i.e.* we may write $P(\theta,\phi) = P(\theta)/2\pi$.

The angular distribution does not quantify the number of products scattered into a particular direction per unit time; it simply reflects the probability that the products are scattered in a particular direction assuming that they are formed in the first place. To quantify the number of products scattered into a particular direction, or more particularly their flux, one needs to define a quantity known as the *differential cross section*. The differential cross section, $\frac{d\sigma}{d\omega}$, characterizes the effective target area of the colliding particles that leads to scattering into a particular angular range, $d\omega$. The angular range is defined in terms of the solid angle ω (with units of steradian), with $d\omega = \sin\theta\, d\theta d\phi$. The angular probability distribution, $P(\theta,\phi)$, is then related to the differential cross section by the equation

$$P(\theta, \phi) = \frac{1}{\sigma}\frac{d\sigma}{d\omega}, \tag{1.6}$$

where $P(\theta,\phi)\sin\theta\, d\theta d\phi$ represents the probability of scattering into solid angles in the range $\theta \to \theta + d\theta$ and $\phi \to \phi + d\phi$. Integration of the differential cross section over all scattering angles yields the so called *integral cross section* discussed already in Section 1.2.2:

$$\sigma = \int_0^{2\pi} \int_0^{\pi} \frac{d\sigma}{d\omega} \sin\theta\, d\theta\, d\phi. \tag{1.7}$$

As we have noted already, the above discussion assumes the reactants to be randomly oriented in space. Chapter 9 discusses, amongst other things, the interesting issue of what happens to the reactivity when reactants are polarized.

Reactions that are dominated by long range attractive forces, which take place preferentially *via* glancing blow collisions at high impact parameters, tend to have large cross sections and show scattering into the *forward* direction. For these processes, little deflection of the atoms occurs during reaction, and the products fly away at small scattering angles in a similar direction to that of the reactants. Conversely, reactions that take place *via* the more repulsive short range region of the PES tend to require more head-on collisions. These have small impact parameters, and lead to *backward* scattering of the reaction products, *i.e.* scattering in the *opposite* direction to that in which the reactants were moving. These simple concepts are developed much more in subsequent chapters, particularly in Chapters 5 and 6.

1.2.4 State-to-state Cross Sections and Rate Constants

We have thus far neglected the fact that atoms and molecules have internal degrees of freedom, associated with electronic, rotational and vibration motion. Atoms and molecules often react with cross sections that vary considerably with initial quantum state. For example, vibrational excitation of a bond that is broken in a reaction often enhances the reactivity more than excitation of a 'spectator' bond which is preserved during reaction. Furthermore, the quantum states of the products of inelastic and reactive scattering processes are often populated in highly specific ways. As we shall see in Chapters 6 and 7, some reactions are so selective in their energy disposal that they can lead, for example, to vibrational population inversions in the products. The dependence of the cross section for a collision on initial and final quantum states is a recurring theme throughout this book, and is a particular focus of Chapter 7. If we write the state-to-state cross section as σ_{if}, where i and f refer to the initial and final states, then the initial state specific cross sections can be obtained simply as a sum over the final states

$$\sigma_i = \sum_f \sigma_{if}. \tag{1.8}$$

To obtain the integral cross section from the initial-state-dependent cross sections one needs to know what the populations are in the initial states, $P(i)$, since the integral cross section is a weighted sum over the initial state specific cross sections

$$\sigma = \sum_i P(i)\sigma_i. \tag{1.9}$$

In the case that the system is at thermal equilibrium, then the populations over the initial states will simply be characterized by the Boltzmann distribution

$$P(i) = \frac{g_i \mathrm{e}^{-\varepsilon_i/k_B T}}{q}, \tag{1.10}$$

where k_B is Boltzmann's constant, q is the appropriate partition function (*e.g.*, for the rotational or vibrational degrees of freedom of the reactants), and g_i and ε_i are the degeneracies and energies of level i.

As already noted, it is also common to quantify the rates of state-specific processes in terms of state specific rate constants. At well-defined relative velocity the relationship between the rate constants and the cross sections is simply given by (see Figure 1.4.)

$$k_{if}(v_\mathrm{rel}) = v_\mathrm{rel}\sigma_{if}(v_\mathrm{rel}). \tag{1.11}$$

In this expression we emphasize that the cross section also depends in general on the relative velocity. Extending simple collision theory, one can think of the state-to-state rate constant as an effective volume swept out by the reactants in specific quantum states and leading to products departing in specific final states.

1.2.5 Thermal Rate Constants

We return finally in this subsection to the thermal rate constant. Thermal rate constants are a primary focus of Chapter 12, which investigates chemistry at very low temperatures. As we have observed, the term *thermal* rate constant implies that the system is at thermal equilibrium, with each state populated according to the Boltzmann distribution law. For a bimolecular reaction, the thermal rate constant can be obtained from the integral cross section by averaging over the Maxwell-Boltzmann distribution of velocities, $f(v_{\mathrm{rel}})$

$$k(T) = \int_0^\infty v_{\mathrm{rel}}\sigma(v_{\mathrm{rel}})f(v_{\mathrm{rel}})\mathrm{d}v_{\mathrm{rel}}, \tag{1.12}$$

where

$$f(v_{\mathrm{rel}}) = \left(\frac{m}{2\pi k_{\mathrm{B}}T}\right)^{1/2} \mathrm{e}^{-mv_{\mathrm{rel}}^2/2k_{\mathrm{B}}T}4\pi v_{\mathrm{rel}}^2. \tag{1.13}$$

It should be clear from the above that the thermal rate constant is a highly averaged quantity, and that to gain the most insight into the dynamics of a reaction it is helpful to avoid as much as possible the blurring effects of this averaging over quantum states and relative velocity. For this reason, in most of the remainder of this book we will be using cross sections to quantify the rates of processes of interest.

1.3 EXPERIMENTAL CONSIDERATIONS

Most of the experiments to be described in these Tutorials are concerned with isolated collisions of molecules, either with other atoms or molecules, with photons, or with surfaces. We need to quantify what we mean by an 'isolated collision', and also to indicate the experimental methods that are most commonly used to study such events.

1.3.1 Single-collision Conditions

One of the most important considerations when carrying out an experiment aimed at probing the dynamics of a reactive collision is the need to isolate the event of interest amongst a perpetual background of other collisions. Collisions generally change the direction of motion, kinetic energy or internal state of a molecule, which presents a serious problem when these are precisely the properties we wish to measure. For this reason, it is vitally important to detect the products of the collision of interest before the information they hold is lost through secondary collisions. When this is achieved, the experiment is said to be carried out under *single-collision conditions*.

There are a number of ways in which we can satisfy the criterion of single-collision conditions. The two general approaches are often referred to in terms of *spatial isolation* and *temporal isolation*, though a given approach is often a combination of the two. Spatial isolation is usually employed in experiments that require the newly formed reaction products to travel some distance to a detector before their properties are measured. In this case we must ensure that the mean free path of the particles (*i.e.* the average distance travelled between collisions) is larger than the distance the particles must travel to the detector. This is achieved by maintaining a very low background pressure inside the experimental apparatus. An example of spatial isolation is found in crossed molecular beam experiments (see Section 1.3.2), in which the background pressure is often kept at less than 10^{-7} Torr. At this pressure, the mean free path may be several hundreds of metres, much longer than the distance the products must travel from the reaction centre to the detector. Temporal isolation is achieved when we ensure that the mean time between collisions is longer than the time required to make a measurement. This condition is also satisfied by the crossed molecular beam example given above, but the term is often used to describe experiments carried out at much higher pressures in which products are detected almost instantaneously after their formation, usually by some kind of spectroscopic technique. An example is a laser pump-probe experiment, in which the first laser initiates reaction, usually by breaking a chemical bond, and the second probes one or more products by a spectroscopic technique such as laser-induced fluorescence (LIF) or resonantly-enhanced multiphoton ionization (REMPI) (see Section 1.3.4 and Study Box 5.2). The mean free path and time between collisions in such an experiment may be relatively short, but the use of pulsed lasers allows the pump and probe laser pulses and pump-probe delay times to be even shorter. In many experiments these times are on the order of nanoseconds, allowing measurements to be made in gases at relatively high pressures (a few tenths of a Torr – see Problem 1). State-of-the-art pump-probe experiments using femtosecond lasers allow detection over such fast time scales that measurements are no longer restricted to the gas phase, and the dynamics of liquids and other condensed phases may be probed (see Chapter 11 and Study Box 7.4).

1.3.2 Crossed Molecular Beams

Product angular and velocity distributions, as introduced in Section 1.2.4, were first measured using the technique of crossed molecular beams. These experiments operate on a very simple principle: two molecular beams containing the reactants are crossed, usually at right angles, reaction occurs at the point of intersection, and products scattered at a particular angle are detected by a mass spectrometer equipped with an electron-impact ion source, which ionizes the neutral products prior to entry into the mass analyzer. The flight time from the crossing region to the detector yields the product velocity, and by stepping the detector through the possible scattering angles, the entire product velocity-angle distribution may be obtained. Such an experiment is shown schematically in Figure 1.6.

The experiment described above is the classic ‘universal’ crossed beam experiment, so called because the detection technique (electron impact ionization) works for any molecule and the experiment therefore has the capacity to measure the scattering distribution of virtually any chemical species, regardless of its spectroscopy (see Section 6.2). However, in modern crossed beam experiments, the moveable mass spectrometer is often replaced with a laser-based detection scheme, as discussed in Chapters 5 and 6. Such schemes provide product quantum state selectivity and improved sensitivity, together with the possibility of measuring product angular momentum polarization effects (see Chapters 5 and 9).

The collision energy in a crossed beam experiment is determined by the speeds of the molecules in the two beams and by the angle at which they cross. Using a simple model in which the thermal energy of the molecules inside the beam source is converted into translational kinetic energy of motion along the beam direction, the terminal velocity of a molecular beam is found to be[3,4]

$$v_{\text{max}} = \left(\frac{2k_B T_0}{m} \frac{\gamma}{\gamma - 1} \right)^{1/2} \tag{1.14}$$

where T_0 is the temperature inside the source, m is the mass of the molecules in the beam, and $\gamma = C_p/C_V$ is the ratio of heat capacities at constant pressure and volume for a particular gas. In the case of gas mixtures, weighted averages of m and γ are used in Eq. (1.14), and a range of reactant speeds may therefore be obtained by 'seeding' the reactant in different inert carrier gases (He, H_2, N_2, Ne, Ar, Kr, and Xe are common choices). This provides a relatively straightforward means by which to vary the collision energy. Whether or not the beam crossing angle can be altered depends on the design of the instrument, but there are several crossed beam experiments in existence for which the collision energy can be controlled in this way (see, for example, refs. [5,6]).

1.3.3 Motion in the Laboratory and Centre-of-mass Frames

The scattering distribution of the products recorded in a reaction dynamics experiment is measured in some fairly arbitrary *laboratory* (LAB) frame, determined by the geometry of the experiment. Extracting the scattering distribution from the experimental data collected in a crossed beam experiment generally requires a LAB-frame to centre-of-mass (CM) frame transformation. Consider the crossed beam experiment shown in Figure 1.6. The fact that the beams cross at right angles in this example is not a necessary requirement for measuring the scattering distribution. We could just as easily have crossed the beams at a different angle, in which case conservation of momentum and energy would have given the measured scattering distribution (in the LAB-frame) an entirely different appearance. We could have carried out a different experiment entirely, in which case the measured scattering distribution would change again. However, no matter what experiment we choose to carry out, we are probing the same chemical process. The scattering distribution provides a 'fingerprint' for a chemical reaction in much the same way as a spectrum provides a 'fingerprint' for a molecule, so there must be some way in which we can compare the scattering distributions from different experiments and obtain chemically meaningful information. As it turns out, all we need to do is to transform the results into the *centre-of-mass* (CM) frame. In this frame, the collision is seen from the viewpoint of an observer travelling along with the centre-of-mass of the system, and the two reactants appear to undergo a collision at the position of their centre-of-mass. This transformation is illustrated for a crossed beam experiment in Figure 1.7.

The CM-frame is independent of experimental geometry, allowing results from different types of experiments to be compared, and also provides a much more intuitive picture of the collision dynamics. It is helpful at this stage to work through some of the details concerning the LAB to CM transformation.

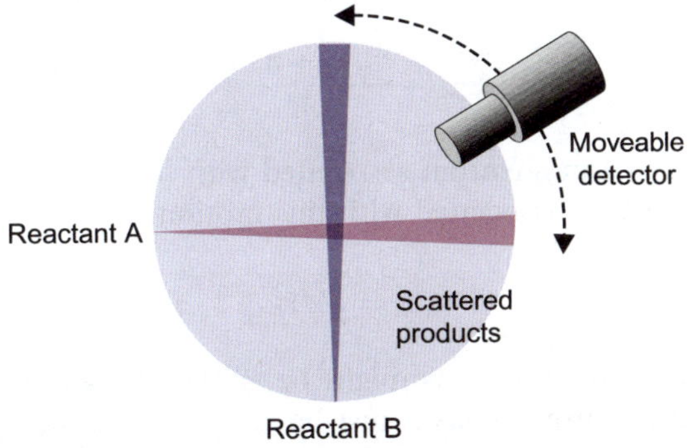

Figure 1.6 Schematic of a crossed beam experiment.

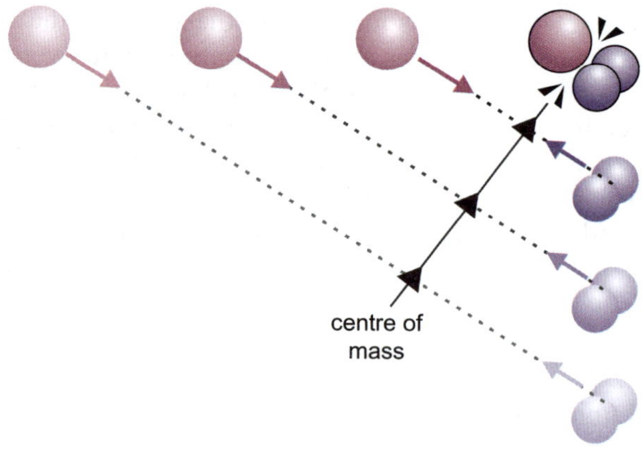

Figure 1.7 Lab to CM-frame transformation: to an observer travelling along with the centre-of-mass, the reactants appear to approach one another from opposite directions. The diagram shows a 'head-on' collision, but as we have seen from Section 1.2.1, this invariably will not be the case; in reality collisions will take place with a range of impact parameters.

Before the collision, the two particles have LAB-frame velocities $\boldsymbol{v}_1$ and $\boldsymbol{v}_2$, and kinetic energies $K_1 = \frac{1}{2}m_1v_1^2$ and $K_2 = \frac{1}{2}m_2v_2^2$, such that the total kinetic energy is

$$K = K_1 + K_2 = \tfrac{1}{2}m_1v_1^2 + \tfrac{1}{2}m_2v_2^2. \tag{1.15}$$

Because the CM-frame is simply the frame in which we are travelling along with the centre-of-mass of the system, all we need to do to determine the velocities of the particles in the CM-frame is to subtract the velocity of the CM from $\boldsymbol{v}_1$ and $\boldsymbol{v}_2$. We will call the resulting CM-frame velocities $\boldsymbol{u}_1$ and $\boldsymbol{u}_2$ to differentiate them from the LAB velocities.

$$\begin{aligned} \boldsymbol{u}_1 &= \boldsymbol{v}_1 - \boldsymbol{v}_{\mathrm{CM}} \\ \boldsymbol{u}_2 &= \boldsymbol{v}_2 - \boldsymbol{v}_{\mathrm{CM}}. \end{aligned} \tag{1.16}$$

We can determine $\boldsymbol{v}_{\mathrm{CM}}$ by using the fact that the total momentum may be written either as the momentum of the centre-of-mass, or as the sum of the momenta of the two individual particles:

$$M\boldsymbol{v}_{\mathrm{CM}} = m_1\boldsymbol{v}_1 + m_2\boldsymbol{v}_2, \tag{1.17}$$

where M is the total mass of the two particles. This rearranges to give

$$\boldsymbol{v}_{\mathrm{CM}} = \frac{m_1\boldsymbol{v}_1 + m_2\boldsymbol{v}_2}{M}. \tag{1.18}$$

In the above, we have defined a momentum associated with the motion of the centre-of-mass. We can also define the kinetic energy associated with this motion.

$$K_{\mathrm{CM}} = \tfrac{1}{2}Mv_{\mathrm{CM}}^2. \tag{1.19}$$

In the absence of an external force (*e.g.*, as might be applied by an external electric or magnetic field), the momentum of the centre-of-mass must be conserved. Hence, the velocity, momentum, and kinetic energy of the centre-of-mass are constant throughout the collision (this is true for any type of collision, whether elastic, inelastic, or reactive). Energy 'tied up' in the motion of the centre-of-mass is therefore not available for the collision. For a reactive collision, this energy does not help to overcome any activation barrier that may be present.

We have now defined the total kinetic energy and the kinetic energy associated with the motion of the centre-of-mass. The remaining kinetic energy is the energy associated with relative motion of the two particles. This energy *is* available for the collision, and consequently, it is often referred to as the *collision energy*, E_c, or sometimes the 'CM-frame kinetic energy'.

$$K_{rel} \equiv E_c = \tfrac{1}{2}\mu v_{rel}^2. \qquad (1.20)$$

Having defined all of the relevant parameters involving the reactants we now turn to the products. The product velocities may be determined by requiring that momentum and kinetic energy (or total energy in the case of an inelastic or reactive collision) are conserved during the collision. Usually it is most straightforward to carry out this calculation in the CM-frame, though the same results are obtained if it is carried out in the LAB-frame;

$$\begin{aligned} m_1\boldsymbol{u}_1 + m_2\boldsymbol{u}_2 &= m'_1\boldsymbol{u}'_1 + m'_2\boldsymbol{u}'_2, \\ \tfrac{1}{2}m_1\boldsymbol{u}_1^2 + \tfrac{1}{2}m_2\boldsymbol{u}_2^2 &= \tfrac{1}{2}m'_1\boldsymbol{u}_1'^2 + \tfrac{1}{2}m'_2\boldsymbol{u}_2'^2. \end{aligned} \qquad (1.21)$$

For elastic and inelastic scattering, the masses of the reactants and products are the same ($m_1 = m'_1$, $m_2 = m'_2$), whilst for reactive scattering the reactant and product masses will generally differ. Note also that in a treatment of inelastic or reactive scattering, the second of the above equations, requiring conservation of energy, would contain contributions from reactant and product internal energies and reaction endo- or exoergicities, as discussed in Section 1.4.2. At this point we have two equations in the two unknowns $\boldsymbol{u}'_1$ and $\boldsymbol{u}'_2$ (the final CM-frame velocities). Remember that in the CM-frame the total momentum is zero both before and after the collision; this simplifies the solution of these equations considerably, since the first equation becomes simply $m'_1\boldsymbol{u}'_1 + m'_2\boldsymbol{u}'_2 = 0$.

Once we have determined the CM-frame velocities of the collision products, we can find the equivalent LAB-frame velocities simply by adding on the velocity of the centre-of-mass (which, as we have already seen, stays constant throughout the collision):

$$\begin{aligned} \boldsymbol{v}'_1 &= \boldsymbol{u}'_1 + \boldsymbol{v}_{CM} \\ \boldsymbol{v}'_2 &= \boldsymbol{u}'_2 + \boldsymbol{v}_{CM}. \end{aligned} \qquad (1.22)$$

Further details concerning the LAB to CM transformation can be found in Chapter 6 and Study Box 6.1, whilst details concerning the mathematics of changing variables of integration are given in Study Box 1.1.

1.3.4 Pump-probe Experiments

The pump-probe experiments to be described here have their origins in the technique of flash photolysis, pioneered by Norrish and Porter in the 1950's and 60's.[7] The basic idea is to excite a molecule with an intense pulse of light, and then monitor one or more of the different processes that the molecule undergoes subsequently. This is generally achieved using a spectroscopic technique, such as absorption or fluorescence. If the excited molecule undergoes dissociation to produce atoms or radicals, these transient species may be studied directly, or they may be allowed to react with other species. In the latter case, either the rate of these secondary reactions or the nature of the products formed might in turn be studied.

STUDY BOX 1.1: CHANGES OF VARIABLES AND JACOBIAN DETERMINANTS

When performing an integral we often need to change the variables of integration, either to simplify the integration or because we would rather work in a different coordinate system from that in which the integral is stated. For example, we might want to transform from Cartesian to spherical polar coordinates, or *vice versa*.

In one dimension, changing variables is a straightforward procedure. Say we have an integral stated in terms of a variable x:

$$\int_a^b f(x)\,\mathrm{d}x, \tag{B1.1.1}$$

and we would like to rewrite it in terms of some new variable, t. So long as we can express our variable x in terms of the new variable t, *i.e.* $x = x(t)$, we can use the chain rule to carry out the transformation, as follows:

$$\int_a^b f(x)\,\mathrm{d}x = \int_c^d f(x(t))\frac{\mathrm{d}x}{\mathrm{d}t}\,\mathrm{d}t. \tag{B1.1.2}$$

Note that we have also changed the limits on our integral to match those appropriate to the new variable, t.

In two or more dimensions, the approach is completely analogous to that outlined above for one dimension. For example, in three dimensions, say we want to transform from the variables (x, y, z) to the new variables (u, v, w). We need to know the functional relationship between our two sets of variables, *i.e.* $x = x(u, v, w)$, $y = y(u, v, w)$, $z = z(u, v, w)$. We then have

$$\int\int\int_R f(x,y)\,\mathrm{d}x\,\mathrm{d}y\,\mathrm{d}z = \int\int\int_{R'} f(x(u,v,w), y(u,v,w), z(u,v,w))\left|\frac{\partial(x,y,z)}{\partial(u,v,w)}\right|\mathrm{d}u\,\mathrm{d}v\,\mathrm{d}w. \tag{B1.1.3}$$

Here R is the region we are integrating over in the (x, y, z) coordinate system, and R' is the same region in the (u, v, w) coordinate system. The quantity inside the straight brackets is known as a *Jacobian determinant* (or often just a Jacobian), and is given in full by

$$\left|\frac{\partial(x,y,z)}{\partial(u,v,w)}\right| = \begin{vmatrix} \frac{\partial x}{\partial u} & \frac{\partial x}{\partial v} & \frac{\partial x}{\partial w} \\ \frac{\partial y}{\partial u} & \frac{\partial y}{\partial v} & \frac{\partial y}{\partial w} \\ \frac{\partial z}{\partial u} & \frac{\partial z}{\partial v} & \frac{\partial z}{\partial w} \end{vmatrix}. \tag{B1.1.4}$$

The Jacobian matrix is a matrix containing all of the partial derivatives of (x, y, z) with respect to (u, v, w), and the Jacobian determinant is the determinant of this matrix. We see that the volume element for the integral in the (u, v, w) coordinate system is simply the Jacobian determinant multiplied by $\mathrm{d}u\,\mathrm{d}v\,\mathrm{d}w$, *i.e.*

$$\mathrm{d}x\,\mathrm{d}y\,\mathrm{d}z = \left|\frac{\partial(x,y,z)}{\partial(u,v,w)}\right|\mathrm{d}u\,\mathrm{d}v\,\mathrm{d}w. \tag{B1.1.5}$$

As an example, consider the transformation from Cartesian to spherical polar coordinates. We have

$$\begin{aligned} x &= r \sin\theta \cos\phi \\ y &= r \sin\theta \sin\phi \\ z &= r \cos\theta. \end{aligned} \tag{B1.1.6}$$

The Jacobian determinant is therefore

$$\left|\frac{\partial(x,y,z)}{\partial(r,\theta,\phi)}\right| = \begin{vmatrix} \frac{\partial x}{\partial r} & \frac{\partial x}{\partial \theta} & \frac{\partial x}{\partial \phi} \\ \frac{\partial y}{\partial r} & \frac{\partial y}{\partial \theta} & \frac{\partial y}{\partial \phi} \\ \frac{\partial z}{\partial r} & \frac{\partial z}{\partial \theta} & \frac{\partial z}{\partial \phi} \end{vmatrix} = \begin{vmatrix} \sin\theta\cos\phi & r\cos\theta\cos\phi & -r\sin\theta\sin\phi \\ \sin\theta\sin\phi & r\cos\theta\sin\phi & -r\sin\theta\cos\phi \\ \cos\theta & -r\sin\theta & 0 \end{vmatrix}. \tag{B1.1.7}$$

Expanding the determinant (left as an exercise for the reader – you will need to make use of the identity $\sin^2 x + \cos^2 x = 1$) leads to the familiar result that $dx\,dy\,dz = r^2 \sin\theta\, dr\, d\theta\, d\phi$.

Claire Vallance

The advent of tuneable lasers (see Study Box 1.2) has opened the way to a whole range of new experimental techniques, many of which employ the basic principles of flash-photolysis experiments. Laser-based studies often rely on a two-laser 'pump-probe' approach, in which the first ('pump') laser initiates reaction and a short time later the second ('probe') laser detects the products using a spectroscopic technique such as LIF or REMPI (see Study Box 5.2). As an example, reactions of fast atoms are often initiated by UV photolysis of a diatomic precursor, with the speed of the atoms being determined by the bond dissociation energy and the laser wavelength. To study reactions of fast chlorine atoms, for example, Cl_2 is often dissociated at 308 or 355 nm, giving Cl atom speeds of 2040 or 1641 m s^{-1}, respectively.*

Since we require single-collision conditions (see Section 1.3.1) for a reaction dynamics experiment, the maximum permissible delay between the pump and probe lasers is determined by the average time between collisions for the molecules in the gas sample. This is typically a few tens of nanoseconds for a gas-phase pump-probe experiment, depending on the pressure. However, much shorter delays are possible. Chemical reactions usually occur over the course of a few tens to a few hundreds of femtoseconds. The development of femtosecond laser sources has opened up an entirely new field of chemistry known as femtochemistry, in which pump-probe experiments on the femtosecond timescale are used to probe chemical reactions in a time-resolved manner. These experiments literally allow the course of the reaction, and the changes in energy level structure as the system evolves from reactants to products, to be probed in real time as the reactive collision occurs. As noted in Section 1.3.1, the extremely short time scales accessible with femtosecond lasers also make it possible to carry out pump-probe measurements on liquids and other condensed phases, as discussed in Chapter 11 and Study Box 7.1.

1.4 GUIDING PRINCIPLES IN DYNAMICAL STUDIES

We have thus far introduced some of the basic machinery and language of molecular reaction dynamics, and given an indication of the ways in which dynamics experiments are generally

*The Cl atom velocity is determined through conservation of energy (taking into account the energy provided by the photon and the bond dissociation energy) and momentum. Problems 1 and 2 of Chapter 8 provide examples of calculations of the recoil velocities in the molecular photodissociation of HCl and Cl_2. Further details about the technique can also be found in Study Boxes 5.2 and 6.3.

STUDY BOX 1.2: LASERS IN REACTION DYNAMICS

Many reaction dynamics experiments rely heavily on the use of lasers, both for initiating reaction and for probing the nascent reaction products. A laser provides a short, intense pulse of light at a precisely defined wavelength. This is ideal for initiating reaction at a well defined time through scission of a specific chemical bond, and there are many examples throughout this book of photon-initiated processes of this type (see, in particular, Chapter 8). Following reaction, a wide range of laser spectroscopy techniques are available for probing the newly-formed products. The most widely-used of these are resonance-enhanced multiphoton ionization (REMPI) and laser-induced fluorescence (LIF), both described in Study Box 5.2. Often, laser initiation of reaction and laser-based product detection are combined in a pump-probe experiment, as described earlier in this chapter.

To understand the properties of laser light and the physical principles underlying laser action, we need to consider the various possible absorption and emission processes for an atom or molecule. We will consider transitions between two states labelled 1 and 2, where 1 is the lower state. Chemists will be most familiar with the processes of *stimulated absorption*, in which absorption of a photon stimulates excitation to a higher energy level, and *spontaneous emission*, in which an excited state spontaneously emits a photon to return the system to a lower state. However, there is a third process, known as *stimulated emission*, in which a photon incident on an excited state molecule stimulates emission to the lower state. The two photons resulting from this process (the initial incident photon and the photon emitted from the molecule), are in phase with each other, which we shall see has important consequences for the properties of laser light. The rates for the three process are determined by the *Einstein coefficients*. Spontaneous emission is a first-order process, with the rate law

$$\frac{\mathrm{d}n_2}{\mathrm{d}t} = A_{21} n_A \qquad \text{Spontaneous emission} \qquad \text{(B1.2.1)}$$

where n_1 and n_2 are the populations of states 1 and 2, and A_{21} is the Einstein A coefficient for the transition. Stimulated absorption and stimulated emission are both second order processes, with the rate laws

$$\frac{\mathrm{d}n_2}{\mathrm{d}t} = B_{21} n_A \rho(v) \qquad \text{Stimulated emission} \qquad \text{(B1.2.2)}$$

and

$$\frac{\mathrm{d}n_1}{\mathrm{d}t} = -B_{21} n_1 \rho(v) \qquad \text{Stimulated absorption} \qquad \text{(B1.2.3)}$$

respectively. In the above, B_{12} and B_{21} are the Einstein B coefficients for stimulated absorption and emission, respectively, and $\rho(v)$ is the radiation density at the frequency ν of the transition.

When 1 and 2 are non-degenerate states, B_{12} and B_{21} are equal. Stimulated absorption and stimulated emission are therefore symmetrical processes with identical cross sections. In other words, the probability of a photon of suitable energy incident on state 1 causing a transition to state 2 is the same as the probability of a photon incident on state 2 causing a transition to state 1. Under equilibrium conditions this means that we can excite at most 50% of molecules to the upper state, since at this point the rates of stimulated absorption and stimulated emission become equal. In the general case, where the states may be non-degenerate, we have

$$\frac{B_{21}}{B_{12}} = \frac{g_1}{g_2} \qquad \text{(B1.2.4)}$$

where the g_i are the degeneracies of the two states. The Einstein A and B coefficients for a general transition between states 1 and 2 are related by

$$\frac{A_{21}}{B_{21}} = \frac{8\pi h \nu^3}{c^3} \tag{B1.2.5}$$

where c is the speed of light and h is Planck's constant.

Laser action depends on creating a *population inversion*, in which the population of the upper state is larger than that of the lower state. This is virtually impossible to achieve in a two-level system, for the reasons outlined above. However, in a three level system, if molecules are excited from the ground state to the highest energy state (in a laser this is known as *pumping*), a population inversion is immediately created between the middle and upper states. A photon emitted spontaneously during a transition from the upper to the middle state can go on to cause stimulated emission in a second molecule. The two photons emitted from the second molecule can stimulate emission in two further molecules, producing four photons in total, and so on. In the presence of a population inversion, the initial photon is rapidly amplified through stimulated emission; in fact, the word 'laser' was originally an acronym for 'Light Amplification by Stimulated Emission of Radiation'. In a three-level system the population inversion is usually maintained by rapid relaxation of the middle state back to the ground state.

The amplification can be increased by placing the emitting species inside an optical cavity consisting of a pair of mirrors. The photons are then reflected back and forth within the cavity, greatly increasing the path length over which stimulated emission can occur. If one mirror is only partially reflecting then a fraction of the light will exit the cavity on each pass as a laser beam. Laser action in a three-level system is shown schematically in Figure 1.8.

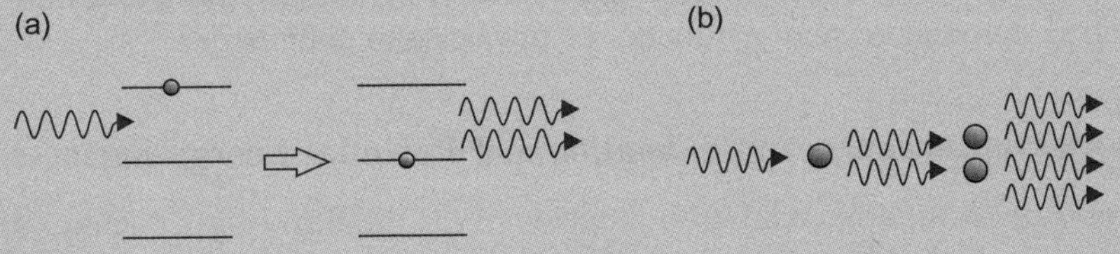

Figure 1.8 (a) Stimulated emission; (b) Light amplification by stimulated emission of radiation.

Laser light has a number of interesting properties. As noted above, the photons generated in stimulated emission are in phase with each other. As a result, the photons emitted from a laser source all have identical phase, and the light is said to be *coherent*. Since the light comes from a single atomic or molecular transition, the photons also all have the same frequency, and so the light is monochromatic. Laser beams are also highly collimated, a result of their generation within an optical cavity. Photons travelling parallel to the cavity axis are preferentially amplified, while photons with a significant off-axis component will eventually 'walk off' the mirrors after a number of reflections and be lost from the cavity, and therefore from the laser beam.

Laser action can be generated in many different materials, leading to the development of a wide variety of lasers, a number of which are used in reaction dynamics studies. Of particular note are tuneable dye lasers, which use a solution of organic dye as the lasing medium. The large number of accessible rotational and vibrational levels mean that a given dye usually emits over a broad range of wavelengths, often several tens of nanometers. By replacing one of the laser cavity mirrors with a grating, a single wavelength may be selected, and the wavelength may be tuned by changing the angle of the grating relative to the laser cavity axis. Optical parametric oscillators (OPOs) are an alternative type of laser that use non-linear optical processes to produce widely tuneable light.

Some of the most commonly-used lasers for reaction dynamics studies are listed below, together with their wavelengths of operation and pulse lengths.

1. Excimer lasers – common operation wavelengths 157 nm, 193 nm, 248 nm, 308 nm; typical pulse length 10–20 ns.
2. Nd:YAG lasers – common operation wavelengths 1064 nm, 532 nm, 355 nm, 266 nm; typical pulse length 5 ns.
3. Dye lasers (usually pumped by an excimer or Nd:YAG laser) – tuneable over a broad range of wavelengths; pulse length determined by pump laser.
4. Optical parametric oscillators (OPOs) – broadly tuneable, pulse lengths determined by pump laser.
5. Ti:sapphire lasers – emit in the range from 650 to 1100 nm; pulse lengths in the femtosecond range.

Non-linear optical processes (see Study Box 11.2), particularly frequency doubling, frequency tripling, and frequency mixing, are often used to generate further wavelengths in the UV and infra-red. For example, a 'workhorse' nanosecond laser system consisting of a dye laser pumped by a Nd:YAG laser and equipped with frequency doubling and tripling units is capable of providing continuously tuneable light over the wavelength range from around 190 to 850 nm. Adding frequency mixing options allow the generation of wavelengths well into the infrared.

Claire Vallance

performed. Next we might ask whether there are any simple guiding principles that can help with the interpretation of the results from such experiments. Fortunately there are, and they are principles which will be familiar to many students of physics and chemistry.

1.4.1 The Born-Oppenheimer Approximation and Potential Energy Surfaces

Perhaps the most important approximation in chemistry is the Born-Oppenheimer approximation, in which the motions of the electrons and nuclei are assumed separable. Because nuclei are much heavier than electrons, they may be treated as approximately stationary on the timescale of electronic motion. Similarly, the electrons respond essentially instantaneously to motions of the nuclei, and are said to *adiabatically* follow the nuclear motion.* This allows the total wavefunction of the system $\Psi(\boldsymbol{r}, \boldsymbol{R})$ to be written in the product form

$$\Psi(\boldsymbol{r}, \boldsymbol{R}) = \chi_{\mathrm{n}}(\boldsymbol{R})\psi_{\mathrm{e}}(\boldsymbol{r}; \boldsymbol{R}). \tag{1.23}$$

Here $\chi_{\mathrm{n}}(\boldsymbol{R})$ is the nuclear wavefunction, describing the nuclear positions in terms of nuclear coordinates $\boldsymbol{R}$, and $\psi_{\mathrm{e}}(\boldsymbol{r}; \boldsymbol{R})$ is the electronic wavefunction, describing the positions of the electrons. The electronic wavefunction depends on the coordinates of the electrons, $\boldsymbol{r}$, and also parametrically on the nuclear coordinates, $\boldsymbol{R}$. As we will see in detail in Chapters 2 and 4, solution of the Schrödinger equation for the motion of the electrons

$$\hat{H}_{\mathrm{e}}\psi_{\mathrm{e}}(\boldsymbol{r}; \boldsymbol{R}) = E(\boldsymbol{R})\, \psi_{\mathrm{e}}(\boldsymbol{r}; \boldsymbol{R}), \tag{1.24}$$

*'Adiabatic' literally translates as 'not passing through', and in the sense used here is in the context of the *adiabatic theorem* from quantum mechanics. This states that a system remains in its instantaneous eigenstate if a given perturbation is acting on it slowly, and if there is an energy gap between the eigenvalue and the eigenvalues of the other states of the system.

yields the total electronic energies, $E(\boldsymbol{R})$ (the electronic energy levels), for a particular nuclear geometry. If calculations are carried out for many nuclear configurations, the $\boldsymbol{R}$ dependence of $E(\boldsymbol{R})$ can be mapped out. For a particular electronic state, $E(\boldsymbol{R})$ is effectively the potential energy that the nuclei experience at a given configuration, and is in fact the familiar potential energy surface of Section 1.1. It is usually given the symbol $V(\boldsymbol{R})$. Note that each electronic state of the system will have a different PES. As we saw qualitatively in Section 1.1, the potential energy surface is of use because it can be used to determine the forces between particles at any given configuration. More quantitatively, the force on the system is simply the negative gradient of the potential. For example, in one dimension we have

$$F(R) = -\frac{\mathrm{d}V(R)}{\mathrm{d}R}, \tag{1.25}$$

whilst in general we may write

$$\boldsymbol{F}(\boldsymbol{R}) = -\nabla V(\boldsymbol{R}). \tag{1.26}$$

Once the PES is known then, classically, one has to solve Newton's equations subject to a given set of initial conditions to obtain the motion of the nuclei, or the 'trajectory', over the surface. Classical methods to determine the dynamics of nuclear motion are discussed further in Chapter 9. To solve the nuclear problem quantum mechanically is quite a complex task, requiring solution of the Schrödinger equation for the motion of the nuclei, and is the main topic for discussion in Chapter 3.

As outlined above, the Born-Oppenheimer approximation leads to the important concept of the electronic state, and of the potential energy surface which defines the variation in electronic energy of a particular electronic state with nuclear coordinates. However, there are occasions when the Born-Oppenheimer approximation fails, and it is important to understand the underlying reasons for such non-Born-Oppenheimer (or *non-adiabatic*) behaviour. This is the main topic of Chapter 4.

1.4.2 Energy and Momentum Conservation and Kinematics

In Section 1.3.3 we touched on the important roles played by the conservation of energy and linear momentum. For inelastic and reactive collisions only the total energy is conserved in general; the partitioning between kinetic and internal energy before and after such a collision is illustrated in Figure 1.9. For a simple A + BC collision, the total energy, E, can be written

$$E = E_{\mathrm{c}} + E_{\mathrm{r}} + E_{\mathrm{v}} = E'_{\mathrm{t}} + E'_{\mathrm{r}} + E'_{\mathrm{v}} - \Delta H^{\ominus}(0\,K). \tag{1.27}$$

where E_{c} is the collision energy (see Section 1.3.3), E_{r} and E_{v} are the rotational and vibrational energies before collision, and E'_{t}, E'_{r} and E'_{v} are the translational, rotational and vibrational energies after collision. The energy available to the products, E_{avl}, is the sum of the product energies $E'_{\mathrm{t}} + E'_{\mathrm{r}} + E'_{\mathrm{v}}$, and one often talks about the fraction of the available energy deposited into a particular degree of freedom. For example, the fraction of the energy released into vibration could be calculated from the equation $f'_{\mathrm{v}} = E'_{\mathrm{v}}/E_{\mathrm{avl}}$.

Another important conserved quantity is angular momentum. We have already noted that in the absence of external forces the orbital angular momentum, ℓ, is conserved in the elastic scattering of structureless particles. More generally, account also needs to be taken of the internal rotational and/or electronic angular momentum of the colliding particles, and it is only the total angular momentum, $\boldsymbol{J}$, which is conserved. For a simple A + BC → AB + C reaction, if we neglect electronic angular momentum of the atoms and molecules, which is generally only a small contribution,

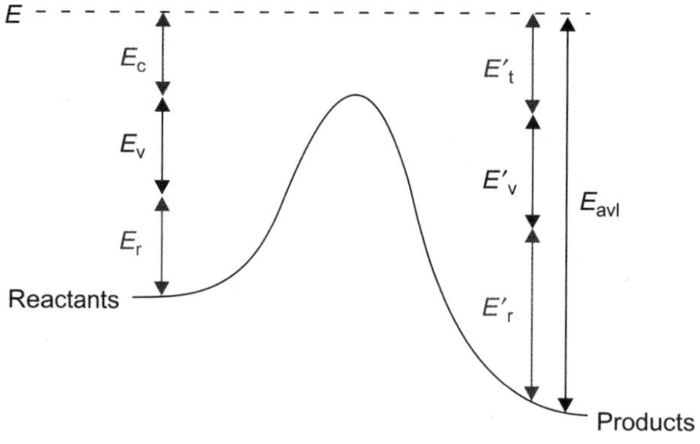

Figure 1.9 The energetics of a simple A + BC reaction.

angular momentum conservation can be written

$$\boldsymbol{J} = \boldsymbol{j}_{\mathrm{BC}} + \boldsymbol{\ell} = \boldsymbol{j}'_{\mathrm{AB}} + \boldsymbol{\ell}', \tag{1.28}$$

where $\boldsymbol{j}_{\mathrm{i}}$ refer to the internal rotational angular momentum of species i, and ℓ and ℓ' are the initial and final orbital angular momenta, respectively.

In many cases the constraints of angular momentum conservation are more restrictive than implied by Eq. (1.28). There are two reasons for this. Firstly, many experiments in reaction dynamics employ molecular beam expansion techniques, and the internal degrees of freedom of molecules entrained in such beams are often cooled down to temperatures as low as a few Kelvin.* Under these conditions, the initial rotational angular momentum can often be neglected and we may write

$$\boldsymbol{J} \simeq \boldsymbol{\ell} = \boldsymbol{j}'_{\mathrm{AB}} + \boldsymbol{\ell}'. \tag{1.29}$$

The mass combination may also play an important role, something which is usually referred to as a *kinematic effect*. If we label heavy and light atoms as $\mathcal{H}$ and $\mathcal{L}$, respectively, then for a light atom transfer reaction, *i.e.*

$$\mathcal{H} + \mathcal{L}\mathcal{H}' \rightarrow \mathcal{H}\mathcal{L} + \mathcal{H}', \tag{1.30}$$

one might expect rather low levels of rotational excitation of the products, because of the low moment of inertia of the '$\mathcal{H}\mathcal{L}$' AB molecule, and hence its wide rotational energy level spacing. Under these circumstances, together with the initial condition of low rotational excitation, angular momentum conservation reduces to

$$\boldsymbol{J} \simeq \boldsymbol{\ell} \simeq \boldsymbol{\ell}'. \tag{1.31}$$

We see that in this case initial orbital angular momentum is channeled preferentially into final orbital angular momentum. Given the link between the orbital angular momentum and relative

*Although one often makes reference to the rotational and vibrational 'temperatures' of molecules in molecular beam expansions, it should be born in mind that the translational, rotational and vibrational degrees of freedom are generally not at thermal equilibrium under such conditions, and the effective temperatures obtained for the various degrees of freedom may not be the same.

velocity (see Eq. (1.1)), it is perhaps not surprising that light atom transfer reactions also often display a propensity to conserve kinetic energy (such that $E_c \simeq E'_t$).

Another class of reaction, that involving a light attacking or departing atom, also displays characteristic angular momentum constraints. In the latter case we have

$$\mathcal{H} + \mathcal{H}'\mathcal{L} \rightarrow \mathcal{H}\mathcal{H}' + \mathcal{L}, \tag{1.32}$$

and the product orbital angular momentum ($|\ell'| = \mu' v'_{rel} b'$) is generally small compared with the AB rotational angular momentum, again due to the much lower moment of inertia involved. The resulting angular momentum constraint is

$$\boldsymbol{J} \simeq \boldsymbol{\ell} \simeq \boldsymbol{j}'_{AB}, \tag{1.33}$$

and initial orbital angular momentum is channeled preferentially into product rotational excitation. This has the practical consequence that a measurement of the rotational population distribution of the product AB molecule can be used to determine the opacity function of the reaction, $P(\ell)$ or $P(b)$.[8]

The role of kinematics in bimolecular chemical reactions is returned to again in several places in these Tutorials, but particularly in Study Box 7.1, which discusses the use of mass weighted coordinates for potential energy surfaces.

1.4.3 Effective Potentials and Centrifugal Barriers

Angular momentum plays another important role in molecular collisions. Associated with the orbital angular momentum is a centrifugal kinetic energy, which at a separation R between the reactants is given classically by*

$$K_{cent} = \frac{|\ell|^2}{2\mu R^2} = \tfrac{1}{2}\mu v_{rel}^2 \frac{b^2}{R^2} = E_c \frac{b^2}{R^2}. \tag{1.34}$$

If the orbital angular momentum is conserved during the collision, then one may think of the reaction as occurring on an *effective potential*

$$V_{eff}(R) = V(R) + K_{cent}, \tag{1.35}$$

which can possess a *centrifugal barrier*, even in the absence of a barrier on the potential energy surface. This is illustrated in Figure 1.10.

Although, the orbital angular momentum is only rigorously conserved in the case of elastic scattering of atoms, the centrifugal barrier remains an important feature of all binary collisions, and will be returned to frequently in this book, particularly so in Chapter 12. Reactants must overcome the barrier in the effective potential if reaction is to occur. This centrifugal barrier may be overcome if the reactants have sufficient kinetic energy along the radial coordinate – known as the *radial kinetic energy*. In addition, quantum mechanically the reactants can tunnel through the barrier in the effective potential. Finally, as we have seen, for collisions between real atoms and molecules the orbital angular momentum may not be conserved, and can be exchanged with other sources of angular momentum (*e.g.*, molecular rotation), such that only the total angular momentum is conserved. This angular momentum exchange provides another means by which the barrier in the effective potential may be overcome.

*The formal derivation of the equations below for atom-atom scattering is given in Chapter 12, Section 12.2.1.

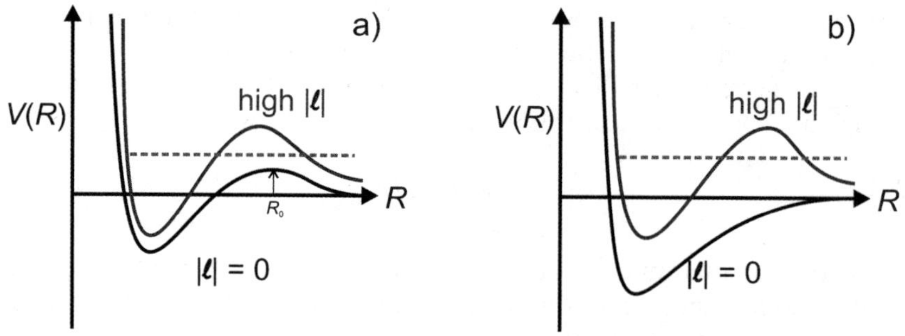

Figure 1.10 The centrifugal barrier for reactions (a) with and (b) without a barrier. Reaction at the energy of the purple dashed line needs to overcome the barrier on the effective potential (blue line), even though, in panel (b), there is no barrier on the radial slice through the potential energy surface. Note that, through the relationship $|\ell| = \mu v_{\text{rel}} b$, collisions with high orbital angular momenta are glancing blow, high impact parameter collisions.

As a simple example of the relevance of the effective potential, consider the *line-of-centres model* of chemical reactions. Imagine two spherical reactants, A and B, undergoing a collision as in Section 1.2.1. Assume that reaction can only take place if the kinetic energy along the line-of-centres, the radial kinetic energy, is greater than zero at the location of the barrier, R_0, *i.e.*

$$\left(\tfrac{1}{2}\mu\dot{R}^2\right)_{R=R_0} \geq 0, \tag{1.36}$$

where $\dot{R} = \frac{dR}{dt}$. In the line-of-centres model we may calculate the radial kinetic energy at any separation R simply as the initial collision energy minus the effective potential, *i.e.*

$$\tfrac{1}{2}\mu\dot{R}^2 = E_c - V_{\text{eff}}(R). \tag{1.37}$$

If the interaction potential, $V(R)$, has a significant barrier (see Figure 1.10(a)), then it is reasonable to assume that the barrier in the effective potential will be located close to that in $V(R)$, *i.e.* we need only consider the barrier in the effective potential at R_0, and we can therefore write (from Eqs. (1.34) and (1.35))

$$V_{\text{eff}}(R_0) = V(R_0) + E_c\frac{b^2}{R_0^2} = E_0 + E_c\frac{b^2}{R_0^2}, \tag{1.38}$$

where E_0 is the barrier height on the potential curve, located at R_0. Substituting Eq. (1.38) into Eq. (1.37) and rearranging yields an equation for the maximum impact parameter at which reaction can occur

$$b_{\text{max}}^2 = R_0^2\left(1 - \frac{E_0}{E_c}\right). \tag{1.39}$$

If we assume that reaction occurs with unit probability provided that the radial kinetic energy at the barrier is greater than or equal to zero, then Eq. (1.4) may be used to estimate the cross section as

$$\sigma = \pi b_{\text{max}}^2 = \pi R_0^2\left(1 - \frac{E_0}{E_c}\right). \tag{1.40}$$

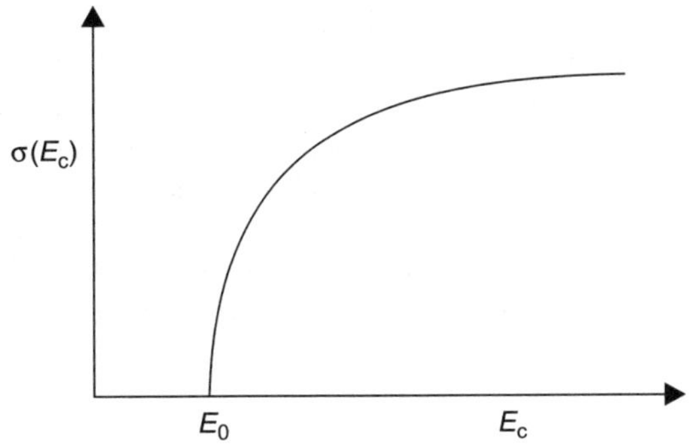

Figure 1.11 The collision energy dependence of the reaction cross section (the excitation function) predicted by the line-of-centres model.

This equation predicts a cross section that increases monotonically with collision energy, as shown in Figure 1.11, with a threshold at the energy of the barrier E_0, and an asymptotic cross section at high energy of πR_0^2. The variation in the cross section with collision energy is usually referred to as an *excitation function*. The physical reason for the rise in the cross section with collision energy is that higher energy collisions have more energy to surmount the centrifugal barrier in the effective potential, and reaction can hence take place over a wider range of impact parameters. The functional form derived from the simple line-of-centres model is qualitatively observed in many reactions with high potential energy barriers (see also Problem 3 of this chapter). As we will see in Chapter 12, the same treatment can be used to model reactions without barriers, which dominate chemistry at low temperatures.

1.4.4 Statistical *versus* Non-statistical Reactions

Another important general factor in determining both the rate of a chemical reaction or energy transfer process, and the utilization and disposal of energy, is whether or not the reaction behaves statistically. At fixed energy, the term *statistical* is taken to mean that each quantum state of the system is equally accessible. Whether or not a reaction behaves statistically often comes down to a question of timescales, and in particular the timescale for the process of interest compared with that required to randomize the energy. The principle mechanism for the latter is *intramolecular vibrational redistribution* (IVR), a process which is discussed in detail in Chapter 7. IVR is responsible for the flow of internal energy around a molecule. If one excites a particular bond in a molecule, by direct absorption of light, for example, then that energy tends to leak out of the bond initially excited due to the coupling between the different vibrational modes of the molecule. If the timescale for reaction is long compared with that for IVR then the internal energy in the system becomes randomized (or 'statistical') over the course of the reaction, and there is usually a clear signature of that randomization in the reaction products, for example, in the population distributions over the product quantum states. On the other hand, if the reaction takes place on a timescale faster than IVR, then it is said to be under dynamical control, and, for example, the shape of the potential energy surface will dictate the disposal of energy in the reaction products, as opposed to the number of accessible product quantum states.

Statistical reactions are usually those which take place on potential energy surfaces with deep wells, such that the reaction can be thought of as occurring *via* an intermediate or long-lived collision complex. The timescale for such reactions can extend for many vibrational periods, and the intermediates may be long-lived even on the timescale of molecular rotation (*i.e.* picoseconds). In the latter case, the angular distribution of the reaction products can provide a useful signature of complex formation, as discussed in Chapter 6.

An important issue addressed in both Chapters 7 and 11 is whether or not it is possible to 'beat' IVR, and to perform state or bond selective chemistry, or even control the outcome of chemical reactions before IVR occurs. In some cases the answer is 'yes': IVR is insufficiently quick on the timescale of the reaction for complete randomization of energy to occur, and in such cases state and bond selective chemistry is to be expected. It is also possible for the experimenter to actively intervene by probing the reaction after a short delay time, before IVR is complete. As we will see in Chapter 11, it is now becoming possible to tailor short pulses of light to cause dissociation of a molecule in a particular way, and this can only be achieved if IVR occurs on a longer timescale than the excitation pulse employed.

1.4.5 Transition States and Transition State Spectroscopy

Transition state theory is an example of a statistical theory of reaction rates, and is widely applied to both bimolecular and unimolecular reactions. Transition state theory is covered in many texts on kinetics and dynamics (see, for example, refs. [2,9–11]), and it will not be a primary focus of this book. The concept of the transition state is nonetheless an important one, particularly for the discussions in Chapters 6 and 7, and it will be touched on repeatedly in the pages that follow.

The transition state of a chemical reaction is defined formally as a dividing surface separating reactants from products (see Figure 1.12). The potential energy often passes through a maximum along one degree of freedom (where the barrier along the reaction coordinate is located), and a minimum along all other directions. This feature corresponds to a saddle point on the PES, and is discussed further in Chapter 2. The location of the saddle point is usually close to where the transition state is placed. The transition state can be thought of qualitatively as a 'bottleneck' through which the reactants must pass in order for them to become products,* and the shape of the PES in this region tends to determine the overall reaction rate. In transition state theory, one

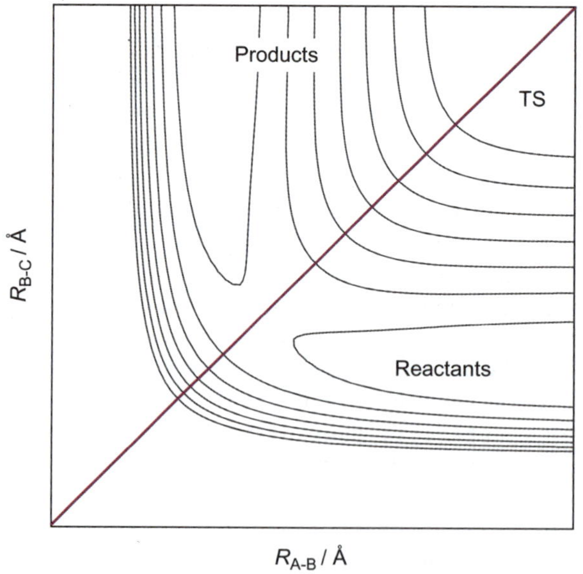

Figure 1.12 The transition state dividing surface (shown in purple) separating reactants from products.

*Because the bottleneck is dynamical in nature, in that it takes account of the rovibrational energy levels in the region of the transition state, the dividing surface is not necessarily best located exactly at the barrier to reaction.[10]

attempts to estimate the rate of reaction by simply considering the properties of the reactants and the transition state alone, without taking account of any other details of the PES. Transition state theory is often referred to as a 'statistical rate theory' because the reaction rate is calculated on the basis of a statistical estimate of the probability of the reactants reaching the transition state. Once the reactants reach the transition state, the reaction is assumed to proceed with unit efficiency.

For the purposes of reaction dynamics, it is often more common to think of a *transition state region* on the PES, rather than a formal dividing surface. This is the region where the reactants are close enough to interact with each other, and consequently where chemical bonds are made and broken. In the majority of this book we will be using the term 'transition state' in this rather looser sense. The reactants only sample the transition state region of the PES for a time on the order of a few tens to a few hundreds of femtoseconds; transition states are amongst the most transient of chemical species, and studying them experimentally presents a huge challenge.

To date, two approaches have been developed that allow transition states to be probed directly. One is the femtosecond laser pump-probe technique pioneered by Zewail,[12] introduced briefly in Section 1.3.4, and discussed more fully in Chapter 11. Femtochemistry experiments of this type allow the reaction to be followed in real time, and earned Zewail the 1999 Nobel Prize in Chemistry. The second approach, developed by Neumark and coworkers at Berkeley,[13] is now widely known as *transition state spectroscopy*. Neumark's technique, illustrated schematically in Figure 1.13(a), takes advantage of the fact that for certain reactions, while the transition state is, by definition, extremely unstable, adding an electron results in a stable negative ion with a nuclear geometry often quite similar to that of the transition state of interest. This has the consequence that a transition state may be prepared by forming the corresponding stable negative ion and then using a UV laser pulse to photodetach an electron to form the desired transition state species. The total energy imparted to the electron is fixed by the wavelength of the laser pulse, and is partitioned between the energy required to reach the final quantum state of the neutral transition state species and the kinetic energy of the ejected electron. The kinetic energy distribution of the ejected electrons therefore mirrors the energy level structure of the transition state, and may be analyzed to reveal its nuclear configuration, vibrational frequencies, and lifetime.

One of the most celebrated examples of transition state spectroscopy in action is summarized in a joint experimental and theoretical study carried out in 1993 on the 'benchmark' reaction $F + H_2 \rightarrow HF + F$.[14] As shown in Figure 1.13(b), the electron energy spectrum following photodetachment from the FH_2^- ion showed strong evidence for a bent transition state rather than the colinear

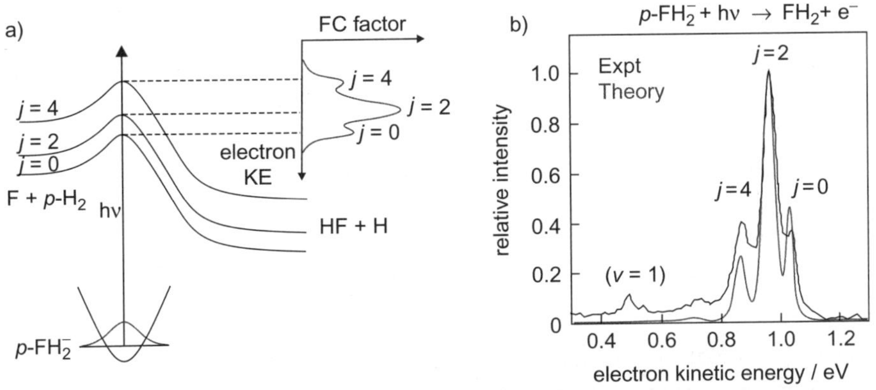

Figure 1.13 Panel (a): schematic of the photoelectron detachment spectroscopy used to probe the transition state region of the $F + H_2$ reaction. Panel (b): the photoelectron spectrum of FH_2^-. Adapted from ref. [14].

transition state that had previously been assumed, shedding new light on the results of many earlier studies. Further experiments which help to reveal the nature of the transition state region are discussed in particular in Chapters 6 and 7.

1.5 SUMMARY

In the preceding sections we have set the scene for the remaining Tutorials on molecular reaction dynamics. We have introduced some of the 'language' of reaction dynamics that you will find in the following pages, and have highlighted a number of the key concepts and themes to be investigated. We hope that the material covered in the following Tutorials will be of use to you in your studies and future research, and that it will inspire some of you to embark on new projects and initiate novel fields of research in the future.

1.6 PROBLEMS

1. (a) Determine the collision frequency of an OH radical in 100 mTorr of Ar. Assume that the collision cross section is 50 Å^2.
   (b) Use Poisson statistics to estimate the fraction of OH radicals that have undergone zero, one, or two collisions at a time delay of 50 ns after their photolytic production (*e.g.*, *via* the photodissociation of hydrogen peroxide).
2. The reaction

$$K + I_2 \rightarrow KI + I$$

was studied at a mean relative velocity of 800 m s^{-1}, with I_2 in thermal equilibrium at 300 K. The reaction cross section was found to be 170 Å^2.
Use the data in the table below to estimate
(a) the total energy available to the products,
(b) the maximum orbital angular momentum quantum number, ℓ_{max}, and
(c) the rotational energy of the KI product if $j' = \ell_{max}$, where j' is the rotational angular momentum quantum number for KI.
Identify any assumptions made in obtaining the estimate in (b).

	I_2	KI
D_0/kJ mol^{-1}	149	319
ω_e/cm^{-1}	214.5	186.5
B_e/cm^{-1}	0.037	0.061

[The mean vibrational energy of the I_2 reactants may be calculated assuming $E_v = hc\omega_e/\left(e^{\theta_v/T} - 1\right)$ with $\theta_v = hc\omega_e/k_B$.]

3. The cross section, $\sigma_r(E_c)$, for the endothermic reaction

$$K + HCl \rightarrow KCl + H$$

increases with the collision energy, E_c, in the following way:

E_c/kJ mol^{-1}	9	15	30	50
$\sigma_r(E_c)/10^{-20}$m^2	0.5	1.25	2.0	2.2

(a) Show that the cross section data are consistent with the line-of-centres model

$$\sigma_r(E_c) = \pi d^2\left(1 - \frac{E_0}{E_c}\right) \qquad E_c \geq E_0$$

and determine the threshold energy, E_0, and the limiting, high collision energy cross section, πd^2.

(b) In terms of the reaction cross section, the thermal rate constant can be written

$$k(T) = \left(\frac{8k_BT}{\pi\mu}\right)^{1/2}\int_{E_0}^{\infty}\frac{E_c}{k_BT}\sigma_r(E_c)\,e^{-(E_c/k_BT)}\frac{dE_c}{k_BT}$$

Use this equation to obtain a line-of-centres expression for $k(T)$. Comment on the result you obtain. You may use the following integral without proof

$$\int_a^{\infty}(x-a)\,e^{-x}dx = e^{-a}$$

4. Explain how the constraints imposed by the conservation of angular momentum influence the disposal of rotational energy in the reaction

$$Ba + HI \rightarrow BaI(v', j') + H$$

This reaction has been studied under crossed molecular beam conditions, at a reactant relative velocity, $v_{rel} = 976\,ms^{-1}$; the rotational state distribution in the product, BaI, was found to peak at the value $j' = 420$. Given the orbital angular momentum of the reactants in this reaction can be written $|\ell| = \mu v_{rel} b$, estimate the most probable impact parameter, b, and the reaction cross section. [Take the masses to be $m_{Ba} = 137.3\,u$, $m_H = 1.0\,u$, and $m_I = 126.9\,u$.]

CHAPTER 2

Potential Energy Surfaces: the Forces of Chemistry

MATTHEW A. ADDICOAT AND MICHAEL A. COLLINS

Research School of Chemistry, Australian National University, Canberra, Australia

2.1 INTRODUCTION

The aim of theoretical chemical dynamics is to understand the mechanisms and calculate the rates of chemical reactions. Why? Hopefully, because then we could understand how reactions occur, devise new reactions, and do chemistry more efficiently. What does "mechanism" mean? We might say we understand the mechanism of a reaction if we know where every atom is at each instant during the molecular rearrangement. More realistically, we might say that we know the most probable positions of the atoms and how these change during the reaction. It is important to remember that the atoms in a molecule are always in motion. Thinking in terms of Newton's laws, we realise that the atoms are moving subject to forces. If the forces are known, we could (in principle) calculate how the atoms move and work out the reaction mechanism. Using some statistical mechanics, we can also calculate the probability that the reaction occurs and what the rate of the reaction would be under conditions of thermal equilibrium.

In this chapter, we will look at how we calculate the forces acting on the atoms for reactions in the gas phase. The task is difficult for two reasons: It takes a fair amount of computer time to calculate the forces on the atoms for just one particular arrangement of the atoms (molecular configuration), and there are an enormous number of different molecular configurations relevant to a chemical reaction. As explained in Chapter 1, the forces of chemistry are derived from a function we call the molecular potential energy surface (PES). We begin with a general description of PESs, and finish with how we calculate a PES in practice.

2.2 THE BORN-OPPENHEIMER APPROXIMATION

For the purposes of chemistry, we can assume that a molecule is composed of electrons and atomic nuclei. The motion of these particles is completely determined by the molecular Hamiltonian, $\hat{H}$.

Tutorials in Molecular Reaction Dynamics
Edited by Mark Brouard and Claire Vallance

Published by the Royal Society of Chemistry, www.rsc.org

This Hamiltonian contains the nuclear kinetic energy, $\hat{T}_N$, the electron kinetic energy, $\hat{T}_e$, the Coulomb attraction of the electrons and nuclei, $\hat{V}_{eN}$, and the electron-electron, $\hat{V}_{ee}$, and nuclear-nuclear, $\hat{V}_{NN}$, Coulomb repulsions. In atomic units (in which $e = \hbar = m_e = 1$ and the unit of length is the Bohr radius) we can write,

$$\begin{aligned}\hat{H} &= \hat{T}_N + \hat{T}_e + \hat{V}_{eN} + \hat{V}_{ee} + \hat{V}_{NN} \\ &= \sum_{\alpha=1}^{N} \frac{\hat{\boldsymbol{P}}_\alpha \cdot \hat{\boldsymbol{P}}_\alpha}{2M_\alpha} + \sum_{i=1}^{N_e} \frac{\hat{\boldsymbol{P}}_i \cdot \hat{\boldsymbol{P}}_i}{2} - \sum_{\alpha=1}^{N} \sum_{i=1}^{N_e} \frac{Z_\alpha}{R_{i\alpha}} + \sum_{i=1}^{N_e} \sum_{j<i}^{N_e} \frac{1}{r_{ij}} \\ &+ \sum_{\alpha=1}^{N} \sum_{\beta<\alpha}^{N} \frac{Z_\alpha Z_\beta}{R_{\alpha\beta}},\end{aligned} \tag{2.1}$$

where there are N atomic nuclei, N_e electrons, M_α denotes the mass and Z_α the charge of nucleus α, $\hat{\boldsymbol{P}}_\alpha$ is the operator for a Cartesian momentum vector for nucleus α, $\hat{\boldsymbol{P}}_i$ is that for the momentum of electron i, $R_{i\alpha}$ is the distance between nucleus α and electron i, r_{ij} is the distance between electrons i and j, and $R_{\alpha\beta}$ is the distance between nuclei α and β. The Coulomb interactions depend only on the Cartesian distances between the particles. The motion of a molecule is then completely described by the time-dependent wavefunctions which obey the Schrödinger equation:

$$i\frac{\partial}{\partial t}\Psi(\boldsymbol{x}, \boldsymbol{X}, t) = \hat{H}\Psi(\boldsymbol{x}, \boldsymbol{X}, t), \tag{2.2}$$

where $\boldsymbol{x}$ and $\boldsymbol{X}$ represent the Cartesian coordinates of all the electrons and nuclei, respectively. Equivalently, we can describe any observable associated with molecular motion in terms of the time-independent wavefunctions, for which

$$\hat{H}\Psi(\boldsymbol{x}, \boldsymbol{X}) = E\Psi(\boldsymbol{x}, \boldsymbol{X}). \tag{2.3}$$

It is important to note that in atomic units the nuclear mass is measured in units of the electron rest mass ($m_e = 9.1 \times 10^{-31}$ kg), so M_α in Eq. (2.1) is a large number, of order 10^3 to 10^5. Because of this, $\hat{T}_N$ is very much smaller than the other four terms in the Hamiltonian, and can often (but not always) be treated as a perturbation. We define the electronic Hamiltonian, $\hat{H}_e$:

$$\hat{H}_e = \hat{T}_e + \hat{V}_{eN} + \hat{V}_{ee} + \hat{V}_{NN}, \tag{2.4}$$

so that

$$\hat{H} = \hat{T}_N + \hat{H}_e \tag{2.5}$$

Anticipating that $\hat{T}_N$ is "small", we ignore the nuclear kinetic energy for the moment, and find the eigenfunctions of $\hat{H}_e$ alone (with fixed $\boldsymbol{X}$):

$$\hat{H}_e\psi_i(\boldsymbol{x}; \boldsymbol{X}) = E_i(\boldsymbol{X})\psi_i(\boldsymbol{x}; \boldsymbol{X}) \quad i = 0, 1, 2, \ldots \tag{2.6}$$

Solving the electronic time-independent Schrödinger Equation is what *ab initio* quantum chemistry is all about. In Eq. (2.6) the nuclear coordinates are fixed, but the electronic energy, E_i, is different for different nuclear positions. Eq. (2.6) is useful because if $\hat{T}_N$ is a perturbation in Eq. 2.5, the wavefunctions, ψ_i ($\boldsymbol{x}$; $\boldsymbol{X}$), form a useful basis set for the total wavefunction:

$$\Psi(\boldsymbol{x}, \boldsymbol{X}) = \sum_{i=0}^{\infty} \chi_i(\boldsymbol{X})\psi_i(\boldsymbol{x}; \boldsymbol{X}). \tag{2.7}$$

The coefficients χ_i must be functions of the nuclear coordinates. To find these coefficients, we insert Eq. (2.7) into the Schrödinger equation of (2.3):

$$(\hat{T}_{\mathrm{N}} + \hat{H}_{\mathrm{e}}) \sum_{i=0}^{\infty} \chi_i(\boldsymbol{X})\psi_i(\boldsymbol{x}; \boldsymbol{X}) = E \sum_{i=0}^{\infty} \chi_i(\boldsymbol{X})\psi_i(\boldsymbol{x}; \boldsymbol{X}). \tag{2.8}$$

Taking the operators inside the sum on the left hand side of this equation, and dropping the variables $\boldsymbol{x}$ and $\boldsymbol{X}$ from the notation of the wavefunctions for clarity, this gives:

$$\begin{aligned}
&\sum_{i=0}^{\infty} \left[\sum_{\alpha=1}^{N} -\frac{1}{2M_\alpha} \nabla_\alpha^2 + \hat{H}_{\mathrm{e}} \right] \chi_i \psi_i = E \sum_{i=0}^{\infty} \chi_i \psi_i \\
&\sum_{i=0}^{\infty} \left\{ \sum_{\alpha=1}^{N} -\frac{1}{2M_\alpha} \nabla_\alpha (\psi_i \nabla_\alpha \chi_i + \chi_i \nabla_\alpha \psi_i) + \chi_i \hat{H}_{\mathrm{e}} \psi_i \right\} = E \sum_{i=0}^{\infty} \chi_i \psi_i \\
&\sum_{i=0}^{\infty} \left\{ \sum_{\alpha=1}^{N} -\frac{1}{2M_\alpha} [\psi_i (\nabla_\alpha^2 \chi_i) + 2(\nabla_\alpha \chi_i)(\nabla_\alpha \psi_i) + \chi_i (\nabla_\alpha^2 \psi_i)] + \chi_i E_i \psi_i \right\} = E \sum_{i=0}^{\infty} \chi_i \psi_i,
\end{aligned} \tag{2.9}$$

where use has been made of the product rule to carry out the differentiation. Now if we premultiply Eq. (2.9) by ψ_j^* and integrate over the electronic coordinates (using the fact that different eigenfunctions are orthogonal) we find:

$$\sum_{\alpha=1}^{N} -\frac{1}{2M_\alpha} \nabla_\alpha^2 \chi_j + E_j(\boldsymbol{X})\chi_j - \sum_{i=0}^{\infty} \sum_{\alpha=1}^{N} \frac{1}{2M_\alpha} \times \{2\langle \psi_j | \nabla_\alpha | \psi_i \rangle (\nabla_\alpha \chi_i) + \langle \psi_j | \nabla_\alpha^2 | \psi_i \rangle \rangle \chi_i \} = E\chi_j. \tag{2.10}$$

The terms in curly brackets in Eq. (2.10) couple the nuclear wavefunction for the jth electronic state to nuclear wavefunctions for all the other electronic states. In the Born-Oppenheimer (BO) approximation, we neglect all these terms. Because the nuclear masses are large relative to the electron mass, this is an accurate approximation, so long as the energy $E_j(\boldsymbol{X})$ is not too close to the energy of the other electronic states (near a degeneracy, terms like $\langle \psi_j | \nabla_\alpha | \psi_i \rangle$ become very large). We will leave the situations where these terms are important to Chapter 4, where processes involving multiple electronic states are discussed. Considering the remaining terms in Eq. 2.10, within the BO approximation, we have for each electronic state:

$$\hat{H}_{\mathrm{nuc}} \chi_j = \left[\sum_{\alpha=1}^{N} -\frac{1}{2M_\alpha} \nabla_\alpha^2 + E_j(\boldsymbol{X}) \right] \chi_j(\boldsymbol{X}) = E\chi_j(\boldsymbol{X}); \tag{2.11}$$

which will be recognized as the nuclear Schrödinger equation. Solutions of Eq. (2.11) yield energy levels for vibrations and rotations, and the continuum wavefunctions that describe a reactive system. The Hamiltonian for the nuclear motion is just the sum of the kinetic energy operator and the potential energy, $E_j(\boldsymbol{X})$, which is the total electronic energy for electronic state j. From now on we drop the j subscript and assume that $E(\boldsymbol{X})$ refers to the energy of any electronic state of interest. Given that $E(\boldsymbol{X})$ is simply the potential energy experienced by the nuclei, it is often given the symbol $V(\boldsymbol{X})$ (as we shall see later in this chapter).

In qualitative terms, we have arrived at Eq. (2.11) because nuclei are thousands of times more massive than electrons, and therefore move much more slowly. It is a reasonable approximation to treat the nuclear and electronic motions separately, such that the nuclei move in an average potential set up by the electrons, while the electrons (effectively instantaneously) move around stationary nuclei. The energy of each electronic state is a function of the positions of the nuclei.

2.2.1 Dynamics

To calculate how the atomic nuclei move, we must solve the nuclear Schrödinger Eq. (2.11) or the corresponding time-dependent equation:

$$i\frac{\partial \chi}{\partial t} = \hat{H}_{\text{nuc}}\chi. \tag{2.12}$$

Chapters 3 and 5 will discuss practical methods for solving these numerically difficult quantum dynamics equations.

In many cases we can learn a great deal by treating the motion over the PES classically. The quantum Hamiltonian on the left hand side of equation (2.11) has a classical equivalent:

$$H_{\text{nuc}} = \sum_{\alpha=1}^{N} \frac{\boldsymbol{P}_\alpha \cdot \boldsymbol{P}_\alpha}{2M_\alpha} + E(\boldsymbol{X}). \tag{2.13}$$

The classical equations of motion of the nuclei are given by (in Newtonian form, $\boldsymbol{F} = M\boldsymbol{a}$):

$$-\frac{\partial E(\boldsymbol{X})}{\partial X_\alpha} = M_\alpha \frac{\mathrm{d}^2 X_\alpha}{\mathrm{d}t^2}, \tag{2.14}$$

where X_α denotes a Cartesian coordinate of nucleus α. The derivatives of the electronic energy, $E(\boldsymbol{X})$, with respect to the nuclear positions, $\boldsymbol{X}$, are the forces of chemistry. To solve the equations of motion for the atoms in order to track their trajectories over the surface, we therefore require both the energy and the gradient of the surface at each point. Remember that the PES, $E(\boldsymbol{X})$, is a continuous surface, though we usually only calculate the electronic energy at discrete geometries. In some cases, *ab initio* calculations provide gradients and second derivatives as well as energies. However, a surface useful for chemical dynamics will normally require energy and gradient information at very many geometries, because the atoms move around a great deal during a reaction.

2.3 COORDINATES

The positions of the N nuclei are completely determined by $3N$ Cartesian coordinates, measured with respect to some laboratory axis system. However, Eq. (2.1) shows that the Hamiltonian only depends on the inter-nuclear distances. Hence, the PES (the average of $\hat{H}$ for an electronic eigenstate) also only depends on the inter-nuclear distances. These distances are independent of the position of the whole molecule, so the PES is the same if the molecule is in Canberra or Oxford. Moreover, these distances are independent of the orientation of the molecule: If we rotate a rigid molecule, its PES does not change. In addition, the interatomic distances and PES are unchanged if the molecule is inverted (*i.e.* if the sign of all the Cartesian coordinates is changed). Three coordinates are required to specify the position of the molecule as a whole (say the sum over the nuclei of each x, y and z coordinate). Three coordinates are also required to specify the orientation of the whole molecule (*e.g.*, three Euler angles). So the PES must be a function of just $3N-6$ *internal* coordinates,* which specify the *shape* of the molecule. What exactly are these $3N-6$ coordinates? This is a surprisingly tough question to answer!

Traditionally, internal coordinates have been chosen to follow the bonding structure of the molecule: bond lengths, covalent bond angles and dihedral angles (for atoms ABCD, this is the angle between the normals to the planes ABC and BCD). However, if chemical reactions are considered, then many different bonding structures are possible. For N atoms, there are $N(N-1)/2$ possible bonds, $N(N-1)(N-2)/3!$ possible bond angles and $N(N-1)(N-2)(N-3)/4!$ possible dihedral angles. This seems to be far too many coordinates. Moreover, if three atoms in a molecule happen to be colinear, all associated dihedral angles become undefined. A fancy way of restating the earlier observations about the motions of the whole molecule is to say that a PES is an *invariant* of the group of translations in three dimensions, of the special orthogonal group, SO(3), of rotations in three dimensions, and of the corresponding orthogonal group, O(3), formed by SO(3) plus inversion.[15] A branch of group theory, called invariant theory, has been used to show that a PES can be expressed purely in terms of the $N(N-1)/2$ atom-atom distances; the bond angles and dihedral angles are not really necessary.[16,17] However, group theory also shows that all of these $N(N-1)/2$ distances may be needed. There is therefore no set of $3N-6$ coordinates that can describe a PES for all possible molecular shapes! To be clear: in the vicinity of almost all molecular shapes, we can choose $3N-6$ atom-atom distances, or combinations of bond lengths, bond and dihedral angles, that exactly describe how the molecular shape changes as the Cartesian coordinates of the atoms change. However, this choice will not be valid everywhere. The "deep" reason for this apparent anomaly is that the space of molecular shapes ("shape space") is a curved Riemannian manifold, not a rectilinear space like the Euclidean space described by the Cartesian coordinates.[18,19] To appreciate that this must be so, we need only remember that if a molecule becomes colinear, the dimension of its shape space increases from $3N-6$ to $3N-5$. If a molecule is only nearly linear, its shape space must be severely curved. Among the physical consequences of the curved nature of molecular shape space is that internal motion (vibration) and orientational motion (rotation) are not dynamically separable. This non-separability is most apparent for nearly colinear molecules. In general then, to describe a PES *globally*, we will need the full set of $N(N-1)/2$ atom-atom distances. If we wish, we can also include bond and dihedral angles in this *redundant* set of coordinates (redundant in the sense that only $3N-6$ are needed in any local region).

Eq. (2.1) also tells us that the PES is infinite if any inter-nuclear distance is zero. Some authors[20,21] have incorporated this property of the PES into the choice of coordinates, using $R_{\alpha\beta}^{-1}$ as a coordinate, rather than the inter-nuclear distance $R_{\alpha\beta}$. Similarly, the PES must become independent of the position of an atom if that atom is infinitely far from all the other atoms. This has suggested using coordinates such as $\exp(-\rho R_{\alpha\beta})$ in place of $R_{\alpha\beta}$.

Once some set of redundant (over-complete) coordinates have been chosen, it is relatively straightforward[21] to find a set of $3N-6$ coordinates which provide a good description of the change in shape of a molecule in the locale of any given structure (except for linear structures). To understand this, let $\{Z_i, i=1,\ldots, N_{\text{red}}\}$ represent a set of redundant internal coordinates, and $\{X_k, k=1,\ldots, 3N\}$ be the Cartesian coordinates. Then the changes in the internal coordinates relative to small changes in the Cartesian coordinates at some structure, $\mathbf{X}_0$, are given by

$$\delta Z_i = \sum_{k=1}^{3N} \left.\frac{\partial Z_i}{\partial X_k}\right|_{\mathbf{X}=\mathbf{X}_0} \delta X_k, \tag{2.15}$$

*In a diatomic or linear molecule only two coordinates are required to specify the orientation of the molecule, so in this case the PES is a function of $3N$–5 internal coordinates.

or in matrix notation,

$$\delta \boldsymbol{Z} = \boldsymbol{B}\delta \boldsymbol{X}. \tag{2.16}$$

The $N_{red} \times 3N$ matrix $\boldsymbol{B}$ (often called a Wilson B-matrix[22]) can always be written in a singular value decomposition:[23]

$$\boldsymbol{B} = \boldsymbol{U}\boldsymbol{\Lambda}\boldsymbol{V}^T, \tag{2.17}$$

where $\boldsymbol{U}$ and $\boldsymbol{V}$ are unitary matrices* and $\boldsymbol{\Lambda}$ is a diagonal matrix of *singular values*†. Note that there are standard computer codes that, given $\boldsymbol{B}$ as input, calculate $\boldsymbol{U}$, $\boldsymbol{\Lambda}$ and $\boldsymbol{V}$. In fact, so long as $\boldsymbol{X}_0$ is not a linear structure, there are exactly $3N-6$ non-zero singular values. Physically, there are only $3N-6$ independent ways that we can change the Cartesian coordinates to change the shape of a molecule. Thus we can define $3N-6$ locally independent internal coordinates, ζ_j, by:

$$\zeta_j = \boldsymbol{\Lambda}_{jj}^{-1} \sum_{i=1}^{N_{red}} U_{ji}^T Z_i \quad i = 1, \ldots, 3N-6. \tag{2.18}$$

To summarize Eqs. (2.16)–(2.18), we could write that $\delta\zeta = \boldsymbol{\Lambda}^{-1}\boldsymbol{U}\delta\boldsymbol{Z} = \boldsymbol{V}^T\delta\boldsymbol{X}$. This means that in some region around $\boldsymbol{X}_0$, the $\{\zeta_i\}$ are an independent set of internal coordinates which correspond to a (unitary) combination of Cartesian coordinates, and which can be used to describe how a molecular shape changes in the vicinity of $\boldsymbol{X}_0$.

2.4 METHODS TO CALCULATE THE ENERGY

Unfortunately for chemists, there exists no strategy to solve the electronic Schrödinger equation exactly for atomic or molecular systems from first principles (*ab initio*). There is, however, a way to approach the ground state wavefunction, $\psi_0(\boldsymbol{x}; \boldsymbol{X})$, systematically. If ψ_{trial} represents a guess or estimate of the exact ground state wavefunction, then the corresponding energy, E_{trial}, is an upper bound for the exact energy:

$$\langle \psi_{\text{trial}} | \hat{H} | \psi_{\text{trial}} \rangle = E_{\text{trial}} \geqslant E_0 = \langle \psi_0 | \hat{H} | \psi_0 \rangle. \tag{2.19}$$

A systematic sequence of more accurate estimates of the wavefunction will produce a converging sequence of estimates for the energy. A detailed review of the systematic *ab initio* methods will not be given here; the reader is instead referred to any of the detailed textbooks devoted to the subject.[24–26] *Ab initio* quantum chemistry is a well-established field, and although it is systematic, a novice wishing to construct a PES would do well to consult an *ab initio* expert to avoid some elementary mistakes. An accurate calculation of the electronic energy, at affordable computational cost, requires careful choice of the basis set and theoretical method employed to approximate E_0 (here we will ignore excited electronic states).

There are a number of popular, commercially available, programs to solve the electronic Schrödinger equation approximately for molecular systems from first principles (*i.e. ab initio*). In nearly all methods, the electronic wavefunction is represented by a linear combination of Gaussian basis functions, centered on the atomic nuclei. These constitute the 'basis set' referred to above. Several Gaussians (with different coefficients) are used to represent efficiently one atomic

*A unitary matrix satisfies the condition $\boldsymbol{U}^*\boldsymbol{U} = \boldsymbol{U}\boldsymbol{U}^* = \boldsymbol{I}$, where $\boldsymbol{I}$ is the identity matrix, and $\boldsymbol{U}^*$ is the conjugate transpose of $\boldsymbol{U}$, *i.e.* the inverse of $\boldsymbol{U}$ is equal to its conjugate transpose, $\boldsymbol{U}^{-1} = \boldsymbol{U}^*$.

†The singular values of a square matrix, $\boldsymbol{A}$, are the square roots of the eigenvalues of $\boldsymbol{A}^*\boldsymbol{A}$.

(hydrogen-like) orbital. In the molecular-orbital-theory approach, linear combinations of these atomic orbitals make up the molecular orbitals. The size and character of the Gaussian basis set has a direct effect on the chemistry that can be described. For example, "diffuse" functions (which decay to zero slowly with distance from the nucleus) are necessary to accurately describe effects such as the molecular polarizability; so-called "polarization functions" are necessary to describe the distortion of electronic density from the typical atomic distributions. The bigger and more "flexible" the basis of Gaussian functions, the more accurate is the description of the electronic wavefunction, but the calculation takes longer. The systematic basis sets developed by Dunning and coworkers provide a reliable approach to establishing convergence of the calculated energy with basis set size. The basis sets developed by Pople and coworkers may be more computationally affordable, but must be chosen with more expert knowledge and care.

The choice of basis set is only the start of the process. We must also choose a method to solve the many-particle second order differential equation that is equation (2.6). The simplest method is called the Hartree-Fock approximation, which assumes that the N_e-electron wavefunction can be composed as the anti-symmetrized product of N_e one-electron wavefunctions or spin orbitals. Anti-symmetrization of the wavefunction means that the wavefunction changes sign if any two electrons are interchanged; this ensures that the Pauli exclusion principle is obeyed (which states that two electrons cannot occupy the same quantum state). In the Hartree-Fock approach, the Coulomb interactions are not treated exactly; rather an electron in a spin orbital interacts with the mean or average field of the other N_e-1 electrons. The mean field itself is determined by the averaged orbitals, so the process is an iterative procedure to produce a self-consistent-field (SCF) estimate of the wavefunction and energy. The distribution of each electron is not correlated exactly with the distribution of the other electrons according to the Coulomb potentials and Eq. (2.6). However, starting with this SCF approximation to Eq. (2.6), the accuracy of the wavefunction and total electronic energy can be improved by accounting for the *electron correlation*. The difference between the Hartree-Fock energy and the exact energy is termed the *correlation energy*.

To study chemical reactions accurately, it is necessary to recover most of this correlation energy. Common methods for including electron correlation effects are as follows. Moller-Plesset (MP*n*) perturbation theory treats electron correlation as a perturbation. The number, *n*, refers to the order of truncation of the perturbation power series expansion. MP2, MP3 and MP4 are commonly used; higher orders are possible but are rarely implemented due to their computational expense. Configuration interaction (CI) methods start from a reference wavefunction provided by a Hartree-Fock or multi-configurational SCF (MCSCF-see below) calculation. The SCF procedure produces many one-electron orbitals, only some of which are occupied by electrons in the final many-electron Hartree-Fock wavefunction. The unoccupied orbitals have higher energies and are referred to as *virtual* orbitals. In a CI calculation, alternative electronic configurations are generated by exciting electrons from the reference occupied orbitals into reference virtual orbitals. The Hartree-Fock reference wavefunction, together with all the alternative configurations form a new basis for ψ_{trial}. The final wavefunction given by a CI calculation is thus a (weighted) sum of anti-symmetrized products of one-electron orbitals. Full CI, which corresponds to including every possible excited configuration, is far too expensive for all but the smallest of molecule/basis set combinations, and so common approximations limit the basis to only singly excited configurations, CI-Singles or CIS, to only doubly excited configurations, CID, or to both singly and doubly excited configurations, CISD. CI methods suffer from the problem that (for a CID calculation) a double excitation in reactant A and a double excitation in reactant B corresponds to a quadruple excitation in AB, which would not be calculated. This problem is termed the *size-consistency* problem. Coupled Cluster (CC) methods also generate excitations from a reference wavefunction. However, unlike CI methods, the formalism of the CC theory guarantees size-consistency. Similar abbreviations for included excitations are used, so for example CCSD includes single and double excitations, CCSD(T) includes single and double excitations and includes the effect of triple excitations using perturbation theory, CCSDTQ includes single, double, triple and quadruple excitations. Finally, Multi-Configuration SCF (MCSCF) improves on SCF by optimising the orbitals for a combination of configurations. A calculation that included all possible electronic configurations would be called a Complete Active Space SCF (CASSCF) calculation. This is prohibitively expensive and so most calculations select a subset of orbitals that electrons may be excited into and do not excite electrons from the core orbitals. Restricted Active Space (RAS) calculations like this often denote the active space in brackets, *i.e.* (m,n) for a calculation placing m electrons in n orbitals. The choice of active space can only be made with knowledge of the chemistry that one is investigating. Like many fields, *ab initio* chemistry is heavily burdened with jargon and notation. It is typical to denote the basis set and type of calculation used with notation like "MP2/aug-cc-pVTZ", which translates as a "second order Moller-Plesset calculation" using an "augmented, correlation consistent polarized valence triple zeta basis set" (we hope that helps).

It is often useful to calculate the energy gradient, ∇E_0, and Hessian (matrix of second derivatives with respect to the nuclear positions). Some *ab initio* methods provide gradients or both gradients and Hessian at modest additional computational cost above that for the energy alone, using so-called *analytic gradient* techniques. Here *analytic* means that the nuclear coordinate derivatives of the approximate equations for the wavefunction and E_0 have been obtained and rearranged in a way that evaluates ∇E_0 using much of the information already generated to calculate E_0. If such methods are not available, then gradients and Hessians can be calculated by finite difference methods. However, this is expensive, since the first derivative of a function of $3N-6$ variables requires $2(3N-6)$ evaluations of the function for a central difference formula. Gradients and Hessians combined require $(3N-6)(3N-5)$ function values.

All of the *ab initio* methods are computationally expensive, limiting the size of systems that can be treated. The computational time for Hartree-Fock calculations scales as N_b^4; MP2 scales as N_b^5, MP4 scales as N_b^7, CCSD scales as N_b^6 and CCSDT scales as N_b^8, where N_b is the number of basis functions. As the generation of a useful PES generally requires such calculations to be carried out at on the order of thousands of molecular geometries, it becomes prohibitively expensive to create a chemically accurate PES for systems with more than a handful of heavy atoms. However, in the authors' experience, some treatment of electron correlation is essential to describe chemical reactions. We will discuss how these *ab initio* methods are employed below when we consider the PES for polyatomic molecules.

2.5 SIMPLE EXAMPLES

2.5.1 Diatomic Molecule

The simplest chemical PES is that for a diatomic molecule. Figure 2.1 shows the MP2/aug-cc-pVTZ PES for hydrogen fluoride.

The PES is a curved function of a single coordinate, obtained by simply repeating the energy calculation for closely spaced values of the internuclear separation, R. We see that there is an energy minimum, E_{min}, and an associated equilibrium value for the bond length (at $R_0 = 0.9218$ Å), a repulsive "wall" at small internuclear separation, and an asymptotic approach toward the energy of the two dissociated atoms as R is increased. The classical "ball and spring" picture of molecular vibration arises from Figure 2.1. Near the minimum, the PES is well approximated by a harmonic potential, a parabola of the form $E(R) = \frac{1}{2}k(R - R_0)^2$, where k is the force constant of the spring. The harmonic vibrational frequency (which is independent of energy) is

$$\omega_0 = 2\pi\nu_0 = \sqrt{\frac{k}{\mu}}, \tag{2.20}$$

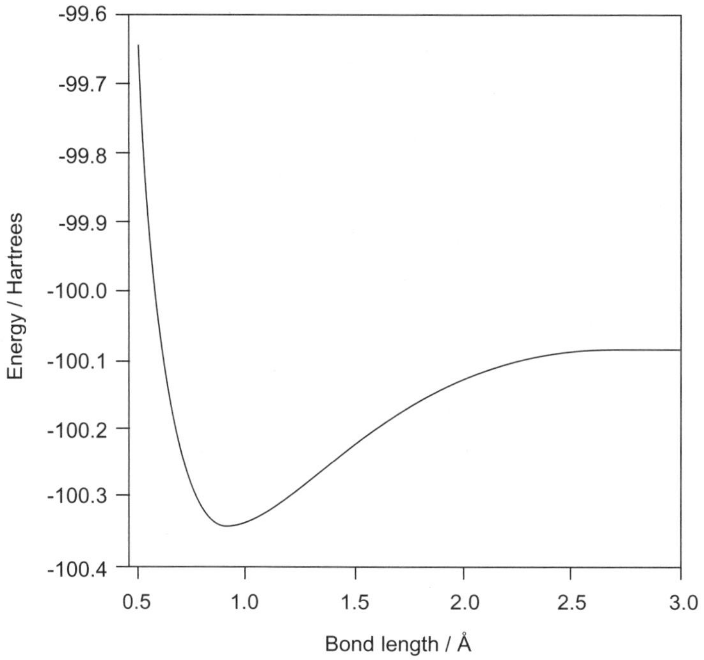

Figure 2.1 The potential energy curve of a stable diatomic molecule, HF.

where μ is the reduced mass of the two atoms. The MP2/aug-cc-pVTZ harmonic frequency of HF is calculated to be 4123 cm^{-1}. The discrete energy levels of the harmonic oscillator near the minimum are:

$$E_n^{vib} = \hbar\omega_0\left(n + \tfrac{1}{2}\right) \quad n = 0, 1, 2, \ldots \tag{2.21}$$

The ground state energy of the molecule is then approximately $E_{min} + \frac{1}{2}\hbar\omega_0$, and the difference in energy between the asymptote, $E(R \rightarrow \infty)$, and $E_{min} + \frac{1}{2}\hbar\omega_0$ is the dissociation energy of the molecule, D_0, measured from the zero-point level.

2.5.2 Constrained Triatomic

A triatomic system has three internal coordinates, and so its PES is already too high-dimensional to plot. However, if we fix one degree of freedom, the PES is only a function of two coordinates and we can draw a representation of it. An important example of a two dimensional PES is that for the colinear collision of an atom with a diatomic molecule, generically referred to as A + BC. In Figure 2.2 below, we have represented the MP2/aug-cc-pVTZ PES for the $H + HF \leftrightarrow H_2 + F$ reaction, where the HH and HF distances are chosen as coordinates and the HHF angle is kept constant (at a convenient angle, which in this case is not 180°).

The energy is represented on the left of Figure 2.2 by a surface plot on a regular grid of HF and HH bond lengths, and as a contour plot on the right of the figure. There are a number of important aspects of such a simple PES that have more general relevance. Several features are evident on this PES. On the far left, where the FH distance is large, we can see that the energy varies with the HH bond length as it does for most diatomic molecules (see Figure 2.1). Moving our eyes right, a succession of these diatomic curves, as the HF bond length is reduced, produces a valley. On the far

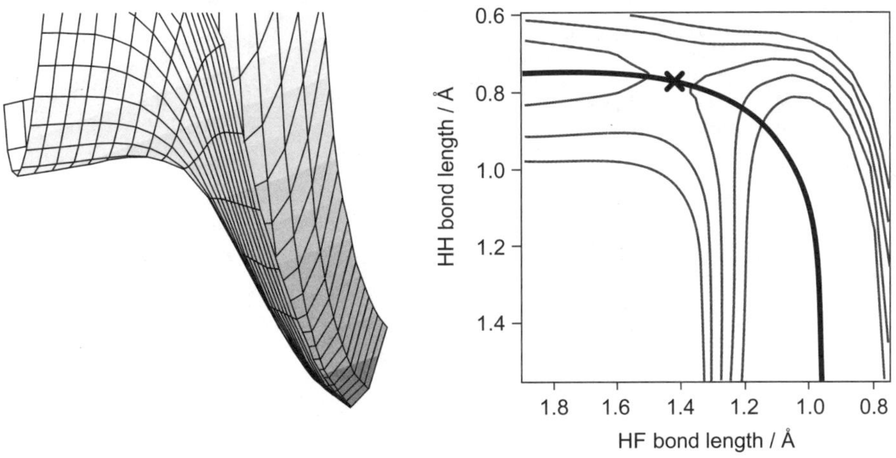

Figure 2.2 A two-dimensional potential energy surface for FH_2. See text for details.

right, where the HH distance is long, we see that the energy varies with the FH bond length as in Figure 2.1. Moving our eyes left, a succession of these diatomic curves, as the HH bond length is reduced, also produces a valley. The valley on the right is deeper than that on the left because H + HF is much lower in energy than HH + F. Looking from left to right we can see there is a single valley (usually called the reaction valley) which rises up to a saddle point before falling steeply down to the product. A saddle point is a *stationary point* (a configuration where the gradient of the PES is zero) for which the matrix of second derivatives has exactly one negative eigenvalue (there the PES is concave up in all directions except one, where it is concave down). The energy has a local maximum along the direction of the valley floor at the saddle point, but is at a minimum in all other directions. The saddle point is traditionally taken to be the location of the *transition state* in the simple *transition state theory* of chemical reaction rates (see Chapter 1 and, for example, refs. [2,10]).

Intuitively, it seems that the lowest energy way to progress from reactants to products would be along the middle of the valley floor, passing through the saddle point (x marks the spot), as indicated by the curved line on the contour plot. This type of line is called a *minimum energy path* (MEP) or *reaction path* (RP), and we will discuss such paths in more detail below. However, it is important to remember that such paths do not indicate the actual path or trajectory that characterizes the motion of a molecule during a reaction. The classical motion of the atoms is governed by Eq. (2.14), where the forces are given by the gradient of the PES. It is not possible to integrate (2.14) by mental arithmetic to see how the atomic positions, $\{X_\alpha(t)\}$, change with time. However, for a simple 2-dimensional PES such as this, it is possible to draw some qualitative conclusions about what effect the PES might have on the molecular motion. Polanyi[27–29] has treated this topic in some detail, and established what are now known as *Polanyi's Rules*. These are outlined in Study Box 2.1. Chapters 3 and 9 provide more details about how dynamical calculations using quantum mechanical and classical trajectory methods are performed.

2.6 POLYATOMIC PESs

For reactions involving polyatomic molecules, the PES is a function of many coordinates and we can no longer represent the surface graphically in any detail. However, many of the simple concepts of the 2-dimensional PES above are still useful. The PES will still be characterized by reactant and product asymptotes, where the surface becomes independent of the coordinates describing the relative positions of the collision fragments. If R denotes the separation of the centres of two

STUDY BOX 2.1: POLANYI'S RULES

With the exception of symmetric reactions (e.g., HF + H → H + FH), the saddle point is unlikely to be located at the mean of the reactant and product geometries, it will therefore be displaced some distance into either the reactant or product valley. Considering the saddle point already noted in Figure 2.2, and remembering that our forward reaction is defined as $H + HF \rightarrow H_2 + F$, it can be seen that the saddle point occurs in the product valley, when the HH bond length is close to the value for isolated H_2. A saddle point occurring in the product region of the PES is termed a "late" barrier, and a saddle point which occurs in the reactant region of the PES is termed an "early" barrier. It follows that a late (product valley) barrier becomes an early (reactant valley) barrier if the reverse reaction is considered.

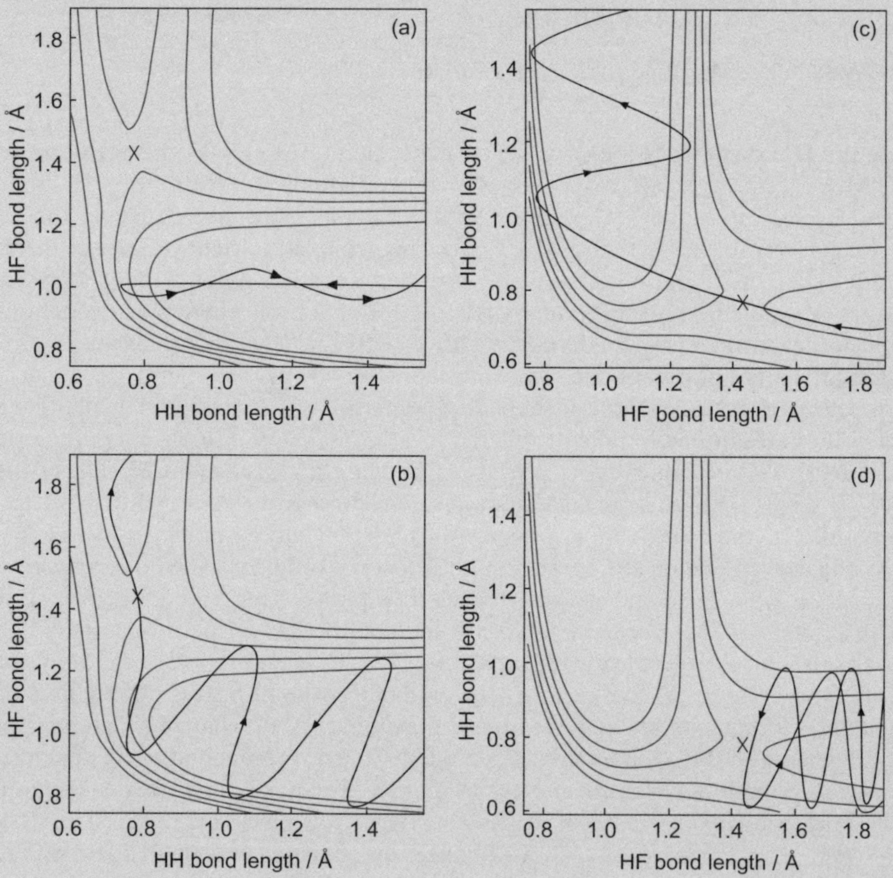

Figure 2.3 Idealized trajectories for H + HF (left panels) and $F + H_2$ (right panels). Note that for both reactions the reactants approach each other from the bottom right of each panel. See text for details.

Returning to the (late barrier) $H + HF \rightarrow H_2 + F$ reaction, suppose that the reactants approach each other with high translational energy, but little vibrational energy in the HF molecule. The path or trajectory followed might look like the dark line in Figure 2.3(a). The path taken will be directed (roughly) parallel with the reactant valley floor and be reflected by the repulsive wall.

Such trajectories will fail to surmount the barrier, and will therefore not lead to reaction. If, at the same total energy, a trajectory had relatively high vibrational energy in the reactants (and low translational energy) then the path followed might look like that in Figure 2.3(b). The path oscillates substantially in the reactant valley as the two reactant fragments approach, and such motion may allow the molecule to direct its energy in the best direction to surmount the barrier. The inverse situation applies to an early barrier as illustrated for the $H_2 + F \rightarrow H + HF$ reaction in Figures 2.3(c) and (d). Translational energy in the reactants carries them smoothly across the early barrier, whereas energy present as vibration is ineffective in crossing the barrier in the reactant valley. The behaviour described qualitatively above, and in Figure 2.3, illustrate what are now commonly referred to as Polanyi's rules.

The idealized trajectories that successfully cross the saddle point suggest that translational excitation in the reactants results in vibrationally excited products and vice-versa. In the case of an early barrier, translationally excited reactants cross the barrier and then are reflected from the wall in the exit (product) valley resulting in motion transverse to the valley; that is, vibration. Conversely, when vibrationally excited reactants surmount a late barrier, the motion over the barrier is parallel to the exit (product) valley, corresponding to relative translation.

The dynamics of the $F + H_2$ reaction is discussed in more detail in Chapter 6. We will see there that in practice, reactions behave in more complicated ways than suggested by the idealized trajectories shown in Figure 2.3. A further discussion of Polanyi's rules, and in particular the use of mass weighted coordinates and skew angles, is also given in Chapter 7 and Study Box 7.1.

Matthew A. Addicoat and Michael A. Collins

fragments (reactants or products), then the way that the PES varies with R at large values of R can often be determined from simple perturbation theory arguments. There are many texts which discuss the electrostatic, inductive and dispersive interactions that govern the PES at long range[30,31] (see also Study Box 12.2). These long range interactions can actually govern the rate of chemical reactions. For example, the long range interaction between an ion and a neutral molecule provides an attractive potential (the energy falls as R decreases), which varies as R^{-4} at long range. As discussed in detail in Chapter 12, this is enough information to determine the probability that the two molecules come in close contact at a given energy or temperature, and since many ion-molecule reactions proceed on contact (without an energy barrier), the long range PES determines the reaction rate.[32]

For polyatomic collisions that involve an energy barrier to reaction, the concepts of a reaction path, a reaction valley and a saddle point on that valley are also still useful. As two large polyatomic species come together, we can think of the $3N-6$ coordinates at long distance as consisting of a coordinate, R, describing their separation, five coordinates that specify the relative orientations of the two fragments, and $3N-12$ internal vibrations of the two species. At closer contact, the five relative orientations are transformed into torsional and angle bending motions in a total of $3N-7$ vibrational motions "orthogonal" to R. The reaction valley walls rise in $3N-7$ directions about a valley floor, and the displacement along the valley floor (R at long range and usually denoted by s in general) is a useful "reaction coordinate" that measures how far the molecular configuration has moved from reactants to products. It is possible (and sometimes useful) to give a quantitative description of the location of the path along the valley floor and to determine the associated reaction coordinate.

To generate the reaction path of Figure 2.2 in a mathematically rigorous manner, we recall that both minima and saddle points have the property that the first derivatives of the energy with respect to all coordinates are zero (that is to say they are stationary points); we distinguish them by means of their second derivatives, which provide information on the curvature of the PES at the stationary point. A minimum has second derivatives that are positive in all directions. A saddle point has positive second derivatives in all bar one coordinate (that in which the system heads downhill to either reactants or products). It is usual to define this latter coordinate as that specifying the directions along the reaction path away from the saddle point. The whole path can be defined in many ways, but to follow Fukui,[33] we begin by writing the mass weighted Cartesian coordinates of the nuclei as the vector $\boldsymbol{q}$:

$$q_k = \sqrt{M_k X_k} \quad k = 1, \ldots, 3N. \tag{2.22}$$

In these mass-weighted Cartesian (MWC) coordinates, the gradient of the PES directly indicates the classical force on each atom. Newton's law of motion, Eq. (2.14), becomes:

$$\frac{\mathrm{d}^2\boldsymbol{q}}{\mathrm{d}t^2} = -\frac{\partial V(\boldsymbol{q})}{\partial \boldsymbol{q}}, \tag{2.23}$$

where $V(\boldsymbol{q})$ is the potential energy, equivalent to the electronic energy $E(\boldsymbol{X})$ of Eqs. (2.11) and (2.13). If we assume that our system experiences infinite friction, we can damp out any oscillatory motion to obtain an equation for motion just along the 'floor' of any valley in the PES. To get this, let us write the coordinates as a function of time, using Eq. (2.23)

$$\boldsymbol{q}(t+\delta t) = \boldsymbol{q}(t) + \delta t\frac{\mathrm{d}\boldsymbol{q}(t)}{\mathrm{d}t} + \frac{1}{2}\delta t^2\frac{\mathrm{d}^2\boldsymbol{q}(t)}{\mathrm{d}t^2}. \tag{2.24}$$

If the friction reduces the velocity, $\frac{\mathrm{d}\boldsymbol{q}(t)}{\mathrm{d}t}$, to zero, then

$$\frac{\boldsymbol{q}(t+\delta t) - \boldsymbol{q}(t)}{\frac{1}{2}\delta t^2} = \frac{\mathrm{d}^2\boldsymbol{q}(t)}{\mathrm{d}t^2} = -\frac{\mathrm{d}V(\boldsymbol{q})}{\mathrm{d}\boldsymbol{q}}. \tag{2.25}$$

Defining a new variable, τ, by $\delta\tau = \delta t^2/2$, we then have

$$\frac{\mathrm{d}\boldsymbol{q}}{\mathrm{d}\tau} = -\frac{\partial V(\boldsymbol{q})}{\partial \boldsymbol{q}}, \tag{2.26}$$

which defines the *path of steepest descent*. From a given saddle point, two paths of steepest descent (forward and reverse directions) can be defined. Concatenation of these two paths gives the *intrinsic reaction path* (IRP). Now, the distance along the intrinsic reaction path is defined to be s:

$$\mathrm{d}s^2 = \|\boldsymbol{q}(\tau+\mathrm{d}\tau) - \boldsymbol{q}(\tau)\|^2 = \left\|\frac{\partial V(\boldsymbol{q})}{\partial \boldsymbol{q}}\right\|^2 \mathrm{d}\tau^2, \tag{2.27}$$

where the double modulus indicates the norm of the vector. Combining equations (2.26) and (2.27), yields:

$$\frac{\mathrm{d}\boldsymbol{q}}{\mathrm{d}s} = -\frac{\partial V(\boldsymbol{q})/\partial \boldsymbol{q}}{\|\partial V(\boldsymbol{q})/\partial \boldsymbol{q}\|}, \tag{2.28}$$

which describes the IRP in terms of the distance s, along the path. The distance, s, is called the *intrinsic reaction coordinate*. By convention, $s = 0$ at the saddle point, $s < 0$ in the reactant valley and $s > 0$ in the product valley. Note that our choice of mass-weighted Cartesian coordinates was actually arbitrary. We could have evaluated a path of steepest descent from the saddle point in any set of coordinates (say some set of internal coordinates). Alternatively, we could choose one coordinate that changes significantly during the reaction, such as a bond length, dihedral angle or distance between two fragments. The coordinate chosen is termed the *distinguished coordinate*. The coordinate is stepped either side of its value at the saddle point to its value at the reactant and product geometries and the energy at each point is determined by a constrained geometry optimization, producing a *distinguished reaction path*. Other options for calculating reaction paths include minimizing a line integral of the energy. This method was originally proposed by Elber and Karplus[34] for use in biological systems, but has been modified and applied to *ab initio* PESs. In this procedure, the reactants and products are fixed first and last points, and the saddle point is a fixed intermediate. The line integral is minimized by optimizing the geometries of intermediate points on the path, the set of optimized geometries then constitute the reaction path. All of these types of reaction path are generally termed *minimum energy paths* (MEPs), and are plausible reaction paths which at least aid our understanding of the qualitative form of the PES. Miller, Handy and Adams[35] derived the complete Hamiltonian for motion in terms of the Fukui reaction coordinate and the coordinates orthogonal to the path. However, it is important to realize that chemical dynamics in a given reaction does not occur on any single reaction path. In particular, a reaction valley will only "confine" a reacting system if the potential energy walls are sufficiently high in comparison with the energy available to the system. Nonetheless, it is *qualitatively* useful to picture a polyatomic PES as a "mountain range" with valleys containing stationary points such as minima and saddle points. These stationary points are "signposts" along the path that connects reactants and products.

Ab initio quantum chemists often present calculations of PESs purely in terms of these signposts. A typical *ab initio* investigation of a chemical reaction starts with identification of the reactant and product structures. This requires minimization of the ground state energy, E_0, with respect to the nuclear positions. Modern *ab initio* program packages perform such minimizations at relatively low cost if analytic gradients are available. The Hessian and harmonic vibrational frequencies are also evaluated. This allows estimation of the reaction enthalpy. These calculations (structure optimization, energy, frequencies and reaction enthalpy) are repeated with a systematic sequence of basis sets and a progression of levels of theory (for example, Hartree-Fock, MP2, MP4, CCSD, CCSD(T)) to establish convergence of the results. There may be experimental data for the frequencies and reaction enthalpy available for comparison. Then, one searches for saddle points that connect reactants and products (there may be more than one saddle point and energy minima intermediate between reactants and products). *Ab initio* program packages can also optimize saddle point structures, if given a sufficiently close initial guess for the structure. A Hessian and frequency calculation is carried out to verify that a saddle point has been found and to evaluate the frequencies at the saddle point. Enough information is now available to carry out a transition-state-theory estimate of the reaction rate, if desired.[37] It is now useful to evaluate the reaction paths, if only to verify that the saddle points and intermediate minima provide a complete set of signposts along the route connecting reactants and products. A distilled representation/simplification of a PES is simply a concatenation of the intrinsic reaction paths connecting all saddle points and intermediate minima between reactants and products. An example of such a representation is shown in Figure 2.4 for the $O(^3P) + H_3^+(^1A_1)$ and $OH^+(^3\Sigma^-) + H_2(^1\Sigma_g^+)$ reactions calculated at the MP2/6-311G(*2d,p*) level. The energy profiles (relative to $O + H_3^+$) along competing minimum energy paths are shown.[36] It is now possible (with additional evaluations of the Hessian along the reaction path) to carry out a variational-transition-state-theory estimate of the reaction rate.[37]

We should emphasize that a basis set and treatment of electron correlation that is adequately accurate at some stationary points on the PES may not be adequate everywhere. An investigation of the reaction paths increases the probability that a globally satisfactory level of *ab initio* quantum chemistry has been determined. For example, the spin unrestricted Moller-Plesset method, UMP2, can be particularly prone to unphysically abrupt changes in the potential energy as a bond is

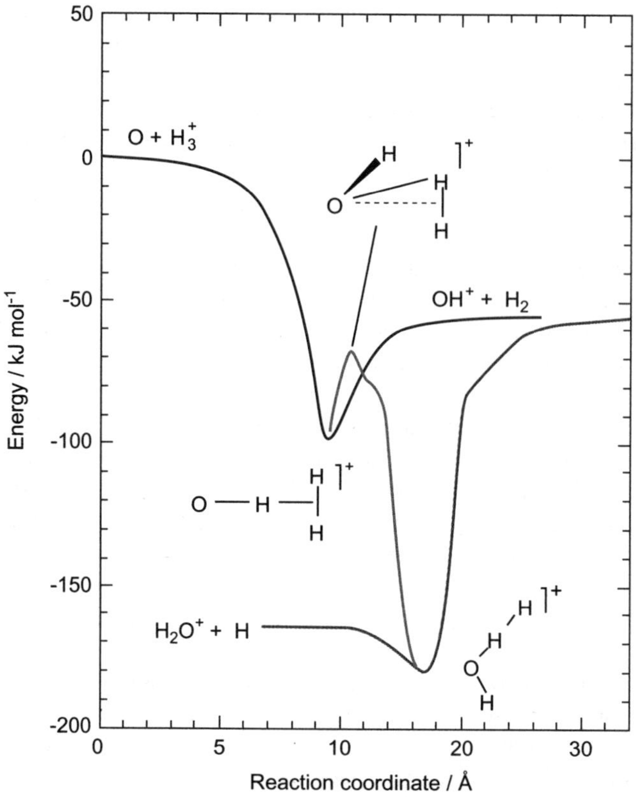

Figure 2.4 Reaction paths for H_3O^+. See text for details. Adapted from ref. 36.

broken (due to so-called *spin contamination**). Such unacceptable features may be discovered when the reaction paths are determined.

However, all of this information is merely the prelude to evaluation of a PES that can be used for a complete classical or quantum dynamical study of the reaction dynamics (see Chapters 3 and 9).

2.7 METHODS FOR CONSTRUCTING PESs

There are two basic approaches to obtaining the data needed in a dynamical calculation: Construct a PES or not. "Not" means that whenever the value of a potential energy or its gradient is required, simply evaluate the quantities *ab initio*. This option has the great virtue of simplicity, and has been used in so-called "on-the-fly" classical trajectory studies of chemical reactions.[38–41] The difficulty is that *ab initio* calculations are very time consuming, and very many different molecular structures are relevant to almost any chemical reaction. Moreover, important quantities such as the barrier height and exothermicity of a reaction can often only be calculated accurately by "high" levels of *ab initio* theory, which accurately account for electron correlation, as noted earlier. Hence, "on-the-fly" calculations can easily become prohibitively time consuming at the necessary level of accuracy. So, we choose to construct a PES. There are then two basic approaches: "functional form fitting" or "interpolation".

*In unrestricted Hartree Fock, the spatial parts of the α and β spin orbitals are not restricted to be the same. This can lead to artificial mixing of different electronic spin states. The resulting solutions are then not eigenvalues of $\hat{S}^2$. See refs. [24–26].

2.7.1 Functional Form Fitting

The oldest established method for constructing a PES is fitting functional forms. Traditionally, the PES is written as a mathematical function of the internal coordinates. This function includes a number of parameters. The values of these parameters are adjusted (fitted) to make the PES agree with *ab initio* calculations of the energy at various molecular structures. As a simple example, we might express the PES for a diatomic molecule, with bond length R, as a Morse function:

$$V_M(R) = D_e\left[e^{-\alpha(R-R_e)} - 1\right]^2, \tag{2.29}$$

where R_e is the equilibrium bond length, and D_e is the dissociation energy measured from the bottom of the well. *Ab initio* quantum chemistry could calculate the equilibrium bond length of the molecule, the energy at equilibrium relative to that of the dissociated atoms, and the second derivative of the energy with respect to the bond length, at equilibrium. The three parameters, R_e, α and D_e, can then be determined so that this functional form reproduces the calculated values exactly. More generally, if we calculated the *ab initio* energy at many different values of R, $\{E_{ab\ initio}[R(n)], n=1,\ldots, N_s\}$, the optimum values of the parameters would be the values that minimize a penalty function, P:

$$P = \sum_{n=1}^{N_s}\{V_M[R(n)] - E_{ab\ initio}[R(n)]\}^2. \tag{2.30}$$

The lower the value of P, the closer is the agreement between the functional form and the *ab initio* values, and the more "accurate" the functional form would be (more about this below). In general, for a polyatomic molecule, the functional form contains very many adjustable parameters, $\{\alpha_i\}$, and is a function of many coordinates, $\{Z_i\}$, but the least-squares penalty function is similar:

$$P = \sum_{n=1}^{N_s}\{V[Z_1(n), Z_2(n),\ldots;\alpha_1,\alpha_2,\ldots] - E_{ab\ initio}[Z_1(n), Z_2(n),\ldots]\}^2. \tag{2.31}$$

Examples of functional forms in many dimensions include the London-Eyring-Polanyi-Sato function for a colinear triatomic molecule, and the functions used for PESs for three and four atom systems developed by many researchers, including Murrell and Varandas and coworkers,[42,43] Schatz,[44] Bowman[45] and many others. Hundreds of such PESs are available through the online POTLIB archive.[46]

There are a few important technical details to note. Firstly, the functional form for the PES is often not a linear function of the parameters (Eq. (2.29) is a simple example demonstrating this fact). Hence, minimizing P is a non-linear minimization problem. In such cases, there are usually many possible sets of parameters, α_i, for which P is a local minimum:

$$\frac{\partial P}{\partial \alpha_i} = 0, \quad \text{for all } \alpha_i. \tag{2.32}$$

Local minima of P may be found using standard numerical methods such as the steepest descent, or conjugate gradient algorithms.[47] More sophisticated methods such as simulated annealing may be employed, but it is usually not possible to be sure the global minimum has been found, implying some uncertainty.

A more basic problem is that the "fit" to the PES cannot be systematically improved using any of these methods. If the number of *ab initio* calculations of the energy, N_s, is larger than the number of adjustable parameters, then the residual (minimum) value of P will not normally be zero: The PES function does not equal the *ab initio* energy, even at the finite set of configurations used. There is no systematic way of modifying the functional form and increasing the number of adjustable parameters so that a zero residual can be obtained, because the shapes of PESs for poly-atomic molecules are very variable and not known *a priori*. Moreover, even if one could dream up a sufficiently flexible functional form for the PES so that the residual was zero, that does not ensure that the functional form is accurate (close to the *ab initio* value) at other geometries not included in P (*i.e.* at points for which *ab initio* calculations have not been performed). Hence, it is very difficult to determine the residual error in the PES function and what effect such errors may have on the dynamics.

Quite recently, a new version of non-linear functional form fitting has been developed which utilizes neural networks.[48,49] The nonlinear parameters include both the architecture of the neural net and the weights used in the nodes. Tens of thousands of molecular structures (with the corresponding *ab initio* energies) form the input from which the neural net "learns" to predict the energy for a given structure. The net can then be tested using a different set of structures and energies to determine the accuracy of these predictions. There is some art involved in choosing the net architecture and in ensuring the net does not suffer from "over-fitting". However, in principle, the complexity of the architecture and the size of the "learning" sample could be systematically increased to produce a sequence of PES of increasing accuracy. Dynamical calculations using this sequence of PESs would provide an indication of the reliability of the PES.

In recent times, PES functions have been derived as linear functions of some of the unknown parameters. This approach utilizes the invariance of PESs to the permutation of indistinguishable nuclei. For example, swapping the positions of two H atoms in methane does not change the energy of the molecule. Eq. (2.1) shows that the energy only depends on the charges of the nuclei at various locations. The set of permutations of indistinguishable nuclei form the Complete Nuclear Permutation (CNP) group. If we choose some set of functions of the internal coordinates, then group theory (applied to the CNP group) defines all the possible polynomial combinations, p_i, of these functions which are invariant to permutations. There may be nonlinear (but usually fixed) parameters involved in the choice of functions, but the PES is represented as a linear combination of the invariant polynomials, with unknown coefficients which must be determined by minimizing P:

$$V[Z_1, Z_2, \ldots; \alpha_1, \alpha_2, \ldots] = \sum_i \alpha_i p_i. \tag{2.33}$$

The PES is guaranteed to have the correct CNP symmetry. The application of invariant theory to PESs was pioneered by Murrell and coworkers,[42,50,51] elaborated by Collins and coworkers,[52,53] and extensively developed and applied recently by Braams and Bowman and coworkers.[54–56] This recent work has benefitted from a computer algebra program for group theory; the number of linearly independent invariant polynomials increases rapidly with the polynomial order considered and without such as automated procedure the algebra associated with deriving all the polynomials is very tedious. The major benefit of this approach is that the minimization of P is simply a problem in linear algebra:

$$\frac{\partial P}{\partial \alpha_i} = \sum_{n=1}^{N_s} 2\left\{\sum_j \alpha_j p_j\big|_{X=X(n)} - E_{ab\ initio}\big|_{X=X(n)}\right\} p_i\big|_{X=X(n)}, \tag{2.34}$$

where the notation '$|_{X=X(n)}$' indicates evaluate at $\boldsymbol{X}=\boldsymbol{X}(n)$. In matrix notation:

$$\frac{\partial P}{\partial \alpha} = \boldsymbol{M}\alpha + \boldsymbol{b}. \tag{2.35}$$

In this equation

$$M_{ij} = 2\sum_{n=1}^{N_s} p_i p_j \big|_{\boldsymbol{X}=\boldsymbol{X}(n)}, \tag{2.36}$$

and

$$b_i = -2\sum_{n=1}^{N_s} p_i E_{ab\ initio} \big|_{\boldsymbol{X}=\boldsymbol{X}(n)}. \tag{2.37}$$

The equation for the best fit PES parameters,

$$\boldsymbol{M}\alpha + \boldsymbol{b} = 0, \tag{2.38}$$

can be solved exactly if the number of samples, N_s, and their distribution in configuration space ensures that $\boldsymbol{M}$ is invertible. Otherwise, one could take the minimal norm solution of Eq. (2.38), found from the singular value decomposition of $\boldsymbol{M}$ (see, for example, ref. [47]). Hence, by comparison with non-linear minimization, the problem of multiple local minima is avoided. The difficulty of systematically improving the accuracy of a functional form remains to some extent, since the number of invariant polynomials is large and changes in large discrete steps when the allowed polynomial order changes.

2.7.2 Interpolation

In this alternative approach to the construction of a PES, the idea is to estimate the *ab initio* energy at any relevant geometry by interpolation from known values of the *ab initio* energy at known configurations. We will refer to the locations of calculated *ab initio* data as *data points*. Interpolation differs from functional form fitting in two fundamental ways. Firstly, an interpolated approximation to a function agrees precisely with the exact values at the data points, whereas form fitting attempts only to minimize any disagreement. Secondly, an interpolation involves no adjustable parameters. For low dimensional systems, such as diatomic and triatomic molecules, this can be achieved simply using well established methods such as cubic spline interpolation. However, simple interpolation methods like splines cannot be applied readily to larger systems. Spline interpolation requires that the data points be arranged on a regular grid in the $3N{-}6$ dimensional space. Even if we ignore difficulties associated with the curved nature of shape space, a regular grid with m points in each dimension contains m^{3N-6} data points in total. The magnitude of the *ab initio* computational task increases exponentially with the dimension of the shape space. There are a number of methods which can potentially avoid this exponential scaling by allowing the data points to be scattered irregularly in an optimal way.

Ho, Rabitz and coworkers introduced and applied a method known as a reproducing kernel Hilbert space (RKHS) interpolation scheme.[57–60] The interpolation for the PES takes the form of an integral equation. Part of the integrand (the 'kernel') in the integral equation can be chosen to ensure that the PES is an interpolation of the *ab initio* data and has the correct asymptotic form (discussed above). The method has thus far only been applied to reactions of triatomic systems, using regular grids of data points, but it may be possible to use scattered data.

Our group has developed and applied an automated method for constructing PESs using *modified Shepard* (MS) *interpolation*.[21,61–63] This type of interpolation is a limiting form of *moving-least-squares* interpolation (more on this below). Only a brief summary is presented here, and the interested reader is referred to the more detailed literature on the MS approach.[64] *Ab initio* calculations provide the energy, energy gradient, and the Hessian at some chosen geometry (data point). This allows us to write a *local* approximation to the PES as a Taylor series to second order in deviations of the structure from that of the data point, using locally well-defined coordinates. Suppose now that these *ab initio* calculations were repeated at many data points scattered throughout shape space. Then we can write the PES globally as a weighted sum of these Taylor expansions:

$$V(\boldsymbol{Z}) = \sum_{g \in \mathrm{CNP}} \sum_{n=1}^{N_{\mathrm{data}}} w[\boldsymbol{Z}; g \circ \boldsymbol{Z}(n)] T[\boldsymbol{Z}; g \circ \boldsymbol{Z}(n)]. \tag{2.39}$$

Here $\boldsymbol{Z}$ represents the internal coordinates, $\boldsymbol{Z}(n)$ is a data point, and $g \circ \boldsymbol{Z}(n)$ represents a data point which is simply a CNP operation on $\boldsymbol{Z}(n)$. The *ab initio* data at $g \circ \boldsymbol{Z}(n)$ is simply a rearranged version of the data calculated at $\boldsymbol{Z}(n)$. Since the data is reproduced at all possible permutations of $\boldsymbol{Z}(n)$, the PES has the correct CNP symmetry.* The quantity $T\,[\boldsymbol{Z};\, g \circ \boldsymbol{Z}(n)]$ represents a Taylor series for the energy, from the data point to the point where the energy is to be evaluated. The weights in Eq. (2.39) are constructed so that $w[\boldsymbol{Z};\, g \circ \boldsymbol{Z}(n)]$ is large if $\boldsymbol{Z}$ is near $g \circ \boldsymbol{Z}(n)$ and small if $\boldsymbol{Z}$ is far from $g \circ \boldsymbol{Z}(n)$. The weights sum to unity and, under some simple constraints, ensure that Eq. (2.39) is an interpolation. The location of the data points is determined iteratively. We can start by placing (say) tens of data points along the MEP, because we expect this reaction valley to be dynamically significant. With this small data set, Eq. (2.39) is inaccurate far from the MEP, but is a global PES nonetheless. At this stage the PES resembles a parabolic valley surrounding a reaction path. Classical trajectories are evaluated on this PES to provide a sample of molecular geometries that correspond to the reaction of interest. One of these geometries is chosen to be a new data point, and added to the data set (once the *ab initio* calculations have been performed). The trajectory sampling, and choosing and adding a new data point, is repeated again and again until the PES is converged. Convergence is made evident by observing when important dynamical observables (*e.g.*, the reaction cross section) do not change significantly as more data is added. This method has been used to obtain converged PESs for systems of up to nine atoms.

Modified Shepard interpolation is a variant of a more general procedure called the interpolating moving least squares (IMLS) method.[65,66] In this approach, the PES is written as a linear function of some unknown coefficients, a_i of basis functions, b_i (usually low order polynomials):

$$V(\boldsymbol{Z}) = \sum_{i=1}^{N_b} a_i(\boldsymbol{Z}) b_i(\boldsymbol{Z}). \tag{2.40}$$

The basis functions are usually second or third order polynomials of the internal coordinates, $\boldsymbol{Z}$. The coefficients, a_i, are optimized every time the PES is evaluated by minimizing the penalty function P:

$$P(\boldsymbol{Z}) = \sum_{i=1}^{N_a} w_n[\boldsymbol{Z}; \boldsymbol{Z}(n)]\{a_i(\boldsymbol{Z}) b_i[\boldsymbol{Z}(n)] - E_{ab\ initio}[\boldsymbol{Z}(n)]\}^2. \tag{2.41}$$

The weight $w_n[\boldsymbol{Z};\, \boldsymbol{Z}(n)]$ is large if $||\boldsymbol{Z}-\boldsymbol{Z}(n)||$ is small. The idea is to make a polynomial approximation to the PES by fitting to the *ab initio* energies of data points which are close to the configuration where the PES is evaluated. The method appears to be sufficiently accurate for classical dynamics studies of reactions, but in this primitive form it is very time consuming to evaluate, as the

*This is called symmetrization by summing over the orbit of the group (an alternative to constructing invariant polynomials).

fitting procedure is repeated every time the PES is evaluated. Gradients and second derivatives of the PES can be evaluated by analytic differentiation of Eq. (2.40). The method provides a means for estimating the optimum location of data points, by estimating the effect of varying the polynomial order on the PES. In recent times,[66] the problem of slow evaluation has been overcome by using the IMLS method to interpolate energies, gradients and second derivatives which form the data needed for a modified Shepard interpolation (which is relatively fast to evaluate in a dynamics calculation).

2.8 OUTLOOK

Over the last ten to fifteen years, there has been a resurgence in the development of methods to construct PESs for chemical reactions and to optimize the application of "on-the-fly" methods. Reliably accurate molecular PESs for reactions involving several atoms have been obtained, and the development and application of exact quantum reactive scattering is no longer restricted by the lack of suitably accurate PESs. Nonetheless, further development of PES construction methods is an active area because the computational cost of *ab initio* PESs is still very substantial. The cost of sufficiently accurate *ab initio* calculations increases very rapidly with the number of electrons in the molecule. Hence, the PES for reactions which involve even a few non-hydrogen atoms may be prohibitively expensive unless more efficient methodology can be derived. Current work is going beyond the Born-Oppenheimer approximation to multi-state PESs and "on-the-fly" dynamics.[67–69]

2.9 PROBLEMS

1. (To answer this question fully requires knowledge of some of the material covered in Chapters 1, and Chapters 5 to 7, including Study Box 7.1.)
   (a) Briefly outline experimental strategies currently available for measuring the partitioning of energy between vibration, rotational, and translation in the products of an exothermic atom transfer reaction

$$A + BC \rightarrow AB(v', j') + C$$

   (b) How may the properties for vibrational and translational energy disposal be influenced by the topography of the potential energy surface over which the reaction proceeds, and by the masses of the atoms involved? Illustrate you answer by reference to the data, determined at 300 K, for the following reactions

	$\langle f_{vib} \rangle$	$\langle f_{trans} \rangle$	Product scattering
$H + Cl_2 \rightarrow HCl(v', j') + Cl$	0.39	0.54	backward
$Cl + HI \rightarrow HCl(v', j') + I$	0.71	0.16	forward

   [$\langle f_{vib} \rangle$ and $\langle f_{trans} \rangle$ are the mean fractions of the total energy disposed into vibration and translation, respectively.]
2. In this question we use a very simple theory of bonding, known as Hückel theory, which ignores the effects of electron repulsion and electron correlation, to obtain information about the relative barrier heights of two reactions. (If you are unfamiliar with Hückel theory, you will need to read about this first in one of the recommended standard texts, for example, refs. [1,70].)

Use Hückel theory and the variation principle to construct the secular equations for the π orbitals of the allyl radical shown below.

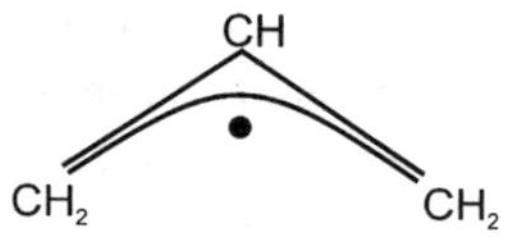

Determine the molecular orbital energies and the linear combinations of atomic orbitals associated with them. Show that the charge density arising from the π electrons is uniform at each carbon atom in the allyl radical. Using Hückel theory to estimate the appropriate barrier heights, suggest the most likely point of attack on the allyl cation, $C_3H_3^+$, by a nucleophilic reactant. You may assume that the σ-bonding framework is unaffected by the attack of the nucleophile.

3. (a) How is a *stationary point* on a PES defined. What differentiates a *saddle point* from an energy minimum.
   (b) What information do we commonly calculate from the Hessian of the PES (a generalization of the second derivative of the energy for a diatomic molecule) at an energy minimum? Can you relate this information to Newton's law for motion close to the energy minimum? What corresponding information could we calculate from the Hessian at a saddle point?
   (c) Saddle points are viewed as 'separating' and 'connecting' the reactant and product valleys on a PES. For many chemical reactions, there are many stationary points on a PES (Figure 2.4 is a simple example). Quantum chemistry programs can determine stationary points on a PES. Given that a saddle point geometry has been found, how would you determine which 'reactants' and 'products' are connected by this saddle point?

CHAPTER 3

Scattering Theory: Predicting the Outcome of Chemical Events

ANTHONY J. H. M. MEIJER[a] AND EVELYN M. GOLDFIELD[b]

[a] Department of Chemistry, University of Sheffield, UK; [b] Department of Chemistry, Wayne State University, Detroit, MI, USA

3.1 INTRODUCTION

This chapter deals with scattering theory and how it is used in practice to predict or explain the outcome of scattering experiments. Two routes are generally taken to obtain this information, classical mechanics or quantum mechanics, although a combination of these two strategies is also possible. This chapter will attempt to focus on the practicalities of running these kinds of scattering calculations rather than on providing an in-depth *ab initio* description of the theory involved. However, to set the scene and introduce notation, we have included a cursory description of the theories of classical and quantum mechanical scattering. For further details we refer to the books listed in Further Reading, but we would like particularly to point out the books by Goldstein on Classical Mechanics[71] and by Zhang on time-dependent quantum mechanics.[72] We have also found the excellent review by Nyman and Yu to be very useful.[73]

3.2 CLASSICAL MECHANICS

3.2.1 Newton's Laws and Conservation Laws

Newtonian or classical mechanics describes the motion of objects according to Newton's three Laws of Motion, as detailed in his Philosophiae Naturalis Principia Mathematica (1687).

The first law is often called the Law of inertia: if no force, $\boldsymbol{F}$, is operating on an object, it is either at standstill or moving with a constant momentum, $\boldsymbol{p}$. This leads to two additional fundamental conservation laws that apply not just to classical mechanics, but also to quantum mechanics (see Section 3.3.2.3). As an aside, these laws are not just of fundamental importance, but are often useful checks on the validity and accuracy of calculations. The first additional conservation law is that if there is no torque, defined as $\boldsymbol{\tau} = \boldsymbol{r} \times \boldsymbol{F}$, where $\boldsymbol{r}$ is the displacement vector from the axis about

Tutorials in Molecular Reaction Dynamics
Edited by Mark Brouard and Claire Vallance

Published by the Royal Society of Chemistry, www.rsc.org

which the torque is being applied, on an object, then the total angular momentum, $\boldsymbol{\ell} = \boldsymbol{r} \times \boldsymbol{p}$, of that object is conserved. For a conservative system, *i.e.* a system in which there are no dissipative forces, such as friction, there is a second additional conservation law, which states that the total energy of the object, *i.e.* the sum of kinetic and potential energy, is also conserved.*

Newton's second law deals with the application of force. A force, $\boldsymbol{F}$, applied to an object induces a change in its momentum, $\boldsymbol{p}$. This is usually expressed as

$$\boldsymbol{F} = m\boldsymbol{a} = \frac{\mathrm{d}\boldsymbol{p}}{\mathrm{d}t}. \tag{3.1}$$

Finally, Newton's third law states that a force exerted by one object on another induces a reaction force of opposite sign ("for every action (or force) there is an equal and opposite reaction"). This law comes in two forms, the weak form and the strong form. The difference between the two is that in the strong form the forces are not just of opposite sign, but also lie along the line connecting the two objects. This is the case for most situations one encounters in scattering. However, the strong form does not apply in cases where, for example, an external magnetic field is present.

3.2.2 Lagrangian & Hamiltonian Mechanics

In the 18th and 19th century much work was done to make Newtonian mechanics more stringent and put it on a more general footing. In particular, the efforts of Joseph-Louis Lagrange and William Rowan Hamilton should be noted, since their reformulations provide the theoretical frameworks that are most commonly used these days and are what is most commonly meant by the term "classical mechanics". Note that these two distinct forms of classical mechanics offer equivalent expressions from slightly different points of view.

In the Lagrangian reformulation of classical mechanics, the evolution in time of a system of particles is given by the Lagrangian $\mathcal{L}(\boldsymbol{q}, \dot{\boldsymbol{q}}, t) \equiv T - V$, where T and V are the kinetic and potential energies, respectively. This Lagrangian depends both on the positions, $\boldsymbol{q}$, of the particles and on their velocities, $\dot{\boldsymbol{q}} = \mathrm{d}\boldsymbol{q}/\mathrm{d}t$. In addition, the Lagrangian can depend explicitly on the time, t. The true trajectory this set of particles takes can be found through the solution of an equation arising from Hamilton's principle (see, for example, chapter 1 of ref. [74] and chapters 1 and 2 of ref. [71])

$$\delta \int \mathcal{L}(\boldsymbol{q}, \dot{\boldsymbol{q}}, t)\mathrm{d}t = 0, \tag{3.2}$$

which states that the variation in the action integral $\int \mathcal{L}(\boldsymbol{q}, \dot{\boldsymbol{q}}, t)\, \mathrm{d}t$ is zero for the true trajectory. This expression holds for any system in which the forces are derived from a single scalar potential function,† which may depend on the positions and velocities of the particles as well as on time (such a system is known as a *monogenic system*).[71] For the case in which the potential only depends on the positions of the particles, the system is also conservative. For almost all scattering problems, Hamilton's principle applies.

Eq. (3.2) can be simplified to Lagrange's Equations of Motion, a set of n second-order differential equations to be solved, where n is the total number of degrees of freedom.

$$\frac{\mathrm{d}}{\mathrm{d}t}\frac{\partial \mathcal{L}}{\partial \dot{q}_i} - \frac{\partial \mathcal{L}}{\partial q_i} = 0 \quad \text{with} \quad i = 1, \ldots, n \tag{3.3}$$

*In a dissipative system, the total energy is still conserved, but not the energy of the object, as some of the energy is transferred to the surroundings.

†A scalar potential has only magnitude at each point in space. This is in contrast to a vector potential, which has a magnitude and a direction at each point

For a conservative system, *i.e.* $\partial\mathcal{L}/\partial t = 0$, Eq. (3.3) can be rewritten as

$$\frac{\mathrm{d}}{\mathrm{d}t}\left(\sum_i \frac{\partial\mathcal{L}}{\partial\dot{q}_i}\dot{q}_i - \mathcal{L}\right) = 0 \tag{3.4}$$

where[71,74]

$$\frac{\partial\mathcal{L}}{\partial\dot{q}_i}\dot{q}_i - \mathcal{L} \equiv H = E \tag{3.5}$$

H is the Hamiltonian of the system, which is equal to the total energy, *i.e.* $E = T + V$. Defining

$$\frac{\partial\mathcal{L}}{\partial\dot{q}_i} = p_i \tag{3.6}$$

where p_i is the conjugate momentum of the particle, we arrive at the following definition for the Hamiltonian

$$H \equiv \sum_i p_i\dot{q}_i - \mathcal{L} \tag{3.7}$$

Applying the variation principle of classical mechanics[71] to H leads to the following set of differential equations which give the evolution of a system in time. Note that for a Cartesian coordinate system, the conjugate momentum and standard linear momentum are exactly the same. However, for general coordinate systems, conjugate momenta may not be clearly related to any common notions of linear or angular momentum.

$$\frac{\partial H}{\partial q_i} = -\dot{p}_i \tag{3.8a}$$

$$\frac{\partial H}{\partial p_i} = \dot{q}_i \tag{3.8b}$$

$$\frac{\partial H}{\partial t} = -\frac{\partial\mathcal{L}}{\partial t} \tag{3.8c}$$

These are the *canonical equations of motion*, a set of 2*n* first order differential equations. They are completely equivalent to the set of *n* second-order differential equations in the Lagrangian formalism. However, the Hamiltonian framework is more popular in molecular dynamics, because of its obvious connections with quantum mechanics. In addition, Eq. (3.7) shows the connection between coordinates and momenta, two quantities that are connected in quantum mechanics through the Heisenberg uncertainty principle.

3.2.3 Example: Scattering in a Central Potential

A depiction of the relevant coordinates for an atom-diatom collision is given in Figure 3.1. This system is used as a guide to illustrate the most important aspects of a classical trajectory calculation. The basic algorithm and setup of the calculations we outline here were first described in ref. [75], and have not really changed since then, even though the analysis of the trajectory calculations has become more sophisticated over the years. For a more complete description than can be given here, see *e.g.*, refs. [75,76].

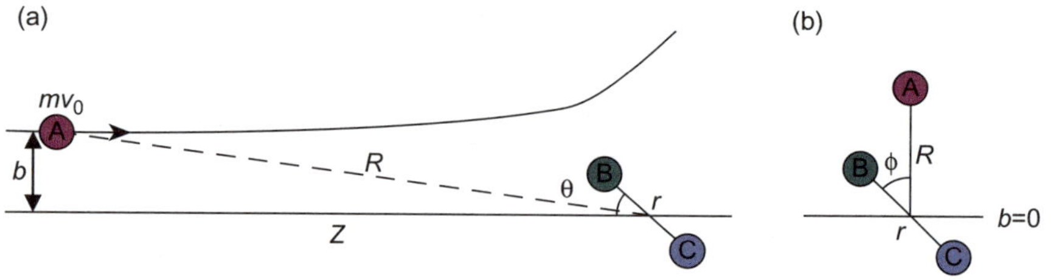

Figure 3.1 Cartoon of a classical trajectory showing the important dynamical variables. Panel (a): Side-on view. Panel (b): Head-on view. Note that the coordinates (R, r, θ) are equivalent to the Jacobi coordinates (R, r, γ). For a three atom system, R, r, and γ define the separation of A from the centre-of-mass of BC, the BC bond length, and the angle between the vectors $\boldsymbol{r}$ and $\boldsymbol{R}$, respectively, as indicated in the figure.

First of all, one needs to determine the most appropriate coordinates to use in a scattering calculation. The centre-of-mass motion of the system can be ignored, since it is a constant of the motion, unless there is an external field (see Chapter 1). For three atom systems there are nine degrees of freedom in total required to specify the positions of the particles. Three degrees of freedom are required to specify the position of the centre-of-mass. Thus, for an atom-diatom collision the total number of coordinates required to define the atomic positions (relative to the centre-of-mass) is six, with six associated conjugate momenta. There are two obvious possibilities for these coordinates. One could use the Cartesian coordinates of A with respect to the centre-of-mass of B, together with the relative Cartesian coordinates of BC.[75–77] Alternatively, one could still use the Cartesian coordinates of A with respect to the centre-of-mass of BC, but in combination with the polar coordinates (r, θ, ϕ) for the BC molecule, as shown in Figure 3.1. The latter choice makes the initial and final matching onto quantum states easier, but does have some drawbacks as well.[78,79]

Once the coordinate system has been decided on, one needs to derive the appropriate Hamiltonian for the chosen coordinate system, and subsequently the equations of motion (Eqs. 3.8), and then allow the system to evolve in time. For the atom-diatom system, the equations of motion are given in ref. [75]. For more elaborate systems, one may want to use, for example, a symmetric top Hamiltonian (see *e.g.*, refs. [80,81] and references therein).

Armed with the equations of motion, one needs to establish the initial conditions for the trajectory. For some coordinates, there is a clear experimentally accessible equivalent. In such a case, the relevant classical coordinate is set equal to the corresponding quantum coordinate. In the atom-diatom case, this means that we can set the initial conjugate momenta of the diatom such that the associated angular momentum is equal to $j(j+1)\hbar^2/r^2$, the quantum rotational angular momentum energy. Finally, the initial position of the incoming atom can be chosen to be in the XZ plane without any loss of generality, since the entire system is invariant with respect to rotation around the Z-axis.

Other parameters cannot be accessed experimentally, and are instead chosen at random from an appropriate distribution function. For example, the impact parameter, b, (see Figure 3.1) needs to be chosen at random (see Chapter 1). A value of $b=0$ corresponds to a head-on collision. Because of the axial symmetry of the system, the probability distribution function for b is actually $b/\sqrt{2\pi}$, so that higher impact parameters are more likely. The initial Z-coordinate is subsequently calculated as $\sqrt{(R_{\text{inf}}^2 - b^2)}$, where R_{inf} is the initial separation between atom and molecule, chosen to be in a region where the interaction potential is negligible. For the molecule, we have to sample the vibrational distribution function for a diatomic. The classical vibrational distribution is different from the corresponding quantum distribution. In the quantum distribution function for the ground state, one is most likely to find the molecule near the equilibrium geometry, whereas in the classical distribution function, it is most likely found near the classical turning points. As an aside, a similar problem holds for the symmetric top molecule.[79–81] Finally, the initial momentum of the incoming atom can be chosen to lie along the Z-axis. Its value will depend on the specific experiment performed. If the atom is generated photolytically from a specific precursor molecule, it will have a specific momentum determined by the photodissociation wavelength, the bond dissociation energy, and the photofragment masses (see Chapter 8). Otherwise, it may be chosen from the Maxwell-Boltzmann distribution at the temperature of the experiment.

After the initial conditions have been chosen, the system is evolved forward in time using the canonical equations of motion derived earlier. Generally, this cannot be done analytically, and a numerical integration scheme is commonly used. There is an extensive literature on this subject, which we will not discuss here. Instead, we refer the interested reader to ref. [47] for more details. The propagation of the system is continued for a length of time until some specified "stop" criterion has been satisfied. In the case of the atom-diatom system, this would generally be that either the distance R has become larger than a certain threshold, indicating reflection, or that r has crossed a threshold, indicating reaction. At this point the trajectory is stored and a new one is started using the procedure outlined above. Thousands of trajectories may be run in this way in order to build up a precise picture of the collision dynamics.

After a specified number of trajectories has been run, all of the calculated data are collated and processed. The total reaction probability may be obtained simply by dividing the number of reactive trajectories by the total number of trajectories run. In the limit of an infinite number of trajectories, this number converges to the total reaction probability.[75] One property that is of particular interest is the opacity function, introduced in Chapter 1, which describes the reaction probability as a function of the impact parameter. For final-state-resolved information a more elaborate procedure needs to be followed. Since classical mechanics does not lead to quantization, continuous values for *e.g.*, the angular momentum of the product are generated by the calculations. To circumvent this problem, some form of "binning" is usually performed. Thus, trajectories are assigned to a certain final quantum state by rounding off the corresponding classical values to match the appropriate quantum properties of the closest state. More details about this so called *quasi-classical trajectory* (QCT) method are given in Chapter 9, whilst the problem of scattering by a central potential is returned to in Chapter 12.

We make three final remarks. Firstly, in *e.g.*, ref. [75] an analytical potential energy surface was available for the $H + H_2$ reaction. However, this is not really necessary. It is perfectly possible to run trajectories and evaluate the electronic energy and its derivatives and Hessians with respect to the geometrical coordinates on the fly (see the discussion at the end of Chapter 2). This technique is called "direct dynamics" and is widely applied (see *e.g.*, refs. [82,83]). Secondly, zero-point energy, a quantum mechanical effect, is not included in classical trajectory calculations. As a result, reactive trajectories are possible that have a final vibrational energy lower than the zero-point energy. Whether these trajectories are classed as reactive or not is a matter of debate (see *e.g.*, refs. [77,84]). Thirdly, classical trajectory calculations, in principle, do not include non-adiabatic coupling, again a quantum-mechanical effect, which means that they follow only a single potential energy surface. Techniques have been developed to include this coupling into classical mechanics calculations, and we refer the interested reader to refs. [85,86] and Chapters 2 and 4 and references therein for more information.

3.3 QUANTUM SCATTERING

3.3.1 Preamble

As with the classical mechanics section above, there is simply too much material to allow us to cover all of quantum mechanics from first principles in this chapter. Instead, we are assuming that the reader has at least a working knowledge of the basics of quantum mechanics. For more in-depth information, we refer to the reading list at the end of these Tutorials, as well as to Chapter 2 of this text.

Quantum mechanics is the ultimate result of the realization that, in particular, processes involving atoms and molecules cannot be explained using classical mechanics. There are two key elements to this. The first is the wave-particle duality (due to de Broglie), which gives a relation between the momentum of a particle and its wavelength,

$$p = \frac{h}{\lambda}. \tag{3.9}$$

By inspection it becomes clear that for light particles (*i.e.* atoms and small molecules) the wavelength is of a similar order of magnitude to the size of the particle. This leads to the second key concept, the Heisenberg uncertainty principle, which for motion along x may be written

$$\Delta x \Delta p \geq \frac{\hbar}{2}. \tag{3.10}$$

This expression states that the uncertainty in the position of a particle times the uncertainty in its momentum must be larger than $\hbar/2$, *i.e.* it is impossible to know precisely both the position and the momentum of a particle to arbitrary precision. The consequence of Eqs. (3.9) and (3.10) is that the theory describing the motion of light particles cannot be deterministic, but must instead be probabilistic. In other words, we can only determine a *probability* that a certain particle is inside a certain volume of space. This probability is expressed through the wave function, ψ, whose square is a probability density.

3.3.2 Fundamental Theory

3.3.2.1 The Schrödinger Equation

The evolution of a classical system of particles is given through Eqs. (3.8). In quantum mechanics the following axiom holds:

Axiom: The time evolution of a system is governed by the time-dependent Schrödinger equation (TDSE):

$$i\hbar \frac{\partial \psi(\boldsymbol{q}, t)}{\partial t} = \hat{H}\psi(\boldsymbol{q}, t), \tag{3.11}$$

where $\psi(\boldsymbol{q}, t)$ is the wave function, which depends on the generalized coordinates of the particles involved, $\boldsymbol{q}$, and $\hat{H}$ is the Hamiltonian operator (see Section 3.3.2.2).

From the time-dependent Schrödinger equation it is straightforward to derive the more widely known time-independent Schrödinger equation, which is often used in electronic structure theory, if we are dealing with a conservative system. In this case the Hamiltonian operator is independent of time. This means we can make the following ansatz for the wave function $\psi(\boldsymbol{q}, t)$.

$$\psi(\boldsymbol{q}, t) = \mathrm{e}^{-iEt/\hbar}\Phi(\boldsymbol{q}) \tag{3.12}$$

Substituting this wave function into Eq. 3.11, the TDSE can now be written as

$$\hat{H}\Phi(\boldsymbol{q}) = E\Phi(\boldsymbol{q}), \tag{3.13}$$

which is the Time-Independent Schrödinger Equation (TISE). For a conservative system both equations are equally applicable and give similar answers. As a result, scattering problems can be solved using either equation.

In the following sections we discuss some TI and TD methods, which can be applied to scattering problems, and also provide pointers to other techniques. However, first we discuss some fundamental aspects of the TDSE and how to obtain the Hamiltonian operator for a general system.

3.3.2.2 Derivation of the Kinetic Energy Operator

All physical information about the system is contained in the Hamiltonian operator, $\hat{H}$. whose expectation value is the total energy of the system, just like the Hamiltonian in classical mechanics is equal to the total energy of the system. Similar to the classical case, the Hamiltonian operator is the sum of the kinetic energy operator and the potential energy operator. The latter is generally a multiplicative operator taken to be defined by the potential energy surface, and in chemical problems comprises the electronic energy as a function of the positions of the nuclei (see Chapter 2 for more details). In order to derive the quantum Hamiltonian from the corresponding classical expression, one needs to take the classical variables and replace them with their quantum mechanical counterparts. Thus, *e.g.*, the position variable x is replaced by the multiplicative operator, $\hat{x}$, and the momentum variable, p_x is replaced by the derivative operator, $\hat{p}_x = -i\hbar\partial/\partial x$, *etc.*

For a simple system, this transformation from a classical to a quantum framework is straightforward. For example, for a particle of mass m in a 1D harmonic potential, the classical Hamiltonian is given as $H = p^2/2m + \frac{1}{2}kx^2$. Thus, the quantum Hamiltonian can be derived to be (dropping the hats on $\hat{x}$)

$$\hat{H} = \frac{-\hbar^2}{2m}\frac{\partial^2}{\partial x^2} + \frac{1}{2}kx^2,$$

which leads to the following expression for the TDSE for a 1D harmonic oscillator:

$$i\hbar\frac{\partial\psi(t)}{\partial t} = \left[\frac{-\hbar^2}{2m}\frac{\partial^2}{\partial x^2} + \frac{1}{2}kx^2\right]\psi(t).$$

For Cartesian coordinate systems, the above can easily be generalized to higher dimensions. However, Cartesian coordinates are not generally appropriate for most scattering systems. Most often, calculations are carried out using coordinates that contain the internal degrees of freedom of one or all of the colliding partners, since this provides the easiest route to obtaining state-resolved information, which can subsequently be compared to experiment. In classical mechanics a form for the kinetic energy is usually easy to find (see *e.g.*, refs. [80,81]). However, the transformation to a quantum mechanical operator can be complex.

There are basically two different ways of obtaining a quantum expression for the kinetic energy from a corresponding classical one. The first method is due to Podolsky.[87–89] In classical mechanics, starting from the Lagrangian, one can write, for a general coordinate system, the kinetic energy as (see, for example, refs. [73,90,91])

$$T = \frac{1}{2}\sum_i^n\sum_j^n p_i \sum_k^n \frac{1}{m_k}\frac{\partial q_i}{\partial x_k}\frac{\partial q_j}{\partial x_k}p_j = \frac{1}{2}\dot{\boldsymbol{q}}^T\mathbb{G}(\boldsymbol{q})\dot{\boldsymbol{q}} = \frac{1}{2}\boldsymbol{p}^T\mathbb{G}^{-1}(\boldsymbol{q})\boldsymbol{p}, \tag{3.14}$$

where $\mathbb{G}(\mathbf{q})$ is the metric tensor,[89,91] and n the total number of degrees of freedom.* $\boldsymbol{q}$ are the generalized coordinates and $\boldsymbol{p}$ their conjugate momenta, defined through Eq. (3.6). These coordinates need to be known as a function of the Cartesian coordinates, $\boldsymbol{x}$. From this expression, the quantum mechanical kinetic energy operator can now be found to be, in the notation of ref. [91]

$$\hat{T} = -\frac{\hbar^2}{2} g^{-\frac{1}{2}} \sum_{ij}^{n} \frac{\partial}{\partial q_i} g^{\frac{1}{2}} (\mathbb{G}^{-1})_{ij} \frac{\partial}{\partial q_j}, \tag{3.15}$$

where g is the determinant of $\mathbb{G}$.[87] Note that for orthogonal coordinates, such as the Jacobi coordinates, defined in the caption to Figure 3.1 and used widely in molecular dynamics, the sum in Eq. (3.15) collapses to just its diagonal elements. See for example refs. [91–93] for an application to four-atom scattering in various coordinate systems and, for example, ref. [90] for an application to the calculation of spectra for weakly-bound van der Waals clusters.

The second method[94,95] starts off with the Laplacian in Cartesian coordinates and transforms this to general curvilinear coordinates *via* a procedure which involves repeated applications of the chain-rule (see also ref. [88] for a review for this approach). This leads to a kinetic energy operator of the form[73]

$$\hat{T} = -\frac{\hbar^2}{2} \left[\sum_{ij}^{n} \{\mathbb{G}\}_{ij}^{-1} \frac{\partial^2}{\partial q_i \partial q_j} + \sum_{ij}^{n} \frac{1}{m_i} \frac{\partial^2 q_j}{\partial x_i^2} \frac{\partial}{\partial q_j} \right]. \tag{3.16}$$

Both methods lead to equivalent expressions for the kinetic energy operator. The major difference is that for operators derived *via* Eq. (3.15) the volume element in any integration is $g^{-1/2}$, whereas in the second method this volume element is 1. If one wants to obtain a form for the kinetic energy operator *via* the first method in which the volume element is 1, one should use the following expression instead of Eq. (3.15)[87,90]

$$\hat{T} = -\frac{\hbar^2}{2} g^{-\frac{1}{4}} \sum_{ij}^{n} \frac{\partial}{\partial q_i} g^{\frac{1}{2}} \{\mathbb{G}\}_{ij}^{-1} \frac{\partial}{\partial q_j} g^{-\frac{1}{4}}. \tag{3.17}$$

3.3.2.3 *Physical Consequences of the TDSE*

The TDSE is perhaps less well-known to the reader than its time-independent counterpart, since the latter is predominantly used in electronic structure theory. However, in molecular dynamics, the time-dependent Schrödinger equation is widely used, because of its obvious attractiveness for the study of systems that evolve in time.

Before we discuss the role of either in molecular dynamics or quantum scattering we indicate a few physical consequences of using the TDSE. We first note that the time dependence of the expectation value for a general operator $\hat{A}$ is given by (see Study Box 3.1)

$$\frac{\partial \langle \hat{A} \rangle}{\partial t} = \frac{i}{\hbar} \langle \psi(\boldsymbol{q}, t) | [\hat{H}, \hat{A}] | \psi(\boldsymbol{q}, t) \rangle + \left\langle \frac{\partial \hat{A}}{\partial t} \right\rangle, \tag{3.18}$$

where $[\hat{H}, \hat{A}] = \hat{H}\hat{A} - \hat{H}\hat{A}$ is the commutator of $\hat{H}$ and $\hat{A}$ (more information on commutators is given in Study Box 3.2). It follows from this expression that $\langle \hat{A} \rangle$ is conserved (a "constant of the motion") if $[\hat{H}, \hat{A}] = 0$ and if $\hat{A}$ is time-independent.

*We draw attention to the differences in notation used in ref. [73] on the one hand and in refs. [90,91] on the other.

STUDY BOX 3.1: TIME-DEPENDENT EXPECTATION VALUE OF $\hat{A}$

To establish how the expectation value of a general operator evolves in time under the influence of the time-dependent Schrödinger Equation, we need a formal expression for $\frac{\partial\langle\hat{A}\rangle}{\partial t}$, which is relatively straightforward. The first step is to recognize that the time-dependence of $\langle\hat{A}\rangle$ is the sum of the time-dependences of each of the constituent components, that is, if $\langle\hat{A}\rangle = \langle\psi|\hat{A}|\psi\rangle$ then using the product rule,

$$\frac{\partial\langle\hat{A}\rangle}{\partial t} = \left\langle\frac{\partial}{\partial t}\psi(t)|\hat{A}|\psi(t)\right\rangle + \left\langle\psi(t)\left|\frac{\partial\hat{A}}{\partial t}\right|\psi(t)\right\rangle + \left\langle\psi(t)|\hat{A}|\frac{\partial}{\partial t}\psi(t)\right\rangle. \qquad \text{(B3.1.1)}$$

Using the expression for the time-dependent Schrödinger Equation in Eq. (3.11), this can be rewritten as

$$\frac{\partial\langle\hat{A}\rangle}{\partial t} = \left\langle\frac{-i}{\hbar}\hat{H}\psi(t)|\hat{A}|\psi(t)\right\rangle + \left\langle\frac{\partial\hat{A}}{\partial t}\right\rangle + \left\langle\psi(t)|\hat{A}|\frac{-i}{\hbar}\hat{H}\psi(t)\right\rangle. \qquad \text{(B3.1.2)}$$

Taking $i/\hbar$ outside the integration and using the fact that $\hat{H}$ is hermitian, this can be rewritten as

$$\frac{\partial\langle\hat{A}\rangle}{\partial t} = \frac{i}{\hbar}\langle\psi(t)|\hat{H}\hat{A}|\psi(t)\rangle - \frac{i}{\hbar}\langle\psi(t)|\hat{A}\hat{H}|\psi(t)\rangle + \left\langle\frac{\partial\hat{A}}{\partial t}\right\rangle, \qquad \text{(B3.1.3)}$$

which is finally trivially rewritten as

$$\frac{\partial\langle\hat{A}\rangle}{\partial t} = \frac{i}{\hbar}\langle\psi(t)|[\hat{H},\hat{A}]|\psi(t)\rangle + \left\langle\frac{\partial\hat{A}}{\partial t}\right\rangle. \qquad \text{(B3.1.4)}$$

Anthony J. H. M. Meijer and Evelyn M. Goldfield

There are two fundamental properties of the TDSE, both of which are easily obtained from Eq. (3.18). The first property we highlight is the obvious fact that the norm of the wave function (*i.e.* the total probability amplitude represented by the wave function) should be conserved; one should not create or destroy matter while propagating a wave function. This is straightforward to prove from Eq. (3.18), if one takes $\hat{A}$ to be the unit operator $\hat{1}$. This operator commutes with the Hamiltonian, *i.e.* $[\hat{H}, \hat{1}] = 0$, and is time-independent. Plugging these facts into Eq. (3.18) leads trivially to the conclusion that the norm of a wave function evolving through the TDSE is conserved. The second property we highlight is that if $\hat{A} = \hat{H}$, we find that

$$\frac{\partial\langle\hat{H}\rangle}{\partial t} = \frac{i}{\hbar}\langle\psi(\boldsymbol{q},t)|[\hat{H},\hat{H}]|\psi(\boldsymbol{q},t)\rangle + \left\langle\frac{\partial\hat{H}}{\partial t}\right\rangle, \qquad (3.19)$$

which is easily rewritten as

$$\frac{\partial E}{\partial t} = \left\langle\frac{\partial\hat{H}}{\partial t}\right\rangle, \qquad (3.20)$$

STUDY BOX 3.2: COMMUTATORS

The order in which operators appear is important. In general, $\hat{A}\hat{B} \neq \hat{B}\hat{A}$. Thus,

$$[\hat{A}, \hat{B}] \equiv \hat{A}\hat{B} - \hat{B}\hat{A} \neq 0. \tag{B3.2.1}$$

$[\hat{A}, \hat{B}]$ is called a commutator (which is an operator in its own right). An example of a commutator is:

$$[\hat{x}, \hat{p}] = i\hbar. \tag{B3.2.2}$$

The proof for Eq. (B3.2.2) is straightforward. Take an arbitrary function $f(x)$ and operate on it with the commutator $[\hat{x}, \hat{p}]$. Then,

$$\begin{aligned} [\hat{x}, \hat{p}]f(x) &= (\hat{x}\hat{p} - \hat{p}\hat{x})f(x) \\ &= x\frac{\hbar}{i}\frac{\partial}{\partial x}f(x) - \frac{\hbar}{i}\frac{\partial}{\partial x}(xf(x)) \\ &= x\frac{\hbar}{i}\frac{\partial}{\partial x}f(x) - \frac{\hbar}{i}f(x) - x\frac{\hbar}{i}\frac{\partial}{\partial x}f(x) \\ &= i\hbar f(x). \end{aligned} \tag{B3.2.3}$$

Given that $f(x)$ is arbitrary, it is clear that Eq. (B3.2.2) holds for any function and is correct.

Generally, commutators have the following properties (note that $[\hat{A}, \hat{A}] = 0$):

$$[\hat{A}, \hat{B}] = -[\hat{B}, \hat{A}] \tag{B3.2.4a}$$

$$[\hat{A}, (\hat{B} + \hat{C})] = [\hat{A}, \hat{B}] + [\hat{A}, \hat{C}] \tag{B3.2.4b}$$

$$[\hat{A}, \hat{B}\hat{C}] = [\hat{A}, \hat{B}]\hat{C} + \hat{B}[\hat{A}, \hat{C}] \tag{B3.2.4c}$$

$$[\hat{A}, [\hat{B}, \hat{C}]] + [\hat{B}, [\hat{C}, \hat{A}]] + [\hat{C}, [\hat{A}, \hat{B}]] = 0, \tag{B3.2.4d}$$

where Eq. (B3.2.4a) is the antisymmetry property, Eq. (B3.2.4b) is the linearity property, and Eq. (B3.2.4d) is Jacobi's identity. Note that if two operators $\hat{A}$ and $\hat{B}$ commute (*i.e.* $[\hat{A}, \hat{B}] = 0$), then an eigenfunction of $\hat{A}$ will also be an eigenfunction of $\hat{B}$.[70]

Classical equivalents of quantum mechanical commutators exist and are called Poisson brackets. These are defined as:[71]

$$[u, v]_{q,p} = \sum_i \frac{\partial u}{\partial q_i}\frac{\partial v}{\partial p_i} - \frac{\partial u}{\partial p_i}\frac{\partial v}{\partial q_i} \tag{B3.2.5}$$

with properties very similar to Eqs. (B3.2.4). For example, the linearity condition in this case is given as:

$$[u + v, w]_{p,q} = [u, w]_{p,q} + [v, w]_{p,q} \tag{B3.2.6}$$

For more details, see ref. [71].

Anthony J. H. M. Meijer and Evelyn M. Goldfield

given that $[\hat{H}, \hat{H}]=0$. So, for a conservative system, in which the Hamiltonian operator is time-independent, we find that the total energy of the system is conserved. This illustrates that even though we are dealing with quantum mechanics here, rather than with classical mechanics as in Section 3.2, the conservation laws posed there apply equally to both theories.

Eq. (3.18) can also be used to show an important connection between classical and quantum mechanics.[96] In particular, Eq. (3.18) can be used to show that the expectation values of the coordinate and momentum operators behave in a classical manner. We show this connection for a 1D system, but the proof is easily extended to more dimensions. For a one-dimensional classical system, a general Hamiltonian is given as

$$H = \frac{p^2}{2m} + V(x). \tag{3.21}$$

Using the canonical equations of motion Eq. (3.8), we find that

$$\frac{\partial H}{\partial x} = \frac{\partial V}{\partial x} = -\dot{p} \tag{3.22a}$$

$$\frac{\partial H}{\partial p} = \frac{p}{m} = \dot{x} \tag{3.22b}$$

$$\frac{\partial H}{\partial t} = 0. \tag{3.22c}$$

For a one-dimensional quantum mechanical system, a general Hamiltonian is given as

$$\hat{H} = \frac{1}{2m}\hat{p}^2 + V(\hat{x}). \tag{3.23}$$

Using Eq. (3.18) we obtain the following expression for the time-dependence of $\langle \hat{x} \rangle$

$$\frac{\partial \langle \hat{x} \rangle}{\partial t} = \frac{i}{\hbar}\langle \psi(x,t)|\big[\hat{H}, \hat{x}\big]|\psi(x,t)\rangle = \frac{i}{2m\hbar}\langle \psi(x,t)|\big[\hat{p}^2, \hat{x}\big]|\psi(x,t)\rangle. \tag{3.24}$$

Note that $[\hat{V}(x), \hat{x}]=0$, since the commutator of $\hat{x}$ with any function of itself is zero. So, given that

$$\big[\hat{p}^2, \hat{x}\big] = \hat{p}[\hat{p}, \hat{x}] + [\hat{p}, \hat{x}]\hat{p} \tag{3.25}$$

and

$$[\hat{p}, \hat{x}] = -i\hbar \tag{3.26}$$

we can rewrite Eq. (3.24) as

$$\frac{\partial \langle \hat{x} \rangle}{\partial t} = \frac{i}{2m\hbar}\langle \psi(x,t)|(-2i\hbar\hat{p})|\psi(x,t)\rangle = \frac{\langle \hat{p} \rangle}{m}. \tag{3.27}$$

The time-dependence of $\langle \hat{p} \rangle$ is given through Eq. (3.18) as

$$\frac{\partial \langle \hat{p} \rangle}{\partial t} = \frac{i}{\hbar}\langle \psi(x,t)|\big[\hat{H}(x), \hat{p}\big]|\psi(x,t)\rangle = \frac{i}{\hbar}\langle \psi(x,t)|[V(x), \hat{p}]|\psi(x,t)\rangle, \tag{3.28}$$

Table 3.1 Comparison of classical equations of motion with the corresponding quantum mechanical equations of motion.

Classical Mechanics	*Quantum Mechanics*
$\frac{\partial x}{\partial t} = \frac{p}{m}$	$\frac{\partial \langle \hat{x} \rangle}{\partial t} = \frac{\langle \hat{p} \rangle}{m}$
$\frac{\partial p}{\partial t} = -\frac{\partial V}{\partial x}$	$\frac{\partial \langle \hat{p} \rangle}{\partial t} = -\left\langle \frac{\partial V(x)}{\partial x} \right\rangle$

where we have used the fact that $[\hat{p}^2, \hat{p}] = 0$, since the commutator of $\hat{p}$ with any function of itself is zero. Thus, using the definition for $\hat{p}$, we can rewrite the right-hand side of Eq. (3.28) as

$$\frac{\partial \langle \hat{p} \rangle}{\partial t} = \frac{i}{\hbar} \left\{ \left\langle \psi(x,t) \middle| V(x) \frac{\hbar}{i} \frac{\partial}{\partial x} \middle| \psi(x,t) \right\rangle - \left\langle \psi(t) \middle| \frac{\hbar}{i} \frac{\partial}{\partial x} V(x) \middle| \psi(x,t) \right\rangle \right\}, \tag{3.29}$$

which upon application of the chain-rule collapses to

$$\frac{\partial \langle \hat{p} \rangle}{\partial t} = -\left\langle \psi(x,t) \middle| \frac{\partial V(x)}{\partial x} \middle| \psi(x,t) \right\rangle = -\left\langle \frac{\partial V(x)}{\partial x} \right\rangle. \tag{3.30}$$

If we now compare the classical and quantum equations of motion, as shown in Table 3.1, it is obvious that the two sets of expressions are similar. In particular, it is clear from the comparison that the expectation values for position and momentum of a quantum particle follow classical-like equations of motion, with instantaneous values for variables in the classical expressions replaced by expectation values in the quantum formulation. This observation is called the Ehrenfest Theorem, and provides an important starting point for many mixed quantum-classical methods.[85]

3.3.2.4 Free Motion

i. Free Particle States

Most chemistry students are intimately familiar with bound quantum systems, such as the "particle-in-a-box" and the harmonic oscillator. In order to study molecular dynamics, however, we must also include unbound motion, in which *at least* one of the variables is *not* constrained by a potential.

The Hamiltonian associated with this free motion is (in 1D):

$$\hat{H}_{\text{free}} = -\frac{\hbar^2}{2m} \frac{\mathrm{d}^2}{\mathrm{d}x^2}. \tag{3.31}$$

The eigenfunctions associated with this Hamiltonian are also eigenfunctions of the momentum operator, since $[\hat{H}_{\text{free}}, \hat{p}] = 0$. So, the following differential equation (in 1D) holds for the eigenvalues and eigenfunctions.

$$\hat{p} u_k(x) = \hbar k u_k(x), \tag{3.32}$$

where $k = p/\hbar$ is the wave vector associated with the eigenfunction $u_k(x)$. Eq. (3.32) leads to the following definition for $u_k(x)$

$$u_k(x) = \frac{1}{\sqrt{2\pi}} \mathrm{e}^{ikx}, \tag{3.33}$$

which we recognize as a "plane wave" or free particle state. The convention is to assume that if $k>0$ the plane wave is moving right, if $k<0$ it is moving left, and if $k=0$ it is a standing wave. $u_k(x)$ is not square integrable, since $\int_{-\infty}^{\infty} u_k^*(x)u_k(x)\mathrm{d}x$ diverges, and is therefore not finite. Also, k is a continuous index, since its value is arbitrary. In addition, $u_k(x)$ does not actually correspond to an observable system. This makes $u_k(x)$ an improper basis in the traditional sense (see Study Box 3.3). However, $u_k(x)$ can still be used as a continuous orthonormal basis for unbound motion, since we can define the wave function $\psi(x,)$ as a linear combination of $u_k(x)$ functions, with coefficients that depend on time. Since k is continuous, the "sum" is in fact replaced by an integral

$$\psi(x,t) = \int_{-\infty}^{\infty} c(k,t)u_k(x)\mathrm{d}k = \frac{1}{\sqrt{2\pi}}\int_{-\infty}^{\infty} c(k,t)\mathrm{e}^{ikx}\mathrm{d}k. \tag{3.34}$$

Thus, $\psi(x, t)$ can be defined to be the Fourier transform of $c(k, t)$. Then, $|c(k, t)|^2$ dk is the probability of finding the particle with a momentum between k and $k+\mathrm{d}k$ at a certain time t. This definition for $\psi(x, t)$ is exploited in the use of Fast Fourier Transforms in the evaluation of $\hat{H}\psi$ in TD methods (see Section 3.3.5.1).

ii. Motion of Gaussian wavepackets

Consider the following (normalized) initial wave function

$$\psi_0 = \psi(x,t=0) = \left(\frac{1}{2\pi\beta^2}\right)^{1/4} \mathrm{e}^{-(x-x_0)^2/4\beta^2}\mathrm{e}^{ik_0x}, \tag{3.35}$$

which is called a Gaussian wavepacket (GWP) (a read of Study Box 3.4 might be helpful at this point). An example of a GWP is given in Figure 3.2.

Here, $k_0=mv_0/\hbar$ is the initial average wavevector of the wavepacket, x_0 is the centre of the wavepacket, and v_0 is the initial velocity. The parameter β is proportional to the width of the wavepacket; β is related to the full width half maximum (FWHM) of the GWP by the relationship FWHM $=4\beta(\ln 2)^{1/2}$.

According to Eq. (3.34) $\psi(x, t)$ is the Fourier transform of $c(k, t)$. Thus, $c(k, t=0)$ is given by the inverse Fourier transform of $\psi(x, t=0)$

$$c(k,t=0) = \frac{1}{\sqrt{2\pi}}\int_{-\infty}^{\infty} \psi(x,t=0)\mathrm{e}^{-ikx}\mathrm{d}x. \tag{3.36}$$

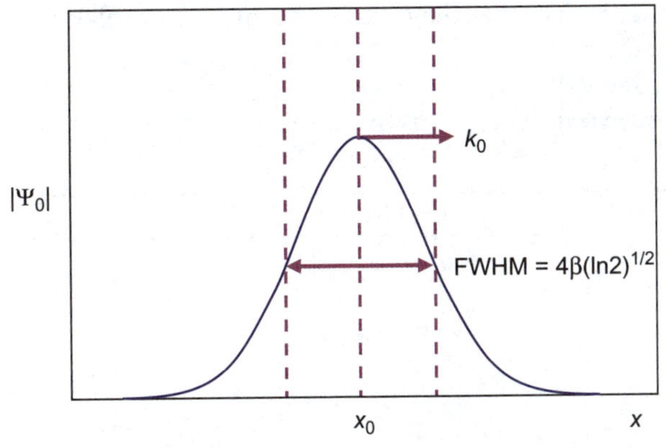

Figure 3.2 Absolute value of a Gaussian Wavepacket as a function of the position along the x-axis. (FWHM = Full Width Half Maximum.)

STUDY BOX 3.3: CONTINUOUS *VERSUS* DISCRETE BASES

Consider an orthonormal, countable set of functions $\{u_i(\mathbf{r})\}$, *i.e.*

$$\int u_i^*(\mathbf{r})u_j(\mathbf{r})\mathrm{d}\mathbf{r} = \langle u_i|u_j\rangle = \delta_{ij}, \quad \text{(B3.3.1)}$$

where δ_{ij} is the "Kronecker delta". It is defined as:

$$\delta_{ij} = \begin{cases} 1 & \text{if} \quad i=j \\ 0 & \text{if} \quad i \neq j \end{cases}. \quad \text{(B3.3.2)}$$

$\{u_i(\mathbf{r})\}$ forms a *basis* for $\psi(\mathbf{r})$ if one can write *uniquely*:

$$\psi(\mathbf{r}) = \sum_i c_i u_i(\mathbf{r}). \quad \text{(B3.3.3)}$$

c_i is defined as $\langle u_i|\psi\rangle$:

$$\langle u_i|\psi\rangle = \langle u_i|\sum_n c_n|u_n\rangle = \sum_n c_n\langle u_i|u_n\rangle = \sum_n c_n\delta_{in} = c_i. \quad \text{(B3.3.4)}$$

This means the following holds:

$$|\psi\rangle = \sum_i c_i|u_i\rangle \overset{B3.3.4}{=} \sum_i \langle u_i|\psi\rangle|u_i\rangle = \sum_i |u_i\rangle\langle u_i|\psi\rangle. \quad \text{(B3.3.5)}$$

Therefore,

$$\sum_i |u_i\rangle\langle u_i| = 1, \quad \text{(B3.3.6)}$$

which is the closure relation or the "resolution of the identity."

In the above discussion the basis functions $u_i(\mathbf{r})$ are orthonormal and the indices labelling them are discrete (*i.e.* integer). As stated in the main text, plain waves can be used as bases for unbound motion, despite the fact that they are not square integrable and despite the fact that the indices $\boldsymbol{k}$ are continuous.

Thus, we can still define orthonormality, the closure relation, and a basis set expansion in a continuous basis. The relevant expressions are given in the table below.

	Discrete Basis	*Continuous Basis*
Orthonormality	$\langle u_n\|u_m\rangle = \delta_{nm}$	$\langle u_k\|u_{k'}\rangle = \delta(k-k')$
Closure relation	$\sum_n \|u_n\rangle\langle u_n\| = 1$	$\int_{-\infty}^{\infty} \|u_k\rangle\langle u_k\|\mathrm{d}k = 1$
Expansion	$\|\psi\rangle = \sum_n c_n\|u_n\rangle$	$\|\psi\rangle = \int_{-\infty}^{\infty} c(k)\|u_k\rangle\mathrm{d}k$

A couple of remarks need to be made here:

1. The delta function $\delta(k-k')$ is an infinitely narrow function centred around k'. It is defined as follows:

$$\delta(k-k') = \begin{cases} \infty & \text{if } k=k' \\ 0 & \text{otherwise,} \end{cases} \tag{B3.3.7}$$

and has the following properties:

$$\begin{aligned} \int_{-\infty}^{\infty} \delta(k-k')\mathrm{d}k &= 1 \\ \int f(k)\delta(k-k')\mathrm{d}k &= f(k') \end{aligned} \tag{B3.3.8}$$

There are a number of ways to represent the delta function:

$$\begin{aligned} \delta(k-k') &= \lim_{\alpha\to\infty}\sqrt{\frac{\alpha}{\pi}}\mathrm{e}^{\alpha(k-k')^2} \quad \text{(Infinitely narrow Gaussian function)} \\ \delta(k-k') &= \frac{1}{2\pi}\int_{-\infty}^{\infty} \mathrm{e}^{-i(k-k')x}\mathrm{d}x \quad \text{(See above table)} \\ \delta(k-k') &= \lim_{n\to\infty}\frac{1}{\pi(k-k')}\sin n(k-k') \quad \text{(Infinitely narrow "sinc" function)} \end{aligned} \tag{B3.3.9}$$

2. Extending the last line of the table with the definition of the plane wave

$$\psi(x) = \int_{-\infty}^{\infty} c(k)u_k(x)\mathrm{d}k = \frac{1}{\sqrt{2\pi}}\int_{-\infty}^{\infty} c(k)\mathrm{e}^{ikx}\mathrm{d}k \tag{B3.3.10}$$

clearly illustrates that $\psi(x)$ is the Fourier transform of $c(k)$.

Anthony J. H. M. Meijer and Evelyn M. Goldfield

Thus, substituting for $\psi(x, t=0)$ from Eq. (3.35), we obtain

$$\begin{aligned} c(k,t=0) &= \frac{1}{\sqrt{2\pi}}\int_{-\infty}^{\infty}\left(\frac{1}{2\pi\beta^2}\right)^{1/4}\mathrm{e}^{-(x-x_0)^2/4\beta^2}\mathrm{e}^{-i(k-k_0)x}\mathrm{d}x \\ &= \left(\frac{2\beta^2}{\pi}\right)^{1/4}\mathrm{e}^{-\beta^2(k-k_0)^2}\mathrm{e}^{-i(k-k_0)x_0}. \end{aligned} \tag{3.37}$$

Eq. (3.37) shows that the Fourier transform of a Gaussian centred around $x=x_0$ in coordinate-space is another Gaussian centred around k_0 in momentum-space.

If one assumes that there is no potential, *i.e.* that the particle undergoes free motion, the time-evolution of the absolute square of the wavepacket is given as[72]

$$|\psi(x,t)|^2 = \left(\frac{1}{2\pi\beta^2(t)}\right)\exp\left[-\frac{(x-x_0-v_0t)^2}{2\beta^2(t)}\right], \tag{3.38}$$

STUDY BOX 3.4: A QUALITATIVE INTRODUCTION TO WAVEPACKETS

A classical particle has a definite position and momentum, both of which may be measured to arbitrary precision. In contrast, a quantum particle is defined by a wavefunction that fills the entire system under consideration, and knowledge of both its position and momentum is limited by the Heisenberg uncertainty principle. How do we reconcile these two wildly divergent descriptions of matter?

If we take a superposition of a number of quantum mechanical wavefunctions, the individual wavefunctions undergo constructive and destructive interference with each other. Constructive interference occurs at a certain position, and the wavefunctions cancel each other out everywhere else. The resulting probability distribution (found by taking the square of the superposition wavefunction) shows a peak in one region of space, and is zero elsewhere, and therefore describes a localized particle. Such a superposition is known as a *wavepacket*. Increasing the number of wavefunctions included in the superposition increases the localization of the particle.

Because the individual wavefunctions making up a wavepacket are time-dependent, the wavepacket is also time-dependent, and the position of the intensity peak in the probability distribution evolves with time. If the wavefunctions making up the superpositions are solutions of the Hamiltonian for a given potential, the motion of the wavepacket will closely resemble that of a classical particle in the same potential. This is a manifestation of the Ehrenfest Theorem, discussed in Section 3.3.2.3.

Experimentally, wavepackets may be prepared by exciting a molecule from its ground state using an ultra-short (usually picosecond or femtosecond) laser pulse. The frequency bandwidth of such a pulse is broad, so that instead of exciting a single quantum state, a coherent superposition of several excited states is created *i.e.* a wavepacket. The ability to prepare quantum mechanical wavepackets has opened up a whole playground of experiments in molecular quantum mechanics, in which the time-dependent rotational, vibrational, and even electronic motion of atoms and molecules may be probed. Some of this work is detailed in Chapter 11.

Claire Vallance

where

$$\beta^2(t) = \beta^2 + \frac{\hbar^2 t^2}{4m^2\beta^2}. \tag{3.39}$$

Thus, a Gaussian wavepacket travelling in a potential-free region remains Gaussian, although it slowly spreads in size, since β is directly proportional to the width of the GWP. Note that this means that the GWP actually becomes narrower with time in momentum-space, since there the width is inversely proportional to β. Note that GWPs are not only used in quantum mechanics, but also form the basis of some semi-classical mixed quantum-classical approaches (see, for example, refs. [97,98] and references therein).

3.3.2.5 *Scattering in TI and TD QM*

The above discussion was completely general. However, for the purposes of this chapter, we are mainly interested in scattering phenomena involving more complex species, than the single particle case considered above. In particular we are interested in studying molecular scattering. Although we can, of course, apply both the time-independent and time-dependent Schrödinger equation to these phenomena, this is most easily done from a TI perspective. See ref. [72] for an equivalent TD approach. Note also that Chapter 12 provides a more extensive discussion of the theory presented here.

In a scattering problem, as mentioned above, one or more coordinates are not constrained by a potential, but some other coordinates may be. For example, in an inelastic collision between an atom and a diatomic, the scattering distance R is not bound, but the vibrational distance r for the diatomic is constrained by a potential. In contrast, for a reactive collision between an atom and a diatomic, both distances will be unbound, corresponding to the two possible product channels. These channels are most naturally described by different coordinate systems, a fact that alerts us even at this early stage to some of the complexities of reactive scattering calculations.

In light of this consideration, it seems natural to divide our Hamiltonian operator into two parts, an operator $\hat{H}_0$ that deals with the parts of the problem at infinite separation, and an operator $\hat{V}$ that represents the interaction potential:

$$\hat{H} = \hat{H}_0 + \hat{V}. \tag{3.40}$$

If there is no potential, the ith eigenfunction for the system is simply a product of a plane wave, $u_{k_i}(\boldsymbol{R})$ at an energy $E = \hbar^2 k_i^2/2m$ with an eigenfunction $\chi_i(\boldsymbol{q}')$ of the internal Hamiltonian $\hat{H}_0$. The internal coordinates $\boldsymbol{q}'$ and the scattering coordinates $\boldsymbol{R}$ together form the complete set of coordinates, $\boldsymbol{q}$, with the motion of the system described by a wave vector $\boldsymbol{k}_i$.

When there is an interaction potential, we must solve the following TISE instead of the free particle equivalent.

$$(\hat{H}_0 + \hat{V})\psi(\boldsymbol{q}', \boldsymbol{R}) = E\psi(\boldsymbol{q}', \boldsymbol{R}), \tag{3.41}$$

where, because of the conservation of total energy, E is equal to the asymptotic energy. A solution of Eq. (3.41) suggests itself immediately, namely

$$\psi_f^{\pm}(\boldsymbol{q}', \boldsymbol{R}) = \chi_i(\boldsymbol{q}')u_k(\boldsymbol{R}) + \frac{1}{E - \hat{H}_0 \pm i\varepsilon}\hat{V}\psi_f^{\pm}(\boldsymbol{q}', \boldsymbol{R}). \tag{3.42}$$

By inspection, it is obvious that this solution has the right properties. If $\hat{V} = 0$, *i.e.* a free-moving system with no interaction between the particles, then $\psi_f^{\pm}(\boldsymbol{q}', \boldsymbol{R})$ is just a product of a plane wave with an asymptotic eigenfunction. Furthermore, if we apply $(E - \hat{H}_0)$ to Eq. (3.42), in the limit of ε going to zero, we obtain the TISE Eq. (3.41). Note that the term $\pm i\varepsilon$ is included in Eq. (3.42) solely to avoid problems with the singular nature of the operator $(E - \hat{H}_0)^{-1}$.

Eq. (3.42) is called the Lippmann-Schwinger Equation. This equation is the starting point for almost all time-independent scattering methods. $\psi_f^{\pm}(\boldsymbol{q}', \boldsymbol{R})$ is a *stationary scattering state*, labelled by the final state of the colliding partners. Its two forms indicated in Eq. (3.42) are each other's complement. $\psi_f^{+}(\boldsymbol{q}', \boldsymbol{R})$ describes the output for a certain incoming stationary wave, whereas $\psi_f^{-}(\boldsymbol{q}', \boldsymbol{R})$ describes the input for a certain outgoing wave. The former expression is more commonly used, since that conforms more closely to the usual experimental setup. For both stationary waves, the precise form in the region where the interaction is strong is not clear *a priori*, but asymptotically, *e.g.*, $\psi_f^{\pm}(\boldsymbol{q}', \boldsymbol{R})$ has the following form[96]

$$\psi_f^{\pm}(\boldsymbol{q}) = \left[\chi_i(\boldsymbol{q}')\mathrm{e}^{i\boldsymbol{k}_i.R} + \frac{\mathrm{e}^{ik_f R}}{R} f_{if}(\boldsymbol{k}_i, \boldsymbol{k}_f)\right], \tag{3.43}$$

whereby the *scattering amplitude*, $f_{if}(\boldsymbol{k}_i, \boldsymbol{k}_f)$, is given as

$$f_{if}(\boldsymbol{k}_i, \boldsymbol{k}_f) = -\frac{m}{2\pi\hbar^2}\langle\chi_i(\boldsymbol{q}')u_{\boldsymbol{k}_i}(\boldsymbol{R})|\hat{V}|\psi_f^+(\boldsymbol{q})\rangle. \tag{3.44}$$

Here, the integral $\langle\chi_i(\boldsymbol{q}')u_{k_i}(\boldsymbol{R})|\hat{V}|\psi_f^+(\boldsymbol{q})\rangle$ is generally referred to as an element of the transition or $\boldsymbol{T}$-matrix. The $\boldsymbol{T}$-matrix (or the $\boldsymbol{S}$-matrix, which is defined as $\boldsymbol{S}=\boldsymbol{I}-2i\pi\boldsymbol{T}$) is generally what is calculated in quantum dynamics calculations.

It is the scattering amplitude, which gives the direct connection with the scattering experiment, since the differential cross section from an initial state i with wave vector $\boldsymbol{k}_i$ to a final state f with wave vector $\boldsymbol{k}_f$, $\mathrm{d}\sigma/\mathrm{d}\omega$ (see Chapter 1), is given as

$$\frac{\mathrm{d}\sigma_{if}}{\mathrm{d}\omega} = |f_{if}(\boldsymbol{k}_i, \boldsymbol{k}_f)|^2. \tag{3.45}$$

From this we can obtain *total* state-to-state cross sections for a particular incoming wave vector $\boldsymbol{k}_i$ by integrating over all directions $\hat{\boldsymbol{k}}_f$. Assuming that all incoming directions are equally possible then allows us integrate over all $\hat{\boldsymbol{k}}_i$ as well. This gives us a final expression for the *total* (or integral) state-to-state cross section as a function of the total energy of the system. Please note that in the two integrations above, the change from $\sigma_{if}(\boldsymbol{k}_i, \boldsymbol{k}_f)$ to $\sigma_{if}(E)$ means that the volume element changes as well. As an aside, we need to note here that the expressions for $\mathrm{d}\sigma_{if}/\mathrm{d}\omega$ turn out not to be practical in computational terms. So, additional approximations and assumptions need to be made to come to expressions that are computationally tractable. In particular, we mention the partial wave expansion, which provides a way to express a plane wave as a sum of Bessel functions of the first kind. This expansion allows one to split the total $\boldsymbol{T}$ matrix into contributions resulting from different orbital angular momenta (for space-fixed coordinates) or different total angular momenta (for body-fixed coordinates). The bottom line is that specific expressions for total (or even differential cross sections) differ, depending on the assumptions made and coordinate systems used. However, they all have in common that the total state-to-state cross section is directly dependent on a sum of squares of the relevant transition matrix elements. See Chapter 12 for one particular derivation.

The form for the stationary scattering state given in Eq. (3.43) invites the following explanation. A stationary scattering state is the sum of a plane wave going through the strong interaction region and a spherical wave, which is the result of the scattering process. This is schematically expressed in Figure 3.3. The correct explanation is more complex than this simple picture; for example, we have so far ignored interference between the incoming wave and the scattered wave.[96] However, if the dimensions of the plane wave coming in are much larger than the region governed by the interaction potential and if the detector is far enough away, the above explanation is correct. See, for example, ref. [96], chapters X and XIX, for a more extensive discussion of this. Note that the above discussion, as well as that in the references quoted, implicitly assumes that there is only one channel, though different product quantum states may be populated within this channel, *i.e.* that the scattering event is inelastic. The complexity of the problem is greatly increased if there is more than one channel, as is the case in a reactive scattering problem.

The Lippmann-Schwinger Equation is exact, provided one can perform the partitioning required in reaching Eq. (3.40). However, it is difficult to solve except for some special cases, given that the scattering state appears in both the left-hand and right-hand sides of Eq. (3.42). Instead, approximations to Eq. (3.42) are generally used. These will be discussed in the next section.

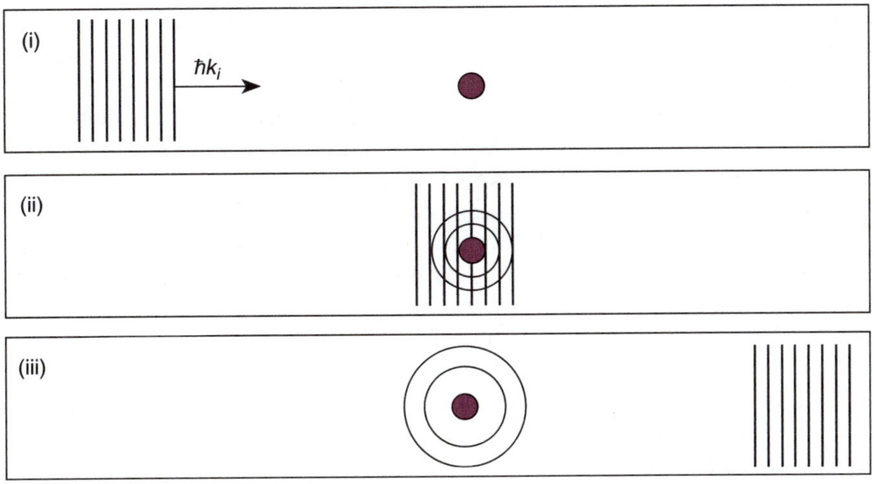

Figure 3.3 Depiction of quantum scattering event (adapted from ref. [96]). A plane wave comes in from the left (panel (i)) and strikes a spherical scatterer (red) (panel (ii)), resulting in a spherical wave being emitted from the scattering centre (panel (iii)).

Finally, we point out that the total cross section $\sigma(E)$ at a given energy E is also interesting, because it is related to the state-resolved reaction rate $k_{if}(T)$ at a given temperature T as

$$k_{if}(T) \propto \int \sigma_{if}(E) \mathrm{e}^{-E/kT} E \mathrm{d}E. \tag{3.46}$$

So, calculating the S-matrix gives a way to calculate reaction rates from first principles in principle. In practice, this will be cumbersome, so short-cuts have been proposed. See, *e.g.*, ref. [99] for one such direct calculation of $k(T)$ without first calculating the cross section.*

3.3.3 Overview of Methods

As mentioned above, the scattering, or S-matrix, is the fundamental quantity that is calculated in any scattering calculation. For a general problem, the elements of the S-matrix are the amplitudes for beginning in some initial state, i, and ending in a final state, f, at a given energy. Thus, the probability of making a transition from state i to state f is given by

$$P_{if} = |S_{if}|^2, \tag{3.47}$$

where S_{if} is an element of the S-matrix. Traditionally, the TD and TI approaches to obtaining S have been distinct and have employed very different methodologies. Recently, however, "time-independent wavepacket" methods[100–105] have been developed that have much in common with time-dependent (TD) methods. Interestingly, these methods have been developed starting from both the TDSE and the TISE.

In standard TI calculations the entire S-matrix is computed at a *particular* energy so that all of the state-to-state reaction probabilities are given in a single calculation and detailed observables such as differential cross sections are easy to obtain. In contrast, TD methods propagate a

*The discussion of the cumulative reaction probability in Chapter 7 provides some background to these alternative methods for calculating $k(T)$.

wavepacket (hence the acronym TDWP, where WP stands for wavepacket) that corresponds to a particular initial state. Thus, a single wavepacket propagation gives rise to one column of the S-matrix at a range of energies.

There are two other advantages to TI methods. TI methods are most suitable for calculations at low energies, such as threshold energies for chemical reactions or energies needed to study very cold systems. Due to finite grid size, the wavepackets used in the TD approach must be absorbed in the asymptotic region to prevent them from reaching the edge of the calculation grid. However, at very low translational energies the wavepacket has a very large de Broglie wave length. To obtain accurate results, the absorbing region must be correspondingly large, with the result that the calculation quickly becomes computationally unfeasible. In contrast, no absorbing regions are needed in a TI approach, since one matches the scattering wave function to the asymptotic wave function (see below). In addition, the wavepacket will move slowly at low energies leading to long propagation times. In a TI approach one calculates a stationary wave, where this is not an issue.

Secondly, TI methods have been formulated using coordinate systems that are natural for describing chemical reactions, such as hyperspherical coordinates.[73] These coordinates are "democratic" in that they treat all asymptotic channels equally and are very well suited to computing state-to-state reaction probabilities. However, TD methods are also used to obtain detailed state-to-state information.[106–108]

On the other hand, because the TI equations involve the entire S-matrix, a large number of coupled rovibrational states are involved in the calculations, leading to a large number of coupled equations that must be solved. In practice, this results in the need to store and manipulate large matrices and also to diagonalize a large matrix hundreds of times. This makes it difficult to scale these methods to systems that contain more than three atoms, thereby limiting their applicability to relatively small molecular systems.

TD methods or wavepacket methods have grown in popularity in recent years due to several factors. There is an intuitive appeal to an initial value problem: select initial conditions, propagate a wavepacket in time and analyze results in the asymptotic region of space. Often the wavepacket itself may give a "picture" of the system in the interaction region that explains the important chemistry. TD approaches are generally easier to understand and to implement than TI methods, which require a more profound grasp of scattering theory. Moreover, they can be more computationally efficient than TI methods and in general scale better for larger systems. Wavepacket methods are often the methods of choice for computing the total reaction probability from an initial state

$$P_i = \sum_{f=1}^{n} |S_{if}|^2, \tag{3.48}$$

where n is the number of open reactive channels at a particular energy and f refers only to reactive channels. As we shall see in Section 3.3.5.4, one can obtain the total reaction probability from a wavepacket propagation without having to compute a detailed S-matrix explicitly.

3.3.4 Time-independent Methods

In TI methods, the TISE, Eq. (3.13) is solved. At least one of the coordinates is unbound, and as a result energy is not discretized. Moreover, in contrast to bound state wavefunctions, the wavefunctions are usually of interest only in so far as they allow the calculation of the S-matrix. There are numerous TI methods, many of which require a rather deep understanding of the complexities of scattering theory. Rather than cataloguing these methods we will illustrate the main features of one of them, R-matrix theory. We will do this using the simple example of colinear nonreactive (inelastic) scattering in which a molecule, BC, initially in a vibrational state i collides with an atom,

A, and is (de)excited to a vibrational state f. We choose to illustrate with inelastic scattering to avoid the additional complexities that arise when the initial and final channels (*i.e.* the connectivity of states i and f) are different. Many of the important features of TI scattering may be illustrated with this simple example.

3.3.4.1 *Asymptotic Form of the Wave Function*

Our system can be described in terms of the two Jacobi coordinates defined in the caption to Figure 3.1: a scattering coordinate R, which is the distance between A and the centre-of-mass of BC, and the vibrational coordinate, r, which is the distance from B to C. In this coordinate system, the Hamiltonian operator for our system takes on a simple form:

$$\hat{H} = -\left(\frac{\hbar^2}{2\mu_R}\frac{\partial^2}{\partial R^2} + \frac{\hbar^2}{2\mu_r}\frac{\partial^2}{\partial r^2}\right) + V(R, r), \tag{3.49}$$

where μ_R and μ_r are the reduced masses of A-BC and BC respectively and $V(R, r)$ is the potential energy.

In the asymptotic (large R) region, the form of the wavefunction is known (note the similarity with Eq. (3.43)):

$$\psi \underset{R\to\infty}{\sim} \left(\frac{\mu_R}{\hbar k_i}\right)^{1/2}\frac{1}{2i}\left(\mathrm{e}^{-ik_iR}\chi_i(r) - \sum_f S_{if}\left(\frac{k_i}{k_f}\right)^{1/2}\mathrm{e}^{ik_fR}\chi_f(r)\right), \tag{3.50}$$

where $\hbar k = \sqrt{2\mu_R E_t'}$ and E_t' is the final translational energy. Finally, χ_i and χ_f are solutions of the vibrational Hamiltonian

$$\hat{H}_v = -\frac{\hbar^2}{2\mu_r}\frac{\partial^2}{\partial r^2} + V(R\to\infty, r). \tag{3.51}$$

Thus, if the asymptotic wavefunctions are known, the S-matrix can be extracted. Note also that following the discussion in Section 3.3.2.5 we can write the Hamiltonian as

$$H = -\frac{1}{2\mu_R}\frac{\partial^2}{\partial R^2} + H_v + V_I(R, r), \tag{3.52}$$

where $V_I(R, r) = V(R, r) - V(R\to\infty, r)$ is the interaction potential in the strong interaction region. When the scattering coordinate, R, is large, the vibrational wavefunctions are uncoupled. However, in the interaction region, the vibrational wavefunctions couple due to the interaction potential. The coupling of the vibrational states in the interaction region as a function of the scattering coordinate determines the S-matrix.

3.3.4.2 *Close Coupling Method*

In the close coupled (CC) method, we expand the scattering wavefunction $\psi(R, r)$ in a complete set of orthonormal, square integrable (L^2) basis functions in each degree of freedom except for the scattering coordinate. Thus in our example, we expand the wavefunction in terms of the asymptotic vibrational states of BC:

$$\psi(R, r) = \sum_{j=1}^{n} C_j(R)\chi_j(r). \tag{3.53}$$

If we substitute this wavefunction into the TISE, multiply on the left by $\chi_{j'}(r)$ and integrate over r, we obtain a set of coupled second-order, linear differential equations:

$$\frac{\hbar^2}{2\mu_R}\frac{\mathrm{d}^2C_{j'}(R)}{\mathrm{d}R^2} = \left(\varepsilon_{j'} - E\delta_{jj'}\right)C_{j'}(R) + \sum_j V_{j'j}(R)C_j(R), \tag{3.54}$$

where ε_j is the vibrational energy of state j, E is the total energy and the elements of the potential energy matrix are given by

$$V_{j'j}(R) = \int \chi_{j'} V_I(R,r)\chi_j \mathrm{d}r. \tag{3.55}$$

These equations can be written as a matrix-vector equation:

$$\frac{\mathrm{d}^2\boldsymbol{C}(R)}{\mathrm{d}R^2} = \frac{2\mu_R}{\hbar^2}[(\boldsymbol{\varepsilon} - E)\boldsymbol{I} + \boldsymbol{V}(R)]\boldsymbol{C} = \boldsymbol{WC}, \tag{3.56}$$

where $\boldsymbol{C}$ is a column vector of length n (where the latter is the total number of BC vibrational states). It is important to note that, due to coupling in the strong interaction region, n will generally be much larger than the number of open vibrational channels at energy E.

3.3.4.3 R-matrix Theory

Our discussion of R-matrix theory is based on the work of Light and coworkers,[109–111] who were the first to use it to solve the CC equations for chemical reactions, and on the excellent review of Nyman and Yu.[73] The basic technique, as laid out by Zvijac and Light,[110] is to divide space into an internal interaction region and an external region in which these interactions have vanished. In the internal region, we expand the wavefunction in an appropriate L^2 basis and solve the TISE by diagonalization as one would do in a bound state problem. The basis must satisfy specified boundary conditions, such that the logarithmic derivative of the wave function,

$$\frac{\mathrm{d}\ln\psi}{\mathrm{d}R} = \frac{\psi'}{\psi}, \tag{3.57}$$

is continuous at the boundary of the region (the primes indicate first derivatives with respect to R). This method is therefore similar to the log-derivative method[112–114] (see also Chapter 12, Section 12.3.3). To obtain the S-matrix, the solution in the internal region is matched to the solution in the external region. For reactive scattering problems, there are several such surfaces in which matching must occur, corresponding to the non-reactive channel and at least one product channel. For inelastic scattering, however, there is generally only one such surface.

The R-matrix, $\mathbb{R}$, relates the translational wavefunction, $\boldsymbol{C}(R)$, to its first derivative,

$$\boldsymbol{C}(R) = \mathbb{R}\boldsymbol{C}'(R) \tag{3.58}$$

at the boundaries of the regions in which the TISE has been solved. If the R-matrix is known at the boundary of the strong interaction and the asymptotic region, the S-matrix may be extracted from it.

Stechel *et al.*[109] take the approach of further dividing the interaction region into a set of M sectors, with sector 1 located near the origin ($R=0$) and sector M located in the asymptotic region of the scattering coordinate, R. Within each sector, the local R-matrix, $\boldsymbol{r}$, is determined. The global R-matrix, $\mathcal{R}$, is determined by assembling the local R-matrices.

The problem of computing local R-matrices may be simplified further by assuming that the potential energy does not vary with R across the width of the sector. At the centre of each sector, the scattering coordinate is $R = R_i$ and the sector width is h_i. One then evaluates the coupling matrix $\boldsymbol{W}$ of Eq. (3.56) at $R = R_i$ and transforms from the asymptotic vibrational basis $\chi(r)$ basis to a new, sector-dependent vibrational basis, $\varphi^{(i)}(r)$. This transformation is used to obtain a set of sector-dependent translational functions, $F_n^{(i)}(R)$, where

$$\varphi_n^{(i)} = \sum_{n'} T_{n'n}^{(i)} \chi_{n'}(r) \tag{3.59a}$$

$$\tilde{\boldsymbol{T}}^{(i)} \boldsymbol{W}(R_i) \boldsymbol{T}^{(i)} = \lambda^2(i) \tag{3.59b}$$

$$F_n^{(i)}(R) = \sum_{n'} T_{n'n}^{(i)} C_{n'}(R). \tag{3.59c}$$

The eigenvalues, $\lambda^2(i)$, are identified as the negative of the kinetic energy available for translation (see Eq. (3.54)). Channels with positive eigenvalues have negative kinetic energy and thus are locally closed channels.

Next, one assembles the translational functions in a column vector in order of increasing kinetic energy and identifies translational functions at the left and right hand boundaries of sector, i, using the following notation:

$$\boldsymbol{F}_L(i) = \boldsymbol{F}^{(i)}\left(R - \tfrac{1}{2}h_i\right) \tag{3.60a}$$

$$\boldsymbol{F}_R(i) = \boldsymbol{F}^{(i)}\left(R + \tfrac{1}{2}h_i\right). \tag{3.60b}$$

Because the transformations in Eq. (3.59) remove the potential coupling in Eq. (3.54), we may write the sector R-matrix, in the locally diagonal representation as

$$\begin{bmatrix} \boldsymbol{F}_L(i) \\ \boldsymbol{F}_R(i) \end{bmatrix} = \begin{bmatrix} \boldsymbol{r}_1^{(i)} & \boldsymbol{r}_2^{(i)} \\ \boldsymbol{r}_3^{(i)} & \boldsymbol{r}_4^{(i)} \end{bmatrix} \begin{bmatrix} -\boldsymbol{F}'_L(i) \\ \boldsymbol{F}'_R(i) \end{bmatrix}, \tag{3.61}$$

where primes again indicate first derivatives and where the elements of $\boldsymbol{r}^{(i)}$ are given by

$$\begin{aligned} \left(r_1^{(i)}\right)_{ij} = \left(r_4^{(i)}\right)_{ij} &= \delta_{ij} \begin{cases} |\lambda_j|^{-1}\coth|h_i\lambda_j|, & \lambda_j^2 > 0 \\ -|\lambda_j|^{-1}\cot|h_i\lambda_j|, & \lambda_j^2 \le 0 \end{cases} \\ \left(r_2^{(i)}\right)_{ij} = \left(r_3^{(i)}\right)_{ij} &= \delta_{ij} \begin{cases} |\lambda_j|^{-1}\operatorname{csch}|h_i\lambda_j|, & \lambda_j^2 > 0 \\ -|\lambda_j|^{-1}\csc|h_i\lambda_j|, & \lambda_j^2 \le 0. \end{cases} \end{aligned} \tag{3.62}$$

A brief derivation of Eq. (3.62) is given in Study Box 3.5.

The wavefunction and its derivative must be continuous at the sector boundaries. The translational functions, $F_n^{(i)}(R)$, are expanded in the vibrational basis appropriate to sector i. Therefore, to ensure continuity between sector i and sector $(i - 1)$, we must transfer from the $\boldsymbol{F}(i - 1)$ representation to the asymptotic representation and back to the $\boldsymbol{F}(i)$ representation:

$$\boldsymbol{F}_R(i-1) = \boldsymbol{Q}(i-1, i)\boldsymbol{F}_L(i) \tag{3.63a}$$

STUDY BOX 3.5: DERIVATION OF LOCAL R-MATRICES

The local R-matrices given in Eq. (3.62) can be derived in a straightforward manner. This derivation is taken from the discussion in Appendix B of Stechel *et al.*[109]

Within each sector the translational functions obey the following equation:

$$\frac{d^2}{dR^2} F_n^{(i)}(R) = \left(\lambda_n^{(i)}\right)^2 F_n^{(i)}(R). \tag{B3.5.1}$$

This simple second order differential equation has two linearly independent solutions

$$\begin{aligned} F_n^{(i)}(R) &= \alpha A_n^{(i)}(R) + \beta B_n^{(i)}(R) \\ F_n'^{(i)}(R) &= \alpha A_n'^{(i)}(R) + \beta B_n'^{(i)}(R), \end{aligned} \tag{B3.5.2}$$

where α and β are arbitrary constants. Since the potential is constant in R across the sector, we can write these solutions as

$$\begin{aligned} A(R) &= \cosh \lambda_m R \\ B(R) &= \sinh \lambda_m R, \end{aligned} \tag{B3.5.3}$$

where $\boldsymbol{\lambda}_{\mathrm{m}}$ may be real or pure imaginary. Note that we have dropped the sector notation. The elements of $\boldsymbol{r}_1$, $\boldsymbol{r}_2$, $\boldsymbol{r}_3$, and $\boldsymbol{r}_4$ form a 2×2 block

$$\begin{bmatrix} F_n(a) \\ F_n(b) \end{bmatrix} = \begin{bmatrix} (r_1)_{nn} & (r_2)_{nn} \\ (r_3)_{nn} & (r_4)_{nn} \end{bmatrix} \begin{bmatrix} -F_n'(a) \\ F_n'(b) \end{bmatrix}, \tag{B3.5.4}$$

where a and b refer to sector boundaries.

The R-matrix is independent of the values of α and β since the relation given in Eq. (B3.5.4) holds for all solutions of Eq. (B3.5.4). Thus we can choose two *different* and convenient boundary conditions, labeled "1" and "2":

$$\begin{aligned} F_1(a) &= 1 \quad F_2(b) = 1 \\ F_1'(a) &= 0 \quad F_2'(b) = 0, \end{aligned} \tag{B3.5.5}$$

where we have dropped the subscript n. Using Eqs. (B3.5.1)–(B3.5.5) the elements of the local R-matrix given in Eq. (3.62) are readily obtained.

Anthony J. H. M. Meijer and Evelyn M. Goldfield

$$\boldsymbol{F}_R'(i-1) = \boldsymbol{Q}(i-1,i)\boldsymbol{F}_L'(i) \tag{3.63b}$$

$$\text{with} \quad \boldsymbol{Q}(i-1,i) = \tilde{\boldsymbol{T}}^{(i-1)}\boldsymbol{T}^{(i)}. \tag{3.63c}$$

The global R-matrix, $\mathcal{R}$, is assembled recursively beginning at the origin and propagating out towards the asymptote. For inelastic scattering, $\mathcal{R}$ is built by relating functions and derivatives at the outer boundary of the last sector so that

$$\boldsymbol{F}_R(i) = \mathbb{R}^{(i)}\boldsymbol{F}_R'(i). \tag{3.64}$$

Note that the global $\mathcal{R}$-matrix has dimension $N\times N$, but that the local R-matrices are of dimension $2N\times 2N$. The recursive equations required for assembling the global R-matrix are

$$\mathbb{R}^{(i)} = \boldsymbol{r}_4^{(i)} - \boldsymbol{r}_3^{(i)} \boldsymbol{Z}^{(i)} \boldsymbol{r}_2^{(i)} \tag{3.65a}$$

$$\boldsymbol{Z}^{(i)} = \left(\boldsymbol{r}_1^{(i)} + \tilde{\boldsymbol{Q}}(i-1, i)\mathbb{R}^{(i-1)}\tilde{\boldsymbol{Q}}(i-1, i)\right)^{-1}, \tag{3.65b}$$

where $\boldsymbol{Q}(i-1, i)$ maps the diagonal basis in sector $(i-1)$ onto that in sector (i). Equations (3.59)–(3.65) were derived by Light and coworkers,[111] and are the result of imposing the continuity of the wavefunction and its derivative at the common boundary of sector (i) and $(i-1)$, as shown in Eq. (3.63).

To begin the recursion, we must have an initial R-matrix, $\mathbb{R}^{(1)}$ In many cases, the interaction potential is highly repulsive at $R=0$. In this case, an appropriate choice is

$$(\mathbb{R}^{(1)})_{jk} = \delta_{jk}|\lambda_j|^{-1}. \tag{3.66}$$

Thus, beginning with the initial $\mathbb{R}$-matrix, $\mathbb{R}^{(1)}$, the sector $\mathbb{R}$-matrices given by Eqs. (3.61) and (3.62) are assembled recursively into the global $\mathcal{R}$-matrix using Eq. (3.65).

The R-matrix is propagated until an asymptotic value of R is reached *i.e.* we are at the right hand side of sector M. We now must convert the translational functions and their derivatives back to the asymptotic basis:

$$\boldsymbol{C}_R(M) = \boldsymbol{T}^{(M)}\boldsymbol{F}_R(M) \tag{3.67a}$$

$$\boldsymbol{C}_R'(M) = \boldsymbol{T}^{(M)}\boldsymbol{F}_R'(M). \tag{3.67b}$$

Combining Eqs. (3.67) with final recursion of Eq. (3.65) gives the equations for the final global $\mathcal{R}$-matrix and the translational wavefunctions in the asymptotic χ basis:

$$\mathbb{R}^{\mathrm{f}} = \boldsymbol{T}^{(m)}\mathbb{R}^{(m)}\tilde{\boldsymbol{T}}^{(m)} \tag{3.68a}$$

$$\boldsymbol{C}_R(M) = \mathbb{R}^{\mathrm{f}}\boldsymbol{C}_R'(M). \tag{3.68b}$$

We can write the asymptotic translational functions and their derivatives in matrix form as

$$\boldsymbol{C} = \boldsymbol{D} - \boldsymbol{OS} \tag{3.69a}$$

$$\boldsymbol{C}' = \boldsymbol{D}' - \boldsymbol{O}'\boldsymbol{S}, \tag{3.69b}$$

where $\boldsymbol{D}$ is a diagonal matrix of open channel incoming asymptotic functions and $\boldsymbol{O}$ is a row vector of outgoing asymptotic functions. Eq. (3.69) is a generalization of Eq. (3.50). Combining Eqs. (3.68) and (3.69) leads to an expression for the S-matrix in terms of the R-matrix

$$\boldsymbol{S} = (\mathbb{R}^{\mathrm{f}}\boldsymbol{O}' - \boldsymbol{O})^{-1}(\mathbb{R}^{\mathrm{f}}\boldsymbol{D}' - \boldsymbol{D}). \tag{3.70}$$

3.3.5 Time-dependent Methods

There are three general steps to a time-dependent calculation describing a reactive process: (i) set up an initial wavepacket; (ii) propagate the initial wavepacket in time until the wavepacket has reached the asymptotic region; and (iii) analyze the wavepacket to determine the quantities of interest. For each of these steps methodological choices need to be made. These choices are often dictated by the particular features of the problem at hand. In this section, we present some of the more commonly used methods.

We will illustrate the discussion by focusing on reactive collisions involving three atoms,

$$A + BC \rightarrow \begin{cases} AB + C \\ AC + B \end{cases}, \tag{3.71}$$

where we no longer confine our system to be colinear. We will represent the problem in reactant-channel Jacobi coordinates (R, r, γ) (see Figure 3.1). While these coordinates are suitable for defining the initial wavepacket, they are far less suitable for describing product states. In order to compute state-resolved reaction probabilities, it is necessary to transform to product-channel Jacobi coordinates at some point in the calculation (this is generally true of both TD and TI methods). If only the overall reaction probability (or integral cross section) is desired, however, it is often the case that one can employ a single set of coordinates throughout the calculation. In unimolecular processes (such as photodissociation) involving only one product channel, one coordinate system can be used effectively to obtain final state information. We work in a body-fixed framework in which the coordinates rotate with the system, and we choose R to be the body-fixed axis. One can also choose to work in a space-fixed system. Within a body-fixed system, one can alternatively choose r to be the body-fixed z axis (or for that matter, any radial coordinate).[115–117]

The three-body Hamiltonian has a fairly simple form in body-fixed Jacobi coordinates:

$$\hat{H} = -\frac{\hbar^2}{2\mu_R}\frac{\partial^2}{\partial R^2} - \frac{\hbar^2}{2\mu_r}\frac{\partial^2}{\partial r^2} + \frac{\left(\hat{J}-\hat{j}\right)^2}{2\mu_R R^2} + \frac{\hat{j}^2}{2\mu_r r^2} + V(R, r, \gamma). \tag{3.72}$$

This Hamiltonian is an extension of the colinear Hamiltonian of Eq. (3.49), with the addition of two terms representing orbital and rotational energy respectively. The quantities $\hat{J}$ and $\hat{j}$ refer to total angular momentum and BC rotational angular momentum operators, respectively. As mentioned above, it is also possible to define a Hamiltonian in space-fixed coordinates, which uses both the orbital angular momentum, given by $\hat{\ell}\hbar = \left(\hat{J} - \hat{j}\right)\hbar$ and the rotational angular momentum, j. However, one should bear in mind, that according to the rules of quantum mechanics only two out of these three angular momenta: total, orbital and rotational, can be specified for a given wavefunction.

3.3.5.1 Representation: Grids and Basis Sets

In order to solve the TDSE, a representation for the wavepacket must be chosen. Both grid and basis set representations are used, and often a mixture of the two is employed. Basis sets are particularly useful for coordinates that do not undergo large changes during the course of the propagation. Thus, a vibrational basis is appropriate to describe the BC motion in an inelastic collision. For bonds that can break, however, it is generally advantageous to discretize the wavefunction on a grid. A basis set of associated Legendre functions is often employed to describe rotational/bending motion, but grids are used for this also. Basis sets that are eigenfunctions of some portion of the Hamiltonian are particularly useful.

As we shall see, propagating a wavepacket involves repeated calculation of the action of the Hamiltonian, $\hat{H} = \hat{T} + \hat{V}$ on the wavepacket. In grid based methods, the action of the potential on the wavefunction, $\hat{V}\psi$, simply involves a multiplication at every point on the grid. The action of the kinetic energy (KE) operator, however, involves derivatives. In Jacobi coordinates, only second derivatives are involved, but in other coordinate systems the KE operator may be more complex and involve both first and second derivatives. For efficient wavepacket calculations it is necessary to have efficient methods of computing these derivatives. We will present three such methods, FFT, DVR and DFFD.

i. Fast Fourier Transform (FFT) Method

The Fast Fourier Transform (FFT) method is based on a simple idea (see also Section 3.3.2.4).[118,119] The Fourier transform (FT) takes the wavepacket from a coordinate representation to the momentum representation. In the momentum representation, the operator $\hat{T}$ is a local operator, *i.e.* only operates on a single grid point, but $\hat{V}$ is not. Thus, $\hat{V}\psi$ is best performed in the coordinate representation, but $\hat{T}\psi$ is best performed in the momentum representation. In a calculation, this means that to obtain $\hat{H}\psi$ the wavefunction is Fourier transformed to the momentum representation, each component is multiplied by $-k^2$ (where k is the wavenumber) and the resulting function is transformed back to the coordinate representation.

On a computer, efficient discrete algorithms, referred to as Fast Fourier transforms (FFTs) are used to effect the FT (see, for example, ref. [47]). FFT methods are based on a very efficient numerical algorithm that scales as $N \log N$, where N is the total number of grid points. However, note that the computational efficiency of particular implementations on a given computer can vary widely.

The mathematics underlying the use of Fourier transforms for computing first and second derivatives can be illustrated using a one dimensional wavepacket $\psi(x)$. The necessary FTs and inverse FTs are:

$$\psi(k) = \frac{1}{\sqrt{2\pi}} \int_{-\infty}^{\infty} \psi(x) e^{-ikx} dx \tag{3.73a}$$

$$\psi(x) = \frac{1}{\sqrt{2\pi}} \int_{-\infty}^{\infty} \psi(k) e^{ikx} dx \tag{3.73b}$$

$$\psi'(x) = \frac{1}{\sqrt{2\pi}} i \int_{-\infty}^{\infty} k\psi(k) e^{ikx} dk \tag{3.73c}$$

$$\psi''(x) = \frac{1}{\sqrt{2\pi}} \int_{-\infty}^{\infty} -k^2 \psi(k) e^{ikx} dk. \tag{3.73d}$$

The Fourier method, being based on an expansion in sines and cosines

$$\exp(ikx) = \cos kx + i \sin kx \tag{3.74}$$

gives the exact derivatives for periodic functions. However, the method can still be applied to wavepackets that are not periodic as long as the wavepacket is zero at the edges of the grid. This condition is met when the potential energy in these regions is high (such as often the case at small values of R and r). The use of absorbing potentials[120] or wavepacket damping[121] to prevent the wavepacket from reaching the edge of the grid also ensures that the condition will be met at grid boundaries where the potential is not high.

ii. Discrete Variable Representation (DVR)

The DVR method (sometimes also known as the pseudospectral method) is widely used to compute the action of a kinetic energy operator in both TD and TI calculations. There are many variations of this method, and we highlight only one of them. Early work of Dickinson and Certain[122] showed the equivalence of a DVR based on orthonormal basis functions on the one hand and Gaussian quadratures on the other. This idea was further developed by Light and coworkers, who applied the method to TI approaches to chemical reactions.[123,124] Since most DVR methods are applied to each coordinate independently, we can illustrate the method using a 1-D TI wavefunction $\psi(x)$ which we expand in a complete set of orthonormal basis functions, $\{\varphi_i(x)\}$, yielding the *finite basis representation* (FBR):

$$\psi(x) = \sum_{i=1}^{N} a_i \varphi_i(x). \tag{3.75}$$

The derivation is the same for the TD case, apart from the fact that the expansion coefficients a_i will be time-dependent in that case. Inverting Eq. (3.75), a_i is obtained in terms of the following integral:

$$a_i = \int \varphi_i^*(x)\psi(x)\mathrm{d}x. \tag{3.76}$$

The DVR method is based on the following approximation. We compute the integral in Eq. (3.76) using an N-point quadrature

$$a_i = \sum_{j=1}^{N} \omega_j \varphi_i(x_j)\psi(x_j), \tag{3.77}$$

where the functions are evaluated at the DVR points, x_j, and ω_j is a weight appropriate to the given quadrature. The DVR points are defined to be the eigenvalues obtained by diagonalizing the coordinate operator in the FBR basis

$$X^{DVR} = Q^\dagger X^{FBR} Q, \tag{3.78}$$

where

$$X_{ij}^{\mathrm{FBR}} = \int \varphi_i^*(x) x \varphi_j(x)\mathrm{d}x. \tag{3.79}$$

It turns out that when the FBR consists of a set of classical orthogonal polynomials such as Hermite or Legendre polynomials, DVR points are just the quadrature points of the appropriate Gaussian quadrature.[122] Note that DVRs can be specified for non-classical polynomials as well, the sinc-DVR[125,126] and potential-optimized DVR (PO-DVR)[127] being prime examples.

The orthonormal set of DVR basis functions is given by

$$\chi_\alpha(x) = \sum_{j=1}^{N} \varphi_j(x) Q_{j\alpha}, \tag{3.80}$$

where the matrix Q has elements

$$Q_{j\alpha} = \omega_\alpha^{1/2} \varphi_j^*(x_\alpha), \tag{3.81}$$

which leads to

$$\omega_\beta^{1/2}\chi_\alpha(x_\beta) = \sum_j \omega_\beta^{1/2}\varphi_j(x_\beta)Q_{j\alpha} = \sum_j Q^*_{j\beta}Q_{j\alpha} = \delta_{\alpha\beta}. \tag{3.82}$$

Thus, the DVR functions have the property that they are zero at all of the DVR points except one.

Thus, to compute the action of the KE operator using the DVR we compute derivatives of the DVR basis functions, using derivatives of the FBR functions and the transformation matrix, **Q**

$$\frac{\partial^2\chi_\alpha}{\partial x^2} = \sum_j \left.\frac{\partial^2\varphi_j(x)}{\partial x^2}\right|_{x_\alpha} Q_{j\alpha}. \tag{3.83}$$

For TD or other iterative calculations that require repeated action of the Hamiltonian on the wavefunction, the KE matrix $\boldsymbol{T}^{\mathrm{DVR}} = \boldsymbol{Q}^\dagger \boldsymbol{T}^{\mathrm{FBR}}\boldsymbol{Q}$ is computed once and used throughout the calculation.

The potential matrix is approximated as a diagonal matrix evaluated at the DVR points, $V_{\alpha\beta} = V(x_\alpha)\delta_{\alpha\beta}$. Note that the DVR method is an approximate method. Its accuracy for a particular calculation is largely determined by the accuracy of the corresponding quadrature to evaluate the potential matrix in the FBR.

iii. Dispersion Fitted Finite Difference

Both the FFT and the DVR matrices share the quality that they are non-local. This means that the expression for the derivative at a grid point x_k involves every other grid point. For example, the KE matrix in the DVR is typically a full matrix with non-zero elements everywhere. For many applications, a more localized method is desirable, partly because such methods are more amenable to parallelization. Finite difference (FD) methods are local methods that have long been used for computing derivatives. Unfortunately, standard finite difference methods are often not accurate enough for chemical dynamics calculation. Recently, finite difference methods that are suitable for such calculations have been developed.[128–130] Here we focus on the dispersion finite difference (DFFD) method of Gray and Goldfield.[128] Another widely used local method, developed by Kouri and coworkers, is the distributed approximating function (DAF) method.[131,132]

Suppose that a wavefunction, $y(x) = \psi(x)$, is defined on an evenly spaced set of grid points on the interval $[a, b]$. Assume the function $y(x)$ is such that $y(a) = y(b) = 0$ and the N grid points do not include the boundary points a and b. Let the vector $\boldsymbol{y} = (y_1, y_2, \ldots, y_N)^{\mathrm{T}}$ denote the corresponding array of function values; a $(2n+1)$-point FD approximation to the second derivative of $y(x)$ can then be written as

$$\boldsymbol{y}'' = \boldsymbol{D}\boldsymbol{y}, \tag{3.84}$$

where $\boldsymbol{D}$ is the banded matrix

$$D = \frac{1}{\Delta^2}\begin{pmatrix} d_0 & d_1 & \cdots & d_n & & & & \\ d_1 & d_0 & d_1 & \cdots & d_n & & & \\ \cdots & d_1 & d_0 & d_1 & \cdots & d_n & & \\ d_n & \cdots & d_1 & d_0 & d_1 & \cdots & d_n & \\ & d_n & \cdots & d_1 & d_0 & d_1 & \cdots & d_n \\ & & d_n & \cdots & d_1 & d_0 & d_1 & \cdots \\ & & & d_n & \cdots & d_1 & d_0 & d_1 \\ & & & & d_n & \cdots & d_1 & d_0 \end{pmatrix}, \tag{3.85}$$

and Δ is the grid spacing. For example, the well-known 3 point Lagrangian FD has $d_0 = -2$ and $d_1 = 1$.

One can show that when $N \gg n$ the eigenvalues of $\boldsymbol{D}$ are given by

$$g(k_j) = \frac{1}{\Delta^2}\left[d_0 + 2\sum_{s=1}^{n} d_s \cos(sk_j\Delta)\right], \tag{3.86}$$

where the wave number $k_j = j\pi/[(N+1)\Delta]$, $j = 1, 2 \ldots, N$. Eq. (3.86) is termed a dispersion relation. Since $p_j = \hbar k_j$ corresponds to a momentum, the quantum view of Eq. (3.86) is that it yields the momentum representation of $\boldsymbol{D}$. However, from basic Fourier transformation theory it can be shown that, in the continuous limit, the wave number representation of $\mathrm{d}^2/\mathrm{d}x^2$ is simply $-k^2$ (see Eq. (3.73d)). This suggests that, when the second derivative is well approximated, the dispersion relation really should be

$$g_{ex}(k_j) = -k_j^2. \tag{3.87}$$

Generally, the error in an FD approximation (or rather, in its dispersion relation, Eq. (3.86)) is very small when k_j is small ($j = 1$) and increases dramatically and monotonically as j increases. In the DFFD approach, the matrix $\boldsymbol{D}$ is computed by fitting Eq. (3.86) to Eq. (3.87) over some range of k_j values that leads to a maximum absolute error less than or equal to some pre-specified value, ε. Instead of requiring that the fit be perfect in the limit of very small k_j, which is the standard Lagrange FD requirement, it is required that the error be no greater than ε across the entire range of k_j up to some $k_{\max}$ determined by ε, where ε is in the range of 10^{-3}–10^{-6}. While the error in the lower k_j limit is not as small as that in the standard Lagrangian approximation, the error for higher k_j is much improved. Since any given problem involves a range of k_j values, superior results can be obtained for comparable grid spacings. The DFFD methods are labeled as $\mathrm{DFFD}_{2n+1}^{-\log(\varepsilon)}$. The elements of $\boldsymbol{D}$ for many of them are tabulated in ref. [128].

3.3.5.2 *Initial Wavepacket*

The specific form of the initial wavepacket is determined by the type of problem that is being solved. For a triatomic reaction, for example, the initial wavepacket may be a product of a Gaussian wavefunction for the scattering coordinate and rovibrational wavefunctions for the diatomic. For reactions involving larger molecules, or for unimolecular reactions, wavefunctions that describe the initial states of the reactants may have to be computed. The initial wavepacket must be expressed in the coordinate system, expressed on the grid and/or expanded in the basis functions that have been chosen for the calculation.

3.3.5.3 *Propagators*

In solving the TDSE, the initial wavepacket is propagated forward in time. There are several propagators that are used for this purpose and we will describe three of them: the split operator, the Chebyshev and the symplectic methods. A most useful discussion of several propagators can be found in ref. [133]. Most propagation methods are based on the exponential form of the TDSE which involves the action of the time evolution operator $\exp(-i\hat{H}\Delta t/\hbar)$ on the wavepacket:

$$\psi(\boldsymbol{q}, t + \Delta t) = \exp\left(-\frac{i\hat{H}(\boldsymbol{q})\Delta t}{\hbar}\right)\psi(\boldsymbol{q}, t), \tag{3.88}$$

which takes the wavepacket from a time t, to a time $t + \Delta t$.

i. Split Operator Method

In the split operator (SOP) method,[134] we rewrite the time evolution operator

$$\exp\left(-\frac{i\hat{H}\Delta t}{\hbar}\right) = \exp\left(-\frac{i(\hat{T}+\hat{V})\Delta t}{\hbar}\right)$$
$$\approx \exp\left(-\frac{i\hat{T}\Delta t}{2\hbar}\right)\exp\left(-\frac{i\hat{V}\Delta t}{\hbar}\right)\exp\left(-\frac{i\hat{T}\Delta t}{2\hbar}\right) + O(\Delta t^3), \tag{3.89}$$

where the approximation in the third term of Eq. (3.89) arises from the fact that we have made the approximation that the operators $\hat{T}$ and $\hat{V}$ commute. The SOP method will be accurate if the time step Δt is small, $\Delta t \ll \pi/V_{max}$, where V_{max} is the maximum in the potential.[134] The method is unitary and stable.

Propagation of one time step Δt occurs as follows:[134,135]

1. The wavepacket is Fourier transformed to momentum space and multiplied by $\exp(-ik^2\Delta t\hbar/4\mu)$.
2. The wavepacket is transformed back to coordinate space and multiplied by $\exp(-iV\Delta t/\hbar)$.
3. The calculations in step 1 are repeated and the wavepacket is Fourier transformed back to coordinate space.

In practice, except for the first and last time steps, the action of the KE operator in step 3 may be combined with the action of the KE operator for the next time step so that only one forward and one backward Fourier transform are required for each time step. There are several variations of the SOP operator, which include switching the roles of $\hat{T}$ and $\hat{V}$ in Eq. (3.89) as well as higher order methods.[72]

ii. Chebyshev Polynomial Method

In contrast to the SOP method, the Chebyshev method is a global method, which means that it permits one to use very large time steps. It is one of the most accurate methods of propagating wavepackets. The time-evolution operator is expanded in a set of Chebyshev polynomials:[135]

$$\exp\left(-\frac{i\hat{H}\Delta t}{\hbar}\right) \approx \sum_{n=0}^{N} a_n \Phi_n\left(-\frac{i\hat{H}\Delta t}{\hbar}\right), \tag{3.90}$$

where Φ_n is the nth complex Chebyshev polynomial. The advantage of Chebyshev polynomials over other possible polynomial expansions is that the maximum error in the approximation is minimal compared to almost all other possible polynomial expansions. Note that the argument to the Chebyshev polynomial is an operator rather than a simple scalar function.

The complex Chebyshev polynomials are defined over the range from $-i$ to i. Thus, one must examine the domain of the operator and scale it to fall within this range. This is accomplished by shifting the Hamiltonian so that its eigenvalues fall into the range from -1 to 1:[135]

$$\hat{H}_{norm} = 2\frac{\hat{H} - \hat{I}\left(\frac{1}{2}\Delta E + E_{min}\right)}{\Delta E}, \tag{3.91}$$

where $\hat{I}$ is the the identity operator, and $\Delta E = E_{max} - E_{min}$, with E_{max} and E_{min} the upper and lower bounds of the eigenvalues of the Hamiltonian operator. These can often be estimated from the potential and the grid, since the discrete representation of the Hamiltonian operator restricts the energy range of the problem.[133] In practice, it is important to estimate E_{max} and E_{min} accurately. If the range is too small, the propagation will be unstable, whereas a large number of polynomials will be required for convergence if the range is too large.

The time-evolution operator is approximated by

$$\exp\left(-\frac{i\hat{H}\Delta t}{\hbar}\right)\approx\exp\left(-\frac{i\bar{E}\Delta t}{\hbar}\right)\sum_{n=0}^{N}a_n\left(\frac{\Delta E\Delta t}{\hbar}\right)\Phi_n\left(-i\hat{H}_{\text{norm}}\right), \tag{3.92}$$

where all of the time dependence is contained in the expansion coefficients, which are given by Bessel functions of the first kind: $a_n=2J_n$, $n>0$ and $a_0=J_0$. The number of expansion terms needed for convergence is determined by the argument to the Bessel functions $\alpha=\Delta E\cdot\Delta t/\hbar$. When $n>\alpha$, the Bessel functions $J_n(\alpha)$ decay exponentially.

The propagation is accomplished by using the Chebyshev recursion relation

$$\varphi_{n+1}=-2i\hat{H}_{\text{norm}}\varphi_n+\varphi_{n-1}, \tag{3.93}$$

with $\varphi_n=\Phi_n\left(-i\hat{H}_{\text{norm}}\right)\psi(t)$, $\varphi_0=\psi(t)$ and $\varphi_1=\hat{H}_{\text{norm}}\psi(t)$. If enough terms are contained in the expansion, the time step can be made arbitrarily large so that only one time step is needed for the entire calculation. In this case, Δt is the final time and $\psi(\mathrm{t})=\psi(t=0)$.

Several of the time independent wavepacket methods[101–105] are also based on Chebyshev expansions.

iii. Symplectic Propagators

Somewhat more recent methods are the symplectic integrators (SI) of Gray and Manolopoulos[136] developed explicitly for wavepacket propagation. These methods are useful because of their extreme simplicity and minimal storage requirements. Like the SOP, these are local methods that require short time steps, but global versions can easily be derived.[136]

The discrete form of the TDSE can be written as (with $\hbar=1$)

$$i\frac{\mathrm{d}\boldsymbol{c}(t)}{\mathrm{d}t}=\boldsymbol{H}\cdot\boldsymbol{c}(t), \tag{3.94}$$

where $\boldsymbol{c}(t)$ is a column vector with N complex components and $\boldsymbol{H}$ is an $N\times N$ Hermitian matrix. If we write $\boldsymbol{c}(t)$ in terms of its real and complex parts

$$\boldsymbol{c}(t)=\frac{1}{\sqrt{2}}(\boldsymbol{q}(t)+i\boldsymbol{p}(t)), \tag{3.95}$$

where $\boldsymbol{q}(t)$ and $\boldsymbol{p}(t)$ are real column vectors, then we can define the quadratic Hamiltonian function as

$$h(\boldsymbol{q},\boldsymbol{p})=\frac{1}{2}\sum_{ij}H_{ij}\left(p_ip_j+q_iq_j\right). \tag{3.96}$$

We arrive at the symplectic form of the TDSE by substituting Eq. (3.95) into Eq. (3.94) and equating the real and imaginary parts:

$$\frac{\mathrm{d}\boldsymbol{q}(t)}{\mathrm{d}t}=\frac{\partial}{\partial\boldsymbol{p}}h(\boldsymbol{q},\boldsymbol{p})=H\cdot\boldsymbol{p}, \tag{3.97a}$$

$$\frac{\mathrm{d}\boldsymbol{p}(t)}{\mathrm{d}t} = -\frac{\partial}{\partial \boldsymbol{q}} h(\boldsymbol{q},\boldsymbol{p}) = -H \cdot \boldsymbol{q}. \tag{3.97b}$$

These equations have the form of Hamiltonian's classical equations of motion (see Section 3.2.2).

In the SI method, we propagate the two real vectors $\boldsymbol{p}$ and $\boldsymbol{q}$. An explicit SI algorithm is given by

$$\boldsymbol{p}_j = \boldsymbol{p}_{j-1} - b_j \Delta t \boldsymbol{H} \boldsymbol{q}_{j-1}, \tag{3.98a}$$

$$\boldsymbol{q}_j = \boldsymbol{q}_{j-1} + a_j \Delta t \boldsymbol{H} \boldsymbol{p}_{j-1}, \quad j = 1, 2, 3, \ldots, m. \tag{3.98b}$$

After m iterations, $\boldsymbol{q}_m$ and $\boldsymbol{p}_m$ are the SI approximations to $\boldsymbol{q}(t+\Delta t)$ and $\boldsymbol{p}(t+\Delta t)$. The coefficients a_j and b_j are chosen to ensure that the integration is accurate to $O(t^n)$. Values of the coefficients for various SI's labeled according to m and n are tabulated in ref. [136].

Eq. (3.98a) is actually programmed as $\boldsymbol{p} \leftarrow \boldsymbol{p} - b_j \Delta t \boldsymbol{H} \boldsymbol{q}$ followed by $\boldsymbol{q} \leftarrow \boldsymbol{q} - a_j \Delta t \boldsymbol{H} \boldsymbol{p}$ for $j = 1, 2, \ldots, m$, which is a sequence of updates that does not require saving any previous information. Thus, the storage requirements for this method are minimal. These methods are most stable if the average energy $\langle \psi | \hat{H} | \psi \rangle$ is close to zero. It is not difficult to adjust the Hamiltonian to ensure that this is the case.

3.3.5.4 *Final State Analysis*

At the end of a time-dependent scattering calculation, one needs to extract the information needed to generate the S-matrix. For this, we need the stationary scattering state, which is the Fourier transform of the time-dependent wave function, *i.e.*

$$\psi^{+}(E) = \frac{1}{2\pi\hbar a(E)} \int_{-\infty}^{\infty} \mathrm{e}^{iEt/\hbar} \psi(\mathrm{t}) \mathrm{d}t, \tag{3.99}$$

where $a(E)$ is equal to the portion of the original wavepacket at that particular energy. As a result, all TD methods that calculate S-matrix elements involve such Fourier Transforms. Two (related) methods will be mentioned here, both used in the context of reactive scattering. The first method calculates S-matrix elements via[137]

$$S_{fi} = \frac{i}{a_{\alpha i}(E)} \sqrt{\frac{k_f}{2\pi\mu_\beta}} \mathrm{e}^{-ik_f R_\beta^\infty} \int_0^\infty A_{fi}^{+}(R_\beta^\infty, t) \mathrm{e}^{iEt/\hbar} \mathrm{d}t, \tag{3.100}$$

where the entrance channel has been labelled α and the exit channel β. The symbols have their usual meaning, where A_{fi}^{+} is the projection of the outgoing wavepacket onto the final states, evaluated at R_β^∞

The second method, which can be shown to give the same result as the first method,[138] concentrates on probabilities rather than on S-matrix elements. It evaluates the flux through an analysis surface at R_β^{s} in the exit channel β *via*

$$P_r(E) = \frac{\hbar}{\mu_\beta} \mathrm{Im} \langle \psi^{+}(E) \Big| \delta\left(R - R_\beta^{s}\right) \frac{\partial}{\partial R} \Big| \psi^{+}(E) \rangle. \tag{3.101}$$

In order to calculate this reaction probability, both the (time-dependent) wave function at this surface and its derivative are calculated and stored. Then, after the calculation has finished, these time-dependent quantities are Fourier-transformed and multiplied to give the energy dependent reaction probability.[139]

3.3.5.5 *Time-dependent Treatment of Photodissociation*

The above discussion mainly focussed on the time-dependent treatment of reactive scattering, *i.e.* reactive collisions. Photodissociation of molecules may be regarded as reactive scattering as well, only we are dealing with a "half-collision" rather than a full collision. Similar methods apply to both problems, and therefore we do not intend to discuss photodissociation further, apart from two comments.

The initial state in photodissociation is determined by the bound state of the molecule before excitation. In general, it will be equal to that bound state multiplied by the transition dipole moment, which is dependent on the length of the bond which is being broken.

The analysis for photodissociation processes is more straightforward than in the case of reactive scattering. In this case the state-resolved reaction cross section is given as the Fourier transform of the auto-correlation function, which is the overlap of the time-dependent wavepacket with the initial wavepacket. For further information about calculations on molecular photodissociation the reader is referred to ref. [140]. A general introduction to photodissociation dynamics is presented in Chapter 8.

3.3.6 Approximation Methods

So far, we have only discussed exact methods for solving the TISE and TDSE, respectively. These methods have been very successful in treating three and four-atom systems. However, in general these methods scale exponentially with the number of dimensions introduced. This means that adding one degree of freedom, in general, makes the calculation one-to-two orders of magnitude more expensive, and every atom adds three degrees of freedom. Therefore, it is not surprising that much effort has been spent on more approximate methods, which can be used to treat larger systems. An adequate discussion of these approximate methods is outside the scope of this chapter, but we will highlight a few important methods here and refer to other textbooks or the literature for more information.

The first set of approximations involve total angular momentum. As mentioned above in Section 3.3.5, calculations are often performed in a body-fixed coordinate system, since this allows us to treat each value of total angular momentum J separately. However, for values of $J>0$ there are $2J+1$ substates to be taken into account. These states are usually labelled by the quantum numbers K or Ω, which are the projections of J and j, the total and rotational angular momentum quantum numbers, on the body-fixed z-axis. The problem is that these substates are coupled through Coriolis coupling. As a result, calculations for $J>0$ are $2J+1$ times larger than calculations for $J=0$. This is sometimes called the angular momentum catastrophe. This catastrophe can be avoided if one uses the power of parallel computers to make $J>0$ calculations tractable.[141–144] In this case no approximations are actually made. In addition, a number of approximate methods have been developed to alleviate this catastrophe.

The first method is to ignore Coriolis coupling and conserve the $\Omega(K)$ quantum number. This is known as the helicity-conserving approximation,[145,146] which is similar in spirit to the coupled-states (CS) approximation,[147,148] and the centrifugal sudden approximation.[149–151] These approximations all decouple various angular momentum terms, their exact form depending on the specific representation chosen for the Hamiltonian. A much more drastic simplification is the Infinite Order Sudden (IOS) approximation.[152,153] In the IOS approximation, not only do we ignore the Coriolis coupling terms, but we also assume that the colliding molecules do not rotate during the collision. Because the orientation angle is assumed to be fixed, the Schrödinger equation can be solved for one orientation at a time. The approximation assumes that the collision time is much faster than a rotational period and is most appropriate for high energy collision processes or collisions involving heavy molecules.[154–156] Finally, another drastic solution is to ignore $J>0$ states altogether and obtain the needed information for the cross sections and reaction rates from the $J=0$ calculation. The J-shifting method is an example of such a method.[157]

A different approach to TD calculations is given by the multi-configurational time-dependent Hartree (MCTDH) method. In this method, the wave function is expanded in a product basis. In addition to the expansion coefficients for the basis functions being time-dependent, the basis functions themselves are as well. The advantage of this is that many fewer basis functions are needed per degree of freedom, so that this method scales much better than the TD calculations outlined above and can be used to tackle larger systems. However, the consequence of this approach is that each degree of freedom is no longer completely coupled to all other degrees of freedom. Instead, each degree of freedom moves in the average field generated by all other degrees of freedom. Essentially, this approach turns the full CI of TDWP methods into a multi-configurational self consistent field (MCSCF) approach. In the limit of enough basis functions and configurations included, MCTDH will be equivalent to standard TD wavepacket approaches.[158,159]

The approximations described so far take all geometric degrees of freedom into account. However, for some reactions, it is clear *a priori* that some degrees of freedom are not really playing a part in the collision, and could be viewed essentially as "spectator-bonds". The obvious approximation to make here is to ignore those bonds altogether. At the next level of sophistication are approaches such as the rotating bond approximation or the semirigid vibrating rotor target model, in which some effect of the "spectator bonds" is still included in the calculation, even though they are not included completely dynamically.[73,127,160]

Finally, the most severe approximations that we will discuss here are classical path methods. These methods form a logical extension of MCTDH type methods in that some degrees of freedom are now treated classically. The classical degrees of freedom move in the average potential of the quantum degrees of freedom and vice versa. Initially, these approximations generally employed the Ehrenfest theorem to obtain the coupling between the two sets of degrees of freedom, but more sophisticated approaches are becoming available (see, *e.g.*, refs. [161–163] for a few examples).

3.3.7 Case study: $H+O_2$

The title reaction is a key reaction in combustion chemistry. It is responsible for chain branching in the oxidation of hydrogen and is the dominant molecular oxygen consuming step in both hydrogen-oxygen and methane-oxygen combustion processes.[164] The reverse reaction is also important in atmospheric chemistry including in ozone destruction. It is an interesting system for quantum scattering for a number of reasons, which will be explained below. As a consequence, several studies concerning the rate coefficients,[165–169] reaction dynamics[139,141,144,170–186] and the reaction intermediate[187–195] are available in the literature. Moreover, there have been numerous experimental studies as well.[170,175,196–212] Note that these references are not exhaustive, and more may be found in, *e.g.*, refs. [141,171,213,214]. Given the large amount of literature on this subject, we will focus on the total cross sections for this reaction and leave it to the reader to explore other aspects in the above-cited literature.

As noted above, there are a number of reasons why the $H+O_2$ reaction has received so much attention. Firstly, there is a wealth of experimental data to compare with the calculations. Both cross sections and reaction rates have been reported in the literature. Secondly, the $H+O_2$ reaction is a tough challenge for quantum scattering calculations for a number of reasons. The reaction contains 2 non-hydrogen ("heavy") atoms, which means that in any grid based method, small grid spacings are needed to converge the calculations. Also, the potential energy surface has a deep well of 2.7 eV, which supports many bound states and resonances. This means that in order to resolve these resonances, long propagation times are needed. An additional complication is that the

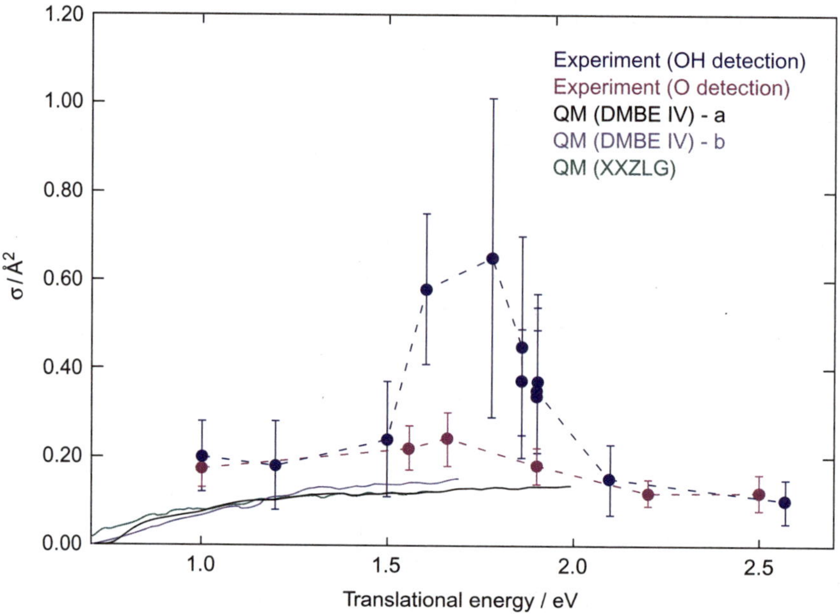

Figure 3.4 Integral cross sections for the $H + O_2$ reaction. The experimental data is obtained through either detection of the OH radical[196–200] or the O atom.[201–205] Lines between experimental points are just there to guide the eye. All theoretical results were obtained through TDWP calculations on the DMBE IV surface ((a)[184] and (b)[178]) and the XXZLG surface.[144]

intermediate metastable HO_2 radical is very floppy, meaning that many angular momentum states will be involved in the reaction. Finally, the potential energy in the O + OH exit channel has an R^{-4} dependence, due to the long range van der Waals forces between the fragments, again meaning large grids and long propagation times. When combined, these factors make the $H + O_2$ reaction a good system on which to test new ideas and new algorithms.

As is clear from Figure 3.4, there are two sets of experimental cross sections, which agree reasonably well for most energies. However, there is one major discrepancy, in that the set of data which measured the reactivity and cross sections *via* monitoring the OH radical product has a large peak at a collision energy of 1.6 eV. This observation in itself became of considerable interest.[175,211] However, later experiments in the group of Wolfrum,[205] which measured the reactivity through observation of the O atom product, concluded that this peak was the result of an experimental artifact.

There are several calculated potential energy surfaces available for this system.[165,168,215–217] The most widely used is the DMBE IV surface[215] due to Pastrana and coworkers, which was published almost 20 years ago. Troe and coworkers designed a global surface[165] on the basis of *ab initio* calculations along the minimum energy path of the $HO_2 \rightarrow H + O_2$ and $HO_2 \rightarrow HO + O$ dissociations.[218] The most recent surface, named XXZLG, by Xu *et al.*[168] has been developed by a cubic spline interpolation of more than 15000 *ab initio* points. Additionally, there is also the early Melius-Blint surface,[217] and the diatomics-in-molecules surface of Kendrick and Pack.[216]

Pack *et al.*[171] were the first to report state-to-state quantum mechanical reaction probabilities for the DMBE IV surface for $J = 0$. They report reaction probabilities for several excited initial rotational states summed over all final states. They also report reaction probabilities from the ground initial state to different product vibrational and rotational states using the CC method (Section 3.3.4.2) with hyperspherical coordinates. Their results were confirmed in 1994 by

Zhang *et al.*[139] and in 1996 by Dai *et al.*[174] who used a time-dependent wavepacket method using the split operator approach to calculate reaction probabilities. It is worth noting that these early results have stood the test of time and are still used as a benchmark for new calculations. However, these calculations were all performed for $J=0$. The importance of angular momentum for this reaction was made clear by Meijer and Goldfield[141,178] for the DMBE IV surface and later confirmed by Lin *et al.*[144] for the XXZLG surface, again using TD wavepacket approaches. Meijer and Goldfield showed that it is $J=15-20$ which contribute most to the final cross section for the DMBE IV surface. For both surfaces both groups also showed that dynamical approximations to deal with $J>0$ are suspect for this particular reaction.[142,144,177]

The XXZLG and DMBE IV surfaces are distinct and yield significantly different probabilities for this reaction. However, on the other hand, the cross sections are very similar. Comparing the theoretical with the experimental results in Figure 3.4 shows that for both surfaces the agreement with experiment is still a long way off. The reasons for this are unclear at present, but two possibilities appear to be major candidates. The first possible reason is that the calculations generally are restricted to just using one initial state, the $v=0, j=1$ state of O_2, which is the rovibrational ground state. The experiments are done at room temperature at which $v=0, j=11$ is the most populated rovibrational state. Some attempts have been made at elucidating the role of molecular rotational angular momentum for this reaction, but no firm conclusions can be drawn yet.[144,174,186,218] The second possibility is that non-adiabatic effects, discussed in detail in Chapter 4, play a role in this collision. There is a conical intersection (see Section 4.2.5) at $E_c=1.6\,\text{eV}$, the effect of which may have been detected experimentally.[175] Additionally, it is possible that interactions between states of different symmetry could pay some role. The $\tilde{X}^2A''$ and $1^2A'$ surfaces are members of a Renner-Teller pair at colinear geometries.* The Renner-Teller intersection occurs at $\approx 0.5\,\text{eV}$ above the energy of the reactants. The $1^2A'$ surface is attractive and correlates to the same product asymptote as the $\tilde{X}^2A''$ surface, although it correlates to excited reactants: $H+O_2\,(^1\Delta_g)$ (see ref. [216]). These reactants are 0.982 eV higher in energy than the ground state reactants.[216] Reaction on the $1A'$ state is slightly exoergic. Therefore, one might expect that it would facilitate reaction if it plays a role *via* the non-adiabatic Renner-Teller interaction. However, if the Renner-Teller interaction plays any role at all, it is not likely to involve direct processes, but rather reactions that proceed *via* HO_2 collision complexes, given that considerable energy must be transferred from the reaction coordinate to the bending modes to reach this linear configuration. Note that the experimental data does not really exclude either explanation for the discrepancy between theory and experiment. It is therefore clear that more data (both experimental and theoretical) is needed to explain this reaction.

Finally, it is interesting to note that such a simple reaction is currently still so poorly understood at a fundamental level and shows such a high level of complexity. Some of this complexity will have to be addressed in the future including the role of rovibrationally excited initial states and non-adiabatic effects. It is hoped that developments in theory as well as in computer algorithms and hardware may make it possible to address them in the future.

3.4 OUTLOOK

Theoretical and computational studies of molecular collisions and, in particular, of reactive molecular collisions have come a long way since the earliest classical calculations on $H+H_2$ by Karplus, Sharma and Porter in 1957[75] and the earliest quantum dynamical calculations on the colinear $H+H_2$ system by McCollough and Wyatt in 1971.[219] Development in this field has been driven by an improvement in algorithms as well as a general increase in the speed of available computer hardware. Coupled with the improved description of potential energy surfaces generated through more accurate electronic structure methods, this means that quantum dynamical

*The Renner-Teller pair form a Π state at linearity, which is split into A′ and A″ components as the molecule bends.

calculations in particular are now at least as accurate as experiment for a number of three-atom systems (see *e.g.*, refs. [220] and [221]). Some four-atom systems have been studied extensively as well, in particular the H_2+OH[139] and $OH+CO$ reactions,[222] since detailed experiments are available for these important combustion reactions.

However, it should be noted that the above sketched situation is not uniformly correct. It should be clear from the discussion in the previous section that, for example, the $H+O_2$ reaction is still very poorly understood at a fundamental level. This reaction (and others like it) shows a surprisingly high level of complexity for a reaction which, after all, only involves three atoms. For this reaction, in particular, the role of rovibrationally excited initial states and non adiabatic effects will need to be addressed in the future. The four-atom calculations mentioned above only looked at total angular momentum $J=0$ and ignored non-adiabatic effects as well.

It can be argued that a significant proportion of the advances made in this field are due to the advances in computer technology, making faster hardware cheaper and therefore more available to a wider range of scientists. However, this improvement in the speed of the hardware (as given by Moore's Law) should not be expected to continue indefinitely, making the further development of new algorithms, in particular those that exploit parallel or multi-core architectures to the fullest, imperative. It should also be borne in mind that quantum dynamics scales exponentially with the size of the problem. This means that each atom added to the system will increase the complexity of the problem by several orders of magnitude. Therefore, treating an entire system rigorously using quantum dynamics will become impossible at some stage. Thus, more approximate methods, such as MCTDH, will need to be developed (further) to increase the maximum size of the problem that can be treated using quantum dynamics. Accurate and detailed experiments on these same systems are therefore critical in order to be able to benchmark and test these methods and to further our understanding of fundamental chemical reactivity.

In the longer term, looking at even larger systems than those being considered here, it is obvious that at some stage it will become impossible to study reactions at the same level of detail as the reactions described above, *i.e.* completely quantum mechanically, nor is it clear that this is completely necessary. Instead, much can be learned by treating only some parts of the system in question quantum mechanically, whereby the rest of the system is either not involved in the dynamics, except perhaps as part of the intermolecular potential, or is treated classically. This would most ideally be done using a hierarchy of methods, which will allow one a smooth transition between treating coordinates quantum mechanically or classically. This is a very active field with many methodologies under development. Examples of these include Feynmann path methods,[223] centroid dynamics,[224] ring-polymer molecular dynamics,[225] semiclassical methods,[226,227] Mp/SOFT,[228] NEO,[229] or Gaussian-MCTDH.[230] Hereby, it should be noted that, for example, the latter method will also allow on-the-fly calculation of the potential energy surface, therefore allowing one therefore to bypass the stage of the development of the PES entirely.

Finally, it is clear that the quantum dynamics community has come a long way since the earliest calculations, but at the same time it is clear there is still a long and exciting way ahead of us to get to grips with chemical reactivity at its most fundamental level.

3.5 PROBLEMS

1. For a Hamiltonian of the form $\hat{H}^2=\hat{p}^2/2\text{m}+V(x)$ calculate the quantities

$$\frac{\partial}{\partial t}\langle x^2\rangle \text{and} \frac{\partial}{\partial t}\langle \hat{p}^2\rangle.$$

Note that the commutator $[\hat{A}, \hat{B}^2]$ can be written as $[\hat{A}, \hat{B}]\hat{B}+\hat{B}[\hat{A}, \hat{B}]$.

2. An H_2 or a D_2 molecule can be idealized as a harmonic oscillator with a force constant $k = 575\,\mathrm{N\,m^{-1}}$. In this idealization the ground state wavefunctions for H_2 and D_2 are Gaussians.

$$\Phi_0(x) = N_0 e^{-\alpha x^2/2},$$

where $\alpha = 2\pi\mu\nu/\hbar$ and $\nu = \frac{1}{2\pi}\sqrt{k/\mu}$, with μ the reduced mass of either H_2 or D_2.
(a) Express the FWHM in terms of μ and k.
(b) What is the FWHM for H_2 and D_2 in m?
Suppose that the potential holding the atoms together is suddenly "removed" and the molecule travels travels through a potential-free region.
(c) Give a expression for the time-dependence of the FWHM.
(d) Calculate the FWHM for both molecules after one vibrational period $T = 1/\nu$.
(e) Calculate the FWHM for both molecules after $t = 1$ second.
3. (a) Show that

$$\frac{1}{\sqrt{2\pi}}\int_{-\infty}^{\infty} e^{-x^2/4\beta^2} e^{-ikx} dx = e^{-\beta^2 k^2}\sqrt{2\beta^2}, \tag{3.102}$$

using the standard integral

$$\frac{1}{\sqrt{2\pi}}\int_{-\infty}^{\infty} e^{-\alpha x^2} dx = \sqrt{\frac{\pi}{\alpha}}. \tag{3.103}$$

(b) Use the result from (a) to prove Eq. (3.37) from Eq. (3.36).
4. Using Eqs. (B3.5.1)–(B3.5.5) of Study Box 3.5, explicitly derive the elements of the local *R*-matrix of Eq. (3.62).
5. Eq. (3.88) gives the relation between $\psi(\boldsymbol{q}, t)$ and $\psi(\boldsymbol{q}, t + \Delta t)$ *via* the time evolution propagator.
(a) Give an expression for the relation between $\psi(t)$ and $\psi(\boldsymbol{q}, t - \Delta t)$.
(b) Use this relation and Eq. VII.18 in to derive an expression for the propagation of $\psi(t - \Delta t)$ to $\psi(t + \Delta t)$.
(c) If you use the above result to propagate a wave function, do the real part and the imaginary part of the wave function have to be propagated simultaneously or can they be propagated separately? Why?
6. Explicitly derive Eqs. (3.97) (a) and (b).

CHAPTER 4

Processes Involving Multiple Potential Energy Surfaces

BERTRAND RETAIL AND ANDREW J. ORR-EWING

School of Chemistry, University of Bristol, UK

4.1 INTRODUCTION

In 1927, Born and Oppenheimer[231] established their famous approximation that nuclear and electronic behaviours in molecules can be treated separately. This so-called adiabatic separation of the motions of electrons and nuclei made possible the introduction of potential energy surfaces (PESs) on which nuclear motions, and therefore chemical reactions, can be considered to take place. The PES describes potential energy changes as a function of nuclear coordinates, and therefore the gradients occurring on a surface quantify the forces acting on the atoms and molecules during a chemical transformation from reactants to products. As we have seen in the preceding chapters, the PES is thus the key notion for a microscopic understanding of chemical reactivity, and lies at the heart of our understanding of molecular reaction dynamics. Most chemical reactions show adiabatic behaviour, which means their dynamics are restricted to occur on a *single* PES. The Born-Oppenheimer (BO) approximation can, however, fail under some circumstances, and this breakdown results in non-adiabatic processes in which nuclear dynamics occur on two or more *coupled* PESs. These non-adiabatic dynamics are the subject of this chapter.

A familiar example of non-adiabatic dynamics is the electron-transfer process commonly known as a harpoon mechanism.[10,11] In the early days of molecular beam studies of reaction dynamics (see Chapter 6), the reactions of alkali metal atoms with halogen-containing molecules, which are classic examples of harpoon mechanisms, were the subject of detailed study. Figure 4.1 shows a simplified diagram of the PESs involved. As the metal atom (M), which has a low ionization energy, approaches to within a critical distance from a molecule such as a dihalogen (X_2) or an alkyl halide (RX), it becomes energetically favourable for an electron to be transferred from the metal to its collision partner. This electron enters an antibonding molecular orbital on the X_2 or RX and encourages dissociation to yield M^+ and X^- (as well as a spectator X atom or R radical) in sufficiently close proximity to form an MX product molecule. The electron hop transfers the reactants from a PES of covalent bonding character onto a PES that is ionic in character, and thus involves a

Tutorials in Molecular Reaction Dynamics
Edited by Mark Brouard and Claire Vallance

Published by the Royal Society of Chemistry, www.rsc.org

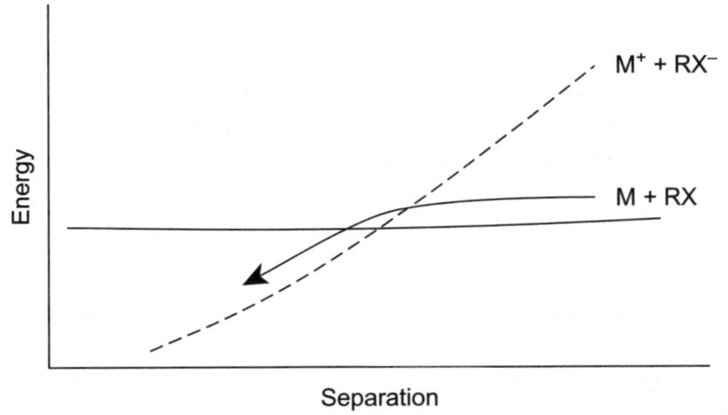

Figure 4.1 A schematic diagram of the intersection of covalent (solid line) and ionic (dashed line) PESs in a charge-transfer mediated reaction.

non-adiabatic transition, which occurs at a point of intersection between the two PESs. The PESs intersect because outside the critical distance for electron transfer, the covalent PES lies lower in energy, but at smaller separations, the ionic PES is energetically favoured due to the Coulombic attraction between the two oppositely charged ions. Harpoon dynamics are well-described in the literature,[10,11] and broader charge-transfer processes have wide-ranging importance in many areas of chemistry and biology. These processes will not be the focus of any further discussion here, but do appear in Problem 3 of this chapter, and in Chapter 6.

The Born-Oppenheimer approximation was first discussed in detail in Chapter 2. The present chapter builds upon the concept of a reaction path on a single, adiabatic PES by examining the typical situations in which these ideas break down. We introduce the important notion of non-adiabatic couplings and their theoretical treatment, and consider both the adiabatic and diabatic representations of PESs. The theoretical concepts are illustrated by examples of reactions in which non-adiabatic couplings and associated dynamics occur. The chapter ends with an overview of current experimental techniques used to probe non-adiabatic reactivity and dynamics in chemical reactions, illustrated by recent experimental discoveries in this evolving field. An in-depth discussion of theoretical treatments of non-adiabatic couplings and scattering dynamics is beyond the scope of the chapter, but several advanced reviews and texts provide further reading in this area.[232–234]

4.2 BREAKDOWN OF THE BORN-OPPENHEIMER APPROXIMATION

The very large difference in the masses of atomic nuclei and electrons enables their motions to be treated separately for many chemical reactions. An electronic Schrödinger equation can thus be solved in a field of static nuclei to obtain electronic energy eigenvalues* U_i and wavefunctions ϕ_i. By repeating this procedure for every possible geometrical arrangement of the (frozen) nuclei, an energy landscape $U_i(\boldsymbol{R})$ can be mapped out as a function of nuclear coordinates $\boldsymbol{R}$; this landscape is also called a PES. As noted in earlier chapters, in the case of a simple atom plus diatomic molecule reaction, three coordinates are necessary to describe the nuclear positions (two bond lengths and one angle). More generally, the surface is parameterized by $3N-6$ coordinates, N being the number of atoms involved in the reaction. The i subscript on the electronic potential energy function is a reminder that several PESs might be obtained, each describing a different electronic state, and having its own associated electronic wavefunction ϕ_i. Nuclear motion on each specific PES is described by a nuclear wavefunction, Ψ_i, as

*Note that the notation for the potential energy and for the wavefunctions is a little different from that employed in Chapters 2 and 3. In this chapter we use U to indicate the adiabatic potential energy and V for the diabatic potential energy. See further below.

discussed in Chapter 3, with both time-independent and time-dependent (wavepacket) formulations being widely used. The BO approximation considers that nuclear motion is restricted to occur on a *single* PES, with the total molecular wavefunction, Φ, approximated as a single product: $\Phi = \Psi\phi$. Under some circumstances, however, this approximation is not valid and several *coupled* PESs might participate in the nuclear dynamics. These circumstances are presented in the following.*

4.2.1 Non-adiabatic Couplings between Adiabatic PESs

To treat the dynamics of a molecular system (such as a chemical reaction) the full Hamiltonian describing the energetics of all the nuclei and electrons is, as shown in Eq. (4.1), usefully separated into contributions from the nuclear kinetic energy operator $\hat{T}_N$ and those from all other terms, which are encapsulated in the electronic Hamiltonian, $\hat{H}_e(\boldsymbol{r}, \boldsymbol{R})$. The latter operator contains all the terms for interactions between the electrons and the nuclei, and also includes the kinetic energy operator for the electrons. Repeating Eq. (2.5), we have

$$\hat{H}(\boldsymbol{r}, \boldsymbol{R}) = \hat{T}_N(\boldsymbol{R}) + \hat{H}_e(\boldsymbol{r}, \boldsymbol{R}). \tag{4.1}$$

Here, $\boldsymbol{R}$ and $\boldsymbol{r}$ denote the coordinates of the nuclei and the electrons respectively. The total wavefunction $\Phi(\boldsymbol{r}, \boldsymbol{R}, t)$ for the time-dependent behaviour of the electrons and nuclei is the solution to the Schrödinger equation:

$$i\hbar\frac{\partial\Phi}{\partial t} = \hat{H}\Phi. \tag{4.2}$$

If we consider one nuclear configuration, described by $\boldsymbol{R}$, we can define the electronic adiabatic wavefunctions $\phi_i(\boldsymbol{r}; \boldsymbol{R})$, which are functions of the electronic coordinates $\boldsymbol{r}$ at fixed $\boldsymbol{R}$, as the solutions of an electronic Schrödinger equation (*c.f.* Eq. (2.6)):

$$\hat{H}_e\phi_i(\boldsymbol{r}; \boldsymbol{R}) = U_i(\boldsymbol{R})\phi_i(\boldsymbol{r}; \boldsymbol{R}), \tag{4.3}$$

where $U_i(\boldsymbol{R})$ is the electronic energy of the ith state (with corresponding electronic wavefunction $\phi_i(\boldsymbol{r}; \boldsymbol{R})$). This equation is obtained by setting the nuclear kinetic energy to zero (as is appropriate for fixed nuclei). The solutions to Eq. (4.3) differ at each set of nuclear coordinates, and the variation of $U_i(\boldsymbol{R})$ with these nuclear coordinates defines the adiabatic potential energy surface for the ith electronic state.

To recognise the possibility of coupling of electronic and nuclear motions, the total wavefunction is expressed as an expansion in the product of the electronic wavefunctions $\phi_i(\boldsymbol{r}; \boldsymbol{R})$ and the nuclear wavefunctions $\Psi_i(\boldsymbol{R}, t)$:

$$\Phi(\boldsymbol{R}, \boldsymbol{r}, t) = \sum_i \Psi_i(\boldsymbol{R}, t)\phi_i(\boldsymbol{r}; \boldsymbol{R}). \tag{4.4}$$

Inserting this expression into the full Schrödinger equation, multiplication through from the left by the complex conjugate of one of the set of orthonormal electronic wavefunctions, $\phi_j(\boldsymbol{r}; \boldsymbol{R})$,* and integration over all electronic coordinates gives the set of equations (see Problem 1):

$$[\hat{T}_N + U_j]\Psi_j - \sum_i \hat{\Lambda}_{ji}\Psi_i = i\hbar\frac{\partial\Psi_j}{\partial t}, \tag{4.5}$$

*Whether or not a non-adiabatic transition occurs is sometimes discussed qualitatively in terms of a Massey parameter.[235] The reader might find it helpful at this point to read the beginning of Section 8.8, where the Massey criterion is introduced in the context of electronic transitions between states of different spin-multiplicity, which is mediated *via* spin-orbit coupling.

where the explicit dependence on coordinates and time have been dropped from the wavefunctions for simplicity.* Eq. (4.5) can conveniently be expressed in matrix form to include all of the electronic states in a single equation (see, for example, reference[70] for matrix formulations of such equations). The nuclear kinetic terms, potential energy functions and coupling elements are then contained within the respective matrices $\hat{T}_\mathrm{N}\boldsymbol{I}$, $\boldsymbol{U}$, and $\boldsymbol{\Lambda}$, where $\boldsymbol{I}$ is an identity matrix of the same dimension as the number of states, and $\boldsymbol{U}$ is a diagonal matrix with elements that are the adiabatic PESs for the different states. The Ψ_j become the rows of a column vector, and the elements of the coupling matrix connect different adiabatic states and are thus non-adiabatic couplings:

$$\hat{\Lambda}_{ji} = \delta_{ji}\hat{T}_\mathrm{N} - \langle\phi_j|\hat{T}_\mathrm{N}|\phi_i\rangle. \tag{4.6}$$

In this expression, δ_{ji} is a Kronecker delta, defined in Study Box 3.3. The non-adiabatic coupling terms involve the action of the nuclear kinetic energy operator on the electronic wavefunctions and their presence leads us to recognize that the motions of the nuclei and electrons are coupled. The nuclear kinetic energy operator is:

$$\hat{T}_\mathrm{N} = -\frac{\hbar^2}{2M}\nabla^2, \tag{4.7}$$

where ∇^2 is the Laplacian operator, involving derivatives with respect to the nuclear coordinates, and M is an appropriate choice of (reduced) mass for the system. Using $\nabla^2 = \underline{\nabla}\cdot\underline{\nabla}$ (with the underlined $\underline{\nabla}$ emphasizing a vector operator), the non-adiabatic coupling term can be re-expressed as (see Problem 2):

$$\hat{\Lambda}_{ji} = \frac{\hbar^2}{2M}(2\boldsymbol{F}_{ji}\cdot\underline{\nabla} + G_{ji}), \tag{4.8}$$

with the derivative coupling vector between different states ($i \neq j$) given by

$$\boldsymbol{F}_{ji} = \langle\phi_j|\underline{\nabla}|\phi_i\rangle = \frac{\langle\phi_j|\underline{\nabla}\hat{H}_\mathrm{e}|\phi_i\rangle}{U_i - U_j}, \tag{4.9}$$

and a scalar second-derivative coupling term

$$G_{ji} = \langle\phi_j|\underline{\nabla}^2|\phi_i\rangle. \tag{4.10}$$

Note that we use bold face for $\boldsymbol{F}_{ji}$ to emphasize that it involves a vector operator. Values of second derivative couplings are typically quite small and can often be neglected; the key couplings will therefore be provided by the first derivative coupling (Eq. (4.9)). The BO approximation results from considering only a single electronic state, so $\hat{\Lambda}_{ji} = 0$ for $i \neq j$. The *group* BO approximation considers non-adiabatic couplings between only a finite subset of the complete set of electronic states and is based on the valid assumption that there is only significant coupling between states that lie close in energy. Note that if two electronic states ϕ_i and ϕ_j are degenerate at some nuclear configuration, the coupling term $\boldsymbol{F}_{ji}$ becomes infinite.

There will be regions of the multidimensional space of possible configurations of the nuclei where the adiabatic electronic wavefunction changes rapidly in character for small changes of the nuclear coordinates. Such locations might include barriers on the potential energy surface or regions in the

*See Eq. (2.9) for the time independent version of Eq. (4.5).

vicinity of avoided crossings or near-degeneracy of two PESs. It is in these regions that the derivative coupling terms will be largest because, as Eq. (4.9) shows, they involve derivatives of the electronic wavefunction with respect to nuclear coordinates. A qualitative interpretation is that, as the dynamics of a chemical reaction take the nuclei through a region where the electronic wavefunction character undergoes a substantial change, the electrons may not be able to adapt to their lowest energy configuration on the adiabatic PES sufficiently quickly to keep up with the nuclear motion. Instead of following the adiabatic pathway corresponding to this optimum electronic configuration, the reaction thus evolves onto another adiabatic electronic PES that is a smoother continuation of the initial electronic configuration.

4.2.2 Diabatic and Adiabatic Representations of the Coupled PESs

A full theoretical treatment of a multi-surface reaction is very challenging when all the derivative couplings of Eqs. (4.9) and (4.10) must be calculated. The couplings change rapidly with nuclear coordinates, and are costly to handle computationally because each term in Eq. (4.9) implies a $3N-6$ dimensional vector coupling. Hence theoreticians prefer to transform the electronic basis set to work with a *diabatic* representation of PESs, and here we make a distinction between diabatic and adiabatic pictures. An adiabatic representation will be obtained by solving Eq. (4.3) for all states using a form of the potential energy matrix that does not have any off-diagonal terms linking different states. Recall that these matrices are functions of the nuclear coordinates $\boldsymbol{R}$ and the adiabatic PESs are the diagonal terms of the PE matrix. The couplings between the different adiabatic PESs are provided by the non-diagonal terms of the nuclear kinetic energy matrix, *i.e.* the second term in Eq. (4.6). In a diabatic representation, the new PE matrix is no longer subject to obeying the BO approximation and is not diagonal; its diagonal elements (the diabatic PESs) are no longer eigenvalues of the full electronic Hamiltonian. The necessary couplings of such a dynamical system of electrons and nuclei are now provided by the off-diagonal parts of the PE matrix, while the nuclear kinetic energy matrix is now diagonal. Study Box 4.1 outlines the mathematics relating adiabatic and diabatic descriptions for two coupled electronic states, and demonstrates the important result that two adiabatic states of the same symmetry cannot cross, whereas crossing is possible for the diabatic states.

Consider an example for the reaction $A+BC \rightarrow AB+C$, with schematic changes in PE as a function of the reaction coordinate displayed in Figure 4.2. As atom A approaches the BC molecule

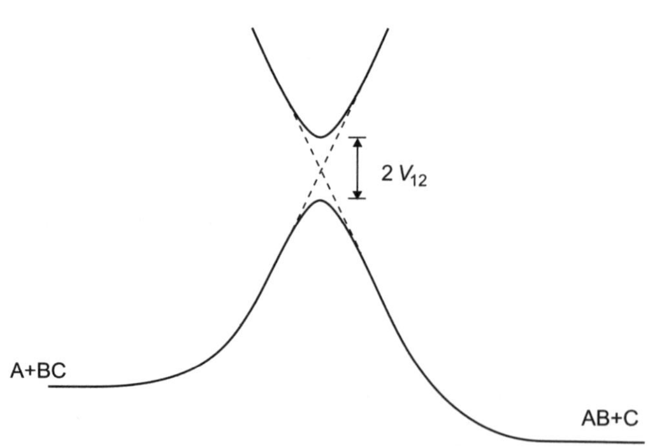

Figure 4.2 A schematic 1D diagram of adiabatic (solid lines) and diabatic (dashed lines) PE surfaces for a chemical reaction $A+BC \rightarrow AB+C$. V_{12} is the coupling between diabatic states at the intersection.

STUDY BOX 4.1: THE COUPLING OF DIABATIC STATES AND THE DERIVATION OF ADIABATIC STATES

Suppose we have two uncoupled, *diabatic* states denoted as $\phi_1^{(0)}$ and $\phi_2^{(0)}$ that are orthonormal eigenstates of a Hamiltonian operator $\hat{H}^{(0)}$, but which are mixed together by an additional weak coupling term ($\hat{V}$) in a Hamiltonian operator $\hat{H} = \hat{H}^{(0)} + \hat{V}$. This new Hamiltonian has eigenstates ϕ of the Schrödinger equation $\hat{H}\phi = E\phi$, which we can consider as our adiabatic states. Each adiabatic eigenstate of $\hat{H}$ can be expressed as a linear combination of the diabatic basis functions:

$$\phi = c_1\phi_1^{(0)} + c_2\phi_2^{(0)}, \tag{B4.1.1}$$

where c_1 and c_2 are constants that describe the amounts of each diabatic state that contribute to the adiabatic state. If this linear combination is inserted into the Schrödinger equation, and the resultant equation is multiplied from the left by either $\phi_1^{(0)}$ or $\phi_2^{(0)}$, and integrated, the well-known outcome is the pair of "secular" equations:[70]

$$\begin{aligned} c_1(H_{11} - E) + c_2H_{12} &= 0 \\ c_1H_{21} + c_2(H_{22} - E) &= 0. \end{aligned} \tag{B4.1.2}$$

Here, $H_{ij} = \langle\phi_i^{(0)}|\hat{H}|\phi_j^{(0)}\rangle$ are the matrix elements of $\hat{H}$, and the integration is over all space. The simultaneous equations have solutions for the energy E that satisfy the quadratic equation

$$E^2 - (H_{11} + H_{22})E + H_{11}H_{22} - H_{12}H_{21} = 0, \tag{B4.1.3}$$

giving two roots that describe the energies of the two adiabatic states resulting from the coupling between the diabatic states:

$$E_{\pm} = \tfrac{1}{2}(H_{11} + H_{22}) \pm \tfrac{1}{2}\sqrt{(H_{11} - H_{22})^2 + 4H_{12}H_{21}}. \tag{B4.1.4}$$

These energies are implicitly a function of the nuclear coordinates of all the constituent atoms in the system because the diabatic wavefunctions, the Hamiltonian operators and the coupling terms all depend on the nuclear geometry (note that $E_{\pm} \equiv U_{\pm}$ in the notation of Section 4.2.1). If we express the full Hamiltonian in terms of its two constituent parts, $\hat{H} = \hat{H}^{(0)} + \hat{V}$, and assume that the coupling term is only non-zero between different diabatic states (so $V_{11} = V_{22} = 0$ but $V_{12} \neq 0$) then $H_{11} = H_{11}^{(0)} = E_1^{(0)}$ (*i.e.* V_1 in the notation of Section 4.2.1) is the energy of the diabatic state $\phi_1^{(0)}$, and likewise for diabatic state $\phi_2^{(0)}$. At a point of degeneracy between the two diabatic states, $E_1^{(0)} = E_2^{(0)}$ and thus $H_{11} = H_{22}$. By inspection of Eq. (B4.1.4), the adiabatic states will then only have the same energy, $E_+ = E_-$, (and thus the adiabatic potentials touch or cross) if $H_{12}H_{21} = |H_{12}|^2 = 0$. The first part of this relationship follows from the Hamiltonian being a Hermitian operator.[70] Note also that because $\phi_1^{(0)}$ and $\phi_2^{(0)}$ are eigenstates of $\hat{H}^{(0)}$, it follows that $H_{12} = V_{12}$, and the separation between the adiabatic states, ϕ_+ and ϕ_-, at the intersection of the two diabatic states is $2V_{12}$. (For further development of this model see Problems 4 and 5 of this chapter.)

The presence of a non-zero coupling term that is totally symmetric affects the two diabatic states if they possess the same symmetry. In a *diatomic molecule*, the only geometric distortion that is possible is an extension or compression of the internuclear distance, which preserves the symmetry. The coupling term cannot therefore become zero on symmetry grounds as a result of a change in the nuclear geometry, so the adiabatic state energies are prevented from being degenerate, and the adiabatic potential energy curves cannot cross (this well-known result is the *non-crossing rule* – see Figure 4.3). If the diabatic states have different symmetries, the adiabatic states can cross unless the coupling term has an appropriate symmetry to mix the two diabatic eigenstates. In this case, the coupling is again not affected by a (symmetric) molecular distortion, so there will not be a distortion-induced crossing of the adiabatic potentials. The more complicated case of a polyatomic molecule is discussed later in this chapter.

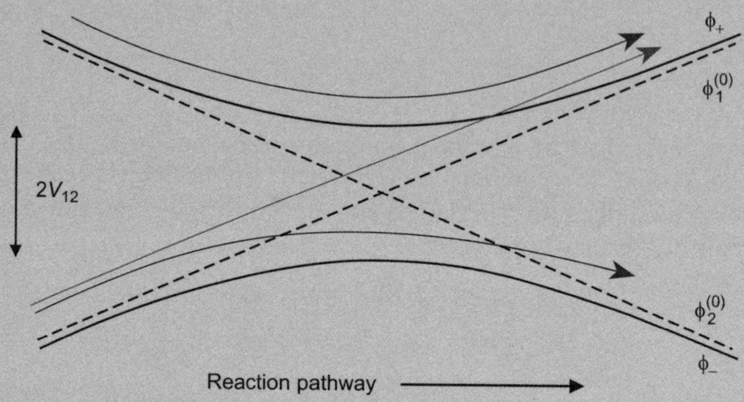

Figure 4.3 Schematic illustration of the crossing of diabatic states ($\phi_1^{(0)}$ and $\phi_2^{(0)}$, dashed lines) and the avoided crossing of the adiabatic states (ϕ_+ and ϕ_-, solid lines) in a region of interaction of two electronic states. The red arrow demonstrates diabatic passage through the interaction region, and the blue arrows are the adiabatic pathways on the lower and upper adiabats.

Bertrand Retail and Andrew J. Orr-Ewing

a bonding configuration starts to be established between the A and B atoms, while at the same time the BC bonding becomes weaker, giving rise to a PE barrier. There are two regions with different electronic character for the reactive system corresponding to the reactant and product sides of the barrier. In the transition state (TS) region, complementary adiabatic and diabatic pictures can be considered. The BO approximation assumes that the electronic structure of the molecular system is able to adapt rapidly compared to any change in nuclear geometry. When the nuclear motion takes the reaction across the barrier region, the electrons thus rearrange to generate a new electronic configuration for each nuclear position (which, in the present case, will be a linear combination of AB and BC bonding characters); the resultant adiabatic PE curves are obtained by diagonalization of the electronic Hamiltonian and avoid crossing at the TS. The validity of such an approximation is limited to how fast the electronic structure can adapt to a change in nuclear geometry and is mathematically parameterized by the derivative coupling operators of Eqs. (4.6), (4.9) and (4.10), some of which are large in magnitude in the region of the barrier. In a diabatic representation of the PESs, the electronic configuration evolves smoothly with changes to the nuclear configuration, retaining predominantly BC or AB bonding character. The two diabatic PE curves of different character can cross (with the adiabatic barrier arising at or near the crossing point). If the BO approximation fails, the electrons retain their configuration (either AB or BC bonding character) and the nuclear trajectory stays on the same diabatic curve at the barrier; in an adiabatic picture, this is equivalent to hopping to the upper adiabatic curve instead of staying on the lower adiabat proceeding to the products.

4.2.3 The Landau-Zener Model for Non-adiabatic Dynamics

If we consider a one dimensional model in which only the motion along the reaction coordinate induces a non-adiabatic jump at the crossing point of the diabatic surfaces, the non-adiabatic transition probability (P) at an avoided crossing of adiabatic PESs can be obtained using the Landau-Zener formula given below in Eq. (4.11),[236]

$$P = \exp\left[-\frac{2\pi V_{12}^2}{\left(\hbar v \frac{\partial(V_2 - V_1)}{\partial R}\right)}\right]. \tag{4.11}$$

Here, v is the velocity along the reaction coordinate, R; $V_1(R)$ and $V_2(R)$ are the two crossing diabatic PESs; and V_{12} is the coupling between them. The derivative in the denominator is the difference in the gradients of the two diabatic PESs at the crossing point. If the coupling is weak, nuclear trajectories will stay on the same diabatic curve and thus the non-adiabatic transition probability will be high. For such a two-state model, the diabatic coupling term is simply related to the energetic difference of the two adiabatic surfaces (U_1 and U_2) at the avoided crossing point, $2V_{12} = U_2 - U_1$.[237] Hence a small value of V_{12} implies a small splitting of the two adiabats and thus strong derivative couplings in the adiabatic representation (Eq. (4.9)) resulting in enhanced non-adiabatic processes.

The Landau-Zener formulation of non-adiabatic transition probabilities has been employed in QCT calculations of dynamics at intersections between PESs for reactions such as $O(^1D) + H_2 \rightarrow OH + H$ (see Section 4.2.5). Here, however, the method is illustrated for a simpler system involving an avoided crossing between two potential energy curves relevant to the photodissociation of the IBr molecule.[238] Figure 4.4 shows diabatic PE curves for the ground and low-lying excited electronic states of the IBr molecule. The $B^3\Pi(0^+)$ and $Y^3\Sigma^-(0^+)$ states are the focus of this discussion because they have the same ($\Omega = 0$) symmetry in Hund's case (c)[239,240] and are coupled, so the adiabatic states avoid a crossing.*

Figure 4.4 also shows $P_{B/Y}$, the probability of flux initially excited to the diabatic B-state potential by photon absorption in the Franck-Condon region transferring to the diabatic Y-state potential and dissociating to $I(^2P_{3/2}) + Br(^2P_{3/2})$ products. In an adiabatic picture, these dynamics would be viewed as occurring *adiabatically* on the lower of the two $\Omega = 0$ potential energy curves; the non-adiabatic transition probability is given by $P = 1 - P_{B/Y}$. The plot compares experimental measurements with quantum mechanical (QM) wavepacket scattering calculations on coupled PE curves, and the outcomes of a Landau-Zener model calculation, and all are in good agreement. As the photon energy is increased, the molecules are initially excited to points higher up the repulsive inner wall of the B-state potential, and thus the dissociating molecules pass through the B/Y crossing region with greater speed. In accord with Eq. (4.11), the non-adiabatic crossing probability increases and the diabatic crossing probability $P_{B/Y}$ decreases.

4.2.4 A Case Study in Non-adiabatic Dynamics – The Ultraviolet Photodissociation of HCl

Photodissociation dynamics will be discussed in detail in Chapter 8, but the example of ultraviolet (UV) photodissociation of HCl, to form H atoms and $Cl(^2P_J)$ atoms in their ground ($J = 3/2$) and spin-orbit excited ($J = 1/2$) states, is presented here as an illustration of some key concepts in

*The general textbooks on spectroscopy listed in Further Reading provide more information on the notation employed. Study Box 4.3 provides an introduction to Hund's coupling cases, and the notation employed here is also discussed further in Section 8.3.

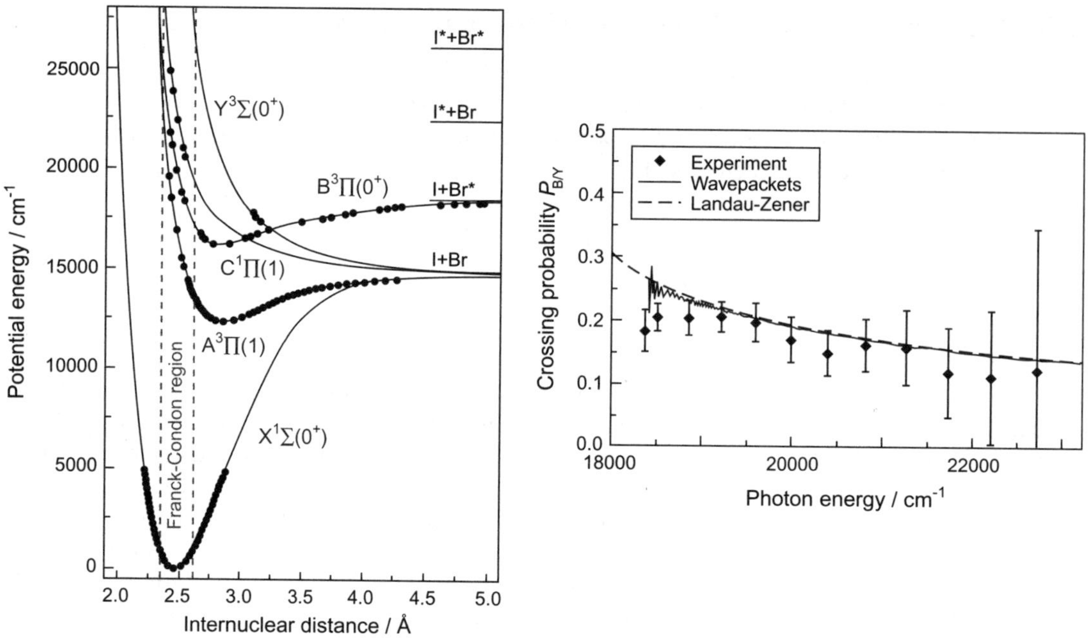

Figure 4.4 Left: Diabatic PE curves for the ground and low lying excited electronic states of IBr. Br and Br* denote, respectively, $Br(^2P_{3/2})$ and $Br(^2P_{1/2})$ and likewise for I and I*. Right: Comparison of crossing probabilities, $P_{B/Y}$, for transitions from the diabatic $B^3\Pi(0^+)$ to the $Y^3\Sigma^-(0^+)$ states obtained from the Landau-Zener model (----), wavepacket scattering calculations (——), and experimental measurements (◆). Reprinted with permission from E. Wrede *et al. J. Phys. Chem.* **114**, 2629 (2001) (ref. [238]), copyright (2001), American Institute of Physics.

non-adiabatic dynamics.[241,242] The ideas developed here are also proving to be significant in the study of non-adiabaticity in bimolecular reactions, as will be illustrated in Section 4.3.

Figure 4.5 shows the ground and low lying adiabatic potential energy curves of HCl relevant to the UV photodissociation dynamics. The states are labelled in a combination of Hund's case (a) and case (c) notations[239,240] in accord with the standard form $^{2S+1}\Lambda(\Omega)$, where S, Λ and Ω are quantum numbers for the total spin angular momentum, the projection of the orbital angular momentum on the internuclear axis, and the projection of the total angular momentum on this axis. In Hund's case (c), for which spin-orbit coupling is strong, Ω is the only good quantum number of these three. In HCl, UV absorption excites the molecules almost exclusively to the $^1\Pi(1)$ state and, in the absence of any rotation-induced couplings, no dissociation can occur by transfer to $\Omega = 0$ states. The focus here will therefore be on states with $\Omega = 1$ only, which are coupled by spin-orbit interactions. These interactions and the effect of their magnitude on non-adiabatic dynamics are discussed further in Section 4.2.6. Despite there being no avoided crossings between these states, non-adiabatic dynamics still occur.

The time-dependent Schrödinger equation describing the evolution of the nuclear wavepacket $\Psi(R, t)$ contains the (time-independent) Hamiltonian:

$$\hat{H}(R) = -\frac{\hbar^2}{2\mu}\frac{d^2}{dR^2} + \hat{V}(R) + \hat{H}_{SO}(R), \tag{4.12}$$

where R is the separation of the H and Cl atoms. The first term is the nuclear kinetic energy operator (with μ the reduced mass), $\hat{V}(R)$ is a diabatic potential energy function for one electronic state, and $\hat{H}_{SO}(R)$ contains the spin-orbit coupling terms which can link different diabatic states. Rotational terms in the Hamiltonian have been omitted. In a matrix formulation of the above

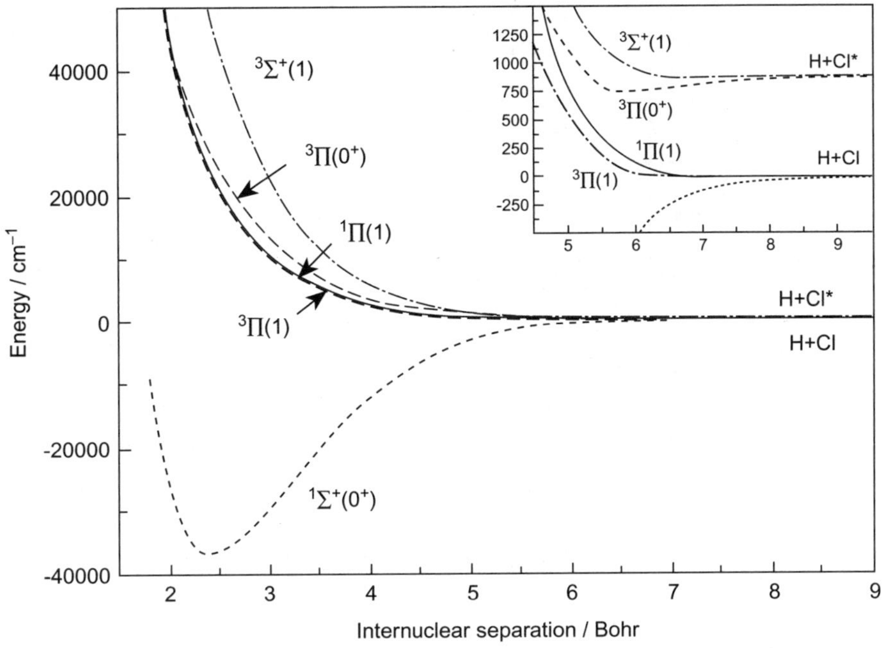

Figure 4.5 Adiabatic PE curves for the ground and low-lying electronically excited states of HCl. The states are labelled with mixed Hund's case (a) and case (c) notation, as described in the text. The inset shows an enlarged view of the long range regions of the PE curves. Cl and Cl* denote, respectively, the $^2P_{3/2}$ and $^2P_{1/2}$ spin-orbit states of the Cl atom, and are split by 882 cm^{-1}. Reprinted with permission from P. M. Regan *et al.*, *J. Chem. Phys.* **112**, 10259 (2000) (Ref. [242]), copyright (2000), American Institute of Physics.

Hamiltonian, incorporating the diabatic representations of the three states of interest, $^1\Pi(1)$, $^3\Pi(1)$ and $^3\Sigma^+(1)$, the electronic part of the Hamiltonian can be expressed as a 3×3 matrix $\hat{V}_e(R) = \hat{V}(R) + \hat{H}_{SO}(R)$. This matrix has diagonal elements that are the sum of the diabatic potential energies and the diagonal matrix elements of the spin-orbit coupling operator; the off-diagonal elements are spin-orbit couplings between different diabatic states. These various terms are described in detail in reference.[241] Diagonalization of $\hat{V}_e(R)$ gives the adiabatic PE curves shown in Figure 4.5. In what follows, diabatic states will be identified as $^1\Pi$, $^3\Pi$ and $^3\Sigma^+$ without the Ω quantum number in parentheses. The adiabatic states are written with the Ω quantum number, preceded by a Hund's case (a) label to identify the dominant component in the Franck-Condon region (*i.e.* at internuclear separations in the range 2 to 3 bohr) as $^1\Pi(1)$, $^3\Pi(1)$ and $^3\Sigma^+(1)$. The adiabatic states are linear combinations of the diabatic states (*c.f.* Eq. (B4.1.1)):

$$\phi_i^{\text{adiab}}(R) = \sum_j c_{ij}(R)\phi_j^{\text{diab}}(R), \tag{4.13}$$

with the summation over the three diabatic states labelled by j, and $i = 1$ to 3 representing the three adiabatic states. The square moduli of the (R-dependent) coefficients, $|c_{ij}|^2$, give the contributions of the three diabatic states to each adiabatic state, and are plotted in Figure 4.6.

There is a marked change in the electronic character of the adiabatic $^1\Pi(1)$ state (panel (b) in Figure 4.6) at $R \approx 4$ bohr; it evolves rapidly from predominantly diabatic $^1\Pi$ character to a nearly equal mixture of the $^1\Pi$ and $^3\Sigma^+$ states. A similar change is evident for the $^3\Sigma^+(1)$ state, which becomes an almost equal mixture of all three diabatic states. Derivative couplings between these

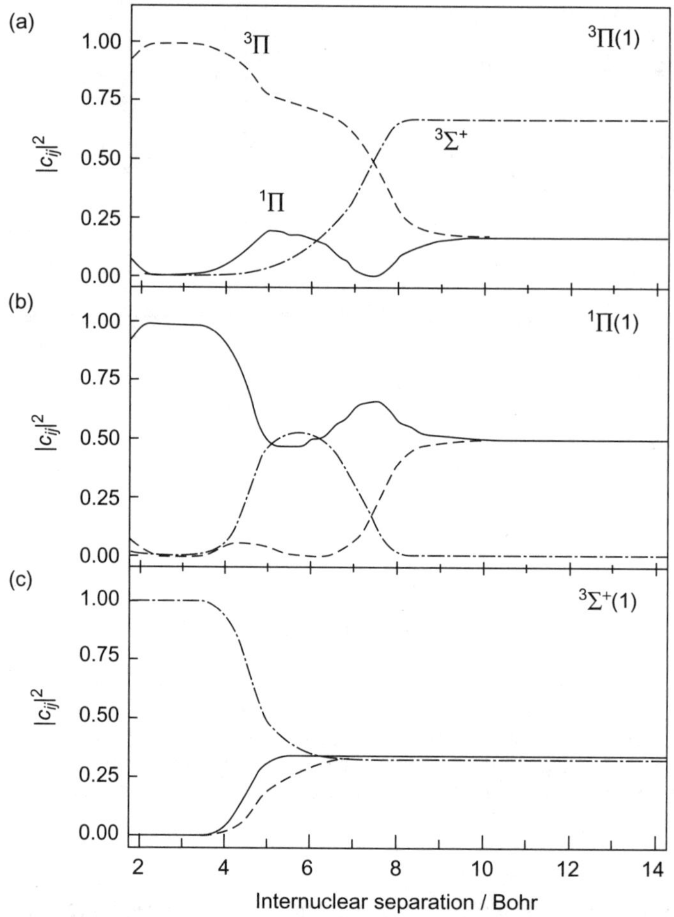

Figure 4.6 Contributions, $|c_{ij}|^2$, of the diabatic states with $\Omega = 1$ to the (a) $^3\Pi(1)$, (b) $^1\Pi(1)$ and (c) $^3\Sigma^+(1)$ adiabatic states of HCl, plotted as a function of internuclear separation. Reprinted with permission from P. M. Regan *et al.* J. Chem. Phys. **112**, 10259, (2000) (ref. [242]). Copyright (2000), American Institute of Physics.

adiabatic states are thus expected to be significant at this separation of the H and Cl atoms. At separations of ~7 to 8 bohr the adiabatic $^1\Pi(1)$ and $^3\Pi(1)$ states undergo further substantial changes in their diabatic state contributions, which may be viewed as the final electronic angular momentum recoupling to form the separated $H(^2S)$ and $Cl(^2P_{3/2})$ atoms. These changes in the electronic character of the adiabatic states should be reflected in non-adiabatic transitions of the nuclear wavepackets during the dissociation caused by significant derivative coupling terms.

The dissociation dynamics have been studied by wavepacket propagation, with calculations initiated by creation of a $t = 0$ wavepacket on the $^1\Pi(1)$ adiabatic PE curve (see Chapter 3 for a discussion of wavepacket propagation techniques). It is mathematically convenient to apply the kinetic energy operator to the wavepacket in the diabatic representation (where there are no kinetic energy couplings between states) for one time step (of 0.024 fs) and to transform into the adiabatic representation to apply the potential energy term (which is diagonal in this representation). Back transformation to the diabatic representation gives the wavepacket propagated over one time step, and may have resulted in some transfer of wavepacket amplitude to another electronic state. This cycle is repeated over numerous time-steps until the asymptotic region ($R > 12$ bohr) is reached, and the wavepackets are

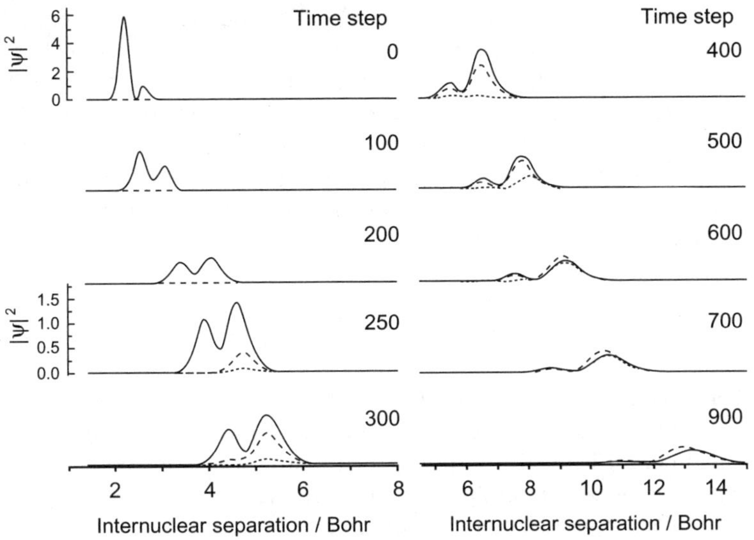

Figure 4.7 Snapshots of adiabatic wavepackets for time steps from $t=0$ to 900 (each time step is 0.024 fs) for the photodissociation of HCl. Wavepackets are identified by their adiabatic electronic state: $^1\Pi(1)$ (——); $^3\Pi(1)$ (----); $^3\Sigma^+(1)$ (— — —). The double-peaked structure of the wavepacket reflects the initial photo-excitation of HCl($X^1\Sigma^+$, $v=1$). Adapted from ref. [242].

then analyzed in the adiabatic representation, *e.g.*, to evaluate branching into the Cl($^2P_{3/2}$) and Cl($^2P_{1/2}$) spin-orbit states for comparison with experimental data. Figure 4.7 shows snapshots of the wavepackets on the $\Omega=1$ adiabatic states at various time steps. The double-peaked structure results from excitation of the $v=1$ vibrational wavefunction of HCl($X^1\Sigma^+(0^+)$) to the $^1\Pi(1)$ state in these particular calculations, which were performed to study vibrationally mediated photodissociation (*i.e.* the photodissociation of a vibrationally excited molecule as opposed to a molecule initially in its ground vibrational state). Transfer of parts of the wavepacket to the $^3\Sigma^+(1)$ state (at $R\sim4$ bohr) and the $^3\Pi(1)$ state (at $R\sim7$ bohr) is evident in the diagram, and is in accord with the discussion above about the changing mixtures of diabatic states comprising each adiabatic state (Figure 4.6).

4.2.5 Multi-dimensional PESs and Conical Intersections

Eq. (4.9) demonstrates that at a point of degeneracy of two adiabatic potential energy surfaces, the non-adiabatic couplings between them become infinite. The arguments presented in Study Box 4.1 demonstrate that for a diatomic molecule, degeneracy or crossing of adiabatic states of the same symmetry is, in general, avoided. Here, we discuss how the non-crossing rule need not apply for a polyatomic system, which can distort along directions other than a totally symmetric bond extension (contrary to the case for a diatomic molecule). Multidimensional seams of *conical intersections* (see below) instead arise between adiabatic PESs and act like (often very efficient) funnels through which the dynamics can take a reactive or photochemical system from one state to another. Conical intersections have been the focus of detailed review articles (*e.g.*, refs. [233] and [234]) and whole volumes of recent books,[232] and their role in chemical reactions, photodissociation dynamics and intramolecular relaxation of excited electronic states (*e.g.*, internal conversion) remains the subject of much ongoing research in systems ranging from small molecules to nucleic acid bases, peptides and proteins. Here, we present only a brief overview of the properties of conical intersections and illustrate their importance in chemical dynamics.

The discussion in Section 4.2.1 concentrated on an adiabatic picture of chemical dynamics, with consideration of the terms that link adiabatic potential energy surfaces and thus can induce non-adiabatic dynamics. This will be the starting point for analysis of conical intersections. In Study Box 4.1, however, we discussed both adiabatic and diabatic pictures of a pair of coupled electronic states, and it will prove convenient in places to consider a diabatic picture as well. The time-dependent Schrödinger equation for dynamics on multiple PESs defined in the adiabatic representation, as expressed in terms of nuclear wavefunctions in Eq. (4.5), can be written in a compact matrix form as follows:

$$[\hat{T}_{\mathrm{N}}\boldsymbol{I} + \boldsymbol{U} - \hat{\boldsymbol{\Lambda}}]\Psi = i\hbar\frac{\partial\Psi}{\partial t}, \tag{4.14}$$

As was noted earlier, $\boldsymbol{I}$ is the identity matrix, $\boldsymbol{U}$ is a diagonal matrix of PESs defined by Eq. (4.3), $\hat{\boldsymbol{\Lambda}}$ is a matrix of non-adiabatic coupling operators, and Ψ is a nuclear wavefunction with components on some or all of the different PESs. In the adiabatic picture, we have a potential energy matrix that is diagonal, and the non-adiabatic couplings, arising through the kinetic energy operator, are contained within the off-diagonal elements of the matrix $\hat{\boldsymbol{\Lambda}}$.

The alternative viewpoint employs a set of diabatic electronic wavefunctions, which are connected to the adiabatic functions by a unitary transformation. The Schrödinger equation is now written as:

$$[\hat{T}_{\mathrm{N}}\boldsymbol{I} + \boldsymbol{V}]\Psi = i\hbar\frac{\partial\Psi}{\partial t}, \tag{4.15}$$

where $\boldsymbol{V}$ is the diabatic potential matrix. The diagonal elements (V_{11} and V_{22} for a two-state system, V_{jj} in general) are the diabatic potential energy surfaces. The off-diagonal matrix elements (*e.g.*, V_{12} and V_{21}, or V_{ij} in general) are the coupling terms, which are local to a region of interaction between coupled states and do not involve derivatives of the electronic wavefunctions with respect to nuclear coordinates. All kinetic energy terms are diagonal (*i.e.* do not connect different diabatic states). Diabatic electronic states generally exhibit smooth variation with changes in the nuclear coordinates, and do not undergo substantial changes in regions of crossing with other states; the orbital character is instead retained.

In polyatomic systems, degeneracy of two adiabatic states ϕ_1 and ϕ_2 can be achieved, whereas it was forbidden for a diatomic molecule (the non-crossing rule, see Study Box 4.1). In a polyatomic molecule there are some configuration(s) of the nuclei for which the electronic Hamiltonian matrix elements satisfy the degeneracy condition $V_{11}=V_{22}$, and the coupling terms V_{12} can become zero for symmetry reasons. The adiabatic states can therefore cross at this geometrical arrangement of the nuclei. A common example of a geometry at which crossings can occur is a high-symmetry arrangement, for which the electronic wavefunctions have different symmetry classifications within the appropriate point group and thus do not interact ($V_{12}=0$). A distortion that reduces the symmetry may cause the symmetry types of the two electronic wavefunctions to become the same in the lower symmetry point group, in which case $V_{12}\neq 0$ and the states can no longer cross in the distorted geometry. The result is a seam of crossing points for nuclear configurations that retain the higher symmetry (but which may differ in, for example, bond lengths, therefore changing the position and energy of the crossing point on the PES), with avoided crossings and no points of degeneracy of the two states for lower symmetry geometries.

In general, we can describe the region of the intersection of two adiabatic PESs in terms of two coordinates. One of the coordinates is the direction along which the nuclei must move in order to bring the two adiabatic states close in energy, and ultimately up to and past a point of degeneracy, while keeping the coupling interaction $V_{12}=0$. This motion maintains the symmetry of the point of intersection to enforce $V_{12}=0$, and is thus totally symmetric; it is known as the *tuning* coordinate because it tunes the energy separation between the two adiabatic states. The second coordinate is known as the *coupling* coordinate and corresponds to a non-totally symmetric motion that reduces the symmetry in such a way as to cause the two states to avoid a crossing (thus removing the degeneracy by inducing $V_{12} \neq 0$). Along the remaining $N-2$ coordinates (where N is the number of degrees of vibrational freedom), there will be an $(N-2)$-dimensional seam of intersections between the adiabatic PESs because motions along these coordinates can take place without lifting the degeneracy of the two states. The orthogonal tuning and coupling coordinates define what is known as the *branching space* of the intersection (also known as the g–h plane[234]). The tuning and coupling coordinates do not necessarily correspond to normal modes of vibration.

The form of the point of intersection between the two adiabatic PESs, when plotted in the two coordinates of the tuning and coupling motions, takes the appearance of a double cone, as illustrated in Figure 4.8, giving rise to the name *conical intersection.* Such conical intersections may arise on symmetry grounds between states of different symmetry (when the coupling between the diabatic states, V_{12} is required to be zero), as mentioned previously. If the two electronic potentials transform as different irreducible representations of the point group of the system, the diabatic coupling term V_{12} is only non-zero for geometries distorted from totally symmetric along the coupling coordinate. The degeneracy condition $V_{11} = V_{22}$ is lifted by motion along the totally symmetric tuning coordinate. The resultant conical intersections are classified as *symmetry induced.* Conversely, when the two adiabatic electronic states have the same symmetry, the states cannot cross at totally symmetric geometries, but an intersection may arise elsewhere at which the conditions for the diabatic potentials $V_{11} = V_{22}$ and V_{12} arise accidentally. Such *accidental* conical intersections are more difficult to locate because the search of configuration space must go beyond the totally symmetric geometries. The two specific motions of the nuclei that lift the degeneracy of the intersection are no longer determined by symmetry.

Degenerate states (with E or T symmetry) give rise to a symmetry-induced conical intersection at high symmetry configurations, resulting in the well-known Jahn-Teller effect[240] in which the molecule undergoes a distortion to lower symmetry to remove the degeneracy. This distortion

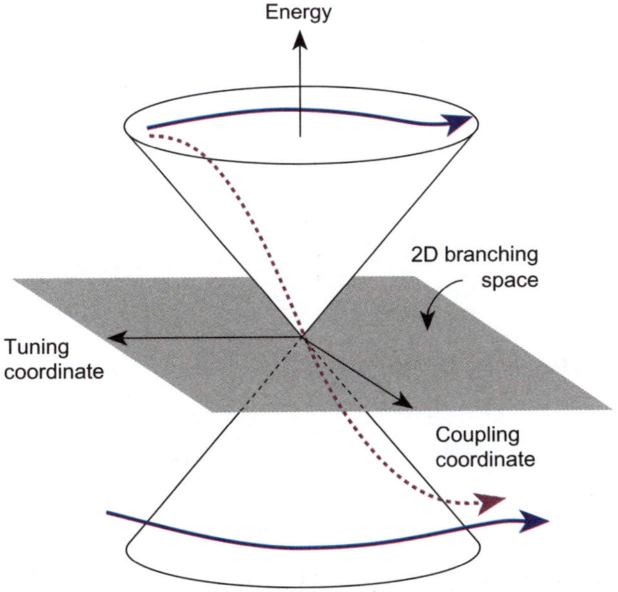

Figure 4.8 A schematic diagram of a conical intersection between two adiabatic potential energy surfaces (illustrated as the upper and lower cones). The blue arrows indicate adiabatic pathways around the conical intersection on the upper and lower adiabatic states; the red arrow is a non-adiabatic pathway between the two states.

corresponds to motion downward in potential energy on the lower of the two PESs, away from a point of conical intersection (and degeneracy) at the undistorted geometry. A much-studied conical intersection of this type arises in the H_3 system[243,244] and features in the dynamics of the reaction $H + H_2 \rightarrow H_2 + H$, which has become a benchmark for comparison of experimental and theoretical studies of chemical reaction dynamics. At the equilateral triangular arrangement of the three H atoms, the system has D_{3h} symmetry, and the lowest two PESs are doubly degenerate with E' symmetry. Distortions resulting from asymmetric extensions of the three H-H distances reduce the symmetry to C_s or C_{2v}, and the two PESs separate in energy, becoming A_1 (lower state) and B_2 (upper state) symmetry types for isosceles (C_{2v}) and A' for scalene (C_s) triangular geometries. Linear triatomic and larger molecules in degenerate electronic states (of Π, Δ... symmetry) experience a similar lifting of the degeneracy with distortion from linearity (mediated by the bending vibration) known as the Renner-Teller effect[240] (see Section 3.3.7). There is a seam of intersection of the resultant two electronic states at linearity, but the topology differs from that of conical intersections.

The conical intersection that arises for $H + H_2 \rightarrow H_2 + H$ and its isotopic variants at the equilateral triangular geometry plays an important role in the scattering dynamics on the lowest adiabatic PES, because it results in a substantial barrier to side-on approach of an H atom to the H_2 molecule. The dynamical consequences of this feature on the PES have been explored in detail in quasi-classical trajectory calculations by Greaves *et al.*[245] The higher lying of the two interacting adiabatic potentials that give rise to the conical intersection is too high in energy (2.7 eV or 260 kJ mol^{-1}) to be accessible at the collision energies of most experimental studies. Dual-surface scattering dynamics are thus not expected for this system under normal circumstances. Nevertheless, the conical intersection has been suggested to play an important role through *geometric phase* effects that are a characteristic of such features. The possible role of the geometric phase is discussed further in Study Box 4.2.

A second example of a reaction that exhibits a conical intersection is $O(^1D) + H_2 \rightarrow OH + H$, for which schematic potential energy curves are shown in Figure 4.9 for both colinear and non-linear geometries. Linear configurations are remote from the minimum energy path on the ground state PES, which passes through the H_2O potential well.[246–248] In a colinear geometry of the three atoms (a configuration with $C_{\infty v}$ symmetry), the approach of the H_2 causes the $O(^1D)$ electronic state to

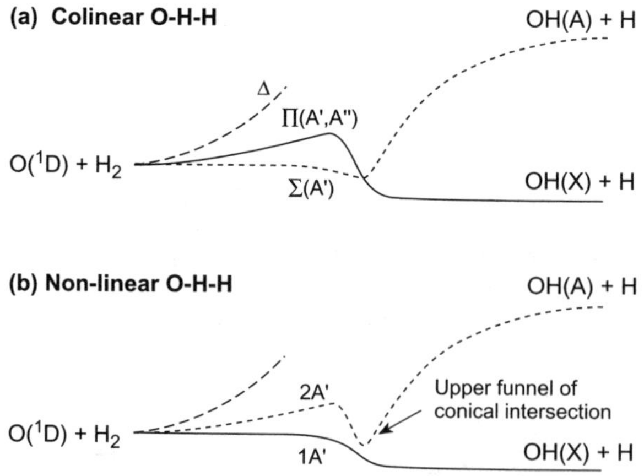

Figure 4.9 Schematic one dimensional cuts of singlet-state PESs for the $O(^1D) + H_2 \rightarrow OH + H$ reaction, correlating to ground state $OH(X^2\Pi)$ and excited state $OH(A^2\Sigma^+)$ radical products. The top diagram (a) is for colinear reaction, and the lower panel (b) for reactions deviating a small amount from linearity. A conical intersection forms between the $1A'$ and $2A'$ states as discussed in the text. The A'' state arising from a component of the Π state is not shown in the lower panel but resembles the Π state in the upper panel. Adapted from ref. [246].

STUDY BOX 4.2: THE GEOMETRIC PHASE AT CONICAL INTERSECTIONS

Herzberg and Longuet-Higgins[252] demonstrated the general result that if nuclear motion on a single adiabatic potential energy surface loops completely around a conical intersection, the adiabatic electronic wavefunction must undergo a change of sign (equivalent to a phase shift of π). As the total wavefunction for the nuclei and electrons must remain single-valued at all points, it therefore follows that the nuclear wavefunction must experience an identical sign change. The change in sign of the electronic wavefunction caused by the nuclei completing one loop (or in general an odd number of loops) around the conical intersection is known as a *geometric* (or *Berry*) phase.[253,254] If a reaction involves a PES with a conical intersection, and the nuclear wavefunction spreads all the way round the region of the intersection (*e.g.*, if the collision energy exceeds all barriers on the minimum energy path around the conical intersection), significant interference effects might thus be expected to arise at points where there is overlap of parts of the wavefunction that have experienced different degrees of encirclement of the intersection (that sum to 2π). The realization that the geometric phase (GP) might have consequences for nuclear scattering dynamics that could be measured has prompted considerable theoretical and experimental research activity.

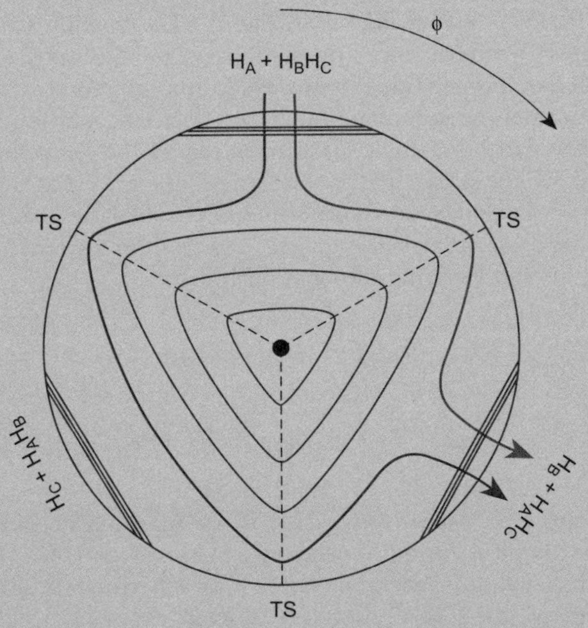

Figure 4.10 A schematic diagram of the PES for the $H+H_2$ reaction; the three H atoms are labelled as A, B and C to distinguish the three possible outcomes. Transition state barriers to H atom exchange are denoted by TS and dashed lines and • marks the location of the conical intersection. The red and blue arrows show direct (over one transition state) and looping (passing over two transitions states) pathways for reaction. The angle of encirclement of the conical intersection is ϕ. Adapted from ref. [220].

A favourite system for study of GP effects has been the reaction $H+H_2 \rightarrow H_2+H$ (and the isotopologues $H+D_2$ and $D+H_2$). In this reaction there is a seam of conical intersections between the ground and first excited adiabatic PESs, which occurs at the equilateral triangular (D_{3h}) arrangement of the three H atoms. Figure 4.10 shows a schematic PES for the reaction illustrating the three possible channels and the location of the conical intersection and barriers to H atom exchange.[220] The lowest energy conical intersection in the seam occurs 2.7 eV above the minimum energy of separated $H+H_2$. Most experimental studies are conducted at energies lower than this, well below the points of intersection of the surfaces, and the reaction can then be considered to occur on a single adiabatic PES. Nevertheless, with barriers to hydrogen exchange of only 0.42 eV, the nuclear wavefunction can extend around the conical intersection with significant amplitude even for the collision energies typical of experimental studies. Considerable controversy over GP effects in the $H+H_2$ reaction and its isotopologues has recently been resolved by fully converged, time-independent QM scattering calculations by Kendrick,[255] time-dependent wavepacket scattering calculations by Juanes-Marcos and Althorpe,[256] and deductions based on topological arguments.[220] The consensus is now that GP effects that are evident in the odd or even partial wave scatterings, associated with different total angular momenta J (see Chapters 3 and 12), cancel out when these odd and even J results are combined to obtain integral or differential cross sections (see Chapters 5 and 6). Thus the effects of the GP are negligible, except perhaps for large J values. This is true regardless of the collision energy. Results of experimental studies of the scattering dynamics at energies as high as 2.4 eV can thus be replicated by QM scattering, or QCT calculations, that make no allowance for GP effects. Other reaction, photodissociation and inelastic scattering systems may, however, hold out better prospects for observation of these consequences of the geometric phase.

The reason that GP effects are negligible in the $H+H_2$ reaction can be understood from Figure 4.10, and relates to interference (or perhaps more precisely, a lack of interference) between the direct (red) and looping (blue) pathways for the reaction $H_A+H_BH_C \rightarrow H_B+H_AH_C$. QM scattering calculations of the square moduli of the scattering amplitudes $f_{direct}(\theta)$ and $f_{loop}(\theta)$ would reveal the different scattering angle (θ) distributions in the centre-of-mass (CM) frame for reactions that proceed by these two pathways (see Chapter 3). Extraction of the scattering amplitudes for these two mechanisms from normal QM calculations is, however, complicated because the two pathways are inextricably mixed. Juanes-Marcos *et al.*,[220] demonstrated that they can be computed instead by using the relationships:

$$f_{direct}(\theta) = \frac{1}{\sqrt{2}}\{f_{NGP}(\theta) + f_{GP}(\theta)\} \tag{B4.2.1}$$

$$f_{loop}(\theta) = \frac{1}{\sqrt{2}}\{f_{NGP}(\theta) - f_{GP}(\theta)\}, \tag{B4.2.2}$$

where $f_{GP}(\theta)$ and $f_{NGP}(\theta)$ are, respectively, scattering amplitudes derived from QM scattering calculations with and without inclusion of geometric phase effects (NGP here stands for neglect of the GP). When such calculations were performed for state-to-state scattering for the $H+H_2$ reaction at a total energy of 2.3 eV, over a restricted range of total angular momentum ($0 \leqslant J \leqslant 10$, corresponding semi-classically to low impact parameters), the scattering amplitudes for the direct and (much lower probability) looping pathways were distributed, respectively, in the backward ($90 \leqslant \theta \leqslant 180°$) and forward ($0 \leqslant \theta \leqslant 90°$) hemispheres, with no overlap. From manipulation of equations (B4.2.1) and (B4.2.2), it is clear that $f_{NGP}(\theta)$ and $f_{GP}(\theta)$ can be expressed as linear combinations of $f_{direct}(\theta)$ and $f_{loop}(\theta)$. The differential cross sections without and with GP effects are obtained from the square moduli of the NGP and GP scattering amplitudes, and will be essentially identical because they differ only in interference terms that are products of $f_{direct}(\theta)$ and $f_{loop}(\theta)$, and which are negligible because the two functions do not overlap. For higher J (corresponding to larger impact parameters) there is some overlap between the scattering amplitudes for direct and loop pathways, and as a result the calculated differential cross sections with and without GP effects show some differences,[256] but these vanish when integrated over all scattering angles to obtain integrated cross sections. These differences remain to be observed in an experiment. The reasons for very different scattering directions for the products of direct and looping reaction paths are revealed by QCT calculations. The direct mechanism involves approach of H_A towards the H_C end of the H_BH_C molecule, with rapid abstraction of H_C. Conversely, the loop pathway corresponds to approach of H_A towards the H_B end (at large impact parameters) followed by insertion between the H_B and H_C atoms and prompt removal of H_C.

Bertrand Retail and Andrew J. Orr-Ewing

split into $^1\Sigma$, $^1\Pi$ and $^1\Delta$ components. The $^1\Delta$ state is repulsive, but the $^1\Pi$ and $^1\Sigma$ states correlate respectively to $OH(X^2\Pi)+H(^2S)$ (ground state) and $OH(A^2\Sigma^+)+H(^2S)$ (electronically excited) products. The $^1\Pi$ state exhibits a small barrier, and thus in the region of approach of the reactants, lies higher in energy than the $^1\Sigma$ state. After this barrier, however, the $^1\Pi$ state drops in energy whereas the $^1\Sigma$ state rises, and the two states cross. If the geometry deviates slightly from colinearity, the symmetry is reduced to C_s, and the Σ state becomes of $^1A'$ symmetry (symmetric to reflection in the plane of the three atoms) while the Π state splits into a $^1A'$ and a $^1A''$ component (the latter being antisymmetric to reflection). The $^1A'(\Sigma)$ and $^1A'(\Pi)$ surfaces can interact, resulting in the $1^1A'$ ground state PES and the $2^1A'$ excited state surface that correlates to electronically excited OH(A) radicals. The $^1A''(\Pi)$ PES (denoted as $1^1A''$) cannot mix with the $^1A'$ surfaces and correlates to OH(X). On the reactant side of the crossing region of the $^1\Sigma$ and $^1\Pi$ surfaces, the $1^1A'$ surface is predominantly the $^1A'(\Sigma)$ PES, and on the product side it is almost entirely of $^1A'(\Pi)$ character. The opposite is the case for the $2^1A'$ surface. There is thus a symmetry induced conical intersection at the crossing because, in linear geometry, the $^1\Sigma$ and $^1\Pi$ PESs can cross (having different symmetry), but a distortion from linearity (the coupling mode) reduces the symmetry to C_s and the $1^1A'$ and $2^1A'$ PESs separate in energy. The tuning mode can be regarded as a colinear motion of the three atoms that brings the $^1\Sigma$ and $^1\Pi$ PESs to a point of degeneracy. Adiabatic reaction dynamics would involve nuclear motion on the $1^1A'$ PES and possibly also the $1^1A''$ PES,[246–249] but any flux that was initially on the $2^1A'$ surface (which is degenerate with the $1^1A'$ surface at large separation of the $O(^1D)$ and H_2) would not contribute to reaction. Non-adiabatic transfer of flux from the $2^1A'$ PES, perhaps at the conical intersection, or elsewhere through derivative couplings, might, however, make a contribution to the reactive scattering. As the $2^1A'$ and $1^1A''$ PESs both derive from the distortion of the $^1\Pi$ state from linearity, they might be expected to have similar topologies in near-linear geometries of approach of the reactants, and thus exhibit similar entrance channel dynamics. Attempts have been made to quantify the extent of non-adiabatic dynamics in this reaction (and the analogous $O(^1D)+HD$ reaction), both by experimental study and quasi-classical trajectory (QCT) and quantum mechanical scattering calculations combined with surface-hopping techniques.[246–248] The surface-hopping methods developed by Tully and coworkers provide a way to incorporate multi-surface dynamics into QCT calculations.[86,250,251] The derivative couplings between PESs for the $O(^1D)+H_2$ reaction are greatest in the vicinity of the seam of conical intersections, but they are also non-negligible in the reactant region where the $1^1A'$ and $2^1A'$ PESs approach degeneracy. The QCT calculations suggested that trajectories that are started on the $2^1A'$ surface at collision energies in excess of the barrier, and then undergo non-adiabatic transitions to the $1^1A'$ PES, play a minor role. The contribution to reaction cross sections from dynamics started on the $2^1A'$ PES is about half that of trajectories initiated on the $1^1A''$ PES, and less than 10% of the total reaction cross section. Both QM and QCT calculations also show the importance of mixing between flux on adiabatic $1^1A'$ and $2^1A'$ surfaces at large $O(^1D)–H_2$ separations. The reaction is discussed further in Section 6.3.1.2.

4.2.6 Vibronic, Coriolis and Spin-orbit Couplings between PESs

The preceding sections concentrated mostly on non-adiabatic couplings and dynamics in the vicinity of crossings of PESs, but non-adiabatic interactions can also occur between PESs that do not cross (as illustrated in Section 4.2.4). Both types of interaction can be mediated by vibrational modes, the rotation of the whole molecular system or by spin-orbit couplings, and this section examines these coupling mechanisms.

The coupling of electronic states by vibrational motion is called vibronic coupling (*vibronic* being a contraction of vibrational and electronic). The types of interaction to be discussed below are sometimes referred to as *weak* vibronic coupling to distinguish the *strong* vibronic couplings that occur at conical intersections (recall that tuning and coupling modes of the molecular system split the degeneracy of the conical intersection, and there is thus a coupling of electronic and vibrational motions). Vibronic couplings will be non-zero only if the product of the symmetries of the two relevant electronic states, ϕ_i and ϕ_j, and the normal mode of vibration, Q, contains the totally symmetric representation of the point group of interest, *i.e.*

$$\Gamma_i \otimes \Gamma_j \otimes \Gamma_Q \supset A. \tag{4.16}$$

If the coupling is symmetry allowed, its magnitude will depend on several factors. In the case of weak couplings, a common situation (most notably in the case of unimolecular reactions such as in photodissociation) is the coupling occurring between a bound and an unbound state. The unbound PES supports a continuum of states for nuclear motion, parameterized by the density of states, $\rho(E)$, which is the number of quantum states within a particular energy range. In the case of vibronic coupling, the transition probability (per unit of time) for the non-adiabatic change between electronic states of the same multiplicity is given by Fermi's Golden Rule:

$$P_{i \to j} = \frac{2\pi}{\hbar} \left| \langle j | \hat{T}_{\mathrm{N}} | i \rangle \right|^2 \rho_j(E). \tag{4.17}$$

An equivalent expression is often given in a diabatic representation.[257] The probability depends on the square of the non-adiabatic coupling matrix element, which couples the adiabatic bound state ϕ_i with the unbound state ϕ_j *via* a perturbation provided by the nuclear kinetic energy operator, as well as on the density of unbound states. The case of couplings between two unbound states is more relevant for bimolecular reactions, but there is no simple analytical expression for the transition probability. It is clear, however, that this probability will be enhanced if the perturbation (the vibrational coupling mode) brings the two coupled states close to resonance. For example, if two adiabatic states are separated by $1000\,\mathrm{cm}^{-1}$, vibrational excitation of a bending mode of frequency close to this value will be a much more efficient coupling mode than a higher-frequency stretching mode, provided the symmetry requirements of Eq. (4.17) are satisfied. This process is the microscopically reversible equivalent of the electronic to vibrational energy transfer processes that occur in many inelastic collisions and contribute to quenching of excited electronic states.

Coupling between electronic states induced by the overall rotational motion of the nuclei is called the rotational or Coriolis coupling. In a bimolecular collision it is a coupling mediated by the nuclear orbital angular momentum, ℓ. As discussed in Chapter 1, classically, the magnitude of ℓ is related to the impact parameter b *via* $|\ell| = \mu b v_{\mathrm{rel}}$, where μ is the reduced mass of the system, and v_{rel} is the magnitude of the relative velocity vector, *i.e.* the relative speed of the two orbiting partners. Reactions involving a stripping mechanism (with large b) will be subject to a greater Coriolis coupling of electronic states than reactions occurring *via* a rebound mechanism (with small b). In bimolecular reactions, Coriolis coupling is generally considered to have only a minor effect in comparison with vibronic couplings. Its (symmetry-based) selection rules differ from those for radial derivative couplings, and can therefore make Coriolis interactions significant in specific cases; for example, in linear configurations of the atoms, with a well defined projection of the electronic angular momentum on the internuclear axis (Ω), Coriolis couplings can mix states with different values of Ω (see Section 3.3.6).

So far, the discussion has neglected the effects of the spins of the electrons. When electron spin is taken into account, however, several terms must be added to the total Hamiltonian to describe further types of interaction; examples are spin-orbit, spin-spin and spin-rotation couplings (which involve coupling of the electron spin angular momentum with, respectively, the electronic orbital angular momentum, the spin angular momenta of other electrons, and rotation of the nuclear framework).[257] In the case of bimolecular reactions, spin-orbit coupling is the most important of these interactions, and is the focus of the subsequent discussion. The phenomenon of spin-orbit coupling arises from the interaction of the intrinsic magnetic moment of an electron (associated with the spin) with the magnetic field generated by its orbital motion, or that of other electrons. In both photodissociation and bimolecular reaction dynamics, non-adiabatic couplings mediated by spin-orbit interactions can arise when either a reactant or product species has non-zero values for its electronic orbital and spin angular momenta. Benchmark reactions for study of the effects of spin-orbit coupling thus include reactions of, or that liberate, halogen atoms (for which $S=1/2$, $L=1$, where S and L are the quantum numbers for the total electron spin and the electronic orbital angular momentum, respectively) or open-shell radicals such as OH or NO (both possessing ground electronic states with $S=1/2$, $\Lambda=1$, where Λ is the quantum number for the projection of the electronic orbital angular momentum onto the internuclear axis).

To understand how the strength of the spin-orbit coupling of an asymptotically open-shell product will affect the nuclear dynamics of the whole reaction, it is instructive to digress again to examine the photodissociation dynamics of the hydrogen halide (HX) molecules, building on the content of Section 4.2.4. HX photodissociation can be thought of as a unimolecular reaction in which the dynamics occur on PE curves corresponding to electronically excited states. These excited states are initially populated in the Franck-Condon region (*i.e.* the range of small internuclear distances defined by the vibrational wavefunction of the stable molecule in its ground $X^1\Sigma^+$ state prior to photoexcitation) by absorption of a UV photon. Figure 4.5 illustrates the important PE curves for HCl photodissociation, and the same patterns of states arise for HF, HBr and HI. The key difference in the HX series in going from HF to HI is the increasing magnitude of the spin-orbit coupling of the halogen atom. Strong spin-orbit interactions in HI make the spin a poor quantum number and there is mixing of (nominally) singlet and triplet states; Ω remains a good quantum number, however, and Hund's case (c) is the best description of angular momentum couplings. In HI, the selection rule for photoexcitation, $\Delta\Omega=0, \pm1$ results in initial population of the $^1\Pi(1)$, $^3\Pi(1)$ and $^3\Pi(0^+)$ states (with the number in parenthesis denoting Ω, and use of a mixed Hund's case (a)/case (c) notation as in Section 4.2.4 to facilitate identification of the states – see also Study Box 4.3). A very different population of states thus occurs when compared to HF or HCl, for which the spin-orbit interaction is weaker (so Λ, S and Σ remain good quantum numbers), a Hund's case (a) description of the states is appropriate, and the $^1\Pi(1)$ state is almost exclusively populated by photo-excitation because of the $\Delta S=0$ selection rule. At longer range, each molecule presents the same set of adiabatic PE curves (see Figure 4.5), the main difference being the energy separation between the two asymptotic channels, leading to $H(^2S_{1/2})+X(^2P_{3/2})$ and $H(^2S_{1/2})+X^*(^2P_{1/2})$, with X and X* the ground and spin-orbit excited states, respectively, of the halogen atom. Note that the adiabatic PE curves shown in Figure 4.5 include the contributions to the potential energy from spin-orbit interactions, and thus X and X* are shown as different energy asymptotes; it is worth emphasizing that having included the spin-orbit interaction in the adiabatic potentials (by diagonalization of the Hamiltonian), the dynamics on these PE curves are not coupled by spin-orbit effects, but the off-diagonal interactions of the nuclear kinetic energy operator can cause non-adiabatic transitions. The difference in energy at large separations depends on the strength of the

STUDY BOX 4.3: ANGULAR MOMENTUM COUPLING AND HUND'S COUPLING CASES

Every quantum mechanical angular momentum has an associated magnetic moment, which can interact with external fields and with other magnetic moments. When dealing with atoms, the angular momenta we are interested in are the electronic orbital and spin angular momenta, and for molecules we must add to these the angular momentum associated with molecular rotation.

Consider a single quantum mechanical magnetic moment interacting with an external magnetic field. The magnetic moment may take one of $2J+1$ orientations relative to the field axis. Before the external field is switched on, all of these orientations have the same energy. After the field is switched on, this single energy level will split into $2J+1$ components, since the energy of interaction between the magnetic moment and the external field depends on their relative orientation. This is the well-known Zeeman effect.

A similar effect occurs when two quantum mechanical magnetic moments interact with each other. The energy levels involved will be split according to the number of possible relative orientations (and therefore interaction energies) between the two magnetic moments, in a phenomenon known as angular momentum coupling. To take a simple classical analogy, it is clear that the interaction energy between two bar magnets depends on their relative orientation, since if the two magnets are placed end-to-end they will either attract or repel each other, depending on whether opposite or like poles are in contact. The number of possible relative orientations for two quantized magnetic moments may be determined from the Clebsch-Gordan series, which gives the possible magnitudes for the vector sum of two quantized vectors $\boldsymbol{J}_1$ and $\boldsymbol{J}_2$. The series takes the form $J_1+J_2, J_1+J_2-1, \ldots, |J_1-J_2|$.

In a molecule, coupling is possible between any of the angular momenta the molecule possesses. This leads to spin-orbit coupling, spin-rotation coupling, spin-spin interactions, and Λ doubling (orbital-rotation coupling). The energy level patterns that arise from all of these interactions depend on the relative magnitudes of the various types of coupling. For linear molecules, Hund classified the energy level patterns according to a number of different *cases*, defined by comparing the splitting caused by the various types of angular momentum coupling, ΔE, with the separation between rotational energy levels in the molecule. The four coupling cases are defined below.[239,262] Note that 'well separated' and 'close' are defined by $\Delta E \gg BJ$ and $\Delta E \ll BJ$, respectively.

Hund's case (a) coupling: Strong coupling of $\boldsymbol{L}$ to the molecular axis (states with different $|\Lambda|$ quantum number are well separated in energy); strong spin-orbit coupling (states with different $|\Omega|$ quantum number are well separated in energy).

Hund's case (b) coupling: Strong coupling of $\boldsymbol{L}$ to the molecular axis (states with different $|\Lambda|$ quantum numbers are well separated in energy); weak spin-orbit coupling (states with different $|\Omega|$ are close in energy).

Hund's case (c) coupling: Coupling of $\boldsymbol{L}$ to the molecular axis is weaker than spin-orbit coupling; strong spin-orbit coupling (states with different $|\Omega|$ are well separated).

Hund's case (d) coupling: Weak coupling of $\boldsymbol{L}$ to the molecular axis (states with different $|\Lambda|$ are close); weak spin-orbit coupling (states with different $|\Omega|$ are close in energy).

Claire Vallance

spin-orbit coupling in the isolated atom (for F atoms the spin-orbit splitting is 404 cm^{-1}, for Cl it is 882 cm^{-1} and for I it is 7603 cm^{-1}). In the case of HI, the PE curves remain well separated in energy, and the coupling of the adiabatic states *via* the nuclear kinetic energy operator is negligible, so the bond dissociation is fully adiabatic.[258] For HF and HCl, however, the large energy separation of states in the Franck-Condon region decreases at extended bond lengths, rendering non-adiabatic couplings operational to promote transfer of reactive flux from one surface to another, as seen in Section 4.2.4. The couplings occur because of the change in electronic character as the bond length extends (so $\partial\phi_e/\partial R \neq 0$),[259] as shown in Figure 4.6. For HF or HCl, the evolution from a Hund's case (a) description in the Franck-Condon region into Hund's case (e) in the asymptotic region (involving recoupling of angular momenta from the molecular to the separated atom limit) implies a significant change in electronic character of adiabatic potentials that lie close in energy, and thus non-negligible first derivative couplings. For HI, this change is insignificant as case (c) is equivalent to case (e) as the bond length increases ($\hat{H}_{SO}$ is the dominant term in $\hat{H}_e$ in both cases) and thus the dynamics are fully adiabatic. The photodissociation dynamics of HBr lie between these two extreme cases because of the intermediate spin-orbit coupling in the Br atom (with spin-orbit splitting of 3685 cm^{-1}).[260] The behaviour of the hydrogen halides should be, at least in principle, transposable to the long range parts of PESs describing chemical reactions involving halogen atoms,[261] as will be discussed further in Section 4.3.

4.3 EXPERIMENTAL PROBES OF NON-ADIABATIC REACTIVITY AND DYNAMICS

The preceding sections have largely concentrated on the theoretical concepts underlying our understanding of reaction dynamics on coupled PESs. A fast-growing body of experimental data is now serving to identify reactions in which non-adiabatic transitions occur, and to quantify the extent to which these non-adiabatic dynamics influence reaction pathways. These experimental data provide important tests of the rigour of theoretical treatments of dynamics on multiple PESs. This section thus presents an introduction to the experimental strategies that are used to probe non-adiabatic reactivity, and discusses ways to unravel the dynamics of these non-Born-Oppenheimer reaction pathways. The key to success lies in quantum-state-specific preparation and/or detection of reactants and products, respectively.

Most measurements of non-adiabatic reactivity have been reported for reactions in which one of the reactants or products has a ground electronic state split by weak spin-orbit couplings, *e.g.*, Cl (882 cm^{-1}) or F (404 cm^{-1}) atoms. This results in closely separated PESs asymptotically leading to each spin-orbit state. One experimental approach is to prepare a specific population of the reactant in its spin-orbit-excited state, and to examine a reaction in which the PESs derived from this excited state do not correlate adiabatically with energetically accessible reaction products (so all adiabatic reaction pathways are closed). Observation of reaction products from the reaction of the spin-orbit-excited reactant is therefore an indicator of non-adiabatic dynamics. More detailed information derives from measurements of quantum-state-specific product distributions. Fingerprints of non-adiabatic reaction mechanisms are often revealed through consideration of energy and momentum conservation. For example, because the excited reactant carries additional internal energy compared to the ground-state reactant, this extra energy should be reflected in the kinetic or internal energies of the products. The reactions of $F(^2P)$ and $Cl(^2P)$ atoms with H_2 serve as useful illustrations. Figure 4.11 shows correlation diagrams for the low lying PESs of these systems.

For the $Cl + H_2$ reaction, different sources of Cl atoms can be exploited to examine the reactivity of the two spin-orbit states: photolytic production using Cl_2 photodissociation at a wavelength of 355 nm (the third harmonic of a Nd:YAG laser) produces almost exclusively $Cl(^2P_{3/2})$,[264] whereas an electrical discharge source results in a mixture of ground $Cl(^2P_{3/2})$, and excited $Cl^*(^2P_{1/2})$ atoms.[265] A comparison of the scattering dynamics forming HCl products (*e.g.*, by probing product quantum state distributions and scattering angles) using each source in turn can thus separate out the contributions to the reaction dynamics from Cl and Cl* atoms. This method was employed by Liu and coworkers to observe the non-adiabatic reactivity of Cl* towards H_2.[265] The product

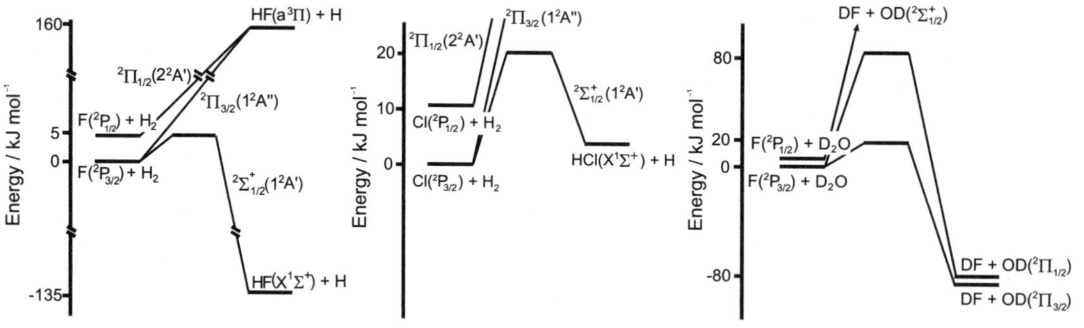

Figure 4.11 Correlation diagrams for the electronic states of the $F(^2P_J)+H_2 \rightarrow HF+H$, $Cl(^2P_J)+H_2 \rightarrow HCl+H$ and $F+D_2O \rightarrow DF+OD$ reactions. Analogous correlation diagrams for reactions of $O(^3P)$ atoms with saturated hydrocarbons are examined in very instructive detail in ref. [263].

differential cross section (DCS) was found to be more backward peaked when using the discharge source, and the product total kinetic energy release distribution had an additional high-energy tail extending to the expected energetic limit for $Cl^* + H_2$. To quantify the Cl* reactivity relative to Cl, knowledge of the exact ratio of Cl* to Cl in the different molecular beam sources is required. This was measured using tuneable vacuum ultra-violet (VUV) radiation in a resonance-enhanced multiphoton ionization (REMPI) detection scheme involving one VUV and one UV photon[266] (see Study Box 5.2). A similar method was used by Yang and coworkers to quantify the collision energy dependence of the relative reactivity of F* and F in the reaction $F+D_2 \rightarrow DF+D$.[267]

Calibration of molecular beam sources is difficult, however, and an alternative way to probe non-adiabatic reactivity is to reverse the direction of reaction, and seek evidence for *formation* of spin-orbit excited products, as has been demonstrated, for example, for $H+HCl \rightarrow H_2+Cl/Cl^*$.[268,269] State-specific, laser-based VUV or UV detection methods (*e.g.*, laser-induced fluorescence (LIF) or REMPI) can distinguish the Cl and Cl* products and quantify the branching ratio, Γ, which is usually evaluated as:

$$\Gamma = \frac{\sigma(\mathrm{Cl}^*)}{\sigma(\mathrm{Cl}) + \sigma(\mathrm{Cl}^*)}. \tag{4.18}$$

Here, σ(Cl) and σ(Cl*) denote the relative cross sections (or calibrated signals) for Cl and Cl* products. A similar strategy was employed by Orr-Ewing and coworkers to quantify the relative rate of non-adiabatic reactivity in the $CH_3+HCl \rightarrow CH_4+Cl/Cl^*$ reaction.[270] UV laser-induced fluorescence (see Study Box 5.2) has also been used successfully to detect excited spin-orbit state OH and OD ($X^2\Pi_{1/2}$) products originating from reactions on coupled PESs in the case of the $H+N_2O \rightarrow OH+N_2$,[271] $H+H_2O \rightarrow OH+H_2$,[272] $O(^3P)+CH_4 \rightarrow OH+CH_3$[273] and $F+D_2O \rightarrow DF+OD$[274] reactions. Figure 4.11 shows the energetics of the adiabatically correlated reaction pathways for the $F+D_2O$ reaction. Under the conditions of the experiment, with a mean collision energy of 21 kJ mol^{-1}, 32% of the OD was formed in the $^2\Pi_{1/2}$ state, despite the ~100 kJ mol^{-1} energetic barrier for the adiabatic reaction pathway to these products.[274,275] Efficient non-adiabatic dynamics in the post-TS region of the ground-state PES must be responsible for the formation of the $OD(^2\Pi_{1/2})$. The measurements were accompanied by *ab initio* electronic structure theory calculations of the derivative couplings between the lowest two adiabatic PESs, which were found to be most significant just after the TS, but to have declined significantly in magnitude by the exit channel region of the PESs.

Once non-adiabatic reactivity has been unequivocally identified in a reaction, and the branching between adiabatic and non-adiabatic pathways established, the reasons for breakdown of the BO approximation can be addressed. In addition to branching data, we thus require dynamical information about the competing adiabatic and non-adiabatic reaction pathways. By combining a single source of both ground and spin-orbit excited reactants with, for example, spectroscopic probing,[276–279] velocity map imaging,[280] or Rydberg atom time-of-flight study of quantum-state resolved products,[267] discrimination is possible between reaction products that arise from adiabatic or non-adiabatic pathways. In the case of $F+D_2$, high-resolution scattering experiments[267] resolved the adiabatic and non-adiabatic product velocity-flux scattering distributions in the centre-of-mass frame, as shown in Figure 4.12, where experimental results are compared with the outcomes of QM scattering calculations. The strong similarity between the experimental results for the F and F* reactions suggests that the non-adiabatic couplings occur in the entrance channel of the PESs, before the system reaches the reaction barrier on the ground surface where the scattering of reaction products is determined. F* atoms are quenched to F atoms early in the course of the reaction, and the subsequent dynamics occur on the lowest energy PES. (Further aspects of the $F+H_2$ reaction are discussed in Chapter 6.)

For the $Cl+H_2$ reaction, the scattering behaviours of the Cl and Cl* pathways were also observed to be similar,[265] but a significant difference was reported in the HCl product rotational distributions.

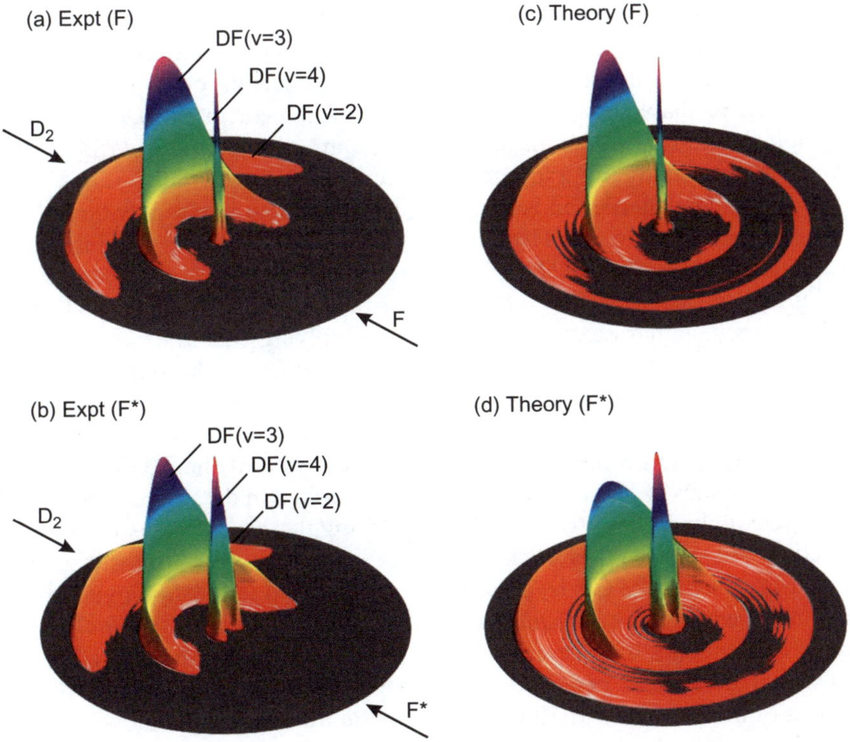

Figure 4.12 A comparison of experimental (left column) and theoretical (right column) D atom product scattering distributions in the CM-frame (shown as fluxes) for the F/F^*+D_2 ($j=0$) reaction at a collision energy of $2.0\,kJ\,mol^{-1}$. The concentric circles correspond to DF formed in different rotational and vibrational states, some of which are labelled. The near quantitative agreement between experiment and theory demonstrates correct treatment of adiabatic and non-adiabatic scattering dynamics on a chemically accurate PES. From L. Che *et al.*, Science **317**, 1061 (2007) (ref. [267]). Reproduced with permission from AAAS.

The experimental data showed a strong non-adiabatic dynamical propensity toward higher rotational excitation of HCl products (for which the most populated rotational level is $j_{HCl}=9$) while the adiabatic route produced HCl with much less rotational excitation (with greatest population in $j_{HCl}\sim 4$). These results suggested that non-adiabatic interactions occur most strongly at bent Cl—H—H geometries. This deduction could account for an increase in the non-adiabatic branching ratio with higher collision energy for the reverse reaction $H+HCl\rightarrow H_2+Cl/Cl^*$ because the ground adiabatic PES favours a colinear minimum energy path, but more bent configurations can be accessed as the available energy increases.[268,269] Initial discrepancies between the experimental findings and theoretical calculations for the $Cl+H_2$ reaction[281,282] were resolved in a comparison of collision energy dependent differential cross sections for the Cl and Cl* reactions from crossed molecular beam experiments and QM scattering calculations incorporating non-adiabatic dynamics.[283] A recent and novel approach to the study of this reaction involved photodetachment of an electron from the ClH_2^- (or ClD_2^-) anion (see Chapter 1), which is a bound species that can be prepared in a molecular beam, coupled with high-resolution velocity map imaging of the velocities of the detached electrons[284] (see Chapters 5 and 6, and Study Box 5.3). Conservation of energy and Franck-Condon arguments for the electron detachment dictate that the electron kinetic energies provide a signature of the internal energies of the neutral ClH_2 (or ClD_2) species formed in the reactant valley of the neutral PES for the $Cl+H_2$ (D_2) reaction, in close proximity to the region where the non-adiabatic dynamics are expected to occur. QM simulations of the structured photoelectron spectra were significantly improved by incorporation of the non-adiabatic couplings between the $^2\Sigma_{1/2}$, $^2\Pi_{3/2}$ and $^2\Pi_{1/2}$ states of the neutral Cl-H_2 system. The same terms had previously been incorporated into the Hamiltonian for the QM scattering calculations of the full dynamics on the coupled PESs. This Hamiltonian includes electrostatic, spin-orbit and Coriolis couplings between the basis set of diabatic states, and the level of agreement of QM simulations with the experimental photoelectron spectra confirms that these are correctly treated in all the theoretical studies. The main non-adiabatic interaction is between the $^2\Sigma_{1/2}$ and $^2\Pi_{1/2}$ states, and arises from spin-orbit interactions. The interaction is most significant close to the weakly bound van der Waals well in the entrance valley of the $Cl+H_2$ PES, but remains a relatively weak interaction.

In the case of the $CH_3+HCl\rightarrow CH_4+Cl$ reaction, Retail *et al.*[285] used velocity-mapped ion imaging techniques to measure product velocity distributions, and found that Cl and Cl* product channels exhibited very similar scattering behaviour, implying that the non-adiabatic couplings opening up the production of Cl* were occurring in the exit channel, after the TS on the ground adiabatic PES. These deductions are consistent with the known regions of non-adiabatic interactions for the $F+H_2$ and $Cl+H_2$ reaction systems, in which the derivative couplings are most significant in the valley lying on the isolated halogen atom side of the TS. They are also strongly reminiscent of the dynamics of HX photodissociation described in detail in Sections 4.2.4 and 4.2.6, in which derivative couplings arise from recoupling of angular momenta as the separated product limit is approached. The outcomes of the CH_3+HCl experiments are consistent with the measured rate coefficients for quenching of Cl* by collisions with CH_4 and its isotopologues, which are larger than the rate coefficients for the H-abstraction reactions.[286] Subsequent investigations of the CH_3+HCl reaction dynamics have revealed the importance of collision energy in promoting non-adiabatic transitions, and have provided evidence for a further role played by low-frequency vibrational modes of methane in mediating non-adiabatic transitions. As shown in Figure 4.13, Cl* production is observed to be higher for $CD_3+HCl\rightarrow CD_3H+Cl/Cl^*$ than for $CH_3+HCl\rightarrow CH_4+HCl$, despite a lower average collision energy.[287] This higher branching is proposed to be a consequence of vibronic coupling induced by the CD_3H bending (ν_3) mode. The CD_3H has bending vibrational frequencies that are close in energy to the spin-orbit splitting of the Cl atom (as shown in the insets to Figure 4.13), and near-resonant vibrational to electronic ($V\rightarrow E$) energy transfer in the exit channel of the ground PES might thus promote Cl* formation (so long as some CD_3H is formed from the reaction with excitation of one or more quanta of the bending mode). The

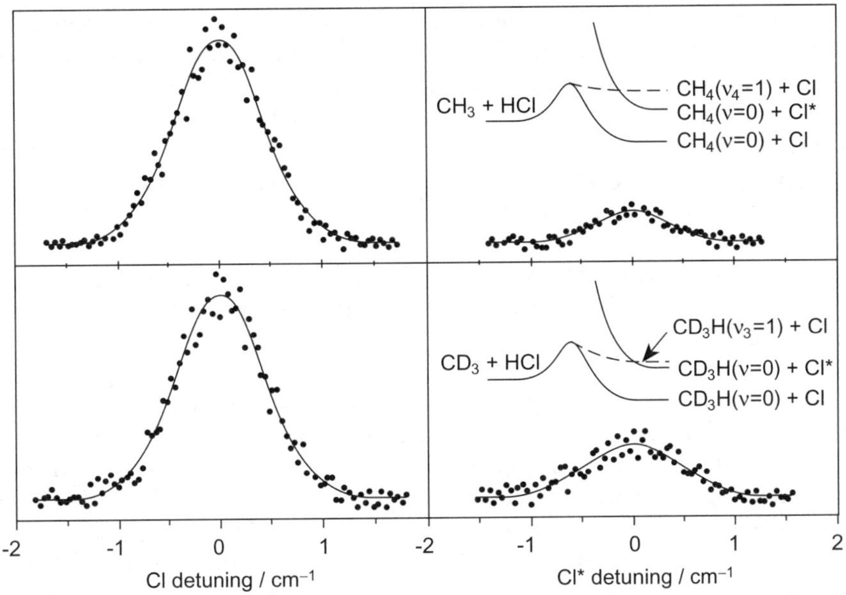

Figure 4.13 Comparison between 2+1 REMPI spectra (•) of the Cl (left) and Cl* (right) products of the $CH_3 + HCl$ (top) and $CD_3 + HCl$ (bottom) reactions, with fits to Gaussian functions (solid lines) used to derive branching ratios with the help of a suitable calibration factor.[270] The horizontal axis displays the detuning of the respective 2-photon resonant transitions and the normalized vertical axis scale is the same for both reactions. Insets: schematic cuts along the reaction coordinates of the relevant *adiabatic* PESs correlating to Cl and Cl* products, with zero-point vibrational energies in the methane co-product (solid lines). The dashed lines represent the same *adiabatic* PES correlating to Cl but with one quantum of vibration in the lowest bending mode of either $CH_4(\nu_4)$ or $CD_3H(\nu_3)$. In the $CD_3H + Cl$ product valley this process is near resonant with the $CD_3H(\nu = 0) + Cl^*$ PES, whereas in the $CH_4 + Cl$ valley, the equivalent surfaces are further apart.

equivalent mode (ν_4) in CH_4 is higher in frequency and thus further from resonance with the adiabatic PESs.

4.4 OUTLOOK

Many of the challenges for future development of experimental, theoretical and computational techniques in the field of molecular reaction dynamics lie in the study of processes on multiple, coupled PESs. As Chapter 8 will discuss, such processes are increasingly the focus of research in photodissociation dynamics, and studies of bimolecular collisions are following suit. This chapter has touched on the complexities of dealing with multi-state PESs and dynamics and the consequences of breakdown of the Born-Oppenheimer approximation, and it is our view that prospects are excellent for developing a deeper understanding in these areas. Opportunities will arise from this greater knowledge, which will be exploited to advantage, for example, in methods for controlling chemical reactivity.

There are many areas of experimental investigation of non-adiabatic dynamics where advances will be made. The collision energy dependence of such dynamics has been examined in only a few bimolecular reactions, yet the non-adiabatic coupling operators are known to depend both on the nuclear kinetic energy and on the energy separation of PESs. Changing the collision energy may allow configurations to be reached that lie away from the minimum energy pathway, but in regions where seams of intersections occur, or, more generally, where derivative couplings are enhanced. The examples provided in earlier sections focused mostly on systems in which spin is conserved, but crossings of singlet and triplet PESs are known to arise in many reactions (*e.g.*, those of $O(^3P)$ and $O(^1D)$ atoms with closed-shell molecules) and spin-orbit couplings can promote singlet-triplet mixings and intersystem crossing. Atom-molecule, radical-molecule and radical-radical reactions exhibiting competing pathways on multiple, low-lying PESs are increasingly being studied by crossed molecular beam techniques (see Chapter 6 for some examples).[288] Measurements of the branching into different product channels and the associated scattering dynamics, combined with electronic structure theory calculations of stable intermediate structures, transition states, and correlations to the various products, are helping to unravel the complicated chemistry of these systems. They remain, however, beyond the current capabilities of reduced or fully dimensional, multi-surface QM and QCT scattering calculations.

A lesson learned from studies of non-adiabatic photodissociation dynamics is that the polarization of the electronic angular momentum of photofragments such as oxygen or chlorine atoms, or OH and other diatomic radicals, depends on interference effects arising from dissociation dynamics *via* competing pathways on coupled PESs[289] (see Chapters 8 and 9). Theoretical and experimental studies of photodissociation have advanced rapidly in this area in recent years, but many of the ideas remain to be translated to bimolecular reactions. If interference effects can influence the dynamics, then perhaps there are prospects for extension of the methods of coherent control by designed laser pulses, as discussed in Chapter 11, to manipulate chemical reactions. Recent experimental studies have succeeded in identifying the coupling modes responsible for conical intersections between PESs in the photodissociation of heteroatom-containing aromatic molecules such as phenol.[290] A likely consequence will be that the effects of vibrational mode excitation on the dynamics of motion through or around such intersections can be examined and understood,[291] providing an alternative mechanism for photochemical control. We are seemingly a long way from applying such methods to bimolecular reactions on PESs coupled by conical intersections or weaker vibronic interactions, but the sophisticated laser, molecular beam and imaging technologies now available to experimentalists may open up such new horizons. Clear-cut experimental demonstrations of geometric phase effects in the nuclear dynamics of bimolecular reactions remain elusive, but their signatures may now be evident in photodissociation dynamics[292] and, with improved theoretical methods for identifying the signatures of the geometric phase, prospects are good for studying examples in chemical reactions.

4.5 PROBLEMS

1. Starting from Eq. (4.1), derive Eq. (4.5). Make use of the expansion of the total wavefunction in nuclear (ψ) and electronic (ϕ) parts:

$$\Phi = \sum_i \psi_i \phi_i$$

where explicit dependence on the coordinates R and r and on time has been dropped to simplify the notation. You will need to employ Eqs. (4.2), (4.3), and (4.6). Hint: the derivation starts by substituting the above expansion into Eq. (4.2), multiplication through by ϕ_j^* and integration over electronic coordinates, making use of orthonormality of the electronic wavefunctions.

2. Starting from Eq. (4.6) and using the definition in Eq. (4.7) of the kinetic energy operator,

$$\hat{T}_N = -\frac{\hbar^2}{2M}\nabla^2$$

derive Eqs. (4.8) and (4.9) using the definitions $F_{ji} = \langle\phi_j|\nabla|\phi_i\rangle$ and $G_{ji} = \langle\phi_j|\nabla^2|\phi_i\rangle$. Hints: to obtain Eq. (4.8), use the relationship for differentiation of the product of wavefunctions $\psi_i\phi_i$ that $\nabla^2\psi_i\phi_i = \phi_i\nabla^2\psi_i + \psi_i\nabla^2\phi_i + 2\nabla\psi_i\nabla\phi_i$; the derivation of the second equality in Eq. (4.9) requires application of the gradient operator ∇ to both sides of the electronic Schrödinger Eq. (4.3).

3. The reaction $Cs + Br_2 \rightarrow CsBr + Br$ occurs *via* a harpoon mechanism in which an electron transfers from Cs to Br_2 at a distance R_c at which the covalent PES for the reaction is crossed by the ionic PES for $Cs^+ + Br_2^-$ ions. At large separations of the Cs and Br_2, the two PESs are separated by an energy corresponding to the difference between the ionization energy of the Cs atom ($I(Cs) = 3.89$ eV) and the electron affinity of the Br_2, ($EA(Br_2) = 2.55$ eV). Estimate a value for R_c assuming that the interaction energy between Cs and Br_2 on the covalent PES varies only very weakly with separation at long range. Hence estimate the cross section for this reaction.
4. Two diabatic states, characterized by the wavefunctions ψ_1 and ψ_2, arise as eigenfunctions of some simple approximate Hamiltonian for a diatomic molecule, and have energies ε_1 and ε_2, respectively. When the full Hamiltonian is considered, a small interaction $\langle\psi_1|\hat{H}|\psi_2\rangle = \Delta$ between these states occurs. By considering trial wavefunctions of the form $\Psi = c_1\psi_1 + c_2\psi_2$, and using the variational principle, find general expressions for the two lowest energy levels (see Study Box 4.1).
5. In the previous question, the energies ε_1 and ε_2 vary with internuclear separation R as $\varepsilon_1 = K(R-4)$, $\varepsilon_2 = -K(R-4)$, whereas $\Delta = K/10$ is a constant. Evaluate and plot the two lowest energy levels at $R = 4.0$, 4.0 ± 0.1, 4.0 ± 0.2, 4.0 ± 0.5, and 4.0 ± 1.0. Plot also the ratio $|c_1/c_2|$ as a function of R for each associated eigenfunction. Comment on the results you obtain.
6. Use the Landau-Zener model (Eq. (4.11)) with the following parameters to calculate the transition probabilities from the $B^3\Pi(0^+)$ to the $Y^3\Sigma^-(0^+)$ state of IBr at selected excitation photon energies in the range shown in Figure 4.4.

Parameters for the IBr B and Y states

Energy at crossing (cm^{-1})	V_{12} (cm^{-1})	$\partial(V_2 - V_1)/\partial R$ (cm^{-1} Å^{-1})
16980	120	6424

7. For colinear approach of an $F(^2P)$ atom to an H_2 molecule, what are the allowed values of the Ω quantum number (for the projection of the total angular momentum on the internuclear axis)? What values of Ω arise for $HF(X^1\Sigma^+) + H$ and $HF(a^3\Pi) + H$ products? Using adiabatic correlation of Ω from reactants through to products, construct a correlation diagram for the adiabatic PESs for colinear reaction of $F(^2P) + H_2 \rightarrow HF + H$ and compare your result to Figure 4.11.
Repeat the analysis for the reaction $F + D_2O \rightarrow DF + OD$ (with the approximation of a colinear reaction).

CHAPTER 5

Elastic and Inelastic Scattering: Energy Transfer in Collisions

DAVID W. CHANDLER[a] AND STEVEN STOLTE[b,c]

[a] Combustion Research Facility, Sandia National Laboratory, Livermore, CA, USA; [b] Laser Centre, Department of Physical Chemistry, Vrije University, Amsterdam, The Netherlands; [c] Institute of Atomic and Molecular Physics, Jilin University, Changchun, China

5.1 INTRODUCTION

Collisional energy transfer is an important and ubiquitous process that works to maintain the world at thermal equilibrium. Details of this process have significant consequences for many important processes in the atmosphere, in combustion environments, non-equilibrium flows such as supersonic molecular beams, and in quantitative laser diagnostics. In the Earth's atmosphere, rotationally and vibrationally inelastic collisions compete with electronic quenching, radiative processes, and chemical reactions in a complex interplay that determines the spatial and temporal concentration of many transient species such as OH. 'Greenhouse gases' such as CO_2 and CH_4 absorb infrared light that is radiated from the Earth, and the collisional redistribution of that energy into states of the molecule that radiate the energy back into space and those that don't dictate the thermal impact of the initial absorption process for heating the atmosphere.

Rotational energy transfer can be a fast and efficient process for populating and depopulating certain quantum states of molecules. The rate at which this occurs has quantum mechanical consequences for the observed spectra originating from those states. The 'energy-time uncertainty relationship' $\Delta E \Delta t \geqslant \hbar$ dictates that the longer a quantum state lives, the sharper the spectral lines which originate from that state, such that a short-lived state will give rise to a broad spectral signature. Measuring these spectral linewidths under different pressure and temperature conditions is an attractive and widely used method for learning about the overall rate of energy transfer from a given rovibronic state.

The nascent product distributions – the rotational, vibrational and electronic state populations – initially formed by chemical reactions or photochemical events are generally non-Boltzmann. Thermal equilibrium is restored between the various degrees of freedom by subsequent elastic and

Tutorials in Molecular Reaction Dynamics
Edited by Mark Brouard and Claire Vallance

Published by the Royal Society of Chemistry, www.rsc.org

inelastic collisional energy transfer processes. This chapter is about the study of these energy transfer processes.

Here we strive to understand the dynamics of energy transfer during a single collision event. We will discuss, only briefly, the study of vibrational to translational energy transfer. Then we will investigate a simple hard sphere model for translational energy transfer between two colliding atoms. A similar model involving a hard shell ellipse will be described which will provide some basic understanding of rotational energy transfer when an atom strikes a diatomic molecule. After discussing these simple models of energy transfer we will discuss modern experiments in this field of study and the theory associated with them. In particular, we will describe a set of experiments that have been carried out on the system of the ground state NO molecule in collision with a rare gas atom.

To study the detailed dynamics of individual collision events one needs to understand the detailed interaction potential between the particles as well as the motion of the particles on that potential energy surface (PES) (see Chapters 2 and 3). As it is difficult and computationally intensive to generate an interaction potential from first principles, potentials have historically been represented by approximate models, some of which involve considerable guesswork. A popular form to describe the interaction between two atoms whose centres are displaced by a distance R is the Lennard-Jones 6–12 potential:

$$V(R) = 4\varepsilon\left[\left(\frac{\sigma}{R}\right)^{12} - \left(\frac{\sigma}{R}\right)^{6}\right]. \tag{5.1}$$

By fitting to experimental data, such as bulk viscosity and speed of sound measurements, one can determine the well depth ε and repulsive core σ parameters. This Lennard-Jones 6–12 form has been shown to be quite successful at representing both the attractive and repulsive forces between two neutral particles. Of course, for any particular atom-diatom system a much more detailed and sophisticated PES may be calculated from first principles (see Chapter 2), and, as we will see in the following discussion, some features of the PES may also be inferred from the results of scattering experiments.

5.2 MODELLING ENERGY TRANSFER PROCESSES

5.2.1 Vibration to Translation Energy Transfer

During a collision, translational, rotational and vibrational energy can be transferred between collision partners, and can also be converted from one degree of freedom to another. In general, vibrational energy transfer happens with a lower probability per collision than rotational energy transfer as a result of the larger energy associated with a vibration than a rotation, and also the period of the vibration relative to the interaction time of the collision. Ehrenfest's adiabatic principle[293] states that if a force capable of causing energy transfer remains essentially constant during a quantized periodic motion (*i.e.* fast vibration and slow interaction) then the disturbance will result only in an adiabatic adaptation of the period of the motion, and no energy will be transferred. However, if the type of energy changing force is large and lasts only a short time compared to the oscillation time of the periodic motion then there will be efficient transfer of quantized energy associated with that periodic motion. In other words, if the periodic motion is a 'blur' (fast vibrational motion and slow interaction time), then there will be little transfer of energy, whilst if the collision is sudden compared to the motion of the molecule then significant energy can be transferred. This explains to a large extent why rotational motion, on a picosecond time scale, is more easily coupled to translational degrees of freedom than a vibration that occurs on a femtosecond time scale. It also follows that the lower the temperature (and, therefore, the slower the collisions) the lower the probability of transferring energy in a collision. This propensity rule breaks down when the energy of the collision is equal to or less than the well depth associated with the van der Waals forces between the particles, which occurs at temperatures below a few degrees Kelvin. The slowest vibrational energy relaxation rate measured to date is that for vibrationally excited N_2 in pure liquid nitrogen (77 K temperature),[294] which has a corresponding lifetime of 100 s. The vibrationally excited nitrogen molecule survives trillions of collisions without transferring its vibrational energy to translational motion. The Ehrenfest theorem also explains the somewhat surprising result that the vibrational relaxation of H_2 is faster than the vibrational relaxation of D_2 at the same temperature, despite H_2 having a larger vibrational energy (4401 cm^{-1}) than D_2 (3115 cm^{-1}). At a given temperature, H_2 moves more quickly and so collisions couple to the vibrational motion better than for the more slowly moving D_2.

Typically, vibrational relaxation is a rare outcome of a thermal collision. It can be understood to occur as a result of a very strong impulsive interaction arising from a direct impact on the repulsive wall of the molecule. The molecular motion therefore changes most rapidly near the distance of closest approach of the collision partners during the collision. A somewhat quantitative model for this case was developed by Landau and Teller.[295] The model assumes a colinear collision of an atom A with a diatomic molecule BC. It ignores the attractive part of the intermolecular potential, and models the repulsive wall with an exponential function that takes into account only the interaction potential between atom A and atom B as a function of their separation R_{AB}:

$$V(r, R) = V_0(r) + C \exp(-\alpha R_{AB}), \tag{5.2}$$

with

$$R_{AB} = R - \left[\frac{m_C}{m_B + m_C}\right] r. \tag{5.3}$$

Here, r denotes the diatomic bond length of BC, and R is the distance between the centre of A and the centre-of-mass (CM) of the BC molecule. The inverse length factor α, which characterizes the steepness of the exponential repulsive wall, and the repulsive potential constant C are determined by a best fit of the repulsive wall of the appropriate Lennard-Jones 6–12 potential to an exponential. The quantity $V_0(r)$ denotes the potential of the isolated BC molecule, *i.e.* the potential when atom A is at a large distance from B. It is usually described with a harmonic oscillator potential, *i.e.* $V_0(r) = -\frac{1}{2}k(r-r_e)^2$, where r_e is the equilibrium bond length, and k is the force constant, such that the vibrational frequency is $\nu = \frac{1}{2\pi}\sqrt{k/\mu_{BC}}$ with $\mu_{BC} = m_B m_C/(m_B + m_C)$. For a non-vibrating diatom, in which $r = r_e$, $V(r_e, R)$ takes the simple form $V(R) = C\exp(-\alpha R)$. Often the probability of energy transfer is calculated by classical time-dependent perturbation theory on the assumption that the probability of vibrational to translational energy transfer is proportional to the square of the Fourier component of the varying force at the vibrational oscillator frequency, very much in accord with Ehrenfest's principle.[9,293,296,297] In the thermal collision regime, the probability of a transition from $v = 1 \rightarrow v' = 0$ is very low because only the incoming momentum that is directed along the vibrating molecular bond couples efficiently to $V_0(r)$. This probability therefore is a maximum for a head-on collision at zero impact parameter. The Landau and Teller result for the probability that a hard sphere collision results a $v = 1 \rightarrow v' = 0$ transition in a thermal sample turns out to be unexpectedly simple:[297]

$$P_{1\rightarrow 0} = \exp\left[-3\left(\frac{2\pi^4 \mu_{A-BC} \nu^2}{\alpha^2 k_B T}\right)^{1/3}\right], \tag{5.4}$$

where μ_{A-BC} is the reduced mass of the collision partners and T the temperature. This simple model is amazingly accurate at predicting the temperature dependence of vibrational energy transfer. When the log of the probability of energy transfer for a particular vibrational relaxation is plotted *versus* $T^{-1/3}$ a straight line is almost always obtained except at very low temperatures. This is called a Landau-Teller plot. The power of $\frac{1}{3}$ arises from the fact that only the projection of the incoming momentum of the incoming atom along the BC molecular axis promotes vibrational relaxation. Additional assumptions of the model can be found in, *e.g.*, ref. [298].

The failure at low temperatures is caused by the formation of dimers and the role played by the attractive van der Waals potential between particles (atoms and molecules). When two approaching particles pass into their so-called van der Waals well their relative velocity will increase as they approach each other, resulting in a somewhat energetic collision despite the low temperature. Dimers become abundant at temperatures for which the average translational energy is comparable to the depth of the attractive well in the interaction potential between the molecules or atoms. In spite of the weakness of its bond, a dimer can live for a long time, and in this regime the assumption of a strong sudden collision on the repulsive wall is no longer applicable.

A somewhat more accurate three-dimensional model, developed by Schwartz, Slawsky and Herzfeld and known as SSH theory,[299] uses a similar exponential potential to model the repulsive wall, but adds a constant that accounts for the attractive forces in order to more accurately model the true potential. This yields more quantitative results than the original Landau-Teller theory. As it is a three-dimensional model, SSH theory is sometimes referred to as a "breathing sphere" model. The interaction potential is given by the expression

$$V(r, R) = V_0(r) + C\exp(-\alpha R_{AB}) - \varepsilon, \qquad (5.5)$$

in which ε denotes the van der Waals well depth.[297] Unfortunately, the probability of a vibrational relaxation in a collision as predicted by the SSH model does not lead to a parameterized expression such as that for the Landau-Teller model (Eq. (5.4)).

For a potential that is anisotropic and depends on r, R and $\hat{\boldsymbol{R}}\cdot\hat{\boldsymbol{r}}$, the transition probability for a vibrational and/or rotational transition can also be estimated from a first-order trajectory method called infinite order time-dependent perturbation theory, otherwise known as the 'sudden approximation' (SA):[300–302]

$$P_{i\to f}(b) = \left|\frac{i}{\hbar}\int_{-\infty}^{\infty}\langle v_i, j_i, m_i|V'(r,\hat{\boldsymbol{r}},\boldsymbol{R}(t))|v_f, j_f, m_f\rangle \exp\left(-i\frac{\Delta E_{\text{inel}}}{\hbar}t\right)\mathrm{d}t\right|^2. \qquad (5.6)$$

In this Equation, the bra $\langle v_i, j_i, m_i|$ denotes the complex conjugate of the rovibrational molecular wave function before the collision, that is completely specified by its vibrational quantum number v_i, its rotational angular momentum number j_i, and its magnetic projection quantum number m_i along a particular space fixed quantization axis. Similarly, the ket $|v_f, j_f, m_f\rangle$ denotes one of the possible rovibrational wave functions into which the molecule will be scattered after a collision with an atom at impact parameter b. ΔE_{inel} is the difference in rovibrational energy between the outgoing and incoming molecular quantum state. The integral is over the classical trajectory $\boldsymbol{R}(t)$ with impact parameter b on the $(\hat{\boldsymbol{R}}\cdot\hat{\boldsymbol{r}})$-averaged isotropic potential, $V_{\text{iso}}(r, R)\equiv\langle V(r, R, \hat{\boldsymbol{R}}\cdot\hat{\boldsymbol{r}})\rangle_{\hat{\boldsymbol{R}}\cdot\hat{\boldsymbol{r}}}$, with $r=r_e$. One defines $t=0$ to occur at the turning point of the trajectory, which coincides with the distance of closest approach between the collision partners. The potential appearing in Eq. (5.6) is then defined as $V'(r, \hat{\boldsymbol{r}}, b, t)\equiv V(r, R, \hat{\boldsymbol{R}}\cdot\hat{\boldsymbol{r}}) - V_{\text{iso}}(r=r_e, R)$. The time-dependent matrix element, $\langle v_i, j_i, m_i|V'(r, \hat{\boldsymbol{r}}, \boldsymbol{R}(t))|v_f, j_f, m_f\rangle$, assumes that the spatial direction of the molecular axis $\hat{\boldsymbol{r}}$ is fixed during the trajectory, and treats the inelastic scattering as a time-dependent perturbation with phase shifts which are first order in the anisotropy of the potential; however the angle between the space-fixed molecular axis and the passing atom remains variable. Often, in addition, the trajectory is approximated by a straight line.[303]

The first-order transition probability, *P*(*b*), can be thought of simply as the square absolute value of the Fourier transform of the time-dependent potential.[304] According to Eq. (5.6) the transition probability is proportional to the square of the magnitude of V' (*i.e.* the strength of the interaction potential between the collision partners), and also depends on the exponential term, which oscillates as a function of time. This oscillatory term is modulated by the pre-exponential factor which can only reach an appreciable magnitude during the collision. The time-dependent exponential factor in Eq. (5.6) requires the conservation of flux; in other words the exponential form of the *S*-matrix is not approximated (see Chapter 3). Since the classical trajectory is only influenced by the isotropic part of the anisotropic potential, $V_{\text{iso}}(r = r_e, R)$, the absolute value of the velocity will be conserved during the collision, *i.e.* $v'_{\text{rel}} = v_{\text{rel}}$, which violates the conservation of total energy for the collision. It also ignores a feedback of the oscillatory mode into the relative motion. The same restriction holds for both the Landau-Teller and the SSH models.[298] Sometimes this setback can be partly counteracted by taking the velocity of the classical trajectory to be equal to $\sqrt{v^2 - \Delta E_{\text{inel}}/2\mu_{\text{A-BC}}}$. Again, also according to Eq. (5.6), the transition probability can only be large if $\tau_{\text{col}}/\tau < 1$, where $\tau = \hbar/\Delta E_{\text{inel}}$ is the effective period of the (de)excited motion of the molecule.

The coupled states (CS) and infinite order sudden (IOS) approximations,[115,148,305,306] discussed briefly in Chapter 3, allow for the feedback between the oscillatory mode and the translational motion. Both treatments assume that $\hat{\boldsymbol{r}} \cdot \hat{\boldsymbol{R}}$ remains fixed during the collision. This implies that an atom that impinges with a certain impact parameter *b* onto the broadside of the molecule ($\hat{\boldsymbol{r}} \cdot \hat{\boldsymbol{R}} = 0$ for large *R*) will still point broadside to the molecule at the turning point of the trajectory. This requirement is contradictory to the inelastic transition favouring the condition $\tau_{\text{col}}/\tau < 1$, which demands that the molecular axis remains essentially fixed in space. In spite of this shortcoming, the IOS and CS approximations do well in their prediction of m_j-averaged and $m_{j'}$-summed integral and differential cross sections, but typically do not do that well in their predictions of the m_j and $m_{j'}$ state dependences.[305]

The transition probability, $P_{i\to f}(b)$, (sometimes called the opacity function – see Chapter 1) varies with impact parameter. In this case the integral cross section follows from (see Chapter 1 and Section 5.2.3),

$$\sigma_{i\to f} = 2\pi \int_0^{b_{\max}} P_{i\to f}(b) b \, \mathrm{d}b. \tag{5.7}$$

Vibrational to vibrational energy transfer[306] has also been treated in a similar manner and the probability of energy transfer is seen to be exponentially dependent upon the energy gap between the two vibrational frequencies. There are several excellent reviews of all the relevant vibrational relaxation experiments and theories.[11,297,298,302,307]

5.2.2 Dynamics of a Single Hard Sphere Collision

Many of the features of elastic and rotationally inelastic collisions between atoms and molecules can be understood classically, with the assumption that both particles can be represented as hard objects. A higher level of understanding is gained by adding quantum mechanical behaviour to the classical calculations, giving rise to, for example, the so-called "Quasi Quantum Treatment" (QQT).[308–311] Understanding the finest details requires a more rigorous approach, employing detailed PESs and quantum scattering calculations (see Chapter 3). Here we will first discuss the classical interaction of a hard sphere with a hard ellipse, and then expand the discussion to realistic potential energy surfaces, and to the classical and quantum mechanical modelling of collisions on these surfaces. In Section 5.3 we will discuss experiments that have accompanied development of the theory of rotational energy transfer. Along with these discussions will be descriptions of experiments that make it possible to isolate many of the interesting phenomena related to rotational energy transfer, such as alignment and orientation of the scattered product and the effect of scattering from the different ends of the molecule.

When two particles approach each other at thermal energies and do not experience strong mutual attraction, as is often the case for two closed shell atoms, then their collision can be approximated as only occurring if they pass within a certain fixed distance from each other. They will only scatter when their distance of closest approach is equal to or less than the sum of their hard sphere/hard ellipse radii. This is because the repulsive wall of an inter-atomic or inter-molecular potential that represents the repulsion of the electron clouds is very steep and behaves much as a hard barrier. In this case, time-wise the interaction is sudden on the repulsive wall of the PES and can be described as a "hard sphere collision". A closed shell diatomic molecule interacting with a closed shell atom can be described similarly by a hard shell model. We will see below that many experimental observations about rotationally inelastic collisions, such as rotational rainbow scattering, can be understood in terms of the collision dynamics of a hard sphere with a rigid ellipse. In fact, much can be learned by treating the collision using the hard sphere approximation. Hard sphere radii are tabulated for most atoms (molecules),[298,307,312] and one can think of these as representing the (average) diameter of the repulsive wall of the interparticle potential energy surface at the energy of the collision.*

5.2.3 Translation to Translation Energy Transfer

Imagine a collision between two spherical species A and B, with radii R_A and R_B, and masses m_A and m_B. At the distance of minimum approach, the centres-of-mass of the two spheres are at a distance $R_S = R_A + R_B$. This two-body problem can be recast in the CM-frame in terms of a point particle with mass $\mu_{AB} = m_A m_B/(m_A + m_B)$, striking a stationary sphere with radius, R_S. The velocity of the point particle is $\boldsymbol{v}_{rel}$, the relative velocity of the original two spheres. The model is illustrated in Figure 5.1.

The distance between the incoming trajectory and the parallel line that passes through the centre of the sphere is called the impact parameter, b, and defines the point of impact (see Chapter 1). The

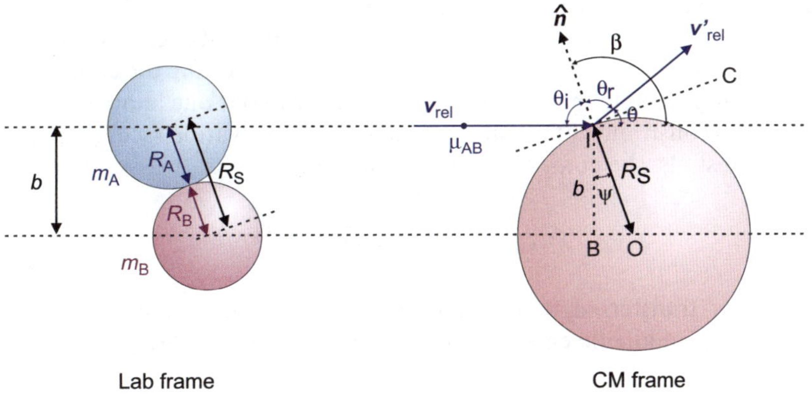

Figure 5.1 Scattering of hard spheres of diameter R_S with relative velocity $\boldsymbol{v}_{rel}$. See text for details.

*The reader might wish to refer to Study Box 5.1 at this point, which gives a general introduction to atom-atom scattering, and, in particular, outlines the concepts of rainbow and glory scattering in such collisions.

impulse to the sphere will be along the radius of the sphere that intersects the point I on the surface. This is how a straight billiard shot is aimed. As the impact of the particle on the sphere is sudden, at point I the velocity component of the particle along the tangent C to the sphere, $\boldsymbol{v}_{\parallel}$, will be conserved, while its component perpendicular to the surface of the sphere along radius IO, $\boldsymbol{v}_{\perp}$, will be reflected perfectly $\boldsymbol{v}'_{\perp} = -\boldsymbol{v}_{\perp}$. Note that primed and unprimed vectors refer to post- and pre-collision velocities, respectively. The particle is deflected specularly with respect to the surface normal, $\hat{\boldsymbol{n}}$, which is directed along IO at an angle β with respect to the incident relative velocity vector, $\boldsymbol{v}_{\mathrm{rel}}$. Specular reflection with respect to $\hat{\boldsymbol{n}}$ means that the angles of incidence, $\theta_{\mathrm{i}} \equiv \pi - \beta$, and reflection, θ_{r}, measured with respect to $\hat{\boldsymbol{n}}$ are equal, *i.e.* $\theta_{\mathrm{i}} = \theta_{\mathrm{r}}$. Because the scattering is specular, it is helpful to resolve the incoming relative velocity vector into two components,

$$\boldsymbol{v}_{\mathrm{rel}} = \boldsymbol{v}_{\parallel} + \boldsymbol{v}_{\perp}. \tag{5.8}$$

The momentum $\boldsymbol{p}_{\perp} = \mu_{\mathrm{AB}}\boldsymbol{v}_{\perp}$ associated with velocity $\boldsymbol{v}_{\perp}$ will be transferred to the particle, while the momentum $\boldsymbol{p}_{\parallel} = \mu_{\mathrm{AB}}\boldsymbol{v}_{\parallel}$ associated with $\boldsymbol{v}_{\parallel}$ will be conserved in the collision. The vector IO is known as the "apse" of the collision. To the stationary sphere it is as if the collision was struck directly along the apse with velocity $\boldsymbol{v}_{\perp}$. Such a 'head on' collision corresponds to an alternative visualization of the collision in the centre-of-mass frame.* Because $\theta_{\mathrm{i}} = \theta_{\mathrm{r}}$, the apse angle β is equal to $(\pi + \theta)/2$, where θ is the scattering angle that the point scatters into in the CM collision frame, *i.e.* the direction of $\boldsymbol{v}'_{\mathrm{rel}}$ with respect to $\boldsymbol{v}_{\mathrm{rel}}$. By definition, the impact parameter $b = \mathrm{IB}$ lies perpendicular to the velocity $\boldsymbol{v}_{\mathrm{rel}}$ of the incoming point particle, and leads to the angle Ψ being equal to $\beta - \pi/2 = \theta/2$. This leads to an equation for the impact parameter in terms of the diameter of the sphere, R_{S}, and laboratory scattering angle, θ,

$$b = R_{\mathrm{S}} \cos(\theta/2). \tag{5.9}$$

For the hard sphere collision, there is a one-to-one correspondence between the scattering angle, θ, and the impact parameter, b, of the collision. This is why it is possible to predict the outcome of a billiard shot. When $b = 0$, direct impact, then $\theta = \pi$, and direct recoil is expected. When $b = R_{\mathrm{S}}$, $\theta = 0$ and no deflection is expected. We will see later that for a hard sphere striking a hard *ellipse* there is no one-to-one correspondence between scattering angle and impact parameter.

At the instant of impact, the part of the initial relative kinetic energy, $E_{\mathrm{t}} \equiv E_{\mathrm{c}} = \frac{1}{2}\mu_{\mathrm{AB}}v_{\mathrm{rel}}^2$, that is associated with $\boldsymbol{v}_{\perp}$, *i.e.* $\frac{1}{2}\mu_{\mathrm{AB}}v_{\perp}^2$, is converted entirely into potential energy V. As noted above, the component of the velocity parallel to tangent C, $\boldsymbol{v}_{\parallel}$, is conserved during the collision. This motion is in fact associated with the orbital motion of the two spheres about one another, and is conserved in this case to satisfy the constraints of angular momentum conservation (see Chapter 1). The energy associated with $\boldsymbol{v}_{\parallel}$ must also be conserved and is equal to:

$$\tfrac{1}{2}\mu_{\mathrm{AB}}v_{\parallel}^2 = \tfrac{1}{2}\mu_{\mathrm{AB}}v_{\mathrm{rel}}^2 \cos^2(\theta/2). \tag{5.10}$$

Therefore, the energy transferred to the stationary particle, $E_{\mathrm{A}\to\mathrm{B}}$, is given by the difference between the initial relative kinetic energy and the conserved kinetic energy:

$$E_{\mathrm{A}\to\mathrm{B}} = V = \tfrac{1}{2}\mu_{\mathrm{AB}}v_{\mathrm{rel}}^2\left[1 - \cos^2(\theta/2)\right]. \tag{5.11}$$

*It should be borne in mind, though, that this visualization needs refining when considering scattering of real particles, for which the attractive part of the potential cannot be ignored (see Study Box 5.1).

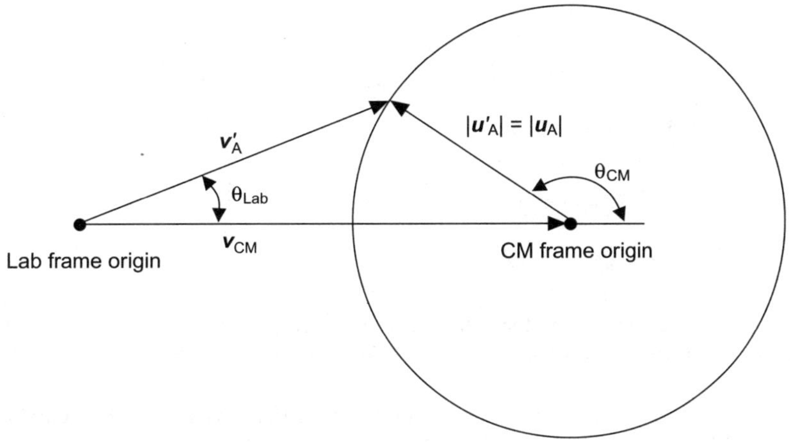

Figure 5.2 The relationship between the velocity of the recoiling particle in both the laboratory and centre-of-mass reference frames.

Substituting the relationship $b = R_S \cos(\theta/2)$ into this equation, we see that the amount of energy transferred is

$$E_{A\to B} = E_c\left[1 - \left(\frac{b}{R_S}\right)^2\right]. \tag{5.12}$$

If $b = 0$, then all of the initial translational energy is transferred. This is the billiard shot for which the queue ball is stationary after the collision. Recall that for $b = 0$ the billiard balls recoil at 180° to each other in the CM-frame. If we measure the recoil angle of the products from the collision, Eqs. (5.9) and (5.12) tell us that for a hard sphere collision we then know the impact parameter, and can determine the amount of translational energy transferred in the collision.

We can now pose the question: what is the amount of energy transferred in a random collision between two hard spheres? To answer this we need to calculate the ratio of the *partial differential cross section* (PDCS) to the total differential cross section (DCS). The partial differential cross section is the DCS for a specific impact parameter, b, or orbital angular momentum, ℓ, but without resolution of the azimuthal scattering angle, ϕ (see Chapter 1). At fixed $\boldsymbol{v}_{rel}$, we are considering a random collision having any possible value of the impact parameter, b. Figure 5.2 shows the relationship between the velocity of particle A in the laboratory (LAB) frame of reference and in the CM-frame of reference. This is important because all of the dynamics must be calculated in the CM-frame and then translated into the LAB-frame (see Chapter 1 and Study Box 6.1 for more discussion of the LAB to CM transformation). After the collision the particle scatters onto a sphere centred on the CM of the system and with dimensions dictated by conservation of energy and momentum. In the LAB reference frame, the scattering (or 'Newton') sphere moves with the velocity of the centre-of-mass, $\boldsymbol{v}_{CM} = (m_A\boldsymbol{v}_A + m_B\boldsymbol{v}_B)/(m_A + m_B)$. Collisions with different impact parameters will scatter in different directions. If the scattering is elastic, such that no energy is put into internal degrees of freedom of the particles, then the product velocity vectors in the CM and LAB-frames are related by the law of cosines* as follows (see Figure 5.2):

$$v_A'^2 = u_A'^2 + v_{CM}^2 - 2u_A' v_{CM}\cos(\pi - \theta_{CM}). \tag{5.13}$$

*Note that Eq. (5.13) is the basis of PHOTOLOC technique, discussed in Study Box 6.3.

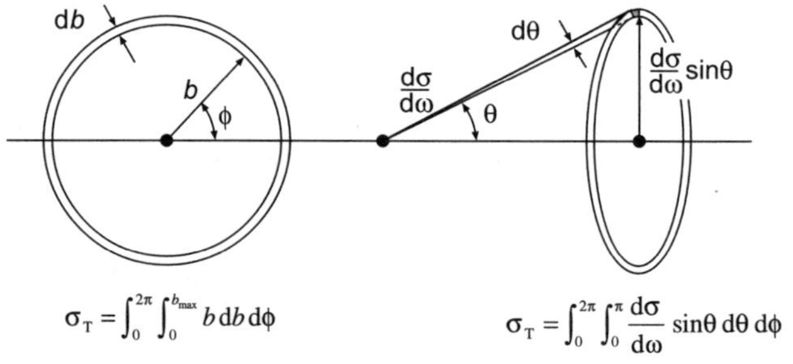

Figure 5.3 The two ways to determine the total DCS of a hard sphere collision are as an integration of the impact parameter over all space or as an integration of the DCS over all angles.

Notice that we use u here to denote velocities with respect to the CM-frame. For elastic scattering one has $u_A = u'_A$ in the CM-frame, whereas in the LAB-frame, $v_A \neq v'_A$. This means that in the LAB-frame elastic collisions can lead to the *translational moderation*, *i.e.* translational cooling, of species with high kinetic energy. If we assume that $v_B = 0$, then the angle θ_{CM} shown in Figure 5.2 becomes equivalent to the centre-of-mass scattering angle θ, *i.e.* $\theta \equiv \theta_{CM}$, and Eq. (5.13), can be transformed to

$$\cos\theta = \frac{(m_A + m_B)}{2\mu_{AB}}\left[\left(\frac{v'_A}{v_A}\right)^2 - \frac{(m_A^2 + m_B^2)}{(m_A + m_B)^2}\right]. \tag{5.14}$$

This results in $v'_A = 0$ if $m_A = m_B$ and also $\cos\theta = -1$, *i.e.* 'backwards scattering' of the products. The condition $v'_A = 0$ illustrates the potential of using collisions between molecules as a means of cooling them translationally, as explored further below towards the end of this subsection (see also Chapter 12).

Next we need to calculate the dependence of the cross section for two hard spheres, A and B with $R_A + R_B = R_S$, on the scattering angle, θ. We do this by noting that we can describe the total scattering cross section in two different ways: either (i) by integrating over all impact parameters; or (ii) by integrating the differential scattering cross section (the scattering cross section per solid angle dω) over all scattering angles. This is possible because, as shown earlier, there is a one-to-one correspondence between impact parameter and scattering angle, Eq. (5.9). Figure 5.3 shows this diagrammatically.

Because the scattering is cylindrically symmetric around the relative velocity vector, the integration of the cross section around the azimuthal spherical angle, ϕ, simply gives rise to a factor of 2π in both cases. The total collision cross section then follows simply from the geometric cross section of a hard sphere $\sigma_T = \pi R_S^2$ (see further below). In the macroscopic world, σ_T relates directly to z_{col}, the number of collisions per unit time of particular molecule A in a gas B of number density n_B, *via* the relationship $z_{col} = v_{rel} n_B \sigma_T$ (see, for example, ref. [11] and references therein). The "collision frequency", z_{col}, is inversely proportional to the mean free path that molecule A travels between two collisions with B. Since a collision with molecule B results in a deflection of A, one obtains σ_T by measuring the attenuation of the intensity $I = n_A v_A$ of a beam of A molecules by a gaseous target (*e.g.*, in a scattering cell or crossed molecular beam) as a function of the density, n_B, of B molecules (see Chapter 6 for more details of the crossed molecular beam technique). An experimental apparatus capable of such a measurement is depicted later in this chapter in Figure 5.11. Designating L_{scat} as the path length (or distance) that the beam of A molecules travel through the region containing B molecules at a number density n_B, one extracts the value of σ_T from:*

$$-\ln\left(\frac{I}{I_0}\right) = \left\langle \frac{v_{rel}\sigma_T(v_{rel})L_{scat}n_B}{v_A} \right\rangle \cong n_B L_{eff}\sigma_T(\langle v_{rel}\rangle)\left\langle \frac{v_{rel}}{v_A} \right\rangle, \tag{5.15}$$

*Eq. (5.15) is analogous to the Beer-Lambert Law for the absorption of light through a medium.

where $L_{\rm eff} = \int n_{\rm B}(L)\mathrm{d}L/n_{\rm B}$ is some effective path length, determined from an average of the number density of B over the interaction region of the experiment. In this equation, I and I_0 are the transmitted and incident beam intensities, respectively. So, one obtains $\sigma_{\rm T}$ by fitting the slope of $\ln(I/I_0)$ as a function of $n_{\rm B}$.[11] Eq. (5.15) applies also to the general case, in which the collision partners are not hard spheres and for which the total collision cross section depends on $v_{\rm rel}$. While many total collision cross section measurements were carried out in the sixties, the theory leading to Eq. (5.15) came some time later.[313] At that time, it was difficult to obtain an accurate value of $n_{\rm B}$ in the scattering cell, within which it is assumed that $L_{\rm eff} \simeq L_{\rm scat}$, and this often restricted the relative accuracy of the absolute value that could be determined for $\sigma_{\rm T}$ to a level of around 15%. About an order of magnitude better accuracy could be achieved in experiments in which the dependence of the relative value of $\sigma_{\rm T}$ upon $v_{\rm rel}$ was explored, since in these measurements the absolute value of $L_{\rm eff}$ and of $n_{\rm B}$ was not as critical.

In reality, molecular scattering cannot be treated exactly using classical mechanics (see Chapter 3 and Study Box 5.1). Diffractive scattering, which strongly contributes at small scattering angles (*i.e.* in the forward direction) for the hard shell type potentials considered here,[314] means that often one has to replace $\sigma_{\rm T}$ by a smaller cross section $\sigma_{\rm incomplete}$, because part of the contribution to $\sigma_{\rm T}$ from scattering into very small angles cannot be detected experimentally. For very small scattering angles, it is rare for the experiments to resolve these products from molecules in the overlapping undeflected incident beam. The simplest form of $\sigma_{\rm incomplete}$ is the partial total cross section, $\sigma_{\rm p}(\theta_{\rm min})$, of all scattering going into an angle $\theta \geqslant \theta_{\rm min}$. In the case of hard spheres, as depicted in Figure 5.3, the one-to-one correspondence between θ and b given by Eq. (5.9) leads to

$$\sigma_{\rm p}(\theta_{\rm min}) = \pi b_{\rm max}^2 = \pi R_{\rm S}^2 \cos^2\left(\tfrac{\theta_{\rm min}}{2}\right) = \pi R_{\rm S}^2 \frac{[1+\cos\theta_{\rm min}]}{2}. \tag{5.16}$$

More details about the A-B intermolecular potential emerge when one measures the scattering process as a function of scattering angle, θ, *i.e.* the number of A-molecules scattered from molecule B, $N_{\rm A-scat}$, into a solid angle, $\mathrm{d}\omega = \sin\theta\,\mathrm{d}\theta\,\mathrm{d}\phi$, per unit time (per second) and per scattering centre, when collisions occur inside a volume element ΔV;

$$\frac{1}{n_{\rm B}}\frac{\mathrm{d}}{\mathrm{d}\omega}\left(\frac{\mathrm{d}N_{\rm A-scat}(\theta,\phi)}{\mathrm{d}t}\right) = \frac{\mathrm{d}\sigma}{\mathrm{d}\omega} v_{\rm rel} n_{\rm A} \Delta V. \tag{5.17}$$

The quantity $v_{\rm rel} n_{\rm A} = I_0$ is called the incident flux density or intensity, defined as the number of A molecules per second per cm^2 that impinge onto the scattering centre of B molecules. By measuring the scattered flux of particle A at a detector for known concentrations and velocities of the colliding particles, one can obtain the DCS per unit solid angle. This is sometimes written more explicitly as $\frac{\mathrm{d}\sigma(\theta,\phi)}{\mathrm{d}\omega}$.

Guided by Figure 5.3, the next goal is to derive $\frac{\mathrm{d}\sigma}{\mathrm{d}\omega}$ from $\sigma_{\rm p}(\theta_{\rm min})$, the partial total cross section given in Eq. (5.16). Inspection of this figure and accompanying equations shows that an infinitesimal decrease in the scattering angle θ corresponds to an infinitesimal increase in $\sigma_{\rm p}(\theta_{\rm min})$, in agreement with Eq. (5.16): *i.e.* $\sigma_{\rm p}(\theta_{\rm min}-\mathrm{d}\theta) > \sigma_{\rm p}(\theta_{\rm min})$. The area of the total cross section that leads to scattering angles between $\theta+\mathrm{d}\theta$ and θ may therefore be written in terms of the partial total cross section as $\sigma_{\rm p}(\theta-\mathrm{d}\theta)-\sigma_{\rm p}(\theta)$.

The total solid angle corresponding to scattering angles between $\theta+\mathrm{d}\theta$ and θ is $2\pi\sin\theta\mathrm{d}\theta$, as also follows from inspection of Figure 5.3. Therefore, we can write

$$\frac{\mathrm{d}\sigma}{\mathrm{d}\omega} \equiv \frac{\sigma_{\rm P}(\theta-\mathrm{d}\theta)-\sigma_{\rm P}(\theta)}{2\pi\sin\theta\,\mathrm{d}\theta} = \frac{1}{2\pi\sin\theta}\frac{-\mathrm{d}\sigma_{\rm P}(\theta)}{\mathrm{d}\theta}. \tag{5.18}$$

STUDY BOX 5.1: AN INTRODUCTION TO RAINBOW AND GLORY SCATTERING

As discussed in both the present chapter and Chapter 1, the direction in which two atomic products of a collision scatter depends on their initial velocity, the impact parameter at which they approach each other, and on the potential energy curve that controls their interaction. To illustrate this, we will consider the scattering behaviour we would expect for two spherical particles (atoms) with an interaction potential of the form shown on the left of Figure 5.4 (*e.g.*, a Lennard-Jones potential).

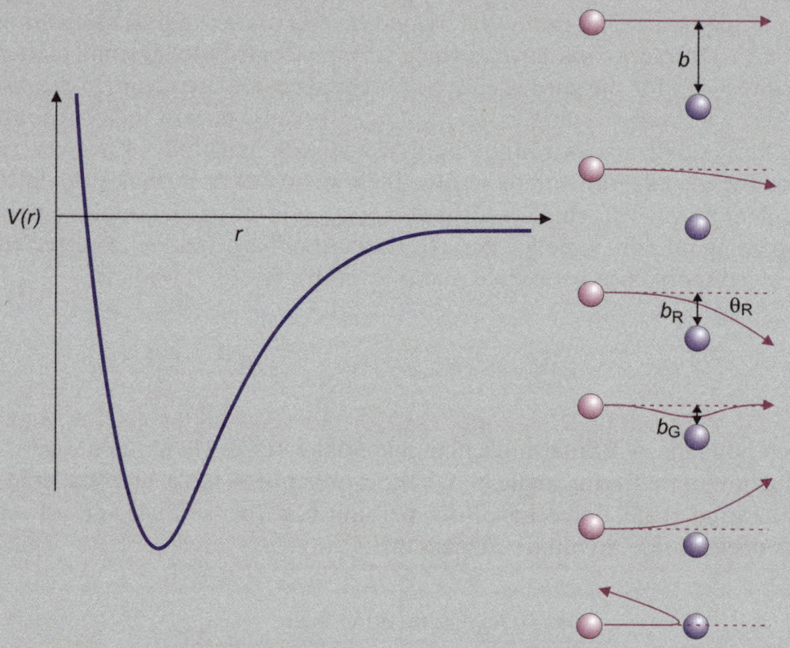

Figure 5.4 Scattering trajectories for particles approaching at a range of impact parameters b (large to small from top to bottom of the figure) under the influence of the interaction potential shown on the left. b_R and b_G denote the impact parameters for rainbow and glory scattering, respectively.

The right hand side of the same figure shows several scattering trajectories for a particle approaching a second stationary particle at various impact parameters. At very large impact parameters, the potential is almost zero and the approaching particle is barely deflected at all. As the impact parameter is reduced, the particles begin to experience the attractive part of the potential, and the approaching particle is deflected towards the stationary particle (a negative deflection). This deflection increases as the impact parameter is reduced, up to a maximum negative deflection known as the rainbow angle θ_R at impact parameter b_R. As the impact parameter is reduced further, the particle initially experiences the attractive part of the potential and is deflected towards the stationary particle, but then at closer separations the repulsive part of the potential starts to act, and the particle is steered away again, reducing the net scattering angle. At a certain impact parameter b_G these two interactions exactly balance each other, and the net deflection is zero.[316] This is known as glory scattering. At still smaller impact parameters the repulsive part of the potential dominates and the deflection angle becomes positive. The dependence of the deflection angle on b is shown in Figure 5.5(a).

Experimentally, the cylindrical symmetry about the relative velocity vector of the colliding particles means that it is only possible to measure the absolute value of the deflection angle. This is plotted in Figure 5.5(b). Also, in an experiment it is not usually possible to control b, and the measured angular scattering distribution is therefore an average over all possible impact parameters. This can lead to some interesting effects. For example, inspection of Figure 5.5(b) shows that for scattering at large angles (*i.e.* backwards scattering) there is a one-to-one correspondence between impact parameter and scattering angle. However, near the rainbow angle three different impact parameters result in scattering into the same angle, making scattering at these angles much more likely and leading to 'rainbow peaks' in the angular scattering distribution.

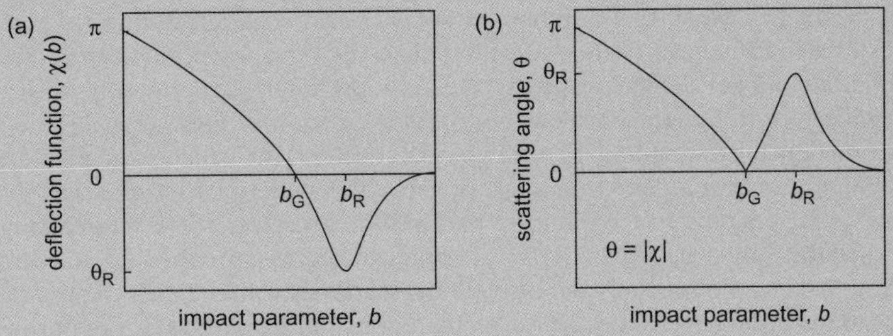

Figure 5.5 (a) Deflection angle χ as a function of impact parameter; (b) Experiments are only sensitive to the absolute value of the deflection angle, $\theta = |\chi|$.

The classical picture of elastic scattering can be made more quantitative by recasting the expression for the differential cross section in terms of the impact parameter. As shown in Figure 5.3, for elastic scattering of atoms, once the impact parameter is defined, the scattering angle is uniquely determined. For elastic scattering it is therefore possible to relate the probability of a collision occurring between impact parameters b and $b + \mathrm{d}b$ and the corresponding probability for elastic scattering into solid angles $\theta + \mathrm{d}\theta$ and $\phi + \mathrm{d}\phi$, *i.e.*

$$b\,\mathrm{d}b\,\mathrm{d}\phi = \frac{\mathrm{d}\sigma}{\mathrm{d}\omega}\sin\theta\,\mathrm{d}\theta\,\mathrm{d}\phi. \tag{B5.1.1}$$

This expression can be rearranged to give an equation for the differential cross section in terms of impact parameter,

$$\frac{\mathrm{d}\sigma}{\mathrm{d}\omega} = \frac{b}{\sin\theta\left|\frac{\mathrm{d}\theta}{\mathrm{d}b}\right|}. \tag{B5.1.2}$$

This reveals that the classical differential cross section tends to infinity when $\sin\theta = 0$ (*i.e.* at 0° or 180°), and when the gradient of the scattering angle *versus* impact parameter plot shown in Figure 5.5 goes to zero, *i.e.* at the angle corresponding to the classical rainbow. One consequence of these discontinuities is that the integral elastic cross section is undefined in classical mechanics. Classically, an elastic collision always occurs, no matter how large the impact parameter and how small the deflection, and, thus, there is no defined maximum impact parameter at which to truncate the integral given in Eq. (5.7).

Fortunately, the discontinuities in the classical elastic differential cross section disappear in quantum mechanics. In quantum mechanics, a maximum orbital angular momentum ℓ_{max}, corresponding to the classical maximum impact parameter, is defined by the Heisenberg Uncertainty Principle: for orbital angular momenta above ℓ_{max} the deflection is so small that to determine it would violate the uncertainty relation $\Delta p_x \Delta x \geq \hbar/2$.[239] Quantum mechanical interference also plays an important role in determining the detailed structure of the scattering distribution. Interference phenomena arise from the de Broglie wave character of the particles involved, and therefore depend on the de Broglie wavelength. The latter is inversely proportional to the momentum associated with the colliding particles, with the proportionality constant being Planck's constant, h:

$$\lambda = \frac{h}{p}. \tag{B5.1.3}$$

In quantum mechanics, when two different pathways lead to scattering into the same region of space, the result is interference. In the present context, the three wavefunctions corresponding to the three different impact parameters near the rainbow angle can interfere with each other, leading to oscillations in the angular distribution in the region of the classical rainbow angle.

The wavefunction corresponding to the 'glory' trajectory (in which attractive and repulsive interactions balance to give a deflection angle of zero) also undergoes interference, this time with wavefunctions corresponding to large impact parameters, for which there is essentially zero interaction and therefore no deflection. Glory scattering arises as an enhanced forward scattering (corresponding to increased transmission through the scattering cell in these experiments) at specific velocities due to quantum interference between different scattering pathways. For particular relative velocities of the colliding particles one gets constructive interference between the forward scattering ($\theta = 0$) trajectories which take place at very high impact parameters and those that impinge at the so called glory impact parameter. At other velocities this interference can become destructive. This creates an oscillation in the transmission signal *versus* velocity, or collision energy, which is known as a Glory. Glory interference is also manifest as rapid oscillations in the differential cross section for elastic scattering at angles close to forward scattering, with the frequency of the oscillations providing information about the location of the well on the potential energy curve.

As we will see later in this chapter, rainbow and glory scattering effects are ubiquitous in many types of scattering processes, including inelastic scattering of molecules.

David W. Chandler and Steven Stolte
Mark Brouard and Claire Vallance

Upon the substitution of Eq. (5.16) into the $\frac{d\sigma_P(\theta)}{d\theta}$ factor of Eq. (5.18) with $\theta_{min} = \theta$ one obtains:

$$\frac{-d\sigma_P(\theta)}{d\theta} = -\frac{1}{2}\pi R_S^2 \frac{d\cos\theta}{d\theta} = \frac{1}{2}\pi R_S^2 \sin\theta, \tag{5.19}$$

which immediately yields:

$$\frac{d\sigma}{d\omega} = \frac{R_S^2}{4}. \tag{5.20}$$

Therefore, the DCS for a hard sphere collision does not depend upon either the velocity or the scattering angle, θ.

The total cross section is found by integrating the DCS over all angles. Substituting the DCS given by Eq. (5.20) into the following equation, we obtain the expected result (see above):

$$\sigma_T = 2\pi \int_0^{\pi} \frac{d\sigma}{d\omega} \sin\theta \, d\theta = \pi R_S^2. \tag{5.21}$$

The total cross section is equal to the area of a particle of radius R_S.* A general equation for the integral cross section of a collision event was given by Eq. (5.7) in terms of the opacity function, $P(b)$, which allows for the variation in the probability of the scattering process (*e.g.*, energy transfer) with impact parameter. For instance, impacts with small b values may lead to vibrational energy transfer with a higher probability than larger impact collisions. For a hard sphere event, $P(b)=1$ from $b=0$ to $b=R_S$ and 0 for all values of $b>R_S$. In this case, Eq. (5.7) yields the expression πR_S^2 for the total (integral) cross section, in agreement with Eq. (5.21). For a rotationally or vibrationally inelastic or reactive scattering event there will in general be some shape to $P(b)$, and the probability of the scattering process will generally be less than unity.

Earlier we derived a relationship between the scattering angle and the amount of energy transferred (see Eq. 5.11). The larger the angle, θ, the more energy transferred. If we now define $\theta_{\min}$ as the smallest scattering angle required to transfer a certain amount of energy, then scattering at larger angles will transfer more energy. The fraction, F, of collisions having $\theta \geqslant \theta_{\min}$ is therefore $F=\sigma_p/\sigma_T=\frac{1}{2}(\cos\theta_{\min}+1)$. In the special case of Eq. (5.14), with $v_B=0$ one obtains:

$$F=\frac{1}{2}\left\{\frac{(m_A+m_B)}{2\mu_{AB}}\left[\left(\frac{v_A'}{v_A}\right)^2-\left(\frac{(m_A^2+m_B^2)}{(m_A+m_B)^2}\right)\right]+1\right\}. \tag{5.22}$$

As $\left(\frac{v_A'}{v_A}\right)^2=\frac{E_A'}{E_A}$, we can substitute this into the above equation, and obtain the expression

$$F=\frac{1}{2}\left\{\frac{(m_A+m_B)}{2\mu_{AB}}\left[\left(\frac{E_A'}{E_A}\right)-\left(\frac{(m_A^2+m_B^2)}{(m_A+m_B)^2}\right)\right]+1\right\}. \tag{5.23}$$

If $m_A=m_B$ then this equation reduces to

$$F=\frac{E_A'}{E_A}. \tag{5.24}$$

Since we have assumed that particle B is initially stationary, $v_B=0$, and particle A is colliding with it, this equation tells us that if $E_A'=0.5E_A$ (*i.e.* at least half of the energy has been transferred to the previously stationary particle B) then the fraction F of collisions that lead to this outcome is half of the total. Half of the collisions cause half of the energy to be transferred. If we want to have a collision that transfers at least 99% of the energy to particle B, then $E_A'=0.01E_A$ and the fraction, F, is then 0.01, so only 1% of the random collisions will accomplish this. This relationship explains why it is not so hard to strike a queue ball while playing billiards and have the queue ball stationary after the collision, assuming a little bit of table friction.

The above result also suggests a mechanism for cooling molecules to very low temperatures. A very low temperature implies a very low velocity in the laboratory frame of reference. It is very difficult to cool molecules below the condensation temperature of helium, about 300 milliKelvin (mK), however, it is possible to make samples of very cold atoms in an apparatus called a magneto optical trap (MOT).[315] Temperatures in the microKelvin (μK) range (*i.e.* a few millionths of a degree Kelvin) are routinely obtained for all the alkali atoms (Li, K, Na Cs, Rb), and for a few other species as well. If a molecule having the same mass collides with the cold atoms, it will have a finite chance of transferring most of its translational energy to the initially cold atom and thereby

*The result for the integral cross section obtained here should be compared with that obtained from the line-of-centres model described in Chapter 1, which shows a collision energy dependence.

come almost to rest in the laboratory frame in the same manner as the billiard ball (for more discussion of 'cold chemistry', see Chapter 12).

As we have noted already, although the hard sphere model above provides much useful insight into molecular collisions, the elastic scattering of atoms cannot be treated exactly as a hard-sphere collision. Study Box 5.1 introduces some of the new features that arise when scattering from a more realistic potential is considered.

5.2.4 Classical Treatment of Translational to Rotational Energy Transfer

Now let us explore the classical mechanics of a hard sphere colliding with a hard ellipse. In this case, rotationally inelastic collisions as well as elastic collisions are allowed. Following either the full three-dimensional treatment of Beck and coworkers,[317,318] or the two-dimensional model of Bosanac and Buck,[319,320] we can investigate two examples, those of an atom striking a homonuclear or a heteronuclear diatomic molecule. In Figure 5.6 we see the relevant vectors for this problem according to the commonly employed model of an ellipsoidally shaped hard shell anisotropic PES. If the diatomic is homonuclear then the CM of the ellipse coincides with the centre of the ellipse, and the displacement δ of the CM of the diatomic to the centre of the ellipse would be zero in Figure 5.6.

Imagine a non-rotating stationary ellipse being struck by a point particle of mass μ, the reduced mass of the collision partners. The particle initially has momentum $\boldsymbol{p}$ and leaves with momentum $\boldsymbol{p}'$ after the collision event. Energy, linear momentum and angular momentum must all be conserved throughout this process. After the collision, the component of the momentum parallel to the surface of the ellipse, $\boldsymbol{p}_{\parallel}$, does not change, only the normal component, $\boldsymbol{p}_{\perp}$, transfers momentum to the ellipse. It is useful to define the kinematic apse of the collision, $\boldsymbol{a}$. This is found by subtracting the incoming momentum vector $\boldsymbol{p}$ from the final momentum vector, $\boldsymbol{p}'$, *i.e.* $\boldsymbol{a}=\boldsymbol{p}'-\boldsymbol{p}$. The kinematic apse in an elastic collision bisects the angle between the incoming and outgoing particle. In Figure 5.6 this implies that $\theta_i=\theta_r$, the angle of incidence and reflection of the particle relative to the surface normal of the ellipse are equal, and the apse is as indicated. As with the case of the collision

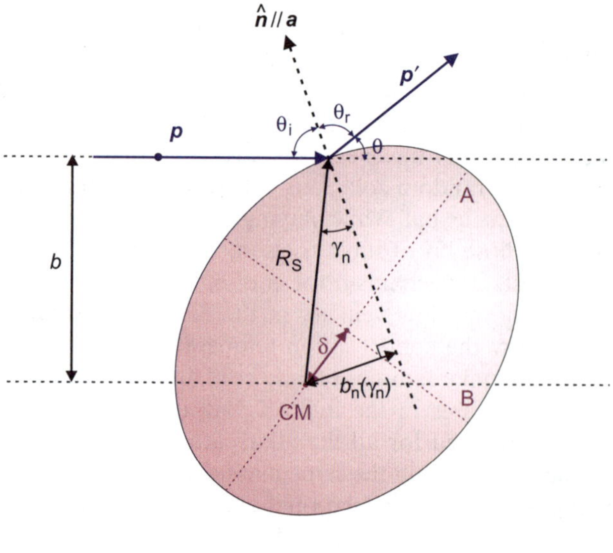

Figure 5.6 A diagram of the geometry of the atom-hard shell ellipsoid in two dimensions.

of rigid spheres, the impact therefore occurs as though a particle of momentum $\boldsymbol{p}_\perp$ undergoes a collision along the direction of the kinematic apse, $\boldsymbol{a}$. The impact parameter, which is defined relative to the CM of the ellipse, is denoted as b. We can also consider an 'effective impact parameter', which determines the torque on the diatomic molecule. This is defined as the distance of closet approach of the apse to the CM of the ellipse, and is denoted $b_n(\gamma_n)$ in Figure 5.6. As indicated, $b_n(\gamma_n)$ is the length of the line that runs from the molecular CM to a perpendicular crossing with the extrapolated surface normal, $\hat{\boldsymbol{n}}$.

We assume that the unit vector along the kinematic apse, $\hat{\boldsymbol{a}} = \frac{\boldsymbol{p}'-\boldsymbol{p}}{|\boldsymbol{p}'-\boldsymbol{p}|} \equiv \hat{\boldsymbol{n}}$, and the momenta $\boldsymbol{p}$ and $\boldsymbol{p}'$ all reside in the plane of the paper. The CM-frame is defined such that $\hat{z}\|\hat{\boldsymbol{p}}$, with the CM x-axis also lying in the $\boldsymbol{p}-\boldsymbol{p}'$ scattering plane. Because of the cylindrical symmetry of a closed-shell diatomic molecule, the impact vector, $\boldsymbol{b}$, and effective impact parameter, $\boldsymbol{b}_n(\gamma_n)$, will reside in the same plane as $\hat{\boldsymbol{n}}$ and $\boldsymbol{r}$.[302,317–327] In general the molecular axis $\boldsymbol{r}$ is not confined to the scattering plane, and therefore $\boldsymbol{b}$ and $\boldsymbol{b}_n(\gamma_n)$ will also not generally lie in the zx plane. However, since the components of $\boldsymbol{p}$ and $\boldsymbol{p}'$ that are not along $\hat{\boldsymbol{a}}=\hat{\boldsymbol{n}}$ remain conserved in a hard-shell collision, a simplified two-dimensional model in which $\hat{\boldsymbol{n}}$ and $\boldsymbol{r}$ are also assumed to reside in the zx plane is very helpful as a means of achieving both a qualitative and a quantitative insight into the dynamics of rotationally inelastic collisions.

As noted, the impulse to the ellipse is imparted as if it had been struck along the kinematic apse $\boldsymbol{a}=\boldsymbol{p}'-\boldsymbol{p}$ by a particle with momentum $\boldsymbol{p}_\perp$. After the collision, $\boldsymbol{p}'_\parallel=\boldsymbol{p}_\parallel$, and from conservation of energy one obtains

$$\frac{p_\perp^2}{2\mu} = \frac{p'^2_\perp}{2\mu} + \frac{j'^2}{2I}. \tag{5.25}$$

Here I is the moment of inertia of the ellipse* and j' is its final (classical) rotational angular momentum. It is common to define an "anisotropy parameter" as $\varepsilon=\mu/I$,[317] and in terms of this, the above equation for energy conservation can be rewritten:

$$p_\perp^2 = p'^2_\perp + \varepsilon j'^2. \tag{5.26}$$

Initially the ellipse is assumed to be non-rotating, $j=0$. Because total angular momentum is conserved in the collision, and we start with a non-rotating ellipse, we also have the relationship,

$$b_n(\gamma_n)p_\perp = -b_n(\gamma_n)p'_\perp + j', \tag{5.27}$$

where the term on the left and the first term on the right of this equation are the magnitudes of the initial and final orbital angular momenta, respectively. Here the minus sign indicates that after the collision the particle is moving away from the ellipse. Combining Eqs. (5.26) and (5.27) and rearranging leads to

$$\frac{\Delta j}{p} = \frac{2b_n(\gamma_n)\cos\theta_i}{1+\varepsilon b_n^2(\gamma_n)}, \tag{5.28}$$

where $p_\perp=p\cos\theta_\mathrm{i}$, and we have used the fact that, if $j=0$, the rotational angular momentum transferred during collision is $\Delta j=j'-0$.

By rewriting Eq. (5.27) as $p'_\perp=j'/b_n(\gamma_n)-p_\perp$, and using the identity $p'_\parallel=p_\parallel=p\sin\theta_\mathrm{i}$, it is possible, after a little rearranging, to obtain the following expression involving the angles of incidence and

*For a molecule, $I=\Sigma_i m_i r_i^2$, where m_i is the mass of atom i, r_i is its separation from the axis of rotation, the centre-of-mass, and the sum is over all atoms in the molecule.

reflection with respect to the surface normal:

$$\tan\theta_r = \frac{p'_\parallel}{p'_\perp} = \frac{p\sin\theta_i}{j'/b_n(\gamma_n) - p_\perp} = \frac{b_n(\gamma_n)\sin\theta_i}{(\Delta j/p) - b_n(\gamma_n)\cos\theta_i}. \tag{5.29}$$

Upon writing Eq. (5.29) as

$$\tan\theta_r = \frac{b_n(\gamma_n)\tan\theta_i}{(\Delta j/p) - b_n(\gamma_n)\cos\theta_i}\cos\theta_i, \tag{5.30}$$

followed by substitution of Eq. (5.28) for $\Delta j/p$, we arrive at a useful and simple relationship between the angles θ_r and θ_i:

$$\tan\theta_r = \frac{1 + \varepsilon b_n^2(\gamma_n)}{1 - \varepsilon b_n^2(\gamma_n)}\tan\theta_i. \tag{5.31}$$

Eq. (5.31) may be related to the scattering angle θ by using the fact that $\theta = \pi - (\theta_i + \theta_r)$, yielding

$$\sin\theta = \frac{2}{1 + \varepsilon b_n^2(\gamma_n)}\cos\theta_i\sin\theta_r. \tag{5.32}$$

Substituting for $\cos\theta_i$ from Eq. (5.28) leads to

$$\sin\theta = \frac{\Delta j/p}{b_n(\gamma_n)}\sin\theta_r, \tag{5.33}$$

which may be rewritten as

$$\sin\theta = \frac{\Delta j/p}{b_n(\gamma_n)}\sqrt{\frac{1}{1/\tan^2\theta_r + 1}}. \tag{5.34}$$

The dependence on θ_r may be eliminated by substituting Eq. (5.29) for $1/\tan^2\theta_r$ and using Eq. (5.28) to eliminate the angular variables $\cos^2\theta_i$ and $\sin^2\theta_i = 1 - \cos^2\theta_i$. This yields an equation that relates the scattering angle, θ, to ε, b_n and $\Delta j/p$:

$$\sin\theta = \frac{\Delta j/p}{2b_n^2(\gamma_n)}\left[\frac{4b_n^2(\gamma_n) - (\Delta j/p)^2\left(1 + \varepsilon b_n^2(\gamma_n)\right)^2}{1 - \varepsilon(\Delta j/p)^2}\right]^{1/2}. \tag{5.35}$$

For small values of the quantity $\varepsilon(\Delta j/p)^2$, Eq. (5.35) contracts to:[320]

$$\sin(\theta/2) = \frac{\Delta j/p}{2b_n(\gamma_n)}\left[1 + \tfrac{1}{4}\varepsilon(\Delta j/p)^2 + \cdots\right]. \tag{5.36}$$

Note, from Figure 5.6, that the actual value of the effective impact parameter b_n does not depend directly on the magnitude of the impact parameter, b, but on the angle γ_n (the polar angle between the surface normal $\hat{\boldsymbol{n}} = \hat{\boldsymbol{a}}$ at the position $\boldsymbol{R}_S$ where the molecular ellipse is hit and the molecular axis $\hat{r}$).

Eq. (5.36) has the important property that it has two roots for every value of Δ_j. Since $b_n(\gamma_n)$ depends solely upon γ_n, for a homonuclear diatomic molecule ($\delta = 0$ with $0 \leqslant \gamma_n \leqslant \pi/2$) there will a particular value of γ_n for which $b_n(\gamma_n)$ reaches a maximal value, *i.e.* one has $\frac{db_n(\gamma_n)}{d\gamma_n} = 0$ at $\gamma_n = \gamma_{n,b_n(\max)}$. Either side of this angle, $b_n(\gamma_n)$ will drop monotonically to zero, as γ_n increases to $\pi/2$ or decreases to 0. For a particular Δj there are therefore two collisions, with different b and γ_n values, that lead to scattering into the same θ. Although we have so far considered only classical models, we should give some thought to the quantum mechanical consequences of the above finding, the main ideas of of which can be found in Study Box 5.1. Multiple pathways to a given scattering angle with the same final rotational state give rise to quantum mechanical interferences, which can be observed in the angular distribution of the scattered molecules. To model these effects, quantum mechanical scattering calculations must generally be performed. In the case of scattering of rigid particles, QQT theory (see Section 5.2.2) provides a simple alternative to a full quantum mechanical (QM) calculation. It is based on a description of the quantum mechanical interference which arises between pathways of different path lengths associated with pairs of trajectories leading to the same outcome.[308–311]

For a homonuclear diatomic molecule, the maximal value of the effective impact parameter $b_n(\gamma_n)$ for an ellipsoidal shell can be shown[320] to be equal to the difference in the major and minor axes of the ellipse, $A-B$. The subsequent substitution of $b_n(\gamma_n) = A-B$ into Eqs. (5.35) and (5.36) implies that for each value of the angular momentum transfer, $\Delta j > 0$, there is a smallest (so called 'classical rainbow') scattering angle θ_{CR} below which that particular transition becomes forbidden, and which corresponds to a maximum in $b_n(\gamma_n)$. The rotational rainbow scattering angle increases with Δj up to a value $\Delta j_{\max}$, at which $\theta_{CR} = \pi$. Under the usual condition that $\varepsilon(\Delta j/p)^2 \ll 1$, Eq. (5.36) can be shown to reduce to:

$$\sin(\theta_{CR}/2) \approx \frac{\Delta j/p}{2(A-B)}. \tag{5.37}$$

When $\theta(\geqslant\theta_{CR}) \to \theta_{CR}$, then $\frac{db_n(\gamma_n)}{d\gamma_n} \to 0$ and from Eqs. (5.36) and (5.37) it follows that if one decreases $\theta \to \theta_{CR}$, then also $\frac{db_n(\gamma_n)}{d\theta} \to 0$. Consequently, the classical DCS will display a singularity (the so called classical rotational rainbow peak) at $\theta = \theta_{CR}$. Measurement of the angular scattering distribution of an atom from a diatomic molecule allows one to determine the scattering angle at which $\frac{d\sigma_{j\to j'}}{d\omega}$ shows its rainbow peak. Substitution of this scattering angle, θ_{CR}, into Eq. (5.37) results in an approximate experimental determination of the difference between A and B for the ellipse, yielding insight into the anisotropy in the repulsive wall around the diatomic molecule. Analytical expressions for the angle γ_n at the rotational rainbow, and also additional material on the ellipsoidal model, can be found in ref. [328].

For a heteronuclear diatomic, scattering from the two different ends of the molecule means that there are two rainbow angles (*i.e.* angles yielding maxima in the scattering intensity) at which $\frac{d\sigma_{0\to j'}}{d\omega} \to \infty$. By measuring the two angles of maximum scattering intensity, the two maximum values of the effective impact parameters b_{n_1} and b_{n_2} for scattering from each end of the molecule under the usual condition of $\varepsilon(\Delta j/p)^2 \ll 1$ follow directly from Eq. (5.36). Having determined the two b_n(max) values, the eccentricity of the ellipse, as well as the off-centre displacement δ of the molecular CM, can be readily calculated from:[320]

$$A - B = (b_{n_1} + b_{n_2})/2, \tag{5.38}$$

and

$$\delta = \frac{(b_{n_1} - b_{n_2})}{[4A/(A+B)]^{1/2}} \approx \frac{1}{\sqrt{2}}(b_{n_1} - b_{n_2})\left[1 + \frac{(A-B)}{4A}\right]. \tag{5.39}$$

Here we have assumed that $A-B \ll A$. It turns out that even if A is not known, if the term in square brackets in Eq. (5.39) is approximated to unity, the equation is still accurate to within a few percent for realistic PESs. Therefore, the measurement of the rainbow scattering angle of atoms from molecules tells us about the shape of the repulsive wall of the diatomic potential. A very nice example of this is provided by the measurements of Houston *et al.*[329,330] on scattering of argon atoms from NO molecules, which will be discussed further in Section 5.3.1.

It is interesting to consider whether there is a preferred direction in which the molecule will rotate as a result of collision, or whether, following a random collision, the diatomic molecule spins equally in all directions? From conservation of angular momentum, and recognizing that the cross product $\boldsymbol{b}_n(\gamma_n)\times\boldsymbol{p}_\perp=\boldsymbol{R}\times\boldsymbol{p}=\boldsymbol{\ell}$ is the orbital angular momentum of the collision partners prior to the collision, we can write $\Delta\boldsymbol{J}_{\text{tot}}=\Delta(\boldsymbol{b}_n(\gamma_n)\times\boldsymbol{p}_\perp)+\Delta\boldsymbol{j}=0$, where $\boldsymbol{J}_{\text{tot}}$ is the total angular momentum, which immediately rearranges to:

$$\Delta\boldsymbol{j} = -\Delta\left(\boldsymbol{b}_n(\gamma_n)\times\boldsymbol{p}_\perp\right). \tag{5.40}$$

This equation states that, because the total angular momentum is conserved, any gain in the rotational angular momentum of the molecule, $\boldsymbol{j}$, must be offset by a loss in orbital angular momentum, $\boldsymbol{\ell}$. At each position, $\boldsymbol{R}_\text{S}$, on the surface of the ellipse at which an impact can take place, the momentum transfer vector, $\Delta\boldsymbol{p}_\perp=\boldsymbol{p}'_\perp-\boldsymbol{p}_\perp$, will always point along the apse axis, $\hat{\boldsymbol{a}}=\hat{\boldsymbol{n}}$. Consequently, the cross product of $\boldsymbol{b}_n(\gamma_n)$ and $\Delta\boldsymbol{p}_\perp$ will always point in a direction perpendicular to the apse axis. In other words, the change in the rotational angular momentum vector $\Delta\boldsymbol{j}$ cannot have a component along the apse axis if it is to cancel the change in orbital angular momentum, $\Delta\boldsymbol{\ell}$.[303,331] This is another way of saying that there is no change in the rotor orientation along the momentum transfer vector $\Delta\boldsymbol{p}_\perp$. Later, in Section 5.3.4.5, we will describe an experiment that allows a measurement of the direction in which the vector $\Delta\boldsymbol{j}$ points after a collision, and will see that it does not always have to lie perfectly along the apse of the collision.

Before we leave this classical hard ellipse model, it is interesting to note that, for a scattering of an atom by a cylindrically symmetric molecule, the ratio $\Delta j/p$ at each scattering angle is independent of the collision energy, E_c. Moreover, only the momentum $\boldsymbol{p}_\perp$ along the surface normal $\hat{\boldsymbol{n}}$ turns out to be relevant, since $\boldsymbol{p}_\parallel$ is conserved during a collision. This causes the azimuthal angle ϕ_n of the molecular axis $\hat{\boldsymbol{r}}$ in Figure 5.6 to be irrelevant for a collision involving a 'single hit' between the molecule and the atom. In the case of a single-hit collision, the two- and three-dimensional treatment of the collision problem lead to identical results. However, the probability for multiple hits occurring during the collision (these are sometimes called chattering collisions, and involve the atom having multiple strikes with the rotating molecule before it moves away) will be enhanced if $\hat{\boldsymbol{r}}$, $\boldsymbol{p}$ and $\boldsymbol{p}'$ all reside in the same plane. In this latter case, a second collision is deemed to occur when $\varepsilon b_n^2(\gamma_n)\geqslant 0.5$.[318] Note that also, when $0.5\leqslant\varepsilon b_n^2(\gamma_n)<1$, Eq. (5.29) still remains valid. The first hit in the collision results in a rebound of the incoming atom from the molecular shell. This brings about an immediate end-over-end rotational motion of $\boldsymbol{r}$ in the plane defined by $\hat{\boldsymbol{n}}$ and $\hat{\boldsymbol{b}}_n(\gamma_n)$. An important question is whether the presence of such chattering collisions might be observable in the measured DCS.

The influence of such possible secondary collisions on the rotational rainbow has been explored by Buck *et al.*, who measured the differential (scattering angle resolved) energy loss spectra in a crossed Xe-CO_2 molecular beam scattering experiment at collision energies between 0.2 eV and 1.6 eV.[332,333] In regard to this system, at scattering angles $\theta\geqslant\pi/2$, Eqs. (5.35) and (5.37) predict that the rotational rainbow ought to occur with a loss of nearly 100% of E_c. This prediction turns out to be in contradiction to the experimental results, in which the peak in the DCS at large $\Delta E_\text{c}=E'_\text{c}-E_\text{c}$ for $\theta\geqslant\pi/2$ remains absent, and the DCS actually decreases with θ. A detailed analysis of individual trajectories run on a partial *ab initio* based PES[327] showed that in the first step of the collision indeed nearly all ΔE_c is transferred to rotational energy. However, before the slow Xe atom has left the interaction zone, it is hit again by the fast rotation of the CO_2 induced in the first step of the energy transfer. Consequently, a large part of the rotational energy acquired by the CO_2 in the first step is re-transferred to translational energy. This results in a decrease in the energy transferred, and completely removes the rotational rainbow structure expected for $\theta \geqslant \pi/2$. Semi-classical coupled state calculations confirmed both the outcome of the experiments and the classical trajectory calculations. For smaller scattering angles, $\theta < \pi/2$, the rainbow angles, as in other cases, corresponded quite well with the predictions of the hard shell model of Eqs. (5.35) and (5.37). The insensitivity of the sudden rotational rainbow to the steepness of the repulsive wall reflects the fact that the collision time is negligibly small compared to the rotational period of the rotor. Apparently, this is not the case for multiple collision rotational "rainbows", for which the collision time is the determining factor.[333,334]

5.3 EXPERIMENTAL STUDIES OF ENERGY TRANSFER

5.3.1 Early Measurements of DCSs and Rotational Rainbows

Two different experimental methods are used to measure DCSs and to observe rotational rainbows. Many classic experiments in the field have employed time-of-flight (TOF) measurements, in which one uses a mass spectrometer at a fixed (laboratory) scattering angle, Θ, to measure the arrival time distribution or mass spectrum (in favourable cases rotationally-resolved) of one of the outgoing collision partners. More recently, many experiments have moved towards using velocity-mapped ion imaging, described in detail later in Section 5.3.4.2 of this chapter (see also Study Box 5.3). In this technique, a laser is used to selectively ionize and detect a particular quantum state, and the velocity (speed and angle) of the ionized scattered product is measured by projecting the resulting ions onto an imaging detector.

In the case that $\boldsymbol{j}'$ may be considered to act as a non-quantized variable (*i.e.* when $|\boldsymbol{j}'|/\hbar \gg 1$, and the angular momentum behaves classically), both methods are considered to be equivalent in terms of the information provided. However, when one resolves the DCS for specific values of the final rotational angular momentum quantum number $j' \approx |\boldsymbol{j}'|/\hbar$ on the order of unity, then one needs to measure the time-of-flight spectra at a very large number of scattering angles so as not to miss the true rotational rainbow maxima for some of the final states. This is not the case with the velocity-mapped ion imaging experiments, as angular distributions for each quantum state are recorded separately. The earliest scattering-angle-resolved studies on the transfer of translational into rotational energy in thermal atom-molecule collisions were carried out on the systems $K + N_2$ and $K + CO$.[317] The velocity spread was 19–17% FWHM for the seeded K-beam, and 18% for the molecular N_2 or CO beam ($T_{rot} \sim 33$ K) and the scattered K-atoms were velocity analyzed by a $\Delta v'/v' = 10\%$ velocity selector before being passed onto an iridium hot wire detector. Subsequently, Bergmann and coworkers[335,336] and, independently, Kinsey, Pritchard and coworkers[337,338] succeeded in employing rotationally quantum-state selective laser depletion techniques to observe the rotationally inelastic, fully-quantum-state-to-quantum-state-resolved laboratory DCS's of, respectively, $\mathrm{He/Ne} + \mathrm{Na_2}(v, j) \longrightarrow \mathrm{He/Ne} + \mathrm{Na_2}(v, j')$ and $\mathrm{Ar} + \mathrm{Na_2}(v, j) \longrightarrow \mathrm{Ar} + \mathrm{Na_2}(v, j')$. The scattered signal into a particular $\mathrm{Na_2}(v', j')$ channel was observed using a rotatable scattering-angle-resolving laser-induced fluorescence (LIF) detector,[339–342] or by the Doppler shift of the probed LIF transition as observed by a fixed LIF detector, which probed the product density at the scattering centre[337,343] (the technique of LIF is discussed more fully in Study Box 5.2). Typically, due to the small value of the rotational constant of Na_2 ($B \approx 0.15\,\mathrm{cm^{-1}}$) even for the rainbow peak at large scattering angle, as expected for $|j' - j| \approx 40$, only a minor fraction of the collision energy could be rotationally inelastic. This restriction made the semi-classical (stationary phase) version of the IOS approximation an elegant tool to analyze and to theoretically predict the observed quantum structures in the DCS's of the state resolved rotationally inelastic scattering of Na_2 with He and Ne. Bergmann and coworkers[335] demonstrated that by modulating the optical pumping of a transition involving a given rovibrational state of Na_2, the population of the state in the beam could be modulated. The

STUDY BOX 5.2: LIF AND REMPI DETECTION SCHEMES

Reaction dynamics experiments generally involve very small numbers of molecules. This is largely a consequence of the fact that such experiments must be carried out under single-collision conditions (see Section 1.3.1), and therefore often at very low pressures. A typical pump-probe experiment, for example, may produce only a few hundred products, or even less, per laser cycle, and these products will usually be present amongst much higher concentrations of other species, such as unused reactants and background gases. Any technique used to detect these products must be extremely sensitive, and also highly specific to the chemical species under study. The two most commonly used techniques that match both of these requirements are laser-induced fluorescence (LIF) and resonance-enhanced multiphoton ionization (REMPI) (see Study Box 1.2 on Lasers in reaction dynamics). The two techniques are illustrated schematically in Figure 5.7.

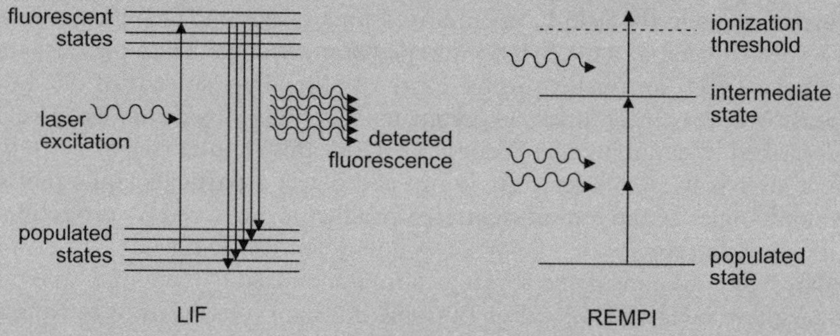

Figure 5.7 Principles of laser-induced fluorescence (LIF) and resonance-enhanced multiphoton ionization (REMPI) detection. See text for details.

Laser-induced fluorescence detection requires that the spectroscopy of the species to be detected is well characterized, and that the species has a fluorescent excited electronic state that is optically accessible from the ground electronic state. The technique is widely used in the detection of small molecules such as OH and NO produced in a pump-probe reaction scheme.

As shown in the figure, nascent products are excited to the upper state, and the photons emitted as the excited state relaxes are detected, with the measured signal being proportional to the original population of the lower state. By scanning the excitation laser over the various accessible rovibronic transitions from the ground state, the rovibrational state populations of the nascent products may be determined. In order to compare the signals from two states, the LIF signals must be corrected to take account of the differing absorption cross sections of the lower state and emission cross sections of the upper state(s) involved in the transitions.

The spectral linewidths of the rovibronic transitions excited in LIF are determined by the velocity distributions of the reaction products through Doppler broadening of the peaks. If the frequency bandwidth of the probe laser is much narrower than the width of the spectral line, then we have *Doppler-resolved LIF*, and the spectral lineshape yields a one-dimensional projection of the product velocity distribution onto the laser propagation axis. If several of these projections are recorded in different experimental geometries (determined by the propagation and polarization directions of the pump and probe lasers), enough information is obtained to reconstruct the entire 3D velocity distribution of the products. Lineshape profiles for a simple diatomic dissociation reaction are shown in Figure 5.8, which illustrates how the lineshapes vary depending on the direction of the probe laser propagation relative to the recoil direction of the photofragments (see Chapter 8 and Study Boxes 8.2 and 8.3). The lineshape profiles may also be analyzed to obtain information on product angular momentum polarization (see Chapters 8 and 9, in particular Study Box 9.1 on detecting angular momentum polarization).

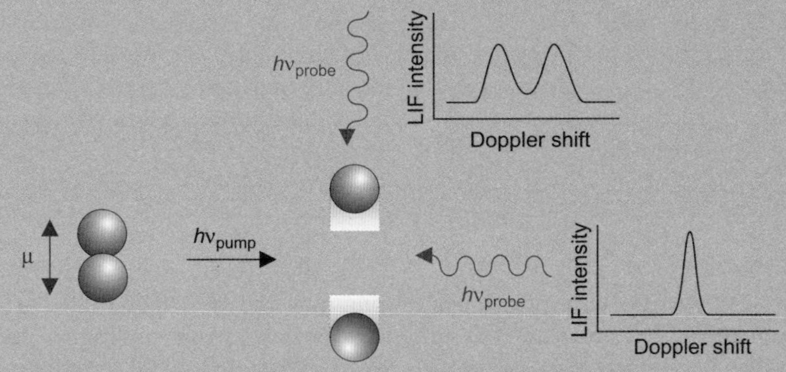

Figure 5.8 Doppler resolved LIF profiles taken in two different experimental geometries for a diatomic molecule dissociating *via* a parallel transition. Each Doppler profile provides a one-dimensional projection of the photofragment velocity distribution along the direction of propagation of the probe laser. When the probe laser radiation propagates along the recoil direction of the photofragments, the line profile is split into two peaks, corresponding to fragments moving parallel or antiparallel to the probe propagation direction.

LIF usually detects neutral product molecules. If the products are ionized then sensitive detection becomes relatively straightforward: particle multipliers capable of detecting single ions have been widely available for several decades, and any ions formed in the experiment may be steered towards the detector by incorporating appropriate electric fields into the instrument. Single photon ionization of most product species of interest is unfeasible, since their ionization energies correspond to soft X-ray photons, which are in rather short supply in most laboratories. However, if two or more photons are used to reach the ionization threshold, standard laser sources emitting in the visible and UV spectral regions may be used. The possibility of multiphoton ionization may seem to fly in the face of one of the most fundamental 'rules' of spectroscopy, namely that in order for a molecule to be excited from state A to state B it must absorb a single photon with an energy that matches the energy separation between the two levels. We are now stating that the same end can be achieved if the molecule absorbs two photons, each having an energy equal to half the energy separation between the two levels of interest. The resolution to this conundrum lies in the existence of entities known as *virtual states*. A virtual state is accessed when a molecule absorbs a photon that does not take it to another of its existing quantum states. Such a state is not a solution of the Hamiltonian for the molecular system, but it is a solution of the molecule + photon system. When the light intensity is extremely high there is a non-zero probability that a molecule will coherently absorb two or more photons simultaneously, and the virtual state can be thought of as the (hypothetical) state accessed after absorption of the first photon. The coherent simultaneous absorption of multiple photons is an example of a non-linear optical technique, more information about which can be found in Study Box 11.2.

Non-resonant multiphoton ionization, in which none of the intermediate levels accessed by successive photon absorptions correspond to real levels of the molecule, is not a particularly efficient process. However, when one of the intermediate states does correspond to an actual state of the molecule, the ionization efficiency is increased enormously, often by several orders of magnitude, and we have resonance-enhanced multiphoton ionization. REMPI schemes are usually labelled in terms of the number of photons required to reach the resonant state and the number of photons required to ionize from the resonant state. So, for example, in a (2+1) REMPI process two photons take the molecule from the ground state to the intermediate state, and a third photon ionizes the molecule from this state.

Efficient REMPI schemes are available for a wide range of molecules, and for this reason the technique is more widely used than LIF detection. Velocity information may be obtained by combining the technique with Doppler-resolved measurements, as with LIF, or using time-of-flight mass spectrometry or velocity-mapped imaging (see Study Box 5.3). A time-of-flight profile provides a one-dimensional projection of the velocity distribution, while velocity-mapped imaging provides a two-dimensional projection of the distribution.

Claire Vallance

population in ($v''=0$, $j''=28$) was modulated by pumping the ($v''=0$, $j''=28$)$\longrightarrow$($v'=6$, $j'=27$) transition between the $X^1\Sigma_g^+$ and the $B^1\Pi_u$ electronic states (corresponding to about 1% of all supersonically cooled Na_2 molecules in the beam). No significant collisional or radiative repopulation of the pumped level was found to occur upstream of the crossing with the target beam. The lifetime of the upper electronic state is short and no electronically excited molecules were able to reach the scattering centre. Scattered Na_2 molecules impinged onto the LIF detector through a narrow collimator at a distance of 5 cm from the scattering centre. The LIF detector was rotated around the scattering centre and probed, upon a proper tuning of its laser frequency, the $v''+\Delta v''$, $j''+\Delta j''$ state population of the scattering angle resolved outgoing Na_2 molecules. In agreement with theoretical predictions, the DCS data for the Na_2-He and Na_2-Ne collision systems showed that for CM-frame scattering angles $\theta > 50°$ nearly all events are inelastic. In a later study, Bergmann and coworkers[339] succeeded in observing both the first, regular rotational rainbow and, at a larger scattering angles, also a secondary 'supernumerary' rotational rainbow. All of these observations were found to be in quantitative agreement with the IOS approximation predictions. These studies also provided an experimental proof of the $|\Delta m_j| \ll j$ (m_j or alignment conserving) propensity rule along the kinematic apse $\boldsymbol{a}=\boldsymbol{p}'-\boldsymbol{p}$ in rotational inelastic differential scattering,[342] as expected from Eq. (5.40).

Another set of beautiful early experiments employing universal detector setups, pioneering the pulsed crossed beam technique (Giese, Gentry and coworkers[344–346]) and improving the continuous wave (CW) supersonic crossed beam technique (Buck and coworkers[347,348]) succeeded in cooling the HD molecules in one of the beams efficiently to the $j=0$ state, which led to a well resolved measurement of the HD-Ne DCS for both inelastic $j=0\rightarrow j'=1$ and elastic $j=0\rightarrow j'=0$ transitions. Shortly thereafter, perfecting the supersonic beam and the time-of-flight spectrum techniques, Faubel *et al.* were able to enhance the energy resolution for rotationally inelastic transitions to $4\,\mathrm{cm}^{-1}$. These improvements resulted in experimental determinations of fully quantum state-to-state resolved DCS's for systems such as He-N_2, CO, O_2 and CH_4.[349] Later on it turned out to be possible even to resolve (though only partly for high j') the DCS for individual rotational state-to-state transitions for a system such as Ar-O_2 up to $\Delta j=18$.[324,350]

Some crossed beam experiments employed a rotatable liquid He-cooled bolometer for product detection.* In conjunction with a HF laser tuned to specific rotational transitions of the $v=0\rightarrow 1$ transition, these experiments provided a determination of the DCS for each final rotational state j'.

*A bolometer is a detector capable of measuring very small variations in temperatures. The bolometer is placed in a cryostat, which allows it to reach temperatures of a few kelvin. When a particle strikes the surface of the bolometer it deposits its energy in the form of heat which can be detected by a thermal sensor. The signal of a rotatable thermal Ge bolometer detector at $T\sim 1.5$ K is proportional to the sum of the kinetic and internal energy of (differentially scattered) molecules that impinge on the detector. At this temperature, the Ge bolometer is sensitive to a minute change of its heat load because it can be operated near the threshold to superconductivity. If a modulated IR laser beam excites only one specific rovibronic transition of the scattered molecule on its way to the bolometer, selective amplification of the resulting ac component of the bolometer signal gives a direct measure of the excited population.

Keil and coworkers[351] employed this method in a supersonic jet-cooled crossed molecular beam study of Ar + HF(j = 1, 2) ⟶ Ar + HF(j'). At the collision energy, E_c, of 120 meV, the highest energetically accessible rotational state was $j' = 6$, and these experiments yielded DCSs for all states up to $j' = 5$. One of the most interesting observations was the presence of an intense forward scattering peak for all j', which decreased in prominence as j' increased. Rotational rainbow structure was also observed for the highly inelastic transitions to $j' = 4$ and $j' = 5$. These unexpected findings were shown to stem from a balance between the attractive and repulsive parts of the potential.[352]

A breakthrough in the experimental determination of rotationally inelastic state-to-state DCSs was brought about by Houston and coworkers,[329,330] and Chandler and coworkers,[353] who built crossed molecular beam machines that were specifically designed to measure both state-to-state rotationally inelastic and reactive DCS's for atom-molecule collision systems. These instruments employed resonantly enhanced multiphoton ionization (REMPI) detection coupled with velocity-mapped ion imaging to characterize the scattering products (see Study Boxes 5.2 and 5.3).

One early study utilizing this technique, carried out by Chandler and coworkers,[353] probed rotational energy transfer in collisions of HCl with Ar atoms, at a collision energy of 538 cm^{-1}. The experiments yielded reactant and product state-resolved DCSs for rotational energy transfer over the full range of CM-frame scattering angles ($\theta = 0$ to 180°), for all energetically allowed collision-induced transitions $j = 0 \rightarrow j' = 1$–6.[353] The measurement was made by crossing a supersonic atomic beam of Ar with a supersonic molecular beam of 5% HCl in Ar. At the scattering centre a laser beam was focused, resonant on a transition of the Q-branch of the $E(^1\Sigma^+) \leftarrow X(^1\Sigma^+)$ system of HCl at a wavelength of ~239 nm. As the laser frequency was tuned through a transition resonant with a ground state rotation level, collision products in that level were ionized and detected using the ion imaging technique (see Study Box 5.3). A typical image, with a maximum count of around 2000 ions in the most intense pixel, was recorded in a time of between 10 and 60 mins, depending on the quantum state being probed. This turns out to be several orders of magnitude faster than the time required by other techniques, which need an order(s) of magnitude longer measuring time and result in a less well resolved and incomplete rotational energy transfer DCSs for a collision system. The experimental results obtained from the study were compared with theory, and the intermolecular potential adjusted to better fit the data.[354–356]

The velocity-mapped ion imaging technique is useful for the determination of DCSs for a particular product quantum state for any molecule that can be quantum state selectively ionized. The scattering process can be inelastic scattering or reactive scattering (see Chapter 6). The technique is also very useful for the measurement of photochemical processes (see Chapter 8),[357] for which information on the dissociation dynamics and energetics is encoded in the velocity of the photofragments.

5.3.2 Molecular Beam Studies of the NO – Rare Gas System

Although the previous discussion of hard ellipse models provides a valuable picture of rotational energy transfer for atom-diatom collisions, it is not sufficient to describe the complex interactions of real molecules. Molecules are not two- or even three-dimensional ellipses, and such simple models can only give one a sense for the dynamics that will occur when a real atom strikes a real molecule. There is one atom-molecule system that has been extensively studied: NO molecules in collision with a rare gas atom. We will use this system to highlight many experimental studies of collisional energy transfer. We will describe a set of experiments and the accompanying theory that has been developed to describe the experiments. These experiments go far beyond the study of energy transfer rates and explore the details of the potential energy surface of a complex diatomic molecule, $NO(X^2\Pi)$.

Collisional energy transfer processes that involve molecules with electronic structure have been particularly interesting to study due to the many possible channels available. In addition to rotational energy transfer, transitions between hyperfine states, lambda-doublet (Λ-doublet) states,* and spin-orbit states are possible. Of particular interest are the dynamics of a molecule in a $^2\Pi$ electronic state colliding with an atom, as this is an example of a molecular collision involving more than one Born-Oppenheimer PES. Due to the accessible and sensitive spectroscopic probes available for NO, the NO electronic, vibrational, rotational and Λ-doublet product state distributions after collision with a rare gas atom have been measured. The total scattering cross section as a function of collision energy has also been measured along with a range of vector quantities. These include the DCS (proportional to the NO product angular distribution) for both the ground and first electronically excited states; and correlated vector quantities such as the rotational angular momentum alignment (which defines the plane in which the NO axis is spinning relative to the relative velocity vector of the collision) and orientation (clockwise or counterclockwise rotation; further information about angular momentum orientation and alignment is given in Chapter 9). In the experiments under consideration, the orientation and alignment were determined for the scattered NO as a function of rotational state and scattering angle. Additionally, the ability of a collision to cause rotational energy transfer as a function of which end of the NO molecule is struck by the Ar atom, the so called steric asymmetry ratio (SA) of the collision, has also been measured.

5.3.2.1 The Ar–NO interaction

To help better appreciate the experiments on inelastic scattering of the NO molecule it is helpful to consider its ground state electronic structure, as well as likely features of the Ar–NO interaction potentials. In concert with this wealth of experimental data, the Ar–NO system has acted as a benchmark for the development of a series of increasingly accurate PESs, and for the development of theoretical scattering techniques for the prediction of dynamics upon these surfaces. Description of the interaction between NO and a rare gas atom is quite complicated, arising, as we have seen, from the fact that the ground state of the NO radical is of $^2\Pi$ symmetry. In order to describe the rare gas-NO system one must construct two angle-dependent intermolecular Born-Oppenheimer adiabatic potential energy surfaces. This is because the reflection operator with respect to the plane of the three nuclei commutes with the electronic Hamiltonian of the rare gas–NO system, with the consequence that the electronic eigenfunction of the system will be symmetric or antisymmetric with respect to reflection in this plane. This leads to the A$'$ symmetry configuration of the Born-Oppenheimer PES, in which the unpaired electron π-orbital of NO points in the plane of the three atoms, and the A$''$ anti-symmetric PES, in which the π-orbital of NO points perpendicular to the nuclear plane. Since the rare gas atom can break the cylindrical symmetry of the diatom, the $V(\mathrm{A}')$ and $V(\mathrm{A}'')$ PESs will not necessarily be degenerate. At small intermolecular separations, R, the Pauli repulsion between the rare gas atom and NO will be larger when the π-orbital lies in the plane of the three atoms, *i.e.* $V(\mathrm{A}') > V(\mathrm{A}'')$. In the limit of large R, when the effect of the Ar atom is very small, the NO open shell rotational wave function is an eigenfunction of both the rigid rotor Hamiltonian, $\hat{H}_{\mathrm{rot}}(\boldsymbol{r})$, and the spin-orbit interaction Hamiltonian $A_{\mathrm{SO}}\boldsymbol{L}\cdot\boldsymbol{S}$ (where A_{SO} is the spin-orbit coupling constant). In the special case of NO for j up to around 10, the spin-orbit splitting is much larger than the rotational energy (*i.e.* $A_{\mathrm{SO}}\boldsymbol{L}\cdot\boldsymbol{S} \gg \hat{H}_{\mathrm{rot}}(\boldsymbol{r})$, so in addition to the electronic orbital angular momentum $\boldsymbol{L}$ of the unfilled π orbital, the electron spin, $\boldsymbol{S}$, is also "tied" to the NO inter-nuclear axis $\boldsymbol{r}$.[239] The NO(X) molecule is then regarded as a prototype of a Hund's case (a) molecule (see Study Box 4.3 for an introduction to Hund's coupling cases). The incoming and outgoing free space NO Hund's case (a) eigenfunctions before and after the collision, which include the azimuthal angle χ of the unpaired π-orbital, can be expressed in terms of symmetric top-like wavefunctions[358]

$$|j, m_j, \bar{\Omega}, \varepsilon\rangle = \left[\frac{2j+1}{16\pi^2}\right]^{1/2}\left[D^{j*}_{m_j,\bar{\Omega}}(\phi,\gamma,\chi) + \varepsilon(-1)^{\bar{\Omega}-\frac{1}{2}}D^{j*}_{m_j,-\bar{\Omega}}(\phi,\gamma,\chi)\right]. \tag{5.41}$$

*For a description of the Λ-doublet levels of NO(XΠ) see ref. [239] and the further discussion below.

Here $\bar{\Omega}(\equiv|\Omega|)$ denotes the absolute value of the sum of the electronic orbital angular momentum projection quantum number $\Lambda=\pm1$ of the unpaired electron in the π-orbital and the projection quantum number $\Sigma=\pm\frac{1}{2}$ of its electron spin, $S=1/2$. Since $\bar{\Omega}=|\Lambda+\Sigma|$, one designates two spin-orbit states $^2\Pi_{1/2}$ with $\bar{\Omega}=1/2$ and $^2\Pi_{3/2}$ with $\bar{\Omega}=3/2$. In the case of NO(X), the origin of the first exited $^2\Pi_{3/2}$ state lies about 120 cm^{-1} above that of its ground state $^2\Pi_{1/2}$. In Eq. (5.41) the symbol $\varepsilon=\pm1$ indicates the so-called parity index of each of the two components (lower or upper) of a Λ-doublet rotational wave function, where the parity p is given by the expression[11]

$$p \equiv (-1)^{j-\varepsilon/2}. \tag{5.42}$$

In contrast to some of the earlier guesses, irrespective of the spin-orbit manifold the upper and lower components of the Λ-doublets of NO(X) have been shown both theoretically[359] and experimentally[360] to carry the definite parity index of $\varepsilon=-1$ and $\varepsilon=1$, respectively. The symmetric top-like rotational wave function of Eq. (5.41) results in a probability distribution for the π-orbital azimuthal angle χ which is cylindrically symmetric. Consequently, the rotational wave function does not match adiabatically to the $V(\mathrm{A}')$ and $V(\mathrm{A}'')$ Born-Oppenheimer PES's. Alexander[358,361] has shown that the electronically adiabatic spin-orbit state conserving rotational inelastic transitions are ruled by a potential of the form:

$$V_{\mathrm{sum}}(R,\gamma) = \tfrac{1}{2}[V(\mathrm{A}'') + V(\mathrm{A}')], \tag{5.43}$$

while the electronically non-adiabatic spin-orbit state changing rotational transitions follow from:

$$V_{\mathrm{dif}}(R,\gamma) = \tfrac{1}{2}[V(\mathrm{A}'') - V(\mathrm{A}')]. \tag{5.44}$$

The dependence of the potential on the Jacobi angle, γ, is then expanded in a series of Wigner rotation functions, $d^k_{n,m}(\cdots)$ (see ref. [239] for a definition of the latter):

$$V_{\mathrm{sum}}(R,\gamma) = \sum_{\lambda=0}^{\lambda_{\max}} V_{\lambda,0}(R)\, d^{\lambda}_{0,0}(\gamma) \qquad V_{\mathrm{dif}}(R,\gamma) = \sum_{\lambda=2}^{\lambda_{\max}} V_{\lambda,2}(R)\, d^{\lambda}_{2,0}(\gamma). \tag{5.45}$$

Without the inclusion of the electronic non-Born-Oppenheimer spatial azimuthal angle coordinate, χ, having only a knowledge of $V_{\mathrm{sum}}(R,\gamma)$ and $V_{\mathrm{dif}}(R,\gamma)$ alone, does not result in the full PES, $V(R,\gamma,\chi)$, of the system. Even in the case that the NO bond axis r is frozen, in order to be able to carry out scattering calculations, a knowledge of the R-dependent potential coupling matrix elements

$$\langle \ell', j', J_{\mathrm{tot}}, M_{\mathrm{tot}}, \bar{\Omega}', \varepsilon' | V(R,\gamma,\chi) | \ell, j, J_{\mathrm{tot}}, M_{\mathrm{tot}}, \bar{\Omega}, \varepsilon \rangle \tag{5.46}$$

is required. In these matrix elements J_{tot}, M_{tot}, ℓ and ℓ' denote the quantum numbers, respectively, for the total angular momentum, its projection onto the quantization axis, and the incoming and outgoing orbital angular moments. Let us consider the electronic part of the potential coupling matrix element between the Hund's case (a) type incoming $|\ell, j, J_{\mathrm{tot}}, M_{\mathrm{tot}}, \bar{\Omega}, \varepsilon\rangle$ and outgoing $|\ell', j', J_{\mathrm{tot}}, M_{\mathrm{tot}}, \bar{\Omega}', \varepsilon'\rangle$ eigenfunctions. The χ dependent term, which reflects the projection of the electronic angular momentum onto the NO axis and carries an eigenvalue Ω, is given by the function $e^{i\Omega\chi}$. The resulting partial coupling matrix elements are of the form $\langle e^{i\Omega'\chi}|V(R,\gamma,\chi)|e^{i\Omega\chi}\rangle$. Since $V(R,\gamma,\chi)$ does not depend on the projection of the spin quantum number onto the NO axis, $\Sigma=\pm\frac{1}{2}$, only the projection quantum number $\Lambda = \pm 1$ of the unpaired π-orbital is of relevance for the collision. This quantum number can be conserved ($\Lambda' - \Lambda = \Omega' - \Omega = 0$), or it can change ($\Lambda' - \Lambda = \Omega' - \Omega = \pm 2$). Next, one can expand the potential as a series

$$V(R, \gamma, \chi) = \sum_{\nu=-\infty}^{+\infty} V_\nu(R, \gamma) e^{i\nu\chi}, \tag{5.47}$$

which leads to a set of partial matrix elements: $\langle e^{i\Omega'\chi} | V_\nu(R, \gamma) | e^{i\nu\chi} | e^{i\Omega\chi} \rangle$, one for each term in the expansion. Therefore, for a transition that conserves $\bar{\Omega}$, one has $\Omega' - \Omega = 0$, and only the $V_0(R, \gamma)$ term remains once the integral over χ in the matrix elements is performed. For a transition $\bar{\Omega} = 1/2 \rightarrow \bar{\Omega}' = 3/2$ or *vice versa*, only terms with $\Omega' - \Omega = \pm 2$ will contribute, and so overall the potential may be reduced to

$$V(R, \gamma, \chi) = V_0(R, \gamma) + V_{+2}(R, \gamma) e^{+2i\chi} + V_{-2}(R, \gamma) e^{-2i\chi}. \tag{5.48}$$

As illustrated in Figure 5.9, at $\chi = \pi/2$ and $\chi = 3\pi/2$ the unpaired electron lobe points perpendicular to the triatomic plane, and for reasons of symmetry one finds $V_{+2}(R, \gamma) = V_{-2}(R, \gamma)$, which results in

$$V(R, \gamma, \chi) = V_0(R, \gamma) + V_{+2}(R, \gamma) \left[e^{+2i\chi} + e^{-2i\chi} \right]. \tag{5.49}$$

Moreover, according to convention, as followed in Figure 5.9, $V(A'')$ corresponds to $\chi = \pi/2$ and $V(A')$ corresponds to $\chi = 0$. So the χ dependence of $V(R, \gamma, \chi)$ is taken into account by:

$$V(R, \gamma, \chi) = V_{\text{sum}}(R, \gamma) - V_{\text{dif}}(R, \gamma) \cos 2\chi. \tag{5.50}$$

More generally, the spin-orbit interaction Hamiltonian, $A_{SO} \boldsymbol{L} \cdot \boldsymbol{S}$, is not strong enough to "tie" $\boldsymbol{S}$ completely to the molecular axis $\boldsymbol{r}$, and the rotational wave functions of the $\Pi_{1/2}$ and $\Pi_{3/2}$ states become mixed. So for NO in higher j-states,[239,361–364] $\bar{\Omega} = 1/2 \rightarrow F_1$ and $\bar{\Omega} = 1/2 \rightarrow F_2$, which leads to the mixed wavefunctions:

$$|j, m_j, F_1, \varepsilon\rangle = a_j |j, m_j, \bar{\Omega} = \tfrac{1}{2}, \varepsilon\rangle + b_j |j, m_j, \bar{\Omega} = \tfrac{3}{2}, \varepsilon\rangle, \tag{5.51}$$

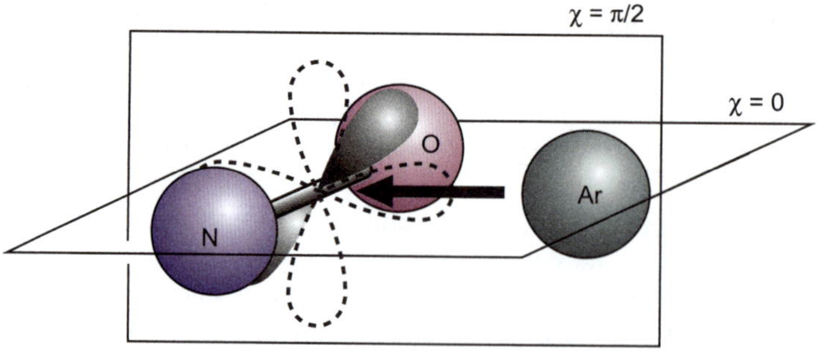

Figure 5.9 A snapshot of an incoming Ar atom in collision with a NO molecule. The orientation of the molecular axis $\boldsymbol{r}_{NO}$ together with the orientation of the π-orbital lobe of the unpaired electron of the NO molecule determine the strength of the intermolecular potential at each fixed value of R. Note that the black arrow (the direction of $\boldsymbol{v}_{rel}$) typically does not reside in the plane of the three atoms.

and

$$|j, m_j, F_2, \varepsilon\rangle = -b_j|j, m_j, \bar{\Omega} = \tfrac{1}{2}, \varepsilon\rangle + a_j|j, m_j, \bar{\Omega} = \tfrac{3}{2}, \varepsilon\rangle, \quad (5.52)$$

with

$$a_j = \sqrt{\frac{X_j + Y_0 - 2}{2X_j}} \quad \text{and} \quad b_j = \sqrt{\frac{X_j - Y_0 + 2}{2X_j}}, \quad (5.53)$$

in which,

$$X_j = \sqrt{4(j+\tfrac{1}{2})^2 + Y_0(Y_0 - 4)} \quad \text{and} \quad Y_0 = \frac{A_0}{B_0}. \quad (5.54)$$

Here $A_0 = 123.13\,\text{cm}^{-1}$ and $B_0 = 1.6961\,\text{cm}^{-1}$ denote respectively the spin-orbit coupling constant (which quantifies the splitting between the $\bar{\Omega} = 1/2$ and $\bar{\Omega} = 3/2$ states) and the rotational constant of NO(X) at $v = 0$. The mixed Hund's case rotational wavefunctions of Eqs. (5.51) and (5.52) carry a probability distribution for the π-orbital azimuthal angle χ around $\boldsymbol{r}$ that is no longer cylindrically symmetric. In addition to the $\varepsilon = \pm 1$ level, one assigns the components of a mixed Hund's case component also with an A′ or A″ label. In the case of $\varepsilon = 1$, the lower Λ-doublet component of the F_1 rotational wavefunction of Eq. (5.51) carries an A′ label, while the lower Λ-doublet component of the F_2 rotational wavefunction of Eq. (5.51) carries an A″ label. Similarly, in the case of $\varepsilon = -1$, the upper Λ-doublet component of the F_1 rotational wave function of Eq. (5.52) carries an A″ label, and the upper Λ-doublet component of the F_2 rotational wavefunction of Eq. (5.52) carries an A′ label. Here, the A′ and A″ labels reflect whether the electronic part of the wave function is respectively symmetric or anti-symmetric with respect to the plane of rotation in the limit of large *j*-quantum numbers,[362–364] as illustrated in Figure 5.10. This high *j*-limit labelling of Λ(A′) and Λ(A″), in which the electronic wave function is respectively symmetric and antisymmetric with respect to reflection of the spatial coordinates, is shown to be responsible for a propensity to favour final NO rotational Λ-doublet states of A″ symmetry upon a collision with a rare gas atom.[362]

The development of accurate potential energy surfaces, along with the appropriate coupling elements between the surfaces, has made Ar–NO one of the most well-characterized scattering systems.[365–367] Accurate quantum mechanical time-independent scattering calculations[368,369] have been performed to predict the results of many of the experimental studies. To go along with these rigorous calculations, classical calculations, as well as QQT and other approximate treatments are being developed, both to explain the experimental results, and to obtain physical insight into the

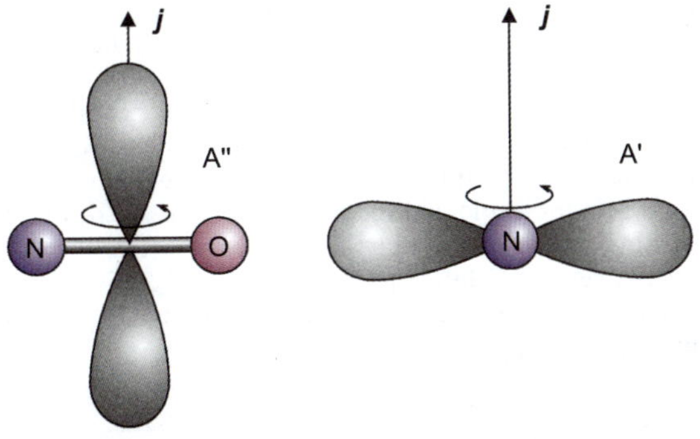

Figure 5.10 Schematic representation of the A″ and A′ symmetry for NO with large *j*.

scattering process. In the remainder of this chapter we highlight several of the most recent experimental studies, and discuss the existing state of the theoretical calculations that have been used to predict the experimental results.

5.3.3 Scalar Measurements

5.3.3.1 Total cross section measurements

One of the first measurements to be made on the NO–Ar system was a determination of the total DCS. Not only was the total DCS measured, but also detected was the difference in the scattering observed when the NO molecule was colliding head on or sideways to the rare gas atom. In addition, glory scattering was also observed (see Study Box 5.1).

The earliest crossed-molecular-beam scattering experiments on the NO–Ar system were performed by Stolte, Reuss and Schwartz in 1973[370] using the apparatus shown in Figure 5.11. A beam of NO molecules was quantum state selected by focusing with an electric hexapole (see Chapter 9 for details), and subsequently aligned along the quantization axis provided by a magnetic field. The molecules could be aligned either parallel or perpendicular to the most probable direction of the relative velocity between the colliding NO molecules and Ar atoms at the scattering centre. The $^2\Pi_{3/2}$, $j=3/2$, $|m_j|=3/2$, $\varepsilon=\varepsilon_u=-1$ state was selected[371,372] in the hexapole. During the adiabatic transition from the hexapole electric field into the uniform magnetic field, this state was observed to transfer into the strongest low magnetic field seeking $j=m_j=\bar{\Omega}=3/2$ state (see Chapter 9).

These experiments measured the relative total collision cross section (not product rotationally-state resolved) for state-selected NO($j=\bar{\Omega}=m_j=3/2$) molecules in two different orientations. The collisions were controlled such that the angular momentum projection along the quantization axis of the magnetic field, $\boldsymbol{B}$, pointed either parallel, $\sigma_{\parallel}$, or perpendicular, $\sigma_{\perp}$, to $\boldsymbol{v}_{\text{rel}}$ (see Figure 5.12). The magnetic field was provided by Helmholtz coils, and the experiments were conducted with a fixed average velocity in the primary beam of 525 ms $^{-1}$.

The expectation value of the actual reactant alignment obtained for the selected $|m_j|=j=\bar{\Omega}=3/2$ quantum state is given by the quantity $\langle P_2(\hat{\boldsymbol{r}}\cdot\hat{\boldsymbol{v}}_{\text{rel}})\rangle=0.2\langle P_2(\hat{\boldsymbol{B}}\cdot\hat{\boldsymbol{v}}_{\text{rel}})\rangle$.* Due to the quality of the

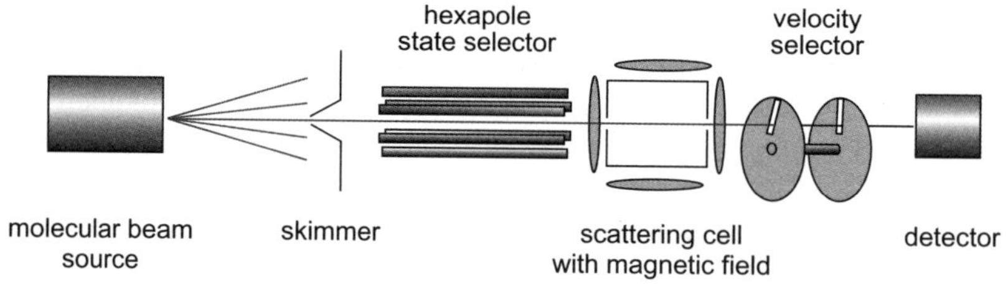

Figure 5.11 Schematic diagram of an apparatus for studying the total DCS as a function of relative velocity, and the dependence of the scattering on reactant orientation. A molecular beam of NO is skimmed and directed into a hexapole state selector. The quantum state selected and spatially oriented NO molecule is directed through an Ar beam, which emerges as an effusive secondary beam from a fused capillary array. Thereafter, the NO beam is passed through a slotted disk velocity selective chopper wheel system (to allow velocity measurements) and upon passing a collimator to give an angular resolution of $\theta\leqslant3.8\times10^{-3}$ rad, the beam is detected by a mass spectrometer detector.

*$\langle P_2(\hat{\boldsymbol{a}}\cdot\hat{\boldsymbol{b}})\rangle$ refers to the expectation value of the second Legendre polynomial in the angle between the unit vectors along $\boldsymbol{a}$ and $\boldsymbol{b}$.

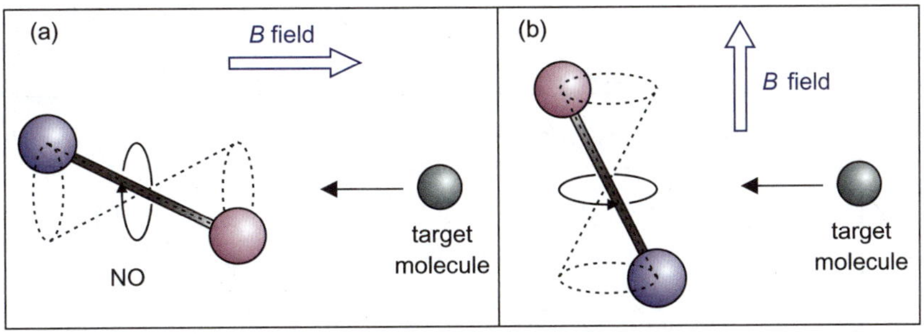

Figure 5.12 Schematic diagram showing the alignment of a hexapole state selected NO molecule in collision with an atom. The nuclear axis can be aligned either (a) parallel or (b) perpendicular to the relative velocity vector of the collision.

data, the authors succeeded in using a more detailed form for the potential to characterize their observations than the usual single parameter dependent anisotropic Lennard-Jones 6–12 model of Eq. (5.1). They added two anisotropy parameters $q_{2,12}$ and $q_{2,6}$, leading to a potential described as

$$V(R) = 4\varepsilon\left[\left(\frac{\sigma}{R}\right)^{12} - \left(\frac{\sigma}{R}\right)^{6}\right] + \varepsilon P_2(\cos\theta)\left[q_{2,12}\left(\frac{\sigma}{R}\right)^{12} - q_{2,6}\left(\frac{\sigma}{R}\right)^{6}\right]. \qquad (5.55)$$

The anisotropy parameters control the magnitude of the angular dependence of the potential through the term in the second Legendre polynomial, $P_2(\cos\theta)$. Here the angle $\theta \equiv \gamma$ is the Jacobi angle between the molecular axis and the vector joining the centres of mass of the colliding particles.

The analysis yielded an anisotropy, A, in the total collision cross section $[A = (\sigma_{\parallel} - \sigma_{\perp})/\bar{\sigma} \backsimeq (\sigma_{\parallel} - \sigma_{\perp})/\sigma_{\parallel} = -(6.5 \pm 1.0) \times 10^{-3}]$, and revealed a larger than expected value of the anisotropy parameter $q_{2,6} = 0.22 \pm 0.02$. This parameter characterizes the long range dispersion contribution to V_{sum}, and may be compared with that expected from the long range dipole-dipole dispersion term, $V_{\text{disp}} = -C_6 R^{-6}[1 + q_{2,6} P_2(\cos\theta)]$ with $q_{2,6} = \Delta\alpha/3\alpha = 0.16$[374] (according to the so called Unsöld approximation), in which $\Delta\alpha = \alpha_{\parallel} - \alpha_{\perp}$ and $\alpha = (\alpha_{\parallel} + 2\alpha_{\perp})/3$ denote the difference of the polarizability along and perpendicular to the molecular symmetry axis and the mean polarizability, respectively. The polarizability values are taken from ref. [374]. Subsequently,[371,375] the measurements of $(\sigma_{\parallel} - \sigma_{\perp})/\bar{\sigma}$, with the orientationally averaged cross section $\bar{\sigma} = (\sigma_{\parallel} + 2\sigma_1)/3$, were extended to investigate the NO–Rg systems, with Rg = Ar, Kr and Xe, over a large range of relative velocities (from 300 m s^{-1} to 2100 m s^{-1}). This enabled the determination of the potential parameters (ε, σ, $q_{2,6}$ and $q_{2,12}$) for the Lennard-Jones 12-6 model, thereby providing very detailed information about the intermolecular forces at play in these systems.

With the apparatus shown in Figure 5.11, several quantities were measured. These included the v_{rel} dependence of the glory oscillations, and of the contributions to the total cross section from large impact parameters, both for aligned and isotropic incoming state selected NO molecules, as illustrated in Figure 5.13. H. Thuis *et al.*[373] succeeded in generating an anisotropic V_{sum} potential capable of reproducing all the observations.

5.3.3.2 Integral cross section measurements

Following these first molecular beam experiments on Ar–NO, which measured the total DCS, and therefore the total amount of scattering averaged over all the possible final rotational states, the group of Andresen[376] carried out a breakthrough experiment in which supersonic molecular beams of Ar and NO were crossed, and laser-induced fluorescence (LIF – see Study Box 5.2)

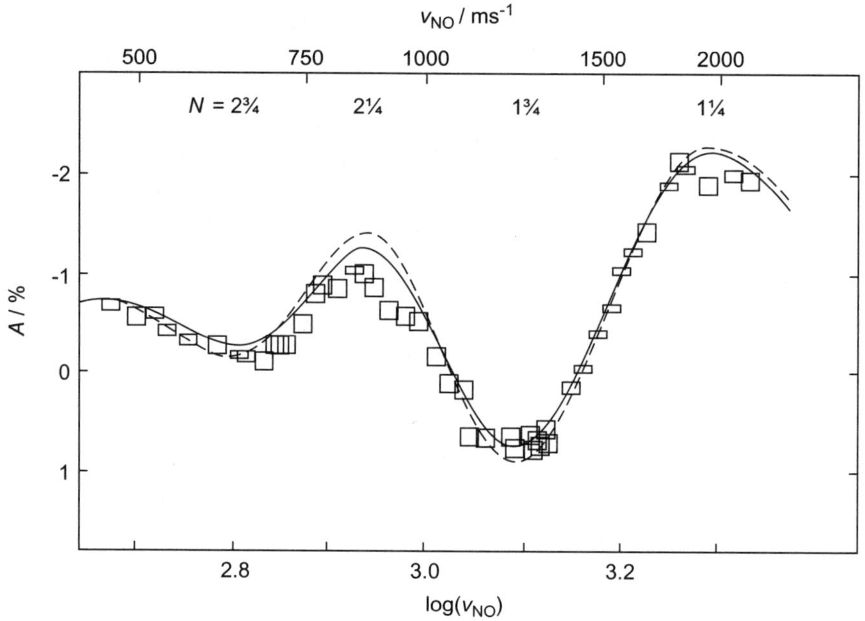

Figure 5.13 The anisotropy in the total collision cross section, $A=(\sigma_{\parallel}-\sigma_{\perp})/\bar{\sigma}$ with $\bar{\sigma}=(\sigma_{\parallel}+2\sigma_{\perp})/3$, *versus* the primary molecular beam velocity, v_{NO}, for the NO-Ar system. The width of each rectangle indicates the uncertainty in the velocity determination (3%); the height displays the statistical error in A. The glory interference phenomena described in Study Box 5.1, that become relevant if $v_{rel}\leq 2^{\frac{1}{6}}\sigma\varepsilon/\hbar$, appear to dominate A. The glory extrema of the total collision cross section of an isotropic potential are indicated with N, as indicated at the top of the figure. Maxima are at $N=1, 2, 3, \ldots$, while minima are at $N=1\frac{1}{2}, 2\frac{1}{2}, 3\frac{1}{2}, \ldots$. The position of the glory extrema follows from $(N-\frac{3}{8}\pi)=a\frac{2\varepsilon\sigma}{\hbar v_{rel}}$, in which $a\simeq\frac{2}{5}2^{\frac{1}{6}}$ depends slightly on the potential model.[327] The extrema of A, which are expected to coincide with the extrema of $\frac{d\sigma_{tot}v_{rel}^{0.4}}{dv_{rel}}$ indeed occur in the figure at $N=1\frac{1}{4}, 1\frac{3}{4}, 2\frac{1}{4}, 2\frac{3}{4}, 3\frac{1}{4}, \ldots$. The solid curves correspond to the best fit obtained with two different potential surfaces (adapted from Thuis *et al.* ref. [373]).

was employed to detect the total amount of product scattered into each final product quantum state. This allowed the integral cross section for each product quantum state to be determined. Figure 5.14 shows the experimental setup used.

These experiments led to the first observation of an "interference" phenomenon in the integral cross section (ICS) (see Figure 5.15). Such an oscillation in the ICS with j' had been predicted theoretically for the rotationally resolved "nearly homonuclear" $^1\Sigma$ molecules, with McCurdy and Miller[377] interpreting the oscillations as arising from interference between the scattering from the two different ends of the NO molecule. This explained the even $\Delta j=j'-j$ propensity observed for the scattering of NO from Ar into $j'-\frac{1}{2}=N'\leqslant 6$: odd rotational states showed less population than their neighbouring even numbered states. The oscillations were generally well reproduced by the exact "close-coupled" (or "coupled channel" (CC)) quantum mechanical scattering calculations* of Alexander, performed using his *ab initio* Ar–NO CEPA(1) and CCSDT PESs.[366,378] These calculations yielded Λ-doublet state-averaged ICSs that confirmed the even N' propensity observed for $N\leqslant 6$.

*See Sections 3.3.4.2 and 12.3.3 for brief introductions to the quantum mechanical scattering methods employed in such calculations.

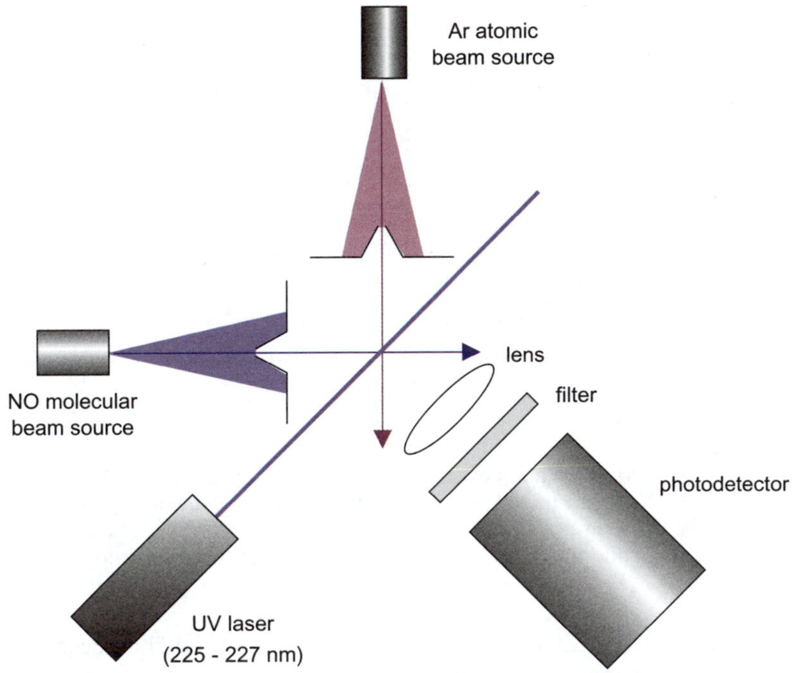

Figure 5.14 Schematic apparatus for measurement of the integral cross section of NO(j = 0.5) molecules colliding with Ar atoms and resulting in population of higher rotational states. The UV laser produces tunable light in the range of 226 nm, which can be tuned to be resonant with a line within the $^2\Sigma^+ \leftarrow {}^2\Pi$ transition. Fluorescence from the upper state is then collected, filtered and recorded.

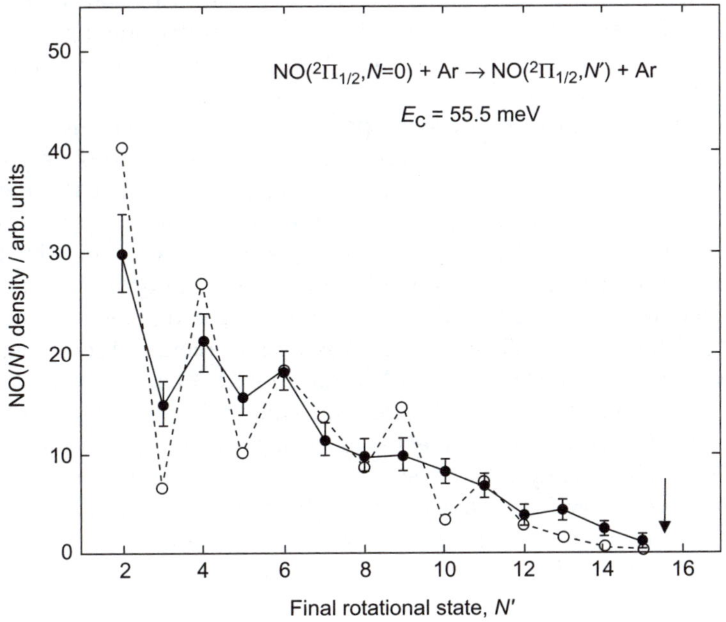

Figure 5.15 NO(N') density for rotational NO($^2\Pi_{1/2}$, N = 0) in collision with Ar. Solid line is the experiment and the dashed line is the theory. Error bars represent the standard deviations from several measurements. The theoretical distribution is normalized to the experimental one at N' = 6. The arrow indicates the energy limit (adapted from Andresen *et al.*[376]).

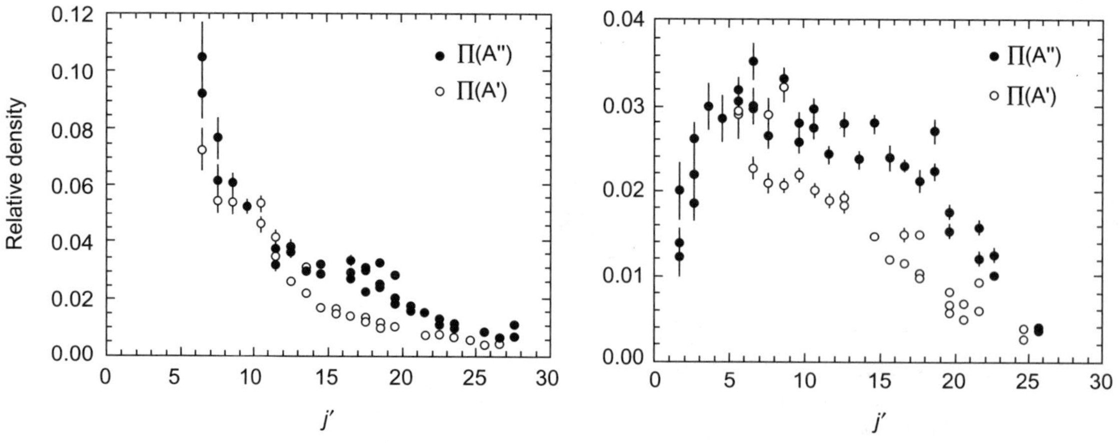

Figure 5.16 Experiments by McBane and coworkers showing observed relative densities for the F_1-F_1 (left panel) and F_1-F_2 (right panel) transitions. Filled circles indicate the $\Pi(A'')$ λ-doublet component and the open circles show $\Pi(A')$. The error bars represent statistical error in individual measurements, while the scatter of points indicates the reproducibility of the results (adapted from Lin *et al.*[379]).

Another scalar quantity of great interest in NO scattering studies has been the propensity to scatter into a particular component of the Λ-doublet of the outgoing rotational states. Using an apparatus very similar to the one shown schematically in Figure 5.14, Lin *et al.*[379] measured the propensity for NO to scatter into the different rotational states as a function of both the final Λ-doublet state and the final fine structure state. At a fixed collision energy of 1775 cm^{-1}, Lin *et al.* obtained the results shown in Figure 5.16. For fine structure conserving collisions (*i.e.* collisions remaining in the $^2\Pi_{1/2}$ state), a small preference for population of states with A″ symmetry is observed for rotational levels above $j = 12$, while for fine structure changing collisions not only are A″ symmetry states more populated in the rotational energy transfer but also an oscillation is seen in the population of these states.

Some levels of theory are able to reproduce this propensity and the oscillation, while others cannot. The theoretically predicted and the observed propensities were found to be in good agreement for scattering into the F_1 spin-orbit manifold (left panel of Figure 5.16). While the approximate CS and the exact CC QM scattering calculations predicted similar ICS's for the spin-orbit state conserving transitions, the CS calculations failed to give accurate ICSs for the spin-orbit changing $F_1 \rightarrow F_2$ transitions (see Chapter 3 for a brief discussion of these approximations). All experimental ICSs for $F_1 \rightarrow F_2$ transitions had to be reduced by a factor of 0.80 to agree overall with theoretical predictions. Even then, for $j' < 10.5$ the theoretically predicted cross sections remained somewhat smaller than the experimentally measured ones for $\bar{\Omega} = 1/2 \rightarrow F_2(A'')$ transitions.[378] This shows that, when observed in detail, rotational energy transfer is complicated and cannot be interpreted quantitatively using simplified or approximate treatments, such as the hard ellipse like model discussed in Section 5.2.4.

In Figure 5.17 the predicted Λ-doublet resolved rotational energy transfer cross sections obtained also using a high level of theory[380] are shown for the He–NO system. In this case, theory predicts a slight propensity to scatter preferentially into A″ symmetry states at $E_c = 1186\ cm^{-1}$. However, the propensity is modest, and remains to be observed experimentally.

5.3.4 Vector Measurements of Rotational Energy Transfer

A vector measurement is one that involves probing the correlation between two or more vectors associated with the scattering process. The most obvious vectors are the relative velocity

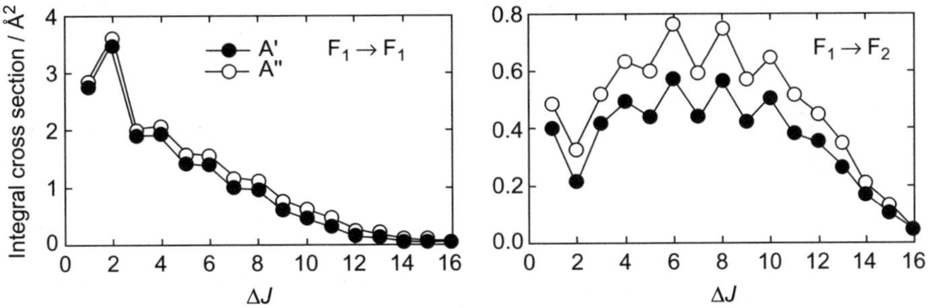

Figure 5.17 Calculations of coupled states integral cross sections for the scattering of NO out of the $j = 0.5$ level of the lower spin-orbit manifold (F_1) as a function of the change in rotational quantum number (Δj) for transitions into levels of the F_1 (left panel) and F_2 (right panel) spin-orbit manifolds at an initial collision energy of $1186\,\text{cm}^{-1}$. The solid and open circles designate, respectively, the sum of the cross sections for the two transitions which populate rotational levels of symmetric (A') reflection symmetry (F_1e or F_2f) and for the two transitions which populate levels of anti-symmetric (A'') reflection symmetry (F_1f or F_2e) (adapted from Yang and Alexander[380]).

vectors of the collision partners. The correlation between the outgoing velocity vector, $\boldsymbol{v}'_{\text{rel}}$, and the incoming, $\boldsymbol{v}_{\text{rel}}$, is quantified by the DCS, proportional to the angular distribution of the scattered products relative to the CM-frame. Other vector quantities are the angular momentum vectors of the molecule before and after collision, $\boldsymbol{j}$ and $\boldsymbol{j}'$. Measurement of the direction of the latter vector relative to the incoming or outgoing relative velocity vectors ($\boldsymbol{v}_{\text{rel}}$ and $\boldsymbol{v}'_{\text{rel}}$, respectively) is described quantum mechanically by the alignment and orientation moments of $\boldsymbol{j}'$ (see Chapter 9 for more details about vector correlations). Alignment may be thought of as a measure of the extent to which the molecule rotates like a frisbee, a wheel, or a propeller as it flies out from the collision. Orientation adds information about whether the molecule is rotating clockwise or counterclockwise. Such measurements often give detailed insight into the forces acting during a collision. Yet another vector is the direction of the inter-nuclear axis of a diatomic molecule, $\boldsymbol{r}$, and how it is oriented relative to the incoming velocity vector. Correlations involving this vector reflect the propensity for collisions from one end or the molecule or the other to yield a specific outcome (*e.g.*, a particularly energy transfer process). Such correlations are usually quantified by a "steric asymmetry parameter", which is very sensitive to the details of the PES.

5.3.4.1 Determination of the steric asymmetry parameter

The value of the steric asymmetry parameter, SA, reflects the dependence of the rotational energy transfer cross section upon which end of a molecule, in this case NO, the collision partner (Ar) strikes. The steric asymmetry parameter is defined by $\text{SA} = (\sigma^{\text{NO}} - \sigma^{\text{ON}})/(\sigma^{\text{NO}} + \sigma^{\text{ON}})$, where σ^{NO} and σ^{ON} denote the cross section for collision with Ar impinging at the N and O ends of NO, respectively. SA takes the limiting values of $\text{SA} = 1$ when $\sigma^{\text{ON}} = 0$, and $\text{SA} = -1$ when $\sigma^{\text{NO}} = 0$, and is zero when $\sigma^{\text{ON}} = \sigma^{\text{NO}}$.

Although this might seem an almost impossible quantity to determine, it is not. As we have seen, by utilizing a hexapole state selector in front of a supersonic molecular beam source of NO one can selectively transmit the low-electric-field-seeking states within the molecular beam (see also Section 5.3.3.1 and Chapter 9). Molecules that reside in a state that increases its Stark energy upon the application of an electric field tend to feel a force towards the region of the lowest potential, which is at the centre of the hexapole field. They can therefore be focused by the field and transported

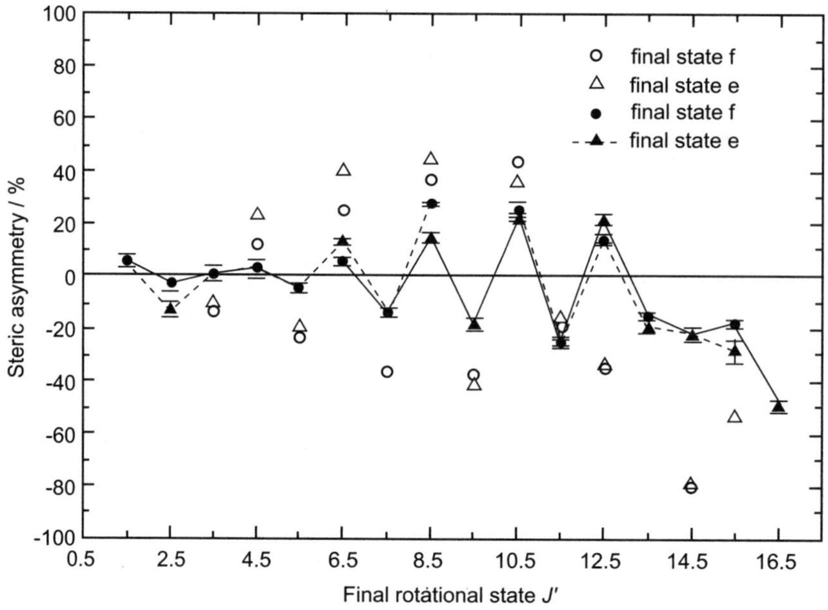

Figure 5.18 Experimental results (filled circles and triangle connected by solid and dashed lines respectively) and theoretical (open circles and triangles). Values of the steric asymmetry of spin-orbit state conserving collisions of Ar and NO are displayed. Note that the steric asymmetry ratio appears to depend strongly on the j' value and only weakly on the parity of the final state. The state with $j' = 15.5$ is the highest energetically allowed (adapted from de Lange *et al.*[382]).

along the axis of the hexapole. In the particular case of NO, it is the molecule that resides in the upper $\varepsilon = -1$ (also called f-) component of the Λ-doublet of the $^2\Pi_{1/2}\, j = 1/2$ rotational ground state that will be focused. In this manner a single quantum state of NO can be transmitted through a hexapole field.[381] This quantum state can then be oriented with a static electric field (the so called orientation field) and in this way the orientation of the molecule can be fixed relative to the velocity vector of the collision with an Ar atom.[382,383] When the electric field axis of the orientation field is pointed such that it coincides with the direction of $\boldsymbol{v}_{\mathrm{rel}}$, then the polarity ($\pm$) of the DC potential generating the orientation field will select the preferred orientation of the NO, *i.e.* the preference for the Ar atom to collide with the N-end (σ^{NO}) or with the O-end (σ^{ON}).

In Figure 5.18 we show data for the steric asymmetry for the NO(X)–Ar system that reveals that rotational levels with even values of $j' - \frac{1}{2}$ are preferred when one side of the molecule is struck, and odd rotational levels are preferred when the other end of the molecule is struck. The remarkable alternation in the sign of the steric asymmetry has been explained as being due to quantum interference arising from scattering off the repulsive parts of the PES from the two halves of the NO molecule.[308]

5.3.4.2 *Differential cross section measurements*

Generalizing the treatment of an elastic hard sphere collision which led to Eqs. (5.16)–(5.21), the DCS for collisions that excite a molecule from initial state i to final state f is the measure of the flux of scattering for that particular transition into a specific solid angle, and is written

$$\frac{\mathrm{d}\sigma_{i\to f}}{\mathrm{d}\omega}, \qquad (5.56)$$

where $d\omega$ represents the solid angle volume element $\sin\theta \, d\theta d\phi$. The units of the DCS are therefore area per solid angle. The integral of the DCS over scattering angle for a particular initial state i and final state f yields the state-resolved ICS discussed in Section 5.3.3.1:

$$\sigma_{i \to f} = \int_0^{2\pi} \int_0^{\pi} \frac{d\sigma_{i \to f}}{d\omega} \sin\theta \, d\theta \, d\phi. \tag{5.57}$$

The angular distribution of the products is simply the DCS normalized to have unit area,

$$P_{i \to f}(\theta, \phi) = \frac{1}{\sigma_{i \to f}} \frac{d\sigma_{i \to f}}{d\omega}. \tag{5.58}$$

Here, $P_{i \to f}(\theta, \phi)$ is the probability of finding the product scattered from state i to state f into a solid angle between ω and $\omega + d\omega$. The DCS is not a function of the azimuthal angle ϕ if the reactants are not oriented in any way, *i.e.* in the CM reference frame, the molecules scatter with cylindrical symmetry about the relative velocity vector of the collision. For a given scattering angle θ as many scatter 'up' as scatter 'down' in the ϕ angle, as was illustrated in Figure 5.3.

The DCS is most commonly measured using the crossed molecular beam technique with a rotatable detector, usually a TOF mass spectrometer (see Chapter 6 for a more detailed account of the crossed molecular beam technique). The apparatus is shown schematically in Figure 5.19. The DCS is obtained by placing the detector at different angles relative to the scattering volume, with the flux of products into the detector determined one angle at a time. This technique has the advantage that detection is "universal" in the sense that it can detect any molecule or atom. It has the disadvantage that it can only determine the flux at one angle at a time, and is not selective to the internal state of the molecule that enters the detector.

Recently, several new laser spectroscopy based methods of determining the DCS of individual quantum states of the products have been developed. One is entirely laser based and relies upon photodissociation of a reactant molecule to define the initial velocity vector. In order to determine the final velocity vector either sub-Doppler[384,385] or time-of-flight spectroscopy[386,387] is performed,

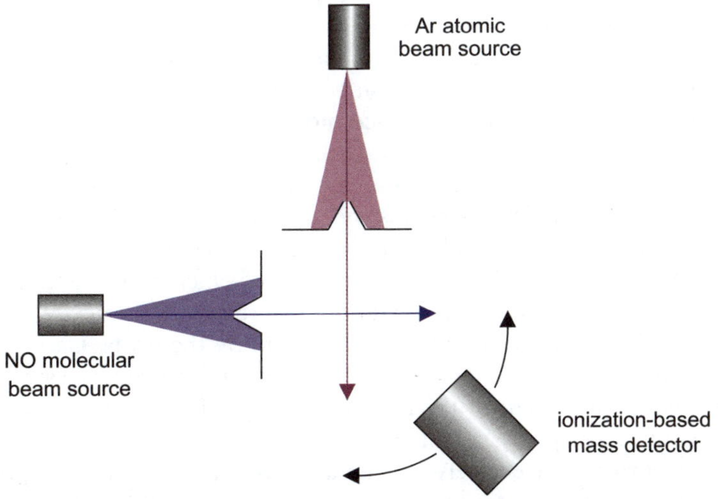

Figure 5.19 Schematic representation of the crossed molecular beam scattering apparatus with mass spectrometer product detection.

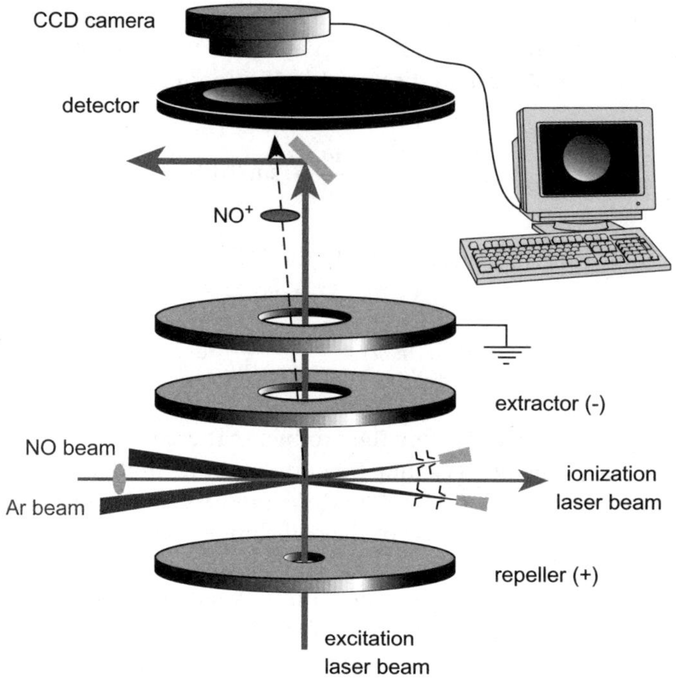

Figure 5.20 Schematic apparatus for the laser-based detection of scattering products from crossed molecular beam scattering of NO from Ar, utilizing REMPI detection of a single rotational state of the NO molecule. See also Study Box 5.3.

or ion imaging[357,388] (see Study Boxes 5.2 and 5.3). This technique is termed PHOTOLOC,[389] and is described briefly in Study Box 6.3.

Another important technique in reaction dynamics, already introduced in Section 5.3.1, is that known as ion imaging,[357,390] and has been used in conjunction with crossed molecular beams to define the initial velocity vector, as shown in Figure 5.20 (see also Study Box 5.3). The imaging technique uses REMPI to selectively ionize a single quantum state of the product molecules within the scattering volume. The more recent variant of ion imaging, velocity-mapped ion imaging, employs the innovation of using lensing ion optics instead of flat field ion extractors.[390] This allows one to record an image that is relatively insensitive to the spatial origin of the ions, and is only dependent upon the velocity of the ion. In crossed molecular beam scattering, where the molecular beams interact at the crossing point for tens of microseconds before product detection is performed by laser ionization, there can be a large spatial spread in the origin of the products. Molecules formed in the early parts of the molecular beams will move distances large in comparison to the laser beam geometry before the probe laser pulse intersects the molecular beam scattering volume. This movement of the scattering product has two effects. Some velocities are detected with greater efficiency than others (those moving slowly in the laboratory reference frame are most easily detected, since they are more likely to remain in volume intersected by the probe laser) and the images are blurred due to the spatial extent of the ions. Velocity mapping eliminates the second of these effects, the blurring due to the spatial extent of the ions, but does nothing to correct the distortion of the images associated with velocity dependent detection efficiency. Due to the velocity dependent detection efficiency, in order to accurately extract a DCS from a velocity-mapped ion image significant modelling of the experimental apparatus is required.[353,391–394] This modelling takes into account what is usually referred to as the 'density-to-flux' transformation. This transformation is discussed further in Study Box 6.5 (see also Study Box 5.4 on Monte Carlo simulation).

STUDY BOX 5.3: VELOCITY-MAPPED ION IMAGING

Principal Features

Velocity-map imaging is an experimental technique that has truly captured the imagination of the reaction dynamics community. The technique allows the spatial distribution of a product species to be visualized directly, yielding the entire three-dimensional angular and velocity distribution (or, strictly speaking, a two-dimensional projection which from which the 3D distribution may be reconstructed) of a photolysis or reaction product in a single digital 'snapshot'. Velocity-map imaging was first developed for the study of photodissociation,[357,390] but is fast becoming the detection method of choice for a wide range of chemical dynamics experiments.

A schematic of a standard velocity-mapped imaging apparatus of the type widely used to study photodissociation processes is shown in Figure 5.21. The parent molecule of interest is expanded in a molecular beam, which is intercepted by a pair of laser beams. The first laser pulse dissociates the parent molecule, and the photofragments fly out from the reaction centre with a velocity distribution determined by the chemical process under study and the polarization of the photolyzing laser pulse. After a short delay of generally a few nanoseconds, the second laser pulse ionizes a chosen quantum state of one of the photofragments, usually via a (1+1) or (2+1) REMPI process (see Study Box 5.2). The ions are then accelerated along a flight tube towards a position sensitive detector by an electric field maintained between a set of three or more ion lenses. Ideally, the molecular beam and photolysis laser beam would cross at a single point in space, so that the spatial distribution of the photofragments at a known point in time after reaction is initiated would yield the velocity distribution directly. However, in practice the two beams cross in a finite interaction volume, which we may expect to cause considerable blurring of the measured distribution. This potential problem is overcome by tuning the electric field such that all ions formed with the same initial velocity strike the same point on the detector, regardless of the position at which they were formed; using these 'velocity mapping' optics, the system behaves as if the ideal point source had been achieved. The field is simultaneously tuned to compress the ion cloud along the axis such that the 3D distribution is 'pancaked' onto the detector, giving the highest possible time resolution and separation of ions with different masses.

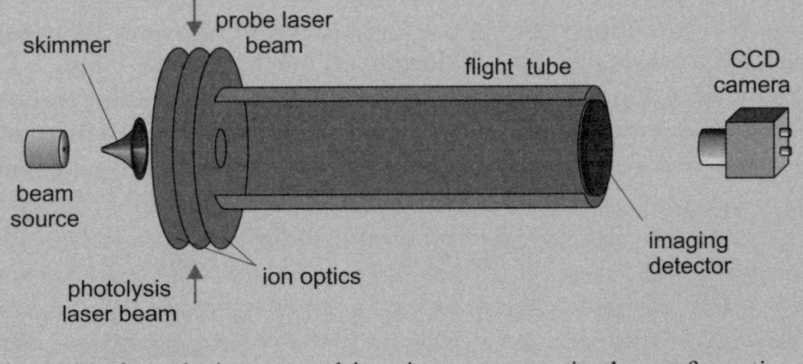

Figure 5.21 Schematic of a velocity-mapped imaging apparatus in the configuration typically used to study molecular photodissociation.

The position sensitive detector consists of a pair of microchannel plates (MCPs) coupled to a phosphor screen. Microchannel plates are very thin resistively coated glass plates arrayed with micron-sized holes or 'channels'. An ion striking the front face of the plates elicits emission of electrons at the entrance to one of the channels. A potential difference maintained between the front and back face of the MCPs accelerates the electrons along a channel, where they are amplified on each collision with the surface, until eventually a burst of around 10^6 electrons is emitted from the back face of the MCPs and accelerated towards the phosphor screen, which emits a flash of light on impact. In this way, the pattern of ions striking the front face of the MCPs is transformed into an optical image at the phosphor screen, which is captured by a CCD camera. Usually we are only interested in imaging one mass at a time, in which case either the MCPs or the CCD camera are gated such that a signal is only recorded at the flight time of the mass of interest. The image captured by the CCD camera is transferred to a PC on each laser pump-probe cycle for processing and data storage.

Each cycle usually generates a few tens to a few hundreds of ions. In order to obtain images with high signal to noise and to gain a clear picture of the scattering distribution, a large number of images are summed to give the final image to be analyzed.

Various real-time image processing algorithms may be used to improve the image resolution. If the incoming images are merely summed as they arrive at the PC, resolution is limited by the fact that each ion strike often ends up illuminating several pixels on the camera, blurring its recorded position. For this reason, each incoming image is often 'event counted'[395] before being included in the sum. For each ion strike, the most intense pixel is located, and a new image is generated in which each of these most intense pixels takes the value 1 and all other pixel values are set to 0. The new images are accumulated over the desired number of pump-probe cycles, with the result that in the final summed image, the pixel values translate directly into the total number of ions detected at that pixel position. In addition to improving image resolution, this procedure also corrects for spatial variations in detector sensitivity, since it is individual events that are summed rather than intensities. Since the pixel resolution of the camera is much lower than the intrinsic resolution of the MCPs (determined by the channel spacing), even higher resolution may be obtained by implementing 'megapixel imaging' algorithms,[396] in which a more sophisticated centroiding algorithm is used for the ion strikes, allowing sub-pixel resolution of the ion position.

Image Analysis

The problem to be addressed in the analysis of velocity-mapped images is how to recover the full 3D scattering distribution from the 2D projection measured in the experiments. For 3D distributions with an axis of cylindrical symmetry, a direct mathematical transform known as the Abel inversion[397] may be used to recover a central 'slice' through the distribution, which contains all the required information (see Figure 5.22). For linearly polarized photolysis the velocity distribution does have an axis of cylindrical symmetry. However, if polarized light is used to probe the photofragments and product angular momentum polarization is significant, the distribution of *detected* fragments will no longer have such an axis, and alternative image analysis schemes must be invoked. These approaches are generally based on forward simulation or basis set fitting.[398,399]

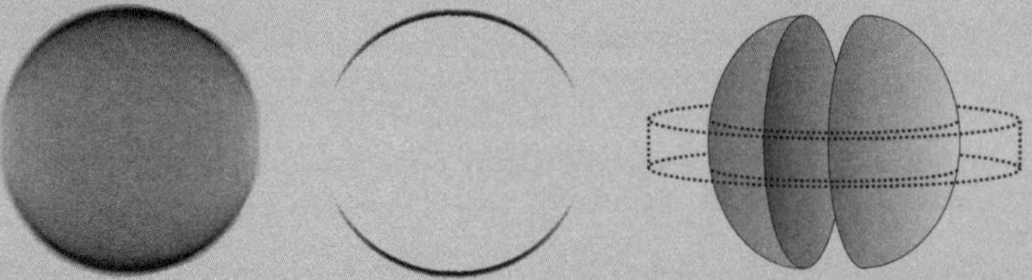

Figure 5.22 The image on the left is a typical 2D 'crushed' image. The Abel inversion 'reinflates' the image and extracts the central slice (centre image). Rotating the slice about the axis reconstructs the 3D distribution. In slice imaging, the detector is time-gated, allowing the central slice to be measured directly, as shown on the right.

Slice Imaging

A variation on the standard velocity-mapped imaging technique allows the 3D scattering distribution to be measured 'slice by slice' over a series of images, rather than in a single measurement of a 2D projection. There are a number of implementations of slice imaging,[400–403] but the basic idea is that the velocity-mapping ion optics are designed to stretch the velocity distribution along the time-of-flight axis, rather than compressing it. The ions then arrive at the detector with a broad flight time distribution, and slices of the distribution may be selected by applying a narrow time gate to the detector to capture a subset of ions in each time-of-flight peak.

Claire Vallance

5.3.4.3 *Quantum state selective DCSs: Velocity map ion imaging*

The group of Houston and coworkers were the first to use the technique of ion imaging[357] to measure the quantum-state-resolved DCS utilizing crossed molecular beam scattering of NO with Ar.[329,330] A molecular beam of NO was scattered from an atomic beam of Ar, and a laser was used to ionize quantum state selectively the NO molecules scattered into high rotational states by collision with Ar atoms. The NO was ionized by using a $(1+1')$ REMPI scheme *via* the A state at 226 nm. Rainbow scattering was observed, and the results could be compared with a two-dimensional hard ellipse model for the scattering. DCSs for both spin-orbit conserving and spin-orbit changing collisions were obtained, and were found to be very similar. Figure 5.23 shows some of the early representative ion images of specific rotational states of the product NO. Earlier, the group of Gentry and Giese[405] had measured the DCSs using the alternative technique of LIF (see Study Box 5.2), and their results generally agreed with the those obtained by the Houston group.

In the $j' = 18.5$ images obtained in the state selected DCS measurements of Houston and coworkers,[330] and shown in Figure 5.23, two rainbows (peaks in the scattering intensity) are observed. From these data the authors were able to use the hard ellipse equations, Eqs. (5.37)–(5.39) to determine both the difference between the A and B axis of the elliptical interaction potential that governs the scattering of Ar from NO, and the off-centre displacement of the NO CM, δ (see Eq. (5.39)). They obtained a value for $A-B$ of 0.32 Å, which compared well with the theoretical value of 0.37 Å.

The work of the Houston and Gentry groups inspired other measurements of the quantum-state-resolved DCS for the NO(X)–Ar system.[406] Many new experiments were stimulated by improvements in ion imaging and other experimental techniques, as well as by the availability of better PESs and advances in the scattering theory used to help interpret the experimental results. As already noted, the improvement to the ion imaging technique that made the largest impact on molecular beam scattering experiments was the introduction of "velocity mapping" of the product ions onto the imaging detector[390] (see above, and Study Box 5.3). A velocity-mapped ion image of NO products in $j' = 7.5$ produced in collisions of NO with Ar is shown in Figure 5.24, and can be compared with the corresponding image (orientation reversed) without velocity mapping, shown in Figure 5.23.

These types of scattering experiments rely upon a supersonic molecular beam to create a cold rotational temperature (on the order of a few degrees Kelvin), and a well defined velocity distribution for the reactants. However, even at these cold temperatures multiple quantum states are often populated in the molecular beam, and this results in the measured DCS being an average over several reactant states. Even if the beam was sufficiently cold that only one rotational state was populated, it would have two almost degenerate Λ-doublet states. Each of these states is predicted to have significantly different scattering characteristics. Recently, the group of Stolte in collaboration with the Chandler group has combined hexapole state selection with velocity-mapped ion

STUDY BOX 5.4: MONTE CARLO SIMULATION TECHNIQUES

The term *simulation* can be defined as a 'technique of performing sampling experiments on a model of a system under study'.[404] Because the sampling process involves the use of random numbers, the method is often referred to as Monte Carlo simulation (this term was first introduced during World War II as a code word for a project involving Monte Carlo simulations on problems related to the atomic bomb).

Computer simulation was once regarded as almost a final resort in the field of data analysis, to be tried only when all else had failed. However, advances in software and processing speed have made simulations an increasingly valuable tool, which is widely used in a large number of disciplines. For cases in which analytic solutions to the equations modelling a system cannot be found, Monte Carlo simulation is often the only option, and consequently it is now one of the most powerful and commonly used techniques for analyzing complex systems. In the context of reaction dynamics, Monte Carlo techniques are widely used to evaluate multi-dimensional integrals over parameters such as the various velocity distributions, laser and particle beam profiles involved in a range of molecular dynamics experiments and theoretical models, in particular, the quasi-classical trajectory method described in detail in Section 9.3.1.1.

In order to carry out a Monte Carlo simulation of a system, a well-defined model is required, parameterized in terms of a finite set of variables. Initialization consists of selection of a large number of parameter sets using a random number generator. 'Trajectories' are then run through the model system for each of the parameter sets, the required data is extracted from each trajectory, and the data from the full set of trajectories is averaged to give the final result. For a simulation containing N sample points, the basic theorem of Monte Carlo integration estimates that the integral of a function f over a multidimensional volume V is given by:[47]

$$\int f\,\mathrm{d}V \approx V\langle f\rangle \pm \sqrt{\frac{\langle f^2\rangle - \langle f\rangle^2}{N}} \tag{B5.4.1}$$

where the terms in angled brackets indicate taking the mean over the N sample points.

$$\langle f\rangle \equiv \frac{1}{N}\sum_{i=1}^{N} f(x_i)$$

$$\langle f^2\rangle \equiv \frac{1}{N}\sum_{i=1}^{N} f^2(x_i)$$

In equation (B5.4.1), the first term, $V\langle f\rangle$, gives the approximate value of the integral, and the second is a one standard deviation error estimate, which gives a rough indication of the probable error in the integral. The key point to take from the expression is that the error in a Monte Carlo integral decreases as $N^{-1/2}$. Checks should always be made that enough sample points have been included in the simulation to ensure that any integrals have converged to their true values (if a simulation is not sufficiently converged then rerunning the simulation with a higher value of N will change the result obtained). A second consideration is the choice of random number generator employed to generate parameter values during the initialization of each trajectory. Random numbers generated by a computer program are usually generated from a sequence, in which the next number is defined precisely by the preceding value through the computer code, and are therefore not really random at all. For this reason they are often called 'pseudo-random' numbers. There are several statistical tests for randomness that a random number generator should pass before being used in a Monte Carlo application, ensuring that the results of the simulation are not dependent on the choice of random number generator.

Claire Vallance

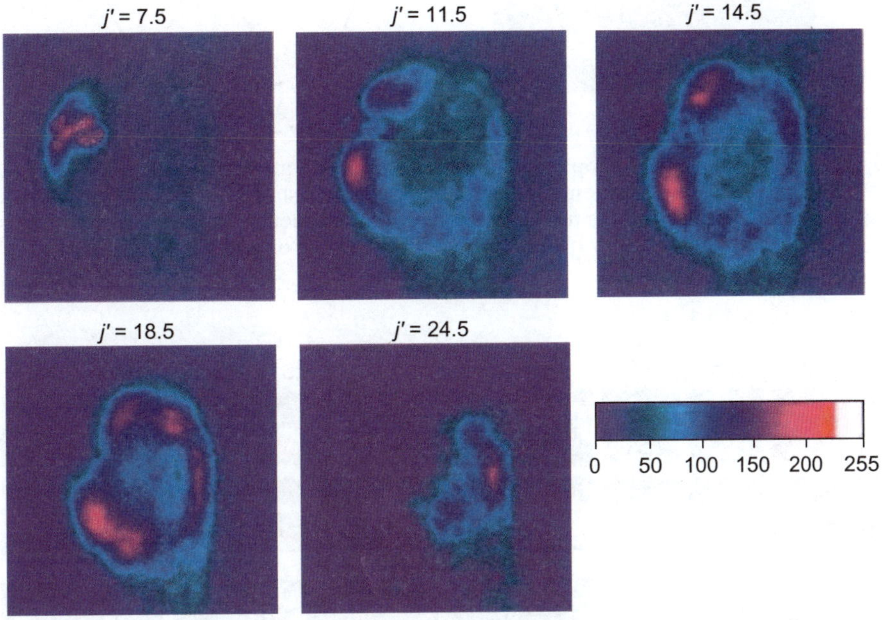

Figure 5.23 Experimental results from the first (spatial) ion imaging measurements of state resolved DCSs for Ar–NO rotationally-inelastic scattering obtained under crossed molecular beam conditions. The data, recorded at a collision energy of $E_c = 1451\ \text{cm}^{-1}$, are shown as false-colour intensity plots. Images of scattered NO are shown for the following state-to-state resolved spin-orbit conserving transitions; $j = 1/2 \rightarrow j' = 7.5$, 11.5, 14.5, 18.5 and 24.5. In these "raw images" one notices the typical shift with j' from favouring forward ($\Theta = 0°$) to backward ($\Theta = 180°$) scattering (adapted from Bontuyan *et al.*[330]).

imaging to make the first measurements of quantum state selective DCSs for scattering from a single quantum state in the NO($^2\Pi_{1/2}$, $j = 0.5$, $\varepsilon = -1$) + He collision system.[407] Some of the results of the study are shown in Figure 5.25. A dramatic change in the DCSs was observed for scattering into upper and lower components of the Λ-doublet states, which could not be explained by the very small energy splitting. An interesting pairing of the shape of the DCSs was observed, depending on the parity of the scattered NO,[408] which cannot be explained by simple classical models such as the hard ellipse model. The scattering at this level of measurement has a very quantum mechanical nature, and the wave-like behaviour of the particles must be taken into account in order to understand the observed dynamics. One theory that has been able to explain this pairing of the DCSs is the Quasi Quantum Treatment (QQT).[308,309] These calculations reveal that the pairing arises from the symmetry properties of the initial and final NO rotational states, and of the potential matrix elements that couple them.

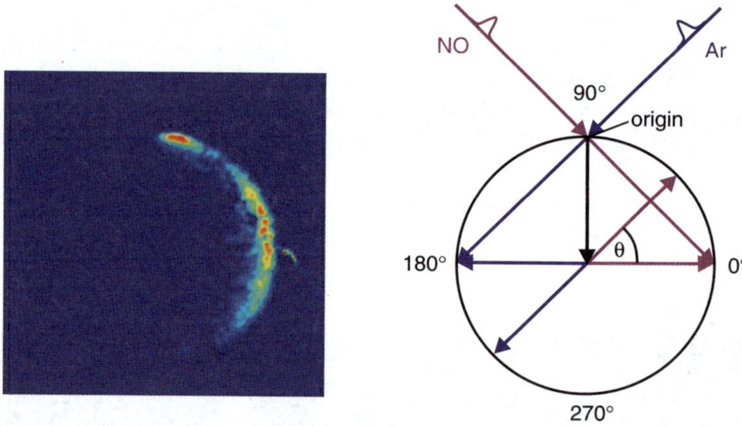

Figure 5.24 Left panel: a velocity-mapped ion image of NO $j'=7.5$ after scattering of NO ($j=0.5$) from Ar. Right panel: the corresponding vector diagram that shows the initial velocity vectors of the molecular beams, and the recoil velocities of the products in both the centre-of-mass (centre of the circle) and laboratory frames (centered around the origin). Red areas are of high scattering intensity and blue low.

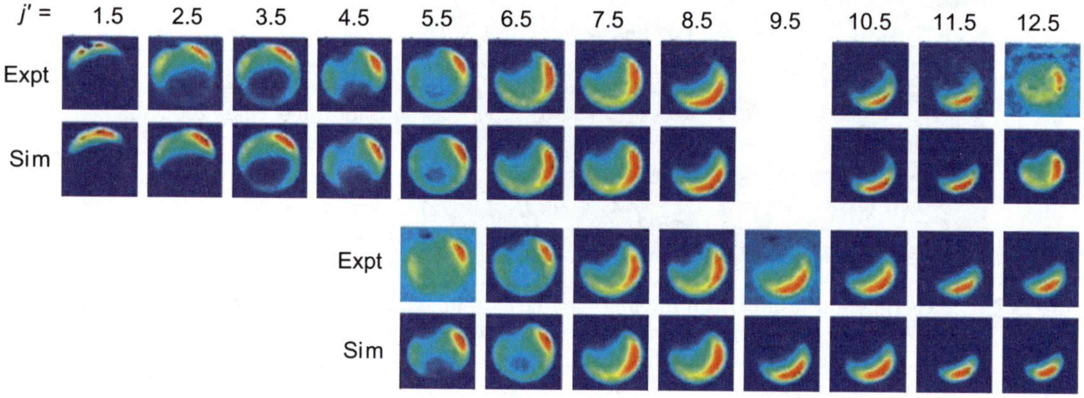

Figure 5.25 False colour intensity plots of experimental and simulated ion images obtained for spin-orbit state conserving rotational transitions resulting from the scattering of a fully state-selected NO($j=1/2$, $\bar{\Omega}=1/2$, $\varepsilon=-1$) beam from a supersonic crossed beam of He atoms at an approximate collision energy of 510 cm^{-1}. The images in the top rows are collected using the $R_{11}+Q_{21}$ branch ($\bar{\Omega}'=1/2$, $\varepsilon'=-1$). The bottom rows show images collected using the $Q_{11}+P_{21}$ branch ($\bar{\Omega}'=1/2$, $\varepsilon'=-1$). The experimental images are raw data, the simulated images are adapted using the optimized extracted DCS and incorporate blurring, velocity correction $T(x,y)$, and an alignment correction. Colour coded as in Figure 5.24 (adapted from Gijsbertsen *et al.*[407]).

5.3.4.4 *Alignment of the rotational angular momentum vector*

As noted at the beginning of Section 5.3.4, in addition to the measurement of the angular distribution of the scattering products as a function of the angular momentum transfer, one can also measure how the product angular momentum vector, $\boldsymbol{j}'$, is directed relative to the scattering angle of the products. This is referred to as measuring the angular momentum polarization of the molecule, specifically its orientation and alignment. One needs a way to describe and quantify the distribution of the angular momentum vector. A common approach is to use "alignment moments". These are moments of a spherical tensor expansion, and are described below.*

We begin by defining the absorption cross section, $I(j', \theta, \phi, \chi_p)$, of the linearly polarized probe light in the laser beam probe frame (PRB). Following Fano and Macek, and R.N. Zare,[239,409–411] $I(j^*, j', \theta, \phi, \chi_p)$ is given by:

$$I(j^*, j', \theta, \phi, \chi_p) = C_{det}\left\{1 - \tfrac{1}{2}h^{(2)}(j^*, j')\left[A_0^{(2)PRB}(j', \theta, \phi, \chi_p) - 3A_{2+}^{(2)PRB}(j', \theta, \phi, \chi_p)\right]\right\}. \quad (5.59)$$

In Eq. (5.59), j' is the rotational quantum number of the scattered NO prior to laser excitation, and j^* denotes that of the resonantly excited rotational level in the $NO(A^2\Sigma^+)$ electronic state. C_{det} is an apparatus sensitivity constant (containing factors for the detector sensitivity and the overlap of the probe laser with the sample, *etc.*) and $h^{(2)}(j^*, j')$ is a linestrength factor that depends on the branch of the probe transition. For pure *R*-branch transitions $h^{(2)}(j'+1, j') = -j'/(2j'+3)$, for pure *P*-branch transitions $h^{(2)}(j'-1, j') = -(j'+1)/(2j'-1)$, and for pure *Q*-branch transitions $h^{(2)}(j', j') = 1$. The quantities $A_0^{(2)PRB}(j', \theta, \phi, \chi_p)$ and $A_{2+}^{(2)PRB}(j', \theta, \phi, \chi_p)$ are the two leading alignment moments of a spherical tensor expansion of the angular distribution of $\mathbf{j}'$, defined in the probe (PRB) laser reference frame.[338] These moments depend on four variables, j', θ, ϕ, and χ_p. As already noted, j' is the final rotational angular momentum quantum number, θ is the deflection or recoil angle of the NO in the scattering plane, and ϕ is the azimuthal deflection angle for the NO. χ_P is the probe polarization angle with respect to the image plane, and is 0° for horizontal polarization and 90° for vertical polarization. The probe laser reference frame is not the frame in which most physical insight can be gained, and therefore most alignment parameters are reported in the "collision frame".† The alignment moments can be rotated from the probe frame to the collision frame using the rotation matrices given by Hertel and Stoll.[412] Below we will show how the alignment moments may be determined by analyzing experimental data and compared to the predictions of theory (see, for example, refs. [413] and [414]).

In Eq. (5.59), the absorption cross section of the linearly polarized probe light depends only on $A_{2+}^{(2)PRB}(j', \theta, \phi, \chi_p)$ and $A_{2+}^{(2)PRB}(j', \theta, \phi, \chi_p)$ and not on any $A_q^{(k)PRB}(j', \theta, \phi, \chi_p)$ moments with odd values of k. The reason for this is that one can only measure alignment with linearly polarized light. Orientation, clockwise *versus* counterclockwise spinning, requires k-odd moments of the distribution to be determined, and as we will see this requires circularly polarized light for detection.

The alignment moments of interest, and their physically realizable limits, are defined as follows in the collision reference frame:[409§]

$$A_0^{(2)col} = \frac{\langle 3\hat{j}_Z^2 - \hat{j}^2\rangle}{j(j+1)} \quad \text{where} \quad -1 \le A_0^{(2)col} \le 2, \quad (5.60)$$

$$A_{1+}^{(2)col} = \frac{\langle \hat{j}_X\hat{j}_Z + \hat{j}_Z\hat{j}_X\rangle}{j(j+1)} \quad \text{where} \quad -1 \le A_{1+}^{(2)col} \le 1, \quad (5.61)$$

---

*The reader might find it helpful to come back to this material after reading Chapter 9. Study Box 9.1 is particularly relevant in providing an introduction to the detection of angular momentum polarization.

†The collision frame is defined such that the Z axis is parallel to $\mathbf{v}_{rel}$, and the ZX-plane coincides with the plane defined by the crossed molecular beams. These lie parallel to the plane of the MCP detector. The polarity of the X-axis is chosen such that the Y-axis points from the centre-of-mass origin directly to the plane of the MCP detector (*i.e.* along the TOF axis). In the probe or detector frame, the Z-axis lies along the direction of propagation of the probe laser radiation, and (for linearly polarized light) the X-axis is defined as that containing the electric vector of the radiation.

§The notation used here is a little different to that found in Chapter 9, but the parameters $A_{q\pm}^{(k)}$col are proportional to the moments $a_{q\pm}^{(k)}$ introduced in that chapter. See also Study Box 8.4.

and

$$A_{2+}^{(2)\text{col}} = \frac{\langle \hat{j}_X^2 - \hat{j}_Y^2 \rangle}{j(j+1)} \quad \text{where} \quad -1 \le A_{2+}^{(2)\text{col}} \le 1. \tag{5.62}$$

When all three alignment moments equal zero, $\boldsymbol{j}'$ has no preferred direction, so its distribution is isotropic. The value of $A_0^{(2)\text{col}}$ relates to the propensity for alignment of $\boldsymbol{j}'$ relative to $\boldsymbol{v}_{\text{NO}}$. When $A_0^{(2)\text{col}}=2$, $\boldsymbol{j}'$ is parallel (↑ or ↓) to $\boldsymbol{v}_{\text{NO}}$. That type of motion may be visualized as the NO spinning like a propeller blade. When $A_0^{(2)\text{col}}=-1$, $\boldsymbol{j}'$ is perpendicular to $\boldsymbol{v}_{\text{NO}}$, *i.e.* NO is spinning either like a frisbee or a wheel. The value of $A_{2+}^{(2)\text{col}}$ reflects the propensity of $\boldsymbol{j}'$ to lie parallel to the X-axis, or to the Y-axis. If $A_{2+}^{(2)\text{col}}=+1$, then $\boldsymbol{j}'$ lies perfectly parallel to the X-axis, while if $A_{2+}^{(2)\text{col}}=-1$, then $\boldsymbol{j}'$ lies exactly parallel to the Y-axis. This moment tells one to what extent the NO molecule prefers spinning like a wheel or a frisbee. Moreover, the value of $A_{1+}^{(2)\text{col}}$ informs about the propensity of $\boldsymbol{j}'$ to be tilted in the XZ plane, with $A_{1+}^{(2)\text{col}}=\pm 1$ indicating a perfect tilt of the $\boldsymbol{j}'$-vector in the direction of the vector along $\boldsymbol{Z}\pm\boldsymbol{X}$. Now that we have considered the interpretation of the alignment moments, we move to their experimental measurement.

5.3.4.5 *Experimental measurement of alignment and orientation*

Alignment of the rotational angular momentum in the scattered NO has been observed and quantified in measurements of the DCS of the NO(X)–Ar system. The experiments utilized polarized lasers to ionize the scattering products. It was observed that the intensity pattern of the ion-images, like those in Figures 5.23, 5.24, and 5.25, were dependent on the polarization of the probe laser beam. As noted above, if $\boldsymbol{j}'$ of the NO molecule is randomly oriented in space, the alignment moments are all zero, and the polarization of the REMPI laser does not affect the measurement.

The groups of Chandler and Cline[415,416] used velocity-mapped ion imaging to measure the scattering angle dependent alignment of the $\boldsymbol{j}'$ vector of NO in the recoil frame of the molecule after its collision with an Ar atom. In these studies, a molecular beam of NO is crossed at 90° with a beam of Ar atoms (see Figure 5.20). In the $(1+1')$ REMPI ionization detection of NO *via* the A state, it was found that the ionization probability of NO (A) upon the absorption of the second laser photon was not dependent on the polarization of this second photon. Therefore, in terms of its sensitivity to alignment and orientation of $\boldsymbol{j}'$ in NO(X), this REMPI scheme is found to behave like a regular one-photon transition.[417] This justifies using the intensity expression in Eq. (5.59), which neglects absorption of the second photon. The azimuthal information intrinsic to imaging detection allowed for the determination of higher-order alignment moments and a more thorough description of the angular momentum alignment.

Figure 5.26 shows the polarization dependent data that were obtained by collecting images with different probe laser polarizations. The second column of the figure shows difference images obtained for the $j'=4.5$, 8.5, 11.5, 12.5 and 15.5 rotational states, obtained by subtracting ion images taken with vertically (V) and horizontally (H) polarized probe laser light (labelled with respect to the image plane). The images have been normalized by the sum of those images (V + H). These difference images, normalized in this manner, do not contain information about the DCS, only the rotational alignment for each scattering angle. This is why the images in the second column of Figure 5.26 are symmetric.

The detected image signal with a vertically polarized laser beam, V (first column of Figure 5.26), or horizontally polarized laser beam, H, $D(j^*, j', \theta, \phi, \chi)$, is given by

$$D(j^*,j',\theta,\phi,\chi) = \int_{v_{\text{NO}}}\int_{v'_{\text{NO}}} A(j^*,j',\theta,\phi,\chi)\frac{\text{d}\sigma_{j'}(\theta)}{\text{d}\omega} f(\theta,\chi,v_{\text{NO}},v'_{\text{NO}})\,\text{d}v_{\text{NO}}\text{d}v'_{\text{NO}}. \tag{5.63}$$

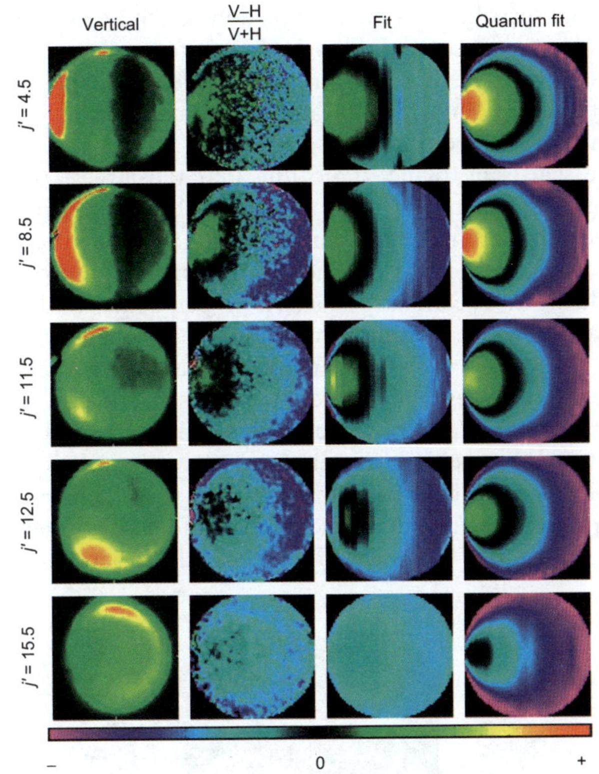

Figure 5.26 NO($\times^2\Pi_{1/2}$, $v' = 0$) ion images following collision with Ar. The images in the first column are taken with vertical (V) laser polarization. The images in the second column are produced by determining (V−H)/(V+H) for each rotational state. The images in the third column are the least-squares fits to the images in the second column from which the values for the alignment moments have been extracted. The images in the first column have the same colour scheme as Figure 5.24. The fourth column is the calculated images taking alignment moments from quantum scattering theory. The colour coding is chosen such that intense red denotes very positive, blue-green corresponds with zero and purple means very negative. Adapted from ref. [416].

Here, $f(\theta, \phi, \boldsymbol{v}_{NO}, \boldsymbol{v}'_{NO})$ is an apparatus function. Integration over a range of initial and final velocities of NO is used to account for the velocity spread in both molecular beams. Since $A(j^*, j', \theta, \phi, \chi_P)$ is the only factor that will vary with laser polarization, to first-order, the (V–H)/(V+H) images separate the effect of angular momentum alignment from the DCS and detection sensitivity in the images. The alignment term, $A(j^*, j', \theta, \phi, \chi_P)$, in Eq. (5.63), is a sum of two terms, since for each pixel, the two azimuthal coordinates above and below the interaction region will project onto the same position of the detector,

$$A(j^*,j',\theta,\phi,\chi) = \frac{I(j^*,j',\theta,\phi,\chi) + I(j^*,j',\theta,-\phi,\chi)}{\sin\theta\ \sin\phi}, \qquad (5.64)$$

in which $I(j^*, j', \theta, \phi, \chi)$ is the absorption cross section of the linearly polarized probe light as defined by Eq. (5.59) in the previous section. From these data, both $A_0^{(2)\text{col}}$ and $A_{2+}^{(2)\text{col}}$ may be determined for each final j' quantum state, and for each scattering angle θ, corresponding to a vertical stripe in the third column of Figure 5.26. The alignment moments may be used to construct three-dimensional distributions of the direction of the angular momentum vector, $\boldsymbol{j}'$, of the scattered NO molecule at each scattering angle, θ, and for each rotational state detected.

Analysis of the images leads to the distributions shown with the images in Figure 5.27. The figure shows the distribution of directions in which the angular momentum vector, $\boldsymbol{j}'$, points relative to the recoil velocity over a range of selected scattering angles. The angular momentum vector is found to point mainly out of the scattering plane, indicating that the largest propensity is for the NO molecules to be rotating in the plane of the scattering in the manner of a spinning frisbee. For forward scattered products the second largest axis is the one pointing along the recoil vector of the NO ($j' = 12.5$) product, indicating some additional propensity for propeller like rotation. For backward scattered products, the second largest axis is perpendicular to the relative velocity vector, but in the plane of the scattering, indicating some additional propensity for wheel like rotation.

Careful inspection of the images shown in Figure 5.27 (and similar images for other product quantum states) also reveals that each of the distributions has a dimple in the side of it. The position of this indentation in the $\boldsymbol{j}'$ distribution corresponds to the apse angle. From our earlier discussion, we know that for a hard ellipse model of the system $\boldsymbol{j}'$ cannot point along the kinematic apse, see Eq. (5.40). This is seen to be approximately the case in the experiments.

In addition to measuring the alignment of the angular momentum vector of the NO molecule after collision with Ar, it is also possible to measure the orientation of the NO molecule

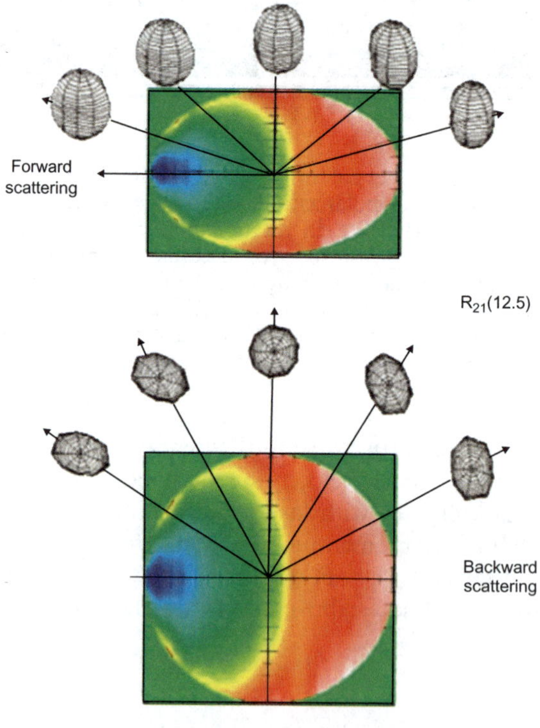

Figure 5.27 Difference image obtained by subtracting two ion images of the $j' = 12.5$ quantum state. One image is taken with polarized light oriented vertically in the laboratory frame and resonant with the NO(A-X) transition near 226 nm. This image is then subtracted from an image taken with the laser polarization rotated 90° to be horizontal in the laboratory frame to obtain the false colored image shown. The $\boldsymbol{j}'$ vector distributions are shown as the line drawings for several scattering angles. All have their largest probability perpendicular to the scattering plane (the NO molecule is rotating in the plane of the scattering, like a frisbee, with highest probability). In these images blue represents high intensity and red low (adapted from ref. [415]).

using right (R) and left (L) handed circularly polarized laser light to ionize the scattered NO molecules.[418] The difference in the images, once again normalized to the sum of the right and left handed images, gives information about the direction of rotation of the molecule. The propagation direction of the laser beam must be such that the right or left-handedness of the laser light coincides with the rotation of the molecule. Depending upon the molecular transition being used in the resonant ionization step, an enhancement will be observed for ionization with one polarization over the other. The amount of enhancement is proportional to the amount of orientation produced in the scattering.

Figure 5.28 shows a set of un-normalized (R-L) difference images for the rotational states, $j' = 6.5$, 8.5, 10.5 and 15.5 formed in the collision. From inspection of the images it is easy to see that the pattern of orientation is different for the different rotational states. The highest rotational state shown in Figure 5.28 is $j' = 15.5$. This quantum state exhibits mainly sideways and backward scattering (see Figure 5.27), and the orientation is very strong. The top of the difference image is entirely of one orientation, corresponding to counter-clockwise rotation of the NO molecule, while the lower half of the image corresponds to clockwise rotation. Inspection of the $j' = 6.5$ difference image shows that the orientation oscillates as a function of scattering angle. This oscillation is quantitatively predicted by time-independent quantum mechanical scattering calculations.[418] The frequency of the oscillations increases with decreasing rotational state, while the intensity of the oscillations decreases with decreasing rotational state. No simple classical model has been agreed upon to account for these orientation-dependent quantum-state-selective rotational energy transfer measurements.[413,414] It is clear that if we are to understand inelastic scattering on the single molecular collision scale we need to understand quantum mechanics as well as classical mechanics.

5.4 OUTLOOK

Collisional energy transfer is an important and ubiquitous process that keeps the world at thermal equilibrium. Depending upon the details of the collisional energy transfer process in question, the rate can vary by many orders of magnitude. Collisional experiments, particularly those using crossed molecular beams, are making it possible to observe directly the detailed outcome of energy transfer processes under well specified conditions. Insight into the factors that control the rates of these events, and the probabilities of collisional kinetic, rotational and vibrational energy transfer, is being provided both from 'simple' models of energy transfer and by essentially exact quantum mechanical scattering calculations on accurate *ab initio* potential energy surfaces.

For reasons of simplicity, most of the molecular beam energy transfer experiments performed to date have explored the rotationally inelastic scattering between an atom and a diatomic molecule at thermal collision energies. Some of the first pioneering experimental and theoretical work in this field was carried out by Toennies and coworkers in 1965. Their work involved the experimental measurement and quantum-mechanical analysis of state-to-state ($|j,m\rangle \rightarrow |j',m'\rangle$) integral rotationally-inelastic cross sections for collisions of TlF with rare gas atoms and other gas molecules.[419,420] In 1979, another benchmark experiment, which inspired numerous theoretical and experimental developments, was the discovery by Beck and coworkers[317,318] of the rotational rainbow in the recoil velocity distributions of K scattered from N_2 and CO, and its dependence upon the CM scattering angle. As shown in this chapter, the IOS approximation solution of this scattering problem, as well as the hard ellipsoid model, helped in the analysis and interpretation of the observed phenomena, and was later extended to non-alkali atom containing scattering systems. In the mid-eighties, it was felt by a number of researchers that rotationally inelastic scattering was essentially a mature field. However, a new ground-breaking direction to this research area was opened up with the theoretical explorations of the collisional energy transfer of open shell (NO)

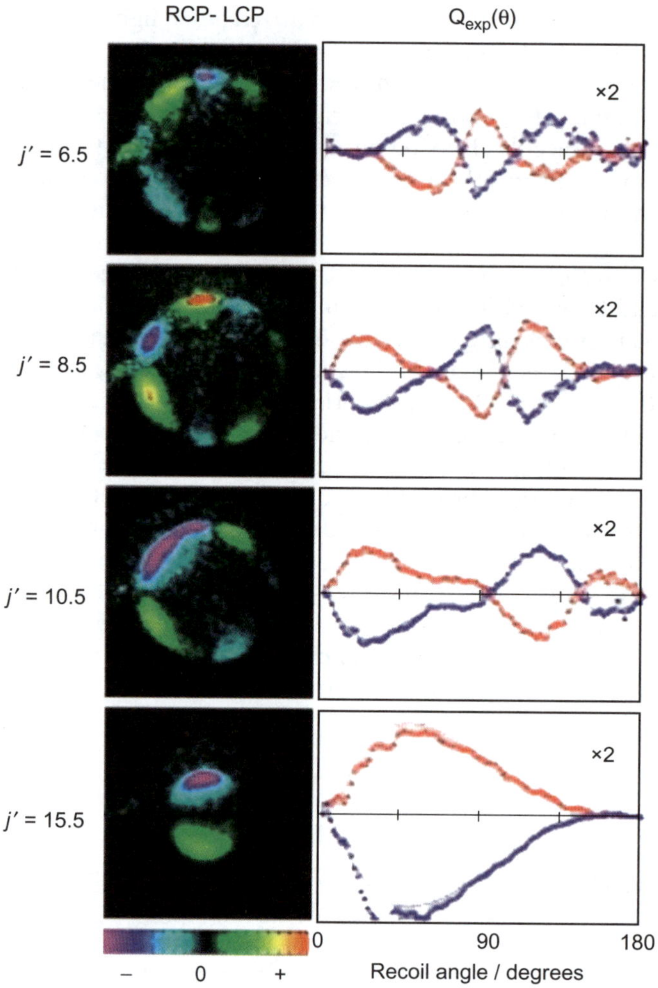

Figure 5.28 Difference image obtained by subtracting two ion images for several final product quantum states. One image is taken with right circularly polarized light and resonant with the NO(A-X) transition near 226 nm. This image is then subtracted from an image taken with the left handed circularly polarized laser light to obtain the false coloured images shown. They have oscillating patterns as a function of scattering angle showing a preference for the molecule to be rotating either clockwise (yellow and red) or counterclockwise (blue and purple) for particular scattering angles. This is strongly dependent upon the quantum state interrogated. The plots in the right panel show the intensity trace as a function of angle for the top and bottom of the images. The colour coding is as in Figure 5.26 (adapted from Lorenz *et al.*[418]).

molecules by Alexander *et al.*[358,361,362] An essential element which helped with the blossoming of the field experimentally was the invention of the ion imaging technique by Chandler and Houston,[357] and the subsequent development of velocity-mapped ion imaging by Eppink and Parker.[390] These techniques made it possible to obtain novel information about vectorial aspects of the scattering process (see Section 5.3.4). At present, such detailed studies on state-to-state collisional energy transfer also include work in which state selected polyatomic molecules are scattered from rare gas atoms and from diatomic hydride molecules. Increasing computational power now makes it possible to calculate accurate *ab initio* PES's for a much larger class of molecule–molecule systems than those mentioned above. However, at thermal collision energies, the number of coupled differential equations that need to be solved to obtain the exact (coupled channel) solution of the rotational inelastic scattering problem for this more extended class of systems easily exceeds a few thousand, and exact dynamical calculations for such scattering problems remains unfeasible at the present time.

State-to-state scattering experiments on rotational energy transfer at collision energies much lower than those sampled under 300 K thermal conditions are now feasible and are receiving increasing attention. Under these conditions, the dynamics tends to be dominated by the anisotropic well instead of the anisotropic hard shell that dominates at higher energies. The study of collisions between molecules at low energies plays a central role in the emerging new research field of cold and ultracold molecules, discussed in detail in Chapter 12. One of the problems for which a practical solution had to be found was the production of (ultra) cold molecules at sufficient densities to be able to study their interactions (see Study Box 12.1). As we will see in Chapter 12, in 1999 Meijer and coworkers[421] invented the Stark deceleration technique that has been shown to yield unprecedented control over both the internal and external degrees of freedom of polar molecules. Stark-decelerated molecular beams allow detailed molecular scattering studies to be undertaken over a very wide range of collision energies.[422]

Alternative methods to yield low collision energies have been developed as well (see Study Box 12.1). As illustrated in Figure 5.24, Elioff *et al.*[423–425] have shown that molecules with sub-Kelvin translational energy can be produced from a single collision between two molecules of comparable mass if they collide with each other from perpendicularly crossed supersonic beams. This method, called kinematic cooling, was applied to the NO-Ar system, and has recently been extended to the 'equal mass' ND_3-Ne system, for which a number low rotational quantum states of ND_3 were brought nearly to rest.[426] Of these the $(J, K) = (2, 2)$ state was theoretically predicted to be the coldest, with $\Delta E_c/k_B = 80\,\mu K$, *i.e.* to be cooled down into the ultra cold (<1 mK) regime.

As, explained in Section 5.2, molecules that are initially made sub-thermal by Stark deceleration or by kinematic cooling, and that reside within a magneto-optical trap filled with ultracold atoms of similar mass, will also result in a further cooling of their translational energy. Such cooled molecules offer an interesting target for the study of energy-transferring collisions in the micro-Kelvin regime, and experiments of this kind are already being pursued.[427] In summary, the field of collisional energy transfer is expected to be a very active area of research for the present and for the coming future, with many new and exciting phenomena to be discovered.

5.5 PROBLEMS

1. Compare the collision systems $H_2(v=1)+H_2(v=0)$ and $D_2(v=1)+D_2(v=0)$ in the case of a pure gas at identical thermal and density conditions.
   (a) Compare the coupling, as predicted by Eq. (5.2), between the vibrational and translational motions of both systems. Use Ehrenfest's adiabatic principle to predict the ratio of the relaxation rates for $v=1$ vibrational-to-translational energy transfer for a pure gas of H_2 molecules and that in a pure gas of D_2 molecules.
   (b) Employ the Landau-Teller result of Eq. (5.4) to calculate this ratio directly.
   (c) Calculate this ratio using Eq. (5.4) in the case that the collider gases, $H_2(v=0)$ and $D_2(v=0)$, are replaced by Ar.
2. (a) Derive Eq. (5.14).
   (b) Derive Eq. (5.16).
   (c) Consider a collision between an incoming hard-sphere-like molecule with an initially stationary atom. Show that to first order the amount of translational energy ΔE_t that is transferred upon collision is at least $\Delta E_t = \frac{1}{2} m_1 v_1^2 (1-F)(1-D^2/4)$ (see Eq. (1.24) from

ref. [427]). Here m_1 and m_2 denote, respectively, the mass of the molecule and the initially stationary atom, and v_1 is the laboratory velocity of the incoming molecule. D corresponds to the mass defect, defined as $(m_1 - m_2)/m_1$, and F is the fraction of all the hard sphere collisions with total collision cross section πR_S^2 that contribute to ΔE_t (see Eq. (5.23)). (Note that this simple equation implies that if one has $D=0$ and $b=0$ all of the initial kinetic energy, $\frac{1}{2}m_1 v_1^2$, is transferred to the atom, and, as follows also directly from Eq. (5.16), this leads to $F=0$.)

(d) In the case of a mass mismatch, $D=10\%$, calculate the maximal fraction of the incoming kinetic energy (*i.e.* at $F=0$ which corresponds to $b=0$ collision) that can be transferred to the atom. Next show that the fraction of collisions $1-F$ that transfer more than 99% of the incoming translational energy of the molecule with $D=10\%$ is about 1%. (Note that if the incoming kinetic energy of the molecules is just low enough that these 1% of molecules become trapped, the other 99% will be ejected from the trap. A value of 1% for the fraction of the incoming molecules to remain in the trap strongly suggests that such trapping experiments are feasible for kinematically slowed HBr that is subsequently trapped in a Rb magneto-optical trap[425] (see Study Box 12.1).)

3. Show that Eqs. (5.26) and (5.27) imply that $\boldsymbol{j} \cdot \hat{\boldsymbol{a}} = \boldsymbol{j}' \cdot \hat{\boldsymbol{a}}$ or $(\boldsymbol{j}' - \boldsymbol{j}) \cdot \hat{\boldsymbol{a}} = \Delta \boldsymbol{j} \cdot \hat{\boldsymbol{a}} = 0$, *i.e.* the projection of the rotational angular momentum onto kinematic apse remains conserved for a hard shell type of collision. (Note that rotationally inelastic state-to-state differential scattering experiments probing the alignment vector, $\boldsymbol{j}'$, of the outgoing molecule for the closed shell Ne-Na_2 system,[344] and for the open shell rare gas-NO system, have been shown to conform to this propensity rule in the quantum mechanical limit.[428–430])
4. Derive Eqs. (5.28) to (5.35), starting from the conservation of energy and angular momentum, Eqs. (5.26) and (5.27).
5. Explain why there is a summation over two terms in Eq. (5.64).

CHAPTER 6

Reactive Scattering: Reactions in Three Dimensions

PIERGIORGIO CASAVECCHIA,[a] KOPIN LIU[b] AND XUEMING YANG[c]

[a] Dipartimento di Chimica, Università degli Studi di Perugia, Via Elce di Sotto 8, 06123-Perugia, Italy; [b] Institute of Atomic and Molecular Sciences, Academia Sinica, No.1, Roosevelt Road, Sec.4, Taipei, 10617, Taiwan; [c] Dalian Institute of Chemical Physics, Chinese Academy of Sciences, 457 Zhongshan Road, Dalian, 116023, P.R. China

6.1 INTRODUCTION

6.1.1 Crossed Molecular Beams

The study of bimolecular reactions has always been at the heart of efforts to understand reaction dynamics, and these processes can be investigated highly effectively under single-collision conditions using crossed molecular beam (CMB) techniques.* The ideal experiment on a bimolecular reaction aims to determine the state-to-state differential cross section (DCS) of the reaction products at well defined relative collision energies, E_c.[9,11] As we have seen in the previous chapter on inelastic scattering, state-to-state integral and differential cross sections provide a very sensitive probe of the underlying forces that drive chemical reactivity, and of the potential energy surface (PES) for reaction. The link between experimental DCSs (and integral cross sections (ICSs)) and the PES is provided by scattering calculations, which may utilize both quantum-mechanical (QM) and quasi-classical-trajectory (QCT) methods. Over the past ten to fifteen years, measurements of state-to-state cross sections, calculations of scattering quantities by QM and QCT methods, and *ab initio* electronic structure computations of the PES have progressed to the level where very accurate comparison can be made between experimental and theoretical results, at least for simple systems involving three atoms. This has deepened significantly our understanding of the dynamics of elementary chemical reactions.

To measure state-to-state DCSs one needs to start from reactants with well defined velocities, $\boldsymbol{v}$ (including approach angle), and internal quantum state populations (including electronic ε, vibrational

*At this point, readers might find it helpful to remind themselves of the material covered in Section 1.3.2.

Tutorials in Molecular Reaction Dynamics
Edited by Mark Brouard and Claire Vallance

Published by the Royal Society of Chemistry, www.rsc.org

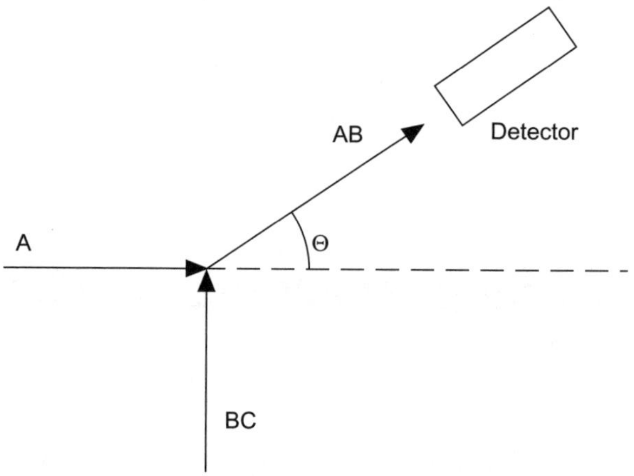

Figure 6.1 Schematic diagram of a crossed-beam experiment on the A + BC → AB + C reaction. Beams of the A and BC reactants cross at a given angle (typically 90°). AB (or C) products are detected in the laboratory frame by a detector that may be rotated in the plane of, and with respect to, the reactant beams.

v, and rotational j states), and to measure the speed and angular distributions of the products:[167]

$$A(\boldsymbol{v}_A, \varepsilon_A) + BC(v_{BC}, \varepsilon_{BC}, v, j) \rightarrow AB(\boldsymbol{v}_{AB}, \varepsilon_{AB}, v', j') + C(\boldsymbol{v}_C, \varepsilon_C). \tag{6.1}$$

(Here we do not consider angular momentum orientation or alignment effects, which are discussed in Chapter 9.) This type of measurement represents a challenging task, not so much for the state selection of the reactants as for the state-resolved detection of the products. In general, DCSs are most directly measured in a crossed-beam configuration, which allows for the best specification of initial conditions. A direct way to measure the DCS is by using a rotatable detector able to probe the velocity of products by time-of-flight (TOF) measurements as a function of the scattering angle, as shown in Figure 6.1. The determination of the kinetic energy distribution (by TOF spectroscopy) of one of the products of a 3-atom bimolecular reaction, such as reaction (6.1), provides a great deal of information on the dynamics, because it also provides information about the undetected product through conservation of energy and linear momentum.[431]

6.1.2 Determining Cross Sections from CMB Experiments

In a crossed molecular beam experiment with a rotatable detector the most common detection scheme employed is electron-impact ionization mass spectroscopy. Using the CMB technique in its 'classic' version (see Section 6.2), product angular and velocity distributions (sometimes referred to as the 'double differential cross section') can be measured at well defined collision energies, E_c. Usually, total DCSs, $\frac{d\sigma}{d\cos\theta}$, summed over all product internal quantum states are determined, though in some kinematically favourable cases vibrationally-resolved DCSs, $\frac{d\sigma(v)}{d\cos\theta}$, can also be obtained if the velocity resolution of the apparatus is sufficiently high. The CMB technique has been central to the investigation of the dynamics of bimolecular reactions over the past 40 years,[432] and has permitted the determination of state-averaged DCSs for an extensive list of reactions. This methodology followed that prescribed in the 1960s during the so called 'alkali era' when, after the first pioneering CMB experiment of Datz and Taylor on the K + HBr reaction, Herschbach and others started a systematic investigation of alkali atom reactions in crossed beams (eventually also using TOF analysis), exploiting the ease of detection of alkali atoms and alkali halide molecules by surface ionization (see, ref. [432] and references therein). These early experiments allowed the main paradigms of reaction dynamics to be established, most notably by Herschbach.[433,434] He investigated a large variety of alkali atom reactions and identified the main reaction mechanisms, such as the 'stripping' and 'rebound' mechanisms, and those involving long-lived complexes (see Section 6.2.1). These paradigms were further rationalized and correlated to features of the underlying reaction PES by the systematic work of Polanyi and coworkers on A + BC type reactions, leading to Polanyi's rules,[27–29] already introduced in Chapter 2 (see also Chapter 7 and Study Box 7.1). Building on this early work, more recent progress in the CMB technique with mass-spectrometric detection, as discussed in Section 6.2, is now permitting investigations of the dynamics of polyatomic reactions exhibiting competing product channels.

When the reaction product is a hydrogen or deuterium atom, the H-Rydberg-atom TOF method may be used for detection.[435–437] This technique was developed during the 1990s, and allows product detection as a function of scattering angle by exciting the H atom product to a high-lying Rydberg state shortly after the reactive collision has taken place, usually within around 50 ns. This 'tagged' atom then undergoes field ionization at the detector, and is detected as an ion. More details about this technique can be found in Section 6.3. Rydberg-tagging has permitted some of the most detailed and accurate measurements of state-resolved and also state-to-state DCSs for a few benchmark three-atom reactions ($H+H_2$, $F+H_2$, $O(^1D)+H_2$, $OH+H_2$), and has recently been extended to O-atom detection.[438,439]

Product detection as a function of scattering angle by laser spectroscopy could in principle be accomplished by means of the laser-induced-fluorescence (LIF) or resonance-enhanced-multi-photon-ionization (REMPI) techniques, described in Study Box 5.2. This approach would allow the measurement of state-specific angular distributions with very high energy and angular resolution. To date, however, despite great advances in LIF, REMPI, and also direct IR absorption techniques, these approaches have been limited by their detection sensitivity. Nevertheless, while the single quantum state product densities resulting from a bimolecular reaction under crossed-beam conditions are generally too low for the available spectroscopic methods to be useful, there are a number of notable exceptions. In selected cases, product detection has been achieved at the collision region using REMPI in combination with either a Doppler-selected ion TOF technique or time-sliced ion velocity-mapped imaging technique. These experiments are discussed in detail in Section 6.4, and a general introduction to velocity map imaging can be found in Study Box 5.3. When possible, this approach has the advantage that it provides the full three-dimensional (3D) state-resolved DCS directly in the CM-frame. Laser-based detection methods can provide a powerful alternative (or complement) to traditional mass-spectroscopy detection methods. However, unlike electron impact ionization based detection, it is not 'universal': the reaction products need to be known, which is often not the case for polyatomic reactions, and their spectroscopy needs to be well characterized.

In addition to vibrationally and also rotationally resolved DCSs, $\frac{d\sigma(v)}{d\cos\theta}$ and $\frac{d\sigma(v,j)}{d\cos\theta}$, high-resolution *pulsed* CMB/laser based techniques also permit determination of total and state-resolved integral cross sections, σ, $\sigma(v)$ and $\sigma(v,j)$, as a function of collision energy, as introduced in Chapter 1. The collision energy may be tuned in a CMB experiment by varying the angle at which the two beams intersect. From all the observables described above, very detailed information on the reaction dynamics can be obtained, especially in cases in which theoretical scattering calculations on *ab-initio* PESs can be performed, thus allowing for direct detailed comparisons between experiment and theory. In this chapter we provide a concise description of the main experimental methods which allow us to map out a three-dimensional view of a bimolecular reactive scattering event, and emphasize the intimate details of chemical reactivity that can be revealed by these studies.

6.2 CMB EXPERIMENTS WITH MASS SPECTROMETRIC DETECTION

A cornerstone in the development of the CMB method, and of the entire field of chemical reaction dynamics, has been the development of CMB instruments equipped with a universal detector,[440]

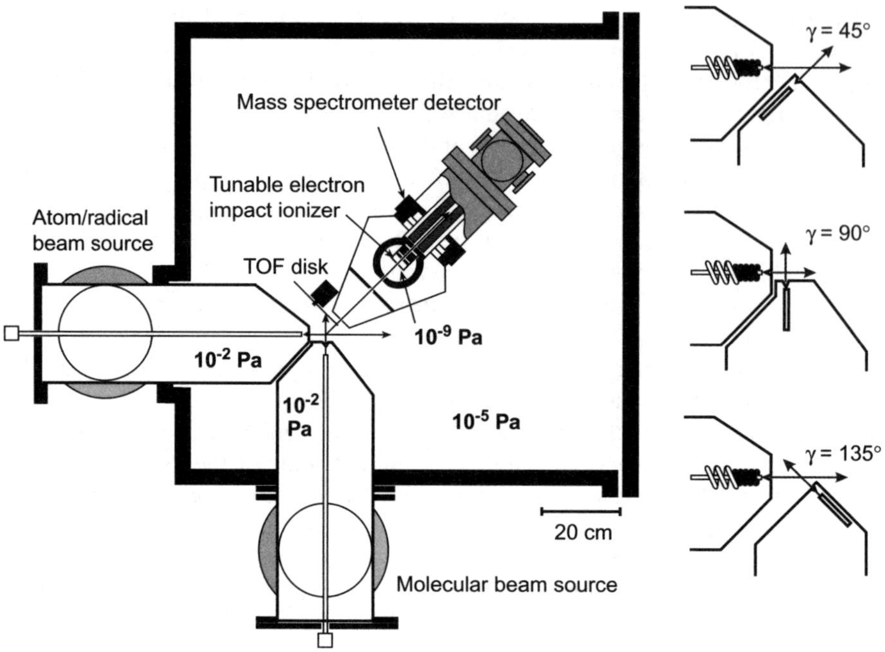

Figure 6.2 Schematic view of a crossed molecular beam apparatus with a rotatable TOF mass spectrometer detector and time-of-flight (TOF) analysis. Three different crossing beam geometries are also shown on the right-hand-side.

such as a mass spectrometer using electron-impact ionization and TOF analysis. The principles of CMB reactive scattering experiments have been discussed at length in a number of reviews[432] and book chapters.[431,441] The method can be outlined as follows. In a typical CMB experiment, as shown schematically in Figures 6.1 and 6.2, beams of atoms and molecules with known, narrow, angular and velocity distributions are crossed, usually at 90°, in a vacuum chamber with a background pressure of 10^{-6}–10^{-7} mbar. The angular distribution and the flight times to the detector of the scattered reaction products are recorded as a function of laboratory (LAB) recoil angle, using the rotatable mass spectrometer detector.[431,440] The low pressure at which the experiments are performed ensures that only single and well-defined collision events are observed. The detector is usually composed of an electron-impact ionizer followed by a quadrupole or magnetic sector mass filter, and an ion counting device, such as a Daly detector or a secondary electron multiplier.[440] A schematic of a universal CMB instrument is shown in Figure 6.2. It should be noted that beam crossing angles different from 90° are also possible (see, for instance, the schematics on the right of Figure 6.2), which makes it easy to vary the collision energy, while otherwise keeping exactly the same beam conditions. The collision energy is given by $E_c = \frac{1}{2}\mu(v_1^2 + v_2^2 - 2v_1v_2\cos\Theta_{12})$, where v_1 and v_2 are the beam velocities in the LAB-frame, μ is the reduced mass of the reactants, and Θ_{12} is the beam intersection angle.

The molecular beams are usually generated by a supersonic expansion of gas from a high-pressure zone behind a nozzle into vacuum. As discussed in Section 1.3.2, molecules (or radicals) are adiabatically cooled during the expansion, so that low rotational temperatures, and hence partial or complete state resolution, are usually achieved. One caveat is that vibrations generally cool less efficiently than rotations. The velocity distribution within the supersonic reactant beams typically has a FWHM (full-width-half-maximum) between a few percent and 30% of the most probable speed, depending on the atomic or molecular species and method of generation in the source. The beams are defined to small angular divergences of typically a few degrees by skimmers and further collimators. The absence of collisions within the beams and the long mean free-path inside the vacuum chambers ensures single collisions between the two reactant species at the collision region, where the two beams intersect each other. In other words, the pressure is low enough that secondary collisions, which would distort the motion of the reactants and products, can be neglected. By measuring the intensity and arrival times of product particles at the detector as a function of scattering angle, the laboratory angular and TOF distributions of the reaction products are determined. The quantities that are measured in CMB experiments with 'continuous' beams* are the product intensity as a function of the LAB-frame scattering angle Θ, usually referred to as the LAB angular distribution, $N(\Theta)$, and the product intensity as a function of Θ and arrival time t, *i.e.* the angle-resolved TOF spectrum, $N(\Theta,t)$.

CMB experiments with continuous mass-spectrometric detection can also be performed using 'pulsed' molecular beams.[442] This is often the case when a transient reactant is more conveniently generated in a pulsed manner, as for instance by laser photolysis or laser ablation of a suitable precursor.[443] In a pulsed beam experiment, one measures only $N(\Theta,t)$ spectra, and from their integration derives the product $N(\Theta)$. In general, however, pulsed crossed beams optimally couple with pulsed laser detection, either in state-resolved spectroscopic modes (see Section 6.4), or in a photoionization/mass-spectrometric mode.[444] The final outcome of a reactive scattering experiment is usually the generation of velocity-flux contour maps of the reaction products for each reaction channel. These show the scattered intensity, $I_{CM}(\theta,u)$, as a function of angle θ and speed u in the CM-frame. Study Box 6.1 provides a discussion of the LAB to CM transformation involved in the data analysis. The product contour map can be regarded as an *image* of the scattering distribution for the reaction in angle-velocity space, and contains much of the available information about the reaction dynamics.

Universal CMB instruments have yielded much fundamental information on primary reaction mechanisms, including product angular distributions, and the partitioning of the total available energy between product internal (vibrational, rotational, etc.) and translational degrees of freedom. The dependence of the reaction cross section on collision energy and reactant orientation has also been probed. For a comprehensive review of reactions studied *via* the CMB technique up until the year 2000, see ref. [432].

While state-averaged DCSs for a reaction product are readily measurable using a universal CMB instrument, the determination of quantum state-selected DCSs is a more challenging problem. A limitation in such experiments is the time-of-flight resolution, which is dictated both by $\Delta l/l$ (the size Δl of the ionization zone relative to the flight distance l) and by the velocity spread of the reactant beams. In CMB experiments using reactant beams generated from continuous supersonic beam sources, beam velocity spreads $\Delta v/v$ are typically 10–30%. There are some special cases: a velocity spread of a few percent can be obtained for beams of molecular hydrogen, and about 5–10% can be achieved for beams of some atomic species. The beam spreads are usually the limiting factor in the attainable resolution, since the detector's intrinsic resolution $\Delta l/l$ is considerably better than 5%.[431,432] Product vibrational,[447,448] and even some partial rotational[449] resolution, has been achieved for the reaction $F+H_2$, in which the vibrational spacing in the HF product is wide enough that the vibrational populations are reflected, through the conservation of energy, as clear peaks in the product velocity distributions. For reactions of the type $X+H_2\rightarrow HX+H$ leading to H atom products, this principle is extended through use of an intrinsically much higher resolution technique, the H-Rydberg atom TOF method mentioned previously and treated in detail in Section 6.3.

*In a 'continuous' beam experiment, time-of-flight (and therefore product speed) information is obtained by placing a slotted chopper in the path to the detector. This crude shutter interrupts the flux of products to the detector, providing a convenient 'zero-time' for the TOF measurements.

STUDY BOX 6.1: THE CLASSIC CMB TECHNIQUE: THE LAB TO CM TRANSFORMATION

As we have seen in Section 1.3.3, and in the present chapter, in a CMB experiment the measurements are carried out in some fairly arbitrary LAB system of coordinates determined by the experimental geometry. The resulting experimental data are referenced to some axis system relevant to the experiment, such as the molecular beam directions. However, for the physical interpretation of the scattering process it is helpful to transform the data (angular, $N(\Theta)$, and velocity, $N(\Theta,v)$ distributions) to a coordinate system that moves with the centre-of-mass (CM) of the collision partners.[431] The reason for this is that in the absence of an external field, the motion of the CM must be conserved throughout the collision, and therefore contributes nothing to the reaction dynamics. The transformation is fairly straightforward, and the relation between LAB and CM fluxes is given by

$$I_{\mathrm{LAB}}(\Theta,v) = I_{\mathrm{CM}}(\theta,u)\frac{v^2}{u^2}, \tag{B6.1.1}$$

where Θ and v are LAB angle and velocity of the scattered products, respectively, and θ and u are the corresponding CM quantities. We see that the scattering intensity observed in the laboratory frame is distorted by the transformation Jacobian v^2/u^2 from that in the CM system (see Study Box 1.1). The velocity vector (or 'Newton') diagram shown in Figure 6.3 illustrates the nature of the LAB to CM transformation for an A + BC reaction, and is generally used for interpreting and discussing scattering results[431] (see also the discussion in Section 1.3.3).

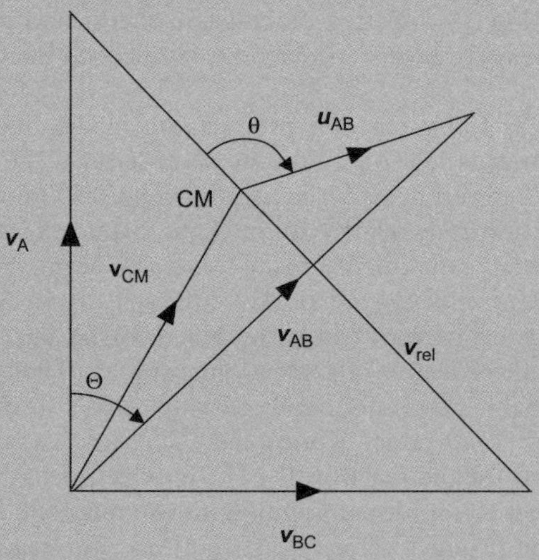

Figure 6.3 The vector (or 'Newton') diagram showing the relationship between LAB and CM velocity vectors and angles for the reaction $A + BC \rightarrow AB + C$. The velocity vector of the CM of the colliding reactants, $\boldsymbol{v}_{\mathrm{CM}}$, allows one to relate the product LAB velocity $\boldsymbol{v}_{\mathrm{AB}}$ and scattering angle Θ to the product CM velocity $\boldsymbol{u}_{\mathrm{AB}}$ and angle θ through the vector addition $\boldsymbol{v}_{\mathrm{AB}} = \boldsymbol{v}_{\mathrm{CM}} + \boldsymbol{u}_{\mathrm{AB}}$. The velocity of the CM may be determined from knowledge of the reactant beam velocities using the expression $\boldsymbol{v}_{\mathrm{CM}} = (m_1\boldsymbol{v}_1 + m_2\boldsymbol{v}_2)/M$ with $M = m_1 + m_2$. The diagram also shows $\boldsymbol{v}_{\mathrm{rel}}$ ($= \boldsymbol{v}_{\mathrm{A}} - \boldsymbol{v}_{\mathrm{BC}}$), the reactant relative velocity vector.

The electron-impact ionization mass spectrometric detector measures the number density of products, $N(\Theta)$, whereas the quantity we want to measure is the product flux. The relation between the actual measured signal (the LAB density) and the CM flux (the quantity we wish to determine) is given by

$$N_{\rm LAB}(\Theta,v) = I_{\rm CM}(\theta,u)\frac{v}{u^2}. \tag{B6.1.2}$$

Further details about this 'density-to-flux' transformation are given in Study Box 6.5. Because of the finite resolution of the experiment, resulting from the finite angular and velocity spread of the reactant beams and the finite angular resolution of the detector, the LAB to CM transformation is not single-valued. For this reason, analysis of the laboratory data usually relies on forward convolution procedures, which simulate the measured signal under the experimental conditions of interest based on trial CM distributions. The trial CM angular and velocity distributions are chosen, averaged over the relevant reactant velocity distributions, and transformed to the LAB-frame for comparison with the experimental distributions. The procedure is then repeated until a satisfactory fit to the observed data is obtained. The CM-frame DCS, $I_{\rm CM}(\theta,u)$, is commonly factorized into the product of the velocity distribution, $P(u)$, or equivalently the product translation energy or *kinetic energy release* distribution, $P(E'_{\rm t})$ and the angular distribution, $P(\theta)$. In the literature, the latter is often referred to as $T(\theta)$, and thus we have:

$$I_{\rm CM}(\theta,u) = T(\theta) \times P(E'_{\rm t}). \tag{B6.1.3}$$

In some cases, the velocity and angular distributions are not independent of each other, and the coupling between the $T(\theta)$ and $P(E'_{\rm t})$ functions needs to be accounted for by using a combined $P(E'_{\rm t},\theta)$ function instead. When multiple reaction channels contribute to the signal at a given mass-to-charge ratio m/z, a more complex situation arises. In these cases a weighted total CM differential cross section reflecting the contributions from the various possible pathways is used in the data analysis of the LAB distributions for a specific m/z:

$$I_{\rm CM}(\theta,E'_{\rm t}) = \sum_i w_i\left[T(\theta) \times P(E'_{\rm t})\right]_i, \tag{B6.1.4}$$

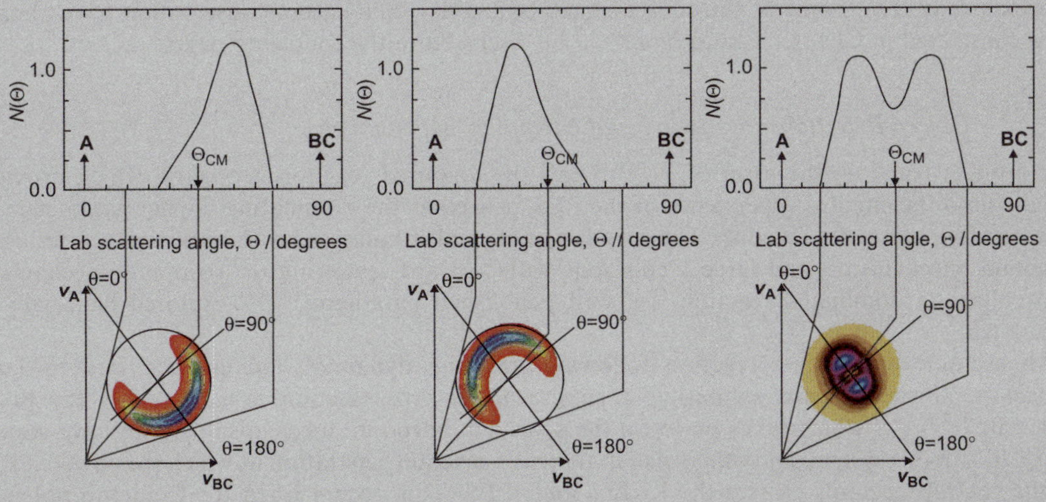

Figure 6.4 Schematic illustration of backward, forward and symmetric scattering of the AB product from an A + BC reaction in the LAB (top) and CM (bottom) reference frame.

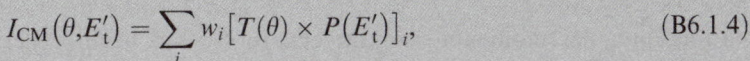

with the parameter w_i representing the relative contribution to the measured signal from the integral cross section for the ith channel (see Section 6.2.3). When the velocity distributions of the reactants are quite narrow, this procedure allows one to recover the CM distributions with high reliability.[431,433,434,445,446]

We conclude this Study Box with a simple illustration of the consequences of the LAB to CM transformation. Figure 6.4 (left and centre panels) depicts typical backward and forward scattering behaviour (with respect to the direction of the A reactant) for a direct A + BC → AB + C reaction. The scattering distributions are shown both in the LAB and CM-frames. In the backward scattering case (left panel), note the backward displacement of the LAB angular distribution, $N(\Theta)$, of the AB product with respect to the position of the CM angle. In the CM-frame, the backward peak in the AB product flux contour map, superimposed on the Newton diagram, with respect to the direction of the A reactant, is clearly seen.

Piergiorgio Casavecchia, Kopin Liu and Xueming Yang

6.2.1 Angular Scattering and Reaction Mechanisms

We have shown in the previous sections how the product angular $T(\theta)$ and translational energy $P(E_t')$ distributions in the CM-frame can be derived from LAB-based measurements. The $T(\theta)$ and $P(E_t')$ functions contain a great deal of information about the dynamics of the process under study; in particular $T(\theta)$ contains detailed information on the reaction mechanism.[431,433,434,445] Elementary reactions may be grouped into two limiting mechanistic categories, namely those following 'direct' (direct scattering) or 'indirect' (complex-forming) mechanisms. These are characterized by distinct $T(\theta)$ distributions. In both types of mechanism, the form of the product angular distribution, $T(\theta)$, is dictated by the angular momentum disposal – the degree of correlation between reactant and product angular momenta – as established by the seminal work of Herschbach during the alkali era.[433,434]

For a generic A + BC reaction such as reaction (6.1), conservation of angular momentum $\boldsymbol{J}$ dictates that the total angular momentum of the reactants and products must be equal: $\boldsymbol{J} = \boldsymbol{J}'$, where $\boldsymbol{J} = \boldsymbol{\ell} + \boldsymbol{j}$ and $\boldsymbol{J}' = \boldsymbol{\ell}' + \boldsymbol{j}'$, with ℓ, ℓ', j, and j' being the reactant and product orbital and rotational angular momenta, respectively. In crossed beam experiments, because of the strong rotational cooling occurring in supersonic beams, j_{BC} is usually small and can be assumed to be approximately zero. Some of the consequences of angular momentum conservation have already been considered in Chapter 1, and more will be discussed in the following pages.

6.2.1.1 Direct Reactions: Rebound and Stripping Mechanisms

For a direct reaction, the form of $T(\theta)$ reveals the favoured reaction geometry, which provides insights into the angular dependence of the PES, as well as the dominating impact parameters, b, that lead to a successful reaction. The correlation of small-b collisions with backward scattering or 'rebound' mechanisms, and large-b collisions with forward scattering or 'stripping' mechanisms in a colinearly dominated reaction is a well established paradigm,[433,434] explored pictorially in Figure 6.5.

An example of a direct reaction following stripping dynamics and characterized by large impact parameters (hence exhibiting a large reactive cross section, $\sigma \approx 200\ \text{Å}^2$) is $K + Br_2 \rightarrow KBr + Br$,[433,434,450] which takes place *via* the so-called 'harpoon' mechanism. As we have seen in Chapter 4, in this system, and others like it, there is a reactant separation at which the covalent PES of the neutral reactants crosses the $K^+Br_2^-$ ionic PES. This occurs when the Coulomb potential energy between the ions matches the difference between the ionization potential of the K atom and the electron affinity of the Br_2 molecule. The crossing point, also known as the Harpoon radius, is

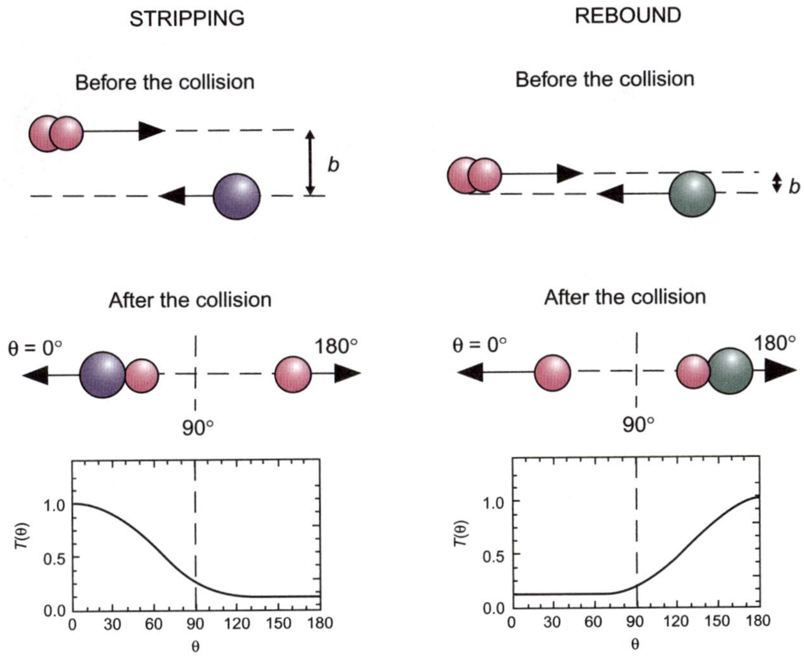

Figure 6.5 Billiard ball (hard-sphere) representation of the rebound and stripping mechanisms in the CM system. The approach direction of the atom is $\theta = 0°$. Left: The atom approaches the diatomic molecule at large impact parameters, b, and strips an atom from it; the molecular product departs in the forward direction ($\theta = 0°$) with respect to the direction of the incoming atom ($\theta = 0°$). The product CM angular distribution will peak at $\theta = 0°$ with the product flux mainly confined in the forward hemisphere, as depicted in bottom left panel. Because b_{max} is large, the reactive integral cross section σ ($= \pi b_{max}^2$) will be large. Right: The atom approaches the diatomic molecule at small impact parameters, b; after the collision the molecular product rebounds in the backward direction ($\theta = 180°$) with respect to the direction of the incoming atom ($\theta = 0°$). The product CM angular distribution $T(\theta)$ will peak at $\theta = 180°$ with the product flux mainly confined in the so-called backward hemisphere, as depicted in the lower right panel. Because in this case b_{max} is small, the reactive integral cross section σ will also be small.

often at quite large separations, and therefore reaction can take place at relatively large impact parameters. The resulting glancing collisions lead to the preference for forward scattered reaction products observed.

Examples of direct reactions following rebound dynamics are $K + CH_3I \rightarrow KI + CH_3$[433,434] and $Cl + H_2 \rightarrow HCl + H$.[451] For the latter, the barrier to reaction is a minimum for a colinear approach of the Cl atom to H_2, as depicted in Figure 6.6(c).[451] The strongly backward peaked HCl product (Figure 6.6(a) and (b)) reflects this highly constrained colinear transition state geometry. These experimental findings and the resulting predictions about the topology of the PES were confirmed by *ab initio* electronic structure calculations of the PES, and by QM as well as QCT scattering calculations on this PES.[432,451]

6.2.1.2 *Indirect Reactions: Long-Lived (Osculating) Complex Mechanism*

As noted above, conservation of angular momentum, and the manner in which angular momentum is disposed of, often has important consequences for the dynamics of chemical reactions. As we have seen, for an $A + BC \rightarrow AB + C$ reaction the total angular momentum is often approximately

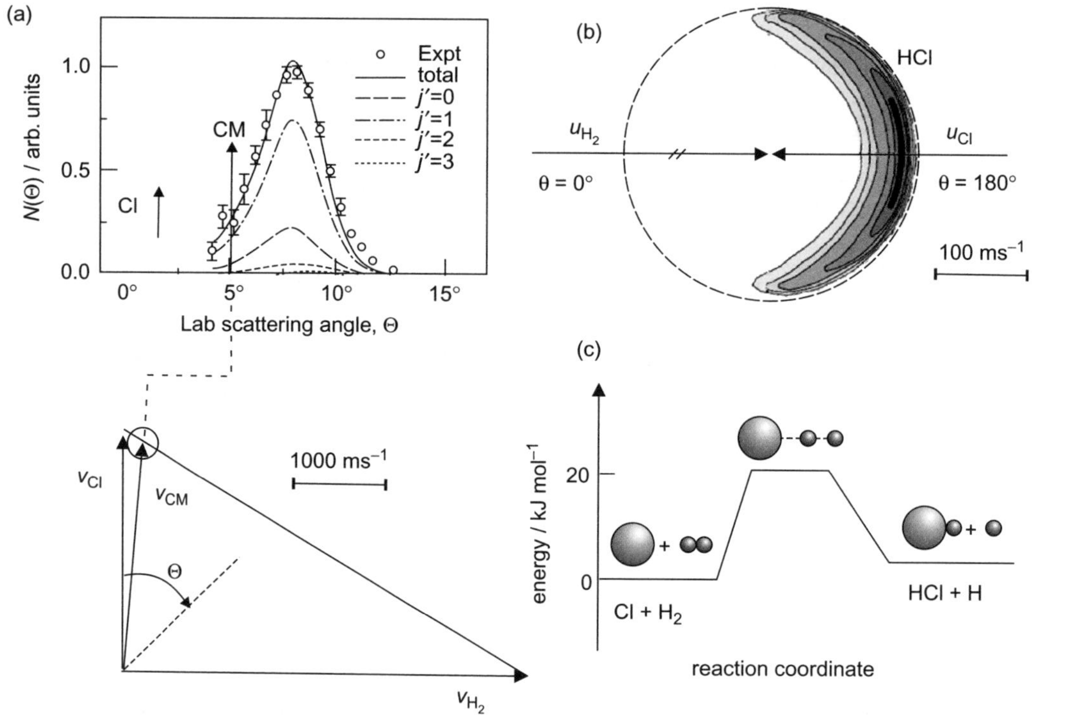

Figure 6.6 (a) The total LAB angular distribution and Newton diagram for the HCl($v'=0$) product from the reaction $Cl + H_2(v=0, j=0, 1, 2, 3) \rightarrow HCl(v'=0) + H$ at $E_c = 24.5\,kJ\,mol^{-1}$, together with simulations based on exact QM calculations on the G3 PES; (b) The HCl($v'=0$) product CM flux contour map: $\theta = 180°$ indicates the backward direction with respect to that of the incoming Cl atom, and u_{H_2} and u_{Cl} are CM velocities; (c) Energy level and correlation diagram (schematic) for the $Cl + H_2 \rightarrow HCl + H$ reaction (adapted from refs. [432] and [451]).

equal to the orbital angular momentum of the reactants, $\boldsymbol{J} \approx \ell$, and this is deposited into orbital and rotational angular momentum of the products, ℓ'_{AB} and $\boldsymbol{j}'_{AB}$.[433,434] If the product receives little rotational excitation then $\boldsymbol{j}'_{AB}$ is small, $\ell' \approx \ell$ and the product relative velocity vector $\boldsymbol{v}'_{rel}$ is constrained to lie in the same plane as the reactant relative velocity $\boldsymbol{v}_{rel}$; that is, we have a 'coplanar' reaction. If such a reaction also involves complex-formation, a backward-forward symmetric angular distribution $T(\theta)$ arises, with a sharp peaking at the poles ($\theta = 0°$ and $180°$). This situation in the CM-frame is schematically depicted in Figure 6.7(a) and (b). Figure 6.4 (bottom-right panel) shows that when the CM product flux contour map has a forward-backward symmetric peaking, the LAB-frame distribution (shown in the upper-right panel) exhibits intensity on both sides of the LAB-frame angle of the CM. Note that in the LAB-frame the form of the CM to LAB Jacobian transformation introduced in Study Box 6.1 means that the backward and forward peaks may not have the same intensity (see Study Box 1.1 for a discussion of the mathematics of Jacobian determinants).

When the angular momentum is constrained such that $\ell' \approx \ell$ the angular distribution takes the form of $1/\sin\theta$, as shown in Figure 6.7(b). This can be explained with the help of the diagrams in Figure 6.7(a). Panel a(i) shows the collision plane for a particular collision event with a well defined direction of the orbital angular momentum. If the fragments were to separate randomly after many rotational periods, then the distribution of products would be uniform in this collision plane. However, in reality, ℓ and ℓ' have cylindrical symmetry around the relative

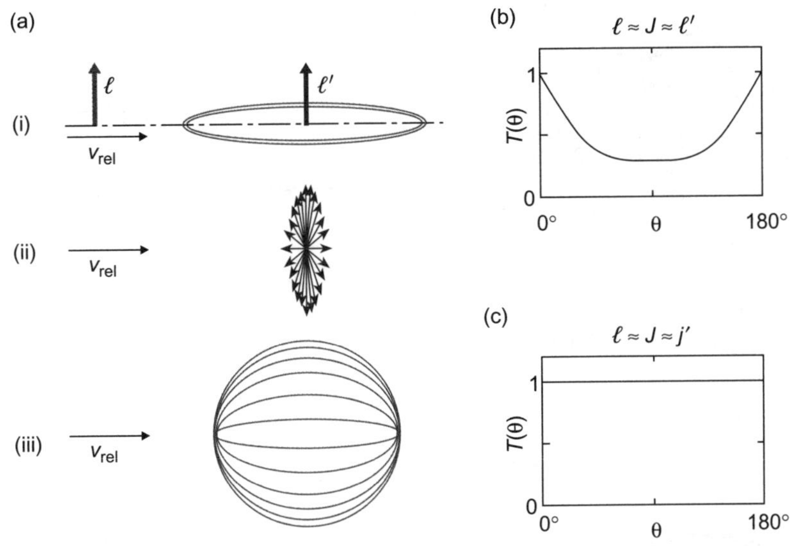

Figure 6.7 For a reaction proceeding through a long-lived complex mechanism, the shape of the backward-forward symmetric angular distribution, $T(\theta)$, reflects the angular momentum disposal. Panel (a) shows the situation when both $\boldsymbol{j}$ and $\boldsymbol{j}'$ are small, and ℓ' is therefore nearly equal to ℓ, and parallel (or antiparallel) to it. This results in a coplanar reaction, as shown in figure a(i). Panels a(ii) and a(iii) illustrate the effect of averaging over the azimuthal ϕ angle, which defines the orientation of the orbital angular momentum and the collision plane of the reaction (see text for details). Panel (b) shows the resulting CM angular distribution, which in this case will have a pronounced ($\propto 1/\sin\theta$) symmetric forward-backward peaking structure (see figure a(iii) and refs. [433] and [434]). In contrast, if the products are rotationally very excited, then $\ell \approx \boldsymbol{j}'$, and the angular distribution will be isotropic (*i.e.* unity over all angular range), as indicated in panel (c).[433]

velocity $\boldsymbol{v}_{rel}$, as illustrated in figure a(ii), and once averaged about the ϕ angle associated with the direction of ℓ about $\boldsymbol{v}_{rel}$, the distribution shows peaks in intensity around the poles, as indicated in panels a(iii) and (b) of Figure 6.7. This argument may be expressed more mathematically in the following way. In the limit that $\ell' \approx \ell$ the probability that the products scatter into angles in the range $\theta \to \theta + d\theta$ and $\phi \to \phi + d\phi$ is independent of θ and ϕ, and thus one can write

$$P(\theta, \phi)d\theta d\phi = \frac{1}{\sigma}\frac{d\sigma}{d\omega}\sin(\theta)d\theta d\phi = N d\theta d\phi, \tag{6.2}$$

where N is a constant.[10] Rearranging the right hand side of this equation directly yields the above mentioned result

$$\frac{1}{\sigma}\frac{d\sigma}{d\omega} = \frac{1}{2\pi^2 \sin(\theta)}. \tag{6.3}$$

The strong $1/\sin\theta$ symmetric behaviour of $T(\theta)$ was first observed by Herschbach and coworkers for the Cs + RbCl reaction.[433,434,452] Another reaction following a long-lived complex mechanism is OH + CO → CO_2 + H. Figure 6.8(a) shows the LAB angular distribution measured at $E_c = 59.0\,\mathrm{kJ\,mol^{-1}}$, together with the Newton diagram for the collision. In (b) the CM CO_2 product

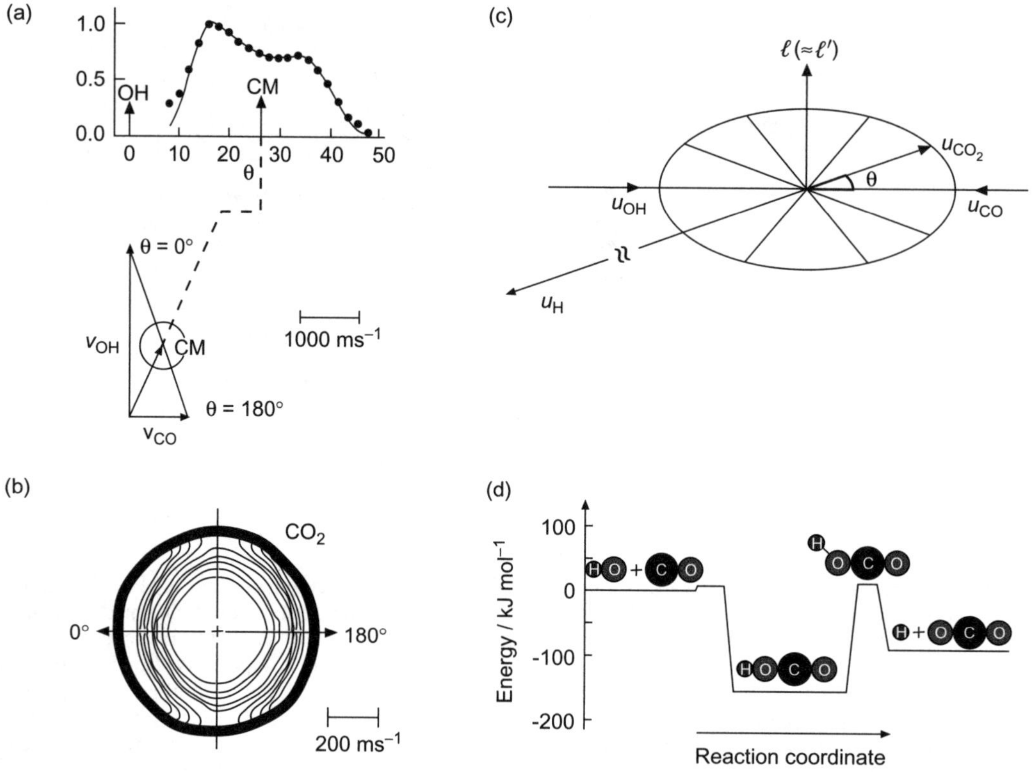

Figure 6.8 (a) CO_2 product LAB angular distribution from the OH + CO reaction at $E_c = 59.0\,\mathrm{kJ\,mol^{-1}}$, together with Newton diagram of the experiment. The circle in the Newton diagram delimits the maximum velocity that the CO_2 product can attain in the CM-frame if all of the available energy is channeled into translation. (b) CM flux contour map of the CO_2 product. It is seen that the CO_2 velocity is very high, peaking at the limits imposed by energy conservation. This corresponds to about 70% of the total available energy being released into translation. (c) Schematic of an OH + CO collision in the CM-frame. CO_2 product velocities are constrained by angular momentum conservation to lie in a plane near that of the reactants. (d) Energy level and correlation diagram (schematic) for the reaction (adapted from refs. [432] and [453]).

flux contour map is depicted. Panel (c) is a schematic representation of the relationship between the vector properties relevant to the collision, and in (d) a schematic of the stationary points on the PES is portrayed.[433,434,453] The topology of the HOCO PES is very different from that of ClH_2 shown in Figure 6.6(c). While in the Cl + H_2 reaction the intermediate complex ClH_2 is unstable (it is a saddle point or transition state on the PES), the HOCO intermediate in the OH + CO reaction has a considerable stability. Also, because $\ell \approx \ell'$ in the HOCO reaction, as Figure 6.8(c) shows, the product CM velocities u_{CO_2} and u_H must lie in a plane nearly identical to that of the incident reactants.

It should be noted that, if the product rotational excitation j'_{AB} is sizeable, then $\ell' < \ell$, and products can scatter out of the plane containing $\boldsymbol{v}_{rel}$. This results in a much less polarized or even isotropic $T(\theta)$ (see Figure 6.7(c)). The above correlations between angular momentum disposal and the shape of the product angular distribution $T(\theta)$ are often important for radical reactions involving polyatomic molecules.[433,434] Here, the shape of $T(\theta)$ can also be related to the geometric structure of the decomposing long-lived complex through its moments of inertia (see refs. [433], [434], and [454] and references therein). According to the 'osculating-complex' model for chemical reactions,[433,434,452] a reaction is said to proceed through a long-lived complex mechanism when the complex lifetime τ is at least $\sim$5 times its rotational period, τ_r. If the complex lives for less than $\sim$5 rotational periods, memory of the initial approach direction of the reactants is not completely scrambled, and $T(\theta)$ becomes asymmetric, with more intensity in the forward direction. The ratio of the backward to forward intensity $T(180°)/T(0°)$ affords an approximate estimate of the complex lifetime through the relation $T(180°)/T(0°) = \exp(-\tau_r/2\tau)$. In the OH + CO reaction, the experimental CO_2 angular distribution shown in Figure 6.8(a) is broad, with products scattered on both sides of the CM angle. At the collision energy of the experiment, the reaction exhibits an osculating-complex behaviour. The forward and backward peaks in the DCS shown in Figure 6.8(b) are of unequal intensity, with $T(180°)/T(0°) = 0.6$, and this corresponds to $\tau/\tau_r \approx 1$.[453]

While the framework outlined so far, based on classical mechanics, explains many features of the dynamics, for a truly accurate description of chemical reactions a QM description is required. The role that quantum effects play in the dynamics of chemical reactions is discussed in Section 6.3.

WORKED PROBLEM (1)

Q: *The bimolecular reaction A + BC → AB + C is investigated in a CMB experiment at a collision energy E_c. The BC molecule is in its ground rovibrational state. Show that the maximum velocity that the AB product can attain in the CM system is*

$$(u_{AB})_{max} = \left[2E'_{t,max}\left(\frac{m_C + m_{AB}}{m_{AB}^2}\right)\right]^{1/2},$$

where $E'_{t,max}$ is the maximum available translational energy (this corresponds to the sum of the collision energy and the reaction exoergicity) and m_{AB} and m_C are the masses of the reactants.

A: From linear momentum conservation, $m_{AB}u_{AB} = -m_C u_C$, and therefore $m_{AB}(u_{AB})_{max} = -m_C(u_C)_{max}$. The maximum translational energy is $E'_{t,max} = \frac{1}{2}m_{AB}(u_{AB})^2_{max} + \frac{1}{2}m_C(u_C)^2_{max}$. From these two relations one obtains the required result by eliminating $(u_C)_{max}$. Note that AB molecules with the maximum velocity are those formed in their ground vibrational and rotational state.

6.2.2 Soft-ionization Detection

As noted previously, the use of electron impact (EI) mass-spectrometric detection renders the CMB method *universal*, *i.e.* applicable in principle to the study of any reaction. Every species may be ionized at the typical electron energy of $\sim$100 eV used in the ionizer which precedes the mass filter of the detector (see Figure 6.2). Since the introduction of the CMB technique in 1969, because of the very low number density of products to be detected in these experiments, the electron energy has generally been matched to the maximum in the ionization efficiency curve, which typically occurs between 60 and 200 eV for most species. This is sometimes referred to as 'hard ionization'.[432,440] Unfortunately ionization at these energies often leads to dissociative ionization, which can represent a serious complication. It may, for instance, lead to a situation in which the signal at a certain *m*/*z* originates from more than one product; this is common for reactions producing organic radicals/molecules because of their tendency to fragment in the ionizer.[431] Only in favourable cases, by exploiting energy and momentum conservation, is it possible to distinguish whether an ion at a given m/z originates from different neutral products (ion fragments coming from a two body fragmentation of the same parent molecule will exhibit identical angular and momentum distributions).[431] Especially troublesome is the case in which interference for product detection results from elastically scattered reactants, since reactant beams are typically very intense and elastic cross sections very large.

An elegant way to overcome this problem is to employ *soft* (*i.e.* non-dissociative) photoionization (PI) detection using tunable VUV radiation in the 5 to 30 eV range, obtained from a third generation synchrotron. This was first demonstrated in the late 1990s by Lee and coworkers at the Advanced Light Source in Berkeley[455] and has been used more recently in Taiwan by Yang and coworkers.[456] Dissociative ionization is avoided by tuning the energy of the ionizing radiation below the dissociative ionization potential of a molecule or radical, so that direct detection of the parent ion is possible. A variety of photodissociation studies have demonstrated the power of *soft* PI by synchrotron radiation.[457–464] Although there have been only a few examples of reactive scattering studies exploiting this approach,[465,466] mainly for sensitivity reasons, recent studies on $O(^3P,^1D) + C_2H_4$,[467,468] and $O(^3P,^1D) + SiH_4$[469] have shown the great potential of this method. On the same theme, single photon PI by a 157 nm F_2 laser (7.9 eV radiation) has recently been used in reactive scattering studies of transition metals, exploiting the low ionization potentials of these metals and their compounds.[444,470] A shortcoming with the VUV PI approach is that absolute PI cross sections are very often not known, and therefore branching ratios cannot easily be estimated.

An attractive, simple and low cost alternative to the use of soft photoionization by synchrotron radiation is based on *soft* electron impact ionization, achieved by using electrons with low, tunable energy.[471–473] Although not affording the same degree of selectivity as VUV synchrotron radiation, the soft EI approach has similar advantages with respect to the dissociative ionization problem; in addition, it offers the possibility of determining branching ratios, since absolute EI cross sections are often known or can be reliably estimated.[474–476] The soft EI approach had not been applied in CMB experiments until recently. This was mainly due to the fact that EI cross sections decrease dramatically towards threshold affording insufficient detection sensitivity. However, by exploiting improvements in instrumentation, which have led to an increased sensitivity of modern CMB instruments relative to their predecessors, it has recently been shown[471–473] that in many cases the simple approach of soft EI permits the achievement of *universal* detection in CMB experiments, and thus the ability to identify all primary reaction products of multichannel polyatomic reactions, determine their branching ratios, and characterize the dynamics of each channel, as briefly discussed next. Undoubtedly, the implementation of soft EI and PI represents a new milestone in CMB studies of reaction dynamics, as well as in photodissociation studies.

6.2.3 Multichannel Polyatomic Reactions

Many elementary reactions of importance in areas of practical interest, such as combustion, atmospheric chemistry, and astrochemistry, involve polyatomic molecules or radicals as reactants and products and are usually characterized by several competing product channels. A major goal in understanding the chemistry of these systems is to identify all possible primary reaction products, characterize their dynamics of formation, and determine the branching ratios into the various channels. In general, it is very difficult to fully characterize multichannel polyatomic reactions at the quantum state-resolved level. Some of the most enlightening experiments carried out to date on such systems have used the CMB technique with soft ionization detection, which allows determination of *state-averaged* DCSs for all product channels, as exemplified below.

6.2.3.1 Example: The Reaction $O(^3P) + C_2H_4$

The reaction of $O(^3P)$ with ethene exhibits six competing energetically allowed pathways:

$$O(^3P) + C_2H_4 \rightarrow H + CH_2CHO \quad \Delta H_0^{\ominus} = -71\,kJ\,mol^{-1} \tag{6.4a}$$

$$\rightarrow H + CH_3CO \quad \Delta H_0^{\ominus} = -114\,kJ\,mol^{-1} \tag{6.4b}$$

$$\rightarrow H_2 + CH_2CO \quad \Delta H_0^{\ominus} = -356\,kJ\,mol^{-1} \tag{6.4c}$$

$$\rightarrow CH_3 + HCO \quad \Delta H_0^{\ominus} = -113\,kJ\,mol^{-1} \tag{6.4d}$$

$$\rightarrow CH_2 + HCHO \quad \Delta H_0^{\ominus} = -29\,kJ\,mol^{-1} \tag{6.4e}$$

$$\rightarrow CH_4 + CO \quad \Delta H_0^{\ominus} = -488\,kJ\,mol^{-1} \tag{6.4f}$$

Of these possible channels, at thermal energies channels 6.4(c), (d) and (f) can only occur *via* intersystem crossing (ISC) from the triplet state to a singlet state PES.

In addition to the combustion of ethene itself, this reaction plays a central role in the overall mechanism for hydrocarbon combustion, and is prototypical for a host of O-atom reactions with alkenes. Many research groups have investigated reaction (6.4) by employing a variety of experimental techniques in different pressure and temperature regimes, and have identified only some of the possible products. The overall rate constant is now well established ($k_{298K} = 7.5 \times 10^{-13}\,cm^3\,molec^{-1}\,s^{-1}$), but the question of the identity of the primary reaction products and their relative importance has been a subject of considerable controversy over the years. Early CMB investigations at $E_c \approx 25\,kJ\,mol^{-1}$ employing hard EI, while confirming the occurrence of channel 6.4(a), the easiest to detect for kinematic reasons,* were unable to explore the other channels. These are much more difficult to detect due to unfavorable kinematics and also to the fact that the expected ion signals from products of channels 6.4(c), (d) and (e) appearing at $m/z = 42$, 15 or 29, and 14 or 30, respectively, coincide with major background peaks and/or with peaks coming from dissociative ionization of CH_2CHO (the most abundant product) and the elastically scattered C_2H_4 reactant. Branching to channel 6.4(f) is thought to be very small. This channel is very hard to study by the CMB technique because one of the two products has the same mass as one of the two reactants and the other is CO, which is a major background mass peak (second only to H_2) in any UHV chamber. This kind of situation is one of the most unfavourable for the CMB technique.

The $O(^3P) + C_2H_4$ reaction is one for which the use of soft EI ionization detection is highly desirable. Using this detection method in a detailed series of angular and velocity distribution measurements at $m/z = 42$, 15, and 14, it has been possible to achieve unambiguous detection of products from each of the five reaction pathways 6.4(a–e), determine their branching ratios, and characterize their dynamics.[473] Figure 6.9 shows the LAB-frame angular distribution for CH_3^+ recorded at $m/z = 15$ using an electron energy of 17 eV. Also shown is the Newton diagram for the collision at a collision energy of $54.0\,kJ\,mol^{-1}$. A crossed molecular beam configuration with $\Theta_{12} = 135°$ was used in these experiments in order to achieve higher angular and especially TOF resolution. Two channels, 6.4(a) and (d), contribute to this signal due to the fragmentation of the primary CH_2CHO product of channel (a) within the mass spectrometer; the relative contributions are disentangled by fitting TOF measurements at selected LAB angles. As an example, see Figure 6.10 (middle panel), in which the TOF spectrum at $m/z = 15$ for $\Theta = 34°$ is shown, together with the fitted contributions from the two channels. Figure 6.10 (top) shows the TOF spectrum recorded at $m/z = 42$, which is used to disentangle the relative contributions from channel 6.4(a), (b), and (c).

*See Section 1.4.2 for a general discussion of kinematic effects, as well as Study Box 7.1.

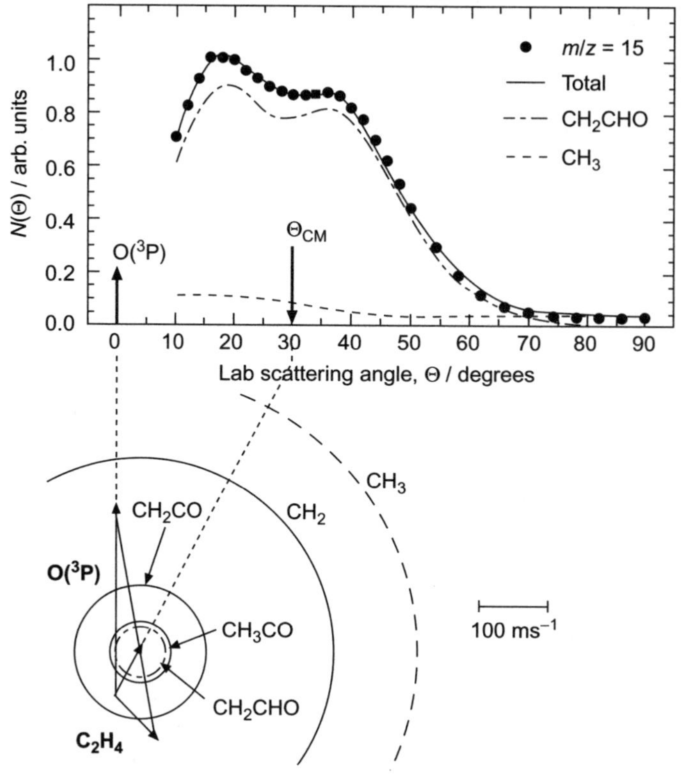

Figure 6.9 LAB angular distribution at $m/z = 15$ (solid circles) from the reaction $O(^3P) + C_2H_4$ at $E_c = 54.0\,kJ\,mol^{-1}$, obtained by using an electron energy of 17 eV, together with the Newton diagram. Error bars are indicated when visible outside the experimental dots. The circles in the Newton diagram delimit the maximum speed that the indicated products can attain on the basis of energy and linear momentum conservation if all of the available energy goes into product translation. The heavy solid line is the total angular distribution calculated from the best-fit product CM translational energy and angular distributions. The separate contributions from the CH_2CHO and CH_3 products from channels 6.4(a) and (d) are shown with dashed-dotted and dashed lines, respectively (adapted from ref. [473]).

The parent ion at $m/z = 43$, CH_2CHO^+, corresponding to one of the main reaction channels, vinoxy radical formation, is not stable and readily fragments in the analyzer, so that measurements of angular and TOF distributions were carried out for the daughter ion at $m/z = 42$.

Figure 6.10 (bottom panel) shows the TOF spectrum recorded at $m/z = 14$, again with an electron energy of 17 eV at $\Theta = 34°$. The three TOF signals shown in Figure 6.10 carry the clear fingerprints of all five product channels. Specifically, the $m/z = 15$ TOF spectrum exhibits a fast peak which is unambiguously due to the CH_3 from the $CH_3 + HCO$ channel (reaction 6.4(d)), and a slower, more intense peak due to dissociative ionization in the ionizer of the vinoxy radical, corresponding to the $CH_2CHO + H$ channel (reaction 6.4(a)). The $m/z = 42$ TOF spectrum exhibits: (i) a dominant peak, analogous to the main peak observed at $m/z = 15$, which is due to dissociative ionization in the ionizer of the vinoxy radical formed in channel (a); (ii) a fast peak, which appears as a shoulder on the main peak and is unambiguously attributed, on the basis of energy and linear momentum conservation, to the ketene product from the channel 6.4(c), $CH_2CO + H_2$; (iii) a small component, peaked at the CM velocity, which is attributed to formation of the acetyl radical from the channel 6.4(b) $CH_3CO + H$ with a very small recoil energy.

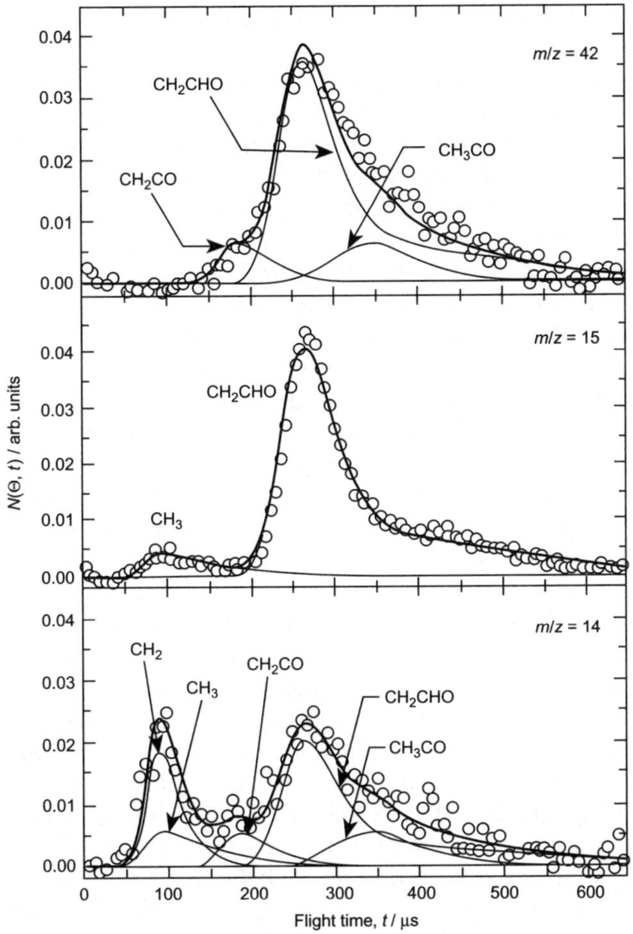

Figure 6.10 TOF spectra at $\Theta = 34°$ for the $O(^3P) + C_2H_4$ reaction at $E_c = 54.0\,kJ\,mol^{-1}$ recorded at $m/z = 42$ (top), 15 (middle) and 14 (bottom) using an electron energy of 17 eV. Open circles are experimental points, while heavy solid lines are the total TOF distributions calculated from the best-fit product CM translational energy and angular distributions for the contributing channels. The contributions from the different products (depicted with different lines) are labelled on the plots. Note that three product channels contribute to the signal detected at $m/z = 42$, two product channels to the signal detected at $m/z = 15$, and five product channels to the signal detected at $m/z = 14$ (adapted from ref. [473]).

The $m/z = 14$ TOF spectrum exhibits contributions from fragmentation of the ketene, vinoxy, and acetyl products, together with a fast peak which can only correspond to methylene formation from channel (e) leading to $CH_2 + HCHO$ (formaldehyde) products. A minor contribution to this fast peak comes also from fragmentation of the CH_3 radical product from channel 6.4(d), but this is very small, since the appearance energy of CH_2^+ from CH_3 dissociative ionization is 15.1 eV and CH_2^+ ions from this process are produced very near threshold at an energy of 17 eV. It is worth noting that the detection of methylene from reaction channel 6.4(e) was only possible because of the use of soft EI. In fact, by using an electron energy of 17 eV it was possible to remove completely the elastic contribution arising from C_2H_4 scattered from the various component species in the O-atom beam. At 60 eV, this signal is about two orders of magnitude larger than the reactive signal, but the electron energy of 17 eV lies below the 18 eV appearance potential of CH_2^+ from dissociative ionization of C_2H_4.

From the measured LAB-frame angular and velocity distributions, product angular and translational energy distributions in the CM system were derived, and the branching ratios between the various competing channels were estimated.[473] It was found that formation of $CH_3 + HCO$ is the major channel (43%), followed by $CH_2CHO + H$ (27%). It was also firmly established that the formation of the molecular products $CH_2CO + H_2$ represents a sizeable channel, accounting for about 13% of the yield, in contrast with the conclusions of kinetic studies. For the first time, it was also shown that a small fraction (1%) of acetyl radicals is formed, and finally, formation of methylene + formaldehyde was observed at the level of 16%, corroborating kinetic investigations. An insight into the overall picture of the dynamics for the various reaction channels can be gained from the 3D contour map representations of the various product fluxes, some of which are shown in Figure 6.11. Most channels in this reaction involve the formation of intermediates with lifetimes comparable to their rotational periods, and many of the observed angular distributions are consistent with the osculating complex model discussed above.

The observation of channels 6.4(b), (c) and (d), which account for about two thirds of the overall reaction yield, can only be rationalized assuming that ISC between triplet and singlet PESs is occurring very efficiently. This is supported by theoretical work using various QM methods and statistical rate theory* on the PESs for both the triplet and singlet electronic states of the $O + C_2H_4$ reaction.[477] Calculated product yields[477] were found to be in excellent agreement with the branching ratio results derived from the CMB experiments.[473] Dynamical QCT calculations with the inclusion of non-adiabatic couplings between the triplet and singlet PESs of C_2H_4O have also recently been carried out; the results show reasonable agreement with the experimental results and contribute to our understanding of the detailed, complex dynamics of this important reaction.[478]

The $O + C_2H_4$ reaction has recently been investigated at a lower collision energy of $E_c = 26.8$ kJ mol^{-1}, also in a pulsed crossed beam experiment, but this time using soft-ionization detection by VUV synchrotron radiation.[467] All carbon-containing products from four exit channels were identified: $CH_2CHO + H$, $CH_2CO + H_2$, $HCO + CH_3$, $CH_2 + HCHO$. A more recent CMB study at $E_c = 12.5$ kJ mol^{-1} employed two different oxygen beams having the same velocity, but containing a factor of twenty different relative concentrations of $O(^1D)$ and $O(^3P)$, and was able to disentangle very clearly the relative reactivity of $O(^3P)$ and $O(^1D)$ with C_2H_4.[468] Although the work has not provided branching ratios, it has revealed the distributions of kinetic energy release of the products, and has also demonstrated the merits of selective photoionization in CMB reactive scattering experiments. Undoubtedly, soft ionization detection by VUV photons or low-energy electrons has opened up new and exciting possibilities in chemical reaction dynamics, enhancing our ability to

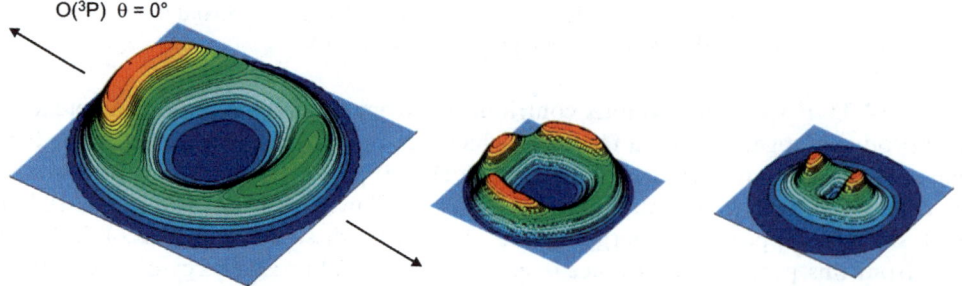

Figure 6.11 Contour maps of products from the $O(^3P) + C_2H_4$ reaction. Left: CH_2CO (ketene), centre: CH_2CHO (vinoxy), right: CH_3CO (acetyl).

*See Section 7.5 for a brief discussion of statistical rate theories.

probe the mechanisms of polyatomic multichannel reactions. Such reactions are of importance in real-world gas-phase chemistry, from combustion to astrochemistry.

6.2.3.2 Example: The Reaction $C(^3P) + C_2H_2$

Often it is of great interest to determine the dynamics and the product branching ratios as a function of E_c over a wide range of collision energies since this allows a much greater region of the PES to be explored than measurements carried out at a single collision energy. Using a CMB apparatus with mass spectrometric detection and featuring molecular beam crossing arrangements with $\Theta_{12} = 45°$, 90° and 135° (as illustrated in Figure 6.2), it is possible to study polyatomic reactions over a wide range of collision energies. For instance, the $C(^3P) + C_2H_2$ reaction, which is of great interest in astrochemistry, exhibits three competing exoergic product channels, two leading to H elimination with formation of linear and cyclic C_3H (labelled *l* and *c* in the scheme below), and the third leading to H_2 elimination with formation of C_3:

$$C(^3P) + C_2H_2\left(\tilde{X}^1\Sigma_g^+\right) \rightarrow l{-}C_3H\left(\tilde{X}^2\Pi_{1/2}\right) + H(^2S_{1/2}) \quad \Delta H_0^{\ominus} = -1.67\,\text{kJ mol}^{-1} \tag{6.5a}$$

$$\rightarrow c-C_3H\left(\tilde{X}^2B_2\right) + H(^2S_{1/2}) \quad \Delta H_0^{\ominus} = -8.8\,\text{kJ mol}^{-1} \tag{6.5b}$$

$$\rightarrow C_3\left(\tilde{X}^1\Sigma_g^+\right) + H_2\left(\tilde{X}^1\Sigma_g^+\right) \quad \Delta H_0^{\ominus} = -105.9\,\text{kJ mol}^{-1} \tag{6.5c}$$

Channel 6.5(c) is spin-forbidden and can only take place *via* ISC from the triplet to the singlet PESs.

Detailed CMB experiments were able to detect the three competing channels, determine their DCSs, and derive the ratio $\sigma_{cyclic}/\sigma_{linear}$ of cross sections for formation of cyclic and linear C_3H, as well as the branching ratio between the H and H_2 elimination pathways. The measurements were made over a very wide range of collision energies, from 3.6 up to about 50 kJ mol^{-1}, as shown in Figure 6.12.[479,480] From the symmetry of the CM angular distributions it was concluded that both cyclic and linear C_3H and C_3 formation from the triplet and the singlet (reached *via* ISC) C_3H_2 intermediates proceed *via* a long-lived complex mechanism,[480] as corroborated by theoretical calculations of the triplet and singlet intermediate lifetimes, which are found to be very long (on the order of many rotational periods, even at high E_c).[481] As a result of the long lifetimes of the triplet intermediate (propargylene), the system has time enough to undergo ISC from triplet to singlet propargylene, which then can readily lead to $C_3 + H_2$ products directly or *via* isomerization to H_2CCC. The reaction mechanism sees the initial addition of the carbon atom to the triple bond of acetylene, with the integral cross section decreasing with increasing E_c,[479] as expected for a barrierless reaction dominated by long range capture forces, as discussed in Chapter 12. The branching ratio $\sigma_{C_3}/(\sigma_{C_3} + \sigma_{C_3H})$ was found to increase with decreasing E_c, as shown in the right panel of Figure 6.12 (similarly to the $\sigma_{cyclic}/\sigma_{linear}$ ratio shown in the left panel of Figure 6.12), as one may reasonably expect from the trend with E_c of the lifetimes of the intermediate C_3H_2 complex(es); in fact, the complex lifetimes increase with decreasing E_c, as does the probability of ISC (see ref. [480]). The decreasing trend of the $\sigma_{C_3}/(\sigma_{C_3} + \sigma_{C_3H})$ ratio with increasing E_c is also confirmed by CMB experiments with pulsed beams, as shown in right panel of Figure 6.12.[482]

In conclusion, CMB experiments with mass-spectrometric detection are able to provide a global picture of the dynamics of both simple (three- and four-atom) and complex (polyatomic) reactions. Although specific examples have been presented above, the approaches described can be readily extended to other polyatomic reactions. However, to probe more deeply into the dynamics of a given reaction, or a given reaction channel, one has to employ state-specific detection of the

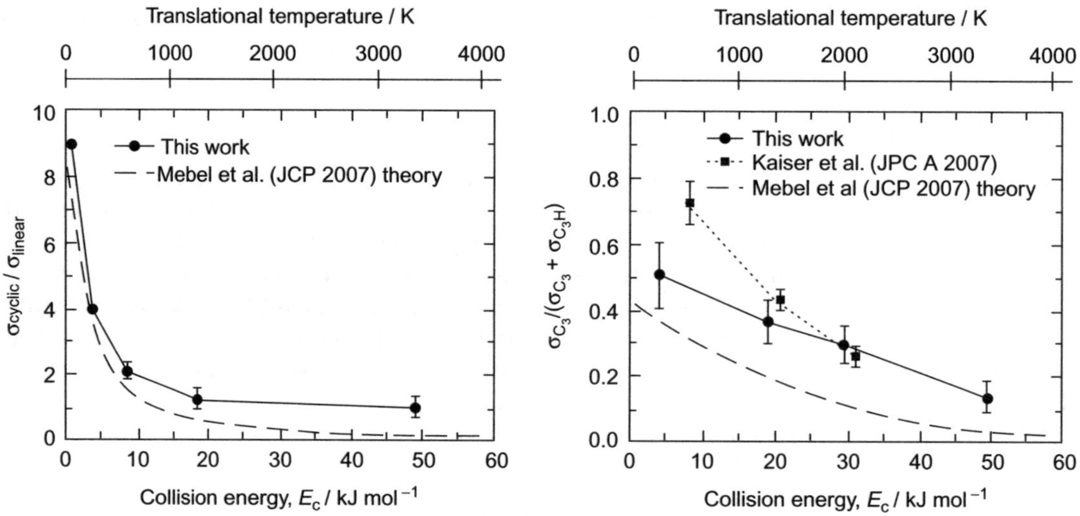

Figure 6.12 Left: Ratio of cross sections for *cyclic-* and *linear*-C_3H formation, $\sigma_{cyclic}/\sigma_{linear}$, in the $C(^3P) + C_2H_2$ reaction as a function of E_c, as derived from CMB studies.[479,480] Right: Branching ratio $\sigma_{C_3}/(\sigma_{C_3} + \sigma_{C_3H})$ as a function of E_c, also derived from the CMB studies.[479,480] In both figures the corresponding translational temperature scale is indicated on the top abscissa. The solid line joining the data points (solid circles) is drawn to guide the eye only. The theoretical data is taken from ref. [481], and the experimental data of Kaiser and coworkers is taken from ref. [482].

products in order to obtain state-resolved DCS. This usually requires the use of laser spectroscopic techniques, which are generally confined to a limited number of relatively simple species. Some of these sophisticated and powerful techniques will now be discussed, and the exquisitely detailed information that can be gained will be emphasized.

6.3 CMB EXPERIMENTS WITH RYDBERG-TAGGING DETECTION

As discussed above, electron impact ionization has proven a very powerful detection technique in studies of reaction dynamics and especially of multichannel reaction dynamics, largely because of its universality. A novel experimental technique, known as H-atom Rydberg tagging, designed to detect specifically H-atom products, was developed in the early 1990s. The technique is described in Study Box 6.2. H-atom Rydberg tagging time-of-flight detection was successfully employed in studies of the dynamics of the important benchmark reaction $H + D_2 \rightarrow HD + H$,[435,436] providing extremely high translational energy resolution and near unit detection sensitivity for the H-atom products. High sensitivity is a crucial factor for the success of any detection method to be coupled to the CMB technique. Figure 6.13 shows a schematic of the setup for this elegant experiment. This method has also been extensively used in studies of the photodissociation dynamics of a variety of molecules, as illustrated by the very high resolution studies of the photodissociation dynamics of H_2O.[483,484]

More recently, the Rydberg tagging technique has been applied, most notably by the group of Yang and coworkers, to the study of the dynamics of a number of important elementary chemical reactions, such as $O(^1D) + H_2$,[485] $H + D_2(HD)$,[486,487] $D + H_2$,[488] and $F + H_2$,[267,489,490] with rotational state resolution. The same technique was also applied to the study of the dynamics of the benchmark four-atom reaction $OH + D_2$ with vibrational state resolution.[491] These high-resolution

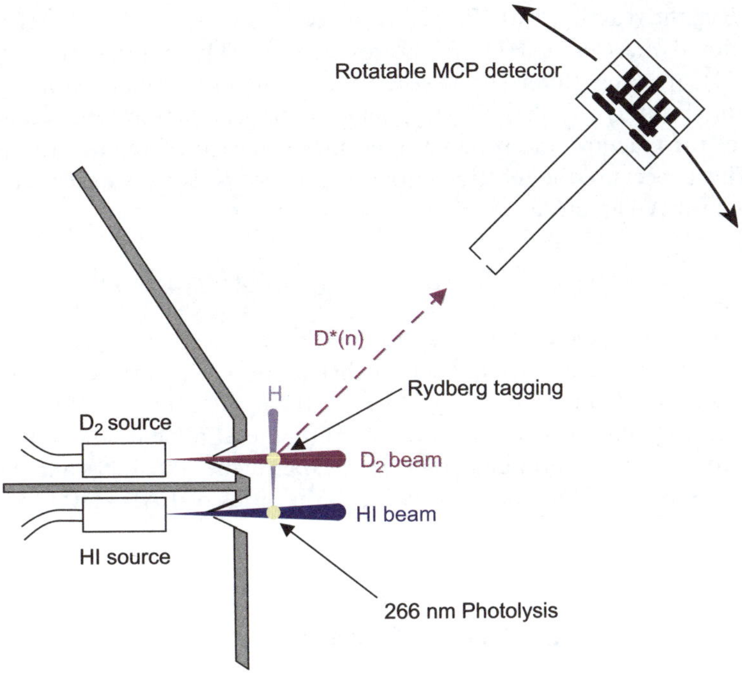

Figure 6.13 The experimental setup for the crossed molecular beams scattering study of the $H + D_2 \rightarrow H_2 + D$ reaction at the collision energy of 1.200 eV (detecting the D atom).

studies allow the probing of the dynamics of these simple reactions at the most fundamental level. In this section, we will give an overview of the state-to-state dynamics of these simple chemical reactions.

WORKED PROBLEM (2)

Q: *Using the crossed molecular beam reaction $H + D_2(j) \rightarrow HD(v', j') + D$ as an example, explain how it is possible to use the H-atom Rydberg tagging technique to determine HD quantum state populations and HD quantum state resolved DCSs.*

A: Because the first excited state of the H-atom (or D-atom) is 10.2 eV above the ground state, which is out of the energetically accessible range for most chemical activations, the H-atom (or D-atom) product formed in chemical reactions is mostly in its electronic ground state. From the translational spectroscopy of the H-atom product as a function of LAB scattering angle, and given sufficient time and angular resolution, one can determine the quantum state distribution of its partner product, as well as the quantum state resolved DCSs.

Since total energy and linear momentum in the reaction process must be conserved, we have

$$E_{\rm int}({\rm H}) + E_{\rm int}({\rm D}_2(j)) + E_{\rm c} + \Delta H = E_{\rm int}({\rm HD}(v',j')) + E_{\rm int}({\rm D}) + E_{\rm t}({\rm HD}) + E_{\rm t}({\rm D})$$

and

$$m_{\rm HD}E_{\rm t}({\rm HD}) = m_{\rm D}E_{\rm t}({\rm D}).$$

In the above, ΔH is the reaction enthalpy, E_c is the collision energy, and $E_t(HD)$ and $E_t(D)$ are the CM translational energies of HD and D, respectively. The internal energies of H and D, $E_{int}(H)$ and $E_{int}(D)$, are equal to zero because they are in their ground electronic states. If D_2 is in the lowest rotational state, $j=0$, then $E_{int}(D_2)$ may also be set equal to zero. (Remember that the internal energy of the reactant and product molecules is measured relative to the energy of the $v/v'=0$ levels. This is because the reaction enthalpy (at zero Kelvin) is measured with respect to these zero point levels.) Therefore,

$$E_{int}(HD(v',j')) = \Delta H + E_c - (1 + m_D/m_{HD})E_t(D).$$

Since the total available energy of the reaction, $\Delta H + E_c$, is known precisely, the measured laboratory D-atom product translational energy distribution ($f(E_t(D))$) can be converted easily into the CM HD product internal energy ($f(E_{int}(HD))$) distribution. With sufficiently high translational energy resolution for the D-atom TOF measurement, the HD quantum state resolved DCS can thus be determined. In a real experiment, the ICSs and DCSs would be determined from a series of TOF spectra obtained at a series of LAB scattering angle.

6.3.1 Quantum Phenomena and Reaction Resonances

Experimental and theoretical studies of simple chemical reactions, especially three or four atom reactions, have played a key role in the fundamental understanding of chemical reaction dynamics. Here we discuss a few simple reaction systems that are prototypes for various different reaction mechanisms, encompassing direct abstraction reactions, insertion reactions and resonance-mediated reactions. All of these processes were studied using the H-atom Rydberg tagging technique. As we will see, because of their simplicity, and also their accessibility to exact quantum treatments, these reactive systems have become important testing grounds for developing an understanding of a number of fundamental concepts in chemistry, such as the role of reactive scattering resonances. Detailed comparison of the results obtained from high-resolution scattering experiments with those predicted by accurate quantum dynamical scattering calculations has contributed greatly to deepening our understanding of chemical reactivity.

6.3.1.1 *Example: $H + H_2 \rightarrow H_2 + H$ – a not so Simple Reaction*

For decades, the $H + H_2$ reaction, sometimes referred to as 'the simplest chemical reaction', has been at the centre of experimental and theoretical dynamics studies. It has been studied using nearly every possible experimental method. The latest milestone in the experimental study of this reaction was reached in 1995 by Welge and coworkers[435,436] using the H-atom Rydberg tagging TOF technique. In this landmark experiment on the $H + D_2$ isotopic variant of the reaction, the H-atom beam was produced by photodissociation of a molecular beam of HI at either 266 nm, 248 nm or 193 nm. As discussed further in Chapter 8, photolysis at the three wavelengths yields H atoms with three different, well defined velocities, providing a means for varying the collision energy. Figure 6.16 shows the product kinetic energy distribution of the D product from the $H + D_2$ reaction at four different collision energies.[496] The remarkable resolution achieved in this study is manifested clearly in the figure, in which the HD products in individual ro-vibrational states are nearly fully resolved.

More recently, the $H + HD \rightarrow H_2 + D$ reaction has also been studied at $E_c = 0.498$ eV[497] and 1.200 eV,[498] with quantitative agreement between experiment and theory being achieved. An interesting oscillation in the product rotational state distribution was observed. This was attributed to the

STUDY BOX 6.2: THE H-ATOM RYDBERG TAGGING TECHNIQUE

The H atom Rydberg tagging technique[435,436] was born of the REMPI method (see Study Box 5.2). The key element of the technique is the efficient $(1+1')$ excitation of the H-atom in the ground state to a high Rydberg state ($n \sim 50$), rather than directly to the ionization continuum as would be the case in REMPI (see Figure 6.14). The first step in the $(1+1')$ scheme is made from the ground $n=1$ state to the $n=2$ intermediate state *via* excitation on the Lyman-α transition. Since high-n Rydberg states of the H-atom are long lived, with lifetimes on the order of milliseconds, it is possible to use this technique to measure the TOF spectra of the H-atom product, which normally has flight times in the region of about 100 μs. Pulsed VUV light at the Lyman-α wavelength is generated using a non-linear optical technique known as difference four-wave mixing, in which an intense beam of nanosecond UV laser radiation is focussed into Kr gas[492] (see Study Boxes 5.2 and 11.2). Because of the efficient production of VUV light the detection efficiency of this method can be almost unity.

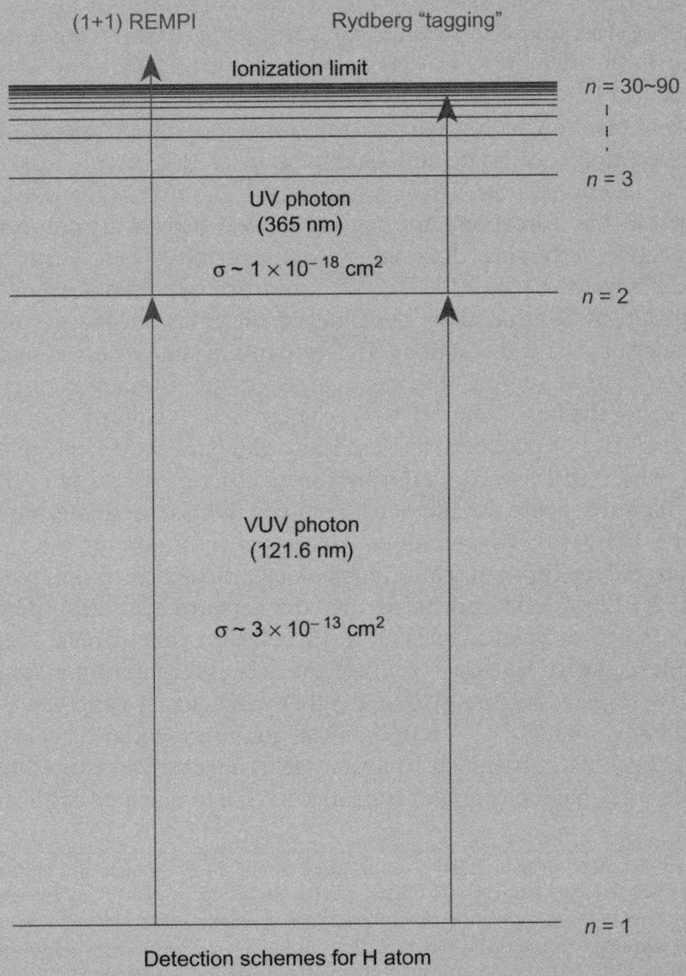

Figure 6.14 Detection scheme for H-atom $(1+1)$ REMPI (left arrows) and for H-atom $(1+1')$ Rydberg tagging technique (right arrows).

The resolution of the technique is also quite remarkable: as high as 0.06% in velocity has been achieved. The origin of this high resolution, compared with that achieved in conventional REMPI-TOF based machines, lies in the fact that the excited Rydberg atoms are neutral species, and therefore do not suffer Coulomb repulsion (see below). As the excited Rydberg atoms drift down the TOF tube, they move apart, so that when they are field ionized at the detector, they experience minimal Coulomb repulsion. In a conventional REMPI experiment, ions are directly generated in the interaction region, and the presence of high number densities of charged species leads to Coulomb repulsion, a perturbation to the ion velocities, and a subsequent degrading of the resolution of the REMPI-TOF instrument. Clearly, however, to fully exploit the high-resolution TOF capabilities of the Rydberg tagging method, reactant beams with narrow velocity distributions are required.[435,436] When this can be achieved, the high sensitivity of the technique make it ideally suitable for high resolution reactive scattering studies of elementary chemical reactions.

Piergiorgio Casavecchia, Kopin Liu and Xueming Yang

nuclear spin statistics* of H_2, which results in even *j* rotational states of H_2 (*i.e. para*-H_2) having a spin statistical weight of one, and odd *j* rotational states (*ortho*-H_2) having a weight of three. This observation is similar to the case of the H + HI reaction studied by Zare and coworkers,[499] in which nuclear spin statistics also have a large effect on the H_2 rotational state distribution.

In a simple chemical reaction with a barrier, the transition state is located close to the energy maximum along the reaction coordinate and usually no discrete structure exists along the reaction coordinate. For some reactions, however, a quasi-bound quantum state *along the reaction coordinate* can be trapped in the transition state region. Such a transiently trapped quantum state is normally called a reaction resonance. Reaction resonances have been a central topic in reaction dynamics for the last several decades,[431,445,500,501] and are explored further in Study Box 6.4. Quantum state-resolved DCSs, and their dependence on collision energy, often provide a key signature of a resonance in chemical reactions. Fully quantum state resolved scattering experiments are therefore a powerful tool for probing resonances in this fundamental system. At collision energies not too far above the barrier, the $H + H_2$ reaction and its isotopic variants display rebound dynamics, with the majority of products scattered into the backward hemisphere. However, Zare and coworkers[500,502] observed forward scattering for a subset of products in the $H + D_2$ reaction, and attributed the forward peaks in the state resolved DCSs to scattering resonances. These experiments used the PHOTOLOC technique, discussed in Study Box 6.3. In a later report, Althorpe *et al.*[221] analyzed the forward scattering peak using exact quantum dynamical calculations based on the BKMP2 PES,[244] and concluded that the forward scattering peaks observed in the DCS for the $H + D_2$ reaction were associated with collision time-delays in the transition state region. Almost simultaneously, Harich *et al.* performed a full quantum state-resolved scattering study of the $H + HD \rightarrow H_2 + D$ reaction at $E_c = 1.200$ eV, and clearly observed a forward scattering peak associated with $H_2(v' = 1)$.[486,497,498] Based on full quantum scattering calculations and further analysis, the forward peak was attributed to a time-delay mechanism caused by slowing-down of the reaction complex as it passes over the top of a barrier associated with a specific 'quantum

*Nuclear spin statistics are described in many textbooks on physical chemistry or quantum mechanics (see, for example, ref. [70]). The possible nuclear spin states of two spin $\frac{1}{2}$ nuclei, as found in H_2, are $\alpha_1\alpha_2$, $\frac{1}{\sqrt{2}}(\alpha_1\beta_2 + \beta_1\alpha_2)$, $\beta_1\beta_2$, and $\frac{1}{\sqrt{2}}(\alpha_1\beta_2 - \beta_1\alpha_2)$. The first three of these are symmetric with respect to exchange of the nuclei, the fourth anti-symmetric. The Pauli Exclusion Principle requires the total wavefunction, Ψ_{tot}, to be antisymmetric with respect to exchange of identical fermions and this leads to a pairing of the symmetric nuclear spin states with odd-*j* (antisymmetric) rotational states, and the antisymmetric nuclear spin states pairing with the *j*-even (symmetric) rotational states. The resulting 3:1 weighting of the odd:even rotational populations is often seen as intensity alternations in rotational spectra, and in this case, in the reaction product rotational state distributions.

STUDY BOX 6.3: PHOTOLOC

The PHOTOLOC technique, described by Brouard *et al.*[385,493] and Zare and coworkers,[386,387] is complementary to the crossed molecular beam method discussed in this chapter, and is based on the laser pump-probe method (see Section 1.3.4). Instead of a crossed molecular beam arrangement, an aligned 'beam' of reactants is generated using polarized laser photolysis of a suitable precursor molecule. In a crossed beam experiment, one of the molecular beams is usually chosen as the reference direction with respect to which the product angular scattering distribution is measured. In photon-initiated reactions, the reference direction is usually chosen as the polarization vector $\boldsymbol{\varepsilon}$ of the pump laser beam. The angular distribution of the photofragments is well defined relative to this direction due to the fact that absorption of a photon requires overlap between $\boldsymbol{\varepsilon}$ and the transition dipole $\boldsymbol{\mu}$ for the electronic transition involved in the dissociation (see Study Boxes 8.2 and 8.3, and Section 8.6). Because the electric vector of the pump laser provides a reference direction, molecular beams are not required in order to measure the angular distribution for a pump-probe experiment; it is perfectly possible to make such measurements using a bulk gas sample, though the LAB to centre-of-mass frame transformation is somewhat less straightforward than in the crossed beam case.

The acronym *PHOTOLOC* denotes a photoinitiated reaction analyzed by the law of cosines,[389] and, as the name suggests, the basis of the method is Eq. (5.13), which we rewrite here in the context of a generic A + BC reaction as

$$v'^2_{AB} = u'^2_{AB} + v^2_{CM} + 2u'_{AB}v_{CM}\cos\theta. \tag{B6.3.1}$$

The relevant vectors are illustrated in Figure 6.15. Species A is assumed to be the photolytically generated reactant, and u'_{AB} and v'_{AB} are the product velocities in the CM and LAB-frames, respectively. v_{cm} is the velocity of the centre-of-mass. The other reactant, BC, is often referred to as the target reactant, and is usually either held at room temperature in a low pressure gas mixture along with the photolytic precursor, or is co-expanded along with the latter in a molecular beam. In both cases, it is often reasonable to assume that the target molecule is stationary relative to the photolytically generated species, A, *i.e.* $v_{BC} \sim 0$, and therefore the angle θ in Eq. (B6.3.1) is the CM scattering angle (see Figure 6.15).

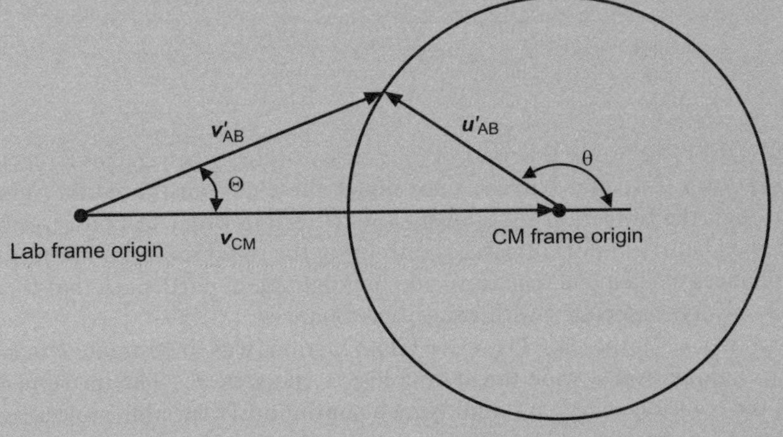

Figure 6.15 Collapsed Newton diagram illustrating the relevant vectors in the PHOTOLOC experiment.[385,386,493]

The principle of the experiment is to determine the CM angular distribution for the reaction of interest through a measurement of the LAB-frame velocity. In particular, if the AB product is probed quantum state selectively, and species C is structureless, such that the probed AB product possesses a single CM speed, u'_{AB}, then the CM angular distribution can be obtained directly from the following equation

$$P(\cos\theta) = P(v'_{AB})\frac{u'_{AB}v_{CM}}{v'_{AB}}. \tag{B6.3.2}$$

The term $\frac{u'_{AB}v_{CM}}{v'_{AB}}$ is a Jacobian that transforms from LAB velocity to CM scattering angle (see Study Box 1.1). In the more general case in which species C is not structureless, or AB is not probed quantum state selectively, then more elaborate procedures are required to extract the CM angular distribution (see, for example, ref. [494], and references therein).

If polarized probe light is used, the experiment also becomes sensitive to polarization of electronic or rotational angular momentum in the reaction products. This is a result of the fact that the transition dipole for the probe transition in correlated with the total angular momentum vector $\boldsymbol{j}$. If $\boldsymbol{j}$ is pointing in a particular direction, then the same is true of the transition dipole. In this case, changing the probe laser polarization will alter the overlap between the transition dipole and the electric vector of the light, altering the detection probability. The dependence of the signal on probe laser polarization may be analyzed in detail in order to determine the angular distribution of $\boldsymbol{j}$.[389,495]

Although the *PHOTOLOC* technique does not generally offer the same angular and energy resolution as the crossed beam method, and the extraction of the full angular distribution can become compromised in certain cases in which the partner to the probed product has internal structure,[494] it can be a useful alternative to conventional methods for certain reactants that are difficult to generate in molecular beams, or for particular systems that might be hard to study in the molecular beam environment due to, for example, kinematic reasons. It has also been widely exploited, particularly by the group of Zare and coworkers (see, for example ref. [387]), to study reactions involving vibrationally excited reagents, which is made possible by virtue of the high reactant number densities achievable by photolytic generation of one of the reactants in a molecular beam co-expansion with the other reactant. Details of some of these types of experiments are presented in Chapter 7.

Claire Vallance and Mark Brouard

bottleneck state' (QBS) (see Study Box 6.4). A time-delay occurs when the total reaction energy is in resonance with the top of such a barrier, since under these circumstances the velocity associated with motion through the transition state must slow to zero in order to conserve the total energy. These studies helped clarify important issues concerning the observable manifestations of quantum mechanical resonances in chemical reactions, and highlighted, in particular, that forward scattering is not necessarily always associated with reactive resonances.

In very recent work,[487] the $H+D_2\rightarrow D+HD$ reaction was investigated using the D-atom Rydberg TOF technique over a wide range of collision energies, E_c. The instrument used allowed the variation of the collision energy by employing a continuously variable molecular beam crossing angle. The collision energy dependent DCSs in the backward direction for the HD($v=0$, $j'=2$) product were measured. As shown in Figure 6.18, oscillations were observed in the energy dependence of the DCSs measured at specific LAB angles, which were apparent over the whole

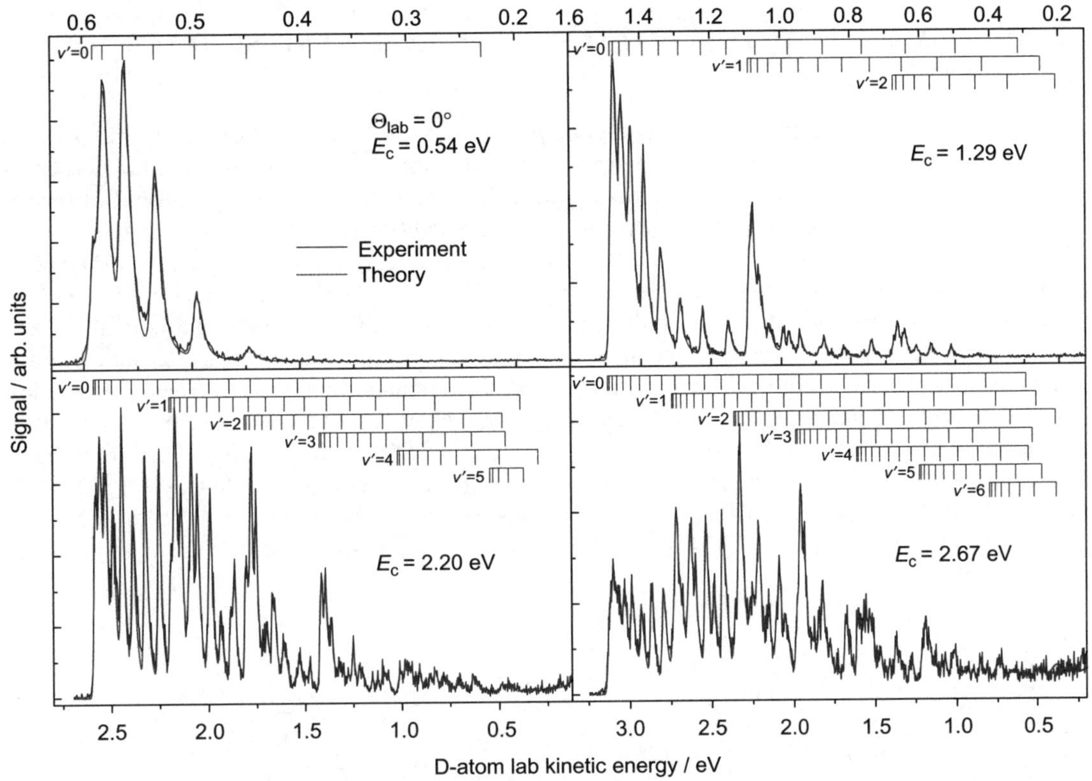

Figure 6.16 The time-of-flight spectra of the D atom product from the $H + D_2 \rightarrow HD + D$ reaction (adapted from ref. [496]). The data were collected at a LAB scattering angle of $\Theta = 0°$, and are shown at four collision energies. The increasing complexity of the TOF spectra with increasing collision energy reflects the population of a wider range of HD coproduct quantum states as the energy is raised.

energy range investigated. Theoretical analysis shows that the oscillations are not resonance peaks associated with QBSs. They are, however, intimately connected to the QBSs. It appears that the energy-dependent oscillations observed in the experiment are the result of the interference of pathways through the network of QBSs. This study demonstrated that quantized bottleneck states can be observed in a bimolecular reaction through high-resolution scattering experiments.

6.3.1.2 *Example: The $O(^1D) + H_2$ Reaction – from Insertion to Abstraction*

A variety of experimental studies have shown that the $O(^1D) + H_2$ reaction is dominated by an insertion reaction mechanism, whereby the diradical $O(^1D)$ inserts into the H–H bond, leading to a highly internally excited H_2O molecule. The reaction is discussed elsewhere in this book in Section 4.2.5 in the context of non-adiabatic effects. Studies of the $O(^1D) + para\text{-}H_2 \rightarrow OH + H$ reaction at $E_c = 5.4\ kJ\ mol^{-1}$ using the H-atom Rydberg TOF technique[437] have provided a detailed picture of the dynamics of this prototypical reaction through measurements of rotationally state resolved integral and differential cross sections. Figure 6.19 shows the H-atom product translational energy distribution at the laboratory angle of $-50°$. All the peaks in the distribution

STUDY BOX 6.4: QUANTUM BOTTLENECK STATES AND SCATTERING RESONANCES

In a typical chemical reaction with an energetic barrier, no discrete bound quantum states exists along the reaction coordinate. However, quantum structures do exist in coordinates perpendicular to the reaction coordinate. These quantum 'states' are normally called quantum bottleneck states (QBSs) because each of them behaves like a dynamical bottleneck in a chemical reaction.[503] The reason for this is that as the reactants come together energy does not flow instantaneously between the reaction coordinate and the internal modes orthogonal to this motion. The orthogonal modes can be thought of as behaving approximately vibrationally adiabatically. Each transition 'state' represents a dynamical barrier to reaction, and reactants are slowed as they approach these barriers. Only in the limit that energy flowed instantaneously between the reaction coordinate and the orthogonal modes would this slowing down not take place. For a single QBS, the energy pathway takes a similar form to the minimum energy pathway, but shifted to a higher energy. Figure 6.17(a) shows a typical energy pathway for a system in a single QBS. For a single QBS model,[504] the reaction probability simply increases as the reaction energy goes above the barrier, as shown in Figure 6.17(b). Even though the picture is quite simple, the dynamics can

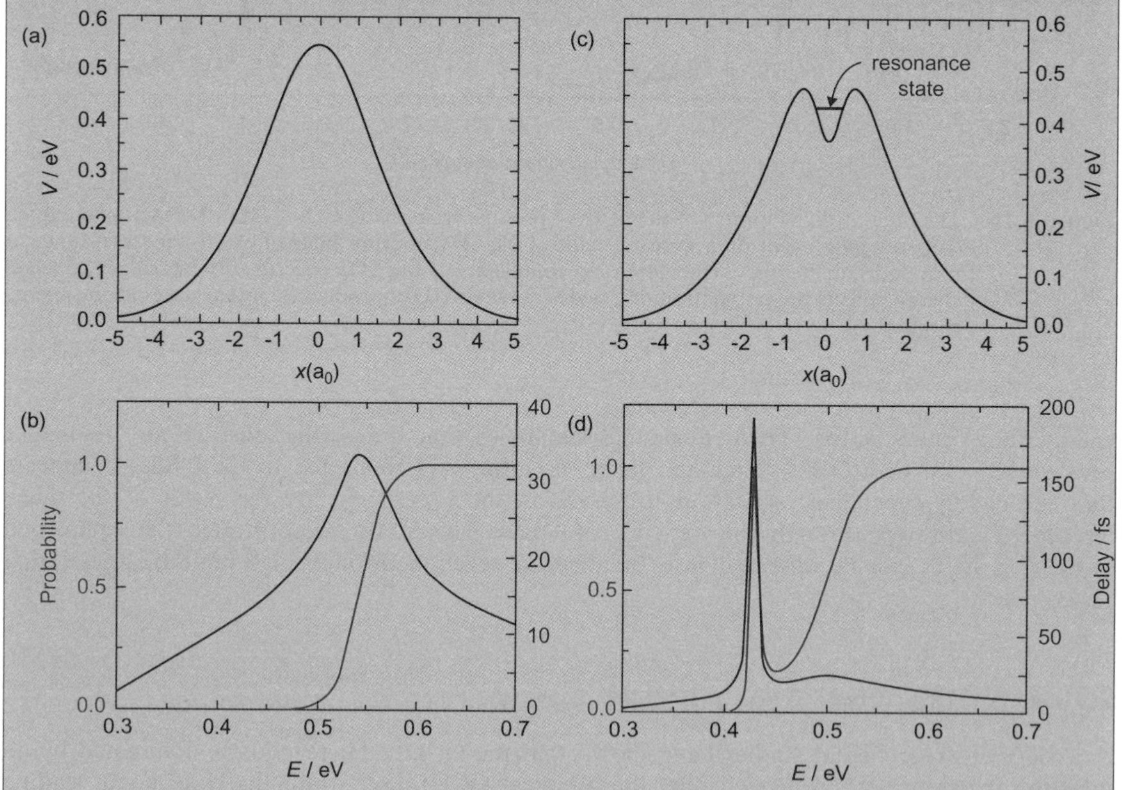

Figure 6.17 One-dimensional views of two model reactions: a reaction with a single QBS pathway and a typical reaction with a dynamical resonance. a) The potential energy along the reaction coordinate for a model reaction with a single QBS pathway; b) The calculated reaction probability and time delay for the model reaction in a); c) The potential energy along the reaction coordinate for a model reaction with a dynamical resonance; d) The calculated reaction probability and time delay for the model reaction in c).

become interesting when the reaction energy is in resonance with the barrier in the QBS (the top of the reaction path). The motion along the reaction coordinate slows as the barrier in the QBS is reached (the velocity at the barrier would reach zero if the reaction energy were resonant with QBS barrier, assuming no coupling or energy exchange with the internal degrees of freedom of the transition state), delaying the reaction by some tens of femtoseconds. It is this delayed mechanism that produces forward scattering products in the $H + D_2(HD)$ reaction. This type of mechanism is considered to be quite general in many chemical reactions.

In special cases, a quasibound quantum state can be transiently trapped in the transition state region, particularly in cases in which there is an energy minimum along the reaction coordinate, as illustrated in Figure 6.17(c). Such a quantum state is often called a dynamical resonance, or reactive scattering resonance. From a simple one-dimensional model of the phenomenon,[504] one can see clearly in Figure 6.17(d) that the resonance state enhances the reaction probability dramatically through resonance-mediated tunnelling. Notice that the tunnelling probability reaches close to unity at an energy close to that of the resonance state. The lifetimes of these transiently trapped states are typically on the order of a few hundred femtoseconds, as also indicated in Figure 6.17(d). At energies close to a resonance state, the dynamics of the reaction, as observed through the DCS, for example, can change dramatically, in part reflecting the different time delays (or transition state lifetimes) in the region of a resonance. Such behaviour has been observed in a number of systems, including the $F + H_2$ reaction, which is discussed later in this chapter. In the limit of a long lived resonance, the DCS at the resonance energy can even show the symmetric forward-backward peaks expected for a long-lived complex.[505]

Piergiorgio Casavecchia, Kopin Liu and Xueming Yang

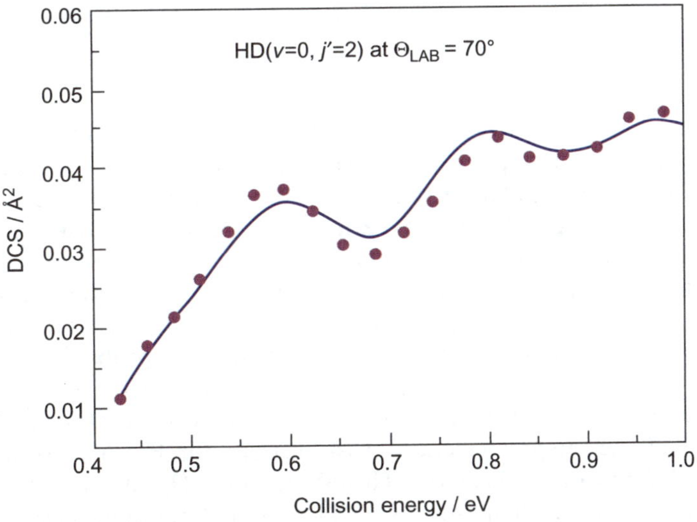

Figure 6.18 The experimental differential cross section (dots) for $H + D_2(v = 0, j = 0) \rightarrow D + HD(v' = 0, j' = 2)$, measured at the laboratory angle of 70°, *versus* the collision energy, E_c. The solid dots are the experimental data points, and the solid curve is the result of the quantum scattering calculation. (Adapted from ref. [487].)

can be assigned, by considering linear momentum and energy conservation, to the OH co-product born in different spin-orbit rovibrational levels. The experimental results show that the angular distributions for each vibrational state of the OH product (averaged over rotational state) are largely forward-backward symmetric. However, for individual rotational states, angular

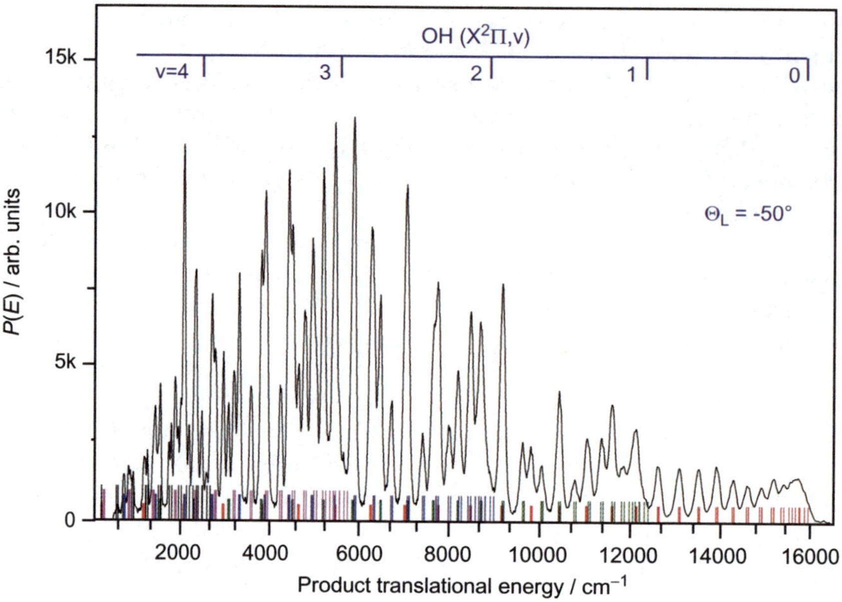

Figure 6.19 The product translational energy distributions, obtained by H-atom Rydberg tagging experiments, at the LAB angle of $-50°$ with respect to the O-atom beam from the $O(^1D) + H_2$ ($v=0$, $j=0$) $\rightarrow$ OH + H reaction. The comb at the top represents the position of the OH(X) vibrational states, while the coloured combs at the bottom show the position of individual OH(X) rovibrational states. The sharp structures reflect the populations in OH co-product rovibrational states (adapted from ref. [437]).

distributions are normally not forward-backward symmetric. In addition to the DCSs, OH product vibration-rotational state population distributions have also been determined, and reveal that the OH(v') vibrational population decreases monotonically as v' increases. Full quantum and QCT dynamics calculations have also been performed for this benchmark system.[506] The theoretical results are generally in good agreement with the experimental data, with the exception that the DCS predicted by the quantum calculations is always much more pronounced than the experimental DCS in a narrow angular range in the forward and backward scattering directions, which points to subtle quantum effects, which are probably not resolvable in the experiments.

The effect of H_2 rotational excitation on the reaction with $O(^1D)$ has also been investigated.[507] The ratio of the integral reaction cross sections for H_2 in $j=0$ and $j=1$ was determined to be $\sigma(j=1)/\sigma(j=0)=0.95\pm0.02$. This shows that H_2 in its rotational ground state, $j=0$, is slightly more reactive than that in its first excited rotation state, $j=1$, in agreement with the quantum dynamics calculations. The effect of H_2 rotational excitation on the OH product state-resolved DCSs has also been investigated. From the translational energy distributions obtained at different scattering angles, the distributions in the backward and forward directions show some noticeable differences for the $H_2(j=0)$ and $H_2(j=1)$ reactions.[508] However, there are no obvious trends in the angular distribution with the OH product state, an observation that is consistent with the relatively random nature of the fragmentation of the intermediate complex.

The Rydberg tagging experiments showed that at a low collision energy of 3.4 kJ mol^{-1}, the $O(^1D) + H_2$ reaction occurs exclusively through an insertion mechanism. A series of studies at higher collision energies using the Doppler-selected TOF technique (see Section 6.4), revealed that while the insertion mechanism dominated below $E_c < 7.5\,\text{kJ mol}^{-1}$, at higher collision energies ($E_c > 7.5\,\text{kJ mol}^{-1}$) an elusive abstraction mechanism channel becomes important[509] (see Section 4.2.5). Recently, Liu *et al.* carried out an experimental study on the $O(^1D) + D_2$ reaction in an effort to clarify this observation further.[510] The experiment was carried out at the collision energies of $8.4\,\text{kJ mol}^{-1}$ and $13.4\,\text{kJ mol}^{-1}$. At the lower collision energy, the experimental results show that $O(^1D) + D_2$ is an insertion type reaction with a forward-backward symmetric product angular distribution. At $E_c = 13.4\,\text{kJ mol}^{-1}$, however, the OD reaction products are clearly more backward scattered with respect to the $O(^1D)$ beam direction than at the lower collision energy. Further analysis of the data shows that the extra backward scattered OD products at $13.4\,\text{kJ mol}^{-1}$ are mainly the rotationally cold OD products in $v' = 4, 5, 6$, which are the product quantum states typically formed in an alternative colinear abstraction mechanism that occurs *via* reaction on an excited electronic state, as suggested by theory[246–248] (see Figure 6.20).

The role of the abstraction pathway in the $O(^1D) + H_2$ has been investigated further using the laser pump-probe technique[249,511] (see Section 1.3.4 and Study Boxes 5.2 and 6.3). In these experiments, OH(X, $v' = 4$) rotational quantum state populations and rotational angular momentum alignment parameters (see Chapter 9) were measured by laser-induced fluorescence (see Study Box 5.2), and found to be consistent with exact QM scattering calculations[511] performed on highly accurate PESs[512,513] only when the abstraction reaction on the excited $1^1A''$ state was included, in addition to the insertion reaction on the ground state surface.

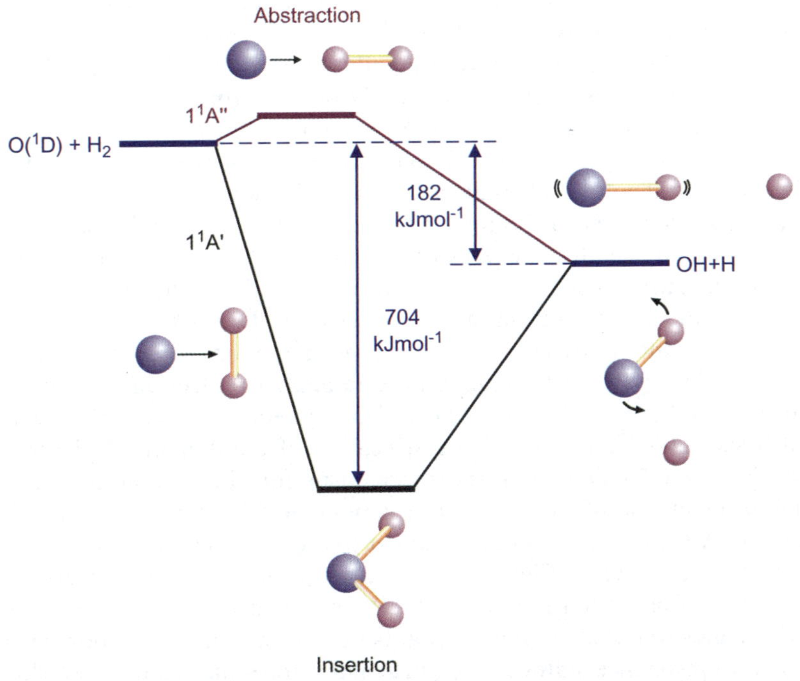

Figure 6.20 Reaction mechanism of the $O(^1D) + H_2 \rightarrow OH + H$ reaction. There are two reaction pathways: insertion and colinear abstraction, with the latter pathway only becoming accessible above the barrier to that reaction. (See refs. [246,248] for details of the theoretical calculations that form the basis for this figure.)

6.3.1.3 Example: The $F+H_2$ Reaction and Reaction Resonances

As we have seen in Section 6.3.1.1 and Study Box 6.3, the existence of reactive scattering resonances is an intriguing issue in the field of reaction dynamics. However, detecting such resonances and studying them has proved to be quite difficult. The most well known example of a reaction exhibiting a dynamical resonance, as opposed to the QBSs described above for $H+H_2$, is the $F+H_2 \rightarrow HF+H$ reaction, which also holds the honour of being one of the most extensively studied elementary reactions. In the 1980s, Lee and coworkers carried out a landmark experiment on the $F+H_2$ reaction using the classic CMB technique with mass-spectrometric detection and TOF analysis, in which vibrational resolution was achieved.[447,448] In this study, an intriguing forward scattering peak for the HF($v'=3$) product was detected and was originally attributed to a reaction resonance. However, both QCT calculations[514] and QM scattering studies[515] on the accurate Stark-Werner PES (SW-PES)[516] concluded that the HF($v'=3$) forward scattering is not in fact likely to be due to a resonance, but arises from quantum mechanical tunnelling through centrifugal barriers which arise for collisions occurring with high orbital angular momenta (*i.e.* at high impact parameters). Although a forward scattered peak was observed in the QCT calculation, its magnitude was much reduced compared with that seen in the QM calculations. The absence of tunnelling in classical mechanics means that the reaction cannot take place at such large orbital angular momentum as in the quantum calculation.

In a recent scattering study of the $F+HD \rightarrow HF+D$ reaction using the Doppler-selected ion TOF technique (see Section 6.4), a step-like increase in the energy dependence of the integral reaction cross section (the so-called excitation function) around 2.0 kJ mol^{-1} was observed,[517] and was again attributed to a reaction resonance. The high resolution and highly sensitive H-atom Rydberg tagging method[489,490,518] was recently employed to perform a fully quantum state-resolved crossed beam scattering study on the $F+H_2$ reaction. As an example of the data obtained, Figure 6.21 shows the TOF spectra of the H-atom product from the reaction at several laboratory scattering directions for a collision energy 2.18 kJ mol^{-1}. The rotational structure of the HF co-product is clearly resolved. From the TOF spectra, rotationally state-resolved DCSs can be derived. Figure 6.22(a) shows 3D contour plots of the state resolved DCS for $F+H_2(j=0)$ obtained in the experiments and from theory (discussed below). A pronounced forward scattering peak in the DCS is observed for the HF($v'=2$) product, which had not been observed previously for this reaction due to the experimental difficulties inherent in accessing this low collision energy range. The fact that the collision energy dependent DCS for HF($v'=2$) scattered in the forward direction shows a clearly identifiable peak at the collision energy of 2.18 kJ mol^{-1},[489,490] provides further evidence for a scattering resonance, since this structure is not expected for the tunnelling mechanism, discussed above, or for the QBS mediated reaction, discussed in Study Box 6.4.

The results of full quantum dynamics calculations based on a newly constructed accurate CCSD(T) PES[518] for the $F(^2P_{3/2})+H_2$ reaction are in almost perfect agreement with the experimental observations (see Figure 6.22(b)). The excellent agreement between theory and experiment provides a solid basis for further theoretical clarification of the dynamical picture of this fundamental reaction. Figure 6.23 illustrates the mechanism for the resonance mediated reaction. Focussing on the product side of the reaction, it is possible to construct a series of vibrationally adiabatic potential (VAP) curves, in a similar manner to the quantum bottleneck states described in Study Box 6.4. These are shown in Figure 6.23 for the various accessible vibrational states of the product HF molecule. Of particular note is the VAP associated with HF($v'=3$), which possesses a significant van der Waals well that supports a number of quasi-bound vibrational states, known as resonance states. Two resonance states are of particular importance as they are close in energy to the reactants, and these states may be assigned the quantum numbers (003) and (103), as shown in the figure, where the three quantum numbers label the van der Waals stretch, the bend, and the HF stretch of the quasi-bound linear complex. The one dimensional wave function for the ground resonance shown in the figure (*i.e.* that labelled (003)) confirms that this state is mainly trapped in

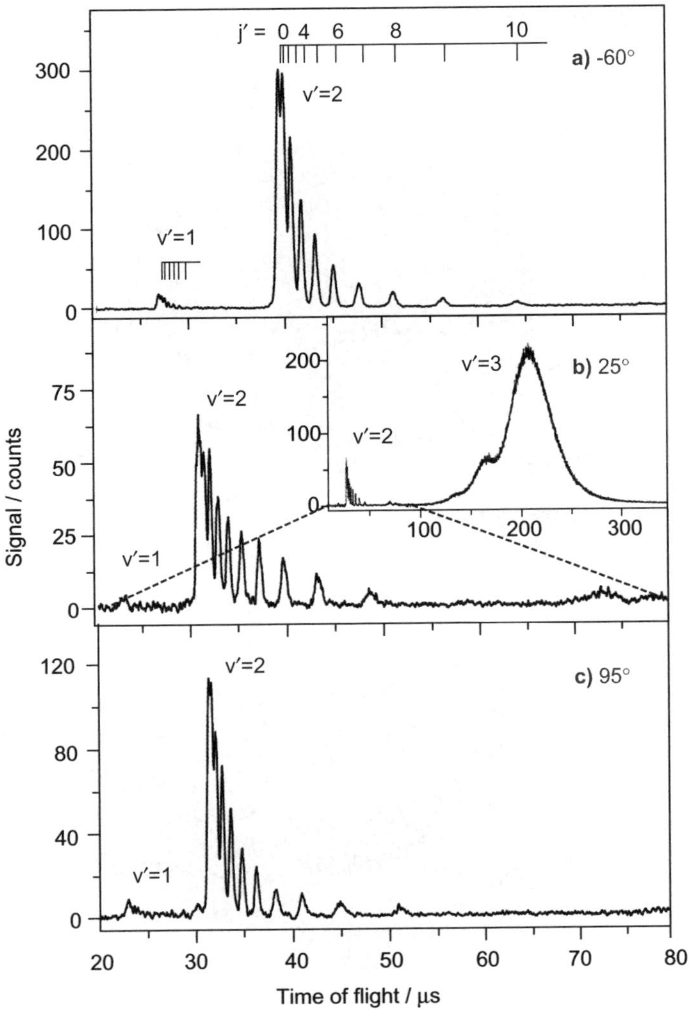

Figure 6.21 Time-of-flight spectra of the H-atom product from the $F + H_2(j = 0) \rightarrow HF + H$ reaction at a series of LAB scattering angles: a) $-60°$ (forward scattering direction); b) 25° (sideways scattering direction); c) 95° (backward scattering direction). (Adapted from refs. [489,490].)

the inner deeper well of the HF($v = 3$)–H VAP with a substantial van der Waals character, whilst the excited resonance wave function is mainly a van der Waals resonance. The resonance mediated reaction occurs first by resonant tunnelling of the reactants through to the resonance states (see Study Box 6.4). Reaction occurs by the coupling of the HF($v' = 3$)–H species to lower vibrational adiabats, particularly that associated with HF($v' = 2$)–H, *via* an intramolecular vibrational redistribution mechanism (see Section 7.4.3). This type of resonance, in which decay is to a different asymptotic channel to the one supporting the resonance, is known as a Feshbach resonance,* and the reaction mechanism, as a Feshbach resonance mediated reaction. The reaction resonance phenomenon has also been clearly observed in the polyatomic $F + CH_4$ reaction[519] (see Section 6.4).

*An alternative decay pathway for a resonance might be to tunnel through a barrier, possibly a centrifugal barrier, in which case dissociation occurs on the same channel as the resonance state. This type of resonance is known as a 'shape' resonance.

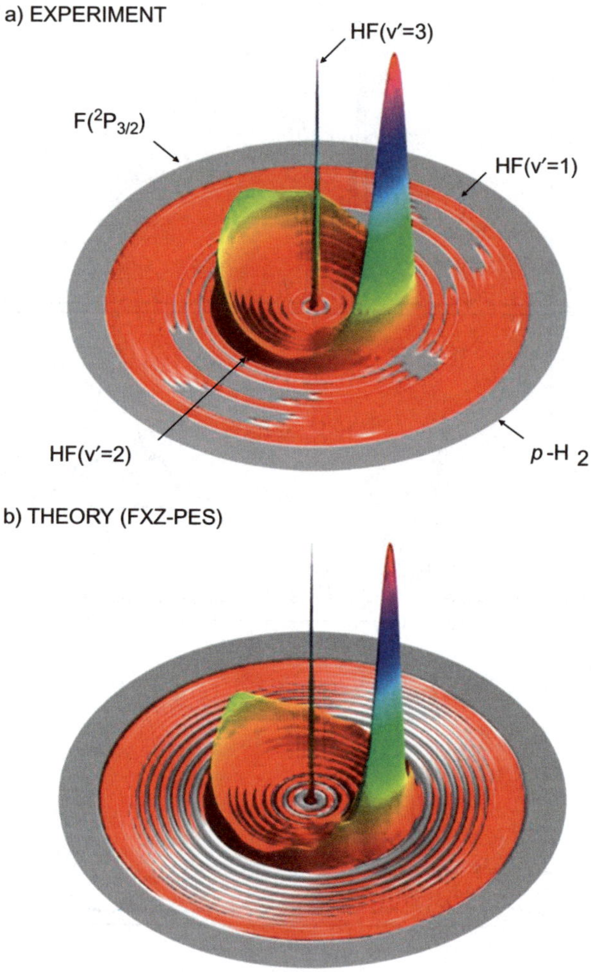

Figure 6.22 The three-dimensional contour plots for the product translational energy and angle distributions for the $F(^2P_{3/2})+H_2(j=0)$ reaction at the collision energy of $2.18\,kJ\,mol^{-1}$: (a) experiment, (b) theory based on a new CCSD(T) potential (FXZ PES). (Adapted from ref. [518].)

6.3.2 The $OH+D_2$ Reaction: Mode Specific Chemistry

For four-atom or larger systems, the challenges for theoretical and experimental state-to-state dynamical studies become much greater because many more internal degrees of freedom (and therefore many more energetically accessible states) are involved. The $OH+H_2 \rightarrow H_2O+H$ reaction has become the most important four-atom benchmark system, since the presence of only one 'heavy' atom makes it amenable to relatively detailed theoretical study. The reverse reaction, $H+H_2O \rightarrow H_2+OH$ and its isotopic variants, has been studied extensively using the *PHOTOLOC* method (see Study Box 6.3),[249] and has also become a prototype reaction for bond selective chemistry (see Section 7.3.2). In the case of the $H+HOD(v) \rightarrow H_2+OD$ or $HD+OH$ reactions, both the rate of reaction and the product channel branching, [OD]/[OH], depend sensitively on the type of mode or bond excitation of the HOD reactant[520,521] (see Section 7.3.3).

A recent study on $OH+D_2 \rightarrow HOD+D$ by Strazisar *et al.*[491] using the D-atom Rydberg tagging method has achieved product vibrational state resolution, the first time this has been achieved in a CMB experiment for a four-atom reaction. Figure 6.24 shows the resulting 3D product flux plot and

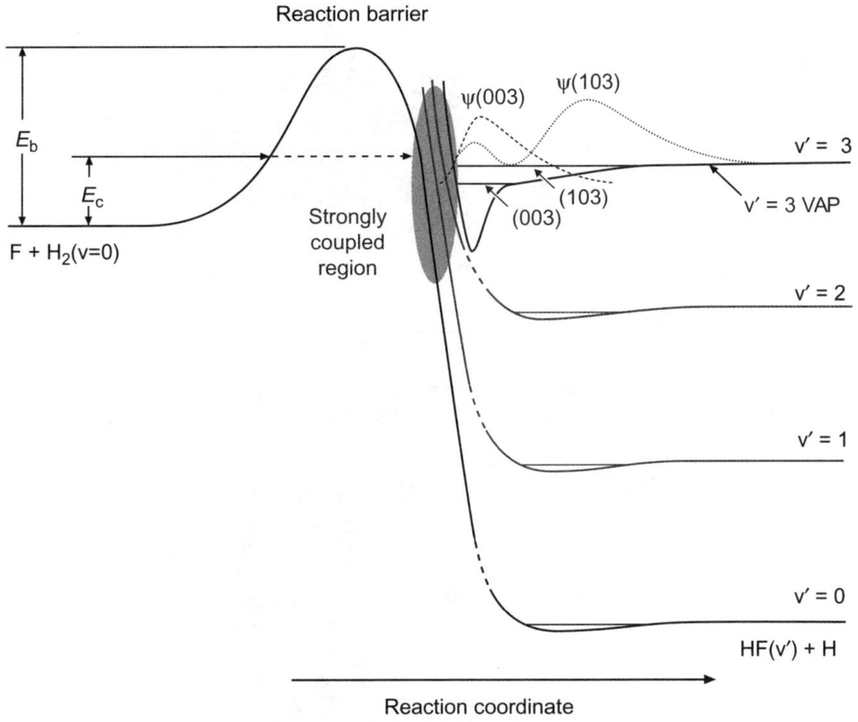

Figure 6.23 The Feshbach resonance mediated mechanism for the $F + H_2$ reaction. Two resonance states are shown trapped in the peculiar HF($v' = 3$)–H vibrationally adiabatic potential (VAP) energy well. The one dimensional VAP curves are traced out from the new CCSD(T) potential surface (FXZ-PES). One dimensional wave functions of the two resonance states, (003) and (103), are also shown. (Adapted from ref. [489].)

surface contour plot.[491] The structure observed in the contour plot can be assigned to the vibrational states of the HOD product. The contour plot shows clearly that the D product is backward scattered with respect to the incoming D_2 molecular beam. By linear momentum conservation, the HOD product is also backward scattered with respect to the OH beam direction, in agreement with earlier results obtained using the classic CMB technique with universal detection.[522] This suggests that the $OH + D_2 \rightarrow HOD + D$ reaction is a direct rebound reaction, as are the $H + H_2$, $Cl + H_2$, and to some extent also the $F + H_2$ reactions. In agreement with quantum scattering predictions,[523] mode-specific reaction dynamics is observed, with vibration in the newly formed oxygen-deuterium bond preferentially excited to $v' = 2$. This result demonstrates that quantum theoretical calculations have progressed to the point where it is now possible to predict energy disposal in four-atom reactions.

The high-resolution H-atom Rydberg tagging TOF technique is expected to provide detailed dynamical information for many other bimolecular and unimolecular reactions leading to an H(D) product. As noted earlier, the same Rydberg tagging concept has recently been extended successfully to the detection of oxygen atoms,[438,439] which may widen the range and scope of this approach to product detection in both reactive scattering and photodissociation studies.

6.4 CMB EXPERIMENTS WITH REMPI DETECTION

Two experimental techniques have recently been developed to measure the CM differential cross section directly under crossed-beam scattering conditions. Both techniques exploit the idea of

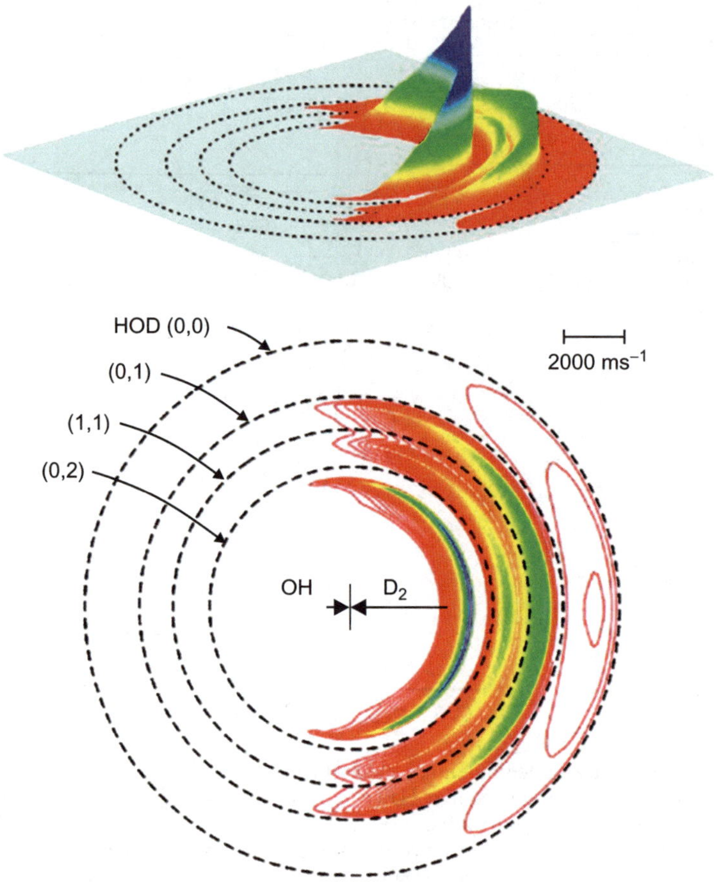

Figure 6.24 The three dimensional and two dimensional contour diagrams for the D atom products of the $OH + D_2 \rightarrow HOD + D$ reaction. The structure in the contour plots arise from the vibrational states of the HOD co-product. (Adapted from ref. [491].)

combining product state-selective detection with a simultaneous measurement of the speed distribution,[401,402,524] and are described in more detail in Study Box 6.5. As we will see, these approaches have permitted a significant advance in the investigation of polyatomic reaction dynamics.

6.4.1 Product Pair Correlations

Going beyond the simple atom plus diatom reactions, most complex reactions yield two *molecular* species as products, rather than an atom and a molecule. Each product will, of course, be formed with its own characteristic nascent internal state distribution, which result from the forces acting during the collision. Conventional measurements of the product state distribution for one species are blind to the state distribution of its co-product; yet, the two products are formed together in a single reactive encounter. Clearly, it is desirable to obtain *coincident* information on the two product state distributions in order to gain deeper insight into how the interaction forces govern the reaction outcomes. Such coincident information may be obtained from quantum state-correlation measurements on the product pair.[525,526] In physical terms, pair-correlation here refers to a measurement of the state distribution of one product when the co-product is coincidently formed in a

STUDY BOX 6.5: DIFFERENTIAL CROSS SECTION MEASUREMENTS IN THE CM-FRAME

The first approach to measuring DCSs in the CM-frame employs a Doppler-selected ion time-of-flight (TOF) technique.[524] This can be regarded as a variant of the 3D velocity imaging method, and exploits high-resolution translational energy spectroscopy by combining three 1D projection techniques in an orthogonal manner to determine the 3D velocity distribution. As we have seen in Chapter 5, the distribution of product velocity vectors arising from a scattering process (including a reaction) has an intrinsic cylindrical symmetry about the reactant relative velocity vector, $\boldsymbol{v}_{rel}$, defined as the CM z-axis. To take advantage of this symmetry, the probe laser propagation axis is set parallel to this initial relative velocity axis. As illustrated in Figure 6.25, one first uses the Doppler-shift technique (see Study Box 5.2) to selectively ionize a subgroup of the products being probed with $v_z \pm \Delta v_z$ in the CM-frame. This requires that the probe laser bandwidth is narrower than the Doppler width of the detection transition. (For this reason, the technique is usually applied to a light species such as the H-atom, since these possess broader Doppler widths.) Rather than detecting all of the resulting ion signal as a single data point, as in the conventional Doppler-shift technique, the velocities of those Doppler-selected ions in the x–y plane are dispersed both temporally (in the y-direction, parallel to the TOF axis) and spatially (in the x-direction). By placing a narrow slit at the space-focusing plane just in front of a microchannel plate (MCP) detector to detect only those ions with $v_x \approx 0$, the v_y-distribution is

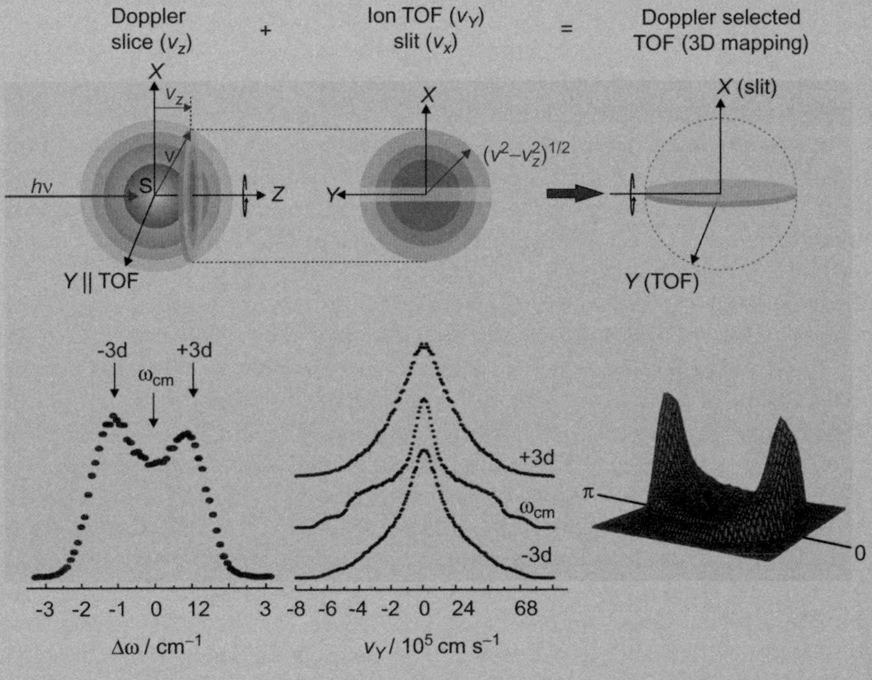

Figure 6.25 Schematic illustration of the basic concept of the Doppler-selected TOF technique. (Adapted from ref. [524]).The hatched slice on the left represents a Doppler-selection of a given v_z. The strip on the Doppler slice (the middle figure) is the 1D v_y-distribution measured under the v_x-restriction of a slit in front of the TOF spectrometer. The combination of many Doppler-selected TOF measurements yields the result shown on the right. The lower figures are the corresponding actual data at each stage for the H-atom product from the reaction of $S(^1D) + H_2$.

then measured through a high-resolution ion TOF velocity spectrometer. Thus, for a given Doppler slice at v_z the measured TOF spectrum corresponds to $S(v_y;\ v_x \approx 0, v_z)$. Due to the cylindrical symmetry around the z-axis mentioned above, there is sufficient information in these data to recover the 'lost information' (*i.e.* that corresponding to large v_x), and for the full 3D velocity distribution to be determined. Since both the Doppler-shift and ion TOF measurements can be readily referred to the CM-frame, and only the products with $v_x \approx 0$ are detected, it can be shown that the quantity measured by the Doppler-selected TOF method is approximately proportional to the probability density distribution in a CM velocity coordinate,

$$S\left(v_y; v_x \approx 0, v_z\right) \propto D\left(v_x, v_y, v_z\right). \tag{B6.5.1}$$

This density distribution as a function of velocity should be invariant to the choice of coordinate used (Cartesian or polar), so expressing the density in terms of the appropriate DCS yields

$$D\left(v_x, v_y, v_z\right) = \frac{\mathrm{d}^3\sigma}{\mathrm{d}v^3} \equiv \frac{\mathrm{d}^3\sigma}{\mathrm{d}v_x \mathrm{d}v_y \mathrm{d}v_z} = \frac{\mathrm{d}^3\sigma}{v^2 \mathrm{d}v \mathrm{d}\omega}. \tag{B6.5.2}$$

The conventional DCS refers to the cross section per solid angle ($\mathrm{d}\omega = \sin\theta \mathrm{d}\theta \mathrm{d}\phi = \mathrm{d}\cos\theta \mathrm{d}\phi$) in the CM-frame polar coordinates. Therefore the double DCS is related to the measured quantity by the simple equation,

$$\frac{\mathrm{d}^2\sigma}{\mathrm{d}v \mathrm{d}\cos\theta} = 2\pi v^2 D\left(v_x, v_y, v_z\right) \propto v^2 S\left(v_y; v_x \approx 0, v_z\right) \tag{B6.5.3}$$

The second approach to the direct measurement of the CM differential cross section is to use the time-sliced velocity-mapped ion imaging technique,[401,402] introduced already in Study Box 5.3. In a similar vein of argument to the above, one can show that Eq. (B6.5.3) also applies to this experimental approach. As with the Doppler-selected TOF technique, the time-sliced ion imaging approach yields directly the product 3D CM velocity distribution, but now with the multiplexed advantage that the two velocity components in the scattering plane are recorded simultaneously.

A correction is usually needed when using either of the two approaches discussed above in a crossed-beam experiment. Laser spectroscopic detection methods (in addition to EI mass spectrometry discussed in Section 6.2) in general measure the number of product particles within the detection volume. In other words, the measured signal corresponds to the product density rather than the flux emerging from the scattering zone. The flux is the number of particles passing through a unit area per unit time, and is proportional to the reaction rate. (Note that the product density here refers to the spatial coordinate, *i.e.* $D(x, y, z)$, and should not be confused with the above $D(v_x, v_y, v_z)$ that refers to the CM velocity space.) A consequence of this is that, because products recoiling at lower LAB velocities will be dispersed in space more slowly than those with higher velocities, their number density will tend to be higher when the probe laser is fired, resulting in a higher detection efficiency. This means that products scattered at different angles in the CM-frame have different detection efficiencies in the LAB-frame, because they have different LAB velocities. As we have already noted in Study Box 6.1, this correction for the LAB velocity-dependent detection efficiency is called the density-to-flux transformation.

Although an approximate correction for this transformation can be made simply by dividing the intensity by the corresponding LAB-frame velocity of the detected species (as was done, for example, in Eq. (B6.1.2)), because of the finite spread of reactant velocities it is usually necessary to perform some form of modelling of the experiment to obtain the CM DCS of interest (using, for example, Monte Carlo simulation techniques – see Study Box 5.4). This modelling usually requires knowledge of the geometry of the overlapping region between the laser beam and the molecular beams, the line-shape of the laser radiation, the velocity distributions of the molecular beams, and the orientation of the laser beam with respect to the molecular beams. Inaccuracies in the modelling of the experimental arrangement will lead to inaccuracies in the DCS that is extracted from the data.

Piergiorgio Casavecchia, Kopin Liu and Xueming Yang
with addition material by *David Chandler and Steven Stolte*

specific state. In terms of probability theory, the product pair correlation represents the joint or conditional probability distributions of the quantum states of the two departing products, $P(v',v)$.

To make the concept more concrete, let us consider the reaction of $F + CH_4 \rightarrow HF(v') + CH_3(v_i)$. The vibrational state distribution of the HF products has been measured by IR chemiluminescence in flow cell experiments and by direct IR laser absorption in a crossed molecular beam experiment. All of the experimental results indicated a highly inverted distribution, peaking at $HF(v'=2)$, with a low degree of rotational excitation. For the methyl radical product, time-resolved diode absorption spectroscopy was used to probe the CH_3 umbrella mode (v_2) excitation in another flow cell experiment. By extrapolating the data to time zero, a vibrational distribution of 0.66:0.24:0.10 was deduced for $v_2 = 0{:}1{:}2$, and a rather cold rotational temperature of 280 K was also found. Combining these independent results leads to the general consensus that the $F + CH_4$ reaction is of the early-barrier type (consistent with the high degree of product vibrational excitation) and that the CH_3 moiety behaves as a pseudo-atom, *i.e.* a spectator whose intramolecular dynamics does not significantly affect the dynamics of the reactive event.

As noted above, however, in any given reactive event the two products HF and CH_3 are always formed as a scattered pair. Hence, for a given HF (or CH_3) state there will be a concomitant state distribution, *i.e.* conditional probability distribution, of the co-product CH_3 (or HF). This conditional probability distribution of CH_3 state will not necessarily be the same when a different HF state is produced, so that independent measurements of the internal state distribution of each product do not necessarily tell the full story. A simple consideration suffices to illustrate this. Shown in Table 6.1 is a hypothetical 4×3 matrix representing the joint probability distribution of the two products from $F + CH_4$.

Each matrix element represents the pair-correlated probability for producing an HF molecule in state v' in a single reactive event when the coincidently formed CH_3 co-product is born in quantum state v_2. The conventional product state distributions, $P(v')$ and $P(v_2)$, are found simply by summing $P(v',v_2)$ over the co-product quantum states, *i.e.* $P(v') = \Sigma_{v2} P(v', v_2)$ and, likewise, for $P(v_2)$. Obviously, it is not possible to retrieve the full $P(v',v_2)$ matrix (of dimension $v' \times v_2 = 4 \times 3 = 12$)

Table 6.1 A hypothetical joint probability matrix to illustrate the concept of the (v',v_2)-correlation distribution of the coincidently formed product pairs from the $F + CH_4 \rightarrow HF\ (v') + CH_3(v_2)$ reaction, where $\Sigma_{v'} \Sigma_{v_2} P(v',v_2) = 1$.

$v' \backslash v_2$	*0*	*1*	*2*	$P(v')$
0	0.03	0.01	0.01	0.05
1	0.12	0.05	0.04	0.21
2	0.41	0.17	0.05	0.63
3	0.08	0.03	0	0.11
$P(v_2)$	0.64	0.26	0.10	1.0

from the information on $P(v')$ and $P(v_2)$ alone (of dimension $v'+v_2=4+3=7$), even if both are available. We require a method for measuring the joint probability distributions directly.

WORKED PROBLEM (3)

Q: *In what situations will the product pair-correlation measurement provide no more dynamical information than the independent determinations of product state distributions of the two molecular products?*

A: It is instructive to consider two limiting cases.[526] If the two products are completely uncorrelated, namely, if the vibrational distribution of HF is independent of the CD_3 state, one has $P(v',v_2)=P(v')P(v_2)$. An obvious example is the spectator case, in which a group of atoms, say the CD_3-moiety, plays little role in the reaction and becomes one of the products. Another dynamically uncorrelated example is the statistical reaction, for which the product pair correlation may not convey much additional information about dynamics, as it can largely be accounted for by statistical models based on energy and angular momentum constraints (see Chapter 7). If, at the other extreme, the vibrations of the two products are strictly correlated, for example $v'=v_2+n$, with n an integer, the joint probability matrix element can then be expressed as $P(v',v_2)=P(v')\delta_{v_2+n,v'}$. In both extreme cases, uncorrelated and totally correlated, it suffices merely to measure the averaged probability $P(v')$ and $P(v_2)$. More often, the situation lies between the two limiting cases, and a direct determination of $P(v',v_2)$ will be more telling.

As it turns out, experimental measurement of $P(v',v_2)$ is in fact rather simple: the method is based purely on conservation of energy and linear momentum in a chemical reaction. Considering the $F+CH_4 \rightarrow HF(v')+CH_3(v_2)$ reaction at collision energy E_c, the conservation of energy leads to

$$E_c - \Delta H_0^{\ominus} = E_{HF(v')} + E_{CH_3(v_2)} + E'_t, \tag{6.6}$$

where for simplicity the rotational energies of the two products are neglected, $\Delta H_0^{\ominus}$ is the change in enthalpy for the reaction ($-133\,\text{kJ mol}^{-1}$ in this case), and E'_t is the total product kinetic energy release, *i.e.*

$$E'_t = \frac{1}{2}\left(m_{HF}u_{HF}^2 + m_{CH_3}u_{CH_3}^2\right), \tag{6.7}$$

with u_{HF} and u_{CH_3} denoting the respective product speeds in the CM-frame. The conservation of linear momentum in the CM-frame requires $m_{HF}u_{HF} + m_{CH_3}u_{CH_3} = 0$. Incorporating momentum conservation into the expression for E'_t one has

$$E'_t = \frac{1}{2}\left(m_{CH_3}\frac{(m_{HF}+m_{CH_3})}{m_{HF}}\right)u_{CH_3}^2. \tag{6.8}$$

Now, if one performs an experiment with a well-defined E_c, which is readily achievable in a crossed-beam experiment, and uses a laser spectroscopic technique to detect one of the products, say CH_3, in a state-selected manner, from Eq. (6.6) one is left with two unknowns, $E_{HF(v')}$ and E'_t. However, if one goes one step further by concurrently measuring the speed u_{CH_3} of the state-tagged $CH_3(v_2)$ at high resolution, then from Eqs. (6.6) and (6.8), the desired state-correlated information of the HF co-product may be obtained directly from the measured CH_3 speed distribution. Repeating the measurements for different tagged states of CH_3 and normalizing them according to the $CH_3(v_2)$

product state distribution then enables the recovery of the full joint probability distribution matrix $P(v', v_2)$.

Illustrated in Figure 6.26 is a set of such correlated distributions for the isotopically analogous reaction $F + CD_4 \rightarrow DF(v') + CD_3(v_2)$ at four different collision energies.[527] These were obtained by the time-sliced ion velocity-mapped imaging technique, discussed in Study Boxes 5.3 and 6.5. Also shown in the background of each plot are the conventional (uncorrelated) vibrational distributions for the two products that are obtained by summing the pair-correlated populations for each fixed vibrational state of the co-products. One notes that $P(v')$ remains highly inverted at all four collision energies, whereas $P(v_2)$ changes from a cold distribution at low E_c to a slightly inverted one at higher E_c values. More significant is the observation of an anti-correlated excitation of the two

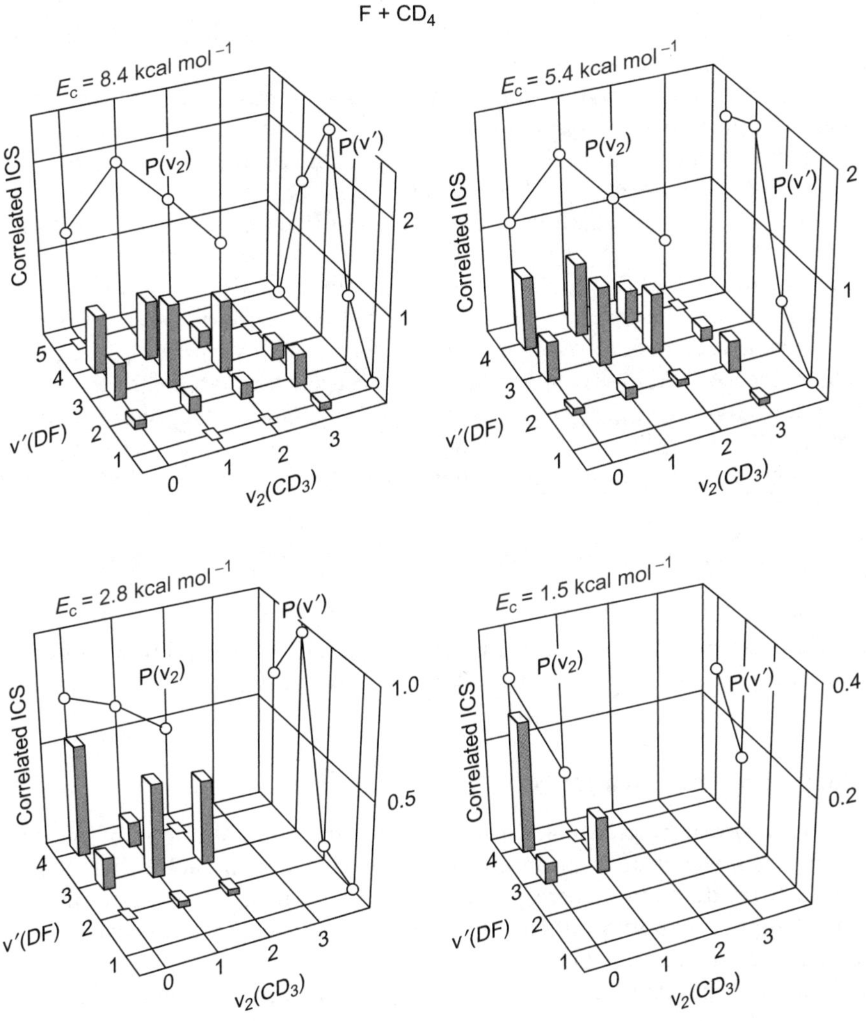

Figure 6.26 3D representation of the correlated integral cross section (CICS) or the correlated state population of product pairs from the reaction of $F + CD_4$ at four collisional energies (1 kcal $mol^{-1} = 4.184\ kJ\ mol^{-1}$). All data are scaled according to the CD_3 vibration state distribution. Summing the pair-correlated populations for each fixed vibration state of CD_3 (or DF) yields the conventional vibration distribution of DF (or CD_3); these are displayed as open circles on the respective background. (Adapted from ref. [527].)

product vibrators, namely, the lower the CD_3 vibrational excitation, the higher the excitation of the DF. This proves beyond any doubt that the CD_3-moiety is *not* a spectator in this reaction – in sharp contrast to the conclusions drawn from the uncorrelated state distribution measurements alluded to earlier (see Worked Problem (4)).

WORKED PROBLEM (4)

Q: *Use kinematic arguments* to provide a qualitative explanation for the anti-correlated excitation of the two product vibrators shown in Figure 6.26.* (The following material is adapted from ref. [528].)

A: For a heavy-light-heavy colinear reaction, the mass-weighted skew angle is small ($\beta = 25.5°$ for the F-D-CD_3 reaction – see Study Box 7.1), and a strong inertial coupling between the two scaled coordinates is anticipated. This in turn leads to a channeling of reactant translational excitation in excess of the barrier to reaction principally into product translation, *i.e.* $E_c \rightarrow E'_t$. For the reactants, the total angular momentum $\boldsymbol{J}$ is almost entirely made up of orbital angular momentum ℓ, since the initial rotational angular momentum of CD_4 in the supersonic molecular beam is negligibly small. Angular momentum conservation therefore dictates that $\boldsymbol{J} \sim \ell = j_{CD_3} + j_{DF} + \ell'$. The reaction of $F + CD_4$ is fast ($k(T) = 1.3 \times 10^{-10}\ \exp(-315/T)\,\mathrm{cm^3\ molec^{-1}\ s^{-1}}$), and *ab initio* calculations suggest a predominantly linear transition state with respect to the F-D-C moiety. As demonstrated for many direct three-atom reactions with a heavy-light-heavy mass combination, ℓ contributes mostly to ℓ', and for a nearly colinear reaction, often $\ell \approx \ell'$, *i.e.* the dynamics is coplanar.[529] For a light-atom transfer reaction, both the initial and final orbital angular momenta (ℓ and ℓ') are mainly carried by the orbital motions of the two heavy species, and thus can be relatively large. The estimated ℓ for the present reaction ranges from about $100\hbar$ to $\gtrsim 200\hbar$ over the energy range shown in Figure 6.26. The probed angular momentum in the CD_3, j_{CD_3}, was small; the angular momentum constraint then leads to small j_{DF}, as found experimentally. Noting energy conservation,

$$E_{\text{total}} = E_c - \Delta H_0^{\ominus} = E_{\text{vib}}(CD_3) + E_{\text{rot}}(CD_3) + E_{\text{vib}}(DF) + E_{\text{rot}}(DF) + E'_t,$$

and recalling that $E_c \rightarrow E'_t$ and that both $E_{\text{rot}}(CD_3)$ and $E_{\text{rot}}(DF)$ (*i.e.* j_{CD_3} and j_{DF}) are small, it follows that the two product vibrational energies ($E_{\text{vib}}(CD_3)$ and $E_{\text{vib}}(DF)$) approximately sum to a constant value of $-\Delta H_0^{\ominus}$, and should therefore exhibit an anti-correlation.

6.4.2 Product Pair-Correlated DCSs

As we have just shown in Section 6.4.1, the quantum state distributions of the two molecular products in a complex chemical reaction are often correlated with each other. Since the relative population of each product state corresponds to the integral of the state-resolved angular distribution over all angles, it should not then be surprising that the state-resolved DCSs of the two products may also be pair-correlated.

Figure 6.27 shows an example of the state pair-correlated angular distribution or the correlated differential cross section (CDCS) for the $F + CD_4 \rightarrow DF(v') + CD_3(0v_200)$† reaction at $E_c = 22.5\ \mathrm{kJ\ mol^{-1}}$.[525] These consist of the flux-velocity contour maps for CD_3 products formed with different degrees of umbrella-mode excitation. This set of data was acquired by the time-sliced

*See Chapter 1 and Study Box 7.1 for an introduction to the role of kinematics in chemical reactions.

†The notation employed here to label the vibrational states of the CD_3 is $(v_1v_2v_3v_4)$, where v_2 is the number of quanta in the umbrella mode, v_2.

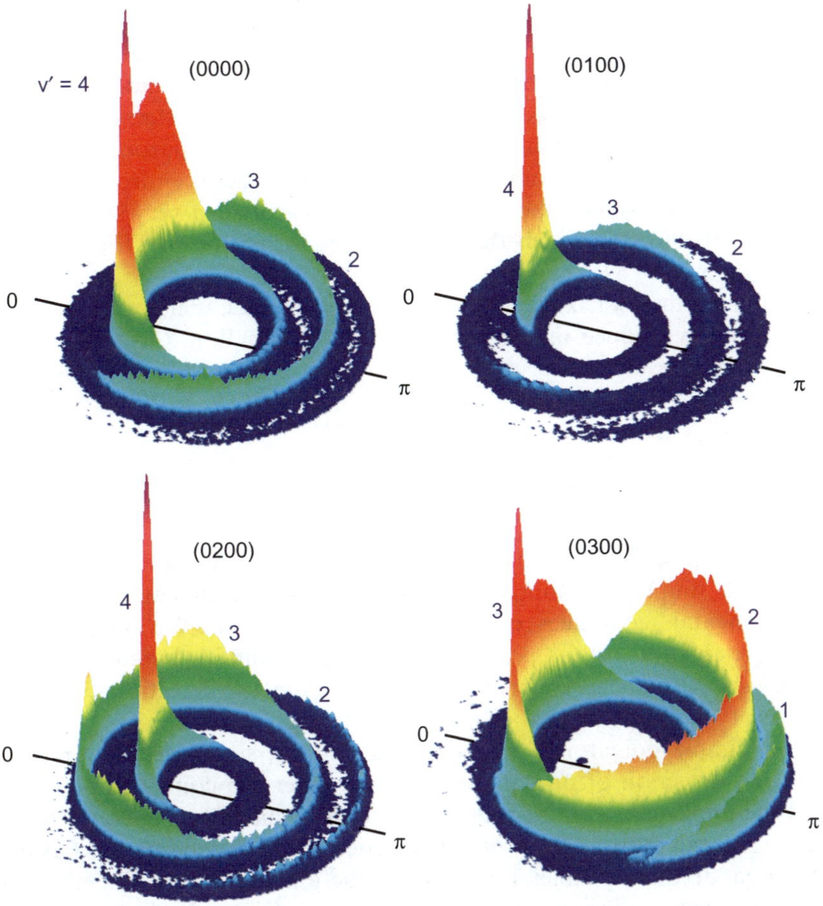

Figure 6.27 State-resolved flux-velocity contour maps of the CD_3 products from the $F + CD_4 \rightarrow DF(v') + CD_3(0v_200)$ at $E_c = 22.5\ kJ\ mol^{-1}$. The pair-correlated DF(v') distributions for a given CD_3 product state are manifested as the nested, ring-like features labelled by the DF vibrational quantum numbers (adapted from ref. [525]). The density-to-flux corrections have been made, and the intensity of each contour has been weighted by u^2 in accordance with conventional representation of the double DCS.[524]

velocity-mapped imaging technique under crossed molecular-beam scattering conditions. Thanks to the high resolution of the experimental technique, the resultant contours feature well-resolved ring structures corresponding to the Newton spheres of the CD_3 products partnered by DF in different quantum states. The energetics of the reaction are well-defined, and the CD_3 products were state-selectivity detected. By conservation of energy and momentum, the maximum velocities of the co-product DF, recoiling from the state-selected CD_3 in different vibrational states, can be calculated and identified unambiguously as the successive rings on each contour. The wide separation of the rings corresponding to the various DF vibrational states indicate unequivocally the low rotational excitations of the DF products (a high degree of rotational excitation in the DF would broaden the peaks for each vibrational state). The angular distribution associated with each ring reflects the preferred scattering direction of the coincidently formed DF states, *i.e.* the pair-correlated DCS.

Although there are some similarities among the results for the four different CD_3 vibrational states, the differences are striking and more interesting. For all CD_3 states, the concomitant DF formed in $v' = 2$ is confined within the backward hemisphere. The gradual protrusion of its angular distribution in the sideways direction with higher excitation of the $CD_3(\nu_2)$ co-product is reminiscent of the usual trend observed for a typical A + BC direct reaction proceeding through a rebound mechanism. The angular distributions for DF($\nu'=3$) are spread over all angles, but the dominant feature shifts progressively from sideways to forward scattering as increasingly more energy is deposited into the umbrella mode of the CD_3 coproduct. In particular, a narrowly-peaked forward feature is quite pronounced for CD_3(0200) and (0300), but is entirely absent for (0000) and (0100). For the DF($\nu'=4$) product, the most prominent feature is a very sharp forward-scattered peak, except for CD_3(0300), for which the concomitant formation of DF($\nu'=4$) is barely open energetically.

The interpretation of these striking CDCSs is complicated. Other experimental observations suggest that a reactive resonance may be at work in this reaction.[519,530] Thus, most likely the observed CDCSs are the result of the subtle interplay between the direct reaction and a resonant scattering pathway.

6.5 OUTLOOK

In this chapter we have described and discussed some of the current state-of-the-art in the study of the dynamics of both simple and complex reactions, with a particular focus on the significant advances in experimental techniques over the last 10–15 years. We have devoted particular attention to the investigation of: (a) polyatomic reactions exhibiting multiple channels, studied using the classic CMB technique with universal mass spectrometric detection; (b) simple three-atom and four-atom prototype reactions, studied using the very high-resolution H-atom Rydberg tagging technique; and (c) simple polyatomic reactions, studied *via* state-selective time-sliced velocity-mapped ion imaging detection of one of the products. Over the course of the chapter, we have emphasized that the implementation of soft ionization by tunable low-energy electrons[531] or tunable VUV synchrotron radiation[468] for product detection in CMB experiments has been central for progress in the area of multichannel polyatomic reaction dynamics. The time-sliced imaging approach also represents a significant advance in the investigation of polyatomic reaction systems, since with sufficient velocity resolution (in both magnitude and direction), the technique allows one to derive state-correlated information on the coproducts in a bimolecular reaction.

Further advances will undoubtedly continue to be made in the development of increasingly sophisticated crossed-beam scattering techniques. For example, the development of table-top lasers capable of producing pulsed light at VUV wavelengths heralds significant contributions from pulsed CMB experiments employing detection *via* pulsed single-photon VUV ionization. F_2 excimer lasers operating at 157 nm (7.9 eV) have been available for some time, and can be used in cases where the ionization potential of the product of interest is particularly low. Single-photon VUV detection using such a laser source has been demonstrated in CMB experiments with both mass spectrometric detection (see ref. [532] and references therein) and ion imaging detection.[533] More recently, Albert and Davis[534] have used intense 9.9 eV laser radiation, generated *via* four-wave mixing in krypton, to study collision complex lifetimes in the reaction of phenyl (C_6H_5) radicals with molecular oxygen. While much of this chapter has focused on crossed-molecular-beam scattering, laser pump-probe experiments in which reactants are co-expanded into a single molecular beam are also providing useful information on polyatomic reaction dynamics (see ref. [535] and references therein), and will continue to do so in the future.

Advances in techniques for studying the kinetics of complex polyatomic reactions provide complementary information to that obtained in CMB experiments. Taatjes and coworkers[536–538] have recently developed a technique that combines pulsed laser photon-initiation and universal detection using quasi-CW synchrotron radiation. The reaction products are recorded in a multiplexed manner, which yields product mass spectra as a function of time after initiation of reaction.

Analysis of the data allows the identification of virtually all of the primary reaction products and their branching ratios, and also yields absolute rate coefficients for each reaction channel.

Following the demonstration that polyatomic multichannel reactions involving an atomic and a hydrocarbon radical may be investigated in crossed-beam experiments with mass spectrometric detection, numerous reactions of this important class are expected to be studied in the near future. In the first instance, the atomic reagents are likely to be include oxygen, nitrogen, carbon and sulphur. Especially interesting are radical-radical reactions that are simple enough to be within the capabilities of current theoretical treatments, both at the level of accurate electronic structure calculations of the PES, and of dynamic calculations by exact quantum and/or quasi-classical trajectory methods.[144] Reactive systems in this category include N + OH, C + OH, O + OH and S + OH.

A new experimental approach has recently been shown to be useful for the study of polyatomic multichannel radical-molecule and radical-radical reactions proceeding *via* the addition-elimination mechanism.[539] A particular isomeric form of a radical intermediate is generated under collisionless conditions through photodissociation of a suitable precursor molecule. The intermediate is chosen such that it is unstable with respect to the reaction coordinate of the bimolecular reaction to be studied. The branching between the ensuing product channels of the energized radical is studied as a function of its internal energy using VUV photoionization with tunable synchrotron radiation. The technique is in some ways reminiscent of the transition state spectroscopy experiments described in Chapter 1, and allows the probing of key portions of the reaction potential energy surface, such as isomerization and dissociation barrier heights.

As we have shown in Section 6.3, Rydberg-tagging TOF spectroscopy of H, D or O atom products facilitates high resolution measurements of product angular and velocity distributions. In the future, we can envisage that O-atom Rydberg tagging[438] will be extended to three-atom reactions involving two heavy atoms, such as $H + O_2$ (a reaction considered to be among the most important in combustion), using the same kind of approach that has so far been limited to three-atom reactions of the type $X + H_2(D_2, HD)$ ($X = H$, F, Cl, $O(^1D)$) (see Section 6.3). We are also hopeful that further improvements in the resolution attainable in H-atom Rydberg tagging experiments will eventually allow rotational resolution to be achieved in the four-atom benchmark reaction $OH + H_2$.

Looking further into the future, one may hope to attempt state-resolved experiments on polyatomic systems to gain further information on the dynamics. Such experiments will almost certainly take the form of CMB experiments with laser spectroscopic detection, possibly coupled to ion-imaging techniques (see ref. [526] and references therein). Although laser spectroscopic detection is limited to those species with known spectroscopy, the highly detailed information obtained in a spectroscopic measurement, particularly when performed in a pair-correlated manner, holds the potential to unveil the fleeting, collective atomic motions in the transition state region of a polyatomic reaction in a way that other techniques are unlikely to achieve.[540] Understanding the transition state region is the key to understanding chemical reactivity. Gaining a better appreciation of the dynamics of the transition state in a few benchmark systems, in particular the ways in which bond-breaking and bond-forming processes occur as a consequence of cooperative motions during atomic rearrangements, will help to build and shape a general conceptual framework for chemical reactivity.

A crucial synergistic contribution to our progress in the understanding of the dynamics of both three-atom and polyatomic multichannel reactions is expected from theory. For instance, there have been great strides in the theory for 'difficult' three-atom reactions such as $H + O_2$, both in the calculation of potential energy surfaces and in quantum dynamics calculations[218] (see Chapter 3), so much so that theory is currently ahead of experiment for this reaction, and more detailed experimental investigations are required. As alluded to above, these will almost certainly become possible in the near future. For polyatomic multichannel reactions, *ab-initio* electronic structure calculations usually provide the stationary points of the relevant reaction PESs. These are useful when attempting to interpret the results of dynamics experiments. However, the prospect of developing full-dimensional PESs for some of these complex systems is even more appealing.[55] Unfortunately, for polyatomic reactions it is not possible to perform rigorous quantum dynamic calculations on accurate multidimensional PESs, and QCT calculations are perhaps the most promising approach to the theoretical treatment of such reactions. Such calculations have already been performed for direct abstraction reactions such as $F + CH_4 \rightarrow HF + CH_3$.[541] Extensions of this approach to multichannel reactions, such as those discussed here, have also started to appear[478] and more are to be expected in the near future.

6.6 PROBLEMS

1. The reaction $C(^3P) + C_2H_2$ at low collision energies proceeds *via* an addition-elimination mechanism whereby formation of an intermediate, C_3H_2, which lives many (>10) rotational periods, is followed by dissociation to $C_3H + H$ products. Knowing also that the C_3H radical product is highly rotationally excited, what do you expect the angular distribution of the C_3H product to look like in the CM reference frame?
2. At low collision energies the reaction $O(^3P) + H_2S \rightarrow HSO + H$ proceeds *via* a direct mechanism on the triplet potential energy surface, while the corresponding reaction of excited oxygen atoms $O(^1D) + H_2S \rightarrow HSO + H$ proceeds *via* a long-lived complex, thyo-peroxide HSOH, on the singlet potential energy surface. What do you expect the angular distribution of the HSO product in the CM-frame of reference to look like in the two cases?
3. With reference to the reaction $O(^1D) + H_2S \rightarrow HSO + H$, what experiments would you perform to estimate the lifetime of the HSOH complex within the *osculating complex model* for chemical reaction? Knowing that at a given collision energy the average lifetime of the HSOH complex is equal to its rotational period, what will be the ratio of the backward to forward intensity of the HSO product CM angular distribution?
4. The figures below show the KI product flux contour plots in the centre-of-mass (CM) frame for the $K + I_2$ (left) and $K + CH_3I$ (right) reactions.

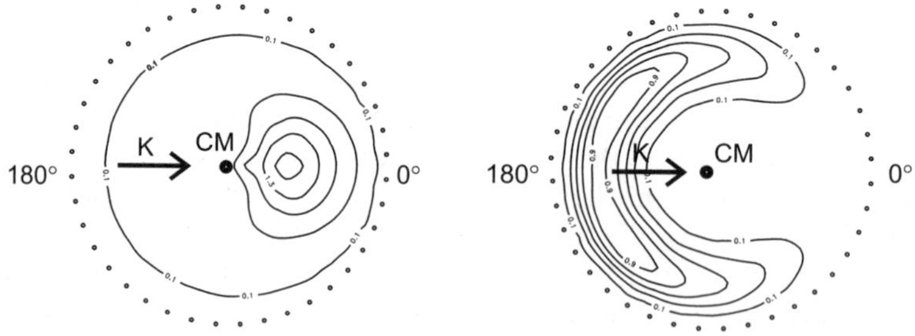

Explain how data like these might be obtained.

What can be learnt about the dynamics of the two reactions from these figures? (The outer rings of dots show the maximum CM velocities of KI in the two reactions.)

Contrast the scattering behaviour observed for the above two reactions with that found for the reaction $S(^1D) + H_2 \rightarrow SH + H$, the differential cross section for which is shown below [adapted from E. J. Rackham, T. Gonzalez-Lezana and D. E. Manolopoulos, *J. Chem. Phys.*, 2003, **119**, 12895].

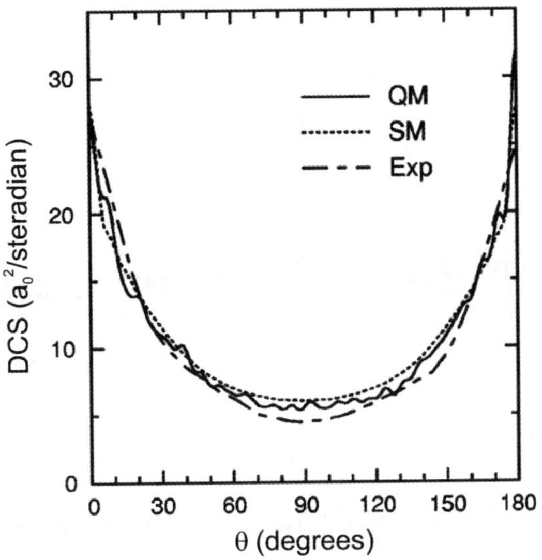

5. As we have seen in this Chapter, OH in its ground electronic state undergoes the reaction

$$OH + D_2 \rightarrow HOD + D,$$

which is isoelectronic with the $F + H_2$ reaction. Figure 6.24 shows plots of the D-atom product flux in the centre-of-mass frame for the above reaction. The experiments were performed under crossed molecular beam conditions, with the D-atom products detected by the H(D)-atom Rydberg tagging method. The dashed rings are labelled according to the number of quanta *m* in the bending mode, and *n* in the OD stretching mode of HOD (*m*, *n*).

(a) What does the above figure suggest about the mechanism of the reaction?

(b) The reaction has also been studied in crossed molecular beam experiments using mass spectrometric detection of the HOD product. Using an appropriate Newton (velocity-vector) diagram for the HOD products, suggest how the product flux contour map of HOD might differ from the one shown in Figure 6.24 for the D atom products. Assume that the reactant beam velocities are the same in the two different experiments.

(c) What spectroscopic experiments might be performed to probe the transition state region of this reaction (or a similar reaction) more directly? (It might be helpful to reread Section 1.4.5, and also refs. [14] and [542] concerning the anion photoelectron detachment spectroscopy of $[FH_2]^-$and $[H_3O^-]$.)

CHAPTER 7

Reactive Scattering: Quantum State-Resolved Chemistry

F. FLEMING CRIM

Department of Chemistry, University of Wisconsin-Madison, Madison, WI, USA

7.1 INTRODUCTION

As we have seen in the previous chapter, a fully quantum-state-resolved reaction is one with all of the quantum numbers of the reactants and products specified. Most generally, the initial quantum numbers identify the electronic, vibrational, and rotational states of the reactants as well as the orbital angular momentum of the collision, and the final quantum numbers identify those same quantities for the products. For example, in the case of the reaction of an atom and a diatomic molecule, both in their electronic ground state, we can write the reaction, as in Chapter 1,

$$A + BC(i) \rightarrow AB(f) + C, \qquad (7.1)$$

where the internal states of the reactants are i and those of the products are f. The reaction cross section for a collision with relative speed v_{rel} is $\sigma_{if}(v_{rel})$ and the corresponding rate constant is $k_{if}(v_{rel}) = v_{rel}\sigma_{if}(v_{rel})$. Averaging this quantity over a thermal distribution of speeds gives the state-to-state rate constant $k_{if}(T) = \langle v_{rel}\sigma_{if}(v_{rel})\rangle$ for molecules in state i with a thermal distribution of relative speeds reacting to form products in state f. Spectroscopic techniques have allowed the preparation of reactants in state i and the detection of products in state f, in some cases approaching the ideal of specifying i and f completely. These experiments show that in many cases the cross section, the distribution of the products among their quantum states, and even the identity of the products depend strongly on the initially prepared state of the reactants. Thus, we know that there are cases in which the identity of the initially prepared state controls the reaction, but, as we shall see, that situation does not apply to all reactions. There are also cases in which the reaction rate is instead independent of the quantum states of the reactants and depends only on the total energy, a situation in which a statistical description of the reaction is particularly useful. In some of these cases, the shape of the potential energy surface, rather than the quantum states of the reactants, determines the distribution of the products among their energetically accessible states.

Tutorials in Molecular Reaction Dynamics
Edited by Mark Brouard and Claire Vallance

Published by the Royal Society of Chemistry, www.rsc.org

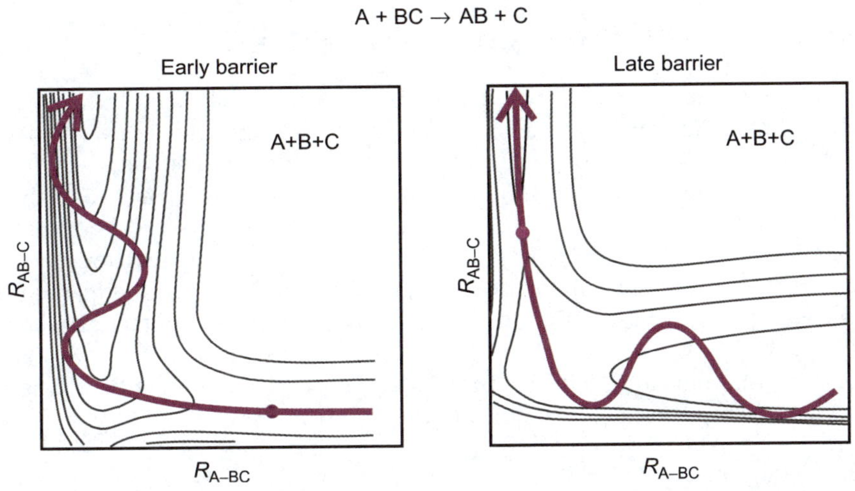

Figure 7.1 Illustration of Polanyi's rules for the disposal and consumption of energy in chemical reactions with barriers that occur early (left) and late (right) along the reaction coordinate.

7.2 MOTION OVER A POTENTIAL ENERGY SURFACE

This chapter examines reactions of molecules in specified quantum states with a particular emphasis on vibrations. Vibration and translation can strongly influence the dynamics of a reaction, and simple descriptions of the motion of a system across the potential energy surface during the course of a reaction illustrate their importance. Figure 7.1 shows the contours for a cut through a potential energy hypersurface along the A-B distance (R_{A-BC}) and the B-C distance (R_{AB-C}) for the reaction, $A + BC \rightarrow AB + C$, where A, B, and C can each be polyatomic species. This cut shows the coordinates for transferring B from C to A, as exemplified by a colinear reaction of an atom with a diatomic molecule, such as the reaction $H + Cl_2 \rightarrow HCl + Cl$ (note that the skew angle, β, is close to 90° for this system – see Study Box 7.1). Careful experimental and computational studies of such reactions are the origin of the *Polanyi rules* that describe the disposal and consumption of energy in a reaction.[27] These have been introduced already in Study Box 2.1, but we revisit these ideas here in a slightly different context.

The key to understanding energy disposal in an exoergic reaction is picturing the geometry and motion of the system during the release of the energy. A reaction with an early barrier, as shown on the left, has its energy maximum located in the entrance valley, where A is approaching BC. The release of the exoergicity of the reaction during the approach deposits energy into vibration of the newly formed A–B bond, as the trajectory sketched in the figure illustrates. The converse is true for a reaction with a late barrier, as illustrated by the surface on the right. The energy release occurs after the trajectory has "turned the corner" and the new products AB and C are separating. Consequently, the energy appears in the relative translation of the products, as illustrated by the trajectory on the right.

Running these illustrative trajectories in reverse provides a simple picture of energy consumption in a reaction. In the reverse reaction of $AB + C$, the surface on the left has a late barrier, and, as the trajectory illustrates, vibrational excitation of AB should preferentially carry the system over the barrier. Similarly, the reverse reaction on the surface on the right has an early barrier, and translation should most effectively promote the reaction. Experiments and trajectory calculations confirm these pictures for reactions of an atom with a diatomic molecule and in some reactions of an atom with a polyatomic molecule. These ideas, based on simple classical mechanics, illustrate the complementary roles of translation and vibration and show how reactions with initial excitation in

STUDY BOX 7.1: KINEMATICS AND SKEW ANGLES

The concept of the motion of nuclei over a potential energy surface is a central one in reaction dynamics. One simple illustration of their use is in Polanyi's rules, which have been discussed in several places in these Tutorials. Although the foundation of these rules lies in rigorous classical trajectory calculations, one often thinks qualitatively about Polanyi's rules, and the motion of nuclei over potential energy surfaces more generally, in terms of motion of a point mass on a three-dimensional potential energy landscape. For a A + BC → AB + C reaction, the appropriate mass of the particle is the reduced mass, $\mu = m_A m_{BC}/M$, where M is the total mass of the system, *i.e.* $M = m_A + m_B + m_C$. It turns out that this picture is only valid if the potential energy surface is drawn with so-called *mass weighted coordinates*.[9–11]

Let us focus on the colinear A + BC reaction. The classical Hamiltonian for the system can be written in terms of simple bond coordinates in the following way

$$H = \frac{1}{2}\left[\mu \dot{R}_{AB}^2 + 2\frac{m_A m_C}{M}\dot{R}_{AB}\dot{R}_{BC} + \mu' \dot{R}_{BC}^2\right] + V(R_{AB}, R_{BC}, \theta = 180°), \tag{B7.1.1}$$

where μ' is the reduced mass of the products, $\mu' = m_{AB}m_C/M$. If we were to represent the potential energy surface for the colinear reaction in terms of bond lengths R_{BC} and R_{AB}, the above equation implies that we would not be able to think of the motion as that of a single particle of mass μ moving over the surface. There are two reasons for this; first the appearance of the cross term $\dot{R}_{AB}\dot{R}_{BC}$ in the kinetic energy, and secondly the different mass factors appearing in the terms involving $\dot{R}_{BC}^2$ and $\dot{R}_{AB}^2$. However, it is possible to transform Eq. (B7.1.1) so that it is written in terms of the Jacobi coordinates R and r defined in Chapters 3 and 12:

$$H = \frac{1}{2}\mu\left[\dot{R}^2 + (\alpha^{\frac{1}{2}}\dot{r})^2\right] + V(R, r, \gamma = 0°), \tag{B7.1.2}$$

where $\alpha = \mu_{BC}/\mu$, with $\mu_{BC} = m_B m_C/m_{BC}$. Note the absence of cross terms in $\dot{R}$ and $\dot{r}$ appearing in this expression. This equation suggests that the picture of a particle of reduced mass μ rolling over a potential energy landscape can be recovered by plotting the surface with the following coordinates

$$\begin{aligned} x &\equiv R = R_{AB} + \frac{m_C}{m_{BC}} R_{BC} \\ y &\equiv \alpha^{\frac{1}{2}} r = \alpha^{\frac{1}{2}} R_{BC}. \end{aligned} \tag{B7.1.3}$$

In the mass weighted (x, y) coordinate system, the bond lengths have coordinates $(R_{AB}, 0)$ and $\left(\frac{m_C}{m_{BC}} R_{BC}, \alpha^{\frac{1}{2}} R_{BC}\right)$. From these coordinates it is readily shown that in the mass weighted system the bond axes are skewed at an angle β, defined by the equation

$$\cos^2\beta = \frac{m_A m_C}{m_{AB} m_{BC}}. \tag{B7.1.4}$$

The effect of the skew angle is illustrated in the left hand panel of Figure 7.2.

The value that the skew angle takes will depend on the masses of the species involved, *i.e.* on the kinematics of the reaction, a topic introduced in Chapter 1. For a reaction involving a light attacking or departing atom, *e.g.*, if $m_A \ll m_B \sim m_C$, then $\cos^2\beta \sim 0$, and the skew angle is approximately equal to 90°. So for reactions involving light attacking or departing atoms, plotting the bond axes at 90° to one another allows the trajectories on the PES to be interpreted in terms of motion of a point mass of reduced mass μ. This is precisely the interpretation we were using in our previous discussions of Polanyi's rules, but now we see that the PES can only be plotted in this way for reactions with this specific kinematics. If we now consider a reaction involving the exchange of a light atom between two heavy atoms, referred to as a light atom transfer reaction, $\cos^2\beta \sim 1$, and the skew angle will be much less than 90°. In this case, to preserve a picture of a ball rolling over a PES, trajectories must be plotted on a surface with a skew angle less than 90°. The panel on the right of Figure 7.2 shows a trajectory on a PES with a relatively small skew angle, and illustrates that although the barrier to reaction occurs quite late in this case, the trajectories are unable to turn the corner sufficiently quickly, and a significant fraction of the energy is released into product vibration. Polanyi's rules would therefore be expected to apply less well for light atom transfer reactions.

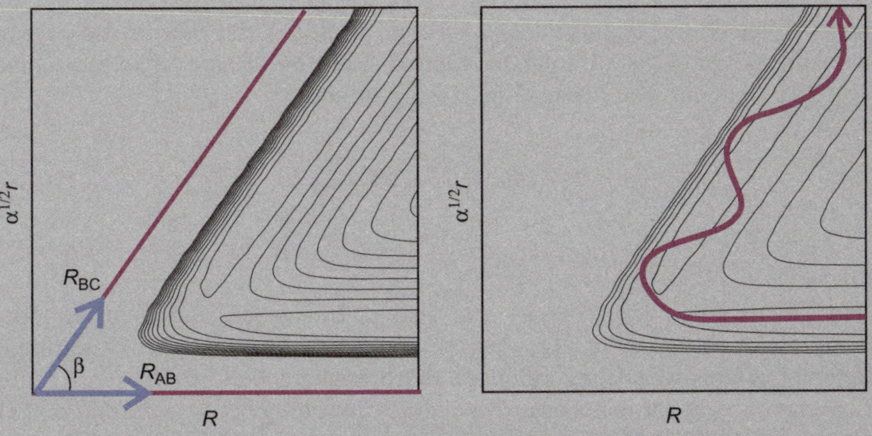

Figure 7.2 Left panel: Illustration of the significance of the skew angle β and the mass weighted coordinate system. Right panel: A schematic trajectory on a surface with a small skew angle. Reactants approach from the bottom right of the figure.

We have already seen in Chapter 1 that a light atom transfer reaction favours the angular momentum constraint $\boldsymbol{J} \simeq \boldsymbol{\ell} \simeq \boldsymbol{\ell}'$, leading to a conservation of orbital angular momentum. Provided the entrance and exit impact parameters for the reaction are similar, such reactions also often display a propensity to conserve kinetic energy, *i.e.* $E_c \simeq E_t'$. In this case, if the reaction is exothermic, then the exothermicity, $-\Delta H^{\ominus}(0\,K)$, tends to be released into the internal degrees of freedom, regardless of the location of the barrier. In terms of the discussion above, as the skew angle is reduced, so the directions of reactant approach and product separation, $\boldsymbol{R}$ and $\boldsymbol{R}'$, on the skewed diagrams tend to become more parallel, and his favours transfer of initial translational motion into that of the products.

Mark Brouard

different degrees of freedom can have very different efficiencies and form products with different excitations.

These classical ideas are more complicated for nonlinear arrangements of the A + BC system, and they can become very complex for reactions of polyatomic reagents. A nonlinear triatomic reactant has three modes of vibration, and the potential energy surface goes from three dimensions for an atom reacting with a diatomic molecule to six dimensions for an atom reacting with a triatomic molecule (see Chapter 2). It is difficult to envision the appropriate cuts through the hypersurface in this case, but, more fundamentally, the distinctions between the different degrees of freedom during the collision are less clear. Even initially preparing a particular bond stretching vibration, as for example the B-C stretch in an ABC molecule, does not ensure that it will be the only vibration excited during the collision. The interaction might cause the energy to flow into other vibrations in the molecule, making the simple correlations described above less appropriate. The failure of these simple pictures points to additional subtleties in reactions with more than one vibrational degree of freedom compared to those in simple three-atom reactions. The limiting case is one in which the outcome of the reaction does not depend at all on the initial state of the reacting molecule. In that case, the system is likely to sample all of the possible states available to it, and one can interpret various features of the reaction using statistical ideas based on the complete redistribution of the available energy, subject to the limits of conservation of energy and angular momentum.

The influence of the initial quantum state on the identity of the products is perhaps the most "chemical" manifestation of a state-to-state reaction. Even in the case of an atom reacting with a diatomic molecule, there are two possible reaction products,

$$\mathrm{A} + \mathrm{BC}(i) \xrightarrow{\sigma_{if}} \mathrm{AB}(f) + \mathrm{C} \tag{7.2a}$$

$$\xrightarrow{\sigma_{if'}} \mathrm{AC}(f') + \mathrm{B}. \tag{7.2b}$$

Each pathway has its own energy barrier and distinct exit valley, and each one potentially has different energy consumption and energy disposal dynamics. Thus, a central question is the extent to which different initial excitations influence the branching between the two channels. For a polyatomic reactant, the question becomes even more interesting because the different channels can correspond to cleaving different bonds in the reactant to form chemically different products, an example of bond-selective chemistry. Reactions in which the barrier location serves as a guide to energy consumption and disposal are "direct", in that the reactants interact for the time it takes the system to cross the energy barrier, but not for long enough to redistribute the energy among all of the available states. In the following, we consider both direct and statistical reactions, as well as the nature of energy flow within an excited molecule.

7.3 STATE-TO-STATE BIMOLECULAR REACTIONS

Our examination of state-to-state reactions focuses on the issues of how excitation influences both the rate of the reaction and the choice of available product pathway. In defining the product pathway we consider both the identities of the products and the quantum states in which they are born. The first questions concern energy consumption: Do molecules prepared in different internal states have different reaction probabilities? Does energy in internal degrees of freedom, such as vibrations, enhance the rate more than the same amount of energy in relative translation? The second question concerns energy disposal and reaction pathways: Do molecules in different internal states react to form products with particular excitations and do different initial excitations lead to different products?

7.3.1 Changing the Reaction Cross Section

One of the first and clearest examples of comparing the effect of vibrational and translational excitation on the reaction of an atom with a diatomic molecule is the study of the reaction of K with vibrationally excited HCl,[543]

$$K + HCl(v = 1) \rightarrow KCl + H. \tag{7.3}$$

The experiment uses infrared light from a chemical laser to prepare HCl in a molecular beam with one quantum of vibrational excitation, and employs surface ionization to detect the KCl product scattered after the HCl beam crosses a beam of K atoms. The 36 kJ mol^{-1} of vibrational energy in the HCl increases the reaction cross section by at least 100 fold over the ground vibrational state, but a comparable amount of energy in relative translation increases the cross section by only a factor of four.[544] The effect of translational excitation on the cross section is well described by a line-of-centres model (see Chapter 1). This behaviour is consistent with a relatively late barrier in this nearly thermoneutral reaction.

The situation is more complicated for reactions involving polyatomic molecules, and a particularly informative example is the reaction of methane with chlorine atoms (see also Chapter 4),

$$Cl + CH_4 \rightarrow HCl + CH_3. \tag{7.4}$$

This reaction has become a prototype for state-specific chemistry in a polyatomic system.[546] The reaction is endothermic by about 8 kJ mol^{-1}, with a reaction barrier of about 15 kJ mol^{-1}, and it has essentially a linear Cl–H–CH_3 geometry at the barrier to reaction. It is most useful to consider isotopically substituted methane in examining bond- and mode-selective reactions, since different outcomes will correspond to isotopically different products which may be identified by spectroscopic or mass spectrometric detection schemes. Figure 7.3 shows the relative cross sections for the reaction of CHD_3 molecules in their ground state, $CHD_3(v=0)$, and with a quantum of CH stretching excitation, $CHD_3(v_{CH}=1)$, as a function of the collision energy. These data were obtained in a crossed molecular beam experiment using infrared laser excitation to prepare the

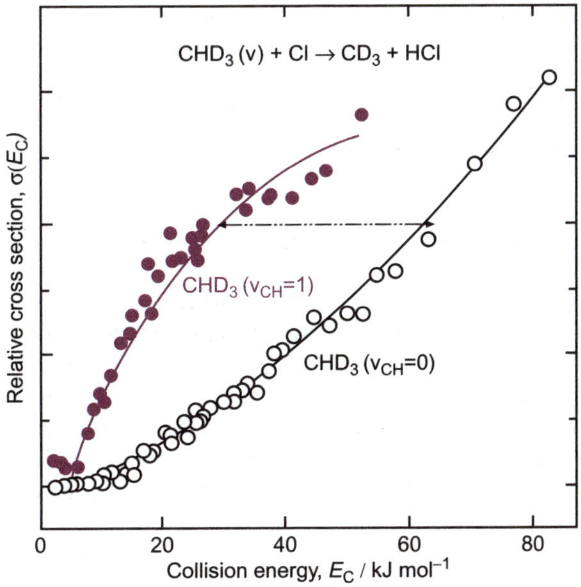

Figure 7.3 The variation in the integral cross sections for the reaction $Cl + CHD_3(v_{CH}) \rightarrow CD_3 + HCl$ with $v_{CH}=0$ and $v_{CH}=1$. Note that although the cross section increases with vibrational excitation of the reactants, the enhancement is about the same as that obtained if the same amount of energy is placed in reactant translation. Adapted from ref. [545].

vibrationally excited CHD_3, and ion imaging to detect the CD_3 products[545] (see Study Box 5.3 and Section 6.4). Both cross sections grow with increasing translational energy, and the cross section for $CHD_3(v_{CH}=1)$ is larger at low translational energy because the molecule contains 36 kJ mol^{-1} of vibrational energy. Thus, these measurements demonstrate that vibrational excitation increases the reaction probability substantially. However, comparing the cross sections for the vibrationally excited and the vibrational ground state molecules also shows that vibration is no more effective than translation in enhancing the reaction probability. The horizontal dotted line in the figure shows that adding 36 kJ mol^{-1} of translational energy to $CHD_3(v=0)$ molecules at a collision energy of about $E_c = 30$ kJ mol^{-1} increases the cross section to the same value as that of molecules with the additional energy in vibration. The simple ideas of Polanyi rules for an atom reacting with a diatomic molecule do not apply to the H-atom abstraction reaction of this polyatomic molecule, for which translation and vibration are comparably effective.

The reaction of another methane isotopomer, CH_3D, with Cl illustrates another subtle aspect of state-resolved reactions. The symmetric C–H stretching vibration (ν_1) and the antisymmetric C–H stretching vibration (ν_4) of CH_3D lie within 100 cm^{-1} of each other, and classically differ only in the phase of their motions. However, $CH_3D(\nu_1)$ molecules are seven times more reactive than $CH_3D(\nu_4)$ molecules.[547] The critical aspect of this striking difference is the change in the initially excited vibration during the collision. The perturbation by the incoming Cl atom changes the vibrational motion differently for the two modes, as shown by a simplified calculation of the evolution of the states as the Cl approaches along one of the C–H bonds. The interaction transforms the symmetric stretching vibration into motion of the C–H bond pointing toward the incoming Cl atom, which favors the reaction. By contrast, it transforms the antisymmetric stretching vibration into motion of atoms away from the incoming atom, a situation that is much less favorable for reaction.

7.3.2 Changing the Reaction Pathways: Populating Different States

The shape of the potential energy surface strongly influences the disposal of energy into the reaction products, and the influence of the initial state i on the population of the product states f can be particularly dramatic for endoergic or thermoneutral reactions. In those cases, we are not concerned with the release of the reaction exoergicity but rather with the fate of the energy in excess of the endoergicity. The reaction of hydrogen, chlorine, or oxygen atoms with vibrationally excited water is an example,

$$X + H_2O(v_1 v_2 v_3) \rightarrow HX(v_{HX}) + OH(v_{OH}). \tag{7.5}$$

The three vibrational quantum numbers of water in the above equation correspond to the symmetric O–H stretching (ν_1), the bending (ν_2), and the antisymmetric O–H stretching (ν_3) normal modes. As the sketch along the reaction coordinate in Figure 7.4 shows, the reaction is endothermic by about 70 kJ mol^{-1} and has a barrier of about 100 kJ mol^{-1}. Experiments using laser light to excite overtone vibrations in water followed by laser-induced fluorescence to observe the OH reaction product have shown that vibrational excitation overcomes the barrier to reaction effectively. For example, four quanta of O–H stretching excitation make the reaction proceed at the gas-kinetic collision rate.[549]

Considering the influence of initial vibrational excitation on the reaction of water and other symmetric molecules is most convenient if we use the *local mode* description of the stretching vibrations in water[548] (for an introduction to local modes, see Study Box 7.2). Because water has two identical O–H bonds, there are no eigenstates of the molecule corresponding to excitation of either one alone. The *normal mode* description uses harmonic oscillators for the collective stretching motions, and in the case of water the two stretching vibrations are the symmetric and antisymmetric

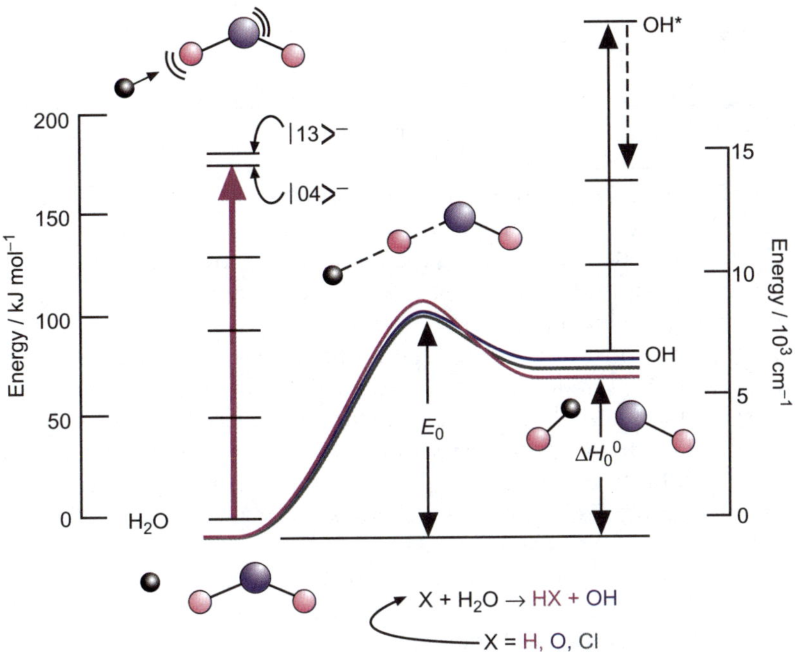

Figure 7.4 Illustration of the energetics for the reaction $X+H_2O \rightarrow HX+OH$. See text for discussion. Adapted from ref. [549].

stretch involving the two O–H bonds. In contrast, the *local mode* description uses anharmonic oscillators corresponding to each of the identical bonds individually. The local mode eigenstates are composed of the appropriate linear combinations of anharmonic oscillators required by the symmetry of the molecule. The normal mode picture is useful because it has the coupling between the two vibrations built in to it, but it does not include the anharmonicity of the oscillators. On the other hand, the local mode picture has the anharmonicity built in, but it lacks the coupling between the bonds. Which description we choose depends on whether the coupling between the bonds or the anharmonicity of the individual bonds is more important. The local mode description is particularly useful for states of oscillators containing light atoms with several quanta of excitation, and we use it in discussing state-selected reactions of water and methane.

The local mode description of a water molecule containing four quanta of O–H stretching excitation designates the states with all four quanta in one bond as

$$|04\rangle^{\pm} = \frac{1}{\sqrt{2}}(|04\rangle \pm |40\rangle), \tag{7.6}$$

where the two quantum numbers designate the stretching excitation of the two individual bonds. The symmetric (+) combination has the same symmetry (A_1) as the symmetric stretching vibration, and the antisymmetric (–) combination has the same symmetry (B_2) as the antisymmetric stretching vibration in the C_{2v} point group of the water molecule.[548] A water molecule with four quanta of excitation could also have one quantum in one of the bonds and three in another, corresponding to the states $|13\rangle^{\pm}$, or, indeed, two quanta in each bond, $|22\rangle$. It is possible to excite either the $|04\rangle^{-}$ or the $|13\rangle^{-}$ state with laser light and detect the $OH(\nu_{OH})$ reaction product with laser-induced fluorescence in order to observe the influence of the initial state on the product state populations. Figure 7.4 shows that the energies of the two states are very similar, and that both lie well above the

STUDY BOX 7.2: LOCAL MODES

Most chemists are very familiar with the normal mode treatment of molecular vibrations. Normal modes provide an excellent description of molecular vibrations for low values of the vibrational quantum number, v, *i.e.* for the harmonic region of the vibrational potential. However, this model often breaks down for highly vibrationally excited molecules. We can demonstrate this by considering a simple example, that of methane highly vibrationally excited in the symmetric stretch mode. According to the normal mode model, as more and more energy is added into the symmetric stretch mode, it is shared equally amongst the vibrations of each C–H bond, and might eventually be expected to lead to dissociation of the molecule, with loss of all four H atoms. Contrast this with the fact that intuitively we know that it is much more likely that once the C–H bond dissociation energy is exceeded, the molecule will lose a single H atom. We would not expect the molecule to be able to contain nearly four times this energy before fragmenting, as we might predict from the normal mode model. So far we have neglected the anharmonicity of the vibrational potential. At high quantum numbers this leads to coupling between different normal modes, but even so, there is no single normal mode of CH_4 in which vibration is localized in a particular C–H bond.

For this reason, the local mode description of molecular vibrations was developed to treat highly vibrationally excited molecules[548]. In this treatment we consider the vibrational motion of the molecule in terms of vibrations of individual bonds ('local modes') rather than collective vibrations ('normal modes') of the whole molecule. Each of these local modes may be treated as a Morse oscillator in much the same way as the normal modes are for low vibrational energies. The overall vibrational motion (and vibrational energy) of the molecule at any energy may be written either as a linear combination of local modes or of normal modes, with, in general, normal modes providing a more intuitive picture of the vibrational motion at low energies, and local modes providing a better description at high energies. Precisely which picture provides the more convenient description of the vibrational motion in a given situation depends on the relative magnitudes of the anharmonicity and inter-mode coupling[548]. If the anharmonicity is substantially larger than the inter-mode coupling, local modes provide the best description, but if the inter-mode coupling dominates, normal modes are better.

Claire Vallance

barrier to reaction, consistent with the ability of four quanta of excitation to make the reaction proceed at the gas kinetic collision rate.

The measurements find that $H_2O(|04\rangle^-)$ reacts with either H or Cl atoms to produce almost exclusively vibrationally unexcited hydroxyl radicals $OH(v_{OH}=0)$ but that reaction of $H_2O(|13\rangle^-)$ produces primarily vibrationally excited radicals $OH(v_{OH}=1)$. The bond with the largest amount of vibrational energy breaks, and the surviving bond retains its initial excitation. In this case, the surviving bond is largely a *spectator* in the reaction. Microscopic reversibility, in which one runs trajectories backwards to predict the ability of different excitations to promote the reverse reaction, suggests that the reverse reaction of HX with OH,

$$HX(v_{HX}) + OH(v_{OH}) \rightarrow X + H_2O(v_1, v_2, v_3), \tag{7.7}$$

should preferentially populate O–H stretching states of water and that vibrational excitation of OH should be less effective than excitation of HX in accelerating the reaction. Indeed, the molecular beam scattering experiments already presented in Section 6.3.2 show that the reaction of OH with

D_2 preferentially forms HOD($v_{OH}=2$).[491] Also, vibrational excitation of the "spectator" OH($v_{OH}=1$) makes little difference in the rate of the reverse reaction of H_2+OH, but excitation of $H_2(v_{H_2}=1)$ increases the rate by a factor of 100.[550–553]

Many of these same ideas about non-reacting bonds being spectators apply to the reaction of Cl with the methane isotopomers, CH_3D, CH_2D_2, and CHD_3. For example, it is possible to prepare CH_3D in either the $|200\rangle$ or the $|110\rangle$ local mode states, where each quantum number represents the excitation in one of the C–H bonds. (Once again appropriate linear combinations produce the A_1 and E symmetry states required in the C_{3v} point group of the molecule.) The state with all of the excitation in one bond ($|200\rangle$) leads primarily to ground vibrational state CH_2D radicals,

$$Cl + CH_3D(|200\rangle) \rightarrow HCl + CH_2D(v_{CH}=0), \tag{7.8}$$

but the state with excitation in two different bonds leads primarily to radicals with C–H stretching excitation,

$$Cl + CH_3D(|110\rangle) \rightarrow HCl + CH_2D(v_{CH}=1). \tag{7.9}$$

The surviving bonds are largely spectators to the reaction, with the initial excitation in the non-reacting bond appearing in that bond in the product.[554] Scattering experiments of the type discussed in Chapter 6, along with other laser based approaches, have explored reactions of the methane isotopomer CHD_3 and have shown that many of the HCl products are vibrationally excited and scattered forward, in the direction of the incoming Cl atom.[545] The spectator picture is not perfect, but it does describe the general features of the energy disposal in these state-to-state reactions.

7.3.3 Changing the Reaction Pathways: Breaking Different Bonds

The examples above involve state-to-state reactions in which preparing a state i steers a reaction to form particular states f preferentially. What about the possibility of steering the reaction to generate chemically different products? Can one choose state i such that it preferentially promotes cleavage of one bond in the molecule? Steering the reaction from a selected initial state to a particular final state is usually referred to as *state-selective chemistry*. In a similar vein, using initial excitation to steer a reaction to cleave one bond preferentially is known as *bond-selective* chemistry. The two are closely related, but differ in that the former involves taking different paths over the same barrier on the PES, while the latter involves taking a path over one barrier in preference to another in order to form a different product.

A conceptually simple means of achieving bond-selective chemistry is to deposit energy into a motion of the system that moves it along the preferred reaction coordinate. However, several features of real molecules complicate this simple view. One is that the molecular eigenstates that one prepares with a laser are often not simple bond motions, but are more complicated combinations of these motions, as described below in Section 7.4. Another is that the interaction of the reactants in a bimolecular reaction can redistribute the energy of the collision partners. Nonetheless, the ability to conduct state-selective reactions in favourable cases suggests that it should also be possible to perform bond-selective reactions. To achieve this goal one needs to identify initial quantum states that lead the system over one barrier in preference to another. Vibrational excitation is a particularly likely candidate because it places energy into relative displacement of the atoms within a molecule, a motion that can have a large component along a preferred reaction coordinate.

A simple example is the reaction of water with H or Cl, described above. One can excite a molecular eigenstate in the water isotopomer HOD either with a large component of O–H stretching or O–D stretching motion. It is possible to identify the bond that breaks by detecting either the OH or OD product using laser-induced fluorescence,

$$X + HOD(\nu_{OH}, \nu_{bend}, \nu_{OD}) \rightarrow HX(\nu'_{HX}) + OD(\nu'_{OD}) \quad (7.10a)$$

$$\rightarrow DX(\nu'_{DX}) + OH(\nu'_{OH}). \quad (7.10b)$$

The reaction is very bond selective. For example, the reaction of H with HOD having four quanta of O–H stretching excitation ($\nu_{OH} = 4$) produces at least a 200-fold excess of OD over OH, signaling the preferential cleavage of the initially excited O–H bond. Similarly, preparing HOD with five quanta of O–D stretching excitation ($\nu_{OD} = 5$), which has about the same total vibrational energy, preferentially cleaves the O–D bond to produce almost solely OH fragments.[555] The high frequencies of the vibrations in water and the weak coupling between the bonds make water a particularly attractive molecule in which to direct the reaction along one reaction channel in preference to another, but methane, which has some of the same characteristics, is an even richer example.

It is possible to steer the reaction of each of the isotopomers of methane, CH_3D, CH_2D_2, and CD_3H, with chlorine to preferentially cleave either a C–H or a C–D bond, depending on the initial excitation.[554,556–558] For example, the reaction of CH_3D with either the C–D or the C–H stretching vibration excited selectively cleaves the excited bond, preferentially passing over one of the two barriers shown schematically in Figure 7.5. In addition, the spectator picture describes the fate of the vibrational energy in the surviving bond. Reaction of CH_3D with both a quantum of C–H stretching and a quantum of C–D stretching excited leaves the bond that survives with a quantum of vibrational excitation,[559]

$$Cl + CH_3D(\nu_{CH} = 1, \nu_{CD} = 1) \rightarrow CH_3(\nu'_{CH} = 1) + DCl \quad (7.11a)$$

$$\rightarrow CH_2D(\nu'_{CD} = 1) + HCl. \quad (7.11b)$$

The energy levels marked on Figure 7.5 show all of the different states for which it is possible to observe state- and bond-selective chemistry for CH_3D. This selectivity is not limited to the bimolecular reactions described here; it is also observed in the dissociative adsorption of methane isotopomers on a Ni surface, for example. In this system, different vibrational modes promote the adsorption preferentially,[560] and vibrational excitation of the C–H bond in CHD_3 promotes its cleavage.[561] Bond-selective chemistry at surfaces is discussed more in Section 10.5.4.

7.4 VIBRATIONAL ENERGY FLOW

The quantum-state-resolved reactions described in the examples above are direct reactions, in which the time taken for the reactants to interact is short compared to that required for vibrational energy to flow around them. This limit is key to state- and bond-selective chemistry, which requires initial preparation and survival of a state that preferentially carries the system over the barrier to the desired products.[562] This requirement leads naturally to considering the nature of states prepared by laser excitation. If every energetically accessible state were populated during the course of the reaction, only the total energy and not the identity of the initially prepared state would matter. It is natural to think in terms of bond excitation, even though, as described below, excitation with a laser may not prepare such states. If one could prepare a vibration of a single bond or even deposit energy selectively into just one part of a molecule, how long would it survive? It is now possible to answer that question using time-resolved (see Chapter 11) or frequency resolved laser spectroscopy (see ref. [563]), but a pioneering gas phase experiment on a chemical activation reaction provided the first direct measure of the timescale.

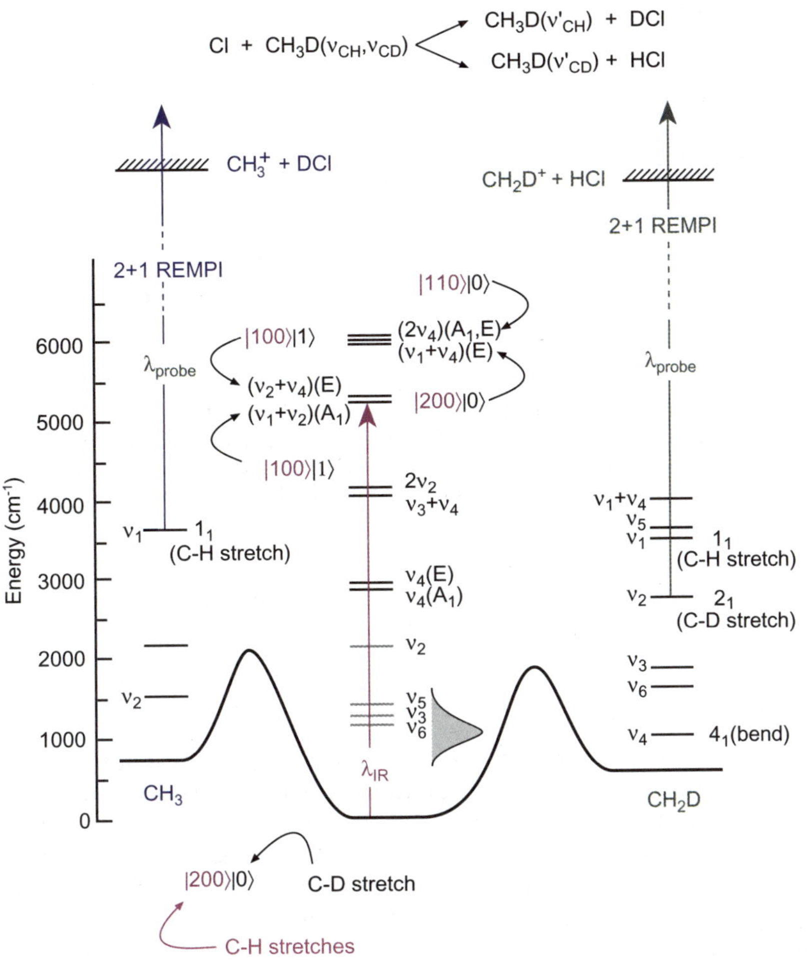

Figure 7.5 Energy profile for the reaction of Cl atoms with $CH_3D(\nu_{CH}, \nu_{CD})$. The pathway to produce CH_3 is shown on the left, while that to produce CH_2D is shown on the right. Adapted from ref. [546].

7.4.1 Chemical Activation: Rabinovitch's Bicycle

The key to that experiment is the addition of photolytically produced deuterated methylene radicals ($^1CD_2$) across the double bond of the cyclopropane derivative shown in Figure 7.6. This exothermic addition forms a bicyclic compound ("Rabinovitch's Bicycle") with the excess energy localized in the newly formed ring. The initially localized energy can cause the adjacent CF_2 to depart, but, if the energy flows to the other end of the molecule, the CF_2 at that end can leave instead. At low pressures, when there is enough time for the energy to flow around the molecule prior to a collision, both processes occur, and fragments from both products appear in the mass spectrum. However, a collision with a buffer gas added to the reaction mixture can stabilize the excited molecule by removing energy. These collisions serve as a "clock" for the energy flow by quenching the excited molecule before the energy can flow to the other end. Indeed, the relative yield of fragments from prompt loss of CF_2 increases sharply when the pressure is high enough that collisions remove the excitation energy before it can flow to the other end of the molecule. Careful analysis of this experiment shows that the energy flow occurs in a few ps in this highly vibrationally excited molecule.[564,565] Because the vibrational period for a 3000 cm^{-1} C–H stretching vibration is about

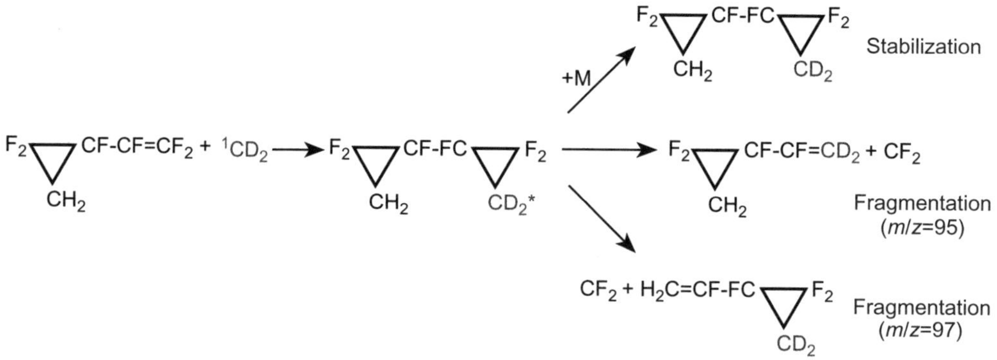

Figure 7.6 Formation of the bicyclic compound, "Rabinovitch's Bicycle",[564,565] which can fragment in two ways, or be stabilized by collision. Collisional stabilization provides the 'clock' by which the timescale for intramolecular redistribution can be evaluated. The two monocyclic products of fragmentation can be distinguished by their different mass spectra, with peaks at the m/z values indicated.

10 fs and that for a 300 cm^{-1} skeletal vibration is about 100 fs, the bonds in a molecule vibrate anywhere from 10 to 100 times during a 1 ps relaxation time.

7.4.2 Laser Excitation

Most studies of state resolved chemistry prepare reactants by laser excitation, making an understanding of the connection between narrow bandwidth excitation of a molecular eigenstate and the flow of energy in a molecule particularly important. It is especially useful to consider the eigenstates in terms of chemically intuitive coordinates, such as bond stretches and bends, in order to see how the composition of the eigenstate reflects the couplings amongst the bonds in a molecule. These are the same couplings that determine the flow of energy out of an initially excited bond into other parts of the molecule, and, as described below, vibrational energy flow and coupling amongst the bonds within a molecule are two sides of the same coin.

A simple description of the structure and coupling of vibrational states illustrates the connection between the couplings amongst the bonds and the flow of vibrational energy within a molecule. Figure 7.7(a) shows the coupling of a bright (optically accessible) zero-order state, $|s\rangle$, to a dark zero-order state, $|l\rangle$, to produce two states $|1\rangle$ and $|2\rangle$, which are eigenstates of the full molecular Hamiltonian. The *zero-order states*, which are eigenstates of some approximate Hamiltonian, are useful because we choose them to describe motions about which we have physical and chemical intuition. For example, the bright zero-order state could be a fundamental C–H stretching vibration, and the dark zero-order state could be two quanta of bending excitation. In this picture, the bright state $|s\rangle$ has an electric dipole transition from the ground state while the dark state $|l\rangle$ does not. The full Hamiltonian of the molecule, expressed using the basis set of the zero-order states, contains terms, such as anharmonicity, that couple the zero-order states to form the molecular eigenstates. Common usage often names the molecular eigenstates in terms of their dominant zero-order state character even though the eigenstate designated as the "symmetric C–H stretch" might have other states mixed into it. In the two-state example in Figure 7.7, we can write the eigenstate $|n\rangle$ as a linear combination of the two zero-order states,

$$|n\rangle = c_s^n|s\rangle + c_l^n|l\rangle. \tag{7.12}$$

The terms in the full Hamiltonian that couple the zero-order states determine the magnitudes of the coefficients c_s^n and c_l^n in the linear combination, and thus the amount of the bright zero-order state

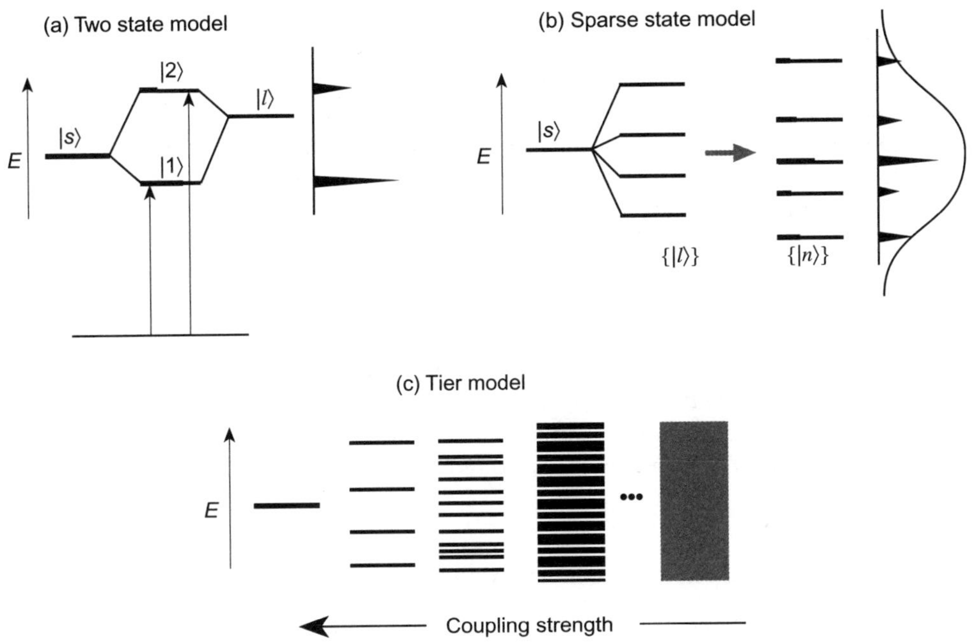

Figure 7.7 Illustration of the mechanism of intramolecular redistribution, in the case of (a) two interacting states, (b) a sparse manifold of states, and (c) the 'tier' model applicable in the limit of a high density of states.

$|s\rangle$ that each of the eigenstates contains. The amount of bright state in each eigenstate in turn determines the intensity of a transition to that eigenstate, as illustrated by the schematic absorption spectrum to the right of the energy level diagram. In that spectrum, the intensity of each transition is proportional to the square of the bright-state coefficient, $|c_s^n|^2$.

Figure 7.7(b) shows a more general situation in which there is a set of dark zero-order states $|l\rangle$ coupled to the bright state to form a set of molecular eigenstates $|n\rangle$. We can write each eigenstate as the corresponding linear combination,

$$|n\rangle = c_s^n|s\rangle + \sum_l c_l^n|l\rangle, \tag{7.13}$$

and the intensities again depend on the amount of the bright state mixed into each eigenstate. A complete analysis of the eigenstate energies and transition intensities in the spectrum predicts the time evolution of the bright state $|s\rangle$, even though the measurement does not excite the bright state.[563] On the other hand, a short pulse of light that contains energies encompassing the entire range of eigenstates sharing the zero-order bright state *does* initially prepare the bright state $|s\rangle$. This spectrally broad pulse creates a bright state by forming a coherent superposition of the molecular eigenstates. Because $|s\rangle$ is not a stationary state of the molecule, its population evolves in time and can have characteristic recurrences, often called quantum beats,[566] the frequencies of which are determined by the energy level spacings (see Problems (1) and (2)). The time evolution reflects the flow of energy in the system subsequent to initial preparation of the bright state and, thus, follows the loss of energy from an initially excited bond. When the number of coupled states is large enough, population leaves the initially excited bright state and does not return during the lifetime of the molecule. In this case, it is possible to describe the energy flow as passing through tiers of zero-order states divided according to the strength of their coupling to the bright state, as Figure 7.7(c) illustrates.

The experiments described in the examples of state-to-state reactions in Section 7.3 use high resolution lasers to prepare molecular eigenstates that are mixtures of a *reactive* bright state, such as a C–H or C–D stretch in the case of CH_3D, and *non-reactive* dark states. The key to the state- and bond-selectivity in those cases is the extent of mixing. Even if several dark states are part of the eigenstate, the reactive bright state component of the eigenstate, dictated by the coefficient c_s^n, determines the reactivity. Thus, an eigenstate with a large C–H stretching component preferentially transfers an H atom to Cl in the reaction, but an eigenstate with a large C–D stretching component transfers a D atom. Even though the eigenstates are not the pure bond stretching states that are part of our chemical intuition, they do contain a dominant state that provides the enhanced reactivity and bond selectivity. In the limit of a completely mixed eigenstate, in which the reactive state is not a greater component of one eigenstate than it is of another, all of the eigenstates have the same reaction probability. As the examples show, there are molecules in which the coupling is limited, the eigenstates are not completely mixed, and state-controlled chemistry is possible.

7.4.3 Intramolecular Vibrational Redistribution: Observing Energy Flow

The couplings that manifest themselves as the state structure and intensities in a high resolution spectrum or as the time-evolution in a time-resolved experiment are responsible for the flow of vibrational energy in an isolated molecule, a process generally known as intramolecular vibrational energy redistribution or intramolecular vibrational relaxation (IVR). This process is a key aspect of describing reactions because the situation is very different depending on whether or not IVR is complete on the timescale of a reaction. In principle, a measurement with high frequency resolution contains the same information as one with high time resolution, since both report on the couplings in the molecule and are related by a Fourier transform.[563] In practice, it is often convenient to use one technique in preference to another depending on the details of the system and the relevant timescales.

One example that provides both time- and frequency-domain data is a "fluorescence depletion experiment", which has been used to observe intramolecular vibrational energy flow in the electronically excited state of the fluorene molecule.[567] Figure 7.8 shows the results of the experiment, in which one 6 ps laser pulse excites the molecule and another identical but time-delayed pulse stimulates emission to the ground state, shutting off (depleting) the fluorescence. The first pulse is short enough to prepare a single bright vibrational state in the electronically excited state. If the initially excited molecule leaves the bright vibrational state before the depleting pulse arrives, there is no decrease in the fluorescence, but if the pulse arrives when there is amplitude in the bright state, the stimulated emission depletes the fluorescence. The traces on the left of the figure show the time evolution for three different levels of vibrational excitation in the excited state, and the traces on the right are the corresponding dispersed emission spectra showing transitions from the excited vibrational state to different levels in the ground state. The bottom traces are for excitation of a vibration lying 834 cm^{-1} above the origin. There are a few sharp transitions in the spectrum on the right and a few clear recurrences (quantum beats) in the time evolution on the left. The presence of a few sharp transitions on the right reflects coupling among only a few zero-order states, and the quantum beats on the left reflect the population flowing back and forth from the initially prepared bright state to those zero-order states. At a higher level of vibrational excitation (1230 cm^{-1}), there are more transitions in the spectrum, which are not completely resolved, and there are, correspondingly, more quantum beats in the time evolution. Finally, at the highest vibrational excitation level shown (1707 cm^{-1}), there is little structure in the spectrum due to the large number of states involved, and essentially no recurrences in the time evolution. The decay is much faster in this

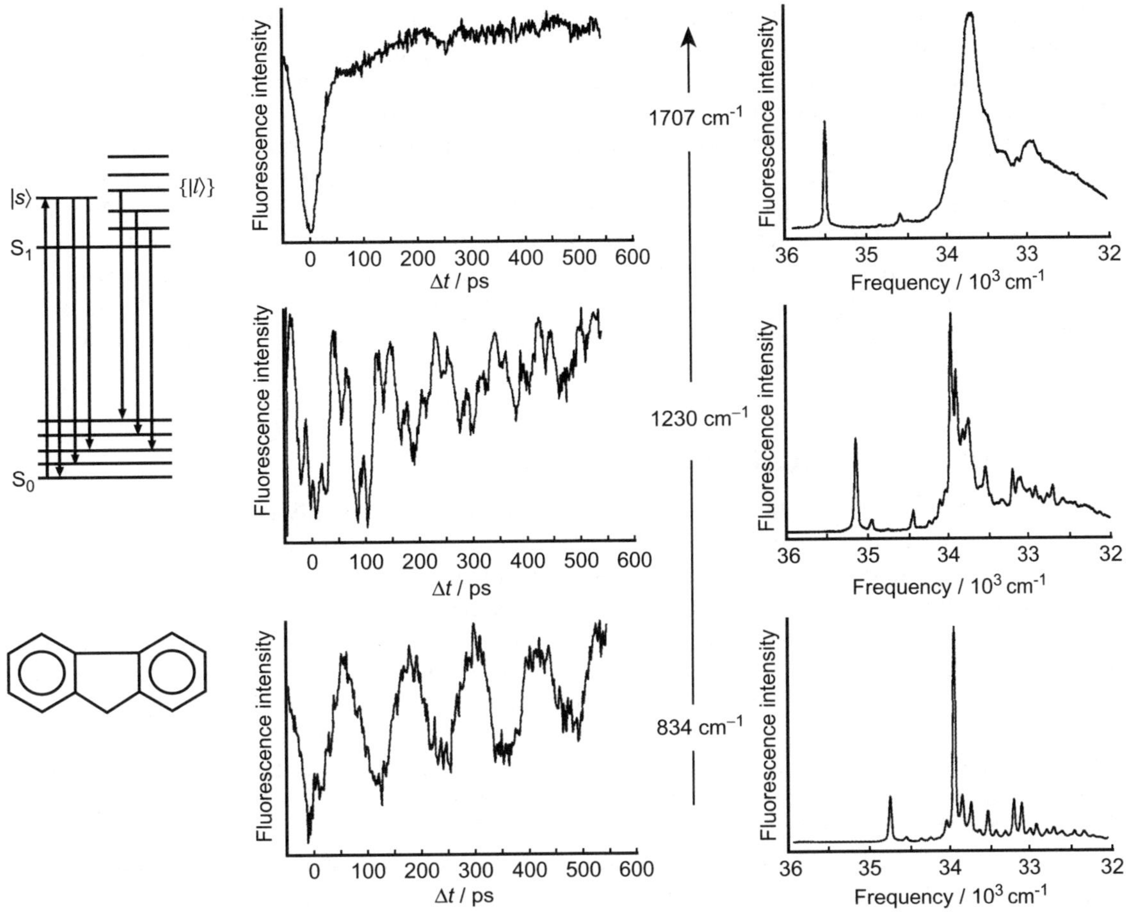

Figure 7.8 Fluorescence depletion experiments on the fluorene molecule (the molecule is shown at the bottom left). The excitation scheme is shown on the left, the middle panel shows the time dependence of the fluorescence depletion signal, while the panels on the right show the dispersed emission spectra. The dispersed fluorescence spectra were measured with an excimer pumped dye laser with a 0.35 cm^{-1} bandwidth. As the internal energy in the excited state increases (in the direction of the arrow), the emission spectra become more complex, and the quantum beat structures in the time-dependent spectra become washed out. Adapted from ref. [567].

regime, reflecting the flow of population out of the prepared vibrationally excited state in less than 20 ps, with no return of the population on the timescales of the experiment.

Other time-domain experiments carried out on electronically excited[568] and ground state molecules[569,570] at different levels of vibrational excitation tell the same story. As the level of vibrational excitation increases, there are more zero-order states nearby to which the bright state can couple, and the energy flow out of the bright state becomes faster. Frequency-domain experiments, in which one analyses the spectra, prove the same point.[563] An important factor in determining the rate of energy flow is the number of quantum states per unit energy, referred to as the density of states, ρ. As we have seen, the density of states is usually a rapidly increasing function of vibrational energy content of the molecule.[571] As a general guideline, vibrational energy flow in a molecule becomes fast when there are on the order of 100 rovibrational states within a 1 cm^{-1} range of energy,[572] so a typical threshold value for extensive energy redistribution corresponds to a density of states $\rho \approx 100$ states/cm^{-1}. There are large variations in this guideline value for reasons that are obvious in Figure 7.7. The energy flow depends on the couplings amongst the levels and on the number of states that are strongly or weakly coupled, as the tier structure in the figure illustrates. Because the details of the state structure are crucial, the density of strongly coupled states does not necessarily grow as steeply as the total density of states as the size of the molecule increases. Energy may flow into states that belong to weakly coupled tiers much more slowly than in flows out of the bright state because it must pass through intervening tiers.[573,574] With these ideas of vibrational energy flow and its relation to the couplings in molecules in mind, we need to consider the opposite limit from that of direct reactions. In direct reactions, we are able to prepare molecules in which the initial states are not highly mixed. How should we describe reactions in which the states *are* extensively mixed or, in time-domain language, in which the flow of energy around the molecule is complete on the timescale of the reaction?

7.5 STATISTICAL REACTIONS

The essential feature of statistical reactions is that the rate depends only on the total energy and angular momentum of the reacting molecules. The concept of motion across the potential energy surface is essential in considering the limit of statistical reactions, just as it is in understanding the direct bond- and state-selective reactions described in Section 7.3. Figure 7.9 shows minimum energy paths across three potential energy surfaces. The first, path (a), is for a bimolecular reaction whose surface has a single maximum, much like those we examined in Section 7.2 for reactions such as $F + H_2$. The second, path (b), is for a bimolecular reaction with a potential energy well along the reaction coordinate, as found in the reaction of $O(^1D) + H_2$, which inserts the oxygen into the H–H bond to form excited water in the course of the reaction (see Chapters 4 and 6). The third path is for a unimolecular isomerization in which an energized molecule passes over a barrier to rearrange, as for example in the isomerization of HCN to HNC. In each of these cases, the reaction rate might or might not vary with the initially prepared quantum state of the system, depending on the time required for reaction compared with that required for energy redistribution in the molecule. For example, in the case of a reaction with a deep well in the PES, it is possible that the time spent in the well is long compared to the IVR time and that the initial states in which the reactants are prepared does not matter. Similarly, the unimolecular isomerization is another likely candidate for extensive state mixing during the reaction.

7.5.1 The Cumulative Reaction Probability

The *cumulative reaction probability*, $N(\varepsilon)$ is a helpful concept in examining the reaction rate constant for a variety of situations.[10,575] It is the sum of the reaction probabilities, $P_i(\varepsilon)$, for each initial

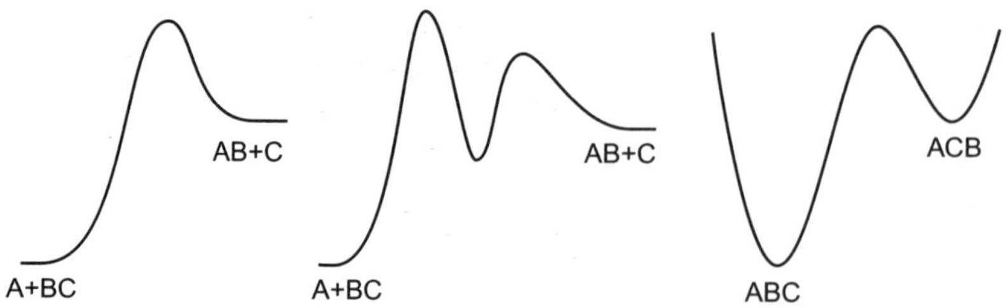

Figure 7.9 Potential energy profiles along the minimum energy path (the reaction coordinate) for a direct reaction with a barrier (left), a reaction with a potential energy well (middle), and a unimolecular isomerization reaction (right).

state i of a system at total energy ε,

$$N(\varepsilon) = \sum_i P_i(\varepsilon) = \sum_n g_n P_n(\varepsilon), \tag{7.14}$$

where the first sum runs over individual quantum states, and the second sum runs over energy levels, each of which have a degeneracy of g_n. The states in the sum include the orbital angular momentum of the collision pair in the case of a bimolecular reaction. The utility of the cumulative reaction probability is that it has a simple relation to the rate constant at the energy ε. This microcanonical rate constant is

$$k(\varepsilon) = \frac{N(\varepsilon)}{h\rho(\varepsilon)}, \tag{7.15}$$

where $\rho(\varepsilon)$ is the total density of states for the system and h is Planck's constant. A convenient means of obtaining $N(\varepsilon)$ that does not involve a complete scattering calculation would be very useful, and in the limit of statistical reactions, transition state theory provides it (see Chapter 1 and further discussion below). In general, however, each different level has a reaction probability that depends on its quantum state, as described in Section 7.3. The vertical arrows on the reaction coordinate in Figure 7.10 show one possible division of the total energy $\varepsilon = \varepsilon_t + \varepsilon_n$ between the relative translational energy ε_t and the internal energy ε_n. If the total energy ε were below the barrier energy ε_0, the classical reaction probability would be zero, but quantum tunnelling would give a finite probability just below the barrier. The sum in the cumulative reaction probability includes all possible divisions of the energy between internal energy and translational energy.

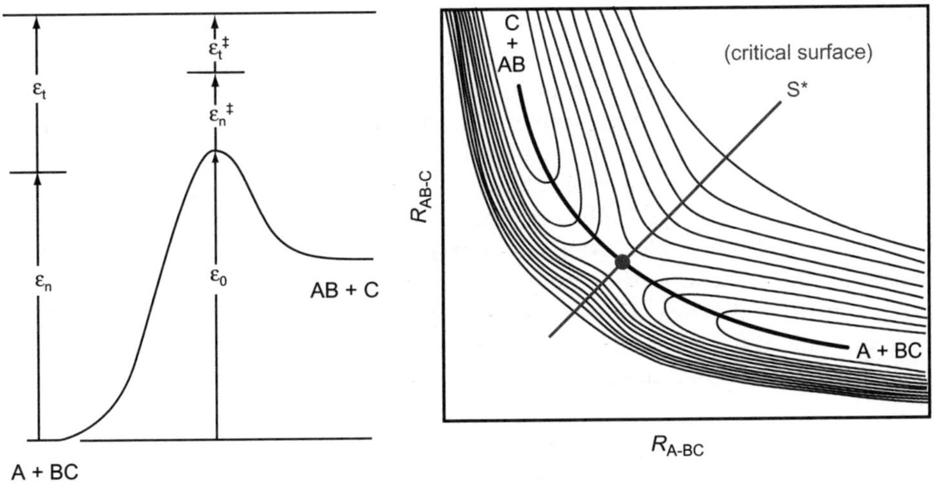

Figure 7.10 The panel on the left shows the potential energy along the reaction path for a simple chemical reaction, with an illustration of the partitioning of energy between translational and internal excitation in the reactants and transition state, which here is located at the barrier. The right panel shows the critical dividing surface which defines the location of the transition state for such a reaction, and separates reactants from products.

7.5.2 The Statistical Assumption: Microcanonical Transition State Theory

A few assumptions lead to a particularly simple form of the cumulative reaction probability. We treat the motion along the reaction coordinate as a *separable, classical* translation that carries the system over the transition state and assume that the system *always goes on to products* once it crosses the transition state. Thus, it is possible at the transition state to divide the available energy, $\varepsilon^{\ddagger}=\varepsilon-\varepsilon_0$, between translation along the reaction coordinate, $\varepsilon_t^{\ddagger}$, and all of the other modes perpendicular to it, $\varepsilon_n^{\ddagger}$. Each of the possible divisions of energy corresponds to a different internal level at the barrier, from which the system goes on to form products. Thus, the reaction probability is $P_n(\varepsilon)=1$ for each of these levels and zero otherwise, a result that makes the calculation of the cumulative reaction probability very straightforward. We just need to count all of the levels above the threshold to the reaction, since each one has unit reaction probability,

$$N(\varepsilon) \rightarrow N^{\ddagger}(\varepsilon) = \sum g_n^{\ddagger}, \tag{7.16}$$

where the sum runs over all levels above the threshold to reaction. In this limit of every state at the barrier reacting with the same (unit) probability, the cumulative reaction probability becomes simply the sum of all the transition states above threshold. The energies of these states depend upon the shape of the potential energy surface at the barrier: the geometry of the molecular system at the barrier determines the rotational constants, and the curvature along the perpendicular coordinates determines the vibrational frequencies. In this transition state theory limit, the microcanonical rate constant is

$$k(\varepsilon) = \frac{N^{\ddagger}(\varepsilon)}{h\rho(\varepsilon)}. \tag{7.17}$$

Knowing the structure at the transition state allows the calculation of the sum of states, $N^{\ddagger}(\varepsilon)$, and knowing the structure of the reactants allows the calculation of the density of states, $\rho(\varepsilon)$. The calculation is particularly useful for unimolecular reactions, such as illustrated in Figure 7.9(c). In that case, the density of states is just that for the energized molecule, and the sum of states is that for the same molecule at the geometry of the transition state, where the total energy available is $\varepsilon^{\ddagger}=\varepsilon-\varepsilon_0$. The microcanonical transition state theory applied to unimolecular reactions is often known as RRKM (Rice-Ramsperger-Kassel-Marcus) theory after the pioneers of this approach.[571,576–578] (See Problem (3) for a further discussion of the statistical treatment of the microcanonical rate constant.) The applicability of statistical theories rests on the assumption of a complete sampling of all energetically available states by both the energized reactants and the molecule at the energy barrier. The first assumption permits use of the total density of states, $\rho(\varepsilon)$, in the expression for the rate constant. If the system sampled only a subset of states, we would have to use a restricted density of states. The second assumption allows us to count all of the levels at the transition state without having to identify ones that preferentially lead to products. The rapid flow of energy in the reacting molecules makes these assumptions realistic and is a crucial requirement for applying statistical theories.

There are several means of obtaining the expression for the microcanonical transition state theory rate constant and several different means of approximating the sum and density of states.[571,576,577,579] The essential physical content of the expression is comparing the number of ways there are of passing through the transition state, $N^{\ddagger}(\varepsilon)$, to the number of ways there are for a molecule of that energy to exist away from the transition state. This view amounts to comparing the number of cells in phase space (see Study Box 7.3) that are on the critical dividing surface between reactants and products, marked as S^* in Figure 7.10, to the total number of cells available to the

STUDY BOX 7.3: PHASE SPACE

Phase space is a multidimensional space that contains all of the possible states of the system under study. Despite the rather unintuitive name, the concept is actually relatively simple. The space has as many dimensions as there are degrees of freedom or parameters in the system, and since we have an axis in phase space for each degree of freedom, any individual state of the system is described by a single unique point in the system's phase space. Phase spaces are used extensively in classical statistical mechanics for describing the motion of an ensemble of systems.

When used in the context of classical scattering, the phase space coordinates are a set of generalized spatial coordinates q_i, one for each particle, and their corresponding generalized momenta p_i. A point in phase space therefore describes the positions and momenta of every particle in the system at a given point in time, and the evolution of the system in time may be represented by its trajectory in phase space.

Phase spaces can also appear in quantum mechanics. Often the coordinates p and q are replaced by Hermitian operators in a quantum mechanical Hilbert space (a space containing the appropriate number of dimensions to describe the quantum system). Each quantum mechanical observable then corresponds to a unique function or distribution in phase space.

One connection between classical and quantal descriptions of phase space is the idea of a 'cell' in phase space. Because the uncertainty principle requires that we do not specify both a position and its conjugate momentum exactly, each point in classical phase space for N particles is actually a small volume element or cell of dimension h^{3N}. Thus, comparing accessible volumes in phase space amounts to counting the corresponding cells, and we speak of the number of cells on the critical dividing surface in statistical theories.

Claire Vallance
with contributions from *F. Fleming Crim*

energized molecule.[579] This point of view emphasizes the statistical nature of the theory. At a single total energy, the probability of the system being in any of the accessible cells in phase space is exactly the same, and the rate of the reaction depends on how many of those cells are on the critical surface with the positive momentum required to carry them across the critical surface.

7.5.3 Quantum State Structure in Unimolecular Reactions

Testing these ideas about statistical reactions of selectively energized molecules is possible using laser excitation to prepare vibrationally excited molecules above the threshold for unimolecular decomposition.[580] One experiment that shows the structured nature of the cumulative reaction probability uses laser excitation to prepare ketene, CH_2CO, in a highly vibrationally excited state, and then monitors its decomposition to singlet methylene and CO,[581]

$$CH_2CO \rightarrow {}^1CH_2(v_1 v_2 v_3, J_{K_a K_c}) + CO(v, J). \quad (7.18)$$

Cooling the ketene in a supersonic jet places it in its few lowest rotational states, and subsequent electronic excitation prepares an excited state with a well-defined initial angular momentum. The excited ketene internally converts to high vibrational levels of the ground electronic state, from which it can decompose to form $^1CH_2$ and CO. With laser-induced fluorescence it is possible to detect individual quantum states of $^1CH_2$, as specified by its three vibrational quantum numbers $(v_1 v_2 v_3)$ and angular momentum quantum numbers $(J_{K_a K_c})$, to monitor the reaction probability. By

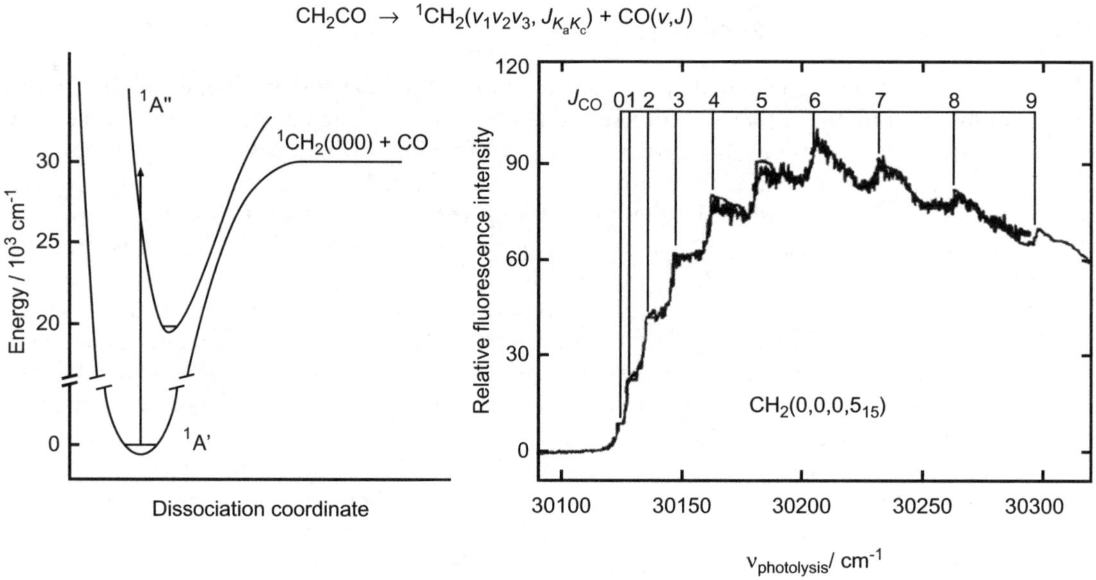

Figure 7.11 Left panel: one-dimensional cuts through the potentials for the ground $\tilde{X}^1A'$ and excited $\tilde{A}^1A''$ states of formaldehyde. Dissociation takes place on the ground state potential following internal conversion from the excited state. Right panel: The intensity of the laser-induced fluorescence signal from $CH_2(000,5_{15})$ as a function of the photolysis laser wavelength. The structure in this spectrum reflects the structure in $N^{\ddagger}(\varepsilon)$. Adapted from ref. [581].

using increasingly energetic excitation photons, it is possible to prepare CH_2CO with increasing amounts of vibrational excitation and observe the resulting products as a function of the excitation energy.

The sketch of the potential along the reaction coordinate in Figure 7.11 shows that the dissociation does not have the sharp barrier in the potential that we used in developing microcanonical transition state theory. At first glance, this situation makes counting the number of ways for the excited molecule to decompose, $N^{\ddagger}(\varepsilon)$, more difficult. The solution to this problem is a limiting case of microcanonical transition state theory known as phase space theory. One counts the number of states at the centrifugal barrier (see Chapters 1 and 12) to the decomposition, which in this case is essentially the number of energetically available product states. The state structure at the "transition state" is identical to that of the products, and the calculation of $N^{\ddagger}(\varepsilon)$ amounts to counting the number of cells in phase space available to the products, a procedure that gives this limiting case its name.

The data in the figure show the signal obtained by detecting the lowest vibrational state of singlet methylene with an angular momentum of $J=5$, $^1CH_2(000,5_{15})$, as a function of the energy deposited into CH_2CO by the photolysis photon. Steps appear in the signal as each new energetically allowed pathway opens. The lowest energy pathway makes $CO(v=0, J=0)$, and the next one makes $CO(v=0, J=1)$. The steps in the signal each reflect a new contribution to $N^{\ddagger}(\varepsilon)$ as each pathway becomes energetically accessible. The solid line running through the data is a phase space theory calculation of the reaction probability. An experiment on the triplet channel, which forms $^3CH_2$ and CO and whose potential has a well-defined transition state, shows similar steps which reflect the structure of the states at that barrier.[582] The conceptually simple notion of counting the ways to get through the transition state is a powerful means of describing many unimolecular reactions, even when studied at the quantum state resolved level.

We might expect this statistical point of view to be inadequate for unimolecular reactions of small molecules at high levels of excitation, and indeed the decomposition of HOCl with an initially excited eigenstate containing six quanta of O–H stretch,

$$HOCl(v_{OH} = 6) \rightarrow OH(v, J) + Cl, \tag{7.19}$$

is much slower that predicted by phase space theory. It also produces OH rotational state populations that disagree with phase space theory.[583–586] Careful vibrational spectroscopy identifies the couplings within the molecule, as described in Section 7.4, and allows the coupling strengths to be extracted in order to predict the time evolution. These weak couplings make the energy flow slower than the unimolecular decomposition rate predicted by phase space theory. The rate of vibrational energy flow within the molecule, rather than statistics, determines the decomposition rate, providing an experimental example of energy flow being slow compared to the decomposition rate. The effects of energy flow being slow compared to unimolecular reaction rate are not necessarily limited to these very small molecules, and calculations suggest that they can also appear in larger systems where the couplings isolate some of the vibrational states.[587,588]

7.6 OUTLOOK

Incisive experiments and detailed theoretical descriptions have provided a sound framework within which to understand quantum-state-resolved reactions. Both technical and conceptual advances promise to increase that understanding further. On the theoretical side, improved computational abilities will make it possible to investigate more complicated systems at a level of detail that matches or exceeds the best measurements. On the experimental side, improved laser technology and electronics permit more highly resolved and sensitive measurements. The combination of scattering techniques, quantum state preparation methods, and state resolved detection schemes, as illustrated in this chapter and in Chapter 6, is providing an extraordinarily fine view of the reactions in experiments where internal states and translational energy are well-controlled.

The possibility of achieving comparable detail for larger molecules with more degrees of freedom is clearly a challenge. One aspect of the problem is obtaining a sufficiently deep understanding of the eigenstates in such systems to exploit them in quantum-state-resolved reactions. As this chapter illustrates, understanding the eigenstates at a chemical level hinges on understanding the couplings in a molecule. A central question is the extent to which increasingly large molecules have their reactive bright states diluted amongst so many dark states that the selective reactivity arising from the bright state disappears. Because this behaviour is just a manifestation of energy flow in a molecule, another approach is to prepare a reactive, non-stationary state, such as the bright state, with a laser pulse that is shorter than the time for energy redistribution. It is even likely that one can tailor a pulse to drive the system along the reaction coordinate, a topic discussed in Chapter 11. This approach again rests on the idea of exciting a motion in the molecule that has a large component along the reaction coordinate. The necessity of operating on the non-stationary state during a time that is short compared to the energy flow time is a challenge, and it is clearly easier to apply such schemes to a unimolecular reaction or a photodissociation than to a bimolecular reaction. The difficulty in the latter comes from bimolecular reactions requiring the encounter of two molecules, with the attendant averaging over orientations and over the time between excitation and collision. A middle ground is to prepare the reactants as a loosely bound complex prior to initiating reaction with a short laser pulse. Thus, the future holds the promise of more sophisticated preparation methods combined with comparably sophisticated detection schemes.

Many of the concepts of quantum-state-resolved chemical dynamics, including not only reaction but also energy flow, apply to processes in liquids[592] (see Study Box 7.4). This environment lacks the simplicity of single collisions and preparation of single eigenstates, but reactions in liquids retain at their core the motion of an energized molecule through a transition state. The competing flow of energy into the surroundings and the modification of the transition state by those

STUDY BOX 7.4: DYNAMICS IN THE CONDENSED PHASE

In Chapter 1 we stated that one of the basic tenets of experimental studies in reaction dynamics is the necessity of isolating the process of interest in the presence of an overwhelming number of background collisions. Such an approach maximizes the dynamical information that can be extracted both from state-of-the-art experiments and from the associated calculations carried out to aid interpretation of the data.

However, much of chemistry does not involve isolated gas-phase molecules, and it is important to consider to what extent the dynamical behaviour observed in gases transfers to processes in condensed phases, such as chemistry at interfaces or in solutions. An introduction to the dynamics of chemical processes at surfaces will be provided in Chapter 10, so in this study box we will focus on reactions in solution.[589–592] The way in which the dynamics of a reaction change in moving from isolated gas-phase molecules to solvated molecules is not easy to predict. In some cases the solvent plays only a small role, perhaps merely perturbing the molecular energy levels slightly, while in others the solvent plays an intimate part in the chemistry.[592]

One of the key differences between gas-phase reactions and those occurring in solution or liquid phases is the timescale of collisions. In a gas at 1 mbar, a typical pressure for a 'bulb' pump-probe experiment, a molecule undergoes a collision about every 100 ns. In molecular beam experiments the pressures involved are much lower, and the time between collisions correspondingly longer, often by orders of magnitude. Nascent reaction products may therefore be probed under single-collision conditions on the nanosecond timescale. In contrast, a molecule in a liquid undergoes a collision approximately every 300 fs, and femtosecond pump-probe experiments must therefore be employed to study the dynamics of molecular processes in solution.[590]

The presence of a solvent modifies the potential energy surface for a chemical reaction by stabilizing or destabilizing the reactants, products, transition state(s), or other structures sampled over the course of the reaction. This may lead to an increase or decrease in the height of an activation barrier for a reaction, with a corresponding change in the reaction rate. These modifications to the PES are generally solvent dependent-a polar molecule will be stabilized to a greater extent by a polar solvent than a non-polar solvent, for example-and the kinetics and dynamics may therefore also depend to a considerable extent on the identity of the solvent. Internal motions of the reacting molecules, particularly vibrational motions, may also be coupled to the solvent, further complicating the dynamics.

The maximum rate at which a reaction may occur in solution is determined by the rate at which reactants diffuse together through the solvent, and is known as the *diffusion controlled* limit. This is the solution-phase equivalent to the *gas-kinetic* limit, though the term 'diffusion controlled' is also often used for reactions in the gas phase. Diffusion and solvent viscosity effects may also influence the reaction rate in more subtle ways. Interactions of the reacting species with the solvent during passage through the transition state can slow down the reaction, and this effect of the solvent viscosity on the reaction is often referred to in terms of 'solvent friction'.

Finally, once the reaction is complete, *solvent cage* effects are often important. This is particularly true for the products of photodissociation, which are generally unstable radical species. In the gas phase there is very little chance of the photofragments encountering each other following dissociation. However, when a molecule is photolyzed in solution, if the nascent photofragments are not formed with sufficient kinetic energy to overcome the solvent-solvent intermolecular forces and 'break out' of the solvent cage in which they are formed, they will quickly re-encounter one another and react in a process known as *geminate recombination*. This often simply regenerates the parent molecule, but in the case of polyatomic photofragments alternative pathways may be available that yield isomers of the original parent molecule.

Dynamics studies in the condensed phase are attracting considerable interest from the reaction dynamics community, and are shedding new light on fundamental processes such as vibrational relaxation, photodissociation, and elementary bimolecular reactions (see ref. [592] and references therein). The ability to probe dynamics in the condensed phase has also opened the way to dynamical studies on much more complicated systems than are accessible to gas-phase experiments. To provide just a single example, work is currently being carried out in the group of Graham Fleming at the University of California at Berkeley to study the photophysics of light harvesting in chlorophyl, a crucial step in photosynthesis.[593]

Claire Vallance

surroundings change the picture, but they do not remove the essential elements. These systems are becoming amenable to studies of the nature of excited molecules and their behaviour in this more complex environment. Studying dynamics in liquids requires high time resolution because of the frequency of interactions, but modern ultrafast laser technology provides that tool. Thus, one of the new opportunities in chemical reaction dynamics is extending detailed studies to new, more complex environments. Obvious examples are liquids and surfaces (see Chapter 10), but more exotic situations, such as helium droplets[594,595] and large clusters, are intriguing possibilities as well. There are already steps being taken in those directions, and there are many more to come.

7.7 PROBLEMS

1. Two reactant molecules A and BC, on collision course with initial velocities $\boldsymbol{w}_{\mathrm{A}}$ and $\boldsymbol{w}_{\mathrm{BC}}$ with respect to their centre-of-mass, react to form AB and C. If the final velocities of the two products are $\boldsymbol{w}'_{\mathrm{AB}}$ and $\boldsymbol{w}'_{\mathrm{C}}$, show that their final translational energy is

$$E_{\mathrm{t}}' = \frac{m_{\mathrm{A}} m_{\mathrm{C}}}{m_{\mathrm{AB}} m_{\mathrm{BC}}} \left(\frac{w_{\mathrm{C}}'}{w_{\mathrm{BC}}} \right)^2 E_{\mathrm{c}} \tag{7.20}$$

   Discuss the dynamical consequences of Eq. (7.20) if the atom C acts as a 'spectator', so that the relative speed of C and BC are equal, $w_{\mathrm{C}}' = w_{\mathrm{BC}}$. Under what conditions would you expect 'spectator' dynamics to convert the initial collision energy, E_{c}, into product translational energy, product internal energy, or to promote collision induced dissociation?
   [The reader might find it helpful to consult Study Box 7.1 before attempting this question.]
2. Let a bright state $|s\rangle$ be described as a linear superposition over zero-order states of a molecule, $|n\rangle$, (*cf.* Eq. (7.12))

$$|s\rangle = \sum_n c_n^s |n\rangle.$$

   The states $|n\rangle$ can be thought of as the first tier of directly coupled states shown in Figure 7.7(b). After excitation of the bright state $|s\rangle$, the time-dependent wave-function, $\Psi_s(t)$, of the system can be written in terms of a linear combination over the true eigenstates of the system (which includes the possibility of coupling to a continuum, or quasi-continuum of states – see Figure 7.7(c))

$$|\Psi_s(t)\rangle = \sum_l c_l^s(t) |l\rangle,$$

where $|l\rangle$ satisfy the time-independent Schrödinger equation, $\hat{H}|l\rangle = E_l|l\rangle$. Show that the time-dependent Schrödinger equation,

$$i\hbar\frac{d\Psi_s(t)}{dt} = \hat{H}\Psi_s(t),$$

is satisfied if the coefficients, $c_l^s(t)$, are given by the expression

$$c_l^s(t) = c_l^s \exp(-iE_l t/\hbar).$$

c_l^s is the coefficient at $t=0$ of the bright state in eigenstate $|l\rangle$, which has eigenvalue E_l. Hence show that the probability of finding the system in state $|s\rangle$ at time t is given by

$$P_s(t) = |\langle s|\Psi_s(t)\rangle|^2 = \left|\sum_l |c_l^s|^2 \exp(-iE_l t/\hbar)\right|^2 \exp(-\gamma t),$$

where the phenomenological decay term $\exp(-\gamma t)$ has been introduced to allow for the decay of the state to the continuum, with rate constant γ.

3. The decays and quantum beats discussed in several cases are relatively easy to calculate from the expression for $P_s(t)$ derived in the previous question.
Consider three eigenstates of energy 3000.000 cm^{-1}, 3000.040 cm^{-1}, and 3000.071 cm^{-1}, and a damping lifetime of 2000 ps. These values mimic the results for an absorption experiment in butyne, and, of course, only the differences in the energy matter.[596]
(a) Using the result from the previous question, show that if damping is ignored ($\gamma = 0$) the time-dependent population in the case of three states can be written

$$P_s(t) = 2\exp(-\gamma t)[|c_1^s|^2|c_2^s|^2 \cos(E_1 - E_2)t/\hbar$$
$$+|c_1^s|^2|c_3^s|^2 \cos(E_1 - E_3)t/\hbar + |c_2^s|^2|c_3^s|^2 \cos(E_2 - E_3)t/\hbar].$$

(b) Plot the time evolution for this system if all three eigenstates contain the *same* amount of bright state character. Plot the time evolution again but with the 3000 cm^{-1} eigenstate having 90% of the bright state character and the other two equal amounts.
(c) Repeat the plots including the damping ($\gamma^{-1} = 2000$ ps). The first case should show a rapid decay almost to zero but the second does not. How does the short time behaviour compare to that of the same system with a damping lifetime of 200 ps? Would you be able to distinguish rapid initial decay (perhaps followed by recurrences) from rapid damping if you observed only the first several hundred picoseconds?
(d) This machinery is useful for examining the effect of adding more states at slightly different energies without having an explicit damping term. One can test how many levels need to be participating for the time evolution to look roughly like a single decay with minor recurrences. Generate a random set of energy levels E_m spanning a range of 0.1 cm^{-1} and a random set of properly normalized coefficients for the levels. Calculate the time evolution over a period of 50 ns for three levels to confirm that your scheme is producing a relatively simple beat pattern such as you obtained above. Obtain the evolution for 5, 15, 25, and 50 levels. (Try it a few times to see how the pattern changes. Note that over some long time this finite system must have recurrences.)
How many levels make the time evolution look like a simple decay? With a maximum energy spread of 0.1 cm^{-1}, what is the fastest decay you can expect? (Consider a two level system with that energy separation.) Are your calculations consistent with this result?

4. One can obtain the classical RRK expression for the rate constant $k(\varepsilon)$ starting with microcanonical transition state theory and using the classical expression for the number of states of a set of s classical oscillators having frequencies ν_i (see, for example, ref. [10]),

$$N(\varepsilon) = \frac{1}{s!}\frac{\varepsilon^s}{\Pi_i h\nu_i}, \tag{7.21}$$

where the product runs over all s oscillators. (The expression $\varepsilon/h\nu_i$ is the maximum number of quanta that can be localized in the i-th oscillator.) The density of states is just $\rho(\varepsilon) = dN(\varepsilon)/d\varepsilon$. Assuming that *only* the frequency of the vibration that becomes the reaction coordinate changes, derive the classical RRK expression

$$k(\varepsilon) = \nu^{\ddagger}\left(\frac{\varepsilon - \varepsilon_0}{\varepsilon}\right)^{s-1}, \tag{7.22}$$

where $\nu^{\ddagger}$ is the frequency along this special coordinate. Sketch the dependence of $k(\varepsilon)$ on ε and s.

What do you think of the assumption that only the one vibration changes frequency?

5. *Trans*-diphenylbutadiene undergoes unimolecular photoisomerization in one of its low lying electronically excited states. The rate constants for isomerization (determined from 'real-time' fluorescence lifetime measurements) vary with energy in the excited butadiene in the following way:

ε/cm^{-1}	2000	3000	4000	5000	6000	7000	8000
$k_2(\varepsilon)/10^{10}$s^{-1}	0.36	1.1	1.9	2.7	3.5	4.1	4.8

(a) Use the data and the expression in question (3) to make a classical RRK estimate of the number of oscillators, s, and the frequency $\nu^{\ddagger}$, given that the critical energy $\varepsilon_0 = 1100\,\text{cm}^{-1}$.

(b) Implicit in the statistical (free energy flow) assumption, upon which RRK theory is based, is that *all* vibrational modes should be active (*i.e.* vibrational energy is assumed to be distributed randomly among all the 3N–6 vibrational modes of the molecule). Using your estimate of $\nu^{\ddagger}$ in (a), calculate $k_2(\varepsilon)$ at a selection of energies given in the table above, assuming all the modes in diphenylbutadiene are active. Comment on the answers you obtain.

CHAPTER 8

Photodissociation Dynamics: the Fragmentation of Molecules by Light

DAVID H. PARKER,[a] ANDRÉ T. J. B. EPPINK[a] AND CLAIRE VALLANCE[b]

[a] Department of Molecular and Laser Physics, Radboud University Nijmegen, The Netherlands;
[b] Department of Chemistry, Chemistry Research Laboratory, University of Oxford, UK

8.1 INTRODUCTION

In this tutorial we describe current, nearly 'perfect' experimental and theoretical studies into the photodissociation of simple diatomic molecules. We celebrate the present status of research in this field and we hope to help you develop at least a qualitative feeling for the dynamics of photodissociation on simple potential energy surfaces. The main models used to understand diatomic molecule photodissociation dynamics are described, and their success and failures are illustrated with the help of several representative molecules. We follow this approach because essentially many of the simple limiting-case models described in this tutorial are employed in all studies of larger molecules, from simple triatomic molecules to biological molecules.* As we show here, there is still much to be learned about small molecule photodissociation.

8.2 PHOTODISSOCIATION AND POTENTIAL ENERGY SURFACES

We study photodissociation in order to understand, at a fundamental level, the behaviour of atoms and molecules as they undergo chemical change brought about by interaction with light.[140,597,598] The breaking of a chemical bond due to absorption of a single photon requires electronic excitation, and photodissociation is just one of many fates that can befall a molecule in an excited electronic state. For a generic system $AB \rightarrow A + B$, including the case that AB is polyatomic, the more commonly observed processes following excitation by a typical high intensity pulsed laser are illustrated in Figure 8.1. One possible fate, the subject of this chapter, is path a), dissociation to

*Study Box 8.1 outlines some of the ways in which photodissociation of a polyatomic molecule differs from that for the diatomic systems discussed in this chapter.

Tutorials in Molecular Reaction Dynamics
Edited by Mark Brouard and Claire Vallance

Published by the Royal Society of Chemistry, www.rsc.org

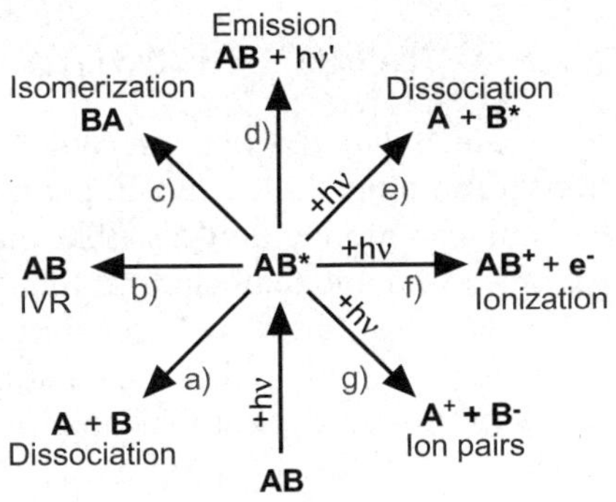

Figure 8.1 Fates of an electronically excited molecule, AB*, after excitation of the ground state, AB, by a photon of energy *hv* in the range of 2–7 eV. One-photon induced processes include: a) dissociation to neutral products A + B; b) intramolecular vibrational redistribution, IVR, or another internal energy transfer process to form a new state $AB^{\dagger}$; c) isomerization to BA; and d) relaxation by emission of a photon *hv*′. Due to the intense laser fields present, absorption of a second or more photons may lead to: e) dissociation, often to higher limits, A + B*; f) ionization $AB^+ + e^-$; and g) ion-pair formation $A^+ + B^-$. With a higher energy photon the processes e)–g) can also be driven directly from the ground state by one-photon excitation.

neutral fragments. Other paths include: b) internal energy transfer processes such as intermolecular vibrational redistribution (IVR), discussed further in Section 7.4.3; c) isomerization; and d) radiative decay back to the ground state. All of these processes must compete with absorption of a second photon from the laser beam used to drive the absorption step. Absorption of more than one photon opens the way to e) dissociation to excited neutral fragments, f) ionization, and other processes such as g) ion-pair formation.

Dissociation itself is a broad term that encompasses many types of events. Some of these are illustrated schematically in Figure 8.2. In all cases illustrated, absorption of a photon of energy *hv* first generates an excited state species. Panel a) in Figure 8.2 shows photoabsorption from a stable ground state to a stable (bound) level of an excited electronic state. Our understanding of the properties of molecules is due in large part to the success of spectroscopy in characterizing this type of bound-bound transition. Absorption may be detected directly using a variety of techniques, including extremely sensitive methods such as cavity ringdown spectroscopy,[599] or may be detected indirectly by monitoring the laser-induced fluorescence (LIF) emitted as the molecule returns to the ground state or some other low-lying state (panel a) in Figure 8.2.

As the photon energy increases, the repulsive wall of the bound upper state is eventually reached. Above the dissociation threshold for the upper state, absorption becomes continuous and direct dissociation (panel b) takes place. Our focus on these bound excited electronic states is narrow – in general there are more unbound (repulsive) electronic states than there are bound states suitable for spectroscopic study. Direct dissociation *via* these unbound states, panel c), can form the main destruction channel for the molecule if absorption is strong at the wavelengths of the available light. An important example is the vacuum ultraviolet (VUV) photodissociation of OH in interstellar space.[600]

A general problem in electronic spectroscopy is that the bound levels of the upper states quickly become unstable with increasing energy (these states are labeled (AB)* in panels (d)–(f) in Figure 8.2) due to coupling with repulsive states. This leads to different forms of predissociation. Panels (d)–(f) illustrate three types of indirect photodissociation.[262] In panel (d), the system is excited to a bound state, whence it cannot undergo direct photodissociation. However, dissociation *is* possible if the molecule undergoes a transition to a nearby state which is repulsive. This process is known as Herzberg type I[262] or electronic predissociation. The radiationless transition involved in

STUDY BOX 8.1: PHOTODISSOCIATION OF POLYATOMIC MOLECULES

The field of photodissociation dynamics has reached the point where we have a very good understanding of the dissociation of diatomic molecules. Experimentally, it is possible to measure photofragment velocity and angular momentum distributions with full reactant and product state selection on both the nanosecond and femtosecond timescales. On the theoretical side, many diatomic molecules are amenable to accurate *ab initio* calculations and in some cases full quantum theoretical scattering calculations, although detailed studies which include the effects of, for example, parent molecular rotation and photofragment electronic angular momentum polarization remain very challenging in practice.

The situation for polyatomic molecules is rather different. Even moving from a diatomic to a triatomic system adds a great deal of complexity. Some of the aspects that need to be considered are:

1. The interaction potential is no longer a simple function of the separation between two atoms, but now depends on two bond lengths and a bond angle.
2. When a diatomic molecule dissociates, the excess energy is shared between electronic excitation of the fragments and their translational kinetic energy. When a triatomic dissociates, we generally also have to consider rotational and vibrational excitation of a molecular (diatomic) fragment, vastly complicating any treatment of energy partitioning during the dissociation process.
3. Excepting the case of linear triatomics, it is unlikely that the transition dipole will lie exactly parallel or perpendicular to the breaking bond, and the transition will generally have both parallel and perpendicular components.
4. When dealing with diatomics, it is relatively straightforward to account for the effects on the dynamics when a long-lived excited state rotates prior to dissociating. However, for a triatomic the vibrational motion of the excited state must be accounted for in addition to rotation of the whole molecule. Both of these effects can lead to deviations from axial recoil (*i.e.* loss of the correlation between the transition dipole $\boldsymbol{\mu}$ for the parent molecule and the velocity $\boldsymbol{v}$ of the photofragment).
5. Full quantum calculations that include the effects of angular momentum polarization are not yet possible on triatomics, and the interpretation of such effects requires recourse to a range of approximate treatments, such as the 'long range' and 'fast dissociation' models.[601]
6. The increase in the dimensionality of the PESs involved in the dissociation opens up the possibility of a range of different intramolecular energy transfer processes such as IVR, internal conversion (IC), and intersystem crossing (ISC). Population transfer *via* conical intersections is also common in polyatomic systems, and the dynamics associated with motion through conical intersections has already been discussed in Chapter 4.

Despite these difficulties, a number of polyatomic photolysis processes have been the subject of detailed studies. For more information about photochemistry and photodissociation of polyatomic species the reader is referred to general textbooks on the subject, including refs. [597,598].

Claire Vallance

this system may be an internal conversion process (between two states of different electronic symmetries or two diabatic states of the same symmetry – see Chapter 4), or an intersystem crossing process. Panel (e) depicts a different predissociation process, in which the excited states species (AB)* lies above the threshold for dissociation on the excited state surface, but there exists a

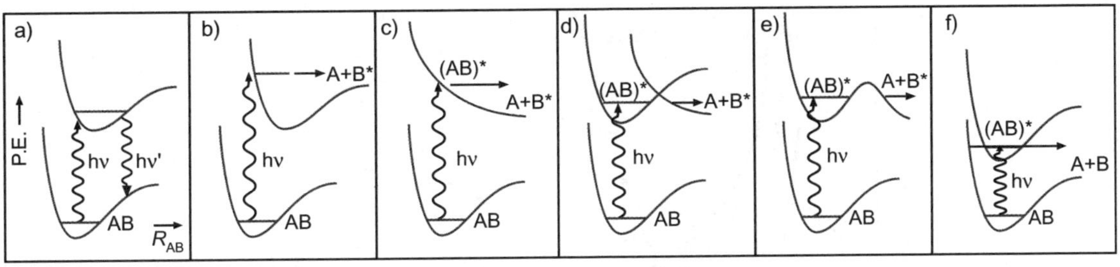

Figure 8.2 a) A transition between two bound electronic states driven by excitation at energy *hν*, and detected by laser-induced fluorescence (LIF) at energy *hν′*. b) A transition to the repulsive wall of a bound state, leading to direct dissociation of AB to products A + B*, where one of the products (B*) can be electronically excited, c) Photo-excitation of a repulsive state, leading to direct dissociation. Panels d)–f) show various predissociation processes from an unstable bound level (AB)*, as described in the text.

barrier. The excited state is quasi-bound and known as a resonance (resonances along reaction pathways are discussed in Chapter 6 and Study Box 6.4). The barrier can be traversed either by tunnelling or by the redistribution of the internal vibrational energy (Figure 8.1(b)), when the species AB is a triatomic or larger molecule. This type of predissociation is known as Herzberg type II[262] or vibrational predissociation. The dissociation lifetime of the excited state species (AB)* is varied. In type I predissociation processes, it is determined by the nature of the electronic coupling between the states involved, and in type II processes by the tunnelling rate or the efficiency of internal energy transfer. Finally, panel (f) of Figure 8.2 illustrates the case in which the excited state species decays, for example *via* spin-conserving internal conversion, IC, to a highly vibrationally and rotationally excited level of the ground electronic quantum state. The system, on the ground state, may lie above the threshold for dissociation and thus break apart to products A + B.

The distinction between the different types of photodissociation illustrated in Figure 8.2 is, in places, somewhat artificial. In general terms, the evolution of a system after photoabsorption is determined by the full set of potential energy surfaces (PESs) involved in the dissociation. Within the framework of the Born-Oppenheimer approximation, the PES(s) for a process may be constructed by solving the Schrödinger equation for each electronic state at all nuclear coordinates. As demonstrated in Chapter 2, in practice, a very good approximation to the surface is obtained by solving the Schrödinger equation at a subset of points and then interpolating these solutions to give a smooth potential energy function. One of the aims of dynamical studies is to understand the connection between the topology of the PESs involved and the observed physical properties of a reaction. An accurate (set of) PES(s) encodes all of the energetics of the process and all of the forces acting on the atomic nuclei for any spatial arrangement of the atoms, and therefore provides a complete theoretical description of the photodissociation event. As we have seen in Chapters 4 to 7, physical observables such as product quantum-state populations, angular scattering distributions and rotational or electronic angular momentum distributions may be predicted by carrying out classical trajectory or quantum scattering calculations over the surfaces. These predictions may be compared directly with experimental data, with the consequence that the results of experiments provide stringent tests both for the accuracy of the calculated PESs and for the dynamical calculations performed on them.

Potential energy surfaces for bound electronic states such as those shown schematically in Figure 8.2(b) may be determined accurately using classical spectroscopy through measurements of line positions, intensities, linewidths, and polarization sensitivity. PESs for repulsive electronic states, such as that shown in Figure 8.2(c), may be determined using translational spectroscopy. This technique uses measurements of one or more photofragment velocities, together with the

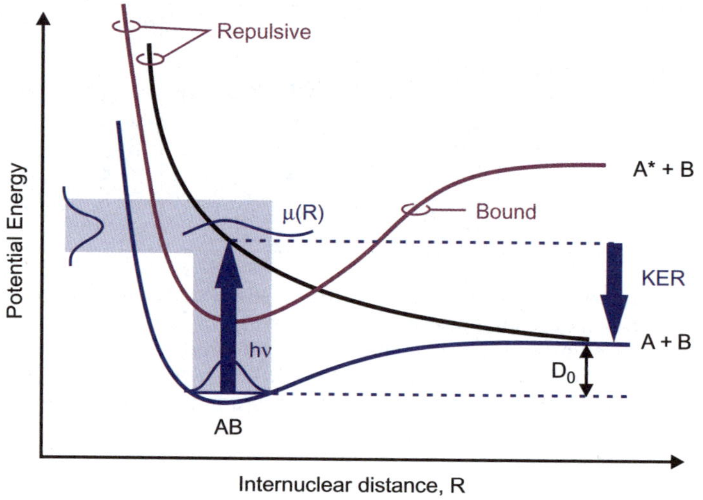

Figure 8.3 Translational spectroscopy: Kinetic energy release, KER, for a photolysis reaction $AB + h\nu \rightarrow A + B$ is determined by measuring the velocity of product A or B. Absorption strength is governed by the projection of the AB wavefunction combined with the transition dipole function $\mu(R)$ (an arbitrary $\mu(R)$ function is shown) onto the repulsive excited state, or the repulsive wall of a bound excited state.

principle of conservation of energy, to determine the kinetic energy release (KER) or photofragment translational energy, E'_t, for a photodissociation process. The energy balance is illustrated in Figure 8.3 for a generic process $AB + h\nu \rightarrow A + B$. The total energy available to the products is simply the photon energy $h\nu$ less the energy D_0 required to break the bond. This is partitioned between internal energy E_{int} and translational kinetic energy, E'_t, in the photofragments, so that $E'_t = h\nu - D_0 - E_{int}$. The way in which the kinetic energy release is partitioned between the two photofragments is determined by conservation of momentum, and therefore depends on the masses M_A, M_B, and M_{AB} of the fragments and the parent molecule: $E_{t,A} = (M_B/M_{AB}) \times E'_t$ and $E_{t,B} = (M_A/M_{AB}) \times E'_t$, where M_A/M_{AB} and M_B/M_{AB} are called the mass partitioning factors. For example, in the dissociation $OH \rightarrow O + H$, the light H atom receives most of the kinetic energy, while for photodissociation of a homonuclear diatomic, such as O_2, the available energy is shared equally between the two product O atoms. We see from the above equations that measurement of one of the product velocities, together with knowledge of the parent and fragment masses, allows the total kinetic energy release to be determined. By measuring the kinetic energy release as a function of the photon energy $h\nu$, the shape of the repulsive curve may be determined. Note that, due to the Franck-Condon principle,* these measurements are limited to mapping out the repulsive curve over the range of internuclear separations sampled by the vibrational wavefunction in the lower state, since the transition probability (and therefore the signal strength) depends on the transition dipole function, which in turn depends on the overlap between vibrational wavefunctions in the lower and upper states.

8.3 AN INTRODUCTION: THE PHOTODISSOCIATION OF Br_2

We begin with the bromine molecule, Br_2, a relatively heavy molecule with a simple electronic structure and with dynamics that should be well described by the Born-Oppenheimer approximation. Recent work by Asano *et al.*,[602] the results of which are shown schematically in Figure 8.4, provides a

*The reader may find it helpful at this stage to turn to Study Box 8.2, which provides an introduction to electric dipole transitions, and includes a discussion of the transition dipole, and the Franck-Condon principle.

STUDY BOX 8.2: AN INTRODUCTION TO ELECTRIC DIPOLE TRANSITIONS

To help understand some of the vector correlations in molecular photodissociation it is first necessary to think about how a photon interacts with a molecule. The formal theory of electric dipole transitions requires the use of time-dependent perturbation theory. This topic is outside the scope of this text, but is covered in many quantum mechanics or spectroscopy text books.[70,239,610] Here we simply use the result. The relevant perturbation to the molecular Hamiltonian when a molecule absorbs plane polarized light *via* an electric dipole transition can be written

$$\hat{H}' \equiv \hat{\mu} = -\boldsymbol{\mu} \cdot \boldsymbol{\varepsilon}, \tag{B8.2.1}$$

where $\boldsymbol{\varepsilon}$ is the electric vector of the light, and the electronic transition dipole moment, $\boldsymbol{\mu}$, can be represented as a sum over all the individual dipoles arising from different charged particles (electrons and nuclei), i, in the molecule

$$\boldsymbol{\mu} = \sum_i q_i \boldsymbol{r}_i, \tag{B8.2.2}$$

where q_i are the individual charges at separations $\boldsymbol{r}_i$ from the centre of mass of the molecule. According to time-dependent perturbation theory, the absorption intensity for a transition between initial state i and final state f is given by the *Fermi Golden Rule*

$$I_{fi} \propto (P_i - P_f)\left|\langle \Psi_f | \hat{H}' | \Psi_i \rangle\right|^2, \tag{B8.2.3}$$

where P_i and P_f are the populations in the initial and final states, respectively, and the wavefunctions include contributions from electronic, vibrational, and rotational degrees of freedom. Eqs. (B8.2.1) and (B8.2.3) imply that the probability of absorbing a photon of plane polarized light will have an angular dependence proportional to $|\hat{\boldsymbol{\mu}} \cdot \hat{\boldsymbol{\varepsilon}}|^2$. Since $|\boldsymbol{\mu} \cdot \boldsymbol{\varepsilon}| = \mu\varepsilon\cos\theta_{\mu\varepsilon}$, the angular dependence will be equivalent to $\cos^2\theta_{\mu\varepsilon}$, where $\theta_{\mu\varepsilon}$ is the angle between $\boldsymbol{\mu}$ and the direction of the electric vector of the linearly polarized light, $\boldsymbol{\varepsilon}$. This aspect of the absorption process is explored further in Section 8.6 and Study Box 8.3.

Let us assume that the molecule is initially randomly distributed in space. Then, neglecting the population term, and assuming that the electric vector of the linearly polarized light lies along the LAB Z-axis, Eq. (B8.2.3) may be rewritten.[239,610]

$$I_{fi} \propto \left|\langle \Psi_f | \hat{\mu}_Z | \Psi_i \rangle\right|^2, \tag{B8.2.4}$$

where $\hat{\mu}_Z$ is the LAB-frame Z component of the dipole moment operator. If we assume that the electronic, vibrational, and rotational motions are separable (*e.g.*, using the Born-Oppenheimer approximation), then we can write the wavefunctions in the product form $\Psi = \Psi_e\Psi_v\Psi_r$. The electronic wavefunction, Ψ_e, depends on the coordinates of the electrons in the molecule, and parametrically on the nuclear coordinates (see Chapter 2); the vibrational wavefunction, Ψ_v, depends solely on the nuclear coordinates; and the rotational wavefunction, Ψ_r, depends (by assumption) solely on the orientation of the molecule in space. It is convenient to express $\hat{\mu}_Z$ in terms of its components in the molecular (body fixed) frame, $\hat{\mu}_j$, with $j = x, y$, or z. The rotation of frames necessary to do this transformation can be performed using Wigner rotation matrices, a discussion of which can be found, for example, in ref. [239]. The result is that the weighting coefficients relating the LAB to the body fixed frame transformation are given by the elements of the direction cosine matrix λ_{Zj}[239,610]

$$\hat{\mu}_Z = \sum_j \lambda_{Zj}\hat{\mu}_j. \tag{B8.2.5}$$

Substituting this equation into Eq. (B8.2.4), and making use of the fact that $\hat{\mu}_j$ depends on the coordinates of the electrons in the body fixed frame, we find

$$I_{fi} \propto \sum_j |\langle \Psi_{f,\mathrm{r}}|\lambda_{Zj}|\Psi_{i,\mathrm{r}}\rangle|^2 |\langle \Psi_{f,\mathrm{e}}\Psi_{f,\mathrm{v}}|\hat{\mu}_j|\Psi_{i,\mathrm{e}}\Psi_{i,\mathrm{v}}\rangle|^2. \tag{B8.2.6}$$

For transitions within a given electronic state we can expand the transition dipole moment $\hat{\mu}_j$ as a Taylor expansion about the equilibrium geometry of the molecule. For a vibrational coordinate Q and electronic state i one can then write

$$\langle \Psi_{i,\mathrm{e}}|\hat{\mu}_j|\Psi_{i,\mathrm{e}}\rangle = \mu_{j,\mathrm{e}} + \left(\frac{\mathrm{d}\mu_j}{\mathrm{d}Q}\right)_\mathrm{e} Q + \cdots, \tag{B8.2.7}$$

where $\mu_{j,\mathrm{e}}$ is the component of the permanent dipole moment of the molecule along molecular frame coordinate j, and $\left(\frac{\mathrm{d}\mu_j}{\mathrm{d}Q}\right)_\mathrm{e}$ is the corresponding first derivative of the dipole moment with respect to coordinate, evaluated at the equilibrium geometry. On substitution of Eq. (B8.2.7) into (B8.2.6) one obtains

$$I_{fi} \propto \sum_j |\langle \Psi_{f,\mathrm{r}}|\lambda_{Zj}|\Psi_{i,\mathrm{r}}\rangle|^2 \left[|\mu_{j,\mathrm{e}}|^2 \delta_{v_f,v_i} + |\langle \Psi_{f,\mathrm{v}}|Q|\Psi_{i,\mathrm{v}}\rangle|^2 \left|\frac{\mathrm{d}\mu_j}{\mathrm{d}Q}\right|_\mathrm{e}^2 + \cdots \right], \tag{B8.2.8}$$

where v_i and v_f are the initial and final vibrational quantum numbers. In the case of a pure rotational transition, the initial and final electronic and vibrational states are both unchanged, and the first term in square brackets on the right side of Eq. (B8.2.8) indicates that the molecule must have a permanent dipole moment in order for a microwave spectrum to be observed. The term $|\langle \Psi_{f,\mathrm{r}}|\lambda_{Zj}|\Psi_{i,\mathrm{r}}\rangle|^2$ provides information about the rotational selection rules and the relative intensities of different rotational transitions. In the case of an IR vibrational transition, the second term in square brackets on the right of Eq. (B8.2.8) reveals that the molecule must exhibit a change in dipole moment with coordinate, Q, *i.e.* an oscillating dipole moment, while the term $|\langle \Psi_{f,\mathrm{v}}|Q|\Psi_{i,\mathrm{v}}\rangle|^2$ provides information about the vibrational selection rules. For harmonic oscillators (and within the linear approximation of the dipole moment function, given in Eq. (B8.2.7)) it is readily shown that this term yields a selection rule $\Delta v = \pm 1$. Higher terms not included here, or vibrational anharmonicity, lead to $\Delta v = \pm 2, \pm 3, \ldots$.

In this chapter we are more concerned with electronic transitions. In this case it is common to assume that the components of the electronic transition moment, $\mathcal{R}_{\mathrm{e}j} = \langle \Psi_{f,\mathrm{e}}|\hat{\mu}_j|\Psi_{i,\mathrm{e}}\rangle$, are independent of nuclear coordinate Q. Within the Born-Oppenheimer approximation one can then write the intensity approximately as

$$I_{fi} \propto \sum_j |\langle \Psi_{f,\mathrm{r}}|\lambda_{Zj}|\Psi_{i,\mathrm{r}}\rangle|^2 |\langle \Psi_{f,\mathrm{v}}|\Psi_{i,\mathrm{v}}\rangle|^2 |\mathcal{R}_{\mathrm{e}j}|^2. \tag{B8.2.9}$$

The intensity of an electronic transition is therefore approximately proportional to a product of three terms, which, on the right side of Eq. (B8.2.9) reflect the rotational, vibrational and electronic parts of the problem, respectively. Specifically, the term $|\langle \Psi_{f,\mathrm{v}}|\Psi_{i,\mathrm{v}}\rangle|^2$ is known as a

Franck-Condon factor, and determines the relative intensities of different vibrational bands in the electronic spectrum. This factor is simply a vibrational overlap integral (where the integral is over nuclear coordinate Q) between vibrational wavefunctions in the two electronic states. The final term in Eq. (B8.2.9) contains information about the relative strength of the electronic transition, and any electronic selection rules that might apply. Eq. (B8.2.9) is quite approximate, but it provides the basis of the Franck-Condon principle, which underpins much of the discussion in this Chapter. Further discussion about this, and more rigorous treatments of absorption spectroscopy and molecular photodissociation can be found in the recommended texts.[70,140,239,610]

Before leaving this study box it is worth noting that the Fermi Golden rule expression for the absorption intensity, equation (B8.2.3), is closely related to a similar expression for the rate of non-radiative transitions in molecules,[70,597,598] as already discussed in Section 4.2.6 (in particular, *cf.* Eqs. (B8.2.3) and (4.17)). Internal conversion and intersystem crossing can also often be treated as time-dependent perturbations, and the analogous expressions for the rates of these processes contain analogous matrix elements, but with different perturbation Hamiltonians, $\hat{H}'$. For intersystem crossing, for example, the perturbation is the spin-orbit interaction, and $\hat{H}'$ would be the spin-orbit Hamiltonian, $\hat{H}_{SO} = A_{SO}\hat{\boldsymbol{L}} \cdot \hat{\boldsymbol{S}}$, where A_{SO} is the spin-orbit coupling constant.

Mark Brouard

good example of the calculation and analysis of potential energy curves involved in the dissociation. The right-hand side of Figure 8.4 shows several of the lowest lying potential energy curves for Br_2, including the X 0_g ground state* (where 0_g is the spectroscopic state label for this curve in the Ω, or Hund's case (c), coupling limit appropriate for molecules in which spin-orbit coupling is strong – see Study Box 4.3). Electric-dipole transitions are allowed between the ground state and the four upper states, A$1_u^{(1)}$, B0_u^+, C$1_u^{(2)}$, and $1_u^{(3)}$. These states are labeled in the figure, and their corresponding state-specific or partial absorption spectra are shown on the left-hand panel. The total absorption spectrum, calculated by summing the predicted partial absorption spectra for each state, agrees well with experimental measurements; this provides a sensitive test of the quality of the *ab initio* calculations.

The potential energy (PE) curves of a diatomic molecule can be built up using correlation diagrams based on symmetry rules developed by Wigner and Witmer,[603] Mulliken,[604] and others, and explained in detail by Herzberg.[262] The approach assumes an infinitely slow (or 'adiabatic') dissociation such that the electronic quantum numbers remain well-defined for all values of the nuclear separation.† We then need to identify the possible electronic states (and their quantum numbers) corresponding to each molecular orbital configuration, and the electronic states that can arise from combining two states of the separated atoms. Finally, we connect molecular and 'separated atom' states with the same quantum numbers, starting with the highest occupied molecular orbital (HOMO) of the molecule and moving on through the lowest unoccupied molecular orbitals (LUMOs). These ideas are illustrated with respect to the Br_2 molecule in the following.

For a molecule composed of high atomic mass atoms, such as Br_2, the quantum number Ω, describing the sum of the projections of $\boldsymbol{L}$ and $\boldsymbol{S}$ onto the molecular axis, is used to characterize each electronic state. The correlation diagram is built up by starting at the lowest energy and connecting Ω states on the molecular side with those on the separated atom side, without allowing

*The states of the bromine molecule are labelled according to Ω (the total electronic angular momentum projection quantum number), with subscripts g and u to denote their inversion symmetry through the centre of the molecule, and +/− superscripts to denote their reflection symmetry in the plane of the molecule. Relevant to the discussion here, there are five electronic states of 1_u symmetry, which are labelled in order of increasing energy as $1_u^{(1)}$ through to $1_u^{(5)}$.

†In the context of photodissociation, an adiabatic process is one that evolves on a single potential energy surface and does not involve crossings to other surfaces. See Chapters 1, 2 and 4 for more discussion.

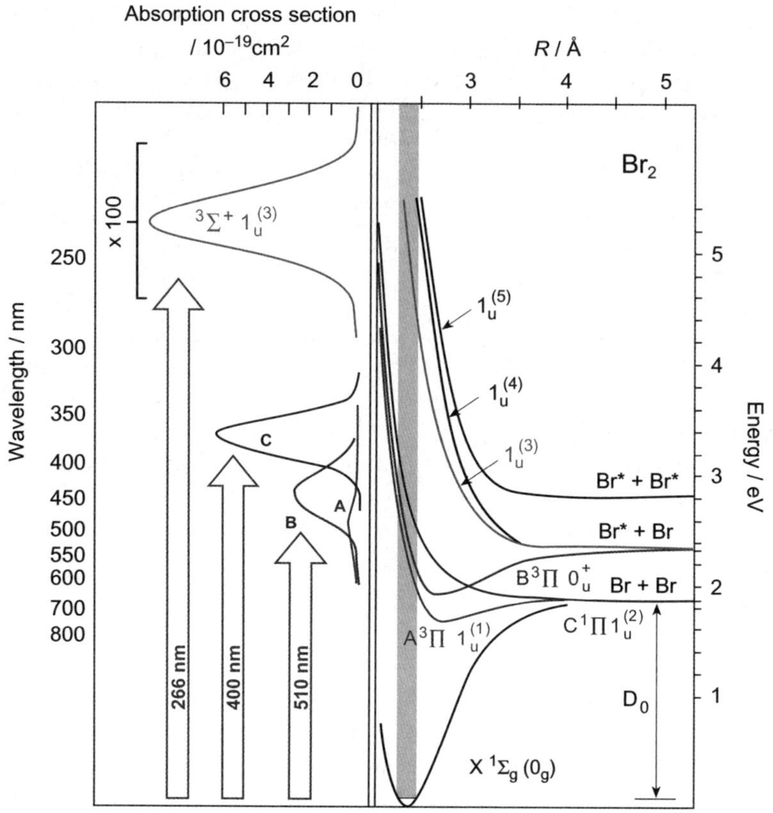

Figure 8.4 Calculated potential energy curves (right side) and their corresponding absorption spectra (left side) for the Br_2 molecule, adapted from the work of Asano *et al.*[602] The region of Franck-Condon overlap with the ground state is marked by a shaded vertical stripe at 2.3 Å.

states of the same Ω to cross. The X 0_g state of Br_2 is the only state possible in the highest occupied molecular orbital (HOMO) configuration $\sigma_g^2\pi_u^4\pi_g^4\sigma_u^0$, and this state is connected in the separated atom limit to the combination of two ground state Br atoms, Br + Br. The second dissociation limit yields one Br atom in its electronic ground state $^2P_{3/2}$, and one in its first spin-orbit excited state, $^2P_{1/2}$; these atomic products are usually written in the shorthand form Br + Br*. In contrast to the HOMO, the $\sigma_g^2\pi_u^4\pi_g^3\sigma_u^1$ LUMO electron configuration of Br_2, corresponding to a $\pi^* \to \sigma^*$ excitation, gives rise to a number of molecular states, namely the A, B and C states. As may be seen in Figure 8.4, the X, A and B states are bound states, while the C state, responsible for the strongest contribution to the absorption spectrum, is a repulsive state. The next excited electron molecular configuration, $\sigma_g^1\pi_u^4\pi_g^4\sigma_u^1$, gives rise to the $1_u^{(3)}$ state. This state has a more strongly repulsive potential energy curve due to the $\sigma \to \sigma^*$ excitation. The A and C states correlate with the lowest Br + Br dissociation limit, while the $B0_u^+$, and $1_u^{(3)}$ states correlate with the Br + Br* limit. More aspects of adiabaticity and state correlations will be discussed as we proceed through this chapter.

The dissociation wavelengths at which the various dissociation limits become accessible are easily predicted from the potential energy curves. D_0, the bond energy measured from the lowest vibrational level, is 1.97 eV for Br_2; thus dissociation to Br + Br becomes possible at energies greater than this, or equivalently, at wavelengths shorter than 628 nm. The spin-orbit (SO) splitting of the Br atom ground state is $E_{SO} = 0.46$ eV, so the Br + Br* and Br* + Br* limits lie at 2.41 eV (511 nm) and 2.87 eV (430 nm), respectively.

Table 8.1 Vibrational periods for the lowest vibrational state, and average rotational periods at 300 K for a range of diatomic molecules.

Molecule	τ_{vib}/fs	τ_{rot} (300 K)/fs
H_2	7.6	148
OH	8.9	266
HCl	11.5	355
HI	14.5	451
O_2	21	959
Cl_2	60	2330
Br_2	105	4050
I_2	155	5960

8.4 FEMTOSECOND PROBES OF PHOTODISSOCIATION DYNAMICS

Suppose we now excite Br_2 to the $1_u^{(3)}$ state using a 266 nm photon. How long would it take for the molecule to dissociate? A rough estimate is found by dividing the distance the Br–Br bond must expand in order to become essentially unbound (this is around 2 Å from Figure 8.4) by the final velocity of the liberated Br atoms. The kinetic energy of the atoms in the Br + Br* limit is $h\nu - D_0 - E_{SO} = 2.2$ eV, corresponding to an individual Br atom kinetic energy of 1.1 eV and a velocity of 1650 m s^{-1} (for ^{79}Br). The dissociation time Δt is thus 2×10^{-10} m/1650 m s^{-1} = 1.2×10^{-13} s or 120 fs (1 fs = 10^{-15} s). We therefore expect the photodissociation process to be finished in around 100 fs. We should compare this timescale to that of electronic excitation (approximately a Bohr period, $\sim10^{-16}$ s) and of molecular vibration and rotation. Table 8.1 lists vibrational and rotational periods in femtoseconds for a range of diatomic molecules. The vibrational periods τ_{vib} have been estimated from the vibrational quantum frequency, $\tau_{vib} \approx \nu_e^{-1}$, and the rotational periods from $\tau_{rot} \approx (h/4cBk_BT)^{1/2}$, with c the speed of light, B the rotational constant in cm^{-1}, k_B the Boltzmann constant, and T the temperature in Kelvin.*

As is evident from Table 8.1, the vibrational period and timescale for Br_2 dissociation at 266 nm are similar, while rotational motion of the parent Br_2 molecule is slow compared to the nuclear motion triggered by photodissociation. In general, parent vibrational excitation can strongly affect the outcome of photodissociation, while the effects of rotational motion are usually limited to the case of dissociation near to threshold. Near threshold, little kinetic energy is available to the photofragments, their velocities are correspondingly low, and rotation and fragment recoil have similar timescales. Photodissociation of rotationally and/or vibrationally state-selected molecules will be discussed later in the text.

Wernet and coworkers[605] have reported a direct measurement of the ultra-short timescale of Br_2 photodissociation.† Their experiments were based on a pump-probe technique (see Section 1.3.4). In the first, 'pump', step, Br_2 is excited by a photon from a 60 fs pulse of 266 nm light. Immediately following excitation, the electronically excited molecule is photoionized with a 23.5 eV, 120 fs pulse of XUV light and the resulting photoelectrons are detected as a function of their translational energy. The XUV pulse has sufficient energy to excite all valence electrons of the molecule, which means that the experiments are able to trace the evolution of the valence electron arrangement in real time as the Br_2 molecules dissociate. The results are shown in Figure 8.5. At time zero the photoelectron spectrum is simply that of Br_2 in the $1_u^{(3)}$ state of the $\sigma_g^1\pi_u^4\pi_g^4\sigma_u^1$ configuration, and

*This expression is based on the classical thermal rotational energy $E_{rot} = \frac{1}{2}I\omega^2 = k_BT$, where I is the moment of inertia and ω the angular velocity, with $\omega = 2\pi\nu$ and ν the rotational frequency in Hz. The rotational period, $\tau_{rot} = 1/\nu$, is therefore given by $\tau_{rot}^2 = \frac{2\pi I}{k_BT}$. On substituting in this expression for the relationship between the moment of inertia and the rotational constant, $I = \frac{h}{8\pi^2cB}$, one obtains the expression given in the main text for τ_{rot}.

†More details about the types of techniques used in these experiments, and about femtochemistry more generally are given in Chapter 11.

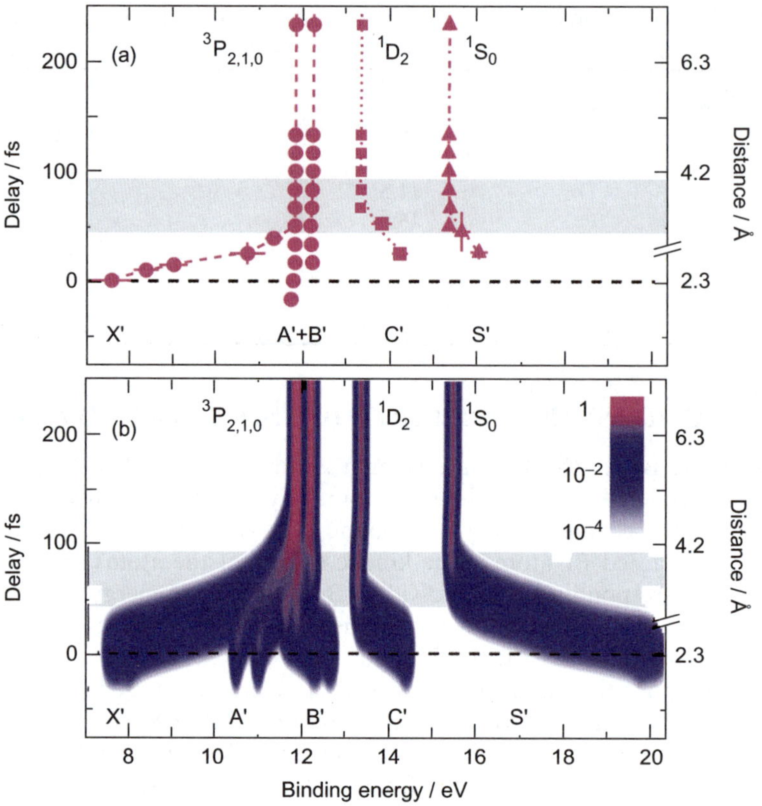

Figure 8.5 Photoelectron energy distributions following the photoexcitation of Br_2 at 266 nm, measured by Wernet and coworkers[605]. Top panel: Measured valence-electron binding energies *versus* delay and distance from the excited molecule (X_0-S_0), through a transition region (grey box) to the free atoms ($^3P_{2;1;0}$, $^1D_2$, $^1S_0$). Data points belonging to the same valence state are connected by lines. (b) Calculated binding energies and photoelectron intensities (logarithmic scale, color-coded intensity scale as indicated) for Fourier-transform-limited probe pulses (20 fs, 0.1 eV). Adapted from ref. [605].

displays the peaks X′, A′, B′, C′ and S′ of the Br_2^+ ion (where the prime indicates that the observed states have an electronically excited core). As the pump-probe delay time is increased, the recorded photoelectron energy distribution directly reveals firstly the rapid rearrangement of electrons in the excited neutral molecule and then the appearance of the evolving free Br(^3P_J) atoms (probed *via* the appearance of the Br^+ ions) all within a time frame of less than 100 femtoseconds.

The work of Wernet and coworkers[605] described above illustrates the power of femtochemistry in directly revealing the evolution of a dissociating molecule, and builds on earlier pioneering studies by Leone and coworkers[606] carried out at 400 nm, which excited the C state (as seen in Figure 8.4). At this point we should briefly consider the resolution available from femtosecond experiments. The time resolution in such an experiment is determined largely by the pulse lengths of the pump and probe lasers. However, the uncertainty principle also plays an important role. To illustrate this fact, consider a hypothetical study in which we would like to make measurements on Br_2 in 0.2 Å steps along the ~2 Å span of the dissociation process. At a dissociation wavelength of 266 nm, yielding Br atoms with recoil speeds of 1650 $m s^{-1}$, this will require a time resolution of around 10 fs. However, for a typical laser pulse-shape, 10 fs corresponds to an energy uncertainty of 1450 cm^{-1} [607] (see Chapter 11), which approaches the spin-orbit splitting of Br and Br*. Furthermore, owing to the similar SO splitting in the Br^+ ion compared with neutral Br, it is not possible to distinguish Br and Br* atomic products in the observed photoelectron kinetic energy distribution, even when using longer pulse lengths. Femtosecond pump-probe studies are powerful and direct probes of the short-time dynamics of photodissociation, but in this case they are unable to tell the whole story.

8.5 FEMTOSECOND DYNAMICS OF THE PHOTODISSOCIATION OF O_2 TO O^+– O^- ION PAIRS

Photodissociation of molecular bromine in the ultraviolet region of the spectrum is described above as an ultra-fast process, occurring on a time-scale that is so short that the energy resolution of the pump-probe laser system is barely sufficient to distinguish the product channels, due to the limits of the uncertainty principle. A more favorable case is photodissociation to ion-pair products (pathway (g) of Figure 8.1). Ion pair states are unusual compared to most electronic states of neutral molecules in that they appear at high energies but can be quite strongly bound. Molecular oxygen, O_2, provides a good example. As shown in Figure 8.6, the $^3\Sigma_u^-(O^+ + O^-)$ ion pair state (dissociation threshold at 17.3 eV) is more strongly bound than the ground electronic state of the molecule, and more strikingly, it remains so over a much larger range of internuclear distances. The long range character makes ion pair states particularly interesting for photodissociation dynamics studies.

Baklanov *et al.*[608] have shown that the $^3\Sigma_u^-(O^+ + O^-)$ ion pair state of O_2 can be reached by stepwise absorption of three 205 nm photons, which are created by frequency quadrupling the output of a Titanium Sapphire laser to produce the 4th harmonic.* If left alone after excitation, the highly localized molecular wavepacket created by the 120 fs UV pulse evolves along the ion pair coordinate, producing equal numbers of O^+atoms in the ^4S state and O^- atoms in the ^2P state. The resulting O^+ and O^- atoms are detected using the velocity-mapped imaging technique,[357,390] described in Study Box 5.3, in which both the speed and angular distribution (with respect to the polarization direction of the linearly polarized laser electric field direction) are recorded. Interpretation of angular distributions induced by multiple photon absorption[609] is more complicated than for those created by one-photon absorption, which will be described in the next section and in Study Box 8.3. O^+ and O^- images are shown as insets in Figure 8.6.

When an intense 815 nm probe pulse (the Ti-Sapphire fundamental) is introduced at selected time delays after the 205 nm excitation pulse, the intense laser radiation causes electron photodetachment from the dissociating O^-–O^+ ion pair state. During this process, the electron is detached primarily from the O^- part of the recoiling ion pair, leaving $O^+ + O$. When this occurs, the O^+ ion is suddenly set free from the attractive force exerted by the negatively charged O^-, and continues its trajectory with a velocity dependent on the O–O distance where detachment took place. The time-dependent velocity distribution for O^+ is shown in the upper panel of Figure 8.7. A vertical cut through the figure yields the velocity distribution obtained from the angle-integrated image for a given 205 nm–815 nm time delay. Signal corresponding to the ion-pair channel is indicated in the figure and simulated using a simple Coulomb attraction potential in the lower panel, where excellent agreement is found between the simulated and measured peak position and width of the energy distribution. The simulation assumes that motion takes place on the flat part of the repulsive curve leading to an electronically excited O*(3s) atom. This curve has the same symmetry as the ion pair state and crosses it (a so-called avoided crossing, discussed in Chapter 4) at both short and long internuclear distances. At the second crossing some of the molecules can escape to the excited atom channels (these excited atoms are directly ionized to form O^+ by the probe laser), producing the horizontal stripes seen in the upper panel of Figure 8.7. From this example we see that for a relatively 'slow' photodissociation process, formation of products can be understood quantitatively using femtosecond pump-probe imaging, essentially along the entire dissociation pathway.

*A read of Study Box 5.2 on the REMPI technique might be helpful at this point.

STUDY BOX 8.3: THE TRANSLATIONAL ANISOTROPY

Here we provide a simple classical derivation for the photofragment angular distribution given in Eq. (8.1).* The LAB-frame angular distribution of a photofragment (or photoelectron) can be expanded in a series of Legendre polynomials, $P_n(\cos\theta)$,

$$P(\theta) = \frac{1}{2}\sum_n a_n[n]P_n(\cos\theta), \tag{B8.3.1}$$

where $[n] \equiv 2n+1$, and θ is the angle between the recoil direction, $\boldsymbol{v}$, and the electric vector of the photolysis light, $\boldsymbol{\varepsilon}$, which we assume here to be linearly polarized (*i.e.* $\cos\,\theta = \hat{\boldsymbol{\varepsilon}}\cdot\hat{\boldsymbol{v}}$). By pre-multiplying both sides of Eq. (B8.3.1) by $P_m(\cos\theta)$, and integrating over $\cos\theta$, one obtains

$$\begin{aligned}\langle P_m(\cos\theta)\rangle &= \int_{-1}^{+1} P_m(\cos\theta)P(\theta)\mathrm{d}\cos\theta \\ &= \frac{1}{2}\sum_n a_n[n]\int_{-1}^{+1} P_m(\cos\theta)P_n(\cos\theta)\mathrm{d}\cos\theta.\end{aligned} \tag{B8.3.2}$$

The Legendre polynomials are orthogonal functions, and the integrals of products of Legendre polynomials can be written

$$\int_{-1}^{+1} P_n(\cos\theta)P_m(\cos\theta)\mathrm{d}\cos\theta = \frac{2}{[n]}\delta_{n,m}, \tag{B8.3.3}$$

such that the expansion coefficients of Eq. (B8.3.1) reduce to

$$a_n = \langle P_n(\cos\theta)\rangle. \tag{B8.3.4}$$

The discussion so far has been entirely in terms of LAB-frame quantities, and what is now needed is to relate the expansion coefficients to quantities in the molecule-fixed frame. This can be achieved by making use of the azimuthally-averaged spherical harmonic addition theorem, which allows us to write[239]

$$\langle P_n(\hat{\boldsymbol{\varepsilon}}\cdot\hat{\boldsymbol{v}})\rangle = \langle P_n(\hat{\boldsymbol{\varepsilon}}\cdot\hat{\boldsymbol{\mu}})\rangle\langle P_n(\hat{\boldsymbol{\mu}}\cdot\hat{\boldsymbol{v}})\rangle. \tag{B8.3.5}$$

This equation is valid provided the distributions of $\boldsymbol{\mu}$ about $\boldsymbol{\varepsilon}$, and $\boldsymbol{v}$ about $\boldsymbol{\mu}$ are uncorrelated. Based on the discussion in Study Box 8.2, the former distribution may be written

$$P(\hat{\boldsymbol{\varepsilon}}\cdot\hat{\boldsymbol{\mu}}) = \tfrac{3}{2}|\hat{\boldsymbol{\varepsilon}}\cdot\hat{\boldsymbol{\mu}}|^2 = \tfrac{1}{2}[1 + 2P_2(\hat{\boldsymbol{\varepsilon}}\cdot\hat{\boldsymbol{\mu}})], \tag{B8.3.6}$$

and therefore

$$\begin{aligned}\langle P_n(\hat{\boldsymbol{\varepsilon}}\cdot\hat{\boldsymbol{\mu}})\rangle &= \langle P_n(\cos\theta_{\varepsilon\mu})\rangle \\ &= \int_{-1}^{+1} P(\cos\theta_{\varepsilon\mu})P_n(\cos\theta_{\varepsilon\mu})\mathrm{d}\cos\theta_{\varepsilon\mu} = \delta_{n,0} \text{ or } \frac{2}{5}\delta_{n,2}.\end{aligned} \tag{B8.3.7}$$

*More rigorous classical or semiclassical derivations can be found refs. [239] and [612], whilst quantum mechanical treatments can be found in refs. [615,616].

Hence, the only terms that survive in the expansion of Eq. (B8.3.1) are those with $n=0$ and $n=2$, and we may write

$$P(\theta) = \tfrac{1}{2}[1 + \beta P_2(\cos\theta)], \tag{B8.3.8}$$

with

$$\beta = 5a_2 = 2\langle P_2(\hat{\boldsymbol{\mu}} \cdot \hat{\boldsymbol{v}})\rangle. \tag{B8.3.9}$$

Eqs. (B8.3.8) and (B8.3.9) reveal that prompt dissociation of a diatomic molecular *via* a parallel transition, for which $\boldsymbol{v} \| \boldsymbol{\mu}$, will have $\beta = +2$, and hence a $\cos^2\theta$ angular distribution of $\boldsymbol{v}$ about $\boldsymbol{\varepsilon}$, as discussed in the main text. On the other hand, prompt dissociation *via* a perpendicular transition, for which $\boldsymbol{v} \perp \boldsymbol{\mu}$, will have $\beta = -1$, leading to a $\sin^2\theta$ angular distribution of $\boldsymbol{v}$ about $\boldsymbol{\varepsilon}$.

Mark Brouard

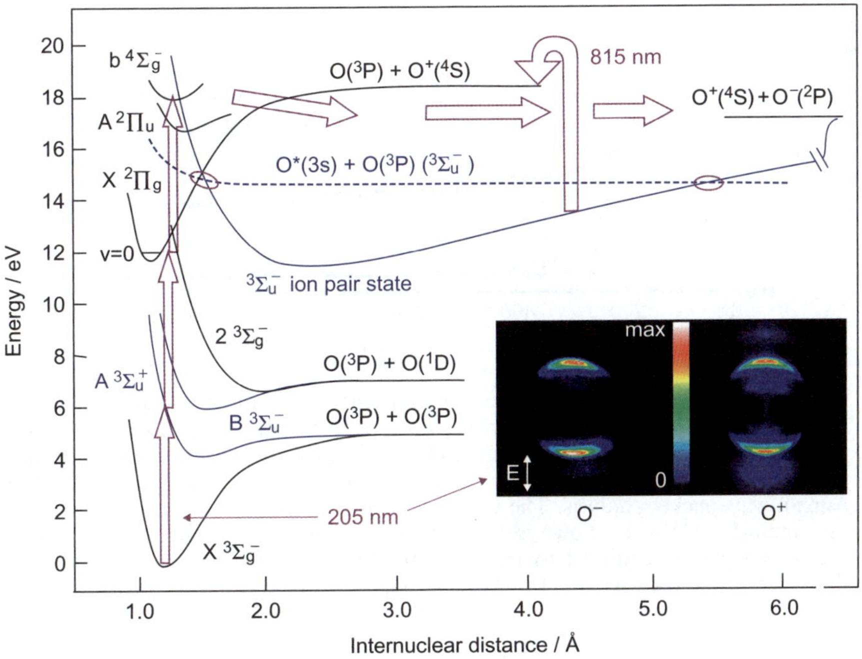

Figure 8.6 Potential energy curves for O_2, and O_2^+ and a possible absorption pathway for the stepwise three photon excitation of O_2 using 205 nm photons.[608] The repulsive wall of the strongly bound $^3\Sigma_u^-$ ion pair state is excited, which results in the production of O^+ and O^- atoms. Velocity-map images of the O^+ and O^- ions produced by 205 nm radiation only are shown as insets, along with a color bar which codes image intensity from low to high signal on a linear scale. The two angular distributions are identical and strongly peaked in the same direction as the electric vector of the 205 nm laser pulse (shown as a double headed arrow). By introducing an intense 815 nm pulse at variable time delays after the 205 nm excitation pulse, the ion pair dissociation can be interrupted by projecting the dissociating molecule onto the first dissociation limit of the O_2^+ ion, producing $O(^3P) + O^+ + e^-$.

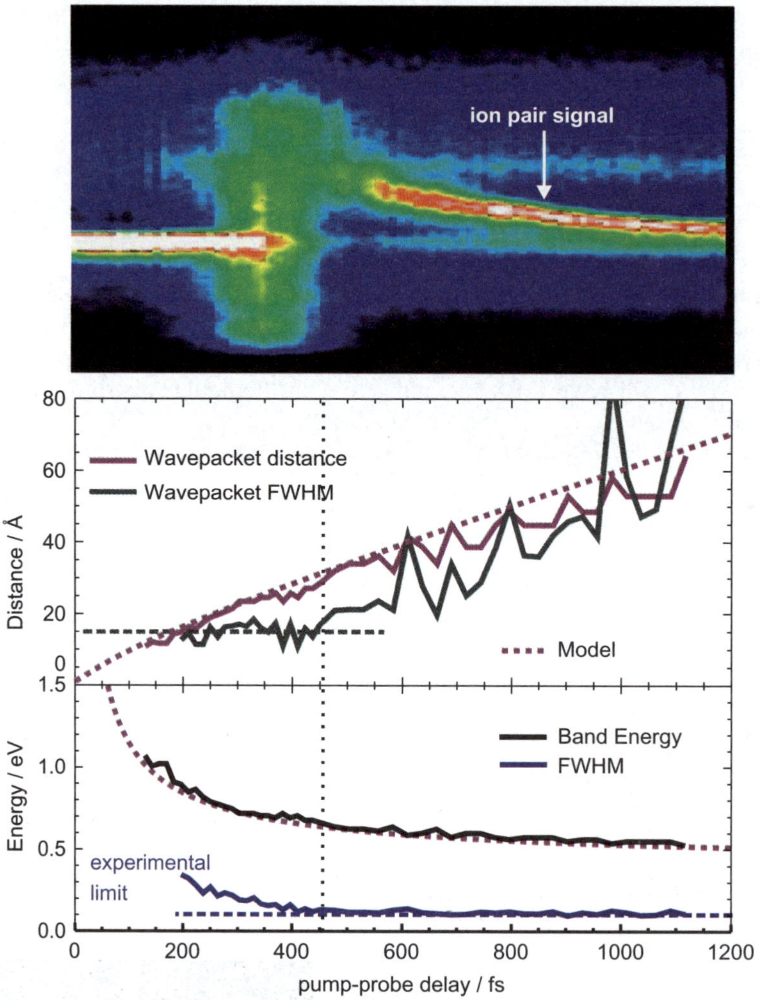

Figure 8.7 Upper panel: O^+ velocity distribution (determined from the angle-integrated ion images) as a function of the 0–1200 fs time delay between the 205 nm pump and 815 nm probe pulses in the ion pair dissociation of O_2.[608] See caption of Figure 8.6 for the color coding. In the lower half of the lower panel the peak and width of the measured O^+ kinetic energy are plotted and simulated using a wavepacket analysis. The employed wavepacket distances and width (full width at half maximum-FWHM) are shown in the upper half of the lower panel. For the period of 0 to 450 fs the wavepacket is assumed to travel on the O*(3s) repulsive curve shown in Figure 8.6, as described in the text. Adapted from ref. [608].

8.6 PHOTOFRAGMENT ANGULAR DISTRIBUTIONS IN Br_2 PHOTODISSOCIATION

In the femtosecond experiment on Br_2 dissociation described in Section 8.4, photoelectron *speeds* were measured using a magnetic bottle spectrometer.[611] The more powerful velocity-mapped[390] ion imaging[357] method employed in the O_2 ion-pair dissociation study described above allows the *velocity* (speed and direction) of any charged particle to be determined. Light is directional (through its propagation and polarization vectors) and molecules are non-spherical, and these two vector properties often result in considerable directionality in the spatial and angular momentum distributions of photo-products. Photofragment angular distributions are characterized experimentally by the spatial anisotropy parameter, β. In the case of the photodissociation of Br_2 by a single photon of linearly polarized light, the measured angular distribution of the Br fragments is presented in the form,[239,612]

$$f(\theta) = \tfrac{1}{2}[1 + \beta P_2(\cos\theta)]. \tag{8.1}$$

In this expression, θ is the angle between the electric vector $\boldsymbol{\varepsilon}$ of the exciting radiation and the velocity $\boldsymbol{v}$ of either Br fragment, both defined in the laboratory frame, in which $\boldsymbol{\varepsilon}$ is parallel to the z axis: the quantity β is the anisotropy parameter defined as $2\langle P_2(\cos\theta_{\mu\varepsilon})\rangle$, with $P_2(\cos\theta) = \frac{1}{2}(3\cos^2\theta - 1)$ being the second-order Legendre polynomial. A simple derivation of Eq. (8.1) is given in Study Box 8.3, from which it will be evident that the anisotropy in the angular distribution arises because the probability of an electric-dipole-allowed excitation process scales as $|\boldsymbol{\mu}\cdot\boldsymbol{\varepsilon}|^2$, where $\boldsymbol{\mu}$ is the transition dipole moment (see Study Box 8.2). Consequently, while the original sample of parent molecules is usually randomly aligned in space, the distribution of photoexcited molecules shows a maximum for those with $\boldsymbol{\mu}$ parallel to $\boldsymbol{\varepsilon}$, and has cylindrical symmetry about $\boldsymbol{\varepsilon}$. As the excited state molecules dissociate, this anisotropic distribution is carried over into the spatial distribution of the photofragments. The β parameter may take values in the range $-1 \leqslant \beta \leqslant 2$. When the transition dipole $\boldsymbol{\mu}$ lies perpendicular to the breaking bond (a 'perpendicular transition', for example $0_g \rightarrow 1_u$ or $\Sigma \rightarrow \Pi$), we have $\beta = -1$ and $f(\theta) = \frac{3}{4}\sin^2\theta$. A parallel transition, such as $0_g \rightarrow 0_u$ or $\Sigma \rightarrow \Sigma$) ($\boldsymbol{\mu}$ parallel to the breaking bond), has $\beta = 2$ and $f(\theta) = \frac{3}{2}\cos^2\theta$. Knowledge of the spatial anisotropy parameter can therefore help elucidate the symmetry of the states populated by photon absorption. This will be illustrated shortly (see also Figure 8.9 and accompanying discussion in the text).

The simultaneous excitation of parallel and perpendicular transitions (*i.e.* a mixed transition, such as that occurring in the region of overlap of the $B\,0_g^+$ and $C1_u^{(2)}$ states of Br_2) gives rise to non-limiting values of β. We note at this point that simultaneous excitation of two states may either be *incoherent*, when there is no fixed phase relationship between the wavefunctions of the two excited states, or *coherent*, in which all excited molecules have the same phase relationship between the two excited state wavefunctions. In laser-based experiments, the coherence properties of the absorbed radiation are often carried over into the excited states formed, and coherent excitation is common. In the most straightforward picture, an incoherent mix of a perpendicular and parallel transition would produce a β value given by $\beta = (x)(2) + (1-x)(-1)$, where x is the fraction of parallel excitation. While this simple approach has been used in most photodissociation studies up to now, it has become clear through more sophisticated measurements of the atomic polarization that coherent interactions between two excited states (Figure 8.8(a)), or between two pathways leading to the same final states, can play a major role in photodissociation. For example, simultaneous optical dipole excitation of two upper states, 1 and 2, from the ground state, g, leads to three terms (see Study Box 8.2 for an introduction to electric dipole transitions),

$$|\langle 1 + 2|\boldsymbol{\mu}\cdot\boldsymbol{\varepsilon}|g\rangle|^2 = |\langle 1|\boldsymbol{\mu}\cdot\boldsymbol{\varepsilon}|g\rangle|^2 + |\langle 2|\boldsymbol{\mu}\cdot\boldsymbol{\varepsilon}|g\rangle|^2 + 2|\langle 1|\boldsymbol{\mu}\cdot\boldsymbol{\varepsilon}|g\rangle\langle 2|\boldsymbol{\mu}\cdot\boldsymbol{\varepsilon}|g\rangle|\cos\Delta\phi, \tag{8.2}$$

with $\boldsymbol{\mu}\cdot\boldsymbol{\varepsilon}$ the component of the dipole operator $\boldsymbol{\mu}$ along the electric vector $\boldsymbol{\varepsilon}$ of the light. The first two terms correspond to individual contributions from states 1 and 2 separately, and the third cross term shows a dependence on the phase shift $\Delta\phi$ between the maxima of the two continuum waves (see Figure 8.8(a)).* This phase shift is dependent on the dissociation wavelength, and Zare and coworkers,[613] among others,[614] have shown that careful product polarization

*The phase shift in scattering theory is discussed in Chapter 12, Section 12.4.3.

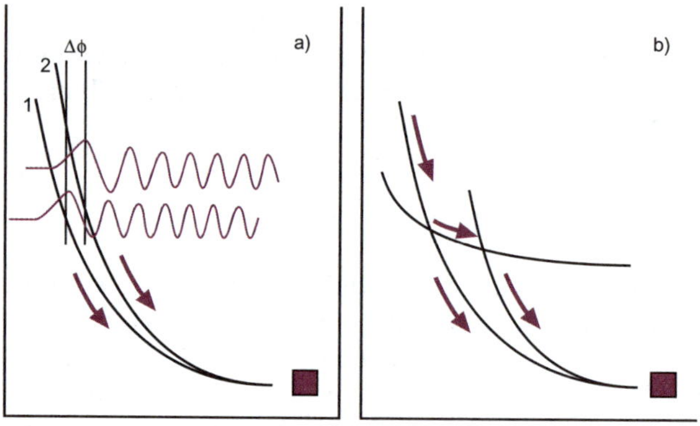

Figure 8.8 a) Interference, which appears at long range, due to simultaneous optical excitation of two upper states. b) Interference due to recrossing of two pathways leading to the same final product states.

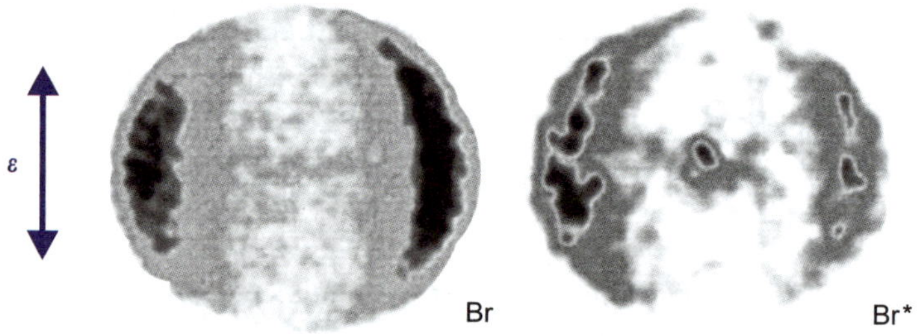

Figure 8.9 Ion images of Br and Br* arising from the photodissociation of Br_2 around 266 nm. Darker areas correspond to higher signal levels. The direction of the laser polarization vector, **ε** is indicated by the arrow. Data from Jee *et al.*[617]

measurements as a function of wavelength can be used to test and improve the accuracy of the potentials involved.

8.6.1 Above-threshold Photodissociation of Br_2 at 266 nm

Figure 8.9 shows two-dimensional raw images of Br and Br* arising from the photodissociation of Br_2 at wavelengths around 266 nm, obtained by Jee *et al.*[617] using the ion imaging method[357] (see Study Box 5.3). What is observed is the two-dimensional projection ('crush') of a sphere of Br^+ ions created by ionizing nascent neutral Br photofragments immediately after their formation. The ion sphere flattens to a disc-shape, with the radius of the disc being directly proportional to the fragment velocity, and the angular pattern, referenced with respect to the direction of the polarization of the dissociation laser, revealing the relationship between the transition dipole and recoil direction for the process.

While the quality of the ion images does not compare with those from the more recent velocity-mapped imaging method,[390] the two images do yield consistent information about the dissociation process. Though Br and Br* atoms are formed in equal numbers, and we should therefore expect their signal intensities to be equal, the Br* image is in fact weaker than the Br image due to the smaller ionization cross section of Br* when using the chosen detection method, namely state-selective (2+1) REMPI (see Study Box 5.2). In both images, most of the signal appears at the outer radius of the images, corresponding to a velocity of $1650\,ms^{-1}$, as predicted above. The fact that only a single 'ring' is observed in the images indicates that all of the products are formed *via* the Br + Br* channel; if some of the products arose from the Br + Br or Br* + Br* channel, the available energy (and therefore the product velocities) would be measurably different, and additional rings would appear in one or both images. Most of the signal appears in a direction perpendicular to that of the dissociation laser polarization axis (shown as a double-headed arrow in the figure). As described in Section 8.6, the polarized dissociation laser 'photo-selects' those Br_2 molecules in the randomly oriented molecular beam whose molecular axis has a significant component along the laser polarization direction. The observed angular distribution is explained when we consider that the electronic excitation is *via* a $0_g \rightarrow 1_u$ transition (see Figure 8.4), for which the transition dipole lies perpendicular to the Br-Br axis. As noted earlier, dissociation is essentially immediate, occurring within 100 fs. Consequently, the Br atom recoil distribution mirrors the axis distribution of the excited Br_2 molecules, giving rise to the observed $\sin^2\theta$ distribution about the laser polarization axis.

From the nanosecond experiment we can thus identify the symmetry of the upper (repulsive) state in the transition and the final product channel (Br + Br*). More detailed experiments for photodissociation at 355 nm[618] also reveal the M_J state distribution for the Br atom, and a 'complete' experiment would measure all of these quantities over the full range of dissociation laser wavelengths (as discussed in more detail in Section 8.7).

8.6.2 Near-threshold Photodissociation of Br_2 at 510 nm

Photodissociation of Br_2 at 510 nm, reported by Wrede *et al.*,[238] yields a very different picture from that at a dissociation wavelength of 266 nm. Figure 8.10 shows that the velocity distribution for $Br(^2P_{3/2})$ atoms peaks at $\sim 60\,ms^{-1}$, yielding a kinetic energy release of 1.5 meV for ^{79}Br. Excitation is thus just above the threshold for Br + Br* production, and occurs primarily to the repulsive wall of the bound $B(^3\Pi_u)\ 0_u^+$ state (Figure 8.2(b)). This state correlates only with the Br + Br* channel, and no dissociation to the lower-lying Br + Br channel is observed. As expected for a $0_u \rightarrow 0_g$ parallel transition, the product angular distribution is characterized by a β value of 2. This limiting β value is seen at the peak of the product kinetic energy distribution, where the signal arises from dissociation of Br_2 molecules in the molecular beam with low rotational energy. However, signal is also seen at higher velocities, in the $90–150\,ms^{-1}$ range, and β for these atomic products is found to decrease from the expected value of 2 down to a close to isotropic value of zero with increasing velocity. This observation is relatively easy to rationalize when we consider the rotational temperature of the molecular beam, which in these experiments was found to be 42 K. For Br_2, this means that a wide range of rotational levels are populated in the beam. The extra internal energy due to rotation is converted to fragment velocity in the photodissociation process, thus the higher rotational states give rise to signal at higher velocities than the lowest rotational states, explaining the appearance of photofragments with velocities in the $90–150\,ms^{-1}$ range. Using both quantum and semi-classical methods, together with accurate potential energy curves for the relevant excited states of Br_2, the authors of ref. [238] successfully modelled the dissociation and showed that this behaviour exhibited by β is due to rotation of the excited Br_2 molecules prior to dissociation. This reduces the correlation between $\boldsymbol{\mu}$ at the instant of photon excitation and $\boldsymbol{v}$, and leads to a corresponding reduction in β, and a clear and direct observation of the breakdown of the axial recoil approximation (see Study Box 8.1). Using this high-velocity resolution approach the authors were also able to determine D_0, the Br_2 bond energy (see Figure 8.4) to high accuracy.

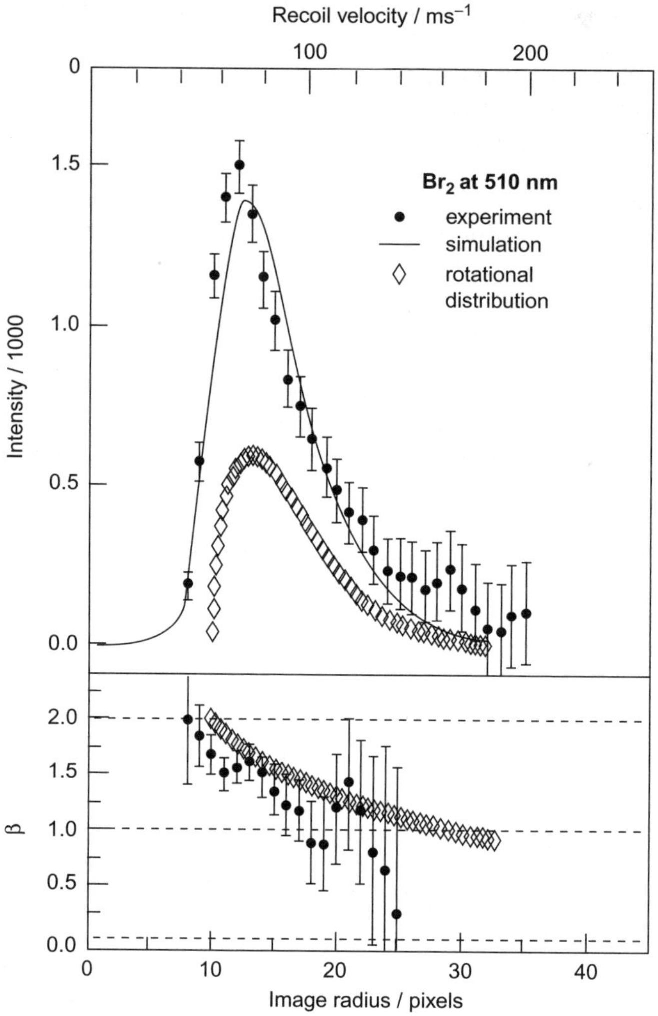

Figure 8.10 Deviations from the axial recoil approximation, as observed in the velocity distribution and β parameter of ^{79}Br atoms from photodissociation of $Br_2(v=0)$ molecules at 510 nm, just above the 511 nm threshold for producing the fragment pair $Br(^2P_{3/2}) + Br^*(^2P_{1/2})$. Also shown is a best-fit simulation based upon the convolution of the rotational state population distribution for Br_2 molecules with $T_{rot} = 42$ K. The bottom panel compares the measured variation of the anisotropy parameter β, with the rotational level dependence predicted by quasi-classical calculations as described in ref. [238].

8.7 ATOMIC PRODUCT POLARIZATION

For ultrafast axial photodissociation, both femtosecond and nanosecond timescale imaging experiments can measure spatial anisotropy and thus determine the direction of the transition dipole relative to the molecular axis. Nanosecond timescale pump-probe experiments on the photodissociation of Br_2 also have the special ability to determine accurately the relative yields of Br and Br* products and the polarization of the angular momentum of these products with respect to the direction of their recoil velocity, $\boldsymbol{v}$. The spatial orientation of an angular momentum vector $\boldsymbol{J}$

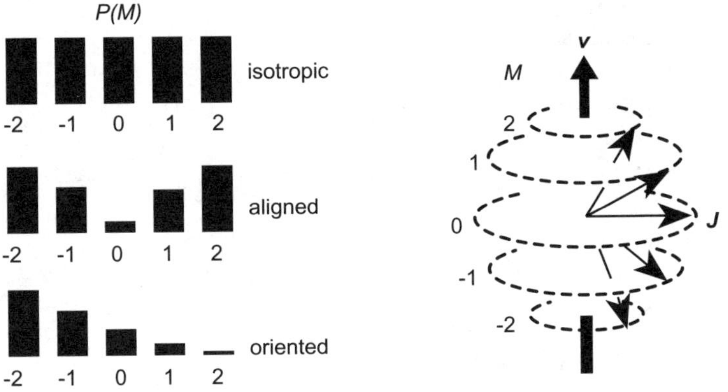

Figure 8.11 Angular momentum polarization for a $J = 2$ atom such as $O(^1D)$.

is described by its projection quantum number M_J (often referred to simply as 'M').* In the molecular frame, the z-axis (or projection axis) is usually taken as the recoil velocity vector, $\boldsymbol{v}$. Atomic polarization therefore reflects an anisotropic M-state distribution, as illustrated in Figure 8.11 for the case of $J = 2$, *e.g.*, an $O(^1D_2)$ atom. In 1968, van Brunt and Zare[619] went as far as to state that "*molecular dissociation can be viewed as a highly efficient atomic M-state selector*".

Polarization of a vector distribution is usually classified in terms of either alignment or orientation (see Chapters 5 and 9). An aligned distribution has the vectors pointing preferentially along an axis, but with equal numbers pointing in the positive and negative direction (such a distribution is often described in terms of 'double headed arrows'), while in an oriented distribution the vectors point preferentially along an axis and also have a preferred direction ('single headed arrows'). The polarized angular momentum distribution reflects the atomic orbital electron densities in the recoiling atom. An isotropic angular momentum distribution, shown at the top of Figure 8.11, results from the case in which all of the M-substates of the atom are equally populated. An aligned distribution arises from a non-statistical distribution of $|M|$, with $\pm M$ levels equally populated, while an oriented distribution arises from a non-statistical distribution of M with unequal populations in the levels with $\pm M$. As an example, photoproducts formed with 100% population in $M = 0$ correspond to an angular momentum distribution in which $\boldsymbol{J}$ is aligned perpendicular to $\boldsymbol{v}$. In the case of Br_2 photodissociation, the angular momentum polarization arises from the population distribution over the $M_J = \pm 1/2$, $\pm 3/2$ levels for atoms formed in the $^2P_{3/2}$ ground state, and the $M_J = \pm 1/2$ levels for atoms formed in the $^2P_{1/2}$ spin-orbit excited state.

A cartoon showing how specific M_J state production could arise, for example, for the $Br_2\ X0_g \rightarrow C(^1\Pi_u)1_u^{(2)}$ $(\sigma_g^2\pi_u^4\pi_g^4\sigma_u^0 \rightarrow \sigma_g^1\pi_u^4\pi_g^4\sigma_u^1)$ transition is shown in Figure 8.12. One of the two spin-paired electrons in the π_g^* orbital of the singlet ground state is excited to the σ_u^* orbital, generating an excited molecular electronic state with $\Omega(AB^*) = \pm 1$. As the molecule dissociates on the excited state, the molecular orbitals 'morph' into atomic orbitals, while conserving the Ω quantum number (the projection of M_J onto the internuclear axis/recoil direction) according to the correlation rules outlined previously. Consideration of the projections of $\boldsymbol{L}$ and $\boldsymbol{S}$ for the atomic products onto the recoil axis, as shown on the right hand side of the figure, reveals preferential population of the $M_J = \pm 1/2$ sublevels, *i.e.* an aligned distribution of $\boldsymbol{J}$ relative to the recoil axis, or in other words, an aligned distribution in the 'molecular frame', in which the z axis is defined to lie parallel to the relative velocity of the photofragments. Because the photofragments have preferred directions of

*Unlike Chapters 5 and 9, we use capital letters here for J and M_J, as a means of emphasizing that we are concerned here with the polarization of the total electronic angular momentum of the atomic photofragments of molecular photodissociation, rather than with the rotational angular momentum polarization of molecular photofragments.

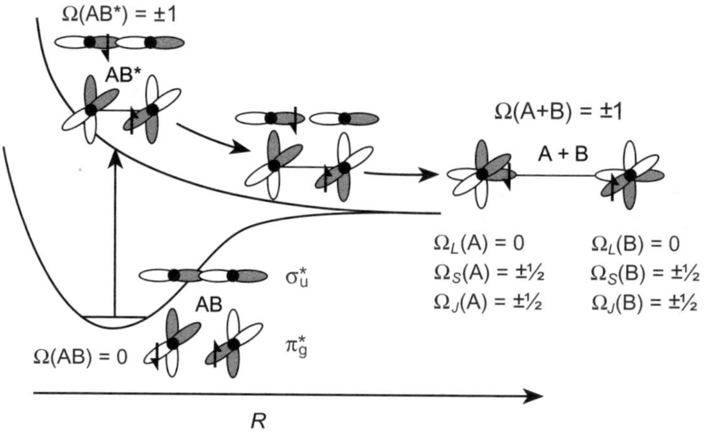

Figure 8.12 Schematic illustration of the origins of atomic photofragment angular momentum polarization, in this case following excitation of a diatomic molecule in a $\pi_g^* \rightarrow \sigma_u^*$ transition. See text for details. Adapted from ref. [620].

recoil relative to the electric vector of the photolysis laser beam, as described previously, their angular momentum is also aligned in the laboratory frame, since the photolysis laser polarization generally defines the direction of Z in this frame.

If polarized light is used in the probe step of a pump-probe scheme, laser-based detection techniques such as laser-induced fluorescence (LIF) or resonance-enhanced multiphoton ionization (REMPI) become sensitive to photofragment angular momentum polarization (see Study Box 9.1). This sensitivity arises from the fact that the transition dipole $\boldsymbol{\mu}_{\text{probe}}$ for the electronic transition involved in the detection step is correlated with the total angular momentum vector $\boldsymbol{J}$. An anisotropic distribution of $\boldsymbol{J}$ therefore implies an anisotropic distribution of $\boldsymbol{\mu}_{\text{probe}}$, and the measured signals will show a dependence on probe laser polarization. This dependence is often dependent on the details of the probe transition chosen. Imaging experiments have proved to be a particularly powerful probe of product polarization.[289,601,620,621] The dependence of the images on probe laser polarization reveals product angular momentum alignment and orientation as a function of both the speed and the recoil angle of the photofragments. This is a case in which highly detailed information may be gleaned on a dissociation process using nanosecond methods.

8.7.1 Atomic Polarization Effects in the Photodissociation of Br_2 and I_2

As the above discussion makes clear, besides the spatial distributions and Br/Br* branching ratio, nanosecond experiments can also provide the final M_J state distributions. Determination of M_J populations generally requires a series of images to be recorded, using different combinations of pump and probe laser polarizations and propagation directions and possibly different REMPI transitions in the probe step. Details of some of the image analysis procedures used to extract the M_J populations from the images may be found in ref. [289].

Photodissociation of bromine at 510 nm takes place *via* the $B0_u^+(^3\Pi_u)$ state, which is known to correlate asymptotically to the atomic M states $|M_A, M_B\rangle$ as

$$|0_u\rangle_B \xrightarrow{R\rightarrow\infty} \tfrac{1}{\sqrt{2}}\left[\left|+\tfrac{1}{2},-\tfrac{1}{2}\right\rangle + \left|-\tfrac{1}{2},+\tfrac{1}{2}\right\rangle\right]. \tag{8.3}$$

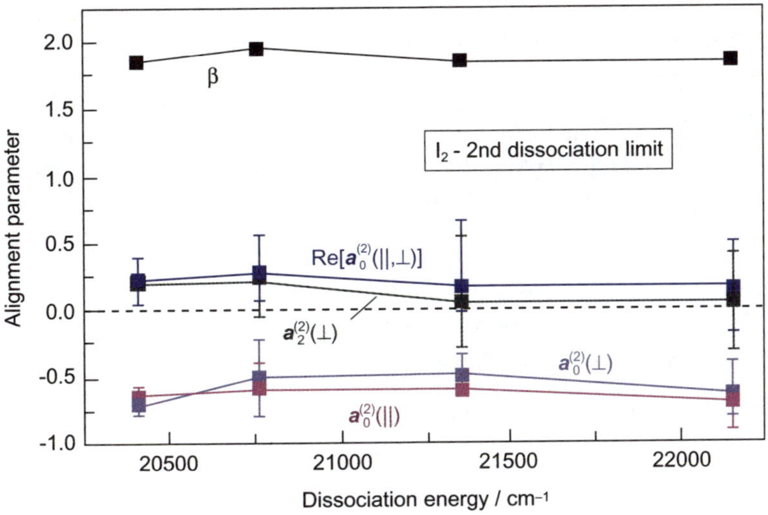

Figure 8.13 I atom polarization properties from the photodissociation of I_2 to the second dissociation limit (I + I*). Data from ref. [622]. The molecular frame polarization parameters $\boldsymbol{a}_q^{(k)}(p)$, are described in Study Box 8.4 and in Chapter 9, and their interpretation is given in the text.

Bromine product atoms are thus produced only in the $|M| = 1/2$ states. The $|M| = 3/2$ states are absent, and the products are thus highly aligned. This correlation holds for all of the dihalogens (F_2, Cl_2, Br_2, and I_2). The potential energy curves for I_2 and Br_2 are quite similar, so we might expect to see similar polarization in the atomic iodine products of I_2 dissociation as we do in the atomic Br photofragments from Br_2 dissociation. Figure 8.13 shows results from a study[622] of the I atom polarization following photodissociation of I_2 over a range of dissociation energies from near threshold (20,000 cm^{-1} for I_2) to $>2000\ cm^{-1}$ above threshold. At these wavelengths, excitation is predominantly to the $B0_u^+(^3\Pi_u)$ state (a parallel transition), with a small contribution (between 2 and 5%) from a perpendicular transition to another state, which then follows a somewhat more complicated pathway to the same products.

The value of β is just slightly less than 2, reflecting the predominantly parallel nature of the transition; the reduction from the limiting value is due to the small contribution to the signal from the perpendicular transition noted above. In Figure 8.13, the polarization of the iodine atoms is expressed in terms of a set of molecular frame polarization parameters, $\boldsymbol{a}_q^{(k)}(p)$, with $p = ||$ and/or $\perp$ (see Study Box 8.4 and Chapter 9). These may be thought of essentially as coefficients in a spherical harmonic expansion of the spatial distribution of $\boldsymbol{J}$, with the symbols in brackets indicating whether the polarization arises from a parallel or perpendicular contribution to the excitation. The value of the $\boldsymbol{a}_2^{(2)}(||)$ parameter is large, at around –0.8, and may be shown to indicate pure $|M| = 1/2$ population. The near-zero value of $\mathrm{Re}[\boldsymbol{a}_1^{(2)}(||,\perp)]$ indicates the lack of any interference between different pathways reaching the same final products. Despite the very small contribution to the signal from the perpendicular pathway, the experiments were sensitive enough also to determine the alignment parameters $\boldsymbol{a}_0^{(2)}(\perp)$ and $\boldsymbol{a}_2^{(2)}(\perp)$ for the products formed through this pathway. These products are also found to be aligned, with essentially pure $|M| = 1/2$ population just as for the products formed through the dominant parallel excitation.

Concentrating on the dominant parallel excitation pathway, we note, importantly, that all of the polarization parameters for this pathway remain essentially constant over the wide range of dissociation energies. This indicates that only a single excited state surface is involved in the dissociation at all wavelengths studied; the dissociation is said to exhibit purely adiabatic behaviour.

STUDY BOX 8.4: POLARIZATION PARAMETERS AND MOLECULAR PHOTODISSOCIATION

There are a number of different formalisms in use for describing angular momentum polarization in the products of molecular photodissociation. In all cases, the alignment or orientation is described in terms of a set of *polarization parameters*. The discussion of polarization in the present chapter uses the $\boldsymbol{a}_q^{(k)}(p)$ parameters developed by Rakitzis and Zare,[624] so we will explain these parameters in a little more detail here.

The $\boldsymbol{a}_q^{(k)}(p)$ parameters are closely related to the polarization parameters discussed in Chapter 9 in the context of bimolecular collisions, and it might be helpful to read Section 9.2.3 prior to reading Section 8.7 of this chapter. In the formalism of Rakitzis and Zare,[624] the starting point is to expand the spatial distribution of the photofragment angular momentum $\boldsymbol{J}$ in the molecular frame as a sum over modified spherical harmonics $C_{kq}(\theta, \phi)$.

$$P(\theta, \phi) = \sum_{k=0}^{2J} \sum_{q=-k}^{k} A_q^{(k)} C_{kq}(\theta, \phi). \tag{B8.4.1}$$

Note that the modified spherical harmonics are related to the spherical harmonics by

$$C_{kq}(\theta, \phi) = \left[\frac{4\pi}{2k+1}\right]^{1/2} Y_{kq}(\theta, \phi). \tag{B8.4.2}$$

The sum goes up to $k = 2J$, where J is the total angular momentum quantum number of the photofragment under study. In Eq. (B8.4.1), θ and ϕ are the polar angles relative to a z axis lying along the photofragment velocity vector $\boldsymbol{v}$ and an x axis lying in the plane of $\boldsymbol{v}$ and the photolysis polarization vector $\boldsymbol{\varepsilon}_{\text{phot}}$. It is therefore valid only for a fixed angle θ_ε between these two vectors. In order to obtain a general expression for the angular momentum polarization that includes the angular scattering distribution of the photofragments, we can break up Eq. (B8.4.1) into a contribution from pure parallel transitions, a contribution from pure perpendicular transitions, and interference terms between the two. To achieve this separation we introduce a set of alignment parameters $\boldsymbol{a}_q^{(k)}(p)$, where $p = \|$ or $\perp$ describes the polarization of the excitation. The $\boldsymbol{a}_q^{(k)}(p)$ parameters are related to the $A_q^{(k)}$ as follows:

$$\begin{aligned}
A_0^{(k)} &= \left[(1+\beta)\cos^2\theta_\varepsilon \boldsymbol{a}_0^{(k)}(\|) + (1-\beta/2)\sin^2\theta_\varepsilon \boldsymbol{a}_0^{(k)}(\perp)\right] \Big/ [1+\beta P_2(\cos\theta_\varepsilon)] \\
A_1^{(k)} &= \sin\theta_\varepsilon \cos\theta_\varepsilon \boldsymbol{a}_1^{(k)}(\|, \perp) \Big/ [1+\beta P_2(\cos\theta_\varepsilon)] \\
A_2^{(k)} &= (1-\beta/2)\sin^2\theta_\varepsilon \boldsymbol{a}_2^{(k)}(\perp) \Big/ [1+\beta P_2(\cos\theta_\varepsilon)],
\end{aligned} \tag{B8.4.3}$$

with $A_q^{(k)} = (-1)^q A_{-q}^{(k)*}$.

The $\boldsymbol{a}_q^{(k)}(\|)$ describe contributions to the alignment from processes involving pure parallel transitions from the ground state and the $\boldsymbol{a}_q^{(k)}(\perp)$ from perpendicular transitions, with the $\boldsymbol{a}_q^{(k)}(\|, \perp)$ arising from interference between the two types of process. For terms up to $k = 2$, the details of this separation are described in ref. [624]. At this point we have an expression for the *molecular frame* angular momentum alignment. To obtain an expression for the measured distribution in the LAB-frame, we need to correct each term in this equation for the detection sensitivity, which will depend on details of the experiment such as the REMPI transition chosen to detect the fragments and the polarization of the probe light. For comparison with experimental data, we also need to transform the expression into the appropriate laboratory frame. As an example, for velocity-mapped imaging experiments the most convenient LAB-frame has z lying along the time-of-flight direction, and x in the plane of z and the photolysis polarization. The details of the molecular frame to LAB transformation are beyond the scope of this Tutorial, but are given elsewhere.[624] Determining the polarization parameters generally consists of measuring the product signal in a number of experimental geometries, and then generating the appropriate LAB-frame expressions for the alignment distribution in the various geometries. These expressions constitute a set of simultaneous equations, which may be solved to extract the alignment parameters.

Other sets of polarization parameters in widespread use in molecular photodissociation studies include the *bipolar moments*,[239] which are based on a semi-classical treatment of polarization effects in molecular photodissociation presented by Dixon.[625] These are closely related to the $f_K(q, q')$ dynamical functions of Siebbeles *et al.*,[614] and the polarization anisotropy parameters of Vasyutinskii and coworkers.[626] The treatment of Siebbeles *et al.*[614] provided the first fully quantum mechanical theory of angular momentum polarization effects in molecular photodissociation. More details of these parameters may be found in the original literature cited above, where the relationships between some of the various parameter sets can also be found (see also, for example, ref. [627]).

Claire Vallance

For molecular bromine, the dissociation channels and the spatial anisotropy of the photofragments determined from the experiments at 266 nm and 510 nm agree well with what we would predict from the calculated potential energy curves. On the face of it this is not surprising owing to the large mass of the Br atoms; nuclear motion should be very slow compared to electronic motion in this molecule, and the adiabatic picture should hold (see Chapters 2 and 4). At a higher level of detail, however, the theoretical analysis[602,623] shows that the $1_u^{(3)}(^3\Pi_u^+)$ state excited at 266 nm crosses the $1_u^{(4)}$ state at an internuclear separation of ~6.5 Å, and following excitation purely to the $1_u^{(3)}$ state, the final products reflect a strong mixing of the $1_u^{(3)}$ and $1_u^{(4)}$ pathways. The mixing of electronic states as the nuclei pass through the 6.5 Å crossing region is an example of a *non-adiabatic process*, as discussed in detail in Chapter 4. Since both pathways correlate to the same Br + Br* dissociation limit, and the M_J distributions expected for adiabatic dissociation *via* these states have not yet been worked out, it is difficult to predict the experimental manifestation of this state mixing. However, analysis of photodissociation *via* the lower electronic states such as the $B0_u^+(^3\Pi_u)$ state shows no curve crossing, and the molecule should indeed follow adiabatic predictions at these dissociation energies.

8.8 THEORETICAL TREATMENT OF PHOTODISSOCIATION

The adiabatic model appears to provide a good description of most aspects of the photodissociation of Br_2, at least at low dissociation energies. However, for photodissociation of Br_2 at higher dissociation energies, and in the photolysis of many other small molecules, the dynamics show considerable deviations from the predictions of this simple model. It is clear that in many cases more than one potential energy curve is involved in the breaking of a chemical bond. Armed with the potential energy curves for Br_2 (Figure 8.4), for example, there are many questions we might ask. Does the molecule stay on the (initially populated) $1_u^{(3)}$ curve throughout the course of the dissociation process, or does it 'jump' to nearby curves on its way towards the separated atom limit? If it does jump, what determines the probability, and how does this depend on the initial excitation energy of the molecule? The probability of eliciting transitions between different spin-orbit states, for example, as the dissociation proceeds through the 'spin-orbit coupling region' linking the molecular states to the atomic states, is often considered in terms of the Massey parameter,[235,628] $\xi_t = \Gamma \Delta E_{SO}/\hbar v$, where Γ is the width of the recoupling zone, ΔE_{SO} is the spin-orbit splitting in the atomic fragment, v is the recoil velocity as it passes through the recoupling region, and ξ_t is known as the translational adiabaticity parameter. According to the Massey criterion, the adiabatic limit will be approached when $\xi_t \gg 1$, *i.e.* when the rate of product separation is very slow (and/or the spin-orbit splitting very large) and therefore the fragmenting molecule is most likely to follow a single, adiabatic surface. Conversely, the situation $\xi_t \ll 1$ corresponds to the sudden, or diabatic, limit. In this case, the rapid traversal of the recoupling zone means that the product fine structure distributions are determined purely by the mapping of the molecular potential energy states onto the asymptotic product states.

Once we have some idea of the crossing probability, can we predict which quantum states, including the value of the atomic M_J quantum number, will be produced as a result of a non-adiabatic process of this type? This is a set of very complex questions, which until recently could not be quantitatively answered for any molecule. However, as we have seen in Chapter 4, for the hydrogen halide molecules HX (X = F, Cl, Br, and I) accurate *ab initio* quantum mechanical calculations have been carried out that follow the dissociation process from start to finish, revealing the evolution of the excited state character along the way.[629]

Hopefully, similar calculations to those on the hydrogen halides will be carried out on a wider range of molecules, including Cl_2 and O_2, in the future. These are challenging calculations, however, and in many cases we must rely instead on highly simplified limiting-case models such as those presented in Figure 8.14 to guide our analysis. The first three panels of this figure show the most popular simple models for photodissociation: adiabatic, sudden, and statistical. All of these models have in common the fact that their predictions are independent of the excitation energy. Panel (d) shows a cartoon representation of an 'actual' process, namely the photodissociation of the HCl molecule as predicted by a fully quantum treatment,[631] and discussed in Chapter 4. Finally, panel (e) describes the transition state model, which is most appropriate for polyatomic molecules (see Chapter 7). Suits and coworkers[630] have observed remarkable deviations ('roaming atom' dynamics) from the simple transition state picture for the photodissociation of formaldehyde.

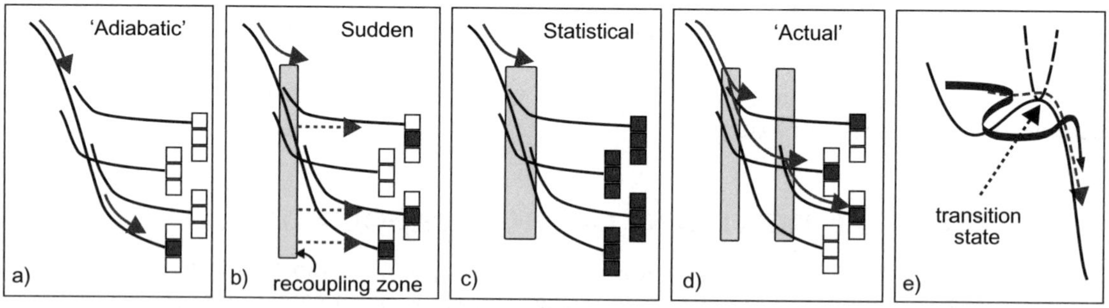

Figure 8.14 Models for photodissociation: a) adiabatic 'curve-following' model; b) Sudden, or diabatic, model, in which the photodissociation flux is projected from one region, the 'recoupling zone', onto all symmetry-allowed final products; c) statistical model, in which the strong coupling limit is reached and all product states are equally populated; d) a cartoon description of the HCl photodissociation dynamics predicted by fully quantum theory; and e) the transition state theory for photodissociation of a polyatomic molecule indicating (dashed) trajectories crossing through the saddle point mode, and (solid line) skirting the transition state in a 'roaming atom' mechanism.[630] The boxes signify possible final state M_J quantum numbers, where, for example, all possible final states are populated in panel 3.

The adiabatic dissociation model, panel (a), forms the basis for the correlation diagrams we have considered so far in our discussion of the photodissociation of Br_2. In the adiabatic model, it is assumed that the photofragments part so slowly with respect to the electronic motion that the system remains in a single electronic state. For Br_2, since the primary excitation is to the $1_u^{(3)}$ state, the adiabatic model predicts that the sole dissociation channel is that leading to Br + Br* products. The sudden or diabatic model, panel (b), approaches the problem from the opposite end of the spectrum: It predicts atomic state distributions in the "sudden limit", in which the molecular states are projected onto the atomic states without allowing for any evolution of the electronic wave function during the dissociation. In the strong coupling or statistical model, panel (c), the states are coupled by the interaction to such an extent that all accessible states are equally populated. Aspects of the adiabatic and sudden models will now be discussed in more detail.

8.8.1 Adiabatic Model of Photodissociation

The adiabatic model was introduced in Section 8.3 for the case of Br_2 in order to describe how a specific molecular electronic state correlates to a specific separated atom limit. Using this model, it is relatively straightforward to predict the product atom Ω (or *L*) quantum numbers for the dissociation of any electronic state of a diatomic molecule. The same approach can also be extended to linear triatomic molecules and to a few highly symmetric polyatomic molecules.[262] We now look in more detail at techniques for calculating the adiabatic potential energy curves for a dissociating molecule.

Because dissociation is initiated by photon absorption, with its specific $\Delta S = 0$ or $\Delta\Omega = 0$ or ± 1 selection rules, it is often useful to include electron spin, a relativistic effect, in the initial quantum mechanical description of the molecule. We begin with the total Hamiltonian for a diatomic molecule, which may be written[257] as, in notation of Chapters 2 and 4,

$$\hat{H} = \left[\hat{H}_{\rm el} + \hat{H}_{\rm SO}\right] + \hat{T}_{\rm N}, \tag{8.4}$$

where $\hat{H}_{\rm el}$ is the electronic part of Hamiltonian, consisting of the electrostatic potential (including electron-electron, electron-nucleus, and nucleus-nucleus terms) and the electronic kinetic energy operators. $\hat{H}_{\rm SO}$ represents the spin-orbit interaction (other relativistic terms are neglected), and $\hat{T}_{\rm N}$ is the nuclear kinetic energy operator. In the simplest treatment, the electronic potential energy curves (within the usual framework of the Born-Oppenheimer approximation), $E_i^{\rm BO}(R)$, where R is the internuclear separation, are found by calculating the eigenvalues of the electronic Hamiltonian $\hat{H}_{\rm el}$ as a function of nuclear separation.

$$\hat{H}_{\rm el}\Phi_i^{\rm BO} = E_i^{\rm BO}(R)\Phi_i^{\rm BO}, \tag{8.5}$$

where

$$\left\langle \Phi_i^{\rm BO} \middle| \hat{H}_{\rm el} \middle| \Phi_j^{\rm BO} \right\rangle = 0 \quad \text{for all } i \neq j, \tag{8.6}$$

and the wavefunctions, $\Phi_i^{\rm BO}$, are the adiabatic electronic wavefunctions obtained within the Born-Oppenheimer approximation. Two curves $E_i^{\rm BO}(R)$ will cross freely unless all of the quantum numbers Λ, Σ, and Ω are shared between the two states. Here, Λ is the projection of the electronic angular momentum on the internuclear axis, Σ the spin projection, and Ω the projection of the total angular momentum (with values $\Omega = |\Lambda \pm \Sigma|$). At short R, the diagonal spin-orbit interaction is approximated to be independent of atomic separation R. In this approximation, the only difference in the potential energy curves for Ω states arising from the same values of Λ and Σ is that they are shifted relative to each other by an R-averaged, Ω-specific spin-orbit energy.

A more complete approach is to use an alternative "relativistic" basis, produced by including the relativistic part of the Hamiltonian when calculating the energy. This yields a new set of potential energy curves $E_i^{\mathrm{rel}}(R)$,

$$\left[\hat{H}_{\mathrm{el}} + \hat{H}_{\mathrm{SO}}\right]\Phi_i^{\mathrm{rel}} = E_i^{\mathrm{rel}}(R)\Phi_i^{\mathrm{rel}}. \tag{8.7}$$

For these states, only Ω is a good quantum number, and the curves $E_i^{\mathrm{rel}}(R)$ will always avoid crossing at intersections of common Ω. Such avoided crossings occur more frequently for relativistic potentials than for pure Born-Oppenheimer potentials. An attractive feature of the relativistic adiabatic model is that a one-to-one mapping of the molecular electronic states to asymptotic (J_{A}, J_{B}) states can easily be constructed, as described earlier for Br_2. Because adiabatic curves with the same Ω never cross, the energy ordering of these states is preserved from short range to the atomic limit. Whether the molecule follows the relativistic adiabatic curves depends on the strength of the spin-orbit coupling. With molecules containing only light atoms, spin-orbit coupling is weak and curve crossing more probable. At long range, the Born-Oppenheimer curves with common Ω are often nearly parallel, and perfectly avoided crossing follows. In the framework of the adiabatic basis, the central question is how the nuclear kinetic energy operator $\hat{T}_{\mathrm{N}}$ couples the adiabatic potentials. Discussions of the various forms of nuclear coupling for the case of O_2 photodissociation can be found in reference[632] (see also the discussion in Chapter 4).

8.8.2 Prediction of Atomic Polarization using the Adiabatic Model

In this section we will consider in a little more detail how to calculate adiabatic potential energy curves for a diatomic, in terms of the form of the Hamiltonians to be used, and the way in which these may be used to predict angular momentum polarization in the atomic fragments.

To recap, anisotropy in molecular photodissociation encompasses both the spatial anisotropy of the photofragments, characterized by the spatial anisotropy parameter β, and the angular momentum polarization, characterized by the M-state distribution of the photofragments. As noted in Section 8.7, M-state selection with respect to a LAB-frame Z-axis arises because the product atoms are polarized with respect to the molecular axis, which in turn is polarized relative to the electric vector of the photolysis radiation, which is fixed in the LAB-frame.

Use of the adiabatic model for predicting the final orbital polarization quantum numbers (M_{A}, M_{B}) of the fragments can be somewhat complex, especially in the case of a photodissociation reaction producing two open-shell fragments

$$\mathrm{AB} + h\nu \rightarrow \mathrm{A}\left({}^{2S_{\mathrm{A}}+1}L_{\mathrm{A}}\right) + \mathrm{B}\left({}^{2S_{\mathrm{B}}+1}L_{\mathrm{B}}\right). \tag{8.8}$$

Br_2 dissociation fits into this category, producing open shell $Br(^2P_{3/2}) + Br(^2P_{1/2})$ atomic products, as do most dissociation processes arising from homolytic cleavage of a chemical bond. To make progress, we use the fact that long range interatomic interactions play a critical role in determining the final atomic polarization arising from a photodissociation process. The long range interactions in a neutral diatomic molecule arise from the electrostatic potential $V(R)$ between the atoms, which may be expressed in terms of a power series in R^{-1}. The most important term in this expansion is usually that describing the quadrupole-quadrupole interaction, proportional to R^{-5}, followed by the van der Waals interaction potential, which depends on R^{-6}. In many cases, we can predict the atomic polarization to a good approximation by considering only the quadrupole-quadrupole term in the Coulomb expansion, though each molecule must be treated on an individual basis when deciding which terms to include.

The general approach is to expand the molecular wavefunction $\Phi_{el}(n, \Omega; R)$ at short R in terms of the atomic fragment $|JM_J\rangle$ states at long R.

$$\Phi_{el}(n,\Omega;R) \overset{R\to\infty}{=} \sum_{M_A M_B} C(n,\Omega,J_A M_A J_B M_B)\hat{A}|J_A M_A\rangle|J_B M_B\rangle. \tag{8.9}$$

In this expression, the $C(n, \Omega, J_A M_A J_B M_B)$ are the expansion coefficients and $\hat{A}$ is an operator that ensures correct antisymmetry with respect to electron exchange. The quantum number n denotes the set of quantum numbers (apart from Ω) defining the molecular electronic state, and the total angular momentum projection quantum number Ω is given by $\Omega = M_A + M_B$. To determine the energies at long range, we set up a Hamiltonian matrix using these wavefunctions together with the Coulomb interaction potential described above, *i.e.* a matrix with matrix elements $\langle\Phi_{el}(n,\Omega;R)|V(R)|\Phi_{el}(n,\Omega;R)\rangle$. Diagonalizing this matrix then yields the potential energy curves and the coefficients $C(\ldots)$, which may be converted to alignment parameters $\boldsymbol{a}_q^{(k)}(p)$, and to M_J state populations (see Chapter 9). These procedures are described in detail by Alexander.[633]

8.8.3 Sudden Model for Photodissociation

In the "sudden" (or diabatic) limit,[140] the repulsive states in the short-range, molecular regime are projected onto the asymptotic atomic limits without accounting for any electronic evolution during the course of the recoil. Rather than considering what happens at each avoided crossing, the initial Hund's case (a) (see Study Box 4.3) basis functions are simply projected onto the atomic basis, with care taken to conserve symmetry labels such as g/u and +/−, as well as the total angular momentum J, its projection Ω, and the total electron spin, S. In many cases, the coefficients of the transformation from the molecular Hund's case (a) basis to the atomic basis may be calculated without any detailed knowledge of the electronic structure.

Two key assumptions underlie the sudden model: (i) the atoms must separate so rapidly that there is no time for the electronic structure to change slowly so as to follow the adiabatic correlation diagram; and (ii) the rearranging of all molecular quantum states happens simultaneously over a relatively narrow range of internuclear distances (this is indicated by the shaded region labeled as the recoupling zone in Figure 8.14(a)). In other words, the non-adiabatic transitions mainly occur in the vicinity of a single point.

Diatomic hydrides such as HCl and OH would appear to be excellent candidates for complying with the sudden model, as the speed of the H-atom recoil is extremely high due to its light mass relative to that of the heavy partner. In the next section the photodissociation dynamics of OH will be described in more detail as a demonstration of the model. At the current level of accuracy for OH photodissociation experiments, all of the observations for direct dissociation agree with predictions of the sudden model. For HCl, however, the quality of the available data, especially the atomic polarization data, is high enough to indicate a significant deviation, illustrating that there are dangers inherent in relying on simple, limiting-case models.

8.8.4 Example: Photodissociation of the OH Radical

8.8.4.1 Background

In contrast to molecular bromine or iodine, photodissociation of the OH free radical is expected to deviate from the adiabatic model. There are two main reasons for this. Firstly, spin-orbit coupling is much weaker for the light O and H atoms, such that the adiabatic energy difference between the states involved is much less than that of molecular bromine. Secondly, owing to the light H atom mass, the nuclear velocity through the transition region is much higher, with the consequence that the Massey parameter is much less than unity, and the molecule is more likely to 'jump' from its initially excited state to other states. The so-called 'sudden' or 'diabatic' model, as described in the previous section, is likely to work better than the adiabatic model. OH is particularly interesting to study from a dynamics point of view because of predissociation (Figure 8.2(d)) of its well-known A-state, the state which is widely used for detection of the molecule by laser-induced fluorescence (Figure 8.2(a)). Relevant potential energy curves for the dissociation, together with their corresponding absorption/dissociation strengths are shown in Figure 8.15.

Several factors make the photodissociation of OH more difficult to study than that of bromine or HCl. As seen in Figure 8.15, strong absorption from ground-state OH is only possible with VUV radiation, which is much more difficult to produce than radiation at UV or visible wavelengths. In addition, OH is a highly reactive free radical which must be made *in situ*. Production of OH always involves co-production of free H and O atoms, which can overwhelm the signal from the dissociation products. Finally, detection of O and H also requires the use of much shorter wavelength (and therefore more difficult to produce) light compared to Br or I atom detection.

OH radicals may be generated by passing a molecular beam of H_2O vapor mixed with argon through a pulsed electric discharge. This method produces rotationally cold but vibrationally hot OH. The low rotational quantum states lead to interesting quantum effects on the β parameter, as described next. The high vibrational temperature of the radicals (~2000 K) makes photodissociation possible when using UV (rather than VUV) laser sources. This is illustrated in Figure 8.15, in which the absorption spectrum of OH in the $v=0$ ground state is compared with OH in the $v=2$ and $v=3$ states, both of which are shown to absorb at wavelengths longer than 200 nm.

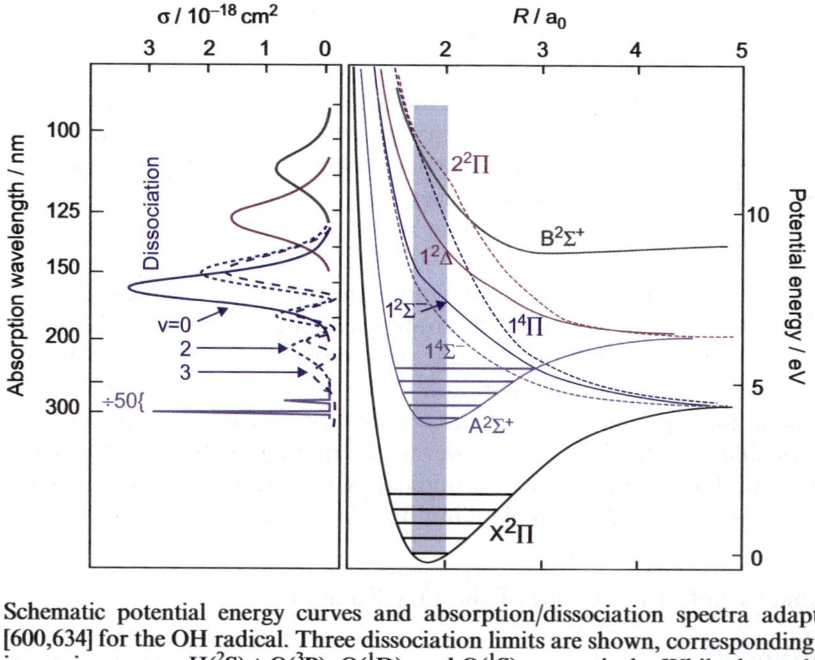

Figure 8.15 Schematic potential energy curves and absorption/dissociation spectra adapted from refs. [600,634] for the OH radical. Three dissociation limits are shown, corresponding to, in order of increasing energy, $H(^2S)+O(^3P)$, $O(^1D)$, and $O(^1S)$, respectively. While the peak absorption of the A state is 50 times stronger than that of the $1^2\Sigma^-$ dissociation continuum, the integrated contribution of A state predissociation is negligible compared to that of direct dissociation in the VUV dissociation region. The Franck-Condon overlap region for $v=0$ is shown as a shaded vertical stripe in the figure. Absorption spectra for transitions to the $1^2\Sigma^-$ state from the ground state $v=0$, 2, and 3 levels are shown. At 200 nm, for example, the photodissociation cross section from $v=2$ is larger than that from $v=3$.

8.8.4.2 *Direct Dissociation of OH* via *the Repulsive* $1^2\Sigma^-$ *State: Sudden Recoil Model*

We consider first the direct photodissociation of OH through the repulsive $1^2\Sigma^-$ state, which is the major destruction channel of OH in diffuse interstellar clouds and in comets.[600] The O–H bond dissociation energy, D_0, of 4.37 eV corresponds to production of $H(^2S)+O(^3P_2)$ atoms, with the spin-orbit-excited $O(^3P_1)$ and $O(^3P_0)$ limits lying 0.020 and 0.028 eV higher in energy, respectively. As the OH $1^2\Sigma^-$ state correlates adiabatically with the $O(^3P_0)+H(^2S)$ limit, at the threshold for photodissociation these are the only products expected. However, as seen in Figure 8.15, Franck-Condon overlap (represented by the shaded vertical stripe) limits absorption from the lower vibrational states of the OH(X) state to excited electronic states lying well above the first dissociation limit. Due to this large excess energy, and also to the low mass and therefore high velocity of the recoiling H atom, OH photodissociation is most likely a non-adiabatic process, and this prediction is confirmed in the measured $O(^3P_J)$ branching ratios for dissociation of vibrationally excited OH and its isotopomer OD at 226 nm and 200 nm, given in Table 8.2. The measured $J=2{:}1{:}0$ branching ratios are close to the 5:3:1 sudden-limit instead of the adiabatic 0:0:1 distribution.

Typical raw $O(^3P_J)$ images for the photodissociation of OD at 200 nm are shown in Figure 8.16, along with the product angular distributions. Note the significant deviations from a $\sin^2\theta$ distribution for $O(^3P_2)$ and $O(^3P_1)$ caused by atomic polarization.

The sudden recoil limit model can also be used to predict the $O(^3P_J)$ atomic polarization; the results of these calculations are shown in Table 8.3. Inspection of the data shows that while the predicted $O(^3P_J)$ $J=2{:}1{:}0$ fine-structure branching ratios are the same as a statistical 5:3:1 pattern, the M_J population distributions are far from statistical. For the $J=1$ products, only the $M_J=\pm1$ states are populated, with no population in $M_J=0$, while for the $J=2$ products only the $M_J=0, \pm1$ states are populated, with no population in $M_J=\pm2$. The sudden recoil model therefore predicts strong O atom polarization for the $O(^3P_1)$ and $O(^3P_2)$ products, which will be detectable in a pump-probe experiment when using a polarized laser in the detection step. As discussed previously, photofragment angular momentum polarization affects the detection sensitivity in REMPI. Because polarized photofragments flying in different directions acquire different detection sensitivities when probed with polarized light, the result is that the *measured* angular distribution depends on the probe polarization. Note that the *actual* angular distribution, *i.e.* the distribution of $\boldsymbol{v}$, is unchanged; the effect of polarization is simply to alter the detection efficiencies for products flying at different angles from the photolysis laser polarization. This effect is often accounted for approximately when fitting experimental data by adding an extra term to the expression for the angular distribution, *i.e.* by using an 'effective' angular distribution $f_{\text{eff}}(\theta)$,[634]

$$f_{\text{eff}}(\theta) = \tfrac{1}{2}[1+\beta_2 P_2(\cos\theta)+\beta_4 P_4(\cos\theta)], \tag{8.10}$$

where $P_4(\cos\theta)$ is the 4th order Legendre polynomial, and β_2 and β_4 are now simply fitting coefficients, from which β and the alignment parameters or M-state populations may be extracted. The value of β_2 determined by fitting the data to the above equation can be markedly different from the

Table 8.2 $O(^3P_J)$ branching ratios for the photodissociation of vibrationally excited OH and OD at 226 and 200 nm.

		J = 2	*J* = 1	*J* = 0
~226 nm	OH	0.59	0.30	0.11
	OD	0.51	0.35	0.14
~200 nm	OH	0.55	0.33	0.13
	OD	0.49	0.35	0.16

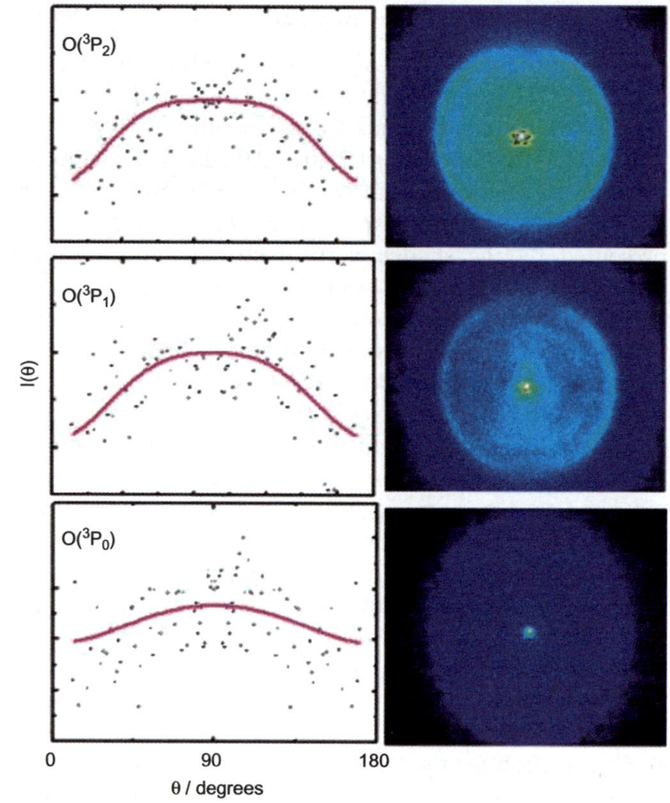

Figure 8.16 Raw velocity-mapped ion images of O(^3P_J, $J=2, 1, 0$) atoms from the photodissociation of OD at 200 nm.[634] The laser polarization direction is as shown in Figure 8.9. The bright spot at the center of the images corresponds to cold O atoms created in the OH discharge source. The outer ring with perpendicular character is from OD, and a weaker signal at smaller radius from background or contaminant OH is also observable. On the left hand side of the figure are angular distributions (fitting parameters in Table 8.3) for the outer ring (O from OD).

Table 8.3 Sudden recoil limit values for O(^3P_J) branching ratios $P(J)$, and $|M_J|$ state populations $P(|M_J|)$ for the photodissociation of OH *via* the $1^2\Sigma^-$ state. Adapted from ref. [634].

J	$P(J)$	$P(\|M_J\|)$			$\beta_2(J)$	$\beta_4(J)$
		$\|M_J\|=0$	$\|M_J\|=1$	$\|M_J\|=2$		
0	1/9	1/9	–	–	−1	
1	3/9	0	1/6	–	−5/7	−2/7
2	5/9	2/9	1/6	0	−25/43	−18/43

'true' β parameter for the dissociation in cases where the atomic alignment is strong. In the case of OH dissociation, fitting data from the O($^3P_0$) product atom, which can have no alignment since M_J can only take the value zero, returns $\beta_2=-1$ and $\beta_4=0$. In this case β_2 is equal to β, and takes the expected value for a perpendicular $\Sigma \leftrightarrows \Pi$ excitation. For the $J=1$ and $J=2$ products, however, we find that $\beta_2>-1$ and β_4 takes significant (negative) values due to the atomic alignment in these products. The sudden-limit predictions from Table 8.3 are compared with the measured values in

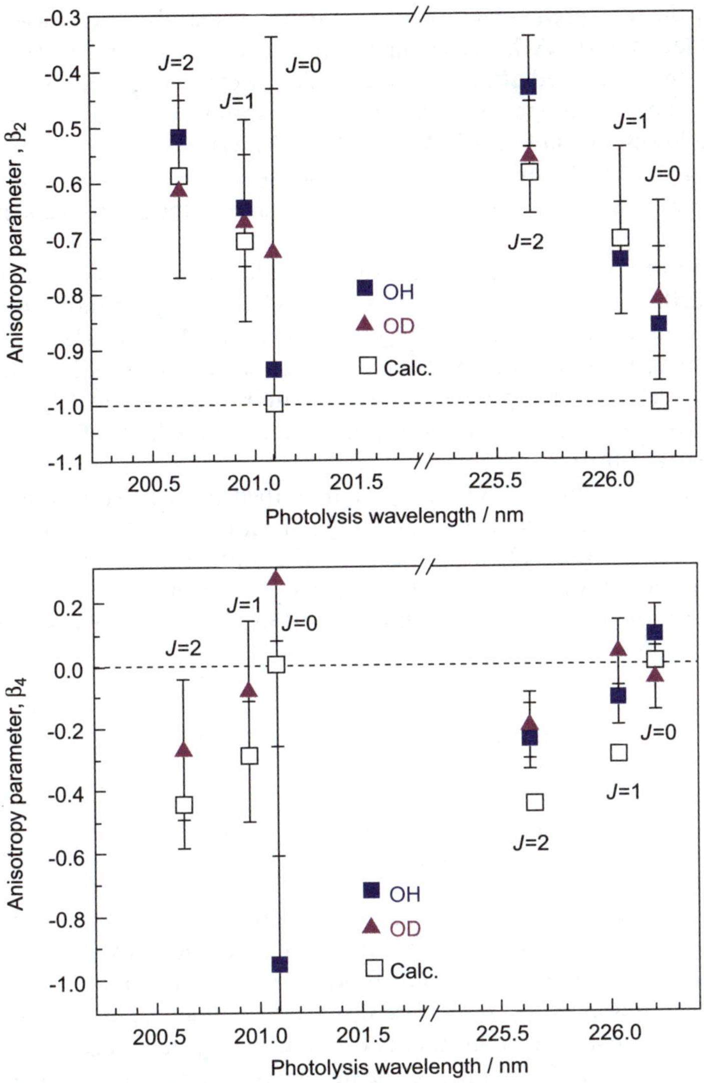

Figure 8.17 Experimental β_2 and β_4 values as a function of wavelength (~226 nm and ~200 nm) for the $J=0$, 1 and 2 fine structure states of the $O(^3P_J)$ atom formed in photodissociation of OH and OD. The experimental data, with error bars, are represented by a filled square for OH and by a filled triangle for OD. Predicted values, calculated in the sudden recoil limit for dissociation *via* the $1^2\Sigma^-$ state, are represented by a open square. Adapted from ref. [634].

Figure 8.17. The model predictions are seen to be in good agreement with the experimental data within the uncertainties of the measurement.

8.8.5 Predissociation of the $A^2\Sigma^+$ state of OH: Interference Effects

Near-threshold direct photodissociation of high rotational states of bromine (Figure 8.10) was found to show significant deviations from the $\beta=2$ value predicted for axial recoil. Large deviations from the limiting β values of $+2$ or -1 can also arise when *low* rotational states of a diatomic undergo dissociation far above threshold but through a slow predissociation mechanism. In this section we describe the slow predissociation of the OH molecule after excitation to the ($v=3$, $N=0$, 1, 2) rovibrational levels of the $A^2\Sigma^+$ electronic state[635] (v and N are the vibrational and rotational quantum numbers in the excited state, *i.e.* N excludes electron and nuclear spin). The discharge source used in the experiments produces a very cold rotational distribution, in our case yielding predominantly the lowest possible level, $N=1$, of the $X(^2\Pi)$ ground electronic state of OH. This level can be excited *via* a P, Q, or R branch transition to the A state, as shown in the spectra in Figure 8.18; the spectral lines are somewhat broadened due to the short ~200 ps lifetime of the A state levels.

The discharge source does a good job of producing state selected parent molecules in their $N=1$ rotational state. Further state selection may be achieved by passing the molecular beam through an inhomogeneous hexapole electric field prior to dissociation. A hexapole is a set of six parallel cylindrical rods whose axes lie on the points of a hexagon; a hexapole field is generated by applying positive and negative electrostatic potentials to alternate rods, and is widely used to achieve state selection in molecular beams of symmetric top molecules. We will briefly outline how this state-selection process works* for OH.

The angular momentum states of OH may be described by the 'basis' wavefunctions $|J\Omega M_J\rangle$, where the quantum number J denotes the total angular momentum (the sum of the electronic and rotational angular momentum), Ω the projection of $\boldsymbol{J}$ onto the internuclear axis, and M_J the projection of $\boldsymbol{J}$ onto the LAB-frame z axis. In order for the wavefunctions to have a definite parity (inversion symmetry), we use the following linear combinations of the basis wavefunctions as our molecular wavefunctions:

$$\Psi_e(0) = \frac{1}{\sqrt{2}}[|J\Omega M_J\rangle + |J\ -\Omega M_J\rangle], \tag{8.11a}$$

$$\Psi_f(0) = \frac{1}{\sqrt{2}}[|J\Omega M_J\rangle - |J\ -\Omega M_J\rangle], \tag{8.11b}$$

where the subscripts e and f on the wavefunctions denote even and odd parity, respectively. The energies of these two states are slightly different due to a phenomenon known as Λ-doubling, which arises because the electronic motion does not follow the nuclear motion exactly (see Chapter 5). When an external field is applied (as is the case inside a hexapole, for example), parity is no longer well defined and rotational states in the two parity subgroups become mixed, leading to coupled wave functions in which the mixing coefficients $a(E)$ and $b(E)$ depend on the strength of the electric field,

$$\Psi_1(E) = a(E)\Psi_e(0) + b(E)\Psi_f(0), \tag{8.12a}$$

$$\Psi_2(E) = -b(E)\Psi_e(0) + a(E)\Psi_f(0). \tag{8.12b}$$

In the low field regime, $\Psi_1(E)$ is dominated by $\Psi_e(0)$ and $\Psi_2(E)$ is dominated by $\Psi_f(0)$. It turns out that in an external field, the states $\Psi_1(E)$ lower their energy by moving to regions of lower field, and the state $\Psi_2(E)$ lower their energy by moving to regions of higher field. A hexapole field is strong near the hexapole rods and zero along the axis, with the result that the $\Psi_2(E)$ states tend to gravitate towards the rods and are lost from the beam, while the $\Psi_1(E)$ states follow sinusoidal trajectories, and are focused to the hexapole axis at certain points. Specific J, Ω, M states may be focused by adjusting the length of the hexapole and/or the magnitude of the electrostatic potentials applied to the rods.

*Further details about hexapole state focussing techniques are given in Chapters 5 and 9.

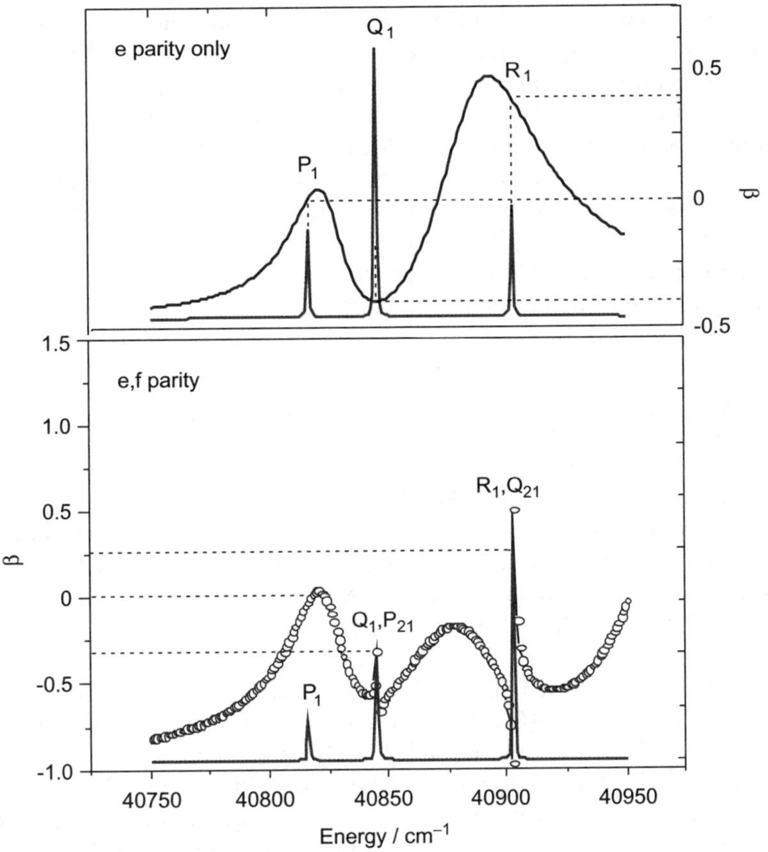

Figure 8.18 Absorption spectrum (purple) and calculated wavelength-dependent β parameters (blue) for the predissociation of OH $A^2\Sigma^+(v=3, N=0, 1, 2)$ e parity states (upper panel) and for mixed e, f parity states (lower panel). Adapted from refs. [634] and [635].

We now return to our consideration of the photodissociation dynamics accompanying a predissociation process, in which dissociation is slow relative to molecular rotation. In particular, can we predict the value of β for dissociation from the $N=0$, 1 and 2 rotational levels of the excited state of OH? The answer turns out to be 'yes'.

Excitation to $N=0$ produces an isotropic state, and we would expect dissociation from this state to yield a β value of zero, assuming that electron spin ($S=1/2$) plays no role in the dissociation. Excitation to $N=1$ or 2 cannot be predicted by a classical model, but a quantum mechanical analysis of this type of process, developed recently by Houston and coworkers,[636] and also by Kutznetsov and Vasyutinskii,[637] has been reasonably successful. Predictions for the wavelength-dependent β parameter from Houston's calculations are shown in Figure 8.18. As described in ref. [636], predissociation from levels with a finite bandwidth causes an interference effect between the different ΔJ transitions similar to that found in a three-slit optical interferometer. Such interference effects are evident in the shape of the β function in the upper panel of Figure 8.18, which shows results for photodissociation from low field seeking states only. At the peak of the P branch (corresponding to dissociation from $N=0$ in the electronically excited state), the predicted value of β is indeed zero. The bottom panel shows the predicted spectra and wavelength-dependent β parameter for the more normal situation with no hexapole state selection, when states of both e and f parity are present in the beam. There are several differences to note. Firstly, P_{21} and Q_{21} absorption bands appear in the spectra, corresponding to transitions starting from $N=1$ states of f parity.* These new transitions are in near overlap with the Q_{11} and R_{11} e-parity branches seen in the top panel for the hexapole state-selected molecular beam. The presence of these bands causes an extra oscillation in β, as seen in the lower panel of Figure 8.18. Experimental measurements yield qualitative agreement with the predictions of the model for OH excited to the $v=3$ vibrational level of the $A(^2\Sigma^+)$ state, and agreement between theory and experiment is near-quantitative for dissociation from a wide range of A(v, N) levels in the related SH radical.

The predissociation process in OH is believed to occur *via* population transfer from low vibrational levels of the A state into the $1^4\Sigma^-$ state, and indeed, the final product branching ratios and M_J distributions are well described by a sudden model assuming direct dissociation from this state.[638]

8.8.6 The 'Bad News': Fully Quantum Analysis of Models for Photodissociation of HX

Our tour of diatomic photodissociation dynamics has nearly come to an end. We have seen for two representative molecules, bromine and OH, that the simple adiabatic model and sudden limit model, respectively, are quite successful in describing the measured properties of the dissociation products. We can thus hope that these limiting case models can be applied to increasingly larger molecules. This should be done with caution, however, especially considering the results of a 'perfect' theoretical analysis of photodissociation described next.

Balint-Kurti and coworkers[629,639–641] have carried out detailed fully quantum calculations for the photodissociation of the hydrogen halides, HX, where X = F, Cl, Br, and I. Lambert *et al.*[259] presented similar calculations for HCl. Potential energy curves and the absorption spectrum for HCl are shown in Figure 8.19. For the case of HCl photodissociation, the sudden model predicts that one-third of the final Cl atom population will be formed in the excited Cl* state and two-thirds of the population in the ground electronic state, while the quantum calculations also predict a branching fraction Cl/Cl* of 0.33. However, the sudden analysis predicts that the net polarization of the H atom is zero, which is in contradiction with the fully quantum analysis.[639]

The sudden model was found to fail due to the second key assumption regarding the existence of a well-defined 'recoupling zone'. Instead of one recoupling zone, two distinct regions of non-adiabatic coupling were found (Figure 8.14(d)), each involving coupling between only two states. A modified two-step sudden approximation may be able to model the main aspects of HCl photodissociation, but the problems uncovered by the fully quantum analysis do not lead to a high confidence in using currently available limiting-case models in general. More details about non-adiabatic effects in the photodissociation of the hydrogen halides can be found in Chapter 4, Section 4.2.4.

8.9 OUTLOOK

At this point it is clear that our ability to probe experimentally the photodissociation dynamics of diatomic molecules is nearing perfection, both on the femtosecond timescale 'during' a photodissociation process and on the long-term timescale where for the final products all possible quantum numbers are fully specified. In the future, it will probably be necessary to carry out fully quantum calculations, which rely completely on accurate PE curves, for a wide range of molecules in order to develop more reliable and intuitive simple models for photodissociation processes. Fortunately, computing power continues to grow and the availability of experimental data will certainly stimulate more of these exact quantum calculations.

*The spectroscopy of OH is more complex than that of normal closed shell diatomics due to the $^2\Pi$ ground state. See ref. [262] for more information.

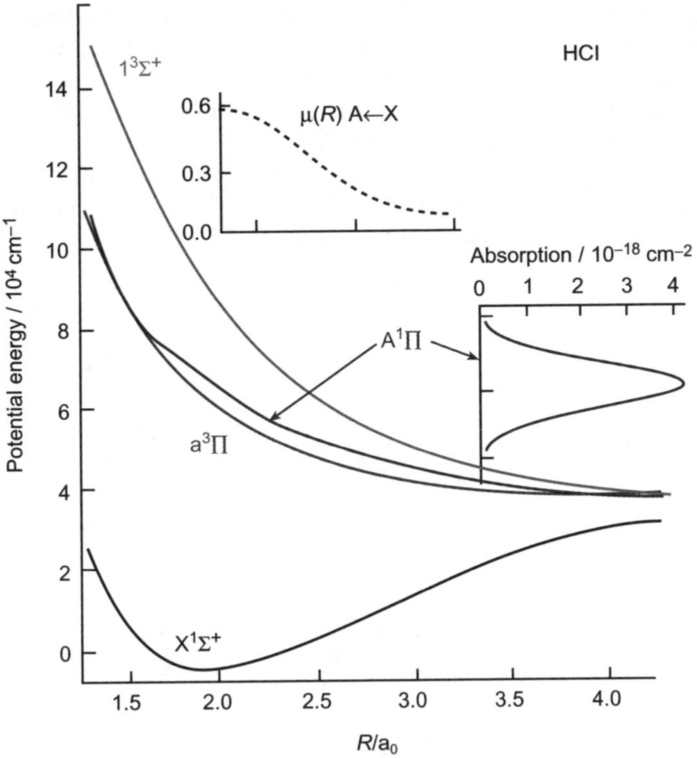

Figure 8.19 Potential energy curves and absorption spectrum for the photo-dissociation of HCl, adapted from ref. [259]. Absorption is dominated by the A←X transition, and the calculated $\mu(R)$ transition dipole function for the A←X transition, in atomic units, is also shown in the figure. The $A^1\Pi$ state correlates with the first $(H(^2S) + Cl(^2P_{3/2}))$ dissociation channel.

8.10 PROBLEMS

1. Velocity-map imaging was used to detect Cl fragments from the photodissociation of molecular chlorine after they had traveled along a 40 cm flight path from the interaction region to the detector. The resulting image is shown below.

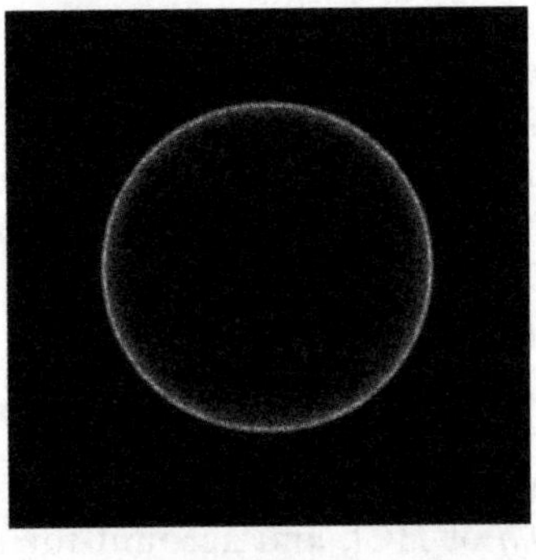

(a) A potential of 3000 V is used to direct the ionized Cl atoms to the detector. What is their flight time? Take the mass of a Cl atom to be 35 g mol^{-1}.

(b) The image appears as a single ring of Cl atoms as a result of conservation of energy and momentum. The outside diameter of the ring is 12.68 mm. What velocity did the Cl atoms acquire as a result of the photodissociation?

(c) The bond dissociation energy of Cl_2 is 243 kJ mol^{-1}. Use conservation of energy to determine the photolysis laser wavelength.

2. The figure below compares an ion imaging, Doppler profile, and TOF measurement of a Newton sphere with $\beta = +2$. The power of 2-D ion imaging is in measuring slow velocities, especially compared to equivalent 1-D techniques.

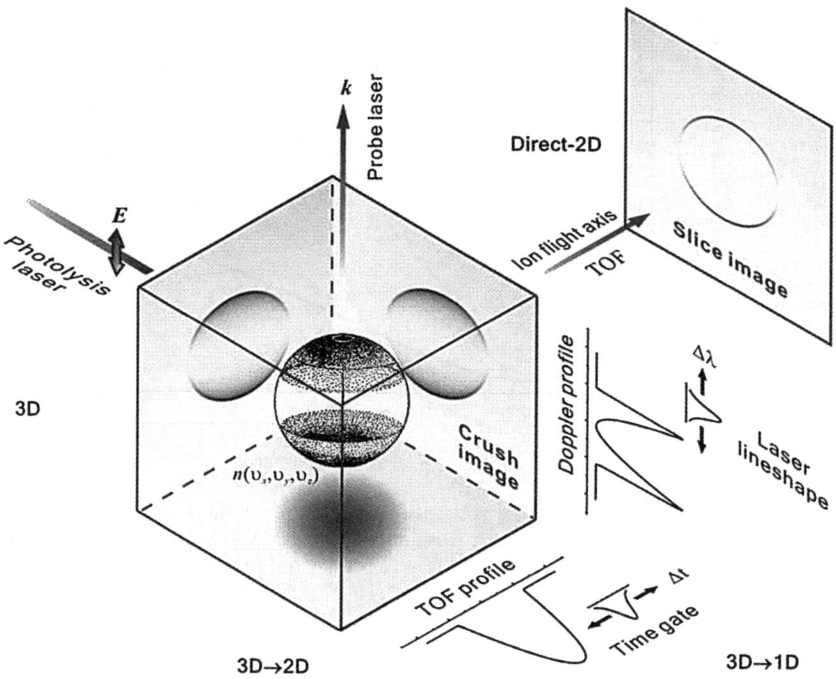

(a) Figure 8.10 of this chapter shows the velocity distribution of Br atoms produced in the 510 nm photolysis of Br_2. Use the velocity corresponding to the peak in the distribution to calculate the laser bandwidth at which one would begin to resolve the 1-D Doppler profile for the Br atoms. In these experiments, Br is detected by (2 + 1) REMPI at 235 nm.

(b) Photoionization of Br by (2 + 1) REMPI at 235 nm creates $Br^+ + e^-$ with 2.5 eV excess kinetic energy. Calculate the recoil velocity (in m s^{-1}) imparted by the ionization, and compare it to the Br velocity resulting from the photolysis process, shown in Figure 8.10.

(c) What velocity resolution might one obtain using the original ion imaging method, in which the ~2 mm beam diameter is also projected onto the detector? Assume in your calculation that the 'ring' in the Br image has radius 10 mm.

3. (a) When HCl, jet-cooled in a molecular beam, is photodissociated at a wavelength of 210 nm, H atoms are observed with two different speeds corresponding to total kinetic energy release (TKER) of H(^{2}S) + Cl(^{2}P) atoms of 10989 cm^{-1} and 11871 cm^{-1}. The ratio of signal intensities for the slower to faster H atoms is 0.75. Both photodissociation channels show anisotropy parameters of $\beta = -1$. Use these data to deduce the bond dissociation energy (D_0) of HCl, and account for the observations of faster and slower channels. What non-adiabatic dynamics must be occurring in the HCl molecule during dissociation? Use the ratio of signal intensities for the two channels to estimate the

probability of a non-adiabatic transition, assuming that the state initially populated in the photoexcitation correlates with spin-orbit ground state products.

(b) If HI is photodissociated at 258 nm, product H atoms are again observed, but with kinetic energies of 6525 cm^{-1} and 14128 cm^{-1}. The faster H atoms again have $\beta = -1$ but, for the slower H atoms, $\beta = +2$. Derive a value of D_0 for HI and explain the observed values of the anisotropy parameter.
[Hint: It might be helpful to take another look at Sections 4.2.4, 4.2.6, and 8.8.6 before attempting this question.]

CHAPTER 9

Stereodynamics: Orientation and Alignment in Chemistry

F. JAVIER AOIZ[a] AND MARCELO P. DE MIRANDA[b]

[a] Departamento de Química Física I, Facultad de Química, Universidad Complutense de Madrid, Spain; [b] School of Chemistry, University of Leeds, UK

9.1 INTRODUCTION

When two molecules collide, the chances that they will react depend on a number of factors. For example:

1. How hard they hit each other, *i.e.* what is the collision energy?
2. How much internal energy they have and how it is distributed, *i.e.* what are their internal (electronic, vibrational, rotational) states?
3. Whether their collision is on-target or off-centre, *i.e.* how large is the impact parameter?
4. The relative directions of their internuclear axes, *i.e.* what are their axial (molecular axis) polarizations?
5. The relative directions of their rotational angular momenta (and therefore planes and senses of rotation), *i.e.* what are their rotational polarizations?
6. The relative directions of other directional quantities, *i.e.* what are, say, the polarizations of the molecular dipole moments and electronic angular momenta?

Among these factors, those numbered 1–3 involve *scalar properties* that are entirely characterized by magnitudes and do not depend on any directions in space. In contrast, factors 4–6 involve *vector properties* characterized not only by magnitudes, but also by spatial directions.

The interplay between vector properties and reactivity is the central concern of studies into stereodynamics. Obvious as it is, the basic realization here is that molecular collisions take place in three-dimensional space. To create a full picture of a molecular collision, one must specify the directions along which reactants approach and products recoil, and also how the reactant and product molecules are positioned, or move, relative to each other.

Tutorials in Molecular Reaction Dynamics
Edited by Mark Brouard and Claire Vallance

Published by the Royal Society of Chemistry, www.rsc.org

The consideration of vector properties opens up new possibilities regarding the analysis and also the manipulation of the dynamics of molecular collisions. From the analytical point of view, the advantage is access to more detailed, less averaged information about the unfolding of the collision dynamics. From the manipulation point of view, vector properties offer a number of parameters with the potential to influence reactivity in a variety of ways.

This chapter describes the basic ideas and methods used to study the stereodynamics of elementary reactions. For simplicity and clarity, the text considers a single type of reaction:

$$\mathrm{A} + \mathrm{BC}(v,j) \xrightarrow{E_c} \mathrm{AB}(v',j') + \mathrm{C}, \tag{9.1}$$

where E_c is the collision energy, v and v' the vibrational quantum numbers of the reactant and product diatomic molecules, and j and j' their rotational angular momentum quantum numbers.

The convention that unprimed symbols refer to reactants and primed symbols to products will be used throughout this chapter. Note also that we are considering the reaction in the energy rather than the time domain, and that electronic and nuclear spin angular momenta are ignored.

By restricting our discussion to reaction (9.1), we are not restricting our discussion of stereodynamical concepts in any significant way. Indeed, consideration of nuclear and electronic angular momenta, of time-domain collision dynamics, or of other types of reactions (involving, say, photons and/or polyatomic molecules), does not require additional concepts; although disregarded in the bulk of the text, such problems are briefly considered in Section 9.5.

The chapter starts with a presentation of the concepts and quantities used to describe the correlations between the directions of the vectors taking part in the stereodynamics of a reaction and the probability that the reaction actually occurs. This is followed by a description of how the reaction properties that quantify these correlations can be calculated or measured, and then by illustrative examples of stereodynamical analysis.

9.2 CONCEPTS AND QUANTITIES

9.2.1 Polarization, Orientation, Alignment

Polarization is a term used when an object (or a set of objects) exhibits anisotropic physical properties; that is, properties that change with spatial direction. For example:

- If a hydrogen atom is in the ground, ^2S state, its electron density is spherically distributed around the nucleus, showing no preference for positions due "up", "down", "left" or "right". The electron position is unpolarized.
- If a beam of unpolarized ground-state hydrogen atoms passes through a Stern-Gerlach apparatus, it splits in two. One beam will contain atoms whose electronic spin is pointing "up", the other will contain atoms whose electronic spin is pointing "down". In each of these two beams, the electronic spins of the H atoms are said to be polarized (though one often simply speaks of "polarized atoms").
- If a diatomic molecule is in the $|j=1,m=0\rangle$ rotational state, the rotational angular momentum of the molecule – the $\boldsymbol{j}$ vector – is most likely to be found lying in a direction perpendicular to the quantization axis, and has zero probability of being found along the quantization axis. The rotational angular momentum of the molecule is polarized.

Polarization comes in many forms. Light, for example, is frequently found with "linear" or "circular" polarization. In stereodynamics, the types of polarization that are most commonly cited are *orientation* and *alignment*.

We say that an object is oriented if it features a physical property whose value is larger along a particular direction (say, $+z$) than along the opposite direction ($-z$). We say that it is aligned if it features a physical property whose value is larger along a particular pair of opposite directions [say, $\pm(x+y)$] than along some direction perpendicular to those [$\pm(x-y)$]. For example:

- If a car travels forward on a straight road, its wheels are aligned but not oriented. If the car moves along z, and y is the vertical axis pointing from the ground to the sky, the wheels are more likely to be found near the yz plane than near planes containing the $\pm x$ direction.
- In the example above, the rotational angular momenta of the car wheels are oriented towards the left of the driver (*i.e.* along the $+x$ direction). These angular momenta will be found to lie along the opposite, $-x$ direction if the car motion is reversed.
- If a car travels at constant speed around an 0-shaped racing car circuit, and if we look at the average over time, then the angular momenta of the car's wheels are oriented with respect to the driver (the angular momenta point towards the driver's left rather than towards than the driver's right), but only aligned with respect to the circuit (on average, the chances of finding the angular momenta along the long or short axes of the circuit are not the same). The wheels are also aligned with respect to the ground (perpendicular rather than parallel to it).
- If a diatomic molecule is in the $|j=1,m=1\rangle$ rotational state, its rotational angular momentum is oriented. It is more likely to be found lying parallel than antiparallel to z, the quantization axis. We can also say that $\boldsymbol{j}$ displays positive orientation with regard to z.
- If a diatomic molecule is in the $|j=1, m=1\rangle$ rotational state, its internuclear axis ($\boldsymbol{r}$) is aligned but not oriented. It is more likely to be found perpendicular than parallel to the $\pm z$ directions. As we will see, we can also say that $\boldsymbol{r}$ displays negative alignment with regard to z.

Note that orientation and alignment can be positive or negative and that they depend on the choice of reference frame. Note also that different properties of a single object can exhibit different polarizations. For instance, the molecule considered in the last two examples had positive $\boldsymbol{j}$ orientation and negative $\boldsymbol{r}$ alignment.

9.2.2 Vector Correlations

Figure 9.1 shows a schematic representation of reaction (9.1) and introduces the vectors one would like to consider when studying the reaction stereodynamics:

- $\boldsymbol{k}$ ($\equiv\boldsymbol{v}_{rel}$) and $\boldsymbol{k}'$, the reactant-approach and product-recoil directions, *i.e.* the relative velocities of the reactants and products;
- $\boldsymbol{\ell}$ and $\boldsymbol{\ell}'$, the orbital angular momenta of reactants and products;
- $\boldsymbol{r}$ and $\boldsymbol{r}'$, the interatomic axes of the reactant and product diatomics;
- $\boldsymbol{j}$ and $\boldsymbol{j}'$, the rotational angular momenta of the reactant and product diatomics.

These vectors are defined in the asymptotic region, where atom-diatom separations are large enough for interactions between them to be negligible. You should study this figure carefully, clarifying in particular why the angular momenta are shown as pointing towards or away from you ($\odot$ or $\otimes$, respectively) and also why the $\boldsymbol{k}$ and $\boldsymbol{k}'$ vectors do not coincide with the blue arrows that are indicative of the trajectories of the atoms and molecules.

The first task of stereodynamical studies is to determine how the reaction probability (or the reaction cross section) depends on the relative directions of these vectors. One must remember, however, that the various vector quantities are not completely independent of each other. For example, the direction of the rotational angular momentum of the BC reactant, $\boldsymbol{j}$, determines the plane and sense of BC rotation, and therefore the directions along which the BC internuclear axis,

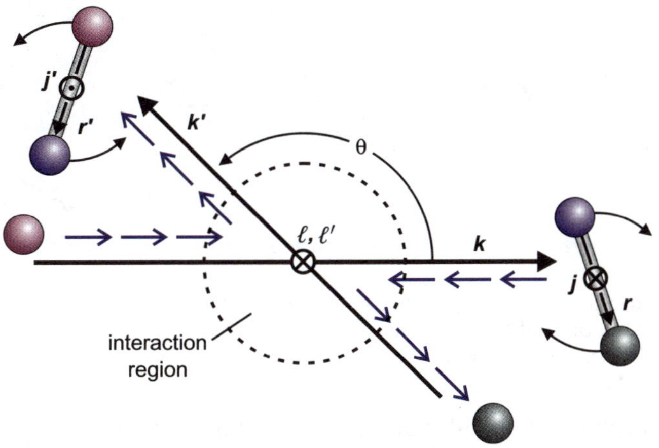

Figure 9.1 Schematic representation of an atom-diatom reaction showing the vectors one would like to consider when studying its stereodynamics. The centre of the interaction region (the region encircled by the dotted line) coincides with the centre-of-mass. The ⊙ and ⊗ symbols indicate vectors perpendicular to the plane of the page and pointing towards or away from the reader, respectively.

r, can lie. The same relation ties ***j'*** to ***r'***, ***ℓ*** to ***k***, and ***ℓ'*** to ***k'***. This must be taken into account when one chooses which vectors to incorporate into the description of the stereodynamics.

9.2.2.1 Which Vectors?

The stereodynamics of reaction (9.1) involves eight vectors that can be separated into four independent pairs, each pair containing vectors that are not independent of each other: ***k-ℓ***, ***r-j***, ***k'-ℓ'***, ***r'-j'***. Given that the directions of the vectors in each pair cannot be independently varied, the best we can do in order to describe the reaction stereodynamics is to take a vector from each pair. The question now is, which four vectors should we pick?

A first factor to consider is that some vectors are more easily observable than others. Experimental specification of the reactant-approach direction, ***k***, is easiest. Determination of the product-recoil direction, ***k'***, is in general second-easiest. In contrast, direct experimental determination of the ***ℓ*** and ***ℓ'*** polarizations is very difficult or even impossible with present-day instrumentation. This tilts the balance towards a description of the reaction stereodynamics in terms of ***k*** and ***k'*** rather than ***ℓ*** and ***ℓ'*** directions, which is indeed the choice made in virtually all of the stereodynamics studies to date.

The situation is not so clear-cut with regard to the ***r-j*** and ***r'-j'*** pairs. Experimentally, there are techniques one can use to specify the ***r*** polarization directly, just as there are techniques one can use to specify the ***j*** polarization[642] (see Section 9.3.2). On the products side, at present only the ***j'*** polarization is directly observable (see Study Box 9.1). But there are other factors one should consider. On the one hand, most chemists would find the description of the collision geometry to be more intuitive if given in terms of internuclear axes. On the other hand, if the *j* quantum number has a well-defined value, descriptions given in terms of rotational angular momenta are bound to contain more information than those provided in terms of internuclear axes: information about the sense of rotation is present in angular momentum but not in internuclear axis distributions, since opposite senses of rotation lead to identical internuclear axis distributions.

The deciding factor here is feasibility. Experimentally, the design of the apparatus determines which vector polarizations one can specify or detect. Theoretically, knowledge of the ***j*** and ***j'*** polarizations allows one to determine the ***r*** and ***r'*** distributions; one can therefore transform a

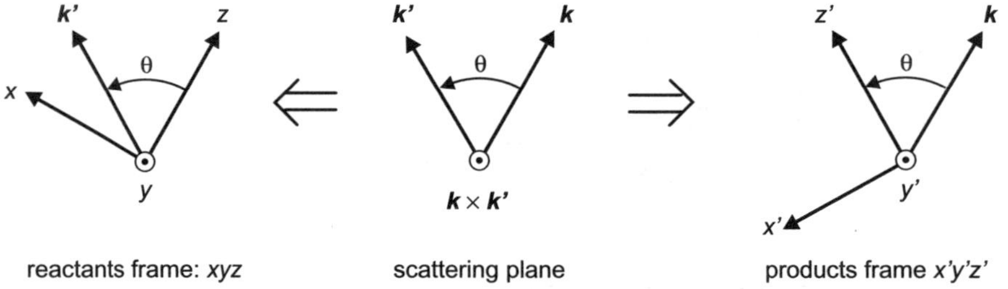

Figure 9.2 The reactant-approach and product-recoil directions, $\boldsymbol{k}$ and $\boldsymbol{k}'$, are used to define the reference frames most commonly used in stereodynamics: the "reactants" and "products" frames. The former is defined by $z = \boldsymbol{k}$ and $y = \boldsymbol{k} \times \boldsymbol{k}'$, the latter by $z' = \boldsymbol{k}'$ and $y' = \boldsymbol{k} \times \boldsymbol{k}'$.

description in terms of rotational angular momenta into a description in terms of internuclear axes. In general, the ideal theoretical approach will involve consideration of both representations of the reaction stereodynamics. We will adopt that approach here, using the $\boldsymbol{j}$-$\boldsymbol{k}$-$\boldsymbol{k}'$-$\boldsymbol{j}'$ and $\boldsymbol{r}$-$\boldsymbol{k}$-$\boldsymbol{k}'$-$\boldsymbol{r}'$ four-vector correlations for the description of the full stereodynamics of reaction (9.1).

There is one more question. What happens if we cannot or do not wish to take one or more of the four vectors into account? As it turns out,[643,644] a vector whose polarization is not observed is mathematically equivalent to a vector that is unpolarized. We shall see how one can take advantage of this to reduce the full description of the reaction stereodynamics (the four-vector correlation) to less detailed descriptions involving only two- or three-vector correlations.

9.2.2.2 Specification of Reference Frames and Vector Directions

We now turn to the question of how to define the angles between the vectors included in the description of the stereodynamics. Although this can be done in many ways, in practice there is little variation between the definitions in use.

Because the reactant-approach and product-recoil directions, $\boldsymbol{k}$ and $\boldsymbol{k}'$, are the most readily measurable vectors, it is natural to use them as starting points. This is also convenient because the plane containing $\boldsymbol{k}$ and $\boldsymbol{k}'$ – the so-called *scattering plane* – and the $\boldsymbol{k} \times \boldsymbol{k}'$ vector perpendicular to it are, respectively, a plane and an axis of symmetry for scattering wavefunctions (this is a consequence of the conservation of parity, see refs. [645] and [646]). The *scattering angle*, θ, is the angle from $\boldsymbol{k}$ to $\boldsymbol{k}'$; its value must lie in the range $0 \leqslant \theta \leqslant \pi$.

Note that, as long as we have only two vectors, the correlation between them is a function of only one angle. This is so because the location of the plane containing the two vectors does not affect how the reacting atoms and molecules are positioned, or move, relative to each other.

This situation changes when one considers more vectors; each new vector requires two additional angles for the specification of its direction. This is usually dealt with by choosing the additional angles to be the polar and azimuthal angles that specify the new vector direction in spherical polar coordinates in a chosen centre-of-mass reference frame. The two commonly used frames, the "reactants" (or xyz) and "products" (or $x'y'z'$) frames, are shown on Figure 9.2.

EXAMPLE

Figure 9.3 shows a possible definition for the five angles that specify the directions of the vectors in the $\boldsymbol{j}$-$\boldsymbol{k}$-$\boldsymbol{k}'$-$\boldsymbol{j}'$ correlation. Note that the angle definitions depend on the choice of reference frame (this example uses the xyz, reactants frame of Figure 9.2) and also on the vector hierarchy, which is the order in which the various vectors are incorporated in the description.

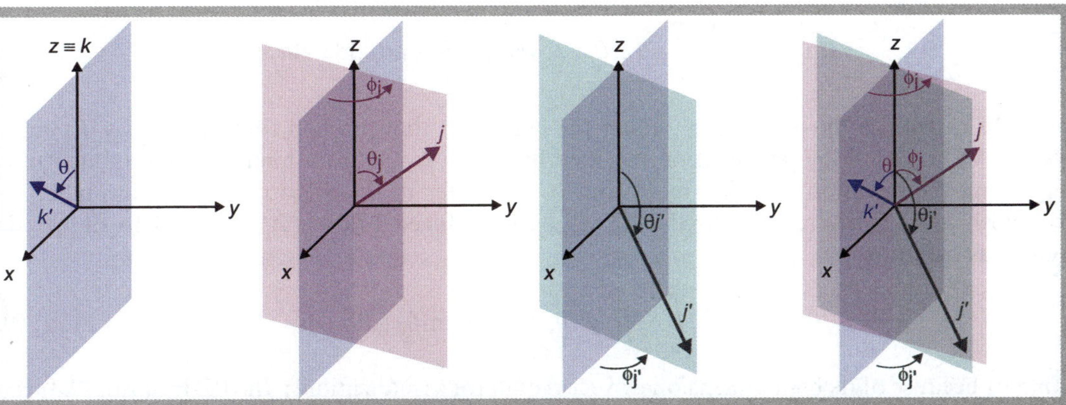

Figure 9.3 A natural hierarchy for the vectors in the ***j-k-k'-j'*** correlation. The first vector, ***k***, is used to define the z axis; the second vector, ***k'***, is used to define the xz plane. All vectors have their directions specified by polar and azimuthal angles referred to the z axis and xz plane. The ***j-k-k'-j'*** four-vector correlation depends on five angles.

9.2.3 Probability Density Functions

In stereodynamics, a *probability density function* (PDF) is a function that ties the angles specifying vector directions to the reaction probability, or cross section. The simplest example is the *product angular distribution*:

$$P(\theta) = \frac{2\pi}{\sigma}\frac{\mathrm{d}\sigma}{\mathrm{d}\omega}, \tag{9.2a}$$

where σ and $\mathrm{d}\sigma/\mathrm{d}\omega$ are the integral and differential cross sections, respectively (see Chapters 1, 5, and 6)*. The constant of proportionality, $2\pi/\sigma$, between the PDF and the differential cross section ensures normalization:

$$\int_0^{\pi} P(\theta)\sin(\theta)\,\mathrm{d}\theta = \frac{2\pi}{\sigma}\int_0^{\pi}\frac{\mathrm{d}\sigma}{\mathrm{d}\omega}\sin(\theta)\,\mathrm{d}\theta = \frac{2\pi}{\sigma}\times\frac{\sigma}{2\pi} = 1. \tag{9.2b}$$

Because it expresses the two-vector, ***k-k'*** correlation, the product angular distribution depends on a single variable: θ, the scattering angle. Note also that $P(\theta)$ is a *conditional* PDF. Its value is not the absolute probability density for product recoil along the direction specified by θ. Instead, what $P(\theta)$ quantifies is this: the probability density, *under the condition that product formation does occur*, for product recoil along the direction specified by θ. Mathematically, the condition just mentioned is imposed by normalization as in Eq. (9.2b).

In what follows we will deal with other PDFs, all of them conditional. In each case we will try to state clearly the normalization used as well as its physical meaning. There is, however, a problem: the terminology can be rather confusing. It will be important that you also do your best to distinguish the various PDFs and associated conditions.

Let us now consider the three-vector, ***k-k'-j'*** correlation. It is quantified by the *normalized θ-dependent **j'** polarization function*:

$$P(\theta, \theta_{j'}, \varphi_{j'}), \tag{9.3a}$$

*Actually, the function of Eq. (9.2a) is $P(\cos\theta)$. It should not be confused with $(2\pi/\sigma)\times(\mathrm{d}\sigma/\mathrm{d}\theta) = (2\pi/\sigma)\times(\mathrm{d}\sigma/\mathrm{d}\omega)\times\sin\theta$. We have used $P(\theta)$ to simplify the notation.

$$\text{with} \int_0^{2\pi}\int_0^{\pi}\int_0^{\pi} P(\theta,\theta_{j'},\varphi_{j'})\sin(\theta)\mathrm{d}\theta\sin(\theta_{j'})\mathrm{d}\theta_{j'}\mathrm{d}\varphi_{j'} = 1. \tag{9.3b}$$

This PDF quantifies the probability, under the condition that product formation does occur, that products will recoil along the direction specified by θ with the AB rotational angular momentum, $\boldsymbol{j}'$, lying along the direction specified by $\theta_{j'}$ and $\varphi_{j'}$ (we refer to the distribution as "normalized" to distinguish it from the renormalized PDF to be introduced below). It is related to the product angular distribution by

$$P(\theta) = \int_0^{2\pi}\int_0^{\pi} P(\theta,\theta_{j'},\varphi_{j'})\sin(\theta_{j'})\,\mathrm{d}\theta_{j'}\,\mathrm{d}\varphi_{j'}. \tag{9.4}$$

In part because of its high dimensionality (too high for visualization!), the PDF of Eq. (9.3) is not terribly handy when what one is looking for is understanding of the whys and hows of the ***k*-*k*′-*j*′** correlation. For this, a more useful PDF is the *renormalized θ-dependent **j**′ polarization function*:

$$P(\theta_{j'},\varphi_{j'}|\theta) = \frac{P(\theta,\theta_{j'},\varphi_{j'})}{P(\theta)}, \tag{9.5a}$$

$$\text{with} \quad \int_0^{2\pi}\int_0^{\pi} P(\theta_{j'},\varphi_{j'}|\theta)\sin(\theta_{j'})\mathrm{d}\theta_{j'}\mathrm{d}\varphi_{j'} = 1. \tag{9.5b}$$

This PDF quantifies the probability, under the condition that products are formed and scattered along the direction specified by θ, that the AB rotational angular momentum, $\boldsymbol{j}'$, will lie along the direction specified by $\theta_{j'}$ and $\varphi_{j'}$.

EXAMPLE

To clarify the meanings of the PDFs defined above, consider the data in the table below, which shows PDF values for particular values of θ, $\theta_{j'}$ and $\varphi_{j'}$. What information is there?

θ	$\theta_{j'}$	$\varphi_{j'}$	$P(\theta, \theta_{j'}, \varphi_{j'})$	$P(\theta)$	$P(\theta_{j'}, \varphi_{j'}\|\theta)$
45°	90°	90°	0.1	0.2	0.5
60°	90°	90°	0.1	0.5	0.2
90°	90°	−90°	0.2	0.5	0.4

The $P(\theta,\theta_{j'},\varphi_{j'})$ column tells us about the ***k*-*k*′-*j*′** correlation; among the outcomes considered, the most likely is that products will recoil along $\theta = 90°$ with $\boldsymbol{j}' \parallel -y$. The other outcomes (product recoil along $\theta = 45$ or $60°$, in either case with $\boldsymbol{j}' \parallel +y$) are equally likely.

The $P(\theta)$ column tells us about the ***k*-*k*′** correlation; among the outcomes considered, the least likely is recoil along $\theta = 45°$. The other outcomes (recoil along $\theta = 60$ or $90°$) are equally likely. Note that here we are considering product recoil regardless of the $\boldsymbol{j}'$ polarization.

The $P(\theta_{j'},\varphi_{j'}|\theta)$ column provides a different view of the ***k*-*k*′-*j*′** correlation; if we compare scattering at $\theta = 45°$ to scattering at $\theta = 60°$, we find that, in relative terms, we are more likely to find $\boldsymbol{j}' \parallel +y$ in the former case. Note that here we are considering the $\boldsymbol{j}'$ polarization at a particular scattering angle regardless of the probability that products will be scattered in that direction.

Careful thought about the full stereodynamical problem shows that the PDFs expressing two-, three- and four-vector correlations can be formulated in a variety of ways. We will not delve into the full range of possible definitions here, but it is important to note that different PDFs provide different perspectives on the dependence of the reaction dynamics on directions in space. As illustrated by the examples and problems that will follow, the convenience of any particular definition depends on the strategy and aims of the study.

9.2.4 Polarization Moments of Probability Density Functions

In Section 9.2.3 we have discussed what PDFs represent, but not their explicit mathematical forms. We shall consider these in this section. We will first see how PDFs can be written in terms of quantitative parameters called polarization moments, and then inspect the shapes that some PDFs take given a particular set of polarization moments.

For definiteness, let us consider a particular PDF: the normalized θ-dependent $\boldsymbol{j}'$ polarization function of Eq. (9.3). This function, $P(\theta,\theta_{j'},\varphi_{j'})$ depends on the angles that specify the recoil and angular momentum directions of the products, $\boldsymbol{k}'$ and $\boldsymbol{j}'$.

Before considering the mathematical form of this PDF, we must examine how it might be determined experimentally. Note the following: experimentally, there is an important distinction between the scattering angle and the angles associated with angular momentum polarization. The former has its value precisely defined: in crossed molecular beam experiments one makes reactants approach along a precisely defined direction and then detects products as they recoil along another, also precisely defined direction. In the case of angular momentum, that does not happen; polarization measurements do not determine particular values for $\theta_{j'}$ or $\varphi_{j'}$. Polarization measurements are carried out with recourse to the interaction of polarized photons with (possibly polarized) molecules; particular photon polarizations lead to quantification of particular components of the molecular polarization (see Study Box 9.1). These components are the *multipoles* of the molecule; the parameters that quantify the contributions of individual multipoles to the overall molecular polarization are the *polarization moments* of the molecule, also known as its *multipole moments.*

Because of this distinction, the theoretical description of the stereodynamical PDFs does not treat the scattering angle in the same way that it treats angles associated with angular momentum polarization. Whereas angles such as $\theta_{j'}$ or $\varphi_{j'}$ are treated as independent PDF variables, the scattering angle is treated as a PDF parameter. This is a subtle distinction, which will hopefully become clearer as we proceed.

9.2.4.1 *Multipolar Expansion*

Back to our PDF. If θ is treated as a parameter, then $P(\theta,\theta_{j'},\varphi_{j'})$ is a function of two independent variables, $\theta_{j'}$ and $\varphi_{j'}$. The independent variables satisfy $0\leqslant\theta_{j'}\leqslant\pi$, $0\leqslant\varphi_{j'}\leqslant 2\pi$, and the value of the integral of the function over $\theta_{j'}$ and $\varphi_{j'}$ can be neither negative nor infinite. Now, it is an established mathematical fact that every mathematical function of this type can be written as a summation (a series) of spherical harmonics:[647]

$$P(\theta,\theta_{j'},\varphi_{j'}) = \sum_{k=0}^{\infty}\sum_{q=-k}^{k} z_{kq}(\theta)\,Y_{kq}(\theta_{j'},\varphi_{j'}), \tag{9.6}$$

where the expansion coefficients, z_{kq}, depend on θ. (In what follows we will use symbols other than $z_{kq}(\theta)$ to denote the expansion coefficients; in each case, the symbol used for the expansion coefficients will depend on the particular PDF to which they correspond.)

Equation (9.6) is an example of a multipolar expansion of a PDF. The spherical harmonics are the multipoles, and the expansion coefficients are the polarization (or multipole) moments. As the spherical harmonics are known functions, in order to fully determine the PDF we only need to determine the numerical values of the polarization moments. These are the dynamical quantities: their values change from one reaction to another and from one PDF to another.

Alas, the relatively simple expression given in Eq. (9.6) is rarely seen as the most convenient form for the multipolar expansion. Worse, there is no unique answer for the question of what exactly *is* the most convenient form of the multipolar expansion. Because of this, different people use different expressions, and there are almost as many forms for the multipolar expansion as there are research groups working on stereodynamics.

Here we will use two forms of multipolar expansion. They are appropriate for a molecule in a definite angular momentum state with angular momentum quantum number j, and incorporate quantum restrictions to polarization.* One describes the polarization of $\boldsymbol{j}$, the other the polarization of $\boldsymbol{r}$ (note that they are separate PDFs, not different representations of a single PDF that simultaneously describes the $\boldsymbol{j}$ and $\boldsymbol{r}$ polarizations). When considering θ-independent polarizations (the corresponding polarization moments are denoted by $a_q^{(k)}$), the PDF describing the polarization of $\boldsymbol{j}$ will be written as

$$P(\theta_j,\varphi_j)=\sum_{k=0}^{2j}\sum_{q=-k}^{k}\frac{2k+1}{4\pi}a_q^{(k)}\langle jj,k0|jj\rangle C_{kq}^*(\theta_j,\varphi_j); \tag{9.7}$$

and the corresponding internuclear axis ($\boldsymbol{r}$) polarization will be written as

$$P(\theta_r,\varphi_r)=\sum_{k=0}^{2j}\sum_{q=-k}^{k}\frac{2k+1}{4\pi}a_q^{(k)}\langle j0,k0|j0\rangle C_{kq}^*(\theta_r,\varphi_r), \tag{9.8}$$

In these equations, $\langle jj,k0|jj\rangle$ and $\langle j0,k0|j0\rangle$ are Clebsch-Gordan (CG) coefficients, and C_{kq}^* is the complex conjugate of a modified spherical harmonic.† Note that the polarization moments that quantify the polarization of $\boldsymbol{j}$ also quantify the polarization of $\boldsymbol{r}$, and that the summation over k only runs up to $k=2j$. The first observation results from the fact that the $\boldsymbol{j}$ and $\boldsymbol{r}$ polarizations are tied to each other, the second from the fact that polarization moments with $k>2j$ cannot contribute to quantum PDFs (see the above footnote concerning the uncertainty principle).

9.2.4.2 *Physical Meanings of Polarization Moments*

We have seen that polarization moments are the measurable parameters that quantify molecular polarizations, but have not dwelled on their physical meanings. Let us do that now, starting with the simple case in which the only non-vanishing polarization moments of the PDFs of Eqs. (9.7) and (9.8) are $a_0^{(0)}$ and $a_0^{(2)}$. In this case, the angular momentum PDF of Eq. (9.7) can be written as

$$\begin{aligned}P(\theta_j,\varphi_j)&=\frac{1}{4\pi}a_0^{(0)}\langle jj,00|jj\rangle C_{00}^*(\theta_j,\varphi_j)+\frac{5}{4\pi}a_0^{(2)}\langle jj,20|jj\rangle C_{20}^*(\theta_j,\varphi_j)\\&=\frac{a_0^{(0)}}{4\pi}\left(C_{00}(\theta_j,\varphi_j)+5\frac{a_0^{(2)}}{a_0^{(0)}}\langle jj,20|jj\rangle C_{20}(\theta_j,\varphi_j)\right),\end{aligned} \tag{9.9}$$

*The uncertainty principle limits the extent to which $\boldsymbol{j}$ (and, consequently, $\boldsymbol{r}$) can be localized along a particular direction. If our PDF were $P(\theta_j,\varphi_j)=\delta(\theta_j-\pi/2)\delta(\varphi_j-\pi/2)$, the $\boldsymbol{j}$ vector would be localized exactly along the y direction, and the values of its x and z components would be known exactly: $j_x=j_z=0$. As the uncertainty principle rules this out, it also rules out the sharply polarized PDF we have considered. In Eqs. (9.7–9.8), this quantum restriction as to how sharply polarized the PDF can be is accounted for by the Clebsch-Gordan coefficients, whose moduli are always smaller than 1 and vanish when $k>2j$. Eq. (9.6), which does not take this quantum restriction into account, is appropriate for a classical, "infinitely polarisable" PDF — hence the involvement of an infinite number of polarization moments.[648]

†A CG coefficient such as $\langle j_1m_1,j_2m_2|JM\rangle$ is a real number that quantifies the probability amplitude that coupling of the angular momentum states $|j_1m_1\rangle$ and $|j_2m_2\rangle$ results in the angular momentum state $|j_1j_2JM\rangle$. Modified spherical harmonics are given by $C_{kq}(\theta,\varphi)=[4\pi/(2k+1)]^{1/2}\,Y_{kq}(\theta,\varphi)$. See refs. [239] and [647].

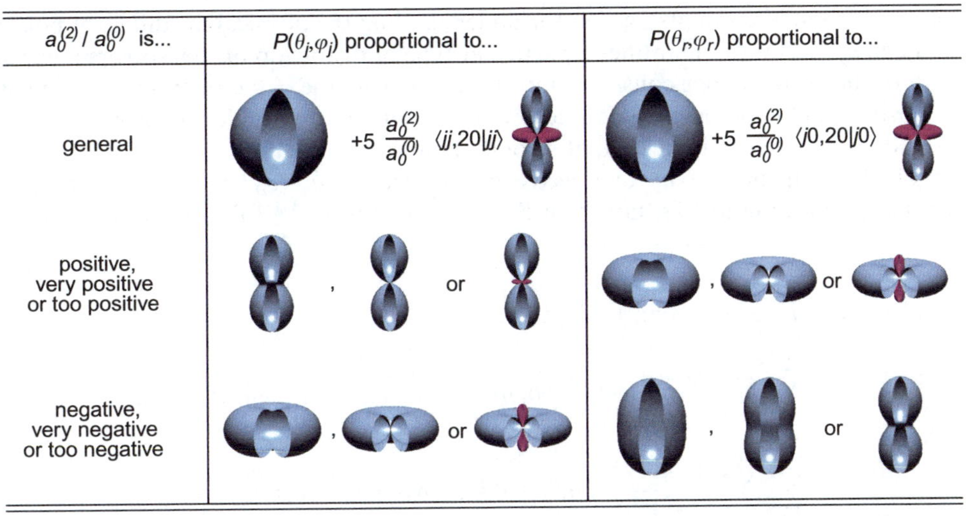

$a_0^{(2)}$ / $a_0^{(0)}$ is...
$P(\theta_j,\varphi_j)$ proportional to...
$P(\theta_r,\varphi_r)$ proportional to...
general
$+5\,\frac{a_0^{(2)}}{a_0^{(0)}}\,\langle jj,20|jj\rangle$
$+5\,\frac{a_0^{(2)}}{a_0^{(0)}}\,\langle j0,20|j0\rangle$
positive,
very positive
or too positive
or
negative,
very negative
or too negative
or

Figure 9.4 Different values of the $a_0^{(2)}/a_0^{(0)}$ ratio lead to different shapes for the angular momentum PDF of Eq. (9.9) and its molecular axis counterpart. Values that are 'too positive' or 'too negative' (*i.e.* lie outside the physical limits of the parameter) lead to PDFs that can take unacceptable, negative values (purple). The functions have been "cut" to facilitate visualization.

where we have used $\langle jm, 00|jm\rangle = 1$ and the fact that spherical harmonics with $q=0$ are real;[647] an analogous expression holds for the molecular axis PDF of Eq. (9.8).

Figure 9.4 gives visual representations of the angular momentum PDF of Eq. (9.9) and its molecular axis counterpart, obtained through use of graphical representations for the C_{00} and C_{20} functions (which the reader will recognize as identical to the s and d_{z^2} orbitals of the hydrogen atom) and PDFs. The functions are positive (negative) where shown in blue (purple).

The top row leaves the value of $a_0^{(2)}/a_0^{(0)}$ unspecified. It indicates that addition (or subtraction) of C_{20} to C_{00} leads to distortion of the spherical, isotropic PDFs one gets when the only non-vanishing polarization moment is $a_0^{(0)}$. If all polarization moments other that $a_0^{(0)}$ are zero, then the rotational angular momentum of the diatomic, $\boldsymbol{j}$, does not show preference for any spatial orientation, and neither does its internuclear axis, $\boldsymbol{r}$. The molecule is unpolarized.

Because the $\langle jj, 20|jj\rangle$ and $\langle j0, 20|j0\rangle$ CG coefficients have different magnitudes and signs (the first is always positive, the second always negative[647]), non-vanishing values for $a_0^{(2)}/a_0^{(0)}$ affect the rotational and axial PDFs in different ways. As can be seen by comparison of the second and third columns of the figure, what happens is this:

- If $a_0^{(2)}/a_0^{(0)}>0$, the rotational PDF is squeezed around the equator and blown near the poles; the $\boldsymbol{j}$ vector is more likely to be found pointing towards either pole, and less likely to be found pointing towards points near the equator. The inverse happens if $a_0^{(2)}/a_0^{(0)}<0$.
- If $a_0^{(2)}/a_0^{(0)}>0$, the axial PDF is squeezed near the poles and blown around the equator; the $\boldsymbol{r}$ vector is more likely to be found pointing towards points near the equator, and less likely to be found pointing towards either pole. The inverse happens if $a_0^{(2)}/a_0^{(0)}<0$.

As you may have guessed, the different implications of the value of $a_0^{(2)}/a_0^{(0)}$ for the shapes of $P(\theta_j,\varphi_j)$ and $P(\theta_r,\varphi_r)$ result from the perpendicularity between $\boldsymbol{j}$ and $\boldsymbol{r}$.

Note also that $a_0^{(2)}/a_0^{(0)}$ values that are 'too positive' or 'too negative' lead to an absurd situation: for some directions, one or both PDFs take negative values. As PDF values are probabilities, this cannot happen. There are, therefore, limits to how positive or how negative $a_0^{(2)}/a_0^{(0)}$ can be, while still giving rise to a physically acceptable PDF.

These are the essential concepts needed for understanding the physical meanings of polarization moments. Each $a_q^{(k)}/a_0^{(0)}$ ratio quantifies a particular type of distortion of the isotropic, unpolarized distribution of the vector under consideration. By comparing the value of this ratio to its limiting (*i.e.* most positive and most negative) values, we can furthermore decide whether the distortion is large or small, and therefore indicative of strong or weak polarization.

A question that remains is why one needs to consider the $a_q^{(k)}/a_0^{(0)}$ ratios rather than just the $a_q^{(k)}$ moments. To understand this, note first that the integral of the PDF of Eq. (9.7) is given by

$$\begin{aligned}
&\int_0^{2\pi}\int_0^{\pi} P(\theta_j,\varphi_j)\sin(\theta_j)\mathrm{d}\theta_j\mathrm{d}\varphi_j \\
&=\sum_{k=0}^{2j}\sum_{q=-k}^{k}\frac{2k+1}{4\pi}a_q^{(k)}\langle jj,k0|jj\rangle\int_0^{2\pi}\int_0^{\pi}C_{kq}^*(\theta_j,\varphi_j)\sin(\theta_j)\,\mathrm{d}\theta_j\mathrm{d}\varphi_j \\
&=\sum_{k=0}^{2j}\sum_{q=-k}^{k}\frac{2k+1}{4\pi}a_q^{(k)}\langle jj,k0|jj\rangle 4\pi\delta_{k0}\delta_{q0}=a_0^{(0)}\langle jj,00|jj\rangle=a_0^{(0)}.
\end{aligned} \tag{9.10}$$

This shows that the probability of finding the $\boldsymbol{j}$ vector pointing along some direction is given by $a_0^{(0)}$, the monopole moment. If the only nonzero moment is $a_0^{(0)}$, then the distribution is isotropic ($\boldsymbol{j}$ does not preferentially lie along any particular direction) and its integral over $\boldsymbol{j}$ directions is given by $a_0^{(0)}$. If other polarization moments (dipole moments have rank $k=1$, quadrupole moments have rank $k=2$, and so on*) also have nonzero values, then the $P(\theta_j,\varphi_j)$ function is distorted (in other words, it becomes anisotropic), but its integral over $\boldsymbol{j}$ directions does not change. It is still given by $a_0^{(0)}$. The conclusion is that the "size" of the PDF – its normalization, which has been discussed in Section 9.2.3 – is determined solely by the monopole moment. To quantify the extent of anisotropy caused by some other multipole moment, one must consider how large, *relative to the monopole*, that multipole moment is; hence the need for division by $a_0^{(0)}$. Note also that often, but not always, the value of the monopole will be set to unity. This is indeed the natural choice for the θ-independent PDF we are considering now, but for θ-dependent PDFs one can choose to set the monopole value to $P(\theta)$ or to unity. The first choice is appropriate for a "normalized" θ-dependent PDF [see Eqs. (9.3) and (9.4)], the second is appropriate for a "renormalized" θ-dependent PDF [see Eq. (9.5)].

We have stated that the $a_q^{(k)}/a_0^{(0)}$ ratios quantify particular types of distortion of an isotropic vector distribution, but detailed inspection of how this works reveals a problem: because spherical harmonics with component $q\neq 0$ are complex, when $q\neq 0$ the polarization moments defined above can take complex values. This leads to some difficulties. After all, we would like to use polarization moments to identify polarization directions in real, three-dimensional space; complex polarization moments are not ideally suited for this task.

9.2.4.3 *Real and Complex Polarization Moments*

As it turns out, it is easy to transform the complex polarization moments, $a_q^{(k)}$, into *real polarization moments*, $a_{q\pm}^{\{k\}}$, of immediate directional significance. The necessary transformations are[412]

$$a_{q+}^{\{k\}}=\frac{1}{\sqrt{2}}\left[(-1)^q a_{+q}^{(k)}+a_{-q}^{(k)}\right]=(-1)^q\sqrt{2}\mathrm{Re}(a_{+q}^{(k)}),\quad q>0, \tag{9.11a}$$

*This terminology comes from the mathematics of the problem. Mathematically, the multipole moments are *tensors* whose *rank* is given by k. The q index identifies the *component* of the tensor.

$$a_{q-}^{\{k\}} = \frac{1}{i\sqrt{2}}\left[(-1)^q a_{+q}^{(k)} - a_{-q}^{(k)}\right] = (-1)^q \sqrt{2}\mathrm{Im}(a_{+q}^{(k)}), \quad q>0, \tag{9.11b}$$

$$a_0^{\{k\}} = a_0^{(k)}, \tag{9.11c}$$

(note the use of curly brackets and the positioning of signs in the notation for the real quantities). Using real polarization moments we can rewrite the PDF of Eq. (9.7) as

$$\begin{aligned} P(\theta_j, \varphi_j) = &\sum_{k=0}^{2j} \frac{2k+1}{4\pi} \langle jj, k0|jj\rangle \\ &\times \left[a_0^{\{k\}} C_0^{\{k\}}(\theta_j, \varphi_j) + \sum_{q=1}^{2j} a_{q+}^{\{k\}} C_{q+}^{\{k\}}(\theta_j, \varphi_j) + a_{q-}^{\{k\}} C_{q-}^{\{k\}}(\theta_j, \varphi_j)\right], \end{aligned} \tag{9.12}$$

where the real spherical harmonics, $C_{q\pm}^{\{k\}}$, are related to their complex counterparts by expressions identical to those of Eq. (9.11)*, except that the $a_{\pm q}^{(k)}$ are substituted by $C_{k\pm q}$. Apart from normalization, the $C_{q\pm}^{\{k\}}$ functions are the familiar angular wavefunctions of the electronic orbitals of the H atom. For example, the angular wavefunctions of p_x, d_{xy} and f_{z^3} orbitals are respectively proportional to $C_{1+}^{\{1\}}$, $C_{2-}^{\{2\}}$, and $C_0^{\{3\}}$.

9.2.5 Density Matrices

Consider a diatomic molecule whose rotational state is described by the wavefunction

$$|\Psi\rangle = \sum_{m=-j}^{j} c_m |jm\rangle, \tag{9.13}$$

where the c_m are complex numbers and the basis states, $|jm\rangle$, are the eigenstates of the $\hat{j}^2$ and $\hat{j}_z$ operators. An alternative representation of this state, very useful in stereodynamics, is given by ρ, the *density matrix* of the molecule.[643] This is the matrix whose elements are given by

$$\langle jm_1|\rho|jm_2\rangle = \langle jm_1|\Psi\rangle\langle\Psi|jm_2\rangle = c_{m_1} c_{m_2}^*, \quad -j \leqslant m_1 \leqslant j, \quad -j \leqslant m_2 \leqslant j. \tag{9.14}$$

There are two particular reasons why density matrices are useful. First, the molecular state is described in a very compact way, involving nothing but a finite set of numbers. This is very convenient for computations and mathematical manipulations.

The second, and most important reason, is that density matrices are more general than wavefunctions. Density matrices can describe not only states like the one of Eq. (9.13) – such states are referred to as *pure*, or *fully coherent*, states – but also *mixed* states that cannot be described by wavefunctions. To see the distinction, let us consider an example.

EXAMPLE

The wavefunctions $f(x) = \sin x$ and $g(x) = \cos x$ describe two states of a certain system. A linear combination of the two, say, $h(x) = f(x) + g(x)$, gives the wavefunction of a third possible system state, with observable properties that differ from those of the states considered initially. This is a

*Notice that to obtain the real spherical harmonics *via* Eq. (9.11) one should use $C_{k\pm q}$ instead of $C_{k\pm q}^*$.

pure, fully coherent state, whose observable properties cannot be directly described in terms of the observable properties of the f and g states from which it was derived. In particular, the new probability density, $|h(x)|^2 = [f(x) + g(x)]^2 = 1 + \sin(2x)$, cannot be directly derived from the original ones, $|f(x)|^2 = \sin^2 x$ and $|g(x)|^2 = \cos^2 x$.

Now suppose that many copies of the system are independently prepared, 70% of them in state f, the rest in state g. Furthermore, suppose that while observing this system we are blind as to whether the observed results stem from f-state or g-state copies of the system. In this case, the observable properties can be directly described as weighted averages of the observable properties of the f and g states. In particular, the new probability density is $0.7|f(x)|^2 + 0.3|g(x)|^2$. In this case, we are dealing with a mixed state. None of the system wavefunctions can lead to the observable probability density, but a density matrix can.

If a system is in a mixed state involving an incoherent superposition of N pure states, its density matrix is the average of the pure-state density matrices:

$$\rho = \sum_{n=1}^{N} w_n \rho_n^{(pure)}, \tag{9.15}$$

where w_n is the statistical weight of the nth state of the incoherent mixture.

We close this section by noting that, if we know the density matrix of a system, we can easily calculate the expectation values of its observable properties, using the formula

$$\langle O \rangle = \frac{\mathrm{Tr}(\rho O)}{\mathrm{Tr}(\rho)}, \tag{9.16}$$

where Tr stands for the trace (the sum of the diagonal elements of a matrix) and O is the matrix representation of the operator associated with the property under consideration.

9.2.6 Polarization Moments of Density Matrices

Just like its PDF, the density matrix of a system can be expanded in a multipolar series. The procedure is similar to the one discussed in Section 9.2.4, except that the multipolar functions used there, the modified spherical harmonics, are replaced by multipolar matrices. As before, one can write the expansion in terms of complex or real multipolar matrices,* which are respectively associated with complex or real polarization moments.

In angular momentum space the multipolar expansion of a density matrix reads

$$\rho = \sum_{k=0}^{2j} \sum_{q=-k}^{k} \frac{2k+1}{2j+1} a_q^{(k)} T_{kq}^{\dagger}, \tag{9.17}$$

where the $T_{kq}^{\dagger}$ are the adjoints of multipolar tensor operators. We define these operators so that their matrix elements in the $|jm\rangle$ representation are

$$\langle jm_2|T_{kq}^{\dagger}|jm_1\rangle = \langle jm_1|T_{kq}|jm_2\rangle^* = \langle jm_1|T_{kq}|jm_2\rangle = \langle jm_2, kq|jm_1.\rangle. \tag{9.18}$$

*The "complex" and "real" qualifiers refer to the eigenvalues rather than to the elements of multipolar matrices. Real multipolar matrices are Hermitian and have real eigenvalues; complex multipolar matrices are not Hermitian and can have complex eigenvalues. As for multipolar matrix elements, they can be complex in either case.

Eq. (9.17) can also be written in terms of real multipolar tensors:

$$\rho = \sum_{k=0}^{2j} \frac{2k+1}{2j+1} \left(a_0^{\{k\}} T_0^{\{k\}} + \sum_{q=1}^{k} a_{q+}^{\{k\}} T_{q+}^{\{k\}} + a_{q-}^{\{k\}} T_{q-}^{\{k\}} \right). \tag{9.19}$$

The matrix elements of the real multipolar tensors are given by

$$\langle jm_1 | T_{q+}^{\{k\}} | jm_2 \rangle = \frac{1}{\sqrt{2}} [(-1)^q \langle jm_2, kq | jm_1 \rangle + \langle jm_2, k-q | jm_1 \rangle], \quad q > 0, \tag{9.20a}$$

$$\langle jm_1 | T_{q-}^{\{k\}} | jm_2 \rangle = \frac{1}{i\sqrt{2}} [(-1)^q \langle jm_2, kq | jm_1 \rangle - \langle jm_2, k-q | jm_1 \rangle], \quad q > 0, \tag{9.20b}$$

$$\langle jm_2 | T_0^{\{k\}} | jm_1 \rangle = \langle jm_2, k0 | jm_1 \rangle. \tag{9.20c}$$

Eqs. (9.17) and (9.18) imply that the elements of the density matrix are given by

$$\begin{aligned} \langle jm_1 | \rho | jm_2 \rangle &= \sum_{k=0}^{2j} \sum_{q=-k}^{k} \frac{2k+1}{2j+1} a_q^{(k)} \langle jm_1 | T_{kq}^{\dagger} | jm_2 \rangle \\ &= \sum_{k=0}^{2j} \sum_{q=-k}^{k} \frac{2k+1}{2j+1} a_q^{(k)} \langle jm_1, kq | jm_2 \rangle. \end{aligned} \tag{9.21}$$

Using formulae for sums of products of Clebsch-Gordan coefficients[239,647] one can invert this formula. The result is an expression for the polarization moments in terms of the elements of the density matrix:

$$\begin{aligned} a_q^{(k)} &= \sum_{m_1=-j}^{j} \sum_{m_2=-j}^{j} \langle jm_1 | \rho | jm_2 \rangle \langle jm_2 | T_{kq} | jm_1 \rangle \\ &= \sum_{m_1=-j}^{j} \sum_{m_2=-j}^{j} \langle jm_1 | \rho | jm_2 \rangle \langle jm_1, kq | jm_2 \rangle \\ &= \mathrm{Tr}(\rho T_{kq}). \end{aligned} \tag{9.22}$$

Using $\langle jm_2, 00 | jm_1 \rangle = \delta_{m_1 m_2}$ we find that, when $k = q = 0$, this equation reduces to

$$a_0^{(0)} = \sum_{m=-j}^{j} \langle jm | \rho | jm \rangle = \mathrm{Tr}(\rho). \tag{9.23}$$

Comparison of Eqs. (9.22) and (9.23) with Eq. (9.16) shows that $a_q^{(k)}/a_q^{(0)} = \langle T_{kq} \rangle$, which in turn implies $a_{q\pm}^{\{k\}}/a_0^{\{0\}} = \langle T_{q\pm}^{\{k\}} \rangle$. Except for a scaling factor (its value is the value of the monopole), the polarization moments are the expectation values of the corresponding multipolar tensor operators.

Note that the PDF expansion of Eq. (9.12) is analogous, but not entirely similar, to the multipolar expansion of Eq. (9.19). For instance, the latter does not involve the $\langle jj, k0 | jj \rangle$ CG coefficient and replaces the 4π denominator by $2j+1$ (see ref. [648] for a discussion of the reasons). This, however, should not lead to confusion regarding the polarization moments, $a_{q\pm}^{\{k\}}$: the polarization moments appearing in Eqs. (9.12) and (9.19) are the same. If calculated under the same circumstances, they agree in meaning and also in value.

9.2.7 Limiting Values of Real Polarization Moments

Figure 9.4 suggests that one might determine the limiting values of real polarization moments by increasing or decreasing a single non-vanishing $a_{q\pm}^{\{k\}}/a_0^{\{0\}}$ ratio until negative PDF values arise. In general, this procedure does not work. The reason is that the limiting values of a given moment need not be associated with a situation in which all other moments with $k>0$ vanish.

The easiest way to determine the limiting values of real polarization moments is through consideration of the eigenvalues of the real multipolar matrices. The key observation is that, except for the $a_0^{\{0\}}$ scale factor (which coincides with Tr(ρ) and, for a pure state, with the square of the norm of its wavefunction, *i.e.* $|\Psi|^2$), $a_{q\pm}^{\{k\}}$ is the expectation value of the $\hat{T}_{q\pm}^{\{k\}}$ operator, as discussed in Section 9.2.6. Now, an expectation value of an operator is a weighted average of the operator's eigenvalues. As such, it can be neither larger than the largest eigenvalue nor smaller than the smallest eigenvalue. It follows from there that the $a_{q\pm}^{\{k\}}/a_0^{\{0\}}$ ratio (that is, the $\langle T_{q\pm}^{\{k\}}\rangle$ expectation value) can never be more positive than the most positive, or more negative than the most negative, of the eigenvalues of the $T_{q\pm}^{\{k\}}$ matrix.

For a compilation of the limiting values of $\langle T_{q\pm}^{\{k\}}\rangle$ when the rank is $k\leqslant 4$ and the angular momentum quantum number is $j\leqslant 5$, see ref. [649].

9.3 EXPERIMENTAL AND THEORETICAL STEREODYNAMICS

In this section we describe how molecules can be experimentally polarized,* and also how the values of reactant and product polarization moments can be extracted from the results of dynamical calculations. When considering product polarization, these are the polarization moments we will be interested in:

- *Normalized polarization-dependent differential cross sections* (PDDCSs). The normalized PDDCSs are the polarization moments of $P(\theta, \theta_{j'}, \varphi_{j'})$ the normalized θ-dependent $\boldsymbol{j}'$ polarization function of Eq. (9.3). This PDF quantifies the $\boldsymbol{k}$-$\boldsymbol{k}'$-$\boldsymbol{j}'$ correlation. Complex and real normalized PDDCSs are respectively represented by $S_{\pm q}^{(k)}(\theta)$ and $S_{q\pm}^{\{k\}}(\theta)$.
- *Renormalized PDDCSs*. These are the polarization moments of the conditional θ-dependent PDF of Eq. (9.5). This PDF also quantifies the $\boldsymbol{k}$-$\boldsymbol{k}'$-$\boldsymbol{j}'$ correlation, and is obtained through division of the PDF of Eq. (9.3) by $P(\theta)=S_0^{(0)}=S_0^{\{0\}}$, see Eq. (9.5a). As a consequence, the complex and real renormalized PDDCSs are respectively given by $S_q^{(k)}(\theta)/S_0^{\{0\}}(\theta)$ and $S_{q\pm}^{\{k\}}(\theta)/S_0^{\{0\}}(\theta)$.
- *Polarization parameters* (PPs). The PPs are the polarization moments of $P(\theta_{j'}, \varphi_{j'})$ the integral $\boldsymbol{j}'$ polarization function:

$$P(\theta_{j'}, \varphi_{j'}) = \int_0^{\pi} P(\theta, \theta_{j'}, \varphi_{j'}) \sin(\theta)\, d\theta, \tag{9.24}$$

where $P(\theta, \theta_{j'}, \varphi_{j'})$ is the normalized PDF of Eq. (9.3). The PDF given by Eq. (9.24) is normalized as per Eq. (9.5b) and quantifies the probability, under the condition that product formation does occur, that the product diatomic will recoil with its angular momentum along the direction specified by $\theta_{j'}$ and $\varphi_{j'}$ irrespective of the scattering direction. By (i) writing down the multipolar expansion for the normalized θ-dependent PDF of Eq. (9.3) and the multipolar expansion for the integral PDF of Eq. (9.24), and (ii) integrating the first expansion over θ, you can show that the complex

*For a description of how molecular polarization can be detected experimentally see Chapters 5 and 8 (Section 8.7) and Study Box 9.1.

PPs [the complex polarization moments of $P(\theta_{j'}, \varphi_{j'})$] are related to the normalized PDDCSs the complex polarization moments of $P(\theta, \theta_{j'}, \varphi_{j'})$ by

$$a_q^{(k)} = \int_0^\pi S_q^{(k)}(\theta)\sin(\theta)\,\mathrm{d}\theta. \tag{9.25}$$

Entirely similar polarization moments will be used to describe the ***k-j-k'*** vector correlation.

We would also like to mention that the rest of this section is quite technical, and could (perhaps should) be skipped on a first reading (in which case proceed to Section 9.4).

9.3.1 Theoretical Methods

Quasi-classical trajectory (QCT) and quantum dynamical calculations are the main theoretical methods used for the calculation of polarization moments. Here we focus on the practical calculation procedures, not on the comparison between classical and quantum descriptions of reaction stereodynamics. For discussions of the latter, see refs. [646,648,650].

9.3.1.1 Quasi-Classical Trajectories

The QCT method has been reviewed many times, and excellent summaries can be found in the literature (see, *e.g.*, refs. [75,651–656]). For atom-diatom reactions, current implementations are very similar to that developed by Karplus *et al.* in the sixties.[75] We will briefly sketch the general methodology before focusing on the calculation of stereodynamical properties.

As we have seen already in Chapter 3, the basic idea is simple: given the initial positions and momenta of the atoms participating in the collision, we want to calculate their *trajectories* (values of nuclear positions and momenta as a function of time) by integration of the classical equations describing the motion of the interacting atoms. This is done until the collision outcome is clear; analysis of the final positions and momenta allows one to decide not only whether the A + BC collision has lead to AB + C products, but also to quantify scalar and vector properties. As no single trajectory can account for the reaction dynamics, many trajectories, with suitably chosen initial conditions, are calculated and then averaged using the Monte Carlo method (see Study Box 5.4). Energy quantization is accounted for in the selection of initial and analysis of final trajectory conditions; it is this assignment of quantum numbers to reactants and products that justifies the "quasi-classical" qualifier. Fitting of the distribution of the various properties of interest is then used to facilitate comparisons with experimental or quantum results.

Equations of Motion and Coordinate System

For three-atom systems, the most efficient trajectory computation procedure is integration of the Hamilton equations (see Chapter 3) using a set of space-fixed frame cartesian coordinates to describe the relative A–BC and internal B–C motions. As the motion of the ABC centre-of-mass is disregarded (the centre-of-mass is taken to be stationary), the Hamilton equations form a system of twelve coupled first-order differential equations;* six describe the relative motion of A and BC, six describe BC vibration and rotation. The values of the constants of motion (the total energy, *E*, and

*As noted already in Chapter 3, to specify the position of three atoms requires $3N=9$ degrees of freedom, but three of these are used to specify the position of the CM. This leaves six degrees of freedom to describe the relative position of the three atoms and the rotation of the system. The same arguments apply for the momenta of the three atoms, yielding twelve equations in all.

the three components of the total angular momentum, $\boldsymbol{J}$) are checked as the trajectory is propagated, and their conservation is used to monitor the accuracy of the integration.

The most convenient space-fixed frame is the one whose Z axis and origin respectively coincide with the reactant-approach direction ($\boldsymbol{k}$) and the BC centre-of-mass. It is also customary to use the initial orbital angular momentum vector ($\boldsymbol{\ell}$) to define the X axis; this implies that atom A is initially placed in the YZ plane.

Numerical Integration

The most time-consuming task in the computation of a trajectory is evaluation of the potential energy and its derivatives. This is of special consequence for two parts of the computer program: the potential energy code; and the differential equations integrator. The former must be fast and accurate; this tips the balance in favour of the use of analytical derivatives even though their programming may require considerable effort. The latter must compromise between robustness and the number of potential energy evaluations; predictor-corrector integrators with fixed or variable step size are usually the most appropriate.

Initial Conditions

The determination of integral and differential cross sections is performed by summing the reaction probabilities over the volume elements of the phase space that contribute to the reaction. Thus cross sections or any other average probability are given by multidimensional integrals over the relevant variables that cover the reactant phase space to be sampled. Therefore, a key issue in the QCT method is the sampling of initial conditions for each trajectory; that is, the determination of the initial values of the coordinates and momenta. Direct selection of initial values for the cartesian (or curvilinear) coordinates and their conjugate moments is not practical. Instead, it is convenient and intuitive to use an indirect method whereby the initial positions and momenta are determined by a set of so-called "collision parameters". For A + BC collisions, they are:

- The impact parameter, b. This is the perpendicular component of the A–BC relative position vector ($\boldsymbol{R}$) with respect to $\boldsymbol{k}$ (and therefore also with the Z axis, which is defined along $\boldsymbol{k}$). The impact parameter of the ith trajectory is usually sampled as $b_i = \xi^{1/2} b_{\max}$, where ξ is a random number satisfying $0 \leqslant \xi \leqslant 1$ and $b_{\max}$ is the largest impact parameter that can lead to reaction. Specification of b implies specification of $|\boldsymbol{\ell}|$. (In this procedure $\boldsymbol{\ell}$ is not quantized, but sampling of discrete, integer values for the ℓ quantum number is also possible.)
- The Euler angles[647] relating the direction of $\boldsymbol{j}$ in the space-fixed frame to its direction in the diatom body-fixed frame. (Alternatively, one can sample the Euler angles of $\boldsymbol{r}$, the BC internuclear axis.) Azimuthal angles are randomly and uniformly chosen in the $[0, 2\pi]$ interval; the cosine of the polar angle is randomly chosen in the $[-1,1]$ interval.
- The phase angle of the BC vibration. This determines the initial BC bond length and whether the bond is being stretched or compressed, and is also randomly and uniformly chosen.
- The vibrational and rotational quantum numbers, v and j. These are given fixed integer values that define the rovibrational state, and therefore the internal energy, E_{vj}, of the BC reactant. The rovibrational energies are those obtained from exact quantum or semiclassical calculations for the BC asymptote of the potential energy surface.
- The collision energy, E_c. This is given a fixed value. (As the total energy $E = E_c + E_{vj}$, one could alternatively fix E instead of E_c.)
- The initial A–BC distance. This is chosen so that the interaction between A and BC is zero (or, strictly speaking, negligible).

Sampling of the randomly-chosen collision parameters is done with the Monte Carlo method (see Study Box 5.4).

Pseudoquantization of Final States

Each trajectory is integrated in time until the resulting chemical species have left the "reaction shell" – that is, until their separation is such that the potential has reached its asymptotic value and the collision products no longer interact. If the trajectory was reactive, the product coordinates and momenta are converted to more directly meaningful quantities such as scattering angle, recoil energy and internal energy.

Pseudoquantization of diatomic states is not difficult. The simplest procedure involves three steps. First, one obtains a real (as opposed to integer) value for the j' quantum number from $|j'|^2 = \hbar^2 j'(j'+1)$. Second, one obtains a real value for the v' quantum number by using a Dunham expansion in $v' + \frac{1}{2}$ and $j'(j'+1)$ for the internal energy of the molecule. Third, one rounds the v' and j' values so calculated to their nearest integers. (More sophisticated methods can be used to overcome problems such as the assignment of levels that are not energetically accessible; in these methods, each trajectory contributes to several product rovibrational states.)

It is important to note the asymmetry between initial and final states. Whereas the specification of v and j is sharp, the specification of v' and j' is not. Individual trajectories comply with microscopic reversibility but the ensemble of trajectories only does so approximately.

Reactivity and Product PDFs

Once a sufficiently large number of trajectories have been calculated, and their initial and final states recorded, one proceeds to compute the PDFs describing various reaction properties. Examples include the opacity function, the product internal energy distribution, and the PDFs we have discussed in Section 9.2.3. Methods based on the use of histograms (binning of the trajectories in "boxes" according to the value of the quantity under consideration) have been widely used, but they are not very convenient for comparisons with continuously-varying experimental or quantum data. The method of PDF expansion in orthogonal polynomials provides a more elegant alternative; expansion in Legendre polynomials, in particular, has proved versatile and efficient. We shall briefly describe this method.

Let x be a variable satisfying $-1 \leqslant x \leqslant 1$, and $P(x)$ the PDF tying the value of x to the reaction probability under the condition that product formation does occur [the normalization condition is $\int_{-1}^{1} P(x)\,\mathrm{d}x = 1$]. This PDF can be expanded in a series of Legendre polynomials:

$$P(x) = \sum_{n=0}^{n_{\max}} \frac{2n+1}{2} a_n \mathcal{P}_n(x), \tag{9.26a}$$

$$a_n = \int_{-1}^{1} P(x)\mathcal{P}_n(x)\,\mathrm{d}x, \tag{9.26b}$$

where $\mathcal{P}_n(x)$ is the Legendre polynomial of degree n, the a_n are the Legendre moments, and we have truncated the expansion at $n = n_{\max}$, with $n_{\max}$ such that the addition of more terms does not change the resulting distribution appreciably.

If we are considering a set of N_r reactive trajectories (this may be, *e.g.*, the set of trajectories ending with products in a specific rovibrational state), the set of values obtained for variable x is $\{x^{(i)}\}$, $i = 1, 2, \ldots, N_r$. The Monte Carlo PDF for this set of values can be formally written as

$$P_{\mathrm{MC}}(x) \approx \frac{1}{N_r} \sum_{i=1}^{N_r} \delta(x - x^{(i)}), \tag{9.27}$$

where $\delta(x-x^{(i)})$ is a delta function* centred on $x^{(i)}$ and the relation becomes an equality in the $N_r \to \infty$ limit. Using Eq. (9.26b) we find that the Legendre moments of $P_{MC}(x)$ are given by

$$a_n = \int_{-1}^{1} P_{MC}(x)\mathcal{P}_n(x)\,dx \approx \frac{1}{N_r}\sum_{i=1}^{N_r}\mathcal{P}_n(x^{(i)}), \tag{9.28}$$

which is the expression that shows how, using trajectory data, we can calculate the Legendre moments. Equation (9.26a) shows how, using the Legendre moments, we can build the PDF.

As an explicit example, consider the PDF of Eq. (9.2a) – the product angular distribution. The variable of this PDF is $x = \cos\theta$. Trajectory data determine it as

$$P(\cos\theta) = \sum_{n=0}^{n_{max}}\frac{2n+1}{2}a_n\mathcal{P}_n(\cos\theta), \quad a_n = \frac{1}{N_r}\sum_{i=1}^{N_r}\mathcal{P}_n(\cos\theta^{(i)}), \tag{9.29}$$

where n_{max} and N_r are parameters of the calculation. Their values must be large enough for the results to be converged.

As every relevant variable (impact parameter, collision energy, etc.) can be transformed into a "reduced" variable in the $[-1,1]$ interval, the method is general. It can be applied to any PDF one can determine from QCT data, and it can be easily extended to multidimensional PDFs.

Polarization Moments

These are determined with a method similar to that of Legendre moments. However, there are complications: the PDFs can depend on more than one variable, and the preferred reference frame is not the space-fixed one used for trajectory propagation.

Let us consider $P(\theta, \theta_{j'}, \varphi_{j'})$, the PDF of Eq. (9.3). We will assume that the θ, $\theta_{j'}$ and $\varphi_{j'}$ angles are defined as in Figure 9.3.

Our first problem is to transform the QCT data (values for the angles specifying the $\boldsymbol{k}'$ and $\boldsymbol{j}'$ directions in the XYZ space-fixed frame) into the data we need (values for the angles specifying the $\boldsymbol{k}'$ and $\boldsymbol{j}'$ directions in the xyz scattering frame).

The definitions used for the XYZ and xyz frames are such that $Z \equiv z \equiv \boldsymbol{k}$. This means that the two frames are related by a rotation around $\boldsymbol{k}$; the angle of the rotation is the dihedral angle specifying the direction of the xz plane in the XYZ frame. Since xz is the plane containing $\boldsymbol{k}$ and $\boldsymbol{k}'$, the appropriate dihedral angle is simply $\Phi_{k'}$, the azimuthal angle of $\boldsymbol{k}'$ in the XYZ frame. This implies: (i) that the polar angles of $\boldsymbol{k}'$ and $\boldsymbol{j}'$ in the xyz frame coincide with their polar angles in the XYZ frame; and (ii) that the azimuthal angles of $\boldsymbol{k}'$ and $\boldsymbol{j}'$ in the xyz frame differ by $\Phi_{k'}$ from their azimuthal angles in the XYZ frame:

	polar angle		*azimuthal angle*	
vector	XYZ	xyz	XYZ	xyz
$\boldsymbol{k}'$	$\Theta_{k'} = \theta$	$\theta_{k'} = \Theta_{k'} = \theta$	$\Phi_{k'}$	$\varphi_{k'} = \Phi_{k'} - \Phi_{k'} = 0$
$\boldsymbol{j}'$	$\Theta_{j'}$	$\theta_{j'} = \Theta_{j'}$	$\Phi_{j'}$	$\varphi_{j'} = \Phi_{j'} - \Phi_{k'}$

Once xyz-frame QCT data are available, we must use them to obtain normalized PDDCSs, the polarization moments of $P(\theta,\theta_{j'},\varphi_{j'})$. Because the PDF depends on three angles, one cannot expand it in terms of a single Legendre polynomial. The appropriate expansion is in terms of a pair of

*See Eqs. (B3.3.7) and (B3.3.8) in Study Box 3.3 for the definition and main properties of the delta function.

modified spherical harmonics. It reads

$$P(\theta, \theta_{j'}, \varphi_{j'}) = \sum_{k=0}^{k_{\max}} \sum_{q=-k}^{k} \sum_{\kappa=|q|}^{\kappa_{\max}} \frac{(2k+1)(2\kappa+1)}{8\pi} s_q^{(\kappa k)} C_{\kappa -q}(\theta, 0) C_{kq}^{*}(\theta_{j'}, \varphi_{j'}), \tag{9.30a}$$

$$s_q^{(\kappa k)} = \int_{-1}^{1} \int_{-1}^{1} \int_{0}^{2\pi} P(\theta, \theta_{j'}, \varphi_{j'}) C_{\kappa -q}(\theta, 0) C_{kq}(\theta_{j'}, \varphi_{j'}) \times \mathrm{d}(\cos\theta)\, \mathrm{d}(\cos\theta_{j'})\, \mathrm{d}\varphi_{j'}, \tag{9.30b}$$

where $k_{\max}$ and $\kappa_{\max}$ play the same role that $n_{\max}$ played in Eq. (9.26a) – they are parameters of the calculation. As we have three angles, we must have three summation indices. As the azimuthal angle of the first spherical harmonic (the azimuthal angle of $\boldsymbol{k}'$ in the *xyz* frame) is zero, this spherical harmonic is real. As the azimuthal angle of $\boldsymbol{j}'$ is defined with respect to the azimuthal angle of $\boldsymbol{k}'$, the components of the two spherical harmonics ($-q$, q) add to zero. This accounts for the fact that an increase of the azimuthal angle of $\boldsymbol{j}'$ by a certain amount is entirely equivalent to a decrease of the azimuthal angle of $\boldsymbol{k}'$ by exactly the same amount.

The analogies with the method of Legendre moments should now be clear. The expansion coefficients are related to the QCT data by

$$s_q^{(\kappa k)} \approx \frac{1}{N_{\mathrm{r}}} \sum_{i=1}^{N_{\mathrm{r}}} C_{\kappa -q}(\theta^{(i)}, 0) C_{kq}(\theta_{j'}^{(i)}, \varphi_{j'}^{(i)}). \tag{9.31}$$

Because of the customary treatment of the scattering angle as a PDF parameter rather than a PDF variable (see Section 9.2.4), Eq. (9.30a) is usually rewritten as*

$$P(\theta, \theta_{j'}, \varphi_{j'}) = \sum_{k=0}^{k_{\max}} \sum_{q=-k}^{k} \frac{2k+1}{4\pi} S_q^{(k)}(\theta) C_{kq}^{*}(\theta_{j'}, \varphi_{j'}), \tag{9.32}$$

where the $S_q^{(k)}(\theta)$ are normalized PDDCSs. They are related to the $s_q^{(\kappa k)}$ coefficients by

$$S_q^{(k)}(\theta) = \sum_{\kappa=|q|}^{\kappa_{\max}} \frac{2\kappa+1}{2} s_q^{(\kappa k)} C_{\kappa -q}(\theta, 0). \tag{9.33}$$

What about other PDFs and their polarization moments? As mentioned at the beginning of Section 9.3, renormalized PDDCSs are obtained from the normalized PDDCSs by division of each of the latter by $S_0^{(0)}(\theta)$. As for polarization parameters, they are obtained through integration of the normalized PDDCS over θ, see Eq. (9.25). Equivalently, the PPs can be directly calculated from the trajectory data. The required formula is

$$a_q^{(k)} = \frac{1}{N_{\mathrm{r}}} \sum_{i=1}^{N_{\mathrm{r}}} C_{kq}(\theta_{j'}^{(i)}, \varphi_{j'}^{(i)}). \tag{9.34}$$

The calculation of reactant polarization moments – PPs and normalized or renormalized PDDCSs – proceeds in an entirely analogous fashion.

*The PDF expansion of Eq. (9.32) is classical and lacks the CG coefficient appearing in its quantum counterpart [Eq. (9.7)]. See ref. [648] for a discussion.

9.3.1.2 Quantum Dynamics

Quantum methods for the description of reaction dynamics have been discussed in Chapter 3. At this point, the important observations are:

- Scattering calculations lead to $\boldsymbol{S}$, the *scattering matrix*. Its elements, $S_{mn} \equiv \langle \Psi_m | S | \Psi_n \rangle$, are the probability amplitudes that collision of reactants in the $|\Psi_n\rangle$ state will lead to formation of products in the $|\Psi_m\rangle$ state.
- $\{|\Psi_n\rangle\}$ is a general asymptotic states basis set. Stereodynamics favours use of the *helicity representation* of the scattering problem.* In this representation, the asymptotic states basis set is $\{|aEJMvj\Omega\rangle|\}$, where a is the channel arrangement label; E is the total energy; J and M are the quantum numbers that determine the magnitude and projection on a space-fixed axis of the total angular momentum vector, $\boldsymbol{J}$; v and j are the vibrational and rotational quantum numbers of the diatom; and Ω is the helicity quantum number, which determines the projection of the diatom's rotational angular momentum (as well as that of the total angular momentum) on the direction of the relative motion.
- Conservation of energy and angular momentum implies that the $\boldsymbol{S}$ elements cannot couple different E, J or M values. Independence of the dynamics with regard to the positions of external observers implies that the $\boldsymbol{S}$ elements cannot depend on M.
- We are only interested in reaction (9.1). The values of E, a, v, j, a', v', j' are all fixed and can be omitted from the notation. We can represent the $\boldsymbol{S}$ elements by $\langle \Omega' | S^J | \Omega \rangle$.
- The *scattering amplitude*, $\langle \Omega' | f(\theta) | \Omega \rangle$, is the probability amplitude, summed over J and M, that collision of reactants with helicity Ω will lead to recoil of products with helicity Ω' along the direction specified by θ. It is related to the $\boldsymbol{S}$ elements by†

$$\langle \Omega' | f(\theta) | \Omega \rangle = \frac{1}{2ik} \sum_{J=0}^{\infty} (2J+1)\, \mathrm{d}^J_{\Omega'\Omega}(\theta) \langle \Omega' | S^J | \Omega \rangle, \tag{9.35}$$

where $k = (2\mu_{\mathrm{A-BC}} E_c)^{1/2}/\hbar$ is the magnitude of the incoming wavevector and $\mathrm{d}^J_{\Omega'\Omega}(\theta)$ is a reduced rotation matrix element.[239,647]

The scattering amplitude is the essential quantity for the calculation of vector properties. As it completely relates the incoming and outgoing states to the scattering angle and to the (differential) reaction probability amplitude, f is the only quantity one needs to define the density matrices of reactants and products.§ Their elements can be written as[646]

$$\langle \Omega_1 | \rho(\theta) | \Omega_2 \rangle = \sum_{\Omega'=-j'}^{j'} \langle \Omega' | f(\theta) | \Omega_1 \rangle^* \langle \Omega' | f(\theta) | \Omega_2 \rangle, \tag{9.36}$$

$$\langle \Omega'_1 | \rho'(\theta) | \Omega'_2 \rangle = \sum_{\Omega=-j}^{j} \langle \Omega'_1 | f(\theta) | \Omega \rangle \langle \Omega'_2 | f(\theta) | \Omega \rangle^*, \tag{9.37}$$

*The reason is that in the helicity representation the state labels are directly related to the vectors we want to correlate: $\boldsymbol{j}$, $\boldsymbol{k}$, $\boldsymbol{k}'$ and $\boldsymbol{j}'$. If the scattering calculations are done in a different representation, one can either transform the $\boldsymbol{S}$ matrix to the helicity representation or else calculate the vector properties in the representation used for the scattering. In general, the first alternative is more convenient.

†See Chapters 3 and 12 for discussions of the scattering matrix, $\boldsymbol{S}$, and the scattering amplitude, $f(\theta)$. Note that $\langle \Omega' | f(\theta) | \Omega \rangle = f_{\Omega'\Omega}(\theta)$ and $\langle \Omega' | S^J | \Omega \rangle = S^J_{\Omega'\Omega}$. Surprising as it may seem, Eqs. (3.44). (9.35), and (12.60) refer the same quantity, only with different representations for the asymptotic states.

§Strictly speaking, we are referring to the *intrinsic* density matrices of reactants and products. See Section 9.4.4.1.

in which case the traces satisfy

$$\mathrm{Tr}[\rho(\theta)] = \mathrm{Tr}[\rho'(\theta)] = \sum_{\Omega'=-j'}^{j'} \sum_{\Omega=-j}^{j} |\langle\Omega'|f(\theta)|\Omega\rangle|^2 = (2j+1)\frac{\mathrm{d}\sigma_{\mathrm{ur}}}{\mathrm{d}\omega}, \tag{9.38}$$

where $\mathrm{d}\sigma_{\mathrm{ur}}/\mathrm{d}\omega$ is the DCS summed over final and averaged over initial helicity (*i.e.* polarization) states. That is, $\mathrm{d}\sigma_{\mathrm{ur}}/\mathrm{d}\omega$ is the DCS that results from reaction of unpolarized reactants.

Note that the matrix elements of Eqs. (9.36) – (9.37) are not specified in the same reference frame; reactant and product helicities are angular momentum projections on $\boldsymbol{k}$ and $\boldsymbol{k}'$, respectively. As a consequence, polarization moments directly calculated from the matrix elements of Eqs. (9.36) and (9.37) will be referred to the xyz and $x'y'z'$ frames on the left and right of Figure 9.2, respectively. While one can carry out stereodynamical analyses using the two frames, rotation of the results for products into the reactant frame (or vice-versa) may be useful. The necessary mathematical expressions for carrying out these transformations are given in the supplementary material.

In general, the density matrices of Eqs. (9.36) – (9.37) involve incoherent superpositions of pure reactant and product states. Disentanglement of the pure states is rather straightforward, but we will not consider these states here.* Instead, we will consider the average PDDCSs of the reaction.

Renormalized PDDCSs are the expectation values of the multipole operators at each scattering angle. To calculate them one can use Eq. (9.16) with $O = T^{\{k\}}_{q\pm}$. In the case of reactants, for example, this leads to

$$\frac{S^{\{k\}}_{q\pm}(\theta)}{S^{\{0\}}_{0}(\theta)} = \frac{\mathrm{Tr}\left[\rho(\theta)T^{\{k\}}_{q\pm}\right]}{\mathrm{Tr}[\rho(\theta)]} = \frac{\sum_{\Omega_1=-j}^{j}\sum_{\Omega_2=-j}^{j}\langle\Omega_1|\rho(\theta)|\Omega_2\rangle\langle\Omega_2|T^{\{k\}}_{q\pm}|\Omega_1\rangle}{(2j+1)\,\mathrm{d}\sigma_{\mathrm{ur}}/\mathrm{d}\omega}. \tag{9.39}$$

The elements of the real multipolar matrices are calculated with Eq. (9.20).

Normalized and renormalized PDDCSs only differ in normalization:

$$S^{\{k\}}_{q\pm}(\theta) = S^{\{0\}}_{0}(\theta) \times \frac{S^{\{k\}}_{q\pm}(\theta)}{S^{\{0\}}_{0}(\theta)} = P_{\mathrm{ur}}(\theta) \times \frac{S^{\{k\}}_{q\pm}(\theta)}{S^{\{0\}}_{0}(\theta)}, \tag{9.40}$$

where $P_{\mathrm{ur}}(\theta)$ is the product angular distribution resulting from reaction of unpolarized reactants†:

$$P_{\mathrm{ur}}(\theta) = \frac{2\pi}{\sigma_{\mathrm{ur}}}\frac{\mathrm{d}\sigma_{\mathrm{ur}}}{\mathrm{d}\omega}, \qquad \sigma_{\mathrm{ur}} = \int_0^{2\pi}\int_0^{\pi}\frac{\mathrm{d}\sigma_{\mathrm{ur}}}{\mathrm{d}\omega}\sin\theta\,\mathrm{d}\theta = \frac{\pi}{k^2(2j+1)}\sum_{J,\Omega,\Omega'}(2J+1)\left|S^{J}_{\Omega'\Omega}\right|^2. \tag{9.41}$$

Polarization parameters – the integral expectation values of the multipole operators – are obtained by integration of the normalized PDDCSs over θ, see Eq. (9.25).

9.3.2 Experimental Production of Polarized Reactants

Production of polarized molecules requires an interaction with external electric or magnetic fields. These fields can be time-dependent or static. The obvious instance of the former is absorption of

*Interested readers should consult ref. [646]. We will make some use of pure reactant and product states in Section 9.4.

†Eq. (9.41) differs from those in the other chapters in more than just notation because of the use here of the helicity representation.

electromagnetic radiation to produce rotationally, rovibrationally or rovibronically excited states, which in most cases leads to angular momentum (and thus internuclear axis) polarization. Alternatively, static fields can interact with molecules *via* their permanent or induced dipole moment.

We start with the preparation of molecules *via* optical pumping (that is, molecular excitation to well defined energy levels using laser techniques), and then consider the interaction of molecules with static electric fields.

9.3.2.1 Optical Methods

It has long been known that when molecules absorb polarized light their rotational angular momentum and internuclear axis become polarized. The advent of lasers, which deliver monochromatic, highly directional, polarized and intense light, has turned light absorption into a particularly convenient approach for molecular polarization.

Let us first consider the absorption of linearly polarized light by closed-shell (that is, Σ-state) diatomics in a well defined v'', j'' rovibrational state. For any given $|jm\rangle$ state, the probability density of finding $\boldsymbol{r}$ pointing along the direction specified by the Θ_r and Φ_r is given by[239]

$$P_{jm}(\Theta_r, \Phi_r) = \left|Y_{jm}(\Theta_r, \Phi_r)\right|^2. \tag{9.42}$$

On the other hand, the probability that the diatomic will indeed be found in the $|jm\rangle$ state after dipole absorption of a quantum of linearly polarized radiation ($j_{\text{ph}} = 1$, $m_{\text{ph}} = 0$) is proportional to the square of the $\langle j''m'', 10|jm\rangle$ CG coefficient. If the diatomic is initially unpolarized, it follows that the probability of finding $\boldsymbol{r}$ along the direction defined by Θ_r and Φ_r is given by the product of these two terms, averaged over the equally populated m'' states.

$$\begin{aligned} P(\Theta_r, \Phi_r) &= \frac{1}{2j''+1} \sum_{m''=-j''}^{j''} \frac{2j+1}{4\pi} \left| \sum_{m=-j}^{j} \langle j''m'', 10|jm\rangle C_{jm}(\Theta_r, \Phi_r) \right|^2 \\ &= \frac{2j+1}{4\pi(2j''+1)} \sum_{m=-j''}^{j''} \langle j''m, 10|jm\rangle^2 \left|C_{jm}(\Theta_r, \Phi_r)\right|^2, \end{aligned} \tag{9.43}$$

where we have used $\langle j''m'',10|jm\rangle \propto \delta_{m''m}$. Using well-known formulae of angular momentum algebra,[647] one can rewrite Eq. (9.43) as[239,657]

$$P(\Theta_r, \Phi_r) = \frac{1}{4\pi}[1 + \mathcal{A}_2(j)\mathcal{P}_2(\cos\Theta_r)], \tag{9.44}$$

where $\mathcal{A}_2(j)$ is a molecular axis alignment parameter that depends on the branch (R or P) of the rotational excitation and also on j''. We see that using linearly polarized light one can align but not orient the molecular axis. We also note that, because the transition amplitude contains a CG coefficient, $\langle j''0, 10|j0\rangle$, that vanishes unless $j''+j$ is odd, closed-shell diatomics cannot undergo Q-branch, $j=j''$ transitions. The $\mathcal{A}_2$ parameter is related to the rotational polarization moment $a_0^{(2)}$:

$$\mathcal{A}_2(j) = 5\langle j0, 20|j0\rangle a_0^{(2)}(j), \tag{9.45}$$

where the factor of five arises from $2k+1=5$.

If the light polarization is circular rather than linear, Eq. (9.43) becomes

$$P(\Theta_r, \Phi_r) = \frac{2j+1}{4\pi(2j''+1)} \sum_{m''=-j''}^{j''} \sum_{m=-j}^{j} \langle j''m'', 1Q|jm\rangle^2 |C_{jm}(\Theta_r, \Phi_r)|^2 \tag{9.46}$$

with $Q = \pm 1$ for right- or left-handed polarization. As in the case of linearly polarized light, it can be shown that: (i) following absorption of a single quantum of light, the internuclear axis can be aligned but not oriented (only the angular momentum can be oriented by absorption of circularly polarized light); and (ii) Q-branch transitions are not allowed. Note also that here the quantization axis is chosen to be in the direction of *propagation* of the radiation, rather than along the polarization axis.

Let us now turn to symmetric tops.[239,657] Their states are characterized by quantum numbers specifying the angular momentum magnitude and its projections on the molecule-fixed figure axis and on a space-fixed axis (respectively, j, k and m).* In terms of the Euler angles specifying the direction of the molecule-fixed reference frame in the space-fixed system of axes (Φ_r, Θ_r, and χ_r), the wavefunction of the $|jkm\rangle$ state is given by

$$\Psi_{jkm}(\Phi_r, \Theta_r, \chi_r) = \sqrt{\frac{2j+1}{4\pi}} D^{j*}_{mk}(\Phi_r, \theta_r, \chi_r), \tag{9.47}$$

where the D^{j*}_{mk} are complex conjugates of rotation matrix elements.[647] Following absorption of one photon of light, the molecular axes are distributed according to[657]

$$\begin{aligned} P(\Theta_r, \Phi_r) = N \sum_{m''=-j''}^{j''} \sum_{m=-j}^{j} \langle j''m'', 1Q|jm\rangle^2 \langle j''k'', 1\Delta k|jk\rangle^2 \\ \times \left[|D^j_{mk}(\Phi_r, \Theta_r, \chi_r)|^2 + |D^j_{m-k}(\Phi_r, \Theta_r, \chi_r)|^2 \right], \end{aligned} \tag{9.48}$$

where N is a normalization constant, and $Q = 0$, or ± 1 for linearly or circularly polarized light, respectively. For parallel bands (when the transition dipole moment lies along the direction of the molecular axis), the selection rules imply: $\Delta j = 0, \pm 1$, $\Delta k = 0$ if $k \neq 0$; and $\Delta j = \pm 1$, $\Delta k = 0$ if $k = 0$. As shown in ref. [657], the above expression reduces to one of the same form as that for a linear top, Eq. (9.44), but with different expressions for the alignment parameters, $\mathcal{A}_2(j, k)$, which now have a dependence on the k quantum number. As in the case of linear tops, the molecular axes can be aligned but not oriented; this is irrespective of whether the light polarization is linear or circular.

Studies of reactions of molecules polarized by optical pumping are relatively scarce. The major hurdle is the fact that laser excitation does not take all molecules to the upper, polarized state. If ground-state species also react, it can be difficult to decide whether the detected products were formed by reactions involving ground-state (and unpolarized) or excited-state (and polarized) reactants.† For this reason, the best candidates for this sort of study are molecules whose ground-state reactivity is negligible in comparison to (or at least substantially smaller than) the reactivity of the excited state.

A second problem is coupling of the rotational angular momentum of the molecule to the nuclear spin ($\boldsymbol{I}$) of its atoms or to external fields. This leads to depolarization[410,643] of the molecule, which can occur on a timescale short enough to be deleterious. Coupling to nuclear spin (hyperfine depolarization), in particular, has been demonstrated to severely reduce the degree of polarization of low rotational states

*The k used here should not be confused with the ones used elsewhere in this chapter to represent the reactants-approach direction or the rank of a polarization moment.

†It is possible, however, to collect data on a shot-to-shot basis, with the polarizing laser alternately on or off; subtraction of the "off" from the "on" data then allows for isolation of reactant polarization effects.

of diatomic molecules such as HF and OH, for which the depolarization timescale is on the order of microseconds.[410,658] To overcome this problem, the optical excitation can be carried out in a homogeneous electric field at the interaction region; the field must be strong enough for the Stark splitting of the *j* manifold to be large in comparison to the hyperfine splitting. This leads to the decoupling of ***j*** and ***I***, which in turn allows the molecular polarization to persist for as long as the field is present.

EXAMPLE

Loesch and coworkers have studied the effect of HF alignment on the dynamics of the Li + HF(v = 1, j = 1) → LiF + H reaction.[659,660] Using light polarized linearly and parallel to the Stark field, they tuned their laser to the $v' = 1 \leftarrow v'' = 0$, $R(0)$, $\Delta m = 0$ transition, and thus produced polarized HF molecules in the $|vjm\rangle = |110\rangle$ state; the resulting molecular alignment moment was $\mathcal{A}_2 = 2$, corresponding to maximum negative ***j*** alignment and maximum positive ***r*** alignment ($a_0^{(2)} = \langle 10, 20|10\rangle = -\sqrt{2/5}$, see Tables 9.3 and 9.4) at a chosen angle with respect to the relative velocity of the crossing Li and HF beams. Variation of the alignment direction was made possible by use of a quadrupolar guiding field.

The measurement was facilitated by the fact that the Li + HF reaction complies with the reactivity requirement mentioned above; HF(v = 1) is about ten times more reactive than HF(v = 0). Product laboratory angular distributions were obtained for unpolarized HF(v = 1, j = 1) and with ***r*** alignment parallel, perpendicular, or at the magic angle with respect to the reactant-approach direction.* Pronounced steric effects were found, with the product quantum state populations and angular distributions varying significantly with alignment of the HF bond axis.[660] A QCT simulation of the experiment was carried out,[661] and the agreement between theory and experiment was excellent. The main conclusions were that (i) differential and state-resolved steric effects were substantial, and (ii) selection of the HF alignment direction allowed for very considerable control of the populations of the various LiF product states.

Molecules can also be polarized by two-photon absorption (or emission). In particular, stimulated Raman pumping can be used to produce vibrationally and/or rotationally excited homonuclear diatomics with a well defined ***j*** polarization; Zare and coworkers have used this technique.[662] Through pumping of the $S(0)$ transition, they were able to produce HD in the $|vjm\rangle = |120\rangle$ state (the quantization axis here is the laser polarization vector, ***E***; when HD is in this state, ***j*** and ***r*** are respectively aligned perpendicular and parallel to ***E***). These polarized molecules were then used to study steric effects in the Cl + HD(v = 1) reaction. Different collision geometries were obtained by specifying different directions for the polarization vectors of the 'pump' and 'Stokes' radiation employed in the stimulated Raman pumping scheme (either both parallel or both perpendicular to the axis of the time-of-flight detector used in the experiment). Again, the results of QCT calculations were found to be in very good agreement with the experimental data. The reaction was found to favour collision geometries with $\boldsymbol{j} \perp \boldsymbol{k}$ and $\boldsymbol{r} \parallel \boldsymbol{k}$.

9.3.2.2 *State Selection and Orientation by Hexapoles*

The orientation of symmetric tops by means of electric field focusing was pioneered in the mid-sixties, when the groups of Bernstein and Brooks and coworkers combined this approach with molecular beam techniques.[663–666] At the time this was an audacious endeavour aimed at disentangling steric effects in chemical reactions at the molecular level. Subsequently the method was enhanced and

*If β denotes the polar angle between the external field and the relative velocity vector, then the magic angle satisfies $\mathcal{P}_2(\cos\beta_{mag}) = 0$, which implies $\cos\beta_{mag} = \sqrt{1/3}$, and therefore $\beta_{mag} \approx 54.74°$. When ***r*** is aligned along this direction, its $k = 2$, $q = 0$ alignment moment with respect to ***k*** vanishes. See ref. [642] for more details.

STUDY BOX 9.1: THE DETECTION OF ANGULAR MOMENTUM ORIENTATION AND ALIGNMENT

The use of polarized light to probe the orientation and alignment of angular momentum has already been mentioned in a number of chapters (see Chapters 5 and 8), and the preceding discussion provides an ideal opportunity to explain the principles of the techniques employed. We start with the expression (9.44) for the bond axis distribution following one photon absorption, generalized to case of a symmetric top molecule[239,657]

$$P(\Theta_r, \Phi_r) = \frac{1}{4\pi}[1 + \mathcal{A}_2(j,k)\mathcal{P}_2(\cos\Theta_r)]. \quad \text{(B9.1.1)}$$

The angle Θ_r in this expression defines the angle that $\boldsymbol{r}$ makes to the LAB Z-axis, which is here defined as the axis of linear polarization of the absorbed radiation. In the case of a parallel transition, Eq. (9.48) can be recast in the above form with the following definitions for the alignment parameter, $\mathcal{A}(j, k)$, with excitation on P, Q and R branches, respectively,[657]

$$\mathcal{A}_2^{\mathrm{P}}(j,k) = \frac{j(j-1) - 3k^2}{j(2j+1)} \quad \text{for } \Delta j = -1, \quad \text{(B9.1.2)}$$

$$\mathcal{A}_2^{\mathrm{Q}}(j,k) = -\frac{j(j+1) - 3k^2}{j(j+1)} \quad \text{for } \Delta j = 0, \quad \text{(B9.1.3)}$$

and

$$\mathcal{A}_2^{\mathrm{R}}(j,k) = \frac{(j+1)(j+2) - 3k^2}{(j+1)(2j+1)} \quad \text{for } \Delta j = +1, \quad \text{(B9.1.4)}$$

where j and k label the initial state in the transition (*i.e.* for the purposes of this study box we have dropped the double prime notation on the j and k quantum numbers). Similar expressions hold in the case of a perpendicular transition.

Assuming $j \gg k$, for a parallel transition in the high j limit the above equations reveal that the alignment of the bond axis tends to $+\frac{1}{2}$ for the P and R transitions, and -1 for Q branch excitation. The same result is obtained in the high j limit with $j \gg k$ for a perpendicular transition. These results suggest the following interpretation, as illustrated in Figure 9.5. For a P or R branch transition in the high j limit, the transition dipole moment $\boldsymbol{\mu}$ lies perpendicular to $\boldsymbol{j}$ (either parallel or perpendicular to the bond axis). Assuming that the electric vector of the linearly polarized light is positioned vertically, in the plane of the page, only those molecules whose transition dipole moments are aligned as shown in the figure can absorb the radiation, due to the dependence of the absorption probability on $|\boldsymbol{\varepsilon} \cdot \boldsymbol{\mu}|^2$, where $\boldsymbol{\varepsilon}$ is the electric vector of the absorbed radiation (see Study Boxes 8.2 and 8.3). As the molecule rotates, the bond axis will always lie perpendicular to $\boldsymbol{j}$, and therefore the distribution of $\boldsymbol{r}$ will be smeared out in space, leading to the results shown in the lower panels in Figure 9.5, which are in accord with the Eqs. (B9.1.2) to (B9.1.4).

The upper panels of this figure illustrate the correlation that exists between the alignment of the transition dipole moment at the instant of excitation and the rotational angular momentum vector, and its dependence on rotational branch. Imagine that $\boldsymbol{j}$ is polarized as shown in the top right panel of Figure 9.5. Plane polarized light aligned vertically in the plane of the page will be able to excite these molecules efficiently if Q branch excitation is employed, but rather weakly if P or R branch excitation is employed. If the polarization of the light was rotated through 90°, so that it was lying out of the plane of the page, the molecule would absorb more strongly on a P or R branch transition than on a Q branch transition. Thus, the polarization of the angular momentum is revealed through a sensitivity of the absorption intensity to probe radiation polarization, or probe transition.

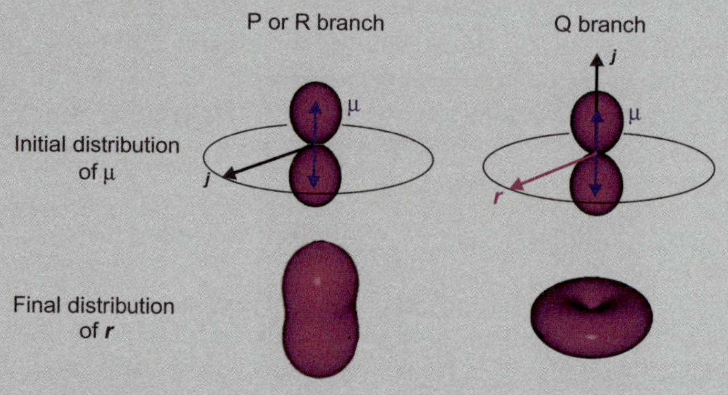

Figure 9.5 Top panels: Schematic of the angular distributions of transition dipole moments excited by linearly polarized light at the moment of excitation, subsequent to excitation on a P or R transition (left) or a Q branch transition (right). The electric vector of the linear polarized light is assumed to be vertical, in the plane of the page. In the left diagram note that $\boldsymbol{j}$ can lie anywhere on the light black circle, similarly for $\boldsymbol{r}$ in the figure on the right. Bottom panels: The final distributions of bond axes, $\boldsymbol{r}$, obtained after averaging over the rotation of the molecule about $\boldsymbol{j}$.

The above arguments are qualitative and more quantitative treatments can be found in refs. [239,409], as well as in Chapters 5 and 8 of the present text. For example, the expression presented in Chapter 5 for the absorption intensity of plane polarized light, given in Eq. (5.59), can be understood qualitatively using the above arguments, provided care is taken to allow for the different reference frames used here and in the previous chapter. In spite of their qualitative nature, the above arguments are quite general and can be readily extended to the probing of products using laser-induced fluorescence or resonantly enhanced multiphoton ionization.

Mark Brouard

improved considerably. In particular, it has been shown to be capable not only of orienting molecules, but also of sometimes achieving a perfect selection of initial molecular states.[667–669]

The technique includes three well defined stages as described in more detail in the following pages: (a) state selection and focusing of reactants by an electrostatic hexapole field; (b) reactant orientation in the collision or interaction zone; and (c) scattering-angle-dependent detection of the collision (or photodissociation) products.

The basis of hexapole state selection is the Stark effect – the shifting of the energy levels of a molecule when it is placed in an electric field. If the field strength and direction are given by the modulus and direction of the electric vector $\mathcal{E}$, the system Hamiltonian is

$$\hat{H} = \hat{H}_0 - \mu_e \mathcal{E} \cos \Theta_\mu, \tag{9.49}$$

where $\hat{H}_0$ is the zero-field molecular Hamiltonian, μ_e is the modulus of the molecular dipole moment, $\boldsymbol{\mu}_e$, and Θ_μ is the angle between $\mathcal{E}$ and $\boldsymbol{\mu}_e$. If the $\boldsymbol{r}$ and $\boldsymbol{\mu}_e$ directions coincide,* then $\Theta_\mu = \Theta_r$.

*Note that here $\boldsymbol{r}$ is the direction of the figure axis of the molecule for symmetric tops (*e.g.*, CH_3F, CH_3Cl, etc.), the figure axis need not be an internuclear axis. In what follows, we will assume that the $\boldsymbol{r}$ and $\boldsymbol{\mu}_e$ directions do coincide. As this convention is not universally adopted, care must be exercised. For instance, the NO dipole moment points from N to O, but its $\boldsymbol{r}$ axis can be defined as pointing from O to N.

If the external field is weak with respect to the molecular potential, the energy can be calculated by perturbation theory, with $\hat{V} = -\mu_e \mathcal{E} \cos \Theta_r$ treated as a perturbation of the zero field Hamiltonian, $\hat{H}_0$.

Let us first assume that the symmetric top is in the $|jkm\rangle$ state. In the coordinate representation, its wavefunction is given by Eq. (9.47). At first order, the perturbation theory correction for the molecular energy is

$$\begin{aligned} W^{(1)} &= -\mu_e \mathcal{E} \langle jmk | D^1_{00}(0, \Theta_r, 0) | jmk \rangle \\ &= -\mu_e \mathcal{E} \langle jmk | \cos \Theta_r | jmk \rangle \approx -\mu_e \mathcal{E} \frac{mk}{j(j+1)}, \end{aligned} \tag{9.50}$$

where we have used $D^1_{00}(0, \Theta_r, 0) = \cos \Theta_r$. The expectation value of $\cos \Theta_r$ is

$$\langle \cos \Theta_r \rangle \approx \frac{mk}{j(j+1)}, \tag{9.51}$$

and vanishes when $k = 0$, which is always the case for closed-shell diatomic molecules; this means that, up to this correction term, these molecules cannot be oriented by a hexapole field.

The second-order perturbation correction to the energy levels of a symmetric top can be written as[239]

$$W^{(2)} = (\mu_e \mathcal{E})^2 \sum_{j' \neq j} \frac{|\langle jkm | D^1_{00}(0, \Theta_r, 0) | j'k'm' \rangle|^2}{E^{(0)}_{j,k} - E^{(0)}_{j',k'}} \delta_{k,k'} \delta_{m,m'}. \tag{9.52}$$

For linear molecules in a Σ state, one has $k = 0$ and zero-order energies given by $E^{(0)}_j = Bj(j+1)$, where B is the rotational constant. With $\Delta m = 0$ and $j' = j + 1$ this results in

$$W^{(2)} = \frac{(\mu_e \mathcal{E})^2}{Bj(j+1)} \frac{j(j+1) - 3m^2}{2(2j-1)(2j+3)}, \tag{9.53}$$

which implies that closed-shell linear molecules can in principle be focussed and oriented *via* their second-order Stark effect. In practice, however, one usually has $\mu_e \mathcal{E} \ll B$, which means that second-order effects are negligible.

A hexapole state selector consists of a hexagonal arrangement of six rods with alternating positive and negative electric charges. The length of the rods is much larger than the distance from the rods to the hexapole axis. In cylindrical coordinates, the expression for the potential inside the hexapole reads

$$V(R, \varphi) = V_0 \left(\frac{R}{R_0} \right)^3 \cos(3\varphi), \tag{9.54}$$

where V_0 is the potential applied to the rods, φ the azimuthal angle about the hexapole axis, and R_0 the inner radius of the hexapole. Therefore, inside the hexapole the magnitude of the electric field is

$$\mathcal{E} = 3 V_0 R \frac{R^2}{R_0^3}, \tag{9.55}$$

and the molecules feel a force that to first order is given by

$$F_{\text{hex}} = -\frac{\partial W^{(1)}}{\partial R} = \mu_e \langle \cos \Theta_r \rangle \frac{\partial \mathcal{E}}{\partial R} = \mu_e \langle \cos \Theta_r \rangle \frac{6 V_0 R}{R_0^3}. \tag{9.56}$$

As the force is linear in R, molecular trajectories in the field are governed by an equation for harmonic motion. The hexapole acts as a lens, focussing molecules with negative $\langle \cos \Theta_r \rangle$ values to the axis of the nodes of a sinusoidal trajectory (these are molecules in *low field-seeking states* – states whose energies decrease with decreasing field strength). Molecules with positive $\langle \cos \Theta_r \rangle$ values (these are molecules in *high field-seeking states* – states whose energies decrease with increasing field strength) are ejected radially out of the field, following exponential trajectories towards the hexapole rods.

Under ideal conditions, the potential applied to the rods in order to focus a particular $|jkm\rangle$ state is related to the expectation value of cos Θ_r by

$$V_0 = \frac{\pi^2 R_0^3}{6L} \frac{M v^2}{\mu_e} \frac{1}{\langle \cos \Theta_r \rangle} = \frac{\pi^2 R_0^3}{6L} \frac{M v^2}{\mu_e} \frac{j(j+1)}{mk}, \tag{9.57}$$

where L is the hexapole length, and M and v are respectively the mass and speed of the molecule. The minimum (or threshold) rod potential at which the molecules can be focussed is given by Eq. (9.57) with $\langle \cos \Theta_r \rangle = -1$. Typical lengths of the hexapole focussing system are 1 m to 2.5 m, and the distance from the molecular beam source to the scattering centre may well be on the order of 3 m; sometimes two or three consecutive hexapoles are employed. Inner hexapole diameters are typically of 7–10 mm, and the diameter of each rod has a similar value. The operating voltages are between 6 and 30 kV, which results in electric fields with strengths larger than 10 kV cm^{-1}. Under these conditions, pulsed-beam hexapole instruments are capable of focussing some 10^{13} molecules cm^{-3}, resulting in a 20–70-fold increase in beam flux due to focussing.

Most instruments of this type use pulsed molecular beams (with gas mixtures of the molecule to be oriented 'seeded' with a light carrier gas, usually He) generated by expanding the gas through a pulsed valve, and pulsed-laser ionization time-of-flight or imaging detection methods. However, some early experiments on closed shell molecules, such as the methyl halides, were performed using quadrupole mass spectrometry detection. In order to achieve population of a single $|jkm\rangle$ state, the beam velocity spread must be small, and its rotational temperature low enough for very few states to be appreciably populated; rotational temperatures of 5–10 K are easily achievable for polyatomic molecules. By varying the fraction of the light carrier gas and/or the source temperature, the velocity of the beam can be selected over a wide range of values. Discharge sources can be used to produce radicals, such as OH, which in its ($^2\Pi$) ground electronic state behaves as symmetric top (with $k \equiv \Lambda$ if the electron spin is neglected, or else with $k \equiv \Omega$ if it is taken into account).[670]

As discussed in Chapter 5, Stolte and coworkers have recently achieved full state selection for NO molecules in the ($^2\Pi$) ground electronic state.[407] The Λ-doublet splitting of the two components of the $j = 1/2$, $\Omega = 1/2$ ground state is so minute (0.01180 cm^{-1}) that, even at the lowest temperatures achievable in a supersonic expansion, the populations in the beam of the two components are equal. However, the two components differ in parity. The lower, positive-parity state turns out to be high-field seeking, and is therefore ejected from the hexapole axis, while the upper, negative-parity state is low-field seeking state and can be focussed. As a result, a molecular beam consisting only of molecules in the $j = 1/2$, $\Omega = 1/2$, negative-parity state could be obtained. Stolte *et al.* took advantage of this to perform the first experiment on fully state-to-state NO collisions.[407] As we have already seen in Chapter 5, they have measured inelastic DCSs for He + NO and Ar + NO collisions in which translational energy was transferred into NO rotation.

Once the molecules have been focussed and state-selected by the hexapole, the orientation is carried out using a guiding field, usually a set of bent rods or parallel field plates, adjacent to the hexapole and overlapping with the scattering centre assembly. Inside the hexapole, the molecules are oriented with respect to the local electric field. However, because the field inside the hexapole is inhomogeneous, the molecules are not oriented in the space fixed (LAB) frame. If the molecules are allowed to pass adiabatically into a region of a uniform field, they will stay in the same quantum state (*i.e.* have the same projection quantum on the new field) and will, therefore, now be oriented in space. The guiding field therefore plays two roles: on the one hand, it preserves the axis of quantization, thus avoiding flips of the selected quantum numbers; on the other, it creates a definite orientation along the reactant-approach direction in the scattering centre. Of course, the direction of the uniform field (this is the axis along which m is defined) can be set either parallel or anti-parallel to $\boldsymbol{k}$.

Once molecular states have been selected and oriented, the $\boldsymbol{r}$ distribution is given by[668,669]

$$P_{jkm}(\cos\Theta_r) = \frac{2j+1}{2}\sum_{n=0}^{2j} c_n(jkm)\mathcal{P}_n(\cos\Theta_r) = \sum_{n=0}^{2j}\left(\frac{2n+1}{2}\langle\mathcal{P}_n\rangle\right)\mathcal{P}_n(\cos\Theta_r). \tag{9.58}$$

The values of the $c_n(jkm)$ coefficients or $\langle\mathcal{P}_n\rangle$ Legendre moments can be readily determined. For low $|jkm\rangle$ states, the actual $\boldsymbol{r}$ distribution differs from the classical one: these states precess about a large average angle with a large range of angles.

Experiments with hexapole focussed and oriented molecules have included photodissociation of oriented molecules, inelastic and reactive scattering in crossed molecular beam arrangements, steric asymmetry in electron impact ionization, and the interaction of oriented molecules with surfaces.[668,669] Except for some particular cases, the molecules were symmetric tops: polyatomics (CH_3I, CF_3I, CF_3Br, t-butyl iodide), open shell diatomics (NO, OH), and linear triatomics in excited bending states (N_2O, OCS).

The first-generation experiments, performed in the late sixties and early seventies, were aimed at the determination of steric effects in reactions of alkali metals with oriented methyl iodide. More refined and quantitative experiments came in the eighties and involved prototypical reactions such as $CH_3I + Rb$ and $CF_3I + K$. In the case of the former, the angular distribution of the RbI product was found to be strongly correlated to the CH_3I orientation; the most significant finding was the experimental confirmation of a substantial "cone of non-reaction".

The second generation of experiments included studies of the $NO + O_3 \rightarrow NO_2^* + O_2$ reaction, in which the NO_2^* chemiluminescence was measured as a function of the NO($j = 3/2$, $k = 3/2$, $m = 3/2$) orientation, and also of the $Ba + N_2O$ reaction with N_2O having a quantum of bending vibration. Experiments have also been carried out to explore the dependence of electron impact ionization cross section on molecular orientation. These experiments were performed on oriented molecular beams of a series of symmetric top species, including CH_3Cl and CCl_3H,[671] and revealed that a distinct preference for ionization exists when the electron approaches the positive end of the molecular dipole. In addition to the experiments already mentioned, there has been a considerable number of applications of the hexapole technique to investigate the steric effect in collisions of oriented molecules with surfaces[672,673] (see Chapter 10). Hexapole state selection techniques have also been recently applied to study the photodissociation of oriented OCS molecules.[674]

The most recent collisional experiments have been directed towards steric effects in inelastic scattering of OH or NO molecules by rare gases and other collision partners,[407,670,675] as discussed in Chapter 5.

9.3.2.3 *Brute Force Orientation: Pendular States*

Hexapole molecular orientation and state selection are restricted to symmetric tops with a permanent dipole moment. In the early 90s, another non-optical method for "heads *versus* tails" molecular orientation was proposed and developed by several groups. It works not only for symmetric tops, but also for Σ-state molecules; all it requires is a permanent molecular electric dipole moment ($\boldsymbol{\mu}_e$) and a strong homogeneous electric field ($\mathcal{E}$) at the reaction volume.

For many years, orientation of polar molecules other than symmetric tops by an electric field was considered impractical. This was because, to first order, the rotational motion of diatomic, linear, or asymmetric top molecules averages out their dipole moment. As a result, they interact with electric fields much more weakly than symmetric tops. For even a slight orientation, extremely high electric fields ("brute force") were deemed necessary. The key parameter is

$$f = \frac{\boldsymbol{\mu}_e \cdot \mathcal{E}}{E_{rot}}, \tag{9.59}$$

where E_{rot} is the rotational energy. Substantial dipole orientation requires $f \geqslant 1$, whereas for molecules at room rotational temperature the f value is typically on the order of 10^{-2} or 10^{-3}. However, the rotational cooling that occurs in supersonic molecular beams changes this situation drastically. Rotational temperatures lower than 5 K can be achieved easily, and the consequence is that one can obtain $f \gg 1$ for many molecules with relatively modest electric fields.

When a polar diatomic enters an electric field, the molecule-field interaction leads to a rearrangement of the $\boldsymbol{r}$ distribution. To understand this, let us first think (quasi)classically and assume that the molecule is in the $j=0$ state. There is no rotational motion, and the internuclear axis can be pointing anywhere. When subject to the electric field, the molecule experiences a torque that results from the interaction of the field with the molecular dipole moment. The molecule starts rotating so that its axis, whose direction coincides with that of $\boldsymbol{\mu}_e$, becomes oriented relative to the field direction. However, because the molecule has acquired (rotational) kinetic energy, its motion does not stop once $\boldsymbol{r}$ becomes antiparallel to $\mathcal{E}$. Instead, the molecule carries on rotating until its kinetic energy is entirely converted again into potential energy. The turning point having been reached, $\boldsymbol{r}$ starts swinging back. The potential is again minimized, and after that another turning point reached, and so on – the motion is that of a spherical pendulum, which can also be seen as a hindered rotation around the $\mathcal{E}$ direction, and such a molecule is said to occupy a *pendular state*. The smaller the swing amplitude, the larger the orientation of $\boldsymbol{\mu}_e$ relative to $\mathcal{E}$. Note also that the initial rotational energy contributes to the swing amplitude: if the molecule is initially rotating, the swing amplitude is larger, and the orientation effect diminishes as the initial j increases. A little calculation shows that this effect is actually very significant. The extent of orientation strongly depends on the initial j value – hence the need for a very low rotational temperature. This implies that only very low rotational states must be populated if this method of orientation is to be successful.

It is not too hard to work out the quantum equations appropriate for linear or symmetric top molecules with permanent dipole moments. For simplicity, let us consider a linear molecule in a Σ vibronic state, and assume that its rotational state in the field-free region is $|jm\rangle$. Let us also assume that the transformation from the field-free to the pendular state is adiabatic. As rotational periods are typically on the order of picoseconds, whereas the time for passage through the fringes of the electric field is typically three orders of magnitudes larger, this is in general a good approximation. For similar reasons, the nuclear spin-molecular rotation coupling can be considered negligible (the hyperfine interaction is small compared to the field-dipole interaction).

The Hamiltonian for a (rigid) dipolar linear molecule in an electric field can be written as

$$\hat{H} = B\left[\frac{\hat{j}^2}{\hbar^2} - w \cos \Theta_\mu\right], \tag{9.60}$$

where B is the rotational constant, w is a dimensionless coupling constant,

$$w = \frac{\mu_e \mathcal{E}}{B}, \tag{9.61}$$

and Θ_μ is the angle between $\boldsymbol{\mu}_e$ and $\mathcal{E}$; if $\boldsymbol{r}$ is defined so as to coincide with $\boldsymbol{\mu}_e$, then $\Theta_\mu = \Theta_r$.

The Hamiltonian of Eq. (9.60) commutes with the $\hat{j}_z$ operator, but not with $\hat{j}^2$. The consequence is that, in the presence of the field, j is not a good quantum number, but m is. The eigenfunctions of the Hamiltonian are

$$\hat{H}|m,n\rangle = W_n(m;w)|m,n\rangle, \tag{9.62}$$

where n is an index whose meaning will soon be discussed, and the eigenvalues W_n depend parametrically on w. We may expand these eigenfunctions in a series of free linear rotor wavefunctions,

$$|m,n\rangle = \sum_{j''=|m|}^{\infty} c_n(j'',m)|j''m\rangle. \tag{9.63}$$

Inserting this series into Eq. (9.62), and then multiplying by $\langle jm|$, an infinite system of linear equations is obtained:

$$c_n(j,m)\left[j(j+1) - \frac{W_n}{B}\right] = w \sum_{j''=|m|}^{\infty} c_n(j'',m)\langle jm|\cos\Theta_\mu|j''m\rangle, \tag{9.64}$$

with $j = |m|, |m|+1, \ldots, \infty$. The only non-vanishing terms within the summation on the right hand side of this equation are those with $j'' = j \pm 1$, and one has

$$\langle jm|\cos\,\Theta_\mu|(j+1)\,m\rangle = \frac{(j+1)}{[(2j+1)\,(2j+3)]^{1/2}}, \tag{9.65a}$$

$$\langle jm|\cos\,\Theta_\mu|(j-1)\,m\rangle = \frac{j}{[(2j-1)\,(2j+1)]^{1/2}}. \tag{9.65b}$$

From Eq. 9.62, the energy of a closed-shell diatomic in an electric field characterized by w is given by the eigenvalue $W_n(m;\, w)$. The n index labels the various solutions of Eq. (9.64) for a given m value and determines the free state $|jm\rangle$ correlating with $|mn\rangle$, where $j = |m| + n - 1$. The infinite system of equations of the type appearing in Eq. (9.64) can be truncated at a finite N value; this value of N corresponds to the number of j levels in a *finite* basis set of free rotor states. Once this is done, the solutions can be found numerically.

Note that Eq. (9.63) implies that the eigenstates of the Hamiltonian Eq. (9.60) are a linear superposition of free rotor states with different j but the same m (for a symmetric top, also with the same k value). These states are the *pendular states* referred to above.[643,676,677] Their internuclear axis distributions are given by*

$$P(\cos\,\Theta_r) = 2\pi|\langle\Theta_r\Phi_r|mn\rangle|^2. \tag{9.66}$$

*The notation $\langle\Theta_r\Phi_r|mn\rangle$ indicates that the state $|mn\rangle$ is written in the coordinate representation given by the angles Θ_r and Φ_r. This notation is equivalent to $\psi_{mn}(\Theta_r,\, \Phi_r)$. For the more familiar case of the wavefunction of the hydrogen atom $\psi_{nlm_l}(r,\, \theta,\, \varphi) \equiv \langle r\theta\varphi|nlm_l\rangle$.

By hybridization of states with different j, the molecular axis can be not only aligned, as in the case of a superposition of $|jm\rangle$ states with the same j and different m values, but also oriented, even for a diatomic molecule. Of course, the extent of orientation strongly depends on the initial rotational states and on the value of the coupling constant w. The larger the value of w, the easier it is to use "brute force" to orient the molecules. Orientation of high rotational states is impractical due to the extremely high electric fields required. Loesch[678] and Friedrich, Herschbach and coworkers[676,677] have shown that the angular distribution of the internuclear axis becomes nearly isotropic for $|mn\rangle$ states correlating with high j values. Consequently, the rotational temperature of the molecular beam must be as low as possible – ideally, one would like to have an initial expansion containing only $j=0$ molecules. It can be shown[679] that the rotational temperature of the molecular beam and the angular distribution of the internuclear axis are approximately related by

$$P(\cos\Theta_r) = \tfrac{1}{2}[1 + \mathcal{A}(T_{\mathrm{rot}}, \mathcal{E})\cos\Theta_r], \tag{9.67}$$

as long as $\mathcal{A}(T_{\mathrm{rot}}, \mathcal{E}) \ll 1$. This parameter is given by

$$\mathcal{A}(T_{\mathrm{rot}}, \mathcal{E}) \approx a\mathcal{E}/T_{\mathrm{rot}}, \tag{9.68}$$

where a is a constant. For CH_3I, for example, with $\mathcal{E}$ in units of kV cm^{-1} and T_{rot} in K, its value is 0.0366.

Loesch and coworkers have used the brute force technique to perform experiments on reactions such as $K+CH_3I$ and $K+ICl$,[678–680] in which cases they measured DCSs and molecular axis polarization-dependent DCSs using a crossed molecular beam apparatus and field strengths up to 20 kV cm^{-1}. The method has been also used for the orientation of asymmetric tops and aromatic molecules.[681]

Friedrich and coworkers have shown that brute force orientation is not necessarily restricted to dipolar molecules (see ref. [682] and references therein). Using a high-power, non-resonant laser in the adiabatic regime (this requires pulse duration $\tau \gg \hbar/B$), one can orient non-polar molecules *via* the interaction of their induced dipole moment with the electric field.

Before closing this section we should note that, at the time of writing, the use of ultrashort (femtosecond) laser pulses to polarize or otherwise manipulate molecules is attracting a great deal of experimental and theoretical interest.[683,684] It is likely that this avenue will lead to new and exciting possibilities regarding the study of reaction and photodissociation dynamics.

9.4 A DETAILED EXAMPLE

We shall now examine in detail an example involving many of the concepts described. It involves simple analytical expressions for product angular distributions and for the wavefunctions of reactants and products, and from this point of view it is quite artificial. From other points of view, however, the example is realistic. In concerns the stereodynamics of the prototype reaction

$$\mathrm{A} + \mathrm{BC}(j=1) \rightarrow \mathrm{AB}(j'=1) + \mathrm{C}. \tag{9.69}$$

The values of E_c, v and v' (respectively, the collision energy, and the initial and final vibrational quantum numbers) are left unspecified because they play no role in the analysis. All of the calculations presented in this section are included in detail in the supplementary material.

We will assume that, perhaps through calculations such as those described in ref. [646], we have found that reaction (9.69) involves three independent transformations from reactant to product states. The reagent and product states in each pathway are pure and therefore fully coherent:*

$$|\Psi_a\rangle = \begin{pmatrix} \frac{1}{\sqrt{2}} \\ 0 \\ \frac{1}{\sqrt{2}} \end{pmatrix} \rightarrow |\Psi'_a\rangle = \begin{pmatrix} \frac{1}{\sqrt{2}} \\ 0 \\ \frac{1}{\sqrt{2}} \end{pmatrix}, \quad P_a(\theta) = \sin^2\left(\frac{\theta}{2}\right), \quad \sigma_a = 1.00\,\text{Å}^2, \tag{9.70a}$$

$$|\Psi_b\rangle = \begin{pmatrix} -\frac{\cos\theta}{\sqrt{2}} \\ ie^{2i\theta}\sin\theta \\ \frac{\cos\theta}{\sqrt{2}} \end{pmatrix} \rightarrow |\Psi'_b\rangle = \begin{pmatrix} -\frac{\cos\theta}{\sqrt{2}} \\ \sin\theta \\ \frac{\cos\theta}{\sqrt{2}} \end{pmatrix}, \quad P_b(\theta) = \frac{3\sin\left(\frac{\theta}{2}\right) + \sin(3\theta)}{4}, \quad \sigma_b = 1.33\,\text{Å}^2, \tag{9.70b}$$

$$|\Psi_c\rangle = \begin{pmatrix} \frac{\sin\theta}{\sqrt{2}} \\ ie^{2i\theta}\cos\theta \\ -\frac{\sin\theta}{\sqrt{2}} \end{pmatrix} \rightarrow |\Psi'_c\rangle = \begin{pmatrix} \frac{\sin\theta}{\sqrt{2}} \\ \cos\theta \\ -\frac{\sin\theta}{\sqrt{2}} \end{pmatrix}, \quad P_c(\theta) = \frac{3\cos\left(\frac{\theta}{2}\right) - \cos(3\theta)}{4}, \quad \sigma_c = 0.01\,\text{Å}^2, \tag{9.70c}$$

where the $P(\theta)$ are angular distributions satisfying Eq. (9.2b), the σ are integral cross sections, and we have used matrix notation to describe the initial and final states. All states are described in the $\{|j=1, m\rangle \equiv |1m\rangle\}$ basis set, which in turn is referred to the *xyz*, reactants frame on the left of Figure 9.2. The matrices have elements representing the contributions to the wavefunctions from the three $|j=1, m\rangle$ components. The basis states' order is that of increasing magnetic quantum number. For instance, the product state of Eq. (9.70c) is

$$|\Psi'_c\rangle = \sin\theta\frac{|1-1\rangle - |11\rangle}{\sqrt{2}} + \cos\theta|10\rangle. \tag{9.71}$$

We shall first consider how we can quantify the reaction stereodynamics. Then we will see how we can use all or part of the quantitative information to rationalize the reaction mechanism(s) and to examine the effect that reactant polarization can have on reactivity.

9.4.1 Quantification of Known Reaction Mechanisms

Let us start with the dynamical information – the polarization moments – and the business of extracting it from the data of Eq. (9.70). We will:

i. Calculate the density matrices of reactants and products.
ii. Calculate the real polarization moments of reactants and products.
iii. Compare the real polarization moments to their limiting values.

We will not present details of all calculations; they can be found in the supplementary material.

*As shown on ref. [646], reaction (9.69) *must* involve three independent transformations. The three reactant (and product) states must be orthogonal, and one of the transformations must involve the asymptotic states shown in Eq. (9.70a). The wavefunctions on Eqs. (9.70b–c) and the angular distributions were all we could vary when designing this example, but even then we had to consider symmetry and orthogonality constraints.

9.4.1.1 Density Matrices

In matrix notation, Eq. (9.14) becomes $\rho = \boldsymbol{c} \cdot \boldsymbol{c}^{\dagger}$, where $\boldsymbol{c}$ is a vector such as those in Eq. (9.70) and $\boldsymbol{c}^{\dagger}$ is its Hermitian transpose. Using this expression one gets

$$\rho_a = \begin{pmatrix} \frac{1}{2} & 0 & \frac{1}{2} \\ 0 & 0 & 0 \\ \frac{1}{2} & 0 & \frac{1}{2} \end{pmatrix}, \qquad \rho'_a = \begin{pmatrix} \frac{1}{2} & 0 & \frac{1}{2} \\ 0 & 0 & 0 \\ \frac{1}{2} & 0 & \frac{1}{2} \end{pmatrix}, \tag{9.72a}$$

$$\rho_b = \begin{pmatrix} f_1(\theta) & -g^*(\theta) & -f_1(\theta) \\ -g(\theta) & 1-2f_1(\theta) & g(\theta) \\ -f_1(\theta) & g^*(\theta) & f_1(\theta) \end{pmatrix}, \quad \rho'_b = \begin{pmatrix} f_1(\theta) & -f_2(\theta) & -f_1(\theta) \\ -f_2(\theta) & 1-2f_1(\theta) & f_2(\theta) \\ -f_1(\theta) & f_2(\theta) & f_1(\theta) \end{pmatrix}, \tag{9.72b}$$

$$\rho_c = \begin{pmatrix} \frac{1}{2}-f_1(\theta) & g^*(\theta) & f_1(\theta)-\frac{1}{2} \\ g(\theta) & 2f_1(\theta) & -g(\theta) \\ f_1(\theta)-\frac{1}{2} & -g^*(\theta) & \frac{1}{2}-f_1(\theta) \end{pmatrix}, \quad \rho'_c = \begin{pmatrix} \frac{1}{2}-f_1(\theta) & f_2(\theta) & f_1(\theta)-\frac{1}{2} \\ f_2(\theta) & 2f_1(\theta) & -f_2(\theta) \\ f_1(\theta)-\frac{1}{2} & -f_2(\theta) & \frac{1}{2}-f_1(\theta) \end{pmatrix}, \tag{9.72c}$$

where $f(\theta)$ and $g(\theta)$ are real and complex functions of the scattering angle:

$$f_1(\theta) = \frac{\cos^2\theta}{2}, \quad f_2(\theta) = \frac{\sin\theta\cos\theta}{\sqrt{2}} = \frac{\sin(2\theta)}{\sqrt{8}}, \quad g(\theta) = ie^{2i\theta} f_2(\theta). \tag{9.73}$$

You can check the results of Eq. (9.72) in several ways: (i) The density matrices must be Hermitian; (ii) because the states we started with were all normalized to unity, the density matrices above must all have $\mathrm{Tr}(\rho) = 1$; (iii) because they represent pure states, they must also satisfy $\mathrm{Tr}(\rho^2) = 1$; and (iv) because we are dealing with orthogonal states, products such as $\rho_a \cdot \rho_b$ and $\rho'_b \cdot \rho'_c$ must lead to the 3×3 matrices whose elements are all zero.[643] You can also check angular momentum conservation: in the $j = j'$ case, at $\theta = 0$ one must invariably have $\rho = \rho'$.

9.4.1.2 Polarization Moments

The easiest way to calculate the real polarization moments of reactants and products is to use Eq. (9.16) with $O = T^{\{k\}}_{q\pm}$; as mentioned in Section 9.2.6, the real polarization moments are related to the expectation values of the real multipolar operators through $z^{\{k\}}_{q\pm} = z^{\{0\}}_0 \langle T^{\{k\}}_{q\pm} \rangle$, where $z^{\{k\}}_{q\pm}$ can stand for normalized PDDCSs [in which case $z^{\{k\}}_{q\pm} \equiv S^{\{k\}}_{q\pm}(\theta)$], for renormalized PDDCSs [in which case $z^{\{k\}}_{q\pm} \equiv S^{\{k\}}_{q\pm}(\theta)/S^{\{0\}}_0(\theta)$], or for PPs [in which case $z^{\{k\}}_{q\pm} \equiv a^{\{k\}}_{q\pm}$].

If we are to use Eq. (9.16), we need the $j = 1$ real multipolar matrices. Using Eq. (9.20) and the values of the CG coefficients,[647] one finds that they are given by

$$T^{\{0\}}_0 = \begin{pmatrix} 1 & 0 & 0 \\ 0 & 1 & 0 \\ 0 & 0 & 1 \end{pmatrix}, \quad T^{\{1\}}_{1-} = \alpha \begin{pmatrix} 0 & i & 0 \\ -i & 0 & i \\ 0 & -i & 0 \end{pmatrix}, \quad T^{\{1\}}_{1+} = \alpha \begin{pmatrix} 0 & 1 & 0 \\ 1 & 0 & 1 \\ 0 & 1 & 0 \end{pmatrix},$$

$$T^{\{1\}}_0 = \beta \begin{pmatrix} -1 & 0 & 0 \\ 0 & 0 & 0 \\ 0 & 0 & 1 \end{pmatrix}, \quad T^{\{2\}}_{2-} = \gamma \begin{pmatrix} 0 & 0 & i \\ 0 & 0 & 0 \\ -i & 0 & 0 \end{pmatrix}, \quad T^{\{2\}}_{2+} = \gamma \begin{pmatrix} 0 & 0 & 1 \\ 0 & 0 & 0 \\ 1 & 0 & 0 \end{pmatrix}, \tag{9.74}$$

$$T^{\{2\}}_{1-} = \delta \begin{pmatrix} 0 & -i & 0 \\ i & 0 & i \\ 0 & -i & 0 \end{pmatrix}, \quad T^{\{2\}}_{1+} = \delta \begin{pmatrix} 0 & -1 & 0 \\ -1 & 0 & 1 \\ 0 & 1 & 0 \end{pmatrix}, \quad T^{\{2\}}_0 = \varepsilon \begin{pmatrix} 1 & 0 & 0 \\ 0 & -2 & 0 \\ 0 & 0 & 1 \end{pmatrix},$$

where $\alpha = 1/2$, $\beta = \sqrt{1/2}$, $\gamma = \sqrt{3/10}$, $\delta = \sqrt{3/20}$ and $\varepsilon = \sqrt{1/10}$.

Now all we need is straightforward calculation; the possibly non-zero θ-dependent $\langle T_{q\pm}^{\{k\}}\rangle$ values for the reactants and products of reactions (9.70a–c) are listed in Table 9.1. Because of the symmetry of the problem, all other expectation values must vanish – see Problems 10–11.

Calculation of the polarization moments requires multiplication of the expectation values of Table 9.1 by $z_0^{\{0\}}$. How do we define the value of this quantity? It depends on what we want the polarization moments, and the PDF associated with them, to represent. The value of $z_0^{\{0\}}$ determines the PDF normalization (*cf.* Section 9.2.4.2), which in turn determines its physical meaning (*cf.* Section 9.2.3).

A useful choice is to set $z_0^{\{0\}}=1$. The polarization moments thus obtained – *renormalized polarization-dependent differential cross sections*, or renormalized PDDCSs – lead to renormalized, θ-dependent PDFs like the one of Eq. (9.5). These PDFs quantify the probability, under the condition that reaction does take place and does lead to scattering along θ, that the molecules will be polarized as prescribed by their polarization moments.

Another useful normalization choice is to set $z_0^{\{0\}}=P(\theta)$. The polarization moments thus obtained – *normalized polarization-dependent differential cross sections*, or normalized PDDCSs – lead to normalized, θ-dependent PDFs like the one of Eq. (9.5). While such PDFs are not particularly useful, by integrating them over θ we obtain PDFs that, under the condition that reaction does take place, quantify the total probability ("total" here means "considering all scattering angles") that the molecules will be polarized as prescribed by their integral polarization moments – their *polarization parameters* (PPs), see Eq. (9.25). The results one obtains by applying this procedure to reactions (9.70a–c) are presented in Table 9.2.

9.4.1.3 *Limiting Values of Polarization Moments*

By themselves, the results listed on Tables 9.1 and 9.2 do not mean much. In order to evaluate whether they are indicative of strong or weak molecular polarization one must know the range of values that each parameter can take.

Table 9.1 Non-zero θ-dependent expectation values, $\langle T_{q\pm}^{\{k\}}\rangle$, of real polarization operators for the asymptotic states of reactions (9.70a–c). These expressions also hold for the renormalized PDDCSs.

State	*k*=0 *q*=0	*k*=1 *q*=1−	*k*=2 *q*=0	*k*=2 *q*=1+	*k*=2 *q*=2+
$\lvert\Psi_a\rangle$, $\lvert\Psi'_a\rangle$	1	0	ε	0	γ
$\lvert\Psi_b\rangle$	1	$4\alpha\mathrm{Im}\,[g(\theta)]$	$\varepsilon[6f_1(\theta)-2]$	$4\delta\mathrm{Re}[g(\theta)]$	$-2\gamma f_1(\theta)$
$\lvert\Psi'_b\rangle$	1	0	$\varepsilon[6f_1(\theta)-2]$	$4\delta f_2(\theta)$	$-2\gamma f_1(\theta)$
$\lvert\Psi_c\rangle$	1	$-4\alpha\mathrm{Im}\,[g(\theta)]$	$\varepsilon[1-6f_1(\theta)]$	$-4\delta\mathrm{Re}[g(\theta)]$	$\gamma[2f_1(\theta)-1]$
$\lvert\Psi'_c\rangle$	1	0	$\varepsilon[1-6f_1(\theta)]$	$-4\delta f_2(\theta)$	$\gamma[2f_1(\theta)-1]$

Table 9.2 Non-zero polarization parameters for the asymptotic states of reactions (9.70a–c).

State	*k*=0 *q*=0	*k*=1 *q*=1−	*k*=2 *q*=0	*k*=2 *q*=1+	*k*=2 *q*=2+
$\lvert\Psi_a\rangle$, $\lvert\Psi'_a\rangle$	1	0.000	0.316	0.000	0.548
$\lvert\Psi_b\rangle$	1	–0.005	–0.241	–0.342	−0.226
$\lvert\Psi'_b\rangle$	1	0.000	–0.241	–0.125	−0.226
$\lvert\Psi_c\rangle$	1	0.065	0.018	0.288	−0.376
$\lvert\Psi'_c\rangle$	1	0.000	0.018	–0.233	−0.376

Table 9.3 Eigenvalues and expectation value ranges of $j=1$ real polarization operators.

	$k=1$ $q=1-$	$k=2$ $q=0$	$k=2$ $q=1+$	$k=2$ $q=2+$
$T_{q\pm}^{\{k\}}$ eigenvalues	$0, \pm\sqrt{\frac{1}{2}}$	$-\sqrt{\frac{2}{5}}, \sqrt{\frac{1}{10}}, \sqrt{\frac{1}{10}}$	$0, \pm\sqrt{\frac{3}{10}}$	$0, \pm\sqrt{\frac{3}{10}}$
$\langle T_{q\pm}^{\{k\}}\rangle$ range	[−0.707, 0.707]	[−0.632, 0.316]	[−0.548, 0.548]	[−0.548, 0.548]

As discussed in Section 9.2.7, this range is determined by the largest and smallest eigenvalues of the corresponding real polarization operator. Now, using the matrix representations of the operators involved [see Eq. (9.74)] one can determine all the necessary eigenvalues and therefore the allowed ranges for the real polarization moments; the results are listed in Table 9.3.

The angular distributions, renormalized PDDCSs and polarization parameters we have obtained are shown in graphical form in Figure 9.6. Note that in the panels showing polarization moments (bottom four rows), the allowed ranges are indicated by the unshaded areas. That is, the borders between the shaded and unshaded areas in each panel indicate the minimum or maximum value the corresponding polarization moment can possibly take; no polarization moment could have a value falling in a shaded part of the panel in which it is shown.

9.4.2 Rationalization of the Complete Set of Stereodynamical Parameters

We now turn to the task of extracting chemical information from the quantitative parameters determined in Section 9.4.1. We will:

i. Directly, but only in part, analyze the polarization moments.
ii. Use the moments to generate stereodynamical portraits, and then analyse the portraits.

Stereodynamical portraits are plots showing PDFs of reactants and products. As we will see, their consideration can make the interpretation of stereodynamical data a great deal easier.

9.4.2.1 Analysis of Polarization Moments

The polarization moments of Tables 9.1–9.2 and Figure 9.6 are the quantitative stereodynamical parameters. Traditionally, they are also the quantities whose physical meanings one turns to when trying to rationalize the stereodynamics of a chemical reaction. However, a point that we wish to make here is that direct analysis of the polarization moments does not provide the easiest route for the elucidation of vector correlations.

The example we are considering is relatively simple. It is described by polarization moments of rank no larger than 2. Using the procedure discussed in Section 9.2.4.2, one finds that the directional meanings of the non-vanishing $\langle T_{q\pm}^{\{k\}}\rangle$ ratios are as given in Table 9.4.

Let us consider mechanism *a*. This is the simplest of the three mechanisms: all renormalized PDDCSs are θ-independent, {1, 1−} and {2, 1+} polarization moments vanish, {2, 0} and {2, 2+} polarization moments take their maximum values, reactant and product polarizations coincide. Here's what we conclude from this:

- As the {1, 1−} moments vanish, there is no angular momentum orientation, and therefore no preference for particular senses of rotation for the reactant or product diatomics.
- $\boldsymbol{j}$ and $\boldsymbol{j}'$ are most likely to be found with positive z alignment (that is, parallel to $\pm z$) and positive x alignment (that is, parallel to $\pm x$).

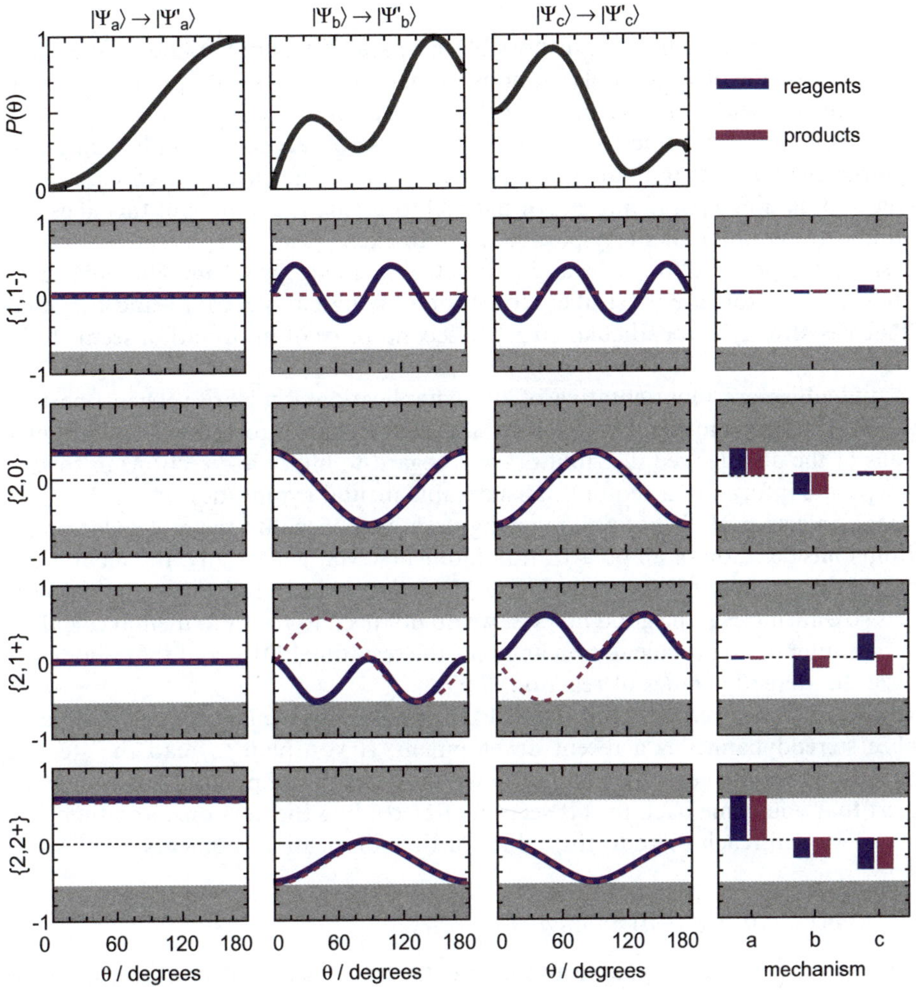

Figure 9.6 Stereodynamical data for reaction (9.70). The three leftmost columns refer to mechanisms *a*–*c* and show quantities that depend on the scattering angle: the angular distribution (top panel) and renormalized real PDDCSs of reactants and products (bottom four panels). The rightmost column shows polarization parameters for all three mechanisms.

Table 9.4 Directional meanings of the non-vanishing $\langle T^{\{k\}}_{q\pm}\rangle$ expectation values.

$k\ q_{\pm}$	$\langle T^{\{k\}}_{q\pm}\rangle$	$\boldsymbol{j}$ (or $\boldsymbol{j}'$): *preference for...*	$\boldsymbol{r}$ (or $\boldsymbol{r}'$): *preference for...*
1 1−	positive	$\boldsymbol{j} \parallel +y$	–
	negative	$\boldsymbol{j} \parallel -y$	–
2 0	positive	$\boldsymbol{j} \parallel \pm z$	$\boldsymbol{r} \perp \pm z$
	negative	$\boldsymbol{j} \perp \pm z$	$\boldsymbol{r} \parallel \pm z$
2 1+	positive	$\boldsymbol{j} \parallel \pm(x+z)$	$\boldsymbol{r} \parallel \pm(x-z)$
	negative	$\boldsymbol{j} \parallel \pm(x-z)$	$\boldsymbol{r} \parallel \pm(x+z)$
2 2+	positive	$\boldsymbol{j} \parallel \pm x$	$\boldsymbol{r} \parallel \pm y$
	negative	$\boldsymbol{j} \parallel \pm y$	$\boldsymbol{r} \parallel \pm x$

- $\boldsymbol{r}$ and $\boldsymbol{r}'$ are most likely to be found with negative z alignment (that is, perpendicular to $\pm z$) and positive y alignment (that is, parallel to $\pm y$).

You may be wondering how $\boldsymbol{j}$ can possibly be parallel to both x and z. Well, it cannot. It is helpful to contrast the observed polarizations to their "opposites": $\boldsymbol{j}$ is preferentially found *parallel rather than perpendicular to* $\pm z$, and *parallel to* $\pm x$ *rather than parallel to* $\pm y$. With a bit of experience, you should be able to recognize that these observations imply that $\boldsymbol{j}$ is preferentially found perpendicular (rather than parallel) to $\pm y$, and cylindrically distributed around it.* The probabilities of finding $\boldsymbol{j}$ parallel to $\pm x$ or parallel to $\pm z$ are the same, but this does not mean that $\boldsymbol{j}$ can be found to be simultaneously parallel to both axes.

Analysis of the polarizations of $\boldsymbol{r}$ and $\boldsymbol{r}'$ involves a similar problem. The polarization moments tell us that these vectors are most likely to be found lying parallel to $\pm y$ and perpendicular to $\pm z$. Given that y is already perpendicular to z, the second piece of information seems to be redundant. But is it?

These difficulties are representative of the complications that arise when one tries to directly analyze polarization moments. Even in very simple cases, one is forced to think about simultaneous distortions of the unpolarized distribution with regard to multiple directions in space. Synthesis of all that separate information requires considerable mental gymnastics.

Let us take a step back. What we are trying to do is to visualize the molecular polarizations that result from consideration of all polarization moments. Mathematically, the mental calculations we are trying to do are those expressed by Eq. (9.12) and its axial polarization counterpart. No wonder the task is so difficult! But the good news is, we do not need to resort to mental calculation. It is easy to use a computer to do the calculation and plot the resulting PDFs, and that is what we shall use to rationalize the stereodynamics of reaction (9.69).

We emphasize, however, that the use of PDF plots (the so-called stereodynamical portraits) to rationalize stereodynamics is a recent development. If you plan to read the stereodynamics literature, you will have to get some familiarity with the analysis of polarization moments. To do that, we suggest that you come back to this section after reading the next one, and then try to relate the conclusions we will reach there to the values of the polarization moments.

9.4.2.2 Analysis of Stereodynamical Portraits

We shall now see how stereodynamical portraits facilitate visualization of vector correlations and, as a consequence, rationalization of a reaction's stereodynamics.

Mechanism a

As we have seen, in this mechanism the polarizations of reactants and products do not change with scattering angle. For the sake of definiteness, however, we have used a particular value of the scattering angle ($\theta = 90°$) in the construction of the stereodynamical portraits of mechanism *a*. They are shown on Figure 9.7 in the following format:

- The top portrait is an "axial portrait". It shows molecular axis polarizations. That is, it displays the spatial distributions of the interatomic axes of the reactant and product diatomics, $\boldsymbol{r}$ and $\boldsymbol{r}'$. These spatial distributions – axial PDFs – were calculated with the molecular axis counterpart of Eq. (9.12) and using the renormalized PDDCSs on the top row of Table 9.1. Reactant and product PDFs are respectively shown in light blue and purple.

*Note that none of our polarization moments quantifies parallel or perpendicular alignment with regard to $\pm y$. Had we used a frame with its z axis where we now have y (that is, along $\boldsymbol{k}\times\boldsymbol{k}'$), the puzzle would not have arisen.

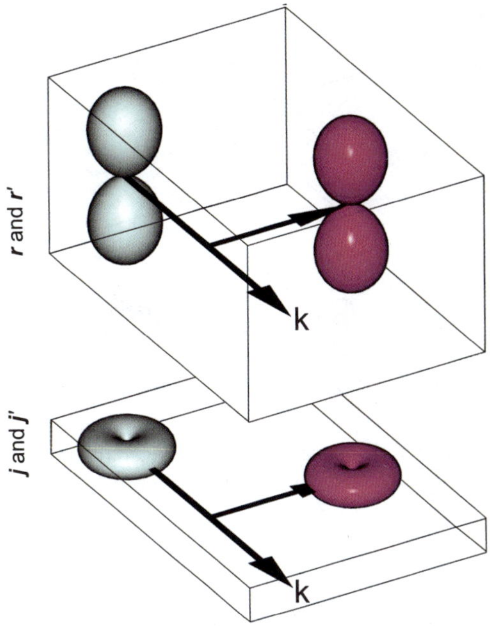

Figure 9.7 Stereodynamical portraits of mechanism *a* at $\theta = 90°$. Top and bottom panels show molecular axis and angular momentum polarizations, respectively.

- The bottom portrait is a "rotational portrait". It shows angular momentum polarizations. That is, it displays the spatial distributions of the rotational angular momenta of the reactant and product diatomics, $\boldsymbol{j}$ and $\boldsymbol{j}'$. These spatial distributions – rotational PDFs – were calculated with Eq. (9.12) and using the renormalized PDDCSs on the top row of Table 9.1. Reactant and product PDFs are respectively shown in light blue and purple.
- Each portrait shows a pair of arrows (whose lengths carry no meaning). The long one is $\boldsymbol{k}$, the reactant-approach direction. The short arrow represents $\boldsymbol{k}'$, the product-recoil direction. The plane containing the two arrows is the scattering plane.
- Reactant polarization is shown at the base of $\boldsymbol{k}$, product polarization at the head of $\boldsymbol{k}'$.
- As the portraits were taken at $\theta = 90°$, the arrows representing $\boldsymbol{k}$ and $\boldsymbol{k}'$ are perpendicular.

Let us now turn to the analysis of these portraits. The first thing to note is that here there is no ambiguity or confusion about the collision geometry. For instance, it is clear that $\boldsymbol{r}$ and $\boldsymbol{r}'$ are aligned along $\pm(\boldsymbol{k} \times \boldsymbol{k}')$, the direction perpendicular to the scattering plane.

Note also that $\boldsymbol{j}$ and $\boldsymbol{j}'$ are aligned [both preferentially perpendicular to $\pm(\boldsymbol{k} \times \boldsymbol{k}')$] but not oriented. As rotational polarizations only carry extra information with regard to axial polarizations when there is angular momentum orientation (see Section 9.2.2.1), here we need only consider the more intuitive axial portraits when trying to rationalize the collision stereodynamics.

So, how can we rationalize the axial portrait of Figure 9.7? What we are looking for is an intuitive description of how the atoms move in the course of the reaction. In other words: can we visualize the collision as a classical trajectory?

If the axial portrait is reduced to its "skeleton", what we have is this: reactants approach in a T- or L-shaped geometry (note that the portrait does not contain information about the impact parameter) and without preference for any particular sense of rotation; product recoil takes place with a similar geometry; the internuclear axes of the reactant and product diatomics are aligned along $\pm(\boldsymbol{k} \times \boldsymbol{k}')$; the picture is the same whatever the scattering angle. How can this happen?

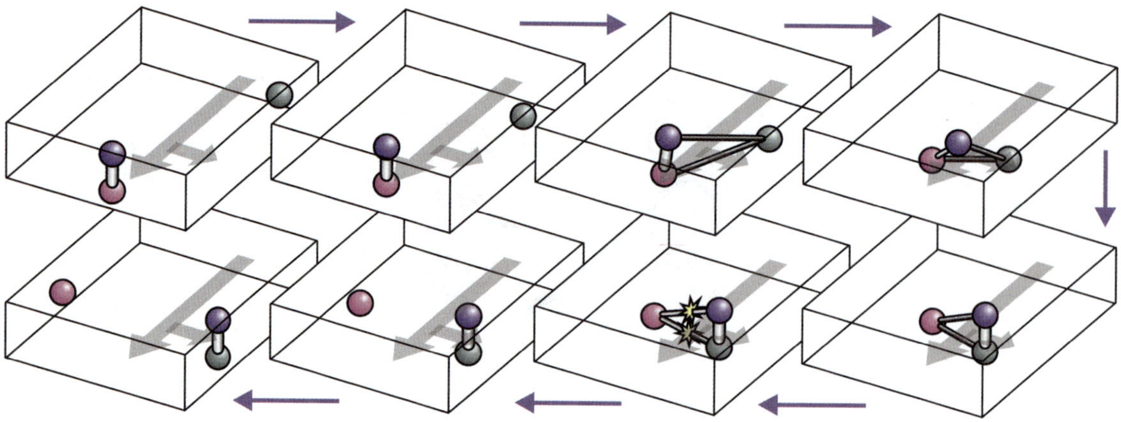

Figure 9.8 Clockwise from top left: a sequence of consecutive snapshots of the classical trajectory used in the rationalization of the stereodynamical portraits of Figure 9.7.

Figure 9.8 shows consecutive snapshots of a classical trajectory that explains the vector correlations of Figure 9.7. This trajectory is not unique; there are other trajectories that can explain the (asymptotic) vector correlations characteristic of mechanism *a*. On the basis of the asymptotic vector properties alone, one cannot decide what particular trajectories are actually representative of the reaction. To do this, one needs additional information. For example, the atomic masses (we have assumed them to be similar), the topology of the potential energy surface (we have assumed it to be consistent with formation of a triangular intermediate), and dynamical information regarding, for instance, the impact parameters contributing to this mechanism. What one *can* be sure of is the correlation between initial and final polarizations: it must be as shown on the stereodynamical portraits of Figure 9.7.

The trajectory progresses clockwise from top left. In each panel, the long and short arrows respectively represent $\boldsymbol{k}$ and $\boldsymbol{k}'$. The trajectory starts as it should: with the BC axis perpendicular to the scattering plane and no rotational orientation. As atom A (green) approaches BC, it pushes away the atom closest to it (C, purple) and attracts the other (B, blue). A triangular intermediate is formed. When the ABC plane becomes parallel to $\boldsymbol{k}'$, the AC and BC bonds are simultaneously broken. This results in the recoil of the AB + C products. As atoms A and B are pushed away from C with equal force, the recoil is such that the AB product is polarized in the manner shown in Figure 9.7. Its axis is perpendicular to the scattering plane and there is no rotational orientation.

Obtaining the level of detail about a reaction that has just been given is one of the holy grails of chemical dynamics. Note that, while we could not obtain all of the details of the collision dynamics from the vector correlations alone, the information about vector correlations is essential.

There is a piece of stereodynamical information we have not yet considered: the product angular distribution. As mentioned earlier, the angular distributions we have associated with each mechanism were chosen somewhat arbitrarily. The one we have chosen for mechanism *a* (one that diminishes as the scattering direction moves from backward to forward) is consistent with the sort of trajectory we have described.

Mechanism b

Here reactant and product polarizations change with θ. For this reason, we need to examine stereodynamical portraits at different scattering angles; the ones we have chosen are $\theta = 60$, 90 and 120°, which are respectively representative of forward, sideways and backward scattering. The portraits are shown on Figure 9.9, with format and colour coding as in Figure 9.7.

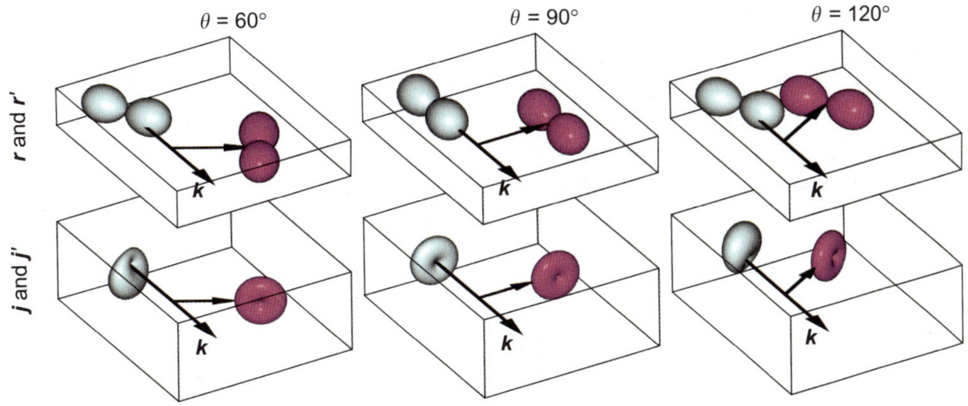

Figure 9.9 Stereodynamical portraits of mechanism b. From left to right, $\theta = 60$, 90 and 120°. Top and bottom panels show molecular axis and angular momentum polarizations, respectively.

Let us first consider backward, $90^\circ \leqslant \theta \leqslant 180^\circ$ scattering. In this region, the axial portraits of mechanism b bear similarities to those of mechanism a: the $\boldsymbol{r}$ and $\boldsymbol{r}'$ polarizations match, and the two vectors are aligned along a direction perpendicular to $\boldsymbol{k}'$. But there is a difference: here $\boldsymbol{r}$ and $\boldsymbol{r}'$ are found to lie preferentially in the scattering plane rather than perpendicular to it.

There are also similarities and differences concerning rotational portraits. As in mechanism a, here there is no product rotational orientation. In contrast to mechanism a, here one does find reactant rotational orientation with regard to $\boldsymbol{k} \times \boldsymbol{k}'$. It is negative at $135^\circ < \theta < 180^\circ$ and positive at $90^\circ < \theta < 135^\circ$ (see the {1, 1−} panel on the second column of Figure 9.6).

The vector correlations found for backward-scattering mechanism b can be rationalized with recourse to trajectories similar to the one shown in Figure 9.8, except that $\boldsymbol{r}$ and $\boldsymbol{r}'$ must lie in the scattering plane; the sketching of such trajectories is left as an exercise. But what is the origin of the reactant rotational orientation, and why is it inverted at $\theta = 135^\circ$?

Figure 9.10 offers an explanation. It shows two sequences of diagrams describing scattering at $\theta = 135^\circ$. Respectively, the top and bottom sequences show how negative and positive $\boldsymbol{j}$ orientation might contribute to $\theta = 135^\circ$ scattering. If the two processes are equally important, there is no net orientation effect. If the top one is dominant (we think this happens at $135^\circ < \theta < 180^\circ$), then there is a preference for $\boldsymbol{j} \| -(\boldsymbol{k} \times \boldsymbol{k}')$. If the bottom one is dominant (which we think happens at $90^\circ < \theta < 135^\circ$), then there is a preference for $\boldsymbol{j} \| +(\boldsymbol{k} \times \boldsymbol{k}')$.

The top sequence is as follows. The first (that is, left) panel shows the coplanar A–BC approach; $\boldsymbol{r}$ is tilted by 45° with regard to approach direction, and the impact parameter is relatively large. As in mechanism a, atoms B and C are respectively attracted towards or pushed away from atom A. This leads to formation of a triangular intermediate that decays *via* simultaneous breaking of its AC and BC bonds; this is shown on the second panel, and is again as in mechanism a. Product recoil is shown on the third panel, and is also as in mechanism a.

The bottom sequence is similar, except that $\boldsymbol{r}$ is initially tilted by −135° and the impact parameter is relatively small.

So, what is the origin of the reactant rotational orientation? As indicated on Figure 9.10, our suggestion is that it facilitates the rearrangement of the positions of atoms B and C.

So much for backward scattering. What about forward scattering? Inspection of the stereodynamical portraits of mechanism b in this region (a snapshot taken at $\theta = 60^\circ$ is shown on Figure 9.9) reveals that the exit stereodynamics is as when scattering is backwards ($\boldsymbol{r}'$ in scattering plane and perpendicular to $\boldsymbol{k}'$, no $\boldsymbol{j}'$ orientation), but the entrance stereodynamics is not: as θ changes

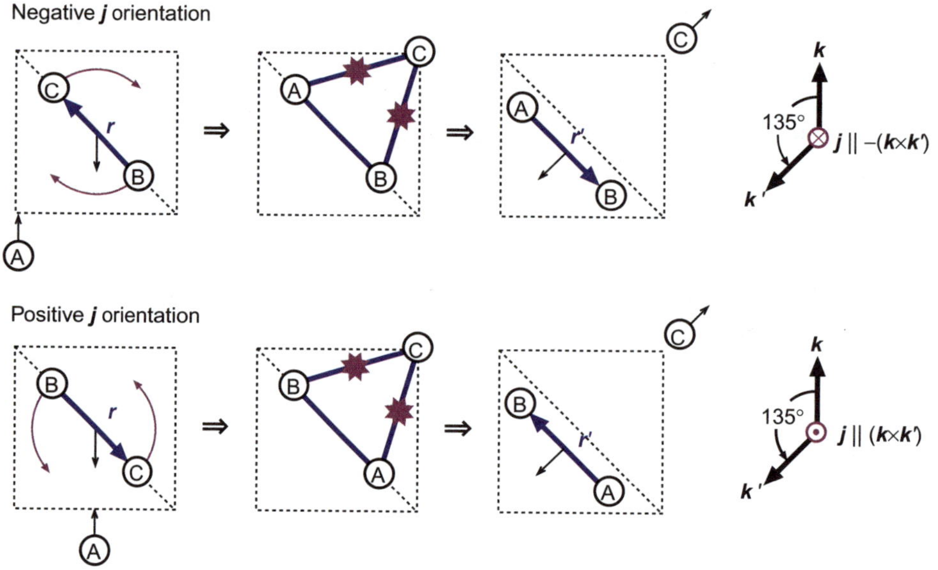

Figure 9.10 The sequence of diagrams on the top, or bottom, panel shows how $\boldsymbol{j}$ orientation along $-(\boldsymbol{k}\times\boldsymbol{k}')$, or along $+(\boldsymbol{k}\times\boldsymbol{k}')$, might contribute to $\theta=135°$ scattering *via* mechanism *b*.

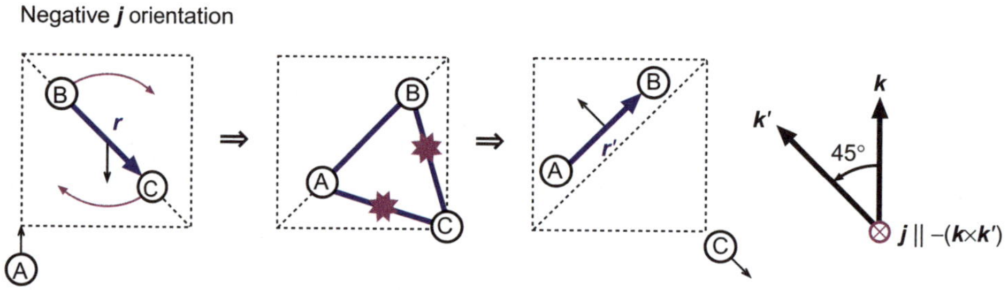

Figure 9.11 This sequence of diagrams shows how $\boldsymbol{j}$ orientation along $-(\boldsymbol{k}\times\boldsymbol{k}')$ might contribute to $\theta=45°$ scattering *via* mechanism *b*. But there is a problem – see text.

from 90° to 45° to 0°, the $\boldsymbol{r}$ alignment direction changes from perpendicular to parallel to perpendicular to $\boldsymbol{k}'$. How can this be rationalized?

Figure 9.11 shows a sequence of diagrams that one might use to rationalize $\theta=45°$ scattering. It is similar to the sequence on the top panel of Figure 9.10, except that the initial positions of atoms B and C have been swapped. The entrance stereodynamics changes as expected (in particular, $45°<\theta<90°$ reactant rotational orientation is accounted for as in the case of backward scattering). Except for the value of the scattering angle, the exit stereodynamics is unchanged.

But there is a problem. Up to this point, we have assumed that the initial reactants interaction involves A–C repulsion and A–B attraction. The trouble is that the diagrams of Figure 9.11 require the opposite. Can this be consistent with the potential energy surface of the reaction? We will assume such a process is indeed possible, but less likely than those leading to backward scattering. This is consistent with the product angular distribution we have selected for mechanism *b*: it favours backward scattering, but the dependence of the scattering probability on θ is more complex than in mechanism *a*. Note also that, using the processes we have considered so far, it is hard to

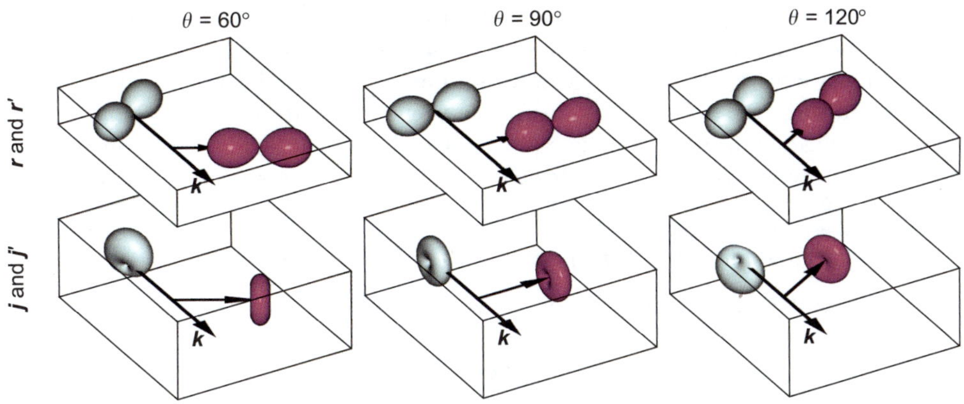

Figure 9.12 As Figure 9.9, but for mechanism *c*.

imagine how reaction (9.69) might lead to $\theta = 0$ scattering; this is why we have associated mechanisms *a* and *b* with angular distributions satisfying $P(0) = 0$.

Mechanisms *a* and *b* must also be degenerate at $\theta = 180°$, when (as at $\theta = 0$) the scattering plane is not defined and therefore one cannot talk about, say, $\boldsymbol{r}'$ alignment perpendicular or parallel to the scattering plane. This is why, having chosen a reference value for the integral cross section of mechanism *a*, we then had to use $\sigma_b = \frac{4}{3}\sigma_a$. This implies that at $\theta = 180°$ the differential cross sections satisfy $\mathrm{d}\sigma_a/\mathrm{d}\omega = \mathrm{d}\sigma_b/\mathrm{d}\omega$, see Eq. (9.2a).

Mechanism c

As mentioned above, mechanisms *a*, *b* and *c* are (must be!) orthogonal. This implies that, having specified reactant and product polarizations in mechanisms *a* and *b*, we have also specified the reactant and product polarizations in mechanism *c*. They are as shown on the stereodynamical portraits of Figure 9.12. The directions along which $\boldsymbol{r}$ and $\boldsymbol{r}'$ are aligned at a particular scattering angle in mechanisms *a*, *b* and *c* are mutually perpendicular, and so are the planes in which $\boldsymbol{j}$ and $\boldsymbol{j}'$ are most likely to be found. In mechanism *c*, the orientation of $\boldsymbol{j}$ is the opposite of that found in mechanism *b*. This is also a consequence of orthogonality (and of the fact that mechanism *a* does not involve $\boldsymbol{j}$ orientation).

The fact that mechanism *c* requires $\boldsymbol{r}' || \boldsymbol{k}'$ suggests that it proceeds *via* a colinear intermediate, in particular when it also requires $\boldsymbol{r} || \boldsymbol{k}'$. This is in stark contrast with what we have found for mechanisms *a* and *b*, and suggests a rather intuitive idea: that reaction mechanisms involving T-shaped or colinear intermediates do not overlap (*i.e.* are orthogonal). It is because of the contrasting character of mechanism *c* that we have associated with it an angular distribution favouring forward (rather than backward, as in the previous cases) scattering. We have also assumed that, if reaction *via* a T-shaped intermediate is likely, reaction *via* a colinear intermediate cannot be. This is why we have chosen the integral cross section of mechanism *c* to be two orders of magnitude smaller than σ_a and σ_b.

9.4.3 Incomplete Information

Collection of complete sets of stereodynamical parameters is often impractical, difficult or impossible. This may be because of experimental limitations (at present this is almost invariably the case, photodissociation processes providing the only possible exceptions), or else because of lack of interest in the role of one or more vectors in the reaction stereodynamics.

In this section we consider reaction (9.69) under two common restrictions that imply that one cannot or need not analyze the ***j-k-k′-j′*** and ***r-k-k′-r′*** four-vector correlations. If reactants are unpolarized (this is equivalent to lack of information about ***r*** and ***j*** polarizations), the stereodynamics is described by the ***k-k′-j′*** and ***k-k′-r′*** three-vector correlations. And if, on top of that, one is only interested in integral data (this is equivalent to lack of information about the scattering direction), then the stereodynamics is described by the ***k-j′*** and ***k-r′*** two-vector correlations.

9.4.3.1 *Unpolarized Reactants:* **k-k′-j′** *and* **k-k′-r′** *Correlations*

To say that an ensemble of particles is unpolarized is to say that, whatever the set of orthogonal polarization states we consider, the polarization states are mixed incoherently and with equal statistical weights – there is no preferred polarization. For example, unpolarized light can be seen as an incoherent mixture of linearly-polarized photons, of which 50% are vertically polarized and 50% are horizontally polarized. It can also be seen as an incoherent mixture of circularly-polarized photons, of which 50% have right-handed polarization and 50% left-handed polarization. These and other descriptions of unpolarized light are entirely equivalent.

In reaction (9.69), the set of orthogonal BC polarization states could be chosen as, say, $\{|jm\rangle|\}\equiv\{|1{-}1\rangle, |10\rangle, |1+1\rangle\}$. Here, however, it will be more convenient to choose it as $\{|\Psi_a\rangle, |\Psi_b\rangle, |\Psi_c\rangle\}$. The reason is that the data in Section 9.4.1 is expressed in terms of these states.

If the three reactant polarization states are superposed incoherently and with equal statistical weights, the density matrix describing unpolarized reactants ("ur") is

$$\rho_{\mathrm{ur}} = \frac{1}{3}\rho_a + \frac{1}{3}\rho_b + \frac{1}{3}\rho_c = \frac{1}{3}\begin{pmatrix} 1 & 0 & 0 \\ 0 & 1 & 0 \\ 0 & 0 & 1 \end{pmatrix}, \tag{9.75}$$

where we have used Eq. (9.72). Except for a normalization factor, the density matrix of an unpolarized molecule is the identity matrix of the appropriate dimension.

What about the products' density matrix? The equally-weighted incoherent superposition of initial states implies an equally-weighted incoherent superposition of mechanisms *a–c*. As the absolute probabilities that products will actually be scattered in a particular direction by each mechanism are given by their differential cross sections (DCSs), one gets

$$\rho'_{\mathrm{ur}} = N\left(\frac{1}{3}\frac{\mathrm{d}\sigma_a}{\mathrm{d}\omega}\rho'_a + \frac{1}{3}\frac{\mathrm{d}\sigma_b}{\mathrm{d}\omega}\rho'_b + \frac{1}{3}\frac{\mathrm{d}\sigma_c}{\mathrm{d}\omega}\rho'_c\right), \tag{9.76}$$

where $N = (\mathrm{d}\sigma_{\mathrm{ur}}/\mathrm{d}\omega)^{-1}$ is a normalizing factor and

$$\frac{\mathrm{d}\sigma_{\mathrm{ur}}}{\mathrm{d}\omega} = \frac{1}{3}\left(\frac{\mathrm{d}\sigma_a}{\mathrm{d}\omega} + \frac{\mathrm{d}\sigma_b}{\mathrm{d}\omega} + \frac{\mathrm{d}\sigma_c}{\mathrm{d}\omega}\right) \tag{9.77}$$

is the DCS that results from the use of unpolarized reactants (that is, the DCS measured in standard crossed-beam experiments).

The product polarization moments can be calculated from ρ'_{ur}, or else from the polarization moments of mechanisms *a–c via* expressions analogous to Eq. (9.76), one for each combination of rank and component (density matrices are replaced by renormalized PDDCSs). The bottom line is this: the stereodynamics associated with unpolarized reactants is an average of the stereodynamics of mechanisms *a–c*. Reactant states are weighted by $w_a = w_b = w_c = 1/3$, product states by $w'_a = (N/3)\mathrm{d}\sigma_a/\mathrm{d}\omega$, $w'_b = (N/3)\,\mathrm{d}\sigma_b/\mathrm{d}\omega$, $\omega'_c = (N/3)\,\mathrm{d}\sigma_c/\mathrm{d}\omega$.

The mathematical expressions for product DCS and PDDCSs can be found in the supplementary material; here we shall only consider plots of the DCS and two of the renormalized PDDCSs – they are shown in Figure 9.13. The average results are depicted as thick black lines, and the data for mechanisms *a–c* as thinner coloured lines. As for stereodynamical portraits, snapshots taken at $\theta = 60$, 90 and 120° are presented in Figure 9.14.

Let us first examine Figure 9.13. Considering the left panel we conclude that the DCS resulting from the use of unpolarized reactants does give considerable insight into the underlying reaction mechanisms: reactivity increases more or less steadily as one moves from forward to backward scattering. The middle panel, however, contains two somewhat misleading features. The first is the eye-catching, fast-varying nature of the curve near $\theta = 0$. This seems to suggest the presence of some curious effect, but it is actually an artifact. None of the underlying mechanisms changes that abruptly. What happens is that the two dominant mechanisms have DCSs that vanish at $\theta = 0$, which means that in the immediate vicinity of that scattering angle the stereodynamical properties are due to the very minor mechanism *c*. Abrupt variations of stereodynamical properties are likely to appear when the average DCS becomes very small; their importance should not be overemphasized.

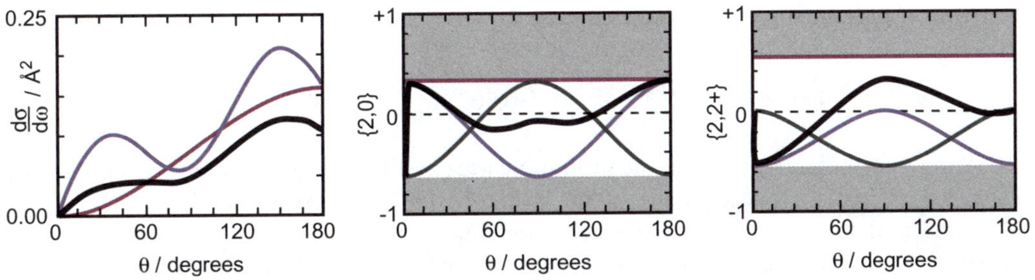

Figure 9.13 DCS (left) and two renormalized PDDCSs that result from the averaging of mechanisms *a* (purple), *b* (blue) and *c* (green). Thick black lines are averages of coloured ones. On the leftmost panel, the DCS of mechanism *c* is too small to be visible.

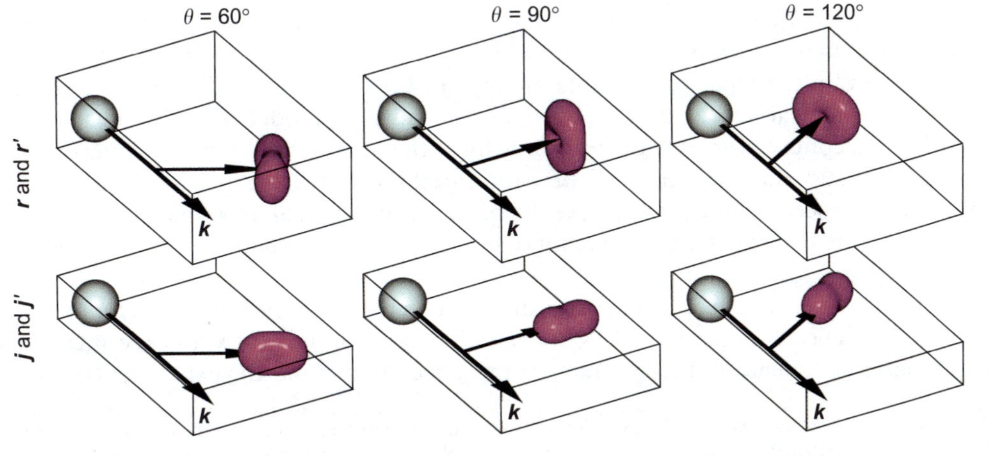

Figure 9.14 As Figure 9.9, but for unpolarized reactants.

The second misleading feature of the middle panel of Figure 9.13 is the suggestion that near $\theta = 90°$ the reaction does not require $\boldsymbol{j}'$ or $\boldsymbol{r}'$ alignment with regard to $\pm\boldsymbol{k}$ (this is what, given our choice of reference frame, the renormalized PDDCS with $k = 2$ and $q = 0$ represents). What happens is that near $\theta = 90°$ the two dominant mechanisms have strong but opposite values for this renormalized PDDCS; the average is not representative of any of them. Contrary to the abrupt PDDCS variation discussed above, this problem cannot be spotted unless the averaging over reaction mechanisms is removed.

Let us now consider the stereodynamical portraits. Note first that the spherical $\boldsymbol{j}$ and $\boldsymbol{r}$ distributions do not represent a $j = 0$, non-rotating BC molecule. Instead, they represent a $j = 1$, rotating but unpolarized molecule. As for the product PDFs, they do capture the essential aspects of the exit stereodynamics: the preference for $\boldsymbol{r}' \perp \pm\boldsymbol{k}'$ and the absence of $\boldsymbol{j}'$ orientation. From this information one may be able to conclude that the exit stereodynamics ensues from break-up of a T-shaped intermediate. Rationalization of the fine details of the portraits, *e.g.*, the variation of PDF shapes with scattering angle, is difficult.

9.4.3.2 *Integral Product Polarization:* k-r′ *and* k-j′ *Correlations*

Using the data of Section 9.4.3.1 one can obtain the normalized product PDDCSs that result from collision of unpolarized reactants. As per Eq. (9.25), their integration over scattering angle yields the polarization parameters (PPs) describing the **k-r′** and **k-j′** vector correlations:*

$$a_0^{\{0\}} = 1.000,\ a_{1-}^{\{1\}} = 0.000,\ a_0^{\{2\}} = -0.002,\ a_{1+}^{\{2\}} = -0.072,\ a_{2+}^{\{2\}} = 0.104.$$

Polarization parameters other than the monopole are small in comparison to their limiting values (see Table 9.3). This is a consequence of the averaging over θ and of the fact that, in general, PDDCS signs and magnitudes change with scattering angle. Although the PPs can be directly illustrative of the underlying mechanisms (in our example this happens in the $a_{1-}^{\{1\}}$ case, for none of the reaction mechanisms involves product rotational orientation), as a rule of thumb one should not expect this to be so. Consider, for example, $a_0^{\{2\}}$. The reason for its practically vanishing value is not irrelevance of $\boldsymbol{r}'$ and $\boldsymbol{j}'$ alignment with regard to $\pm\boldsymbol{k}$, but rather the fact that this type of alignment is strongly dependent on mechanism and scattering angle, *cf.* the middle panel of Figure 9.13.

The message here is this: one should be careful when using PPs, or stereodynamical portraits based on them, to rationalize stereodynamics. For interpretation purposes, PPs are most useful when close to their limiting values (the ambiguity just discussed can then be ruled out) or when complementary information is available. Fortunately, complementary information (kinematic constraints, shapes of potential energy surfaces, additional data from quasi-classical or quantum dynamical calculations, approximate collision models, *etc.*) *is* often available. This provides a context for PP analysis in which the number of possible interpretations is largely reduced, and the usefulness of PPs for interpretation purposes increases accordingly. In our (in this aspect, artificial) example there is no contextual information and PP values are of limited use. In real cases, contextual information is available and PP values can be decisive.

We should also note that, like other polarization moments, PPs can be very useful for validation purposes. If theoretically-determined PPs are found to be in agreement with experimental results, one should have confidence in the accuracy of the calculations (and the experiments).

*Strictly speaking, the **k-r′** and **k-j′** correlations depend only on PPs with component $q = 0$. As the other PPs describe polarization with respect to the x and/or y axes, they cannot play any role when one has no knowledge about the $\boldsymbol{k}'$ direction. See ref. [642] for a detailed discussion.

9.4.4 Manipulation of Collision (stereo) Dynamics

In this section we deal with a question that is almost, but not exactly, the inverse of the one considered in Section 9.4.3. The question is this: if we disregard product polarization, what is the influence of reactant polarization on reactivity? In other words: how can we use reactant polarization to manipulate the integral and differential cross sections of reaction (9.69)?

This question is actually a little more subtle than that of Section 9.4.3. It requires us to consider the distinction between intrinsic, extrinsic and observable reaction properties.

9.4.4.1 Extrinsic, Intrinsic and Observable Reaction Properties

We normally think of chemical reactions as an entirely observable process: reactant molecules approach, interact and are transformed into products that then recoil. However, deeper thought reveals that this observable process actually involves distinct components.[642,646,649]

The first component involves *extrinsic reaction properties*, which are independent of the chemical transformation itself. When we carry out an experiment, or simulate a molecular collision, *we* – not the reaction dynamics – determine how reactant molecules approach. In particular, *we* – not the reaction dynamics – select how the reactant molecules are polarized. For instance, we may carry out an experiment in which we prepare reactants without any polarization. That lack of reactant polarization is an extrinsic property, which we have chosen when designing the experiment. It is entirely independent of the reaction dynamics.

What the reaction dynamics itself does is to transform reactants into products. The relation between reactant states, product states and reaction cross section is determined by *intrinsic reaction properties*. This second component of the observable process is not something we can control. Rather, it is characteristic of the chemical transformation itself. For instance, the reaction may favour T-shaped collision geometries rather than colinear collision geometries (cross sections will be large in the former case, small in the latter). Also, the reaction may lead preferentially to backward or forward product scattering depending on whether the reactant-approach geometry is T-shaped or colinear. The correlations between approach geometry, reaction cross section and recoil geometry are intrinsic properties. They are what we want to probe in our experiment, not what we have specified when designing it.

Loosely speaking, intrinsic properties describe "what the reaction wants", whereas extrinsic properties describe "what the reaction gets". As for observable properties (*e.g.*, the cross section and recoil geometry we will actually observe once we have extrinsically selected a particular approach geometry, and the reaction has intrinsically transformed reactants into products), they result from a compromise between what the reaction wants and what it gets. If we give to it reactants approaching in a T-shaped geometry (that is what the reaction wants), then we observe backward scattering and a large cross section. If, instead, we give to it reactants approaching in a colinear geometry, then we observe forward scattering with a small cross section.

Experimentally, intrinsic reaction properties cannot be directly observed, but they can be inferred from the relation between observable and extrinsic properties. Theoretically, intrinsic reaction properties can be directly examined, and this is what we have focused on up to here.

The question now is: what equation relates the observable DCS of reaction (9.69) to the extrinsic reactants state? The answer (see the supplementary material and Problem 14) is

$$\frac{d\sigma}{d\omega} = \mathrm{Tr}(\rho_i \rho_e) = \frac{\sigma_{ur}}{2\pi} \sum_{k=0}^{2j} \sum_{q=-k}^{k} (2k+1) \left[S_q^{(k)}(\theta) \right]^* A_q^{(k)}, \tag{9.78}$$

where ρ_e and $A_q^{(k)}$ are, respectively, the extrinsic density matrix and complex polarization moments of the reactants (normalized so that $\mathrm{Tr}(\rho_\mathrm{e}) = A_0^{(0)} = 1$),

$$\rho_\mathrm{i} = \frac{\mathrm{d}\sigma_a}{\mathrm{d}\omega}\rho_a + \frac{\mathrm{d}\sigma_b}{\mathrm{d}\omega}\rho_b + \frac{\mathrm{d}\sigma_c}{\mathrm{d}\omega}\rho_c, \tag{9.79}$$

is the intrinsic reactants density matrix, and the

$$S_q^{(k)}(\theta) = \frac{\sigma_a S_q^{(k)}(\theta; a) + \sigma_b S_q^{(k)}(\theta; b) + \sigma_c S_q^{(k)}(\theta; c)}{(2j+1)\sigma_\mathrm{ur}}, \tag{9.80}$$

are the complex intrinsic normalized PDDCSs of the reactants. Note that ρ_i and $S_q^{(k)}(\theta)$ result from incoherent superposition of the (intrinsic) reactant density matrices and normalized PDDCSs associated with each independent reaction mechanism.

Using Eq. (9.78) it can be shown that in Section 9.4.3.1 we have implicitly assumed $\rho_\mathrm{e} = \rho_\mathrm{ur}$, where ρ_ur is the "unpolarized reactants" density matrix of Eq. (9.75). The proof is left to the reader as an exercise.

9.4.4.2 *Control of Reaction Cross Sections*

We can now examine the extent to which reactant polarization can be used to increase or decrease the integral and differential cross sections of reaction (9.69).

First, note the following: because the intrinsic reactant states (the states described by ρ_a, ρ_b and ρ_c) are orthogonal, no extrinsic reactant state can have a large overlap with all of them. The larger the overlap of the extrinsic reactant state with (say) ρ_a, the smaller its overlap with ρ_b and ρ_c. This immediately fixes the extent to which one can manipulate the reaction DCS: whatever the value of the scattering angle, the observable DCS cannot be larger (or smaller) than the largest (or smallest) of the intrinsic DCSs. The observable DCS will be maximum or minimum when the extrinsic reactant state coincides (that is, has unit overlap) with the intrinsic reactant state associated with maximum or minimum intrinsic DCS at the chosen θ value.

Another important observation is that, because extrinsic reactant states are chosen extrinsically – that is, before reactants collide – they cannot change with θ. As intrinsic reactant states generally do change with θ, it is unlikely that any extrinsic polarization can lead to maximum or minimum DCS at all scattering angles.

Let us now consider two specific examples, involving the following extrinsic reactant states:

$$|\Psi_\mathrm{I}\rangle = |1,0\rangle \Rightarrow \rho_\mathrm{e,I} = \begin{pmatrix} 0 & 0 & 0 \\ 0 & 1 & 0 \\ 0 & 0 & 0 \end{pmatrix}, \quad |\Psi_\mathrm{II}\rangle = |1,1\rangle \Rightarrow \rho_\mathrm{e,II} = \begin{pmatrix} 0 & 0 & 0 \\ 0 & 0 & 0 \\ 0 & 0 & 1 \end{pmatrix}. \tag{9.81}$$

Using Eqs. (9.2a), (9.70), (9.72) and (9.78) one gets

$$\begin{aligned}\frac{\mathrm{d}\sigma_I}{\mathrm{d}\omega} &= \frac{\mathrm{d}\sigma_a}{\mathrm{d}\omega}\times 0 + \frac{\mathrm{d}\sigma_b}{\mathrm{d}\omega}\times(1-\cos^2\theta) + \frac{\mathrm{d}\sigma_c}{\mathrm{d}\omega}\times\cos^2\theta \\ &= \frac{1\text{Å}^2}{2\pi}\left[\frac{3\sin(\frac{\theta}{2}) + \sin(3\theta)}{3}(1-\cos^2\theta) + \frac{3\cos(\frac{\theta}{2}) - \cos(3\theta)}{400}\cos^2\theta\right],\end{aligned} \tag{9.82}$$

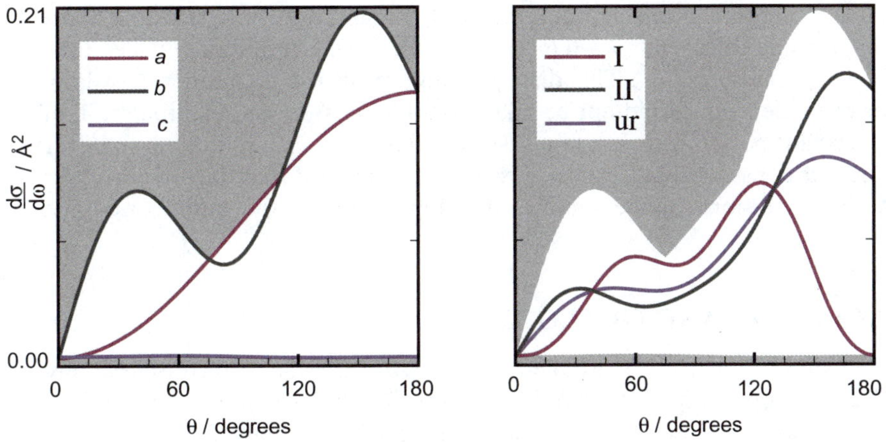

Figure 9.15 Intrinsic (left) and observable (right) DCSs of reaction (9.69). So that the minimum intrinsic DCS is visible, the vertical axes start slightly below 0.

and, with a similar calculation,

$$\frac{d\sigma_{II}}{d\omega} = \frac{1\text{Å}^2}{2\pi}\left[\frac{\sin^2(\frac{\theta}{2})}{2} + \frac{3\sin(\frac{\theta}{2}) + \sin(3\theta)}{6}\cos^2\theta + \frac{3\cos(\frac{\theta}{2}) - \cos(3\theta)}{800}(1 - \cos^2\theta)\right]. \tag{9.83}$$

(You should try to get these results yourself. Once you have done that, you can compare your calculations with those we present in the supplementary material.)

The intrinsic and observable DCSs of reaction (9.69) are plotted on Figure 9.15. The left panel shows the relationship between the intrinsic DCSs and the limits of DCS manipulation – whatever the extrinsic state of reactants, it cannot lead to DCSs penetrating the shaded regions of the figure. The right panel shows the observable DCSs that result from extrinsic preparation of reactants in states I, II, or without any polarization (ur). We see that the extrinsic polarizations we have selected do not do particularly well in minimizing or maximizing the DCS, except when the scattering angle is $\theta = 0$ or 180°. However, they are quite efficient in changing the product angular distribution. While preparation of reactants in state I enhances sideways scattering, preparation of reactants in state II enhances backward and forward scattering.

Let us now consider the manipulation of integral cross sections. Integration of Eq. (9.78) over scattering directions (see Problem 15) leads to

$$\sigma = \sigma_{ur}\sum_{k=0}^{2j}(2k+1)a_0^{\{k\}}A_0^{\{k\}}, \qquad a_0^{\{k\}} = \frac{\sigma_a a_0^{\{k\}}(a) + \sigma_b a_0^{\{k\}}(b) + \sigma_c a_0^{\{k\}}(c)}{(2j+1)\sigma_{ur}}, \tag{9.84}$$

where the $a_0^{\{k\}}$ are intrinsic reactant PPs obtained through incoherent superposition of the intrinsic reactant PPs associated with mechanisms *a–c*.

Using Eq. (9.84) and the data on Eq. (9.70) and Table 9.2 one obtains

$$\sigma\Big/\text{Å}^2 = 0.781 - 0.009A_0^{\{2\}}, \tag{9.85}$$

an expression that shows that, in the case we are considering, the sensitivity of the ICS to reactant polarization is very small. If reactants are extrinsically unpolarized (this implies $A_0^{\{2\}} = 0$), we have

$\sigma_{ur} = 0.781 \text{Å}^2$. Using maximum or minimum $\boldsymbol{j}$ alignment along the reactant approach direction (these imply $A_0^{\{2\}} = 0.316$ or $A_0^{\{2\}} = -0.632$, see Table 9.3), one obtains values that hardly differ: respectively, 0.778 and 0.787 Å^2. This disappointing result is a consequence of how we have constructed our example, and should not be taken as an indication that manipulation of integral cross sections *via* reactant polarization is a hopeless task. There are examples to the contrary. Indeed, one such example was considered in Chapter 5, where it was found that the integral cross section of the inelastic Ar-NO collision can be significantly dependent on the initial orientation of the NO molecule.

9.5 CONCLUSION AND OUTLOOK

In the introduction to this chapter we mentioned that restriction of the discussion to atom-diatom reactions did not imply any significant restriction of the discussion of general stereodynamical concepts. Before closing the chapter we need to justify that statement.

From the stereodynamical point of view, what the study of other types of reaction requires is consideration of additional angular momenta. For example, nuclear spin and electronic angular momenta. How does including these affect stereodynamical analyses?

The key observation here is that, using angular momentum coupling techniques,[239,647] one can regard every bimolecular reaction as a process of this type:

$$[\mathrm{A} + \mathrm{B}](\{n_i\}, j, \Omega) \rightarrow [\mathrm{C} + \mathrm{D}](\{n_i'\}, j', \Omega'), \tag{9.86}$$

where A, B, C and D can be anything (atoms, diatomic molecules, polyatomic molecules, photons, electrons), j and j' are the quantum numbers associated with the magnitude of the total *internal* angular momenta of reactants and products,

$$\boldsymbol{j} = \boldsymbol{J} - \boldsymbol{\ell}, \qquad \boldsymbol{j}' = \boldsymbol{J} - \boldsymbol{\ell}',$$

Ω and Ω' are the reactants and products helicities (that is, projections of $\boldsymbol{j}$ on $\boldsymbol{k}$ and of $\boldsymbol{j}'$ on $\boldsymbol{k}'$), and $\{n_i\}$ and $\{n_i'\}$ are the sets of all other quantum numbers necessary to specify reactant and product states. In a state-to-state reaction these other quantum numbers are fixed and, like v and v' in the case of reaction (9.1), do not play any role in the stereodynamical analysis.

The bottom line is this: if dynamical calculations can be carried out, or polarization measurements made, one can always use them to analyze the reaction stereodynamics. The stereodynamics of every bimolecular reaction is completely described by the $\boldsymbol{j}$-$\boldsymbol{k}$-$\boldsymbol{k}'$-$\boldsymbol{j}'$ correlation.

At present, the main barrier to the development of stereodynamics is the scarcity of detailed comparisons between theory and experiment. Theoretical dynamics (not *stereo*dynamics!) is hampered by the difficulty of scattering calculations. Furthermore, the theoretically accessible reactions can turn out to be inconvenient for experimental studies. As for experimental studies of stereodynamics, they can be very difficult and expensive to perform.

Photodissociation reactions are the possible exceptions. Here a number of detailed measurements and calculations have been performed and compared;[685] a major recent advance was the advent of "complete" photodissociation experiments leading to the measurement of all fragment polarization moments[686] (see Chapter 8).

On the full-collisions side, an important development is expected in the near future; namely crossed-beam experiments involving polarized atoms.[687,688] As theoretical calculations have also progressed in that direction,[279] this avenue may soon lead to detailed comparisons between theoretical and experimental stereodynamics. In the case of inelastic scattering, we have already seen examples in Chapter 5 in which the initial bond-axis has been polarized, and in which the angular momentum polarization of the scattered species has been determined. Some examples of angular momentum polarization effects in bimolecular reactions have also been presented in Chapter 6.

We should also mention the existence of theoretical stereodynamics methods that do not involve polarization moments. They include the nearside-farside and other semi-classical techniques of DCS analysis[689] and the direct inspection of collision and recoil geometries, in particular through consideration of preferred attack angles[690] or through the use of the discrete variable[691] or stereo-directed[692,693] representations of asymptotic states.

Finally, a few words on the *understanding* of reaction stereodynamics. Although one can, at least in principle, determine the stereodynamical parameters of whatever reactive collision one wishes, rationalization of its stereodynamics is a different matter. At present, rationalization strategies invariably involve attempts at visualizing the reactive processes through representative classical trajectories. Intuitive as it may be, this approach is limited: firstly, because chemical reactions can involve effects that cannot be seen in classical trajectories; and secondly, because of a problem that becomes obvious as soon as one considers reactions involving two simultaneously polarisable species; namely that three dimensions may not be enough for a complete view of the stereodynamics.

Alternative approaches have just started emerging. They involve topological descriptions of dynamical effects[694] or the consideration of quantum information/quantum geometry aspects of reaction (stereo)dynamics.[695] These approaches, which at present are virtually unexplored, promise to lead to deep and fresh insights into the underlying structures and the unfolding of chemical reactions. It would be terrific if this book could stimulate its readers to work on such fascinating topics.

9.6 PROBLEMS

Further problems can be found in the online supplementary material.

1. The xyz and $x'y'z'$ reference frames of Figure 9.2 are respectively biased towards reactants and products. Can you think of an unbiased body-fixed reference frame?
2. The vector hierarchy used in Figure 9.3 was $\boldsymbol{k} \rightarrow \boldsymbol{k}' \rightarrow \boldsymbol{j}, \boldsymbol{j}'$. Redraw the figure using a different hierarchy, $\boldsymbol{k} \rightarrow \boldsymbol{j} \rightarrow \boldsymbol{k}', \boldsymbol{j}'$.
3. Using theoretical methods, the authors of ref. [694] have associated a non-adiabatic effect in the dynamics of the $H + H_2$ reaction with product scattering into positive or negative "deflection angles", see Figure 5 of that paper. Experimentally, positive and negative deflection angles cannot be distinguished – they correspond to the same scattering angle. Reactant or product polarization measurements, however, might allow experimental observation of the distinct reaction mechanisms. How? *Hint:* see Problem 2.
4. When reaction (9.69) proceeds *via* mechanism *a*, the real PPs of reactants and products are those on the top row of Table 9.2. What are the value of their complex PPs?
5. A diatomic molecule is in a $j = 1$ rotational state. Calculate its density matrix, knowing that its non-vanishing complex polarization moments are $z_0^{(0)} = 1$, $z_0^{(2)} = \sqrt{1/10}$, $z_{+2}^{(2)} = i\sqrt{3/20}$, $z_{-2}^{(2)} = -i\sqrt{3/20}$. Is this molecule polarized? If so, how?
6. Using the data below, obtain mathematical expressions for the axial and rotational PDFs of the molecule of Problem 5. From the mathematical expressions, can you tell how the molecule is polarized?

$$C_0^{\{0\}}(\Theta, \Phi) = 1, \quad C_{2-}^{\{2\}}(\Theta, \Phi) = (3/4)^{1/2} \sin^2(\Theta) \sin(2\Phi),$$
$$\langle 11, 00|11\rangle = 1, \quad \langle 11, 20|11\rangle = \sqrt{1/10}, \quad \langle 10, 00|10\rangle = 1, \quad \langle 10, 20|10\rangle = -\sqrt{2/5}.$$

7. In the trajectory of Figure 9.8 the internuclear axes of the reactant and product diatomics are oriented. But we have seen that, besides being incompatible with the axial portrait of Figure 9.12, this cannot happen. Show that indeed it does not. *Hint:* draw a new trajectory using reflection through the scattering plane. Can the new trajectory be more or less likely than the original?
8. Use Eqs. (9.18) and (9.22) to show that the complex polarization moments of a density matrix satisfy

$$a_q^{(k)} = (-1)^q \left[a_{-q}^{(k)}\right]^*.$$

Hint: Use the fact that the density matrix is Hermitian (that is, $\langle jm_1|\rho|jm_2\rangle = \langle jm_2|\rho|jm_1\rangle^*$), and the formula[239,647]

$$\langle jm_1, kq|jm_2\rangle = (-1)^q \langle jm_2, k-q|jm_1\rangle.$$

9. Chemical reactions conserve parity. This implies[645] that the elements of helicity-representation scattering matrices satisfy $\langle\Omega'|S^J|\Omega\rangle = \langle-\Omega'|S^J|-\Omega\rangle$, which in turn implies that the density matrix of Eq. (9.36) satisfies

$$\langle\Omega_1|\rho(\theta)|\Omega_2\rangle = (-1)^{\Omega_2-\Omega_1}\langle-\Omega_1|\rho(\theta)|-\Omega_2\rangle.$$

Use this property to show that the polarization moments of $\rho(\theta)$ satisfy

$$S_q^{(k)}(\theta) = (-1)^{k+q} S_{-q}^{(k)}(\theta).$$

Hint: Use Eq. (9.18) and the formula[239,647]

$$(-1)^{\Omega_2-\Omega_1}\langle j\Omega_1, kq|j\Omega_2\rangle = (-1)^{k+q}\langle j-\Omega_1, k-q|j-\Omega_2\rangle.$$

10. Combining the results of Problems 8 and 9, and noting that the symmetry properties obtained there are independent of whether the polarization moments are normalized PDDCSs, renormalized PDDCSs or PPs, we find that the intrinsic polarization moments of chemical reactions must satisfy

$$z_q^{(k)} = (-1)^q \left[z_{-q}^{(k)}\right]^* = (-1)^{k+q} z_{-q}^{(k)},$$

where $z_q^{(k)}$ stands for any of those polarization moments. Use these equalities to prove the following:
(a) If k is even, then the complex polarization moments are actually real.
(b) If k is even, then $z_{q-}^{\{k\}}$ polarization moments vanish.
(c) If k is odd, then the complex polarization moments are actually pure imaginary.
(d) If k is odd, then $z_0^{\{k\}}$ and $z_{q+}^{\{k\}}$ polarization moments vanish.
11. Conservation of parity also implies that the intrinsic rotational PDFs of chemical reactions must satisfy

$$P(\theta_j, \varphi_j) = P(\pi - \theta_j, \pi - \varphi_j).$$

Using this symmetry property and Eq. (9.32) – the classical expansion of rotational PDFs – show that the classical normalized PDDCSs satisfy

$$S_{q}^{(k)}(\theta) = (-1)^k \left[S_{q}^{(k)}(\theta)\right]^*,$$

and therefore that they also follow the rules specified in items (a–d) of Problem 10. *Hint:* Use the following formula:[647]

$$C_{kq}(\pi - \theta, \pi - \varphi) = (-1)^k C^*_{kq}(\theta, \varphi).$$

12. Using the space-fixed frame described in Section 9.3.1.1, sketch an approach geometry one might observe at the start of a classical trajectory. How do the positions of the atoms depend on the collision parameters?
13. Show that the $|\Psi_{\rm II}\rangle$ state of Eq. (9.81) is such that $A_0^{\{1\}} \neq 0$. Taking in consideration the conditions for intrinsic PP values determined in Problem 10, decide how the non-vanishing extrinsic $A_0^{\{1\}}$ will affect the reaction stereodynamics.
14. Prove the second equality of Eq. (9.78). *Hint:* First, rewrite Eq. (9.79) using, for each mechanism, $d\sigma/d\omega = (\sigma/2\pi)P(\theta)$. Next, expand $\mathrm{Tr}(\rho_i\rho_e)$ using Eq. (9.21) twice; note that the expansion coefficients of $P_a(\theta)\rho_a$ are the normalized PDDCSs of mechanism *a*, and similarly for mechanisms *b* and *c*. Finally, use this formula:[647]

$$\sum_{m_1=-j}^{j} \sum_{m_2=-j}^{j} \frac{2k+1}{2j+1} \langle jm_1, KQ | jm_2 \rangle \langle jm_1, kq | jm_2 \rangle = \delta_{kK}\delta_{qQ}.$$

15. Derive Eq. (9.84) from Eq. (9.78). *Hint:* Integration over θ transforms the complex normalized PDDCSs into complex PPs, see Eq. (9.25); integration over φ averages the PPs with $q \neq 0$ to zero; the remaining PPs have $q = 0$ and are therefore real, see Eq. (9.11c).
16. This problem involves fairly long calculations requiring the values of CG coefficients and rotation matrix elements; the supplementary material contains a Maple® worksheet enabling automatic performance of all calculations. As the states and conditions involved can be changed by the user, this problem could be used for a class tutorial, with each student doing different but entirely analogous calculations.
    As discussed in Chapter 12, an ultracold reaction proceeds only *via* the partial wave associated with $\ell = 0$. A consequence of this is that the scattering matrix elements are independent of reactant helicity.[696] As the $\boldsymbol{S}$ elements also satisfy $\langle \Omega' | S^J | \Omega \rangle = \langle -\Omega' | S^J | -\Omega \rangle$ (see Problem 9), for small j and j' values there are few $\boldsymbol{S}$ elements one needs to consider, and fewer still whose values can be independently varied. The problem consists in choosing arbitrary values for the problem parameters and then analyzing the resulting stereodynamics.
    a) Specify the values of j and j'. (The data in the worksheet allows for calculations involving $j \leqslant 5$ and $j' \leqslant 5$.)
    b) Noting that the $\ell = 0$ condition implies $\boldsymbol{j} = \boldsymbol{J}$, decide what $\boldsymbol{S}$ elements can be independently varied, and then assign an arbitrary complex value to each of them. (The worksheet will check whether your specification allows for a full, unique specification of the required $\boldsymbol{S}$ elements, and then scale them so that $\sigma_{\rm ur} = 10\,\mathrm{\AA}^2$.)
    c) Choose a reference frame for the stereodynamical analysis.
    d) Follow the instructions in the worksheet to calculate the following:
        i. DCS of the reaction involving unpolarized reactants, $d\sigma_{\rm ur}/d\omega$.
        ii. Intrinsic density matrices of reactants and products.

iii. Intrinsic real renormalized PDDCS of reactants and products and the associated stereodynamical portraits.
iv. Intrinsic real PPs of reactants and products and the associated portraits.

e) From this data, rationalize the stereodynamics and comment on the possibilities of ICS and DCS control *via* reactant polarization. *Hint:* Think about the consequences of the $\ell = 0$ restriction for the approach direction from the viewpoint of reactive collisions, and also about the consequences of $\boldsymbol{j} = \boldsymbol{J}$ and total angular momentum conservation.
f) Specify a set of directions for $\boldsymbol{r}$ alignment, and then use the worksheet to calculate the resulting ICSs and DCSs. Are the results consistent with your answer to item (e)?
g) Use the worksheet to disentangle the independent reaction mechanisms and calculate the stereodynamical properties of each of them.
h) Rationalize the stereodynamics of the various independent mechanisms. *Hint:* Rather than thinking about the mechanisms in terms of classical trajectories, think about angular momentum conservation.

CHAPTER 10

Surface Scattering: Molecular Collisions at Interfaces

ANDREW HODGSON AND GEORGE DARLING

Surface Science Research Centre, Department of Chemistry, University of Liverpool, UK

10.1 INTRODUCTION

This chapter gives a brief introduction to molecular scattering and reaction dynamics at well-defined surfaces. Many of the considerations applicable to gas-surface scattering are similar to those in the gas phase, but the surface does induce several important differences, most notably in the extended, periodic nature of the scattering potential and in the presence of a thermal heat bath with low energy phonon (vibrational) and electronic (electron-hole pair) excitations.

Molecular reaction dynamics have been studied for many different types of interface, ranging from well-ordered single crystal surfaces of metals, semiconductors and insulators, through adsorbate covered surfaces and self assembled monolayers to liquid surfaces. Studying these surfaces poses a range of different problems, particularly in characterizing and understanding reaction dynamics when the surface is inhomogeneous. On insulators this inhomogeneity occurs because surfaces (particularly oxides) typically have a rather high defect density, while adsorbates and self-assembled monolayers generally have reduced order and liquids are intrinsically disordered. Here we will focus our examples on the most well-characterized systems, which primarily means well-ordered single crystal metal surfaces. These systems have been the focus of considerable study, not least because of the importance of reaction at metal surfaces for catalysis. By restricting the material covered so closely we necessarily exclude much work that is not represented elsewhere in this volume, providing only a flavour of what can be achieved. For a more extensive review of surface dynamics we refer the reader to a recent book,[697] while more specific reviews can be found in refs. [698]–[708].

Tutorials in Molecular Reaction Dynamics
Edited by Mark Brouard and Claire Vallance

Published by the Royal Society of Chemistry, www.rsc.org

10.1.1 Overview of Different Gas-Surface Scattering Channels

When an atom or small molecule scatters from a surface, a number of different outcomes are possible, some of which are illustrated schematically in Figure 10.1 for a diatomic molecule. Depending on the mass, structure and energy of the incident atom or molecule, and the physical and chemical nature of the surface, different scattering channels will predominate. The most pronounced difference occurs between light species, such as $H_2(D_2)$, He and (to an extent) Ne, which scatter elastically from the surface, and all other, heavier species, which undergo inelastic scattering. For a closed shell species such as He, which has no chemical interaction with the surface, the incident beam scatters elastically from the electron density well above the first layer, producing diffraction peaks that reflect the periodicity of the surface. Since close packed metal surfaces have a very flat electron density far from the top layer, courtesy of their delocalized electrons, most of the intensity goes into the specular beam, but more open faces (*e.g.*, face centred cubic (110) surface) and insulators have a much larger corrugation and show strong diffraction. If an attractive potential exists between the atom and the surface, the atom or molecule may scatter into a bound state, creating a selective adsorption resonance. In principle, open shell species such as H and D may also diffract, but chemical interaction with the surface (typically with interaction energies $\geqslant$2.5 eV for transition metal surfaces) generally leads to high energy, inelastic collisions. Diffraction peaks are broadened by disorder, while the fraction of molecules scattered coherently in the elastic beam decreases rapidly with increasing collision energy and surface temperature, and as the mass of the gas becomes comparable to the mass of the surface atoms. The variation of the elastically scattered fraction with surface temperature is known as the Debye-Waller effect.* Inelastic scattering distributions for light species (He), measured by time of flight techniques, show single phonon loss peaks in the scattered distributions and have been used to investigate the phonon bands of surfaces and to measure vibrational spectra of adsorbed molecules. In the case of $H_2(D_2)$, scattering may also excite rotational transitions within the molecule. When there is a strong electronic interaction between $H_2(D_2)$ and the surface, for example if the bond starts to stretch as the molecule begins to dissociate, distortion of the rovibrational eigenstates during collision can give rise to translational to rotational/vibrational energy exchange, as well as an increased corrugation and diffraction. If the barrier to dissociation is energetically accessible, dissociation will occur in the course of the collision, on a timescale of order 10^{-14} s.

Because H_2 and D_2 excite phonons only relatively weakly, and have a simple electronic structure, these two molecules have become the "fruit flies" of gas-surface reaction dynamics. Provided

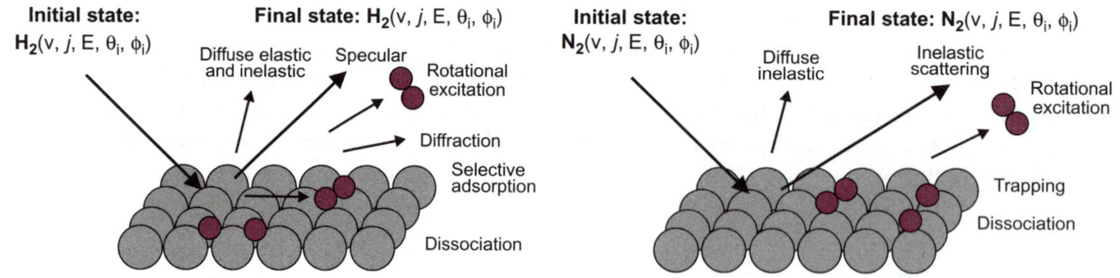

Figure 10.1 Schematic showing the different possible outcomes for thermal scattering of a 'light' molecule such as H_2 (left) or a 'heavy' molecule (*e.g.*, N_2 or suchlike, right) from a periodic surface.

*The Debye-Waller effect is the reduction in the coherent (or Bragg) scattering caused by the thermal motion of the atoms in the solid (in the present context, on its surface). Incoherent/inelastic scattering adds a diffuse background to the elastic diffraction signal.

inelastic excitation of the surface is ignored, the gas-surface interaction can be modelled as a static potential, allowing much of the theoretical machinery developed to describe gas phase scattering to be applied directly to gas-surface scattering, as will be explained in Section 10.2. In contrast, heavier species, with masses more comparable to those of the surface atoms, interact strongly with the surface and scatter inelastically, causing multi-phonon excitation of the surface. In this case, the scattered molecules may undergo rotational excitation arising simply from the impulsive collision with the surface (*e.g.*, N_2), or from a chemical interaction with the surface, which generally makes the potential more anisotropic (*e.g.*, NO). Vibrational excitation of the scattered molecule may also occur, but requires a change in bond length as the molecule interacts with the surface. For heavier, more polarisable molecules, the attractive van der Waals (dispersion or physisorption) interaction creates a long range attractive well into which molecules can be trapped after exchanging energy with the surface. In addition to this site independent physisorption well, short-range chemical interaction with the surface will result in the formation of localized molecular chemisorption states, with well-defined sites and geometries on the surface. These states are not present for $H_2(D_2)$ and provide states into which heavier molecules can scatter and be trapped (Section 10.3). Dissociation in these systems generally occurs *via* a trapping-dissociation mechanism, where the molecule initially becomes trapped in the physisorption well, before transferring into a molecular state and then being thermally dissociated (Section 10.4). Once the coverage of an adsorbate on the surface is sufficient, reaction and desorption will form new products (Section 10.6).

For highly reactive systems, with very different electron affinities or ionization potentials, reaction may occur *via* long range electron transfer, often producing electronically excited products or ions, while several further possibilities can be envisaged at liquid surfaces, including ionization and incorporation into the liquid surface. Space precludes further discussion of these processes here.

10.1.2 Importance of Building an Atomic-Level Understanding of these Channels

Because of the central importance of surface reactions for catalysis, a great deal of interest centres on how molecules dissociate at surfaces. Typically, this is the event that initiates catalytic reaction to form new products, and so understanding such reactions requires a detailed atomistic model of the different channels described above. In particular, we would like to develop a detailed understanding of the way in which the relative importance and outcome of different channels depends on the gas phase molecule and surface in question, so that we can predict (either qualitatively or using computational models) the scattering behaviour that will be important in different circumstances.

In principle, surface reactions can occur *via* either an Eley-Rideal or a Langmuir-Hinshelwood mechanism. In the Eley-Ridel mechanism an atom or molecule from the gas phase reacts directly with a surface species before the gas phase species is thermalized. This channel is significant only when reaction occurs with a high cross section, for example in highly exothermic reactions, particularly those mediated by long range electron transfer. In this case reaction to form new products occurs on the timescale of a single collision, often producing electronically excited species as the electron transfer process follows a non-adiabatic path (see Chapter 4). A variant of the Eley-Rideal mechanism, for reactions with a lower cross section, is for the gas phase species to scatter across the surface, reacting with an adsorbed molecule to form the product before the reactant becomes fully accommodated to the surface heat bath. This mechanism increases the reaction cross section and is particularly important for reactions involving H(D), for which energy exchange to the surface is inefficient.

The vast majority of surface reactions are complex multi-step thermal processes, occurring *via* a Langmuir-Hinshelwood mechanism in which both reactants are fully accommodated to the surface before reaction. However, this does not mean that dynamics are unimportant in describing these reactions, simply that the reaction must usually be broken down into a series of steps whose dynamics and/or kinetics need to be understood separately. Generally, dynamical considerations play a key role in steps that transfer a species from or to the gas phase, whereas the intermediate processes that occur on the surface, such as diffusion, rearrangement, reaction, etc., can be modelled kinetically. This is not to say that dynamical issues are completely irrelevant in these processes, indeed they may be rate limiting, just that they are very difficult to follow and lie outside the limited scope of this chapter. Thus, the dynamical processes important for surface reactions are the initial accommodation (Section 10.3) or dissociation (Sections 10.4 and 10.5) of the reactants on the surface and then the final reaction to desorb new products into the gas phase (Section 10.6).

The adsorption and desorption steps are related; in a scattering picture, recombinative desorption is the time reverse of dissociation, and the two processes can be related *via* detailed balance (Section 10.6). The key theme linking the description of surface reaction dynamics and the gas phase remains the idea of describing motion on a potential energy surface, the validity and accuracy of which is discussed in Section 10.2.

10.1.3 A Brief Comparison with Gas Phase Scattering

Many of the experimental techniques used to investigate surface reaction dynamics originate from gas phase studies, notably molecular beam scattering to study dissociation barriers and energy transfer, and laser spectroscopic measurements to determine energy disposal during surface reaction or scattering (see Chapters 5 to 7). All of the usual state preparation techniques can be applied to the incident gas molecules, including initial state selection, alignment, orientation, etc., and to analyzing the scattered products (see Figure 10.1). These techniques potentially afford a remarkably complete experimental picture of simple surface reactions.[703,704] However, control of the surface is more limited. The surface can be made as ordered as practical, typically by taking a single crystal surface or an ordered adsorbate film, to constrain the collision geometry at the surface, but there always remains the question of the surface thermal heat bath and the distribution of active sites. Phonon or electron-hole pair excitation/destruction is intrinsic to surface reactions, yet is often excluded from theoretical models simply because of the difficulty of describing quantitatively the coupling to the substrate.[709] Since thermal surface reactions greatly favour low energy, thermal channels rather than high energy processes, low energy, low cross section processes are often more important to reaction than high energy, high cross section channels of the type generally studied in dynamical experiments and calculations. Finally, it is worth remembering that surface reactivity is sometimes dominated by minority sites (*e.g.*, the step edge sites in the catalytic dissociation of N_2 on Ru surfaces), or the island edge sites between two reacting phases (*e.g.*, in the catalytic reaction between H_2/O_2 over Pt). Obviously, experiments can be designed deliberately to study these processes, but only if it is recognized that minority sites must be considered.

10.2 THE MOLECULE-SURFACE POTENTIAL ENERGY SURFACE

Much of the general discussion of gas-surface potential energy surfaces (PESs) will be familiar from gas phase systems (see Chapter 2). Terms such as 'diabatic' and 'adiabatic' have the same meaning (see Chapter 4), but there are also significant differences, not least in the computational techniques used to calculate the PES.

When a molecule or atom first approaches a surface there is an attractive interaction due to the long range van der Waals dispersion forces. This decreases into the vacuum as $\sim 1/Z^3$, where Z is the distance of the molecule above the surface. When Z reduces to $\sim 3.5 - 3$ Å, the molecule and surface begin to interact chemically as their electronic wavefunctions overlap. On many surfaces this initial chemical interaction will be repulsive (antibonding), and the combination of this "Pauli" repulsion and the van der Waals attraction results in a physisorption well some distance from the surface. At this distance the detailed surface structure is only weakly felt (at least on flat, *i.e.* not stepped, surfaces), and the molecule behaves in large measure as though it were free to move in the X, Y plane. For small molecules, the van der Waals interaction is weak, producing only shallow surface physisorption wells; for He it is on the order of a few meV, for H_2 on unreactive metal surfaces it is ~30 meV and it can rise to a few hundred meV for larger molecules, such as C_2H_4, on a noble metal. For larger, more polarisable molecules, the physisorption interaction can be substantial, scaling roughly with the total number of atoms in the molecule. Since physisorption involves no orbital rehybridization, the molecular properties are largely indistinguishable from the gas phase. Chemisorption involves much larger energies, often on the order of several eV. Of course, there is a large grey area between these two extremes, where the adsorption energy might be appreciable but we cannot see significant covalent bonding. 'Physisorption' energies of polar molecules on ionic surfaces can be reasonably high because they are largely determined by Coulomb forces between the ions and the partial charges on the molecule, yet there is no clear covalent component to the bond.

We start with the total Hamiltonian describing the interaction of all of the electrons and nuclei in the system,

$$\hat{H} = \hat{T}_{\mathrm{N}} + \hat{T}_{\mathrm{e}} + \hat{V}_{\mathrm{N}}(\boldsymbol{R}) + \hat{V}_{\mathrm{e}}(\boldsymbol{r}) + \hat{V}_{\mathrm{N-e}}(\boldsymbol{R}, \boldsymbol{r}), \tag{10.1}$$

where the first two terms represent the kinetic energies of the nuclei ($\hat{T}_{\mathrm{N}}$) and the electrons ($\hat{T}_{\mathrm{e}}$) with position vectors $\boldsymbol{R}$ and $\boldsymbol{r}$ respectively, the third and fourth terms are the nuclear-nuclear and electron-electron Coulomb repulsion and the final term is the Coulomb attraction between electrons and nuclei. Solving the Schrödinger equation for the motion of all the nuclei and electrons involved in the reaction is challenging for a few atoms and electrons; for a semi-infinite solid it is clearly an impossible task, and so approximations are required. Following the same approach as in the gas-phase (see Chapters 2 to 4), the wavefunction can be written in terms of products of functions, ψ_n, dependent only on nuclear coordinates, and functions, u_n, dependent on both electron and nuclear coordinates,

$$\Psi(\boldsymbol{R}, \boldsymbol{r}, t) = \sum_n \psi_n(\boldsymbol{R}, t) u_n(\boldsymbol{R}; \boldsymbol{r}). \tag{10.2}$$

The Schrödinger equation becomes a set of coupled equations for the nuclear wavefunctions,* ψ_n,

$$i\frac{\partial \psi_n}{\partial t} = (\hat{T}_{\mathrm{N}} + \hat{V}_{\mathrm{N}})\psi_n + \sum_m \langle u_n | \hat{T}_{\mathrm{e}} + \hat{V}_{\mathrm{e}} + \hat{V}_{\mathrm{N-e}} | u_m \rangle \psi_m + \sum_m c_{nm}\psi_m. \tag{10.3}$$

The first term describes the nuclear motion on the inter-nuclear potential, the second gives the electronic energy levels, while the last term couples the electronic and nuclear motions through the $\boldsymbol{R}$ dependence of the electronic functions u_n:

$$c_{nm} = -\sum_i \frac{1}{M_i}\left[\langle u_n | \nabla_{R_i} | u_m \rangle \nabla_{R_i} + \frac{1}{2}\langle u_n | \nabla^2_{R_i} | u_m \rangle\right], \tag{10.4}$$

where M_i and $\boldsymbol{R}_i$ are the mass and position vector for nucleus i. In principle, we could employ either diabatic or adiabatic representations, dependent on the choice of the electronic wavefunctions (see Chapters 2 and 4). However, in practice we can only work with the adiabatic representation, in which the electronic part of the wavefunctions is obtained from an eigenvalue equation involving the nuclear positions only parametrically through the $\boldsymbol{R}$ dependence of the electron-nucleus

*Note that apart from notation Eqs. (10.3) and (10.4) essentially repeat Eqs. (4.5) and (4.8) of Chapter 4.

attraction, *i.e.* we find electronic states for *fixed* nuclear positions,

$$\left[\hat{T}_{\mathrm{e}} + \hat{V}_{\mathrm{N}}(\boldsymbol{R}) + \hat{V}_{\mathrm{e}}(\boldsymbol{r}) + \hat{V}_{\mathrm{N-e}}(\boldsymbol{R},\boldsymbol{r})\right] u_n(\boldsymbol{R};\boldsymbol{r}) = E_n(\boldsymbol{R}) u_n(\boldsymbol{R};\boldsymbol{r}). \tag{10.5}$$

As discussed in Chapter 2, the eigenvalue $E_n(\boldsymbol{R})$, which includes the inter-nuclear repulsion, is a point at position $\boldsymbol{R}$ on the potential energy surface.

There are of course many PESs, each for a different electronic configuration. However, we only have means to compute the ground-state reliably. Because the surface is semi-infinite, this precludes using the full arsenal of quantum chemistry calculations to obtain ground and excited states with arbitrary accuracy; instead, the PES is usually computed using methods from solid-state physics. Density Functional Theory (DFT) is the only easily applicable and reasonably tractable approximation to the full many-body treatment in the solid state.* Since the theorems of standard DFT are applicable only to the electronic ground-state, this is all we can obtain. Although there is no systematic way to improve the accuracy of DFT, it has nevertheless proved accurate enough to have revolutionized the study of gas-surface dynamics, providing unprecedented information on the ground-state PES, and the way in which it varies with surface site, molecular orientation, etc., information that was unobtainable from any other approach. Despite these strengths, it is worth noting that DFT (in present implementations) does not give reliable energies for physisorption interactions and in fact behaves badly at large distances from the surface.

The most common geometric setup for a surface calculation is known as the slab geometry, with supercell boundary conditions, as shown schematically in Figure 10.2. The semi-infinite solid is replaced with a finite number of layers (the slab) of the substrate. Surprisingly few layers, *e.g.*, five, can give a good representation of a metal surface. Parallel to the surface the system has to be periodic. Rather than a single molecule, we therefore have a periodic array of molecules, which will accurately describe adsorption of an isolated molecule if the repeat distance (*i.e.* the size of the supercell) is large enough. In most implementations, the system is also periodic along the surface normal, and the slab has to be separated from its periodic image by an empty region known as the vacuum gap. The slab geometry is indicated in Figure 10.2, and must be chosen with a supercell sufficiently large as to ensure minimal interaction across the vacuum gap, or between adsorbates in adjacent supercells.

The first full DFT calculations of (partial) molecule-surface interaction PESs immediately informed the field. Because the electron density of the close-packed (low index) surface of a noble metal is essentially flat and featureless, it was assumed that molecular interaction potentials would vary weakly with surface site, irrespective of the molecular species. Instead, DFT shows considerable variation in potential with surface site, even for H_2 on a Cu(111) surface (see Study Box 10.1 for an explanation of the indices), as shown in Figure 10.3. These "elbow potentials" are essentially the same as their gas-phase counterparts describing atom exchange reactions, except that the ordinate is the molecule-surface distance, and the abscissa is the intra-molecular bond length. The main point to note is the difference in shape and energetics at the different sites: the elbow is sharper for approach atop a surface atom than at a bridge between them, and the barrier is higher in the atop approach. These differences mean that the surface is corrugated, implying that there exist gradients in the PES that will push the molecules around, and will steer them towards energetically favourable sites. In the literature 'steering' has become almost synonymous with attraction into a potential well, but of course any gradient in the potential will give rise to steering. The corrugation also gives rise to a site dependence of reactivity, although this is not quite as simple as the active site often invoked in the catalysis literature, since different quantum states of the molecule can have different site preferences for reactivity, as we will discuss in Section 10.5.

*A discussion of density functional theory lies outside the scope of this text, but interested readers might wish to look at ref. [70] for an accessible introduction.

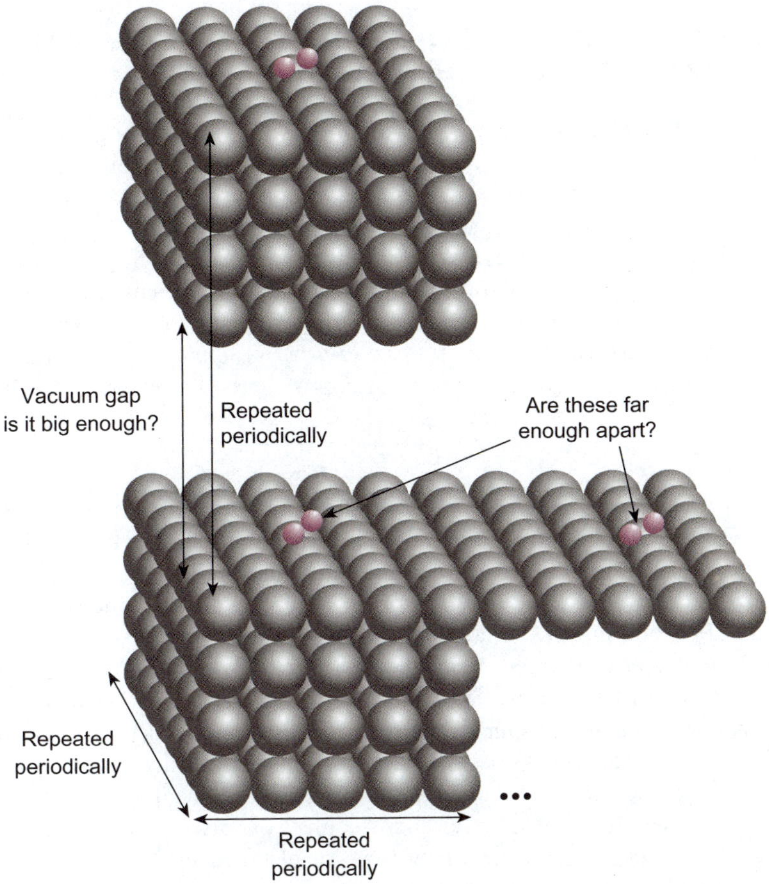

Figure 10.2 Using density functional codes developed for solid-state calculations, it is common to replace the semi-infinite surface with a periodically repeated unit, the supercell. The solid is replaced by a finite number of layers repeated normal to the surface, with a free-space region to represent the vacuum, creating a periodic 3-D supercell. The interaction between an isolated, localized species and the surface can then be calculated within this supercell. Reliable results require the supercell to be sufficiently large, and hence the periodically-repeating adsorbates sufficiently far apart, that any interaction between molecules in neighbouring supercells is negligible.

The chemical nature of the surface will clearly affect the bonding and hence the shape of the PES. For example, when a hydrogen molecule approaches a Cu surface, the interaction between the delocalized metal *s*-states and the molecular σ_g-orbitals gives rise to Pauli repulsion, and a barrier to dissociation. In contrast, for metals with partially filled d-bands, the Pauli repulsion can be overcome by bonding between molecular and localized *d*-orbitals. For small atoms arising from molecular dissociation, the most common adsorption site on a metal surface is the hollow site, where the adsorbate interacts with many surface atoms in a delocalized metallic bond. On a semiconductor or insulator surface, we expect much more localized, classic covalent bonding. For example, H adsorbed on Si bonds with a single Si atom, forming SiH or SiH_2 species, and surface reconstructions can be understood using simple molecular orbital models.

While DFT cannot give us reliable information beyond the ground-state PES, it is obviously desirable to understand excited states. If the substrate is an insulator, for example an ionic solid such as a metal oxide, then we can use a different representation of the surface in order to employ

STUDY BOX 10.1: LATTICE VECTORS, MILLER INDICES, AND RECIPROCAL SPACE

We can think of a surface as a cut through a plane of a solid. The arrangement of atoms at the surface must therefore depend on the crystal structure of the solid, and also on the plane through which we make the cut. For this reason (amongst others) we need a concise way in which to describe crystallographic planes within solids.

Crystal structures are described in terms of a unit cell, specified by the vectors $\boldsymbol{a}$, $\boldsymbol{b}$ and $\boldsymbol{c}$, along axes a, b and c, that define the translational repeat unit of the lattice. Unless the crystal has a cubic structure, these axes will not in general correspond to a set of Cartesian axes within the crystal. A direction or location of a point in a crystal is defined by a lattice vector, which describes the distance from the origin in terms of components along the unit cell axes,

$$\boldsymbol{r} = u\boldsymbol{a} + v\boldsymbol{b} + w\boldsymbol{c} \quad \text{(B10.1.1)}$$

and is written as $[uvw]$. A crystallographic plane is described by its Miller indices (hkl) which specify the inverse of the distance from the unit cell origin at which the plane cuts the a, b and c axes. The indexes, h, k and l are chosen as integers, such that the plane cuts the axes at distances a/h, b/k and c/l from the origin, the plane (hkl) being perpendicular to the vector $h\boldsymbol{a}+k\boldsymbol{b}+l\boldsymbol{c}$.

Note that the above definitions relate to a description of the crystal in *real space* (also sometimes called *direct space*). Much of the theory of crystals, including their electronic structure, vibrational (phonon) structure and diffraction behaviour, relies on an alternative description in terms of a *reciprocal space*. The reciprocal space of a crystal lattice consists of an infinite periodic three-dimensional array of points whose spacings are inversely proportional to the distances between planes in the real space lattice. These reciprocal space lattice vectors, $\boldsymbol{a}'$, $\boldsymbol{b}'$, $\boldsymbol{c}'$, have dimensions of $(\text{length})^{-1}$, which matches the units of wave vectors describing excitations of the crystal, such as a lattice vibration (photon), or electron momentum, and represent the Bragg diffraction condition. Just as any time-dependent property may be described in terms of a Fourier sum in the frequency domain, the spatial properties of a crystal can be described as a sum of components in this reciprocal space.

The surface reciprocal lattice vectors also define the allowed momentum change parallel to the surface during coherent elastic scattering. If vectors $\boldsymbol{a}'$, $\boldsymbol{b}'$ define the surface reciprocal net, the allowed momentum change parallel to the surface is $\boldsymbol{G}_{hk}$, where $\boldsymbol{G}_{hk},=h\boldsymbol{a}'+k\boldsymbol{b}'$, leading to an array of diffraction peaks (h, k). Further discussion of the reciprocal lattice is beyond the scope of this book, but interested readers can find more details in any textbook on crystallography, surface science and solid state physics, *e.g.*, ref. [711].

Claire Vallance
with contributions from *Andrew Hodgson and George Darling*

quantum chemical techniques. Unlike for a metal, the electronic states are localized, and so we can use a cluster model of the surface to capture the bonding and energetics accurately. To account for the rest of the surface, we can embed this cluster in a field of point charges for which the Madelung energy* can readily be computed, as illustrated in Figure 10.4.

*The Madelung energy due to one ion in a crystal as a result of its interactions with all other ions (treated as point charges) is $E = \frac{z^2e^2}{4\pi\varepsilon_0 r_0}M$, where M is the Madelung constant, a geometrical factor which depends on the arrangement of the ions in the crystal, z is the charge on the ion, and r_0 is the nearest neighbour separation.

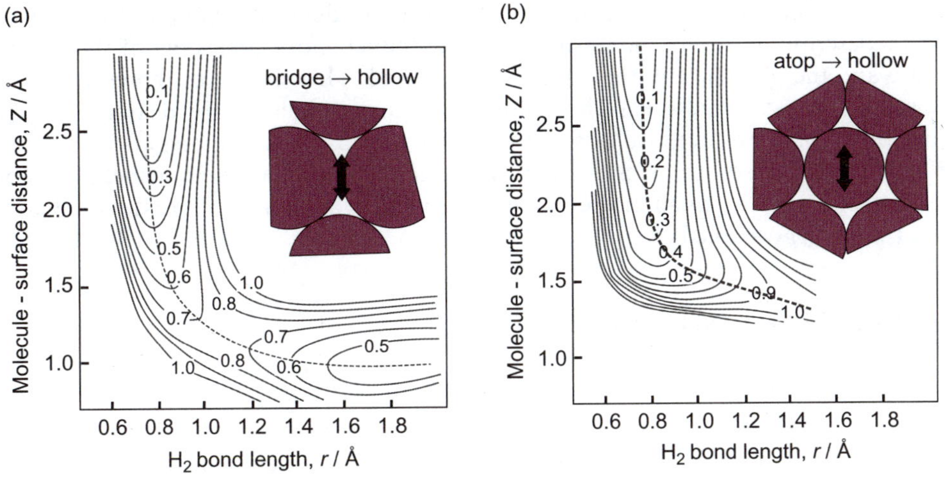

Figure 10.3 Slices of the PES for H_2/Cu(111) taken at two different surface sites as a function of the molecule-surface distance, Z, and H–H bond length, r. The molecules are oriented parallel to the surface. The bridge site has the lowest dissociation barrier on this surface when the atoms dissociate into the hollow sites. On the atop site, the barrier is $\sim$0.2 eV higher, and the bend of the elbow is much more abrupt, adapted from ref. [710].

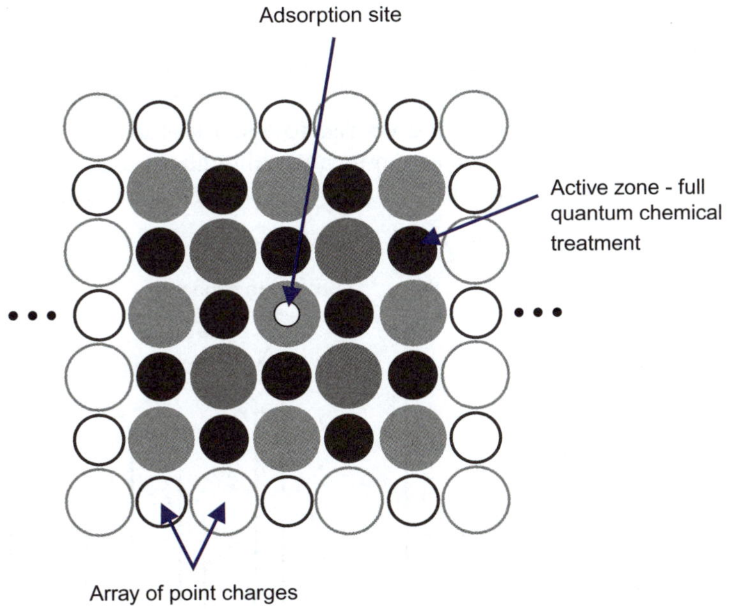

Figure 10.4 The more localized bonding on insulator or semiconductor systems can often be replaced by a finite cluster to which we can apply a full quantum chemistry treatment. The electrostatic field from the remainder of the crystal can be approximated by embedding the cluster in an array of point charges for which the Madelung energy can be included analytically.

Whereas the energies of the ground and excited states of molecules on insulators can be expected to be reasonably well separated, this is not so for metals, which have no gap between their occupied and unoccupied electronic states. In the literature, these excited states (or excitations) are often called 'electron-hole pairs'. In essence, this describes the technical approach to dealing with excitations in a continuum of electronic states. The fermionic description of the electrons is replaced by a bosonic description, where the bosons are a combined entity consisting of the electron in its excited state and the 'hole' left in the band it has vacated. This change of representation is really only a rewriting of the Hamiltonian, but allows us to use techniques for dealing with quantized lattice vibrations (phonons are also bosons) to derive approximate expressions for energy loss to electron-hole pairs. These approaches show that electronic excitation in a metallic substrate is particularly strong when a vacant molecular level, termed an affinity level, crosses the Fermi energy to become (partially) occupied.[712]

10.3 SCATTERING AND TRAPPING

10.3.1 Elastic Scattering

A PES which varies with surface site will cause the scattering molecules to change direction, *i.e.* the forces at the surface will transfer momentum normal to the surface into momentum parallel to the surface. The amount by which the molecules change direction depends on the corrugation of the potential and on the site where the molecule hits. Striking at the point of inflection where the slope of the PES is greatest gives the largest deflection in trajectory, as shown in Figure 10.5. Classically, this leads to a peak in the scattering cross section at a particular scattering angle, known as the rainbow angle (see Study Box 5.1). In contrast to the simple picture shown in Figure 10.5, in reality the peak will be smeared out by energy loss to the substrate, and when we take into account the full two-dimensional nature of the surface and the internal degrees of freedom of the molecule.

In a quantum description of the molecule there is no rainbow singularity, but for initial momentum $\boldsymbol{K}$ parallel to the surface, the final parallel momentum is restricted by the periodicity of the surface to the set of vectors $\{\boldsymbol{K}+\boldsymbol{G}_{hk}\}$, where $\boldsymbol{G}_{hk}$ are the surface reciprocal lattice vectors (see Study Box 10.1). Thus, quantum scattering gives rise to a series of distinct diffraction peaks. Diffraction is a coherent, elastic quantum phenomenon, and although the angular distribution obtained from classical mechanics often appears similar to the intensity envelope of the diffraction peaks, in

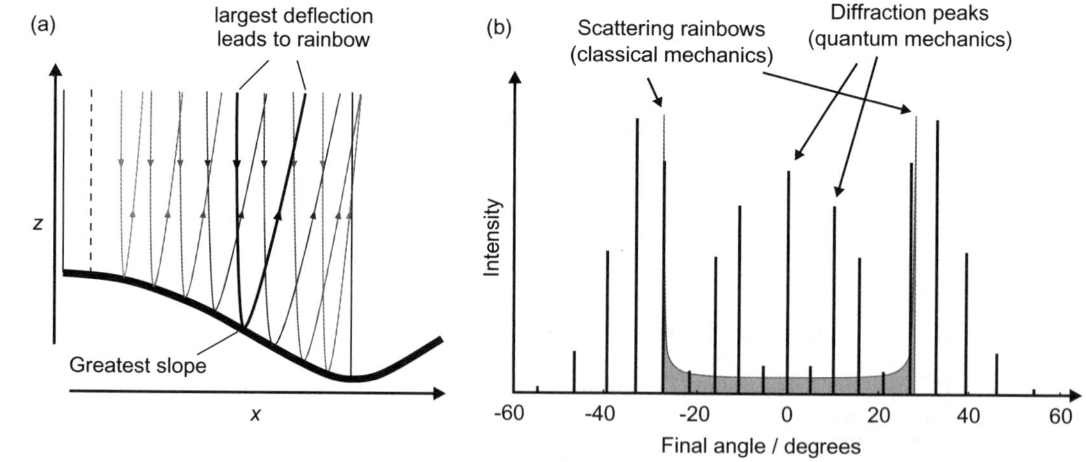

Figure 10.5 Rainbow scattering. (a) Trajectories incident on a simple corrugated two-dimensional model PES, plotted in the X, Z plane, where X is a surface coordinate and Z is the distance from the surface. The trajectories striking the points of inflection where the slope is greatest experience the largest deflection and lead to a rainbow (singularity) in the scattering distribution. (b) The scattering probability as a function of the scattering angle for the model in (a). The classical results show a rainbow, whereas a quantum calculation gives precise diffraction peaks.

reality there will be oscillations in intensity which reflect the detailed structure of the surface and cannot be accounted for classically, very much as in gas phase elastic scattering (see Chapter 5).

The restriction on the diffraction is, of course, that energy must be conserved, so that the highest momentum state from the $\{\boldsymbol{K}+\boldsymbol{G}_{hk}\}$ diffraction set cannot have energy higher than the initial incidence energy. States for which the parallel momentum is too high are closed states for direct scattering back into the gas-phase, but we can access a subset of these closed diffraction states by trapping the molecules into the physisorption well. Motion normal to the surface is exchanged for motion parallel to the surface, as in diffraction, but the translational kinetic energy in the trapped molecules is greater than the initial total energy, E_i. The excess is balanced by the (negative) binding energy, ε_n, of the nth bound-state of the physisorption well, *viz.*,

$$E_i = \frac{1}{2M}\left|\boldsymbol{K} + \boldsymbol{G}_{hk}\right|^2 + \varepsilon_n, \tag{10.6}$$

where the first term gives the translational kinetic energy of the trapped molecules of mass M, and initial momentum $\boldsymbol{K}$ parallel to the surface. In essence, the molecules are diffracting into a channel that is closed far from the surface, and simultaneously dropping into one of the bound states of the physisorption potential. Selective adsorption appears in experimental results as dips in the reflectivity at energies and angles satisfying Eq. (10.6), as shown in Figure 10.6. In principle, selective adsorption can be completely elastic (*i.e.* no energy is lost to the substrate), the trapping is reversed

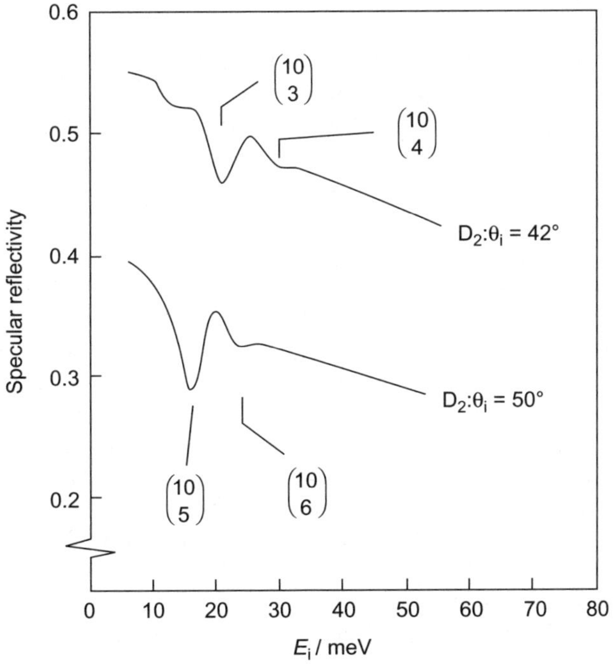

Figure 10.6 Specular reflectivity as a function of incidence energy for H_2 and D_2 scattered from a Cu(100) surface. The intensity decreases overall with increasing energy because the probability of scattering into other channels increases, as does the energy loss to the substrate. Selective adsorption resonances appear as dips in the reflected intensity as the diffraction condition is met, and the precise energy and angle dependence allows one to determine the diffraction state (h, k) (in the examples shown this is the (10) diffraction beam, see Study Box 10.1) and the bound state vibrational level (lower number) into which the molecules are trapped (adapted from Andersson *et al.*[713]).

by diffraction back into open channels of lower parallel momentum, allowing the molecule to escape from the surface.

Selective adsorption can also be achieved by excitation of molecular rotation during the collision with the surface. In this case the increase in rotational energy of the molecule is compensated by the molecule dropping into a bound state of the physisorption well. In general, selective adsorption resonances involve excitation of a combination of molecular rotations and parallel momentum, and will occur for particular energies, angles of incidence and rotational states satisfying:

$$E_i = \frac{1}{2M}|\boldsymbol{K} + \boldsymbol{G}_{hk}|^2 + BJ(J+1) + \varepsilon_n, \tag{10.7}$$

where B is the rotational constant of the molecule and J the final rotational state (assuming an initial state of $J=0$ for simplicity). The bound-state energies are very sensitive to the precise shape and energies of the physisorption well and this makes selective adsorption an extremely accurate probe of the PES far from the surface, the region where density functional theory produces inaccurate results for the energy. Heavier molecules interact too strongly with the substrate vibrations for sharp elastic scattering features to be resolved.

10.3.2 Inelastic Scattering

Motion of the surface atoms also plays a role in the scattering. At the simplest level, molecular energy may be lost to (or possibly gained from) the surface vibrations. A simple binary collision model, often called the cube model, gives sensible trends.[714] The model assumes that the molecule is in collision with an effective surface 'atom', which is initially stationary. From conservation of energy and momentum in the collision, assuming a sudden collision with a hard wall PES, we get the convenient Baulé estimate for the translational energy loss, ΔE, to the substrate

$$\Delta E = \frac{4\mu}{(1+\mu)^2}(E_i + W), \tag{10.8}$$

where E_i is the initial translational energy of the projectile towards the surface, W is the depth of the physisorption well (approximated by a square well) and μ is the ratio M/M_s of the molecular mass, M, of the scattered molecule to the surface atom mass, M_s. The Baulé formula tells us that the energy transferred from the molecular motion into the surface will depend on how well matched the masses are and how hard the molecule hits the surface, *i.e.* on the kinetic energy, E_i. The model can be extended to include the initial thermal motion of the surface atom by considering a range of relative approach speeds and thermally averaging. The appropriate value for M_s depends on how close the molecule gets to the surface atoms, on the stiffness of the surface, and on the degree of corrugation of the PES. If the interaction is really with a large number of surface atoms, as might be the case for delocalized bonding, M_s will be large, while if the bonding is localized and covalent, M_s will be closer to the mass of a single surface atom. If the surface is very soft, it will easily distort, with atoms moving relatively independently, and this will also lead to a smaller value for M_s. On metals, M_s is often estimated as ~ 3 times the mass of a surface atom.

If the incident energy drops below a critical value of

$$E_i = \frac{4\mu W}{(1-\mu)^2}, \tag{10.9}$$

then $\Delta E > E_i$, and the molecule becomes trapped into the physisorption well. For a weakly corrugated surface, where the coupling between motion parallel and normal to the surface is weak, the energy exchange will be determined by the normal component of the motion. Consequently, for a fixed incident energy, we would expect the trapping to increase as the incident angle of the molecules increases away from normal. Of course, trapped molecules will eventually return to the gas-phase once they acquire thermal energy at least comparable to the well depth. They can be easily distinguished from molecules that scatter promptly, but with a loss of energy to the surface, since they will desorb with a completely different energy and angular distribution. The promptly scattering molecules will have an energy less than, but related to, the incident energy; this is known as direct-inelastic (DI) scattering. For DI scattering the angular distribution will be related to the incident angle, typically slightly further from the normal because the energy loss to the substrate will largely come from the normal momentum component. In contrast, the trapping-desorbing (TD) component will desorb in a cosine-type angular distribution peaked about the surface normal, with an energy distribution related to the surface temperature, as shown in Figure 10.7. This is discussed in more detail in Section 10.6.

In addition to changing the translational energy of the molecule, scattering from a surface can also lead to changes in the internal energy. The extent to which this occurs depends on the details of the PES. We will take NO on metal surfaces as an example. NO generally favours chemisorbing with the N end pointing down towards the surface, with the opposite O-down orientation either

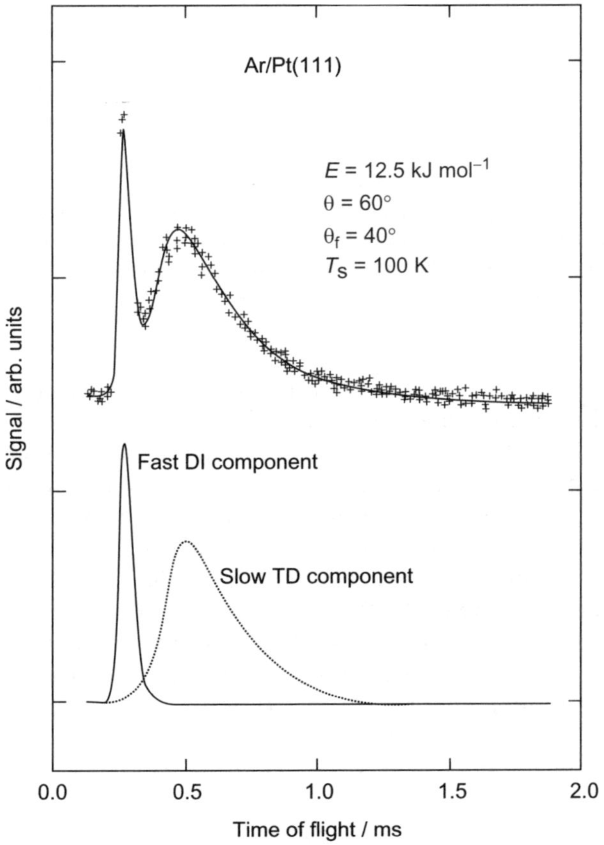

Figure 10.7 Top trace: Experimental time-of-flight data (points) for Ar scattered from a Pt(111) surface. The bottom trace shows a breakdown into a fast direct-inelastic component (scattering only with slight energy loss to substrate vibrations) and a slower trapping-desorption component from molecules that are thermally equilibrated with the surface, (adapted from ref. [715]).

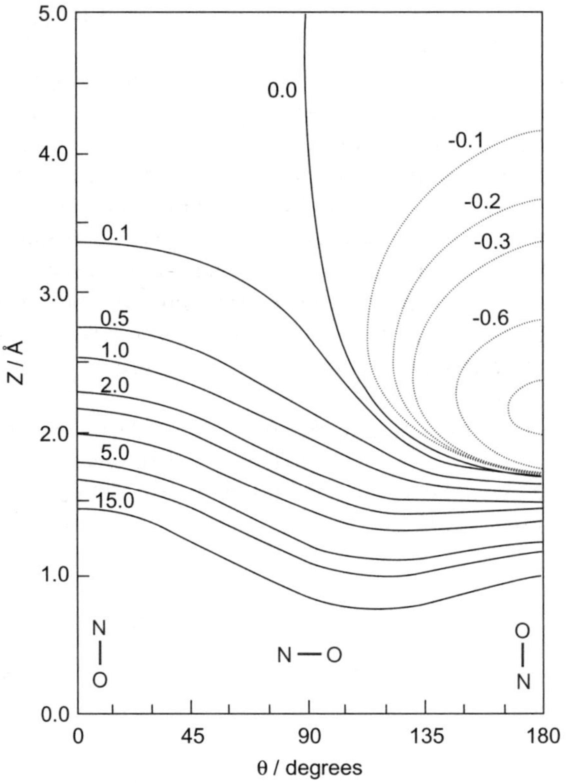

Figure 10.8 A PES for NO interacting with a single site on a metal surface as a function of molecule-surface distance and molecular orientation. The bonding favours the N-down geometry, for which there is a shallow, ~0.75 eV, chemisorption well, whereas for O down the PES is repulsive (adapted from ref. [716]).

more weakly bound or repulsive, as illustrated in Figure 10.8. A molecule moving on this PES will be pulled towards the N-down orientation, *i.e.* it will steer to the right of the PES plot shown in Figure 10.8. Assuming the molecule is not trapped at the surface, it should emerge back into the gas-phase. The steering motion in the molecular orientation coordinate is of course a molecular rotation, so the molecule will emerge with translational energy converted to rotational energy.

Looking carefully at the positive energy contours in Figure 10.8, we see that they are remarkably similar in shape to the energy contour shown in Figure 10.5. There is a particular molecular orientation for which the curvature of the PES is a maximum. Classically, a molecule striking this point of the PES will experience a maximal deflection in the orientation coordinate, *i.e.* it will experience the largest amount of rotational excitation. This gives a rainbow peak in the rotational distribution of the scattered molecules, which can be seen in systems for which the chemisorption well is not too deep.[717] Rotational excitation induced by scattering from the repulsive part of the potential should occur for all molecules, including small molecules like H_2 that have a rather spherical electron density, but in this case the rotational effects are much greater for HD, in which the centre-of-mass is displaced towards the D end.

Depending on the details of the PES, molecules scattering from a surface can also undergo vibrational excitation/de-excitation. This requires the barrier in the potential to occur after the molecular bond has begun to stretch; that is, the barrier is late (*cf.* Polanyi rules for vibrational effects in dissociation, discussed in Section 10.5, and see also Study Box 2.1 and Chapter 7).

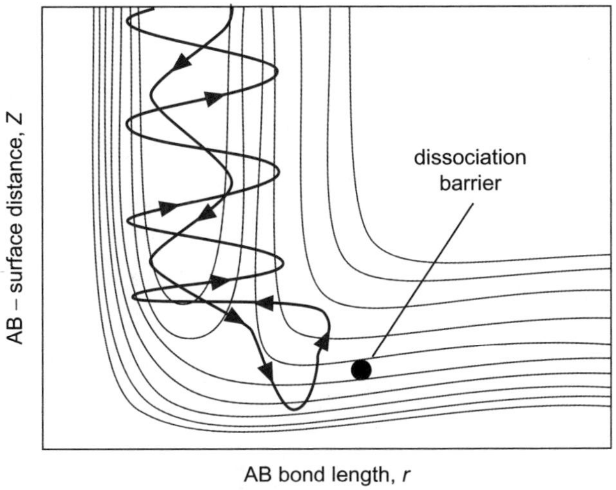

Figure 10.9 A classical trajectory showing a molecule scattering from a low vibrational state to a high vibrational state, with loss of translational energy to vibrations. This trajectory is reversible and a slow, vibrationally hot molecule can be converted into a fast vibrationally cold molecule on scattering from this PES.

Figure 10.9 shows a 'typical' classical trajectory scattering from an elbow potential into an excited vibrational state. Motion in the bend of the elbow couples the translational motion with the vibration, giving translation-vibration inter-conversion. If we neglect energy exchange with the surface, excitation and de-excitation trajectories are simply time-reversed images of each other. Therefore, if we observe vibrational excitation in the scattering of fast, vibrationally cold molecules, then we should also be able to detect fast, vibrationally cold molecules leaving the surface when we scatter slow, vibrationally hot molecules.

The conversions between translational and internal motions considered so far are driven by the PES, but the surface itself can also contribute directly. For example, in scattering even of H_2 from a Cu surface, it is observed that the rotational excitation has a roughly Arrhenius behaviour, at least over the range of surface temperatures studied. Although this temperature dependence is not completely understood, it is clear that it can arise in very simple models in which the PES, such as that shown in Figure 10.3, is mechanically coupled to an oscillator. We simply include another coordinate in the PES: $V(Z) \rightarrow V(Z-y)$, where y is a coordinate describing the surface oscillator. Although this is a crude model of surface motion, for instance it takes no account of the possibility that the topography of the PES changes because the surface is thermally distorted, it does qualitatively reproduce the experimental findings. A much more complicated behaviour appears to occur for NO scattering from noble metals. Again, the excitation of the internal mode (in this case vibrations) shows a roughly Arrhenius dependence on surface temperature, but while mechanical coupling to a surface oscillator can again qualitatively reproduce this finding, detailed modelling indicates that coupling to the electron-hole pairs of the metal is important. The basic idea is that the vibration of the molecule causes the affinity level to sweep constantly across the Fermi level. The fluctuating population in this level couples to the electronic excitations; as the surface temperature increases, so does the population of the electron-hole pair states, thus increasing the coupling, and driving vibrational excitation. Support for this idea comes from the scattering of highly vibrationally excited NO molecules from Au surfaces that have been "precovered" with a small number of Cs atoms in order to lower the work function (*i.e.* the ionization energy of the metal surface).[718] It is found that the very rapid vibrational motion couples so strongly to the electronic states of the metal that electrons are emitted into the vacuum, with a probability increasing with initial vibrational state, as shown in Figure 10.10.

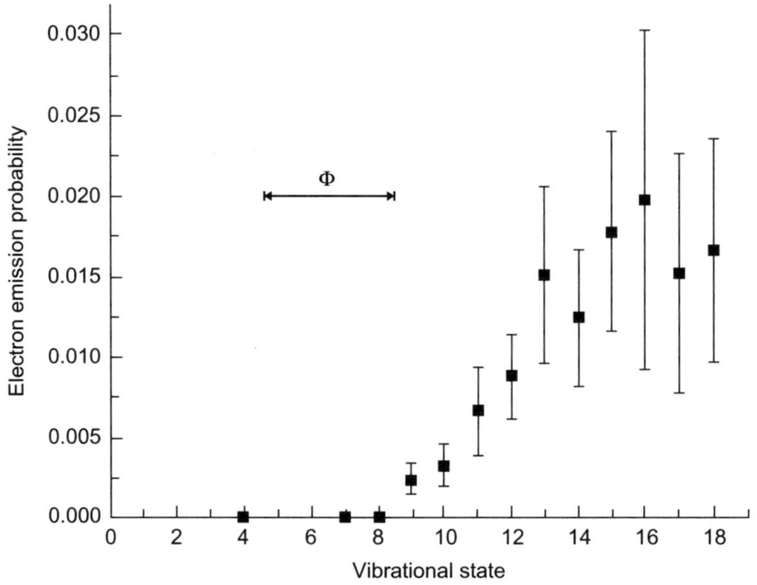

Figure 10.10 Probability of electron emission during scattering of vibrationally excited NO molecules from an Au surface precovered with Cs to lower the work function, Φ. The horizontal bar indicates roughly the number of vibrational states the work function spans (adapted from ref. [718]).

10.4 NON-ACTIVATED DISSOCIATION

The goal of much of gas-surface dynamics research has been to study the fundamental dynamics that lead to dissociation and subsequent surface reaction. Dissociation probabilities have been measured for many reactions, either by measuring the reflectivity of the active surface, or directly by measuring the amount of the dissociation product on the surface (using spectroscopy or a titration technique), or from state and energy resolved measurements of associative desorption by assuming detailed balance (see Section 10.6). There is a close link between the dynamics of scattering and the dynamics of dissociation; we could consider the scattered molecules as those that have simply failed to dissociate. The route to dissociation takes the molecule through the same energy regions of the PES it would traverse when scattering, so the shape of the PES influences both the induced coupling between translations, rotations and vibrations of the molecule and the dissociation probability.

In general terms, we can classify dissociation into three paradigms: direct dissociation, trapping or precursor mediated (*i.e.* indirect) dissociation, and steering-dominated dissociation. In direct dissociation, the molecule breaks apart if it has enough energy to get over a barrier. Increasing the molecular energy should lead to more dissociation as more molecules are then able to overcome the dissociation barrier. In trapping-mediated dissociation, all molecules that dissociate have first been trapped intact in a molecularly adsorbed state. The absorbed molecules explore the surface until they find a reactive site where they can dissociate, or until they obtain enough energy to overcome a dissociation barrier. As the translational energy of a molecule increases, it becomes harder for it to lose enough energy to the surface to be trapped; consequently the signature of an indirect reaction is a decrease in dissociation with increasing E_i. In the final case of steering-dominated dissociation, strong attractive forces pull the molecule into particular surface sites and orientations where non-activated dissociation can occur. The steering forces re-direct the momentum of the molecule, and are therefore less effective at high energy, giving a decrease in dissociation probability as the initial energy increases.

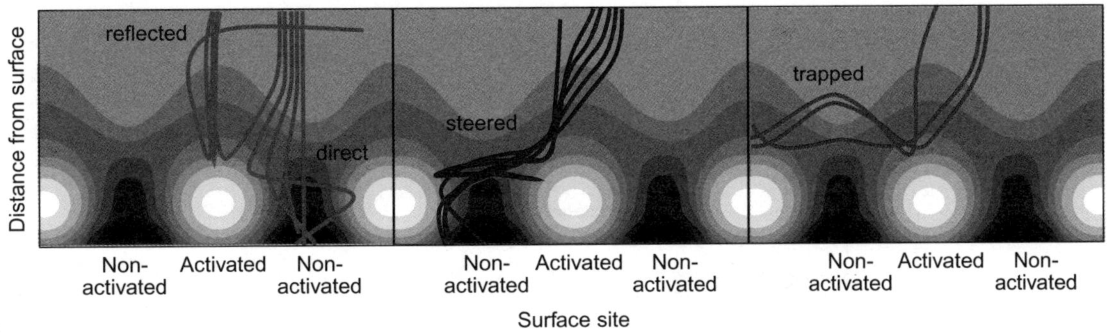

Figure 10.11 Classical trajectories for molecules incident on a surface shown as a function of the lateral site on the surface and the distance from the surface. The potential was chosen such that some sites have an attractive well followed by a barrier to dissociation, as indicated by the shaded regions, while other regions have no barrier to dissociation. The motion of the molecule from one region to another across the surface is indicated by representative trajectories. The left-hand panel shows dissociation or reflection occurring directly at the surface site the molecules initially strike, while the middle panel shows molecules being steered across the surface and dissociating at a nearby site that has no barrier, while the right-hand panel shows trajectories leading to trapping on the surface (adapted from ref. [719]).

In reality, dissociation dynamics cannot be entirely and uniquely categorized in this fashion, and many systems will show a combination of effects. For instance, accessing the molecularly chemisorbed precursor state could be an activated process, leading to an increase in dissociation with molecular energy as the molecules become increasingly likely to overcome the barrier to the precursor state. Steering and trapping are also closely linked and occur together in many simple systems, because the existence of attractive steering forces implies the presence of adsorption wells in the PES into which molecules can trap before finding the dissociation site. For a reaction that is non-activated, we can expect to see all of these effects because of the variation in PES with surface site and molecular orientation. Figure 10.11 shows classical trajectories projected onto two dimensions for a model of hydrogen dissociation on a transition metal. At some sites dissociation is completely non-activated and the potential energy decreases smoothly as the molecule approaches the surface. Molecules striking here simply dissociate as indicated by the direct channel label. At other sites, the chemical interaction is initially favourable when the molecule approaches the surface, but then ultimately unfavourable, producing a barrier to dissociation at this site. Molecules in the vicinity of this barrier are steered into the well, but cannot dissociate there. In this example, the fate of these molecules is either to steer over to the next non-activated dissociation site, or to become trapped at the surface (or scatter back to the gas-phase). As a function of energy, the dissociation probability in this type of model can show a decrease or an increase, depending on the detailed shapes and relative depths of wells and heights of barriers. Figure 10.12 shows experimental data for the N_2/W(100) system. At low energy trapping dominates, but at increasing energy the trapping (and probably steering) is ineffective and we see direct activated dissociation occurring at other sites on the surface.[720]

Increasing the angle of incidence of the molecular beam will reduce the momentum normal to the surface and, following the Baulé model, we expect more trapping and thus more dissociation. The angular dependence of molecular sticking and dissociation has often been quantified in terms of the 'energy scaling'

$$S(E_i, \theta) = S(E_i \cos^n\theta, 0), \tag{10.10}$$

in which, for incident energy E_i and angle θ, the scaling exponent is n. Put simply, if we plot the sticking at all angles of incidence on the same graph, we will find that only when plotted against

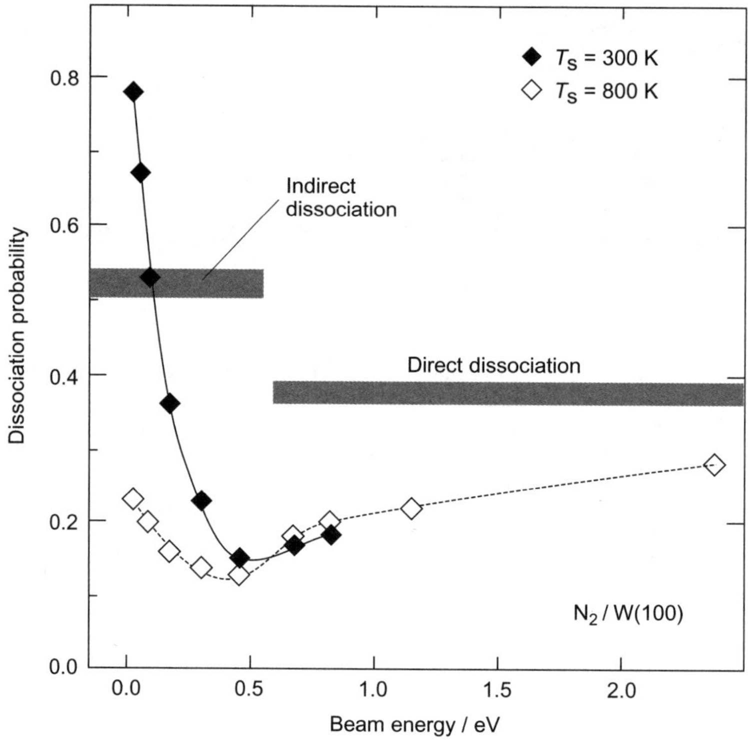

Figure 10.12 Dissociative chemisorption of N_2 on W(100) at 300 K and 800 K, adapted from ref. [720].

$E \cos^n\theta$ do the points all lie on the same curve. If the PES is flat (no X, Y dependence) then only the normal component of momentum influences sticking and dissociation and the scaling exponent would be $n = 2$, commonly referred to as 'normal energy scaling'. For trapping on a corrugated surface, the angular dependence is less pronounced than on a flat surface and the exponent generally ranges between 0 (*total energy scaling*, *i.e.* the dissociation is angle independent) and 2.[721] Plotted against the total translational energy of the molecule, the dissociation probability at normal incidence is greater than that at off-normal incidence. If the dissociation probability at off-normal incidence was greater than that at normal incidence when plotted against total energy, the value of n would be negative. In such a case the accommodation of the parallel momentum would be more important in the trapping process, which could occur on a very highly corrugated PES.

Precursor mediated dissociation should also show a pronounced surface temperature dependence. If all of the trapped molecules can eventually find a reactive site and dissociate when the surface is cold, then increasing the surface temperature will cause a fraction of the molecules to acquire enough energy to desorb, reducing the dissociation probability, as shown in the low energy regime of Figure 10.12. Heating the surface causes a fraction of the molecules to acquire enough energy to desorb rather than dissociate. Experimentally, a decrease in dissociation with increasing surface temperature is generally considered to be a clear signature of precursor mediated dissociation. However, if the molecule is trapped for only a short time in the precursor state, it need not be completely equilibrated with the surface, and the initial rovibronic excitation of the molecule can promote dissociation. When trapped for an appreciable time, the molecules will be thermally equilibrated and can diffuse across the surface to find a favourable dissociation site, possibly a defect or step site if these are more reactive than the terrace.

10.5 DIRECT DISSOCIATION

In this section we will concentrate on direct activated dissociation at surfaces. Whereas non-activated dissociation is accompanied by steering and trapping, activated dissociation is dominated by the shape of the PES in the energy region around the dissociation barrier. In dynamics experiments, the translational energy of the molecule is a parameter we can tune using conventional molecular beam techniques. As we increase the energy, we find that more molecules can overcome higher activation barriers at less favourable reaction sites on the surface and the dissociation probability increases. This is often discussed in terms of a 'range of barrier heights' located at different surface sites (*cf.* Figure 10.3) and different molecular orientations.

10.5.1 Early *versus* Late Barriers – Polanyi Rules

The role of molecular vibration depends on the topography of the PES in the region of the 'elbow' where the molecular bond starts to stretch. Using a Polanyi-rule argument similar to that used to explain vibrational and translational energy release in the gas phase (see Study Box 2.1 and Chapter 7), we can classify the PES as having an 'early barrier' if the dissociation barrier occurs before significant bond extension has taken place, or as having a 'late barrier' if the molecular bond stretches before the barrier maximum, as indicated in Figure 10.13. For an early barrier, the molecular vibration does not assist reaction, while for a late barrier it does, since the barrier effectively lies more in the molecular vibration coordinate than the translational (Z) coordinate. Of course this is a generalization, the molecular bond can soften, *i.e.* the vibrational frequency can decrease, as the molecule approaches the surface, which can lead to vibrationally enhanced reaction even in a non-activated system.[722]

Although the basic idea of vibrationally enhanced reactivity is exactly the same as in the gas-phase, what sets the gas-surface system apart is that the reactivity can be very different at different sites on the surface. As indicated in Figure 10.3, for H_2 on a Cu(111) surface, the lowest barrier corresponds to a bridge-to-hollow dissociation pathway. The barrier is late at this site, so reaction will show vibrational enhancement, but the bend of the elbow is not so sharp that there will be vibrational excitation in the backscattered fraction from this site.[710] However, for the atop-to-hollow pathway the barrier is higher and the bend in the elbow much sharper. As a result,

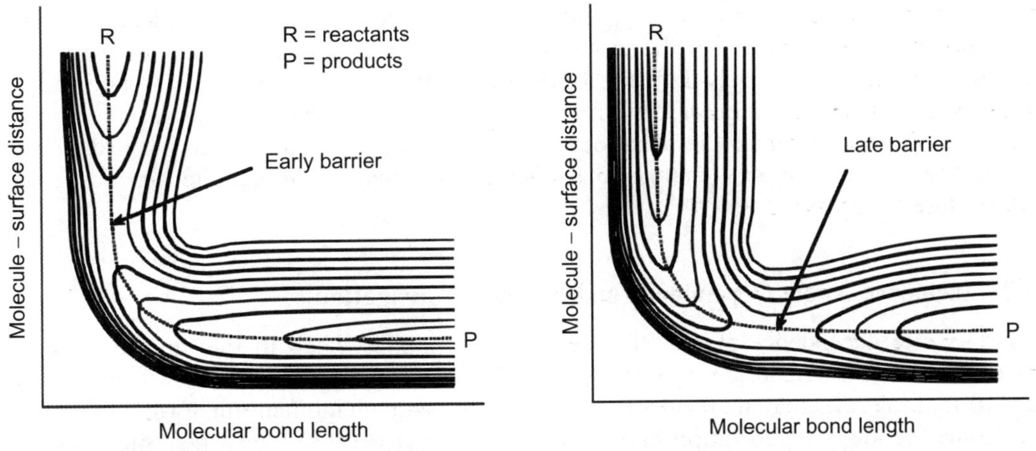

Figure 10.13 PES for early and late dissociation barriers. Reactants approach from the top of the figures. Reaction at such surfaces will follow Polanyi's rules from gas-phase dynamics, with reaction enhanced by vibration for the late barrier PES, but not for the early barrier PES.

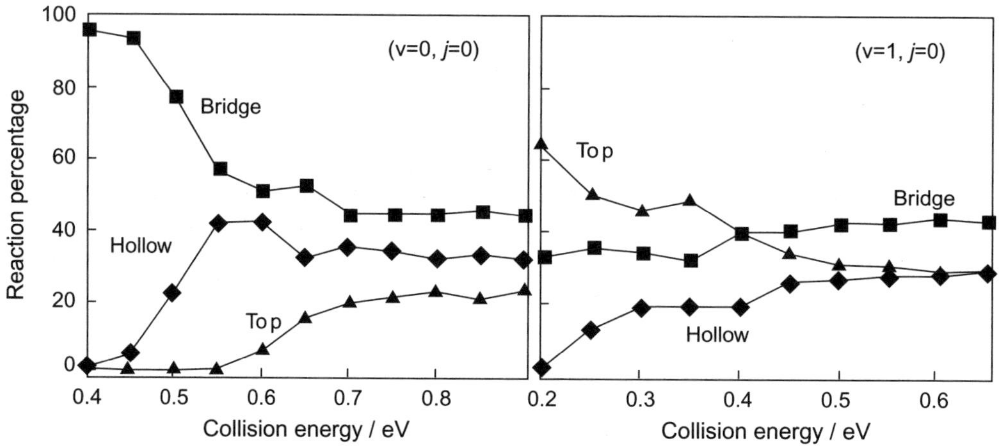

Figure 10.14 Surface site resolved reaction percentages for H_2 on Cu(100) determined by (quasi) classical molecular dynamics simulation. Vibrationally cold molecules dissociate primarily at the bridge site at low energies, whereas vibrationally hot molecules react mostly at the atop site (after ref. [723]).

dissociation at this site requires much higher energy and the backscattered molecules can gain vibrational excitation. Dissociation is dominated by one surface site (the bridge site) whereas vibrational excitation occurs predominantly at the atop site. The thresholds for vibrational excitation and vibrationally enhanced dissociation are actually linked if both happen at the same site, but not if they are spatially separated, and experimental measurements of these thresholds strongly suggest the latter. Site separation does not just occur for reaction or scattering; there is actually a different site preference for reaction of molecules incident in different quantum states. Figure 10.14 shows the percentage of dissociation as a function of molecular translational (collision) energy computed to occur at different surface sites for H_2 molecules incident on a Cu(100) surface in either the vibrational ground-state, or the first excited state.[723] As for the Cu(111) surface, the bridge site is the location of the lowest barrier. Although the barrier is late, and there is significant vibrational enhancement of the reaction, the elbow is not particularly sharply bent (*cf.* Figure 10.3). In contrast, at the atop site the barrier is higher, so reaction of the ground-state molecules is much less favoured, but the elbow is very sharply bent (as at the atop site for Cu(111)), leading to a very strong vibrational enhancement of the reaction at this site. This is such a dominant effect that the pattern of reactivity is completely different for the two quantum states: for the $v=0$ state bridge-site dissociation dominates at low energy, but for the $v=1$ state, over 60% of the dissociation occurs at the atop site. This has significant implications if we wish to make kinetic models of the dissociation, since the 'transition state' can change to a completely different surface site depending on the quantum state of the molecules.[724]

10.5.2 Incidence Angle Dependence and Surface Corrugation

The surface site dependence of the PES should also manifest itself in the angular dependence of the dissociation. If we consider the simple Gaussian barrier PES shown in the left-hand panel of Figure 10.15, molecules that are incident at an angle with enough momentum normal to the surface to overcome the lowest dissociation barrier will fail to react because the surface site with the low barrier lies in the shadow of adjacent sites with higher barriers, *i.e.* above a certain angle of incidence molecules cannot strike the surface sites where the barrier to reaction is lowest. The shadowed region will increase in size as the angle increases, and consequently, when we plot the

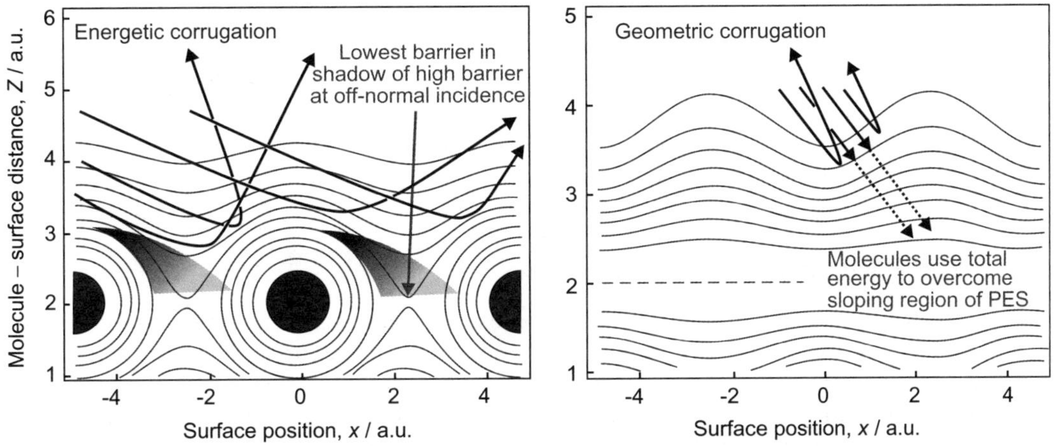

Figure 10.15 Simple models of corrugation (*i.e.* surface site dependence) of the PES. In energetic corrugation the magnitude of the dissociation barrier is site dependent and dissociation at off-normal incidence is lowered by shadowing. If the position or (in the case shown on the right) width of the barrier changes (but not the height) with location, the corrugation is termed geometric and molecules can attack the sloping region of the PES. Distances are expressed in atomic units (a.u.). Adapted from ref. [725].

dissociation probability against *normal energy* (*i.e.* against $E_i \cos^2\theta$–the energy associated with momentum normal to the surface), the probability is higher for normal incidence, decreasing as the molecules strike the surface more obliquely. Corrugation of the magnitude of the dissociation barrier is termed *energetic corrugation.*

Energetic corrugation is not the only change we can make to the PES; the location of the barrier can also vary from site to site. For instance, if the molecule hits an atop site, we might expect the barrier to be further from the surface than at a hollow site, because the atom "sticks out" more from the surface than the hollow (of course this depends on the precise nature of the bonding), alternatively the barrier might simply be broader in the molecule-surface coordinate at one site than another. In both of these cases we will see the same effect, namely that the molecules approaching the surface at off-normal angles of incidence attack the sloping region of the PES to which they are locally normal. Using the total energy to overcome the barrier, these molecules will dissociate more readily than molecules that strike a flat surface because more of the energy is available to overcome the barrier. Plotting the dissociation on the corrugated surface against energy normal to the surface, the probability is higher at off-normal incidence than at normal incidence. This PES corrugation is termed *geometric corrugation* since it was first observed in a PES in which the distance of the barrier above the surface was varied.[725]

10.5.3 Molecular Rotations

Rotation influences dissociation by shadowing effects and by stretching the bond, exactly similar to the behaviour observed in gas-phase dynamics. Shadowing, in this context orientational hindrance, occurs because there is a favourable orientation of the molecule for dissociation to occur. For example, dissociation of most diatomic molecules will favour a geometry in which the molecule lies flat on the surface with the atoms directed towards their favoured chemisorption site. If an orientation is unfavoured, we simply mean that the dissociation barrier is higher, so a model for orientational hindrance looks much like the geometric corrugation picture in Figure 10.15, but with molecular orientation in place of *X*. Motion in the orientation coordinate is of course rotation, so based on the analogy between the two systems, the dissociation should decrease with increasing rotational state. Counteracting orientational hindrance is a centrifugal effect. As the molecule approaches the top of a late barrier its bond stretches. In consequence the moment of inertia increases, causing a decrease in the rotational energies. This decrease in energy feeds into motion over the barrier, and so with greater rotational energy we should see more dissociation. The later the barrier, the more pronounced is this effect.

10.5.4 Mode/Bond Specific Surface Chemistry

While it was quickly established that internal molecular excitation can assist bond dissociation for H_2 at surfaces (Figure 10.14), it has been less clear if reactivity of heavier molecules is mode specific, or if instead the vibrational energy is pooled during the surface collision and assists dissociation only in a statistical manner (see Chapter 7). The system for which there is most evidence is methane dissociation, which is a commercially important process and has been studied on various transition metal surfaces, notably Pt and Ni. Molecular beam experiments show that dissociation is strongly activated, both by translational excitation of the methane and by thermal excitation of the surface.[727] This has led to statistical models of the dissociation process. By choosing a limited number of surface and molecular modes in which the collision energy is pooled, and then applying statistical theory to calculate a reaction rate, these models can reproduce some of the experimental behaviour.[728] Such a model predicts that there will be no mode specific chemistry – it does not matter which vibrational or translational mode is excited. Several recent studies have addressed this question by optically pumping overtone bands and monitoring the change in the methane dissociation probability; one such is shown in Figure 10.16. On both Ni(111) and Pt(111) surfaces, exciting the $2\nu_3$ overtone of the anti-symmetric stretching band promotes dissociation of methane, but with different efficacies on the two different surfaces.[726] In fact, other studies show that putting similar amounts of energy into different vibrational bands results in different amounts of dissociation, implying that the vibrational information is not scrambled and the dissociation dynamics depend on the original mode excited. The complex, multi-dimensional character of the potential for dissociation in this case make this a challenging system to understand in detail, but these results do hold out the possibility of isotope specific surface chemistry and isotope enrichment based on mode specific surface chemistry.

10.5.5 Excitations of the Surface and Dissociation

The simple cube model introduced in Section 10.3.2 can also be adapted to explain the effect of surface temperature (vibration) on the molecular dissociation probability. This model predicts that, near the dissociation threshold, reaction is enhanced when the surface atom (cube) is moving towards the molecule, since the increased collision energy enables a molecule to get over an energy barrier and dissociate. The opposite is true when the cube is moving away from the approaching molecule, reducing the dissociation probability compared to that expected for a stationary surface in the absence of any vibrational motion. Coupling the cube to a harmonic oscillator and Boltzmann averaging the dissociation probabilities, we find that the dissociation has a roughly Arrhenius dependence on temperature. Modelling indicates that the 'activation energy' extracted from this has little to do with the energy of the dissociation barrier in a direct reaction. In addition to any dynamical effects, the dissociation barrier and PES may itself be locally modified by thermal excitation of the surface (*e.g.*, in a sudden approximation), while the surface temperature may also play a role when the reaction is influenced by excitation of electron-hole pairs in the surface, as the thermal population of electron-hole pair states will influence how charge transfers in and out of the molecular affinity level.

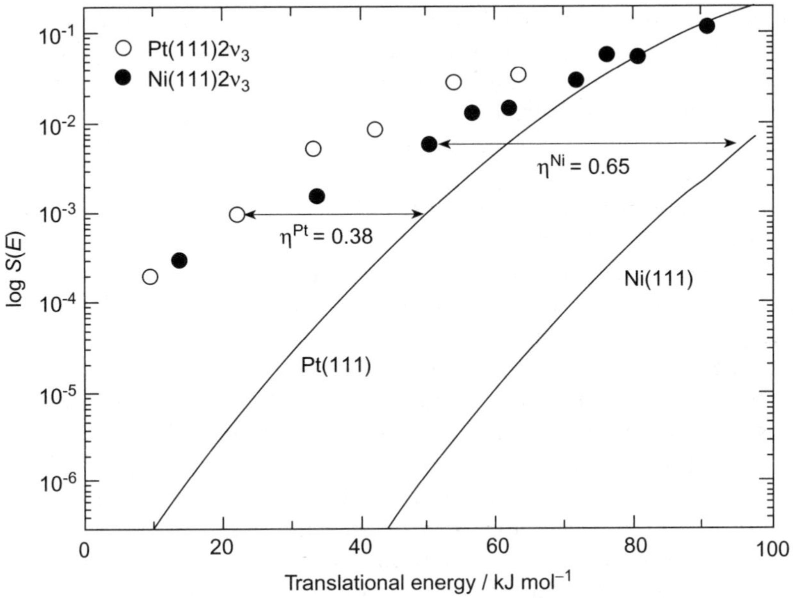

Figure 10.16 Dissociation probability for methane on Pt(111) and Ni(111) after optical excitation of the $2\nu_3$ overtone of the anti-symmetric stretching band. The dissociation of methane in thermal beam experiments is shown by the solid lines while the results of optical pumping are indicated by the points. The vibrational efficacy η is a measure of the relative efficiency of this vibrational mode in promoting dissociation compared to putting the same amount of energy into translational excitation. Adapted from Bisson *et al.*[726]

10.5.6 Non-Adiabatic Effects in Molecular Dissociation Reactions

We have concentrated largely on effects that can be explained within an electronically adiabatic picture, since this is all that we can sensibly obtain from DFT calculations. However, there are many reactions in which the excited states of the molecule could play a role[729] (strictly speaking, the excitation of electron-hole pairs in a metal surface is also non-adiabatic). For example, dissociation of O_2 on Cs surfaces is so strongly non-adiabatic that light and electron emission can be observed during the reaction.[730] In many cases this secondary particle emission cannot be observed, yet there are indications that the reaction may not be determined only by the electronic ground state. The reaction of O_2 with the close packed (111) surface of Al is one such example. Molecular beam experiments indicate that dissociation is activated, with a probability that increases very rapidly with molecular energy, yet DFT calculations indicate that the reaction is non-activated at many surface sites and orientations. This complete failure of any of the standard implementations of DFT to produce an approximately consistent PES has led many to look for non-adiabatic contributions to the dynamics (*e.g.*, ref. [731]). Certainly it is reasonable to expect non-adiabatic effects in such a system. O_2 in the gas-phase is a spin triplet in its electronic ground-state, and as it approaches a metal it can acquire charge to become an O_2^- or even an O_2^{2-} ion (which should spontaneously dissociate), or the spin could simply wash out and the molecule become a singlet. Detailed modelling and spin-constrained DFT calculations indicate that such behaviour could give rise to the observed dissociation barriers.[732]

10.6 RECOMBINATIVE DESORPTION

One of the approaches that has been developed to investigate gas-surface reaction dynamics is to measure how the energy released during a surface reaction is disposed in the products. This type of experiment can be thought of as a "half-scattering" event, similar to the idea of product state analysis after photodissociation, but in this case the transition state for recombination is populated thermally, typically by atomic recombination. Product state analysis can be achieved by detecting molecules as they desorb from the surface using either an angle sensitive detector, such as a mass spectrometer, to determine the angular distribution of the products, or a state specific probe, such as laser-induced fluorescence (LIF), resonance enhanced multiphoton ionization (REMPI) or IR chemiluminescence to determine the energy disposal into the internal, vibrational and rotational coordinates (see Study Box 5.2). In many cases these techniques provide product velocity distributions, which are particularly important in understanding the energetics of recombination, and in a few cases also the rotational alignment of the product angular momentum (see Chapter 9). Product state energy disposal measurements are particularly informative when the recombination dynamics lead to an energy release into specific modes of the product, *e.g.*, when the dynamics are repulsive from the transition state for recombination/reaction and lead directly to desorption, depositing a substantial amount of energy ($\gg k_B T$) into the products. However, even when the energy disposal is less specific, dynamical information can be obtained by looking at the deviation of the product state distributions from a thermal distribution; this is explained in more detail in Section 10.6.1. Product state measurements have been made primarily for H/D recombination, notably from Cu, Ag and Si surfaces, but the technique has also been used to probe N_2 and CO_2 formation in several surface reactions. More experimental detail and examples can be found in other reviews.[701]

Control over the initial conditions is limited to preparing the surface with a known temperature, morphology and coverage of reactants. Since the surface excitation cannot be probed directly, little is known about the processes which populate the transition state. As the products scatter away from the transition state, energy is released into different coordinates, including the surface, depending on the shape of the molecule surface potential. Product state distributions can be calculated using simple models for the energy partitioning into different modes, or by detailed scattering calculations. As ever in such calculations, one must decide how many coordinates need to be included and whether the surface can be treated as rigid, or if energy transfer needs to be considered.

10.6.1 Detailed Balance

For thermal Langmuir-Hinshelwood type reactions, several studies have shown that the product final state distributions are sensitive to the temperature of the experiment, and in some cases to the surface coverage. Any discussion of the recombination dynamics therefore needs to consider how surface temperature affects the energy disposal, and to recognize that the total excitation of the adsorbate/surface is thermal and not well defined (since we cannot know how much energy is partitioned back into the surface during desorption). This thermal aspect of recombination can sometimes be turned to advantage, since under appropriate conditions the dynamics of the reverse, dissociative chemisorption, process can be inferred from desorption measurements using the principle of detailed balance.[733] Taking the explicit example of a diatomic molecule A_2 dissociating on a surface with a temperature T and surface coverage Θ,

$$A_2(g) \leftrightharpoons 2A(\text{ad}), \tag{10.11}$$

the system will be at equilibrium for a particular pressure of $A_2(g)$. Taking the usual definitions* for the sticking probability, S, and the desorption flux distribution, P, then at equilibrium the energy

*The sticking probability is the fraction of molecules that collide with the surface and dissociate to remain trapped on the surface. The desorption distribution $P(E, \theta, v, j; T, \Theta)$ is the relative flux of molecules desorbing with and energy E, at an angle θ to the surface normal into a particular final state (v, j) from a surface with an adsorbate coverage (Θ) and surface temperature (T).

distributions in desorption are related to the sticking probability by,

$$P(E,\theta,v,j;T,\Theta) \propto E\mathrm{e}^{-E/k_\mathrm{B}T}\mathrm{e}^{-\varepsilon_{v,J}/k_\mathrm{B}T}S(E,\theta,v,j;T,\Theta)\cos\theta, \tag{10.12}$$

where $P(E, \theta, v, j; T, \Theta)$ is the translational energy (E) distribution for products desorbing at an angle θ to the surface normal, in quantum state (v, J) with an internal energy $\varepsilon_{v,\,J}$, from a surface with a coverage Θ and temperature T. $S(E, \theta, v, j; T, \Theta)$ is the sticking coefficient for $A_2(E, \theta, v, J)$ under the same conditions of surface temperature T and coverage Θ. This relationship allows desorption measurements to be related directly to detailed sticking probabilities.

Deriving predictions for the energy dependent sticking functions $S(E)$ for individual rovibrational states (v, J) from measurements of the product translational energy distributions, $P(E, \theta, v, j; T, \Theta)$, has several benefits. Firstly, it can provide information about the effect of rotation, vibration and alignment on the dissociation probability, information which frequently could not be obtained directly in an adsorption experiment. Secondly, the product state distribution represents a thermally weighted distribution of the sticking function, that is to say it represents directly the distribution of molecules that would adsorb from a thermal gas at temperature T. If thermal dissociation occurs preferentially at particular sites, for example steps or islands edges, then this will be properly represented in the desorption distributions. In contrast, adsorption experiments typically measure channels with high cross section and may not reveal low energy channels that have a low probability but are statistically highly weighted in the thermal distribution. Finally, considering sticking functions derived from detailed balance, rather than the intrinsically temperature dependent product state distributions, allows us to remove the thermal sampling implicit in desorption measurements and to obtain dynamical information directly.

Relating this detailed balance description to a microscopic picture of adsorption and desorption is not so trivial as the above discussion suggests, since we meet the usual problem in applying time reversal to a reaction, that the energy of the system is not properly conserved between the quantities we would like to equate. In this case, adsorption of a molecule leads to dissociation and phonon creation, whereas desorption arises from the annihilation of phonons in a surface originally at temperature T. While this is a serious problem in the case of a gas phase molecular reaction, where the vibrational excitation is localized, at a surface the phonons are delocalized and scatter away from the adsorbate, leaving the surface at the same temperature as before. Although the comparison of adsorption and desorption has been criticized, time reversal appears to be an over-restrictive criteria and several studies have concluded that the two processes can be directly related even under non-equilibrium conditions.[734] Experimentally, support for this application of detailed balance comes both from measurements of the trapping-desorption of rare gases at metal surfaces and also from the adsorption-desorption of hydrogen on surfaces.

Based on Eq. (10.12), it is possible to calculate how product distributions for recombination P depend on the sticking functions S, or *vice versa*, by making a suitable integration over the coordinates that are not selected. While S is not usually known in sufficient detail to calculate complete desorption distributions, $P(E, v, j; T)$, it is often possible to predict the general behaviour seen in desorption from the form of $S(E)$, or from the vibrational state dependence $S(E, v)$ of sticking, as measured in molecular beam adsorption measurements. This is illustrated below for two of the more common measurements which can be made during desorption, namely translational energy distributions and product angular distributions.

If the energy dependence of the sticking probability $S(E)$ for dissociative chemisorption is known (Sections 10.4 and 10.5), then the translational energy distribution in re-combinative desorption, $P(E)$, can be predicted directly. Assuming that the adsorption (sticking) probability $S(E)$ is measured for a beam incident along the surface normal, $\theta_i = 0$, at low surface coverage ($\Theta \simeq 0$) and is independent of quantum state, then the translational energy release in this

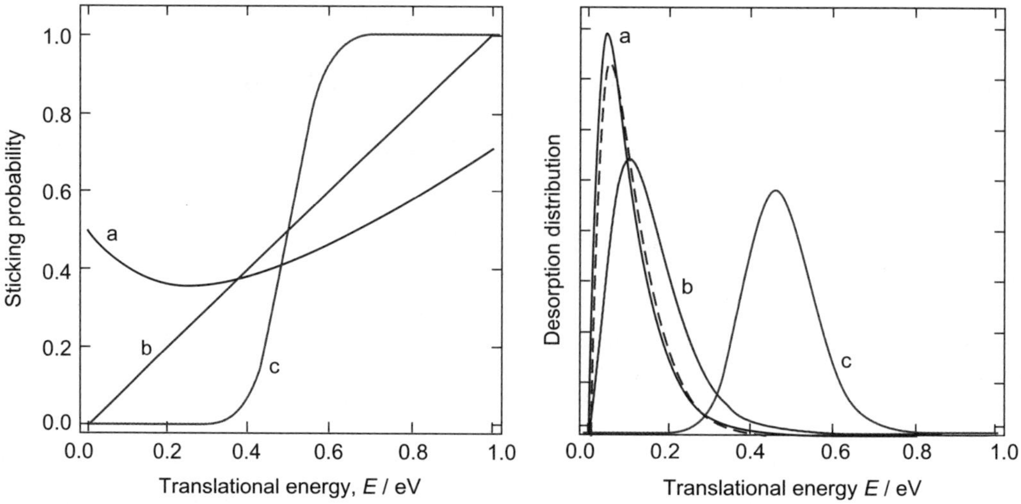

Figure 10.17 Examples of the energy distributions in desorption $P(E)$ (right frame) expected for different generic types of sticking behaviour $S(E)$ (left frame) similar to those seen for H_2 dissociation on different transition metal surfaces. Three types of adsorption behaviour are shown, (a) a weak energy dependence with S decreasing initially with increasing energy as steering/trapping is inhibited, and then showing an activated behaviour at higher energy (*e.g.*, Pd(111)), (b) activated adsorption with no threshold energy (*e.g.*, Pt(111)) and (c) activated adsorption with a clear threshold for dissociation (*e.g.*, Cu(111)). The corresponding energy distributions for the recombinative desorption process are shown on the right, with the thermal (Knudsen) distribution shown by the dotted line. Adapted from ref. [701].

direction is given by

$$P(E; T, \Theta) \propto E\mathrm{e}^{-E/k_B T} S(E; T, \Theta). \tag{10.13}$$

If we further assume that the sticking function does not depend significantly on temperature and surface coverage, then

$$P(E, T) \propto E\mathrm{e}^{-E/k_B T} S(E), \tag{10.14}$$

and we can predict the desorption distributions directly.

Several limiting cases can be imagined based on different behaviour for the dissociation probability $S(E)$. Examples are shown in Figure 10.17. If there is no energy dependence, $S(E) = \text{constant}$, then the desorption distribution will be approximately thermal, $P(E, T) \propto E\exp(-E/k_B T)$. If adsorption is hindered by increasing translational energy, as it is when adsorption is mediated by energy transfer to the surface, then $S(E)$ decreases with E and the energy distributions found in desorption will be sub-thermal. Perhaps the most useful situation is when there is an activation barrier to adsorption, so that $S(E)$ increases abruptly with energy above some critical threshold, as is the case for activated dissociation of H_2 on inert noble metal surfaces. In that case, the desorption distribution is shifted to high energy, reflecting the energy released into repulsion of H_2 from the surface as the atoms recombine and scatter from the surface.

10.6.2 One Dimensional Models for Activated Adsorption-Desorption

For a planar one dimensional system in which dissociation is activated, adsorption will be insensitive to the energy parallel to the surface and must obey a normal energy scaling, $S(E, \theta) = S(E \cos^2\theta)$. If we further assume that sticking shows a sharp threshold, below which dissociation is inefficient, then $S(E)$ can be approximated by a step function at some threshold $E = E_0$. This approximation will only be reasonable in practice when the sticking threshold is sufficiently sharp compared to the thermal distribution that desorption is dominated by energies greater than E_0, *i.e.* the energy width of the increase in $S(E)$ (Figure 10.17) is comparable to or smaller than $k_B T$. The translational energy distribution can then be integrated, giving a simple expression for the mean translational energy release along the surface normal,

$$\langle E_z \rangle = \frac{E_0^2 + 2E_0 k_B T + 2(k_B T)^2}{(E_0 + k_B T)}. \tag{10.15}$$

For a threshold $E_0 = 0$ the expression reduces to $\langle E_z \rangle = 2k_B T$, the correct result for a thermal distribution.

Similarly, the angular distribution of the desorption products can be found by integrating across the translational energy distribution

$$P(\theta, T) \propto \int_0^\infty E e^{-E/k_B T} \cos\theta \, S(E, \theta; T) \, dE. \tag{10.16}$$

For the 1D model, where energy is released only into motion along the surface normal, this gives an analytic result,

$$P(\theta, T) = \frac{E_0 + k_B T \cos^2\theta}{(E_0 + k_B T)\cos\theta} e^{-E_0 \tan^2\theta / k_B T}. \tag{10.17}$$

This expression was originally derived by van Willigen,[735] and can be used to relate the angular distribution directly to the barrier height, E_0. The angular distribution of the desorption products reflects the relative energy released into product motion parallel and perpendicular to the surface during desorption. $P(\theta)$ will broaden with increasing temperature, but will become sharper as E_0 is increased and the repulsion along the surface normal becomes larger. For a threshold $E_0 = 0$ the expression reduces to the thermal result, $P(\theta) = \cos\theta$.

10.6.3 Weakly Coupled Systems (Hydrogen Recombination)

Fully state resolved measurements of hydrogen recombination have been made from several metal surfaces,[701] and from Si.[737] Figure 10.18 shows an example of the translational energy released into recombination of D_2 from Ag(111) and the sticking functions that can be derived from this by detailed balance. These results show that the vibrationally excited species has a lower translational energy requirement for dissociation, vibrational excitation contributing to assist dissociation, as expected for a late barrier system (Section 10.5). Information on the influence of rotation on the recombination dynamics can be obtained by measuring either the translational energy release as a function of rotational state,[738] or the rotational alignment of the products,[739] as shown in Figure 10.19 for D-atom recombination from Cu(111). The rotational alignment is dependent on the translational energy released, since with increasing energy molecules can more easily overcome the dissociation barriers in less favourable orientations. Product state measurements have been recorded for thermal surface recombination of H on Ni(111) and as a result of H resurfacing from

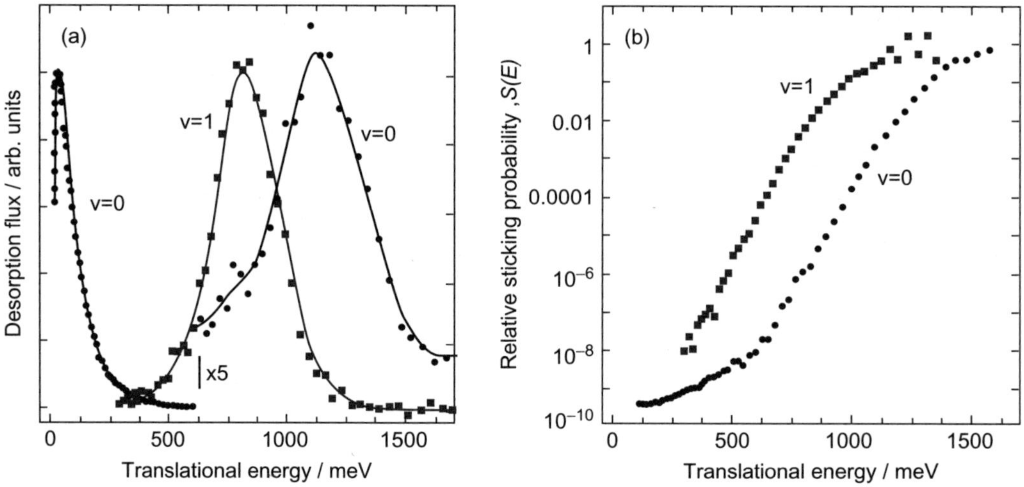

Figure 10.18 (a) Translational energy distributions for D_2 ($v = 0$, $J = 2$) (circles) and ($v = 1$, $J = 2$) (squares) after recombination on Ag(111). The energy distribution from the ground state is bimodal with a maxima at low and high energy (expanded ×5). (b) Predicted sticking functions for these two states by application of detailed balance to the data in (a). Figure adapted from the data of ref. [736].

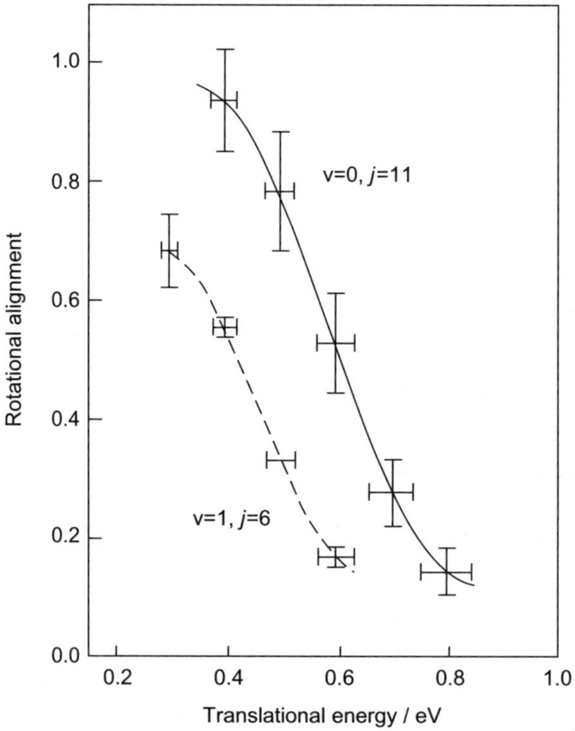

Figure 10.19 Measured rotational alignment for D_2 molecules desorbing from Cu(111). The alignment of the desorbing molecules is positive, indicating a J lies preferentially along the surface normal, often referred to as a "helicopter" rotation. The alignment is largest at low energy, where the molecules just get over the barrier to adsorption-desorption, and falls at higher energy where there is less constraint on the molecular geometry that leads to reaction. Adapted from Hou *et al.*[739]

metastable sites within the bulk (a species implicated in some catalytic hydrogenation reactions). This reveals the different dynamics associated with H recombination on the flat terraces, at steps and following H resurfacing, to give a rather complete, coverage dependent picture of adsorption/desorption.[740]

10.6.4 Heavy Molecules

Product state distributions have also been measured for several reactions that lead to heavy products, for example N_2 formation has been studied in angle resolved quadrupole mass spectrometry measurements[741,742] and by REMPI,[743] OH formation has been probed using LIF,[744] while CO_2 formation has been followed by IR chemiluminescence.[745] As an example, Figure 10.20 shows the angular distributions measured for N_2 formed by recombination on three different surfaces of different reactivity to N. Once closed shell N_2 is formed by recombination at the transition state it is repelled from the surface, gaining translational energy along the surface normal to form a sharply peaked angular distribution. The translational energy distribution (and the internal state distributions) have been measured by REMPI detection of the N_2 and confirm the correlation between the mean energy release and the focusing of the angular distribution along the surface normal expected from a 1D model. State resolved measurements for Cu(111) indicate that the energy release occurs specifically into translational excitation, with rotational levels being approximately thermally populated, and vibration having more excitation than rotation.[743] On other surfaces, the product state angular distributions can be much more complex. For example, N_2O decomposition on several face centred cubic (110) surfaces leads to "inclined" desorption, with N_2 being desorbed into two lobes directed at approximately 45° to the surface normal.[746] This behaviour is attributed

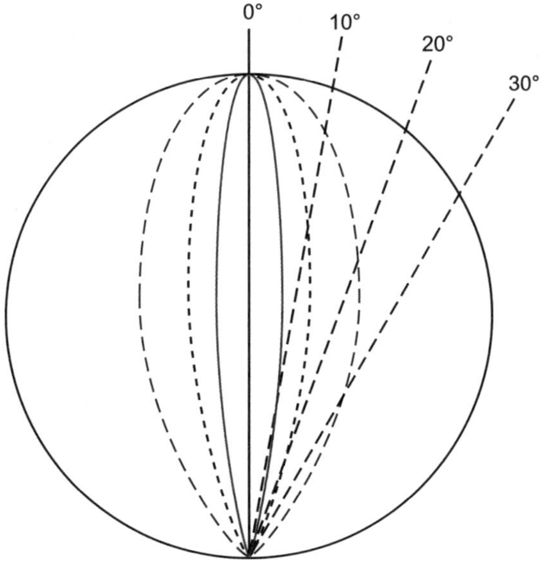

Figure 10.20 Angular distribution $P(\theta)$ of N_2 formed by recombination of N atoms from different metal surfaces after fitting to a $P(\theta) = \cos^n\theta$ distribution. The plots are normalized to the flux along the surface normal (0°) and should be scaled by a factor of $(n+1)$ to compare the fluxes directly. The outer circle shows a statistical ($\cos\theta$) distribution, with increasingly sharply focused angular distributions measured from Ru(0001) ($\cos^7\theta$), Cu(111) ($\cos^{28}\theta$) and Ag(111) ($\cos^{75}\theta$). As expected, experiments also show the translational energy released into N_2 motion increases from Ru(0001) to Cu(111) and Ag(111).[701] Adapted from ref. [701].

to a planar N_2O transition state for dissociation, and indicates the specific nature of the product energy release and dissociation dynamics as N_2 is repelled from the O and from the metal surface.

10.7 OUTLOOK

Many of the areas in which surface dynamics is being developed lie on the fringes of this brief review. The problem of describing non-adiabatic systems and electronic excitation during scattering has been alluded to, and remains an important and active area of research, as does charge transfer,[708] which we have not discussed in any detail. Surface photochemistry,[706] bond specific surface chemistry,[561] and time resolved measurements of surface dynamics[747] are all areas in which new developments have been reported recently, and which will stimulate a better understanding of the molecular and electronic processes underpinning gas-surface reaction dynamics.

10.8 PROBLEMS

1. Using conservation of energy and momentum perpendicular to the surface, derive the classical expression for the energy exchange ΔE (Eq. (10.8)) when a gas phase species, mass m_g, strikes a surface cube, of mass m_s, where $\mu = m_g/m_s$, with a translational energy E_i along the surface normal.
2. Assuming a physisorption well can be represented by an attractive well of depth W followed by a repulsive hard wall, show that the classical condition for trapping is given by $E_i = 4\mu W/(1-\mu)^2$ (Eq. (10.9)).
3. Assuming that a molecule can be adsorbed only when its translational energy towards the surface is greater than some critical value E_0, *i.e.* $S=0$ for $E < E_0$ and $S=1$ for $E \geqslant E_0$, derive an expression for the average energy release along the surface normal $\langle E_z \rangle$ (Eq. (10.15)). Hence find the limiting behaviour for a sticking threshold $E_0 = 0$ or $E_0 \gg k_B T$. Hint: use Eq. (10.13) to calculate $\langle EP(E) \rangle / \langle P(E) \rangle$.
4. Assuming the adsorption probability $S(E, \theta)$ is unity above some critical normal energy $E_0 \cos^2\theta$, derive the expression for the anticipated angular distribution $P(\theta, T)$ (Eq. (10.17)). Hint: this requires integrating Eq. (10.16) with appropriate limits for $E(\theta)$.

CHAPTER 11

Femtochemistry and the Control of Chemical Reactivity

HELEN H. FIELDING AND ABIGAIL D. G. NUNN

Department of Chemistry, University College London, U.K.

11.1 INTRODUCTION

The typical vibrational period of a covalent bond in a molecule is on the order of tens of femtoseconds (fs), where 1 fs = 0.000 000 000 000 001 s or 1 billionth of a millionth of a second. If we wish to observe these ultrafast processes we need to probe them on an ultrafast timescale; this became possible with the advent of femtosecond lasers in the late 1980s.* Femtosecond laser experiments generally employ pump-probe techniques in which the pump laser pulse either initiates a chemical reaction or creates a non-stationary state known as a wavepacket.† A probe pulse is then fired at a series of well-timed intervals after the pump pulse and is configured to generate an observable that will provide information about the evolving system. Precisely what dynamical information can be retrieved from a pump-probe scheme is highly dependent on the excitation scheme and the observation method, and so a great deal of effort is directed towards developing innovative detection schemes to map out as much of the dynamical pathway as possible.

The application of femtosecond lasers to probe molecular dynamics was pioneered by Ahmed Zewail. His experiments were the first to record a "movie" of the progress of a chemical reaction by using sequenced spectroscopic signatures of the nuclear motion, and the significance of his work was recognized by the award of the 1999 Nobel Prize in Chemistry.[749] The concept of controlling the outcome of a photochemical reaction using light is closely related to the concept of the real-time observation of femtochemistry. In an experiment using two femtosecond laser pulses, Zewail and coworkers were able to increase the yield of the XeI product in the bimolecular reaction $Xe + I_2 \rightarrow XeI + I$.[748] The experiment is illustrated in Figure 11.1. The yield of XeI, as measured by a chemiluminescence signal, was found to be dependent on the timing of the two laser pulses. To maximize the reaction yield, the second laser pulse, the *control pulse*, needed to arrive at the

* Readers might find it helpful at this point to refresh there memories of Study Box 1.2 on Lasers in reaction dynamics.
† Wavepackets are discussed in Section 11.4.1 of this chapter, but also in Chapter 3 in the context of quantum mechanical scattering calculations. Chapter 1 provides an introduction to pump-probe methods.

Tutorials in Molecular Reaction Dynamics
Edited by Mark Brouard and Claire Vallance

Published by the Royal Society of Chemistry, www.rsc.org

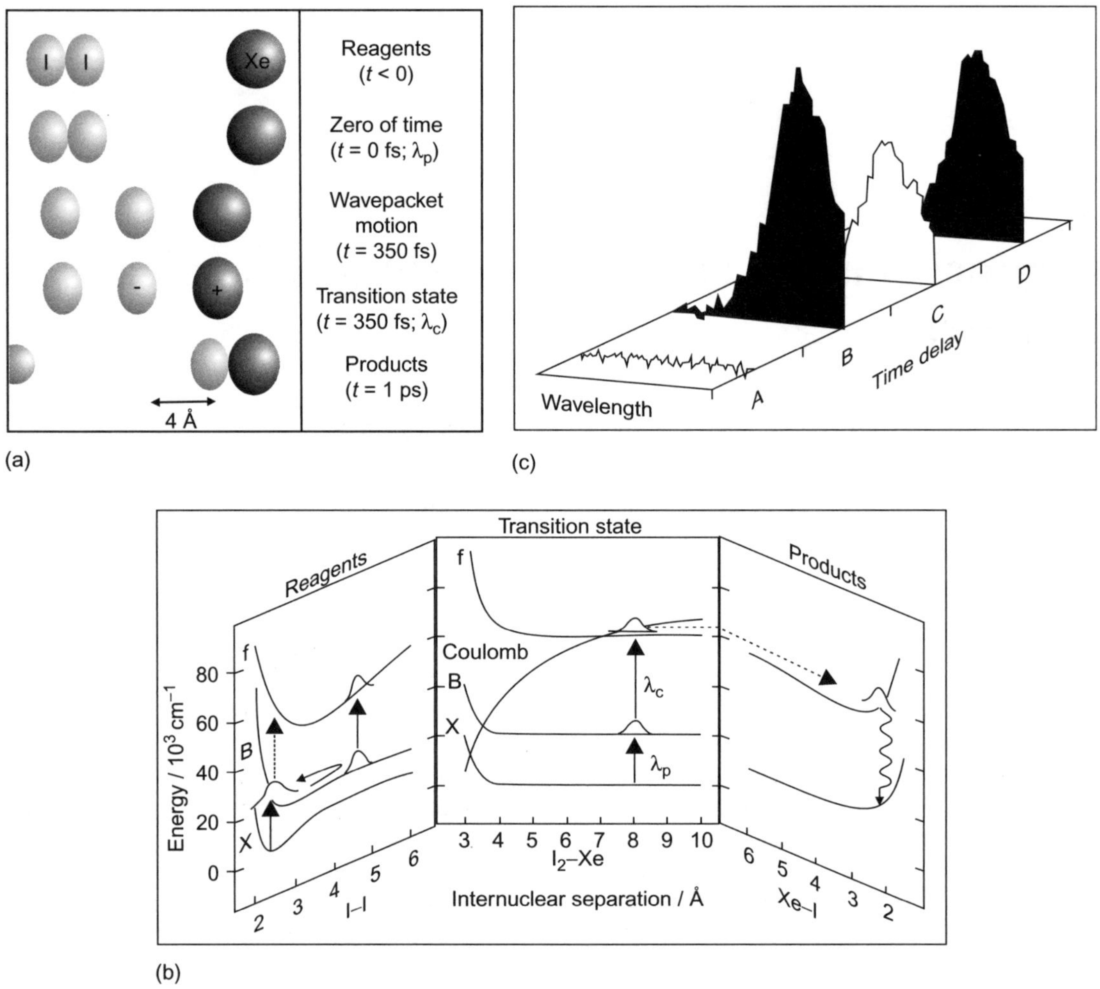

Figure 11.1 Scheme (a) illustrates the timescales for the progression of the controlled reaction of Xe and I_2 with femtosecond laser pulses. The panels in (b) illustrate the mechanisms of two possible control schemes of the reaction. The left hand panel shows the I-I coordinate at large Xe-I distances, and a possible control scheme that depends on timing the second pulse to meet an I_2 vibrational wavepacket at the Franck-Condon window to take the molecule to the ion-pair state, f, from where a harpoon reaction is possible after collision with Xe (see Chapter 4). The central panel shows the I_2-Xe coordinate and an alternative reaction mechanism in which the control pulse must be timed to match the wavepacket motion of the I_2-Xe collision pair, which is oscillating on the shallow I_2-Xe potential: at the correct internuclear separation, the molecule is in the harpooning region and absorption of the control pulse takes it to the ion-pair state. The right panel illustrates the product formation and chemiluminescence on the Xe-I coordinate, well-separated from the other I atom. The detected chemiluminescence of Xe-I at 253 nm was used as a measure of product yield, and was recorded as a function of the delay between pump and control laser pulses. The chemiluminescence spectra at various pump-control delays is shown in (c), where the delays labelled A, B, C and D correspond to a negative time, zero-of-time, 350 fs and 700 fs respectively. Adapted from ref. [748].

molecule at a particular time to be absorbed by the molecule at the correct phase in the motion of a vibrational wavepacket that was initially excited by the first pulse, the pump pulse.

There has been an explosion in activity in the field of femtochemistry during the last decade, and a wide variety of problems have been investigated, ranging in complexity from simple bond breaking in diatomic molecules to intramolecular processes such as internal conversion, isomerization and electron transfer in larger molecules.* There are a number of excellent reviews describing these scientific achievements (for example, see ref. [750]). The purpose of this chapter is to explain the tools, principles and techniques of ultrafast spectroscopy and femtochemistry. In particular, we will provide an overview of femtosecond laser technology and pump-probe schemes.

As ultrafast laser technology adapts and improves, new opportunities for probing and controlling molecular dynamics and reactions present themselves. An understanding of the principles of current femtosecond laser technology is a vital tool for carrying out any experiment in femtochemistry, and so in Section 11.2, we explain some key concepts associated with this type of laser. In Section 11.3 we discuss the counterpart of the ultrafast laser in a femtochemistry experiment: the various possible ways in which ultrafast processes may be spectroscopically recorded. In Sections 11.4 and 11.5, the discussion throughout is lead by some of the most important experiments in the published femtochemistry literature; Section 11.4 is about observation of ultrafast processes, and Section 11.5 is about their control.

11.2 FEMTOSECOND LASERS

Femtosecond laser technology is a research field in itself, and there are many excellent reviews available, see for example ref. [751]. This section is intended to provide a basic introduction to femtosecond lasers and how they may be applied to spectroscopic and dynamical studies of molecules. In a typical pump-probe molecular dynamics experiment, the pump and probe femtosecond laser pulses will have different wavelengths, although they are derived from a single femtosecond laser oscillator. A schematic diagram of the optical layout of a typical two-colour experiment is presented in Figure 11.2. In what follows, we will track the optical pathway from the initial generation of infrared femtosecond pulses in a femtosecond oscillator, through a femtosecond amplifier, to the optical parametric amplifiers (OPAs) which generate new wavelengths tuneable across the far infrared, visible and ultraviolet regions of the spectrum. On the way, we describe some of the important details of the optical processes involved, and at the end we describe some of the diagnostic tools required to characterize femtosecond light pulses.

11.2.1 Femtosecond Laser Oscillators

As outlined in Study Box 1.2, a laser cavity in its simplest form consists of a gain medium enclosed between two cavity mirrors.[752] The medium is 'pumped' in order to achieve a population inversion,

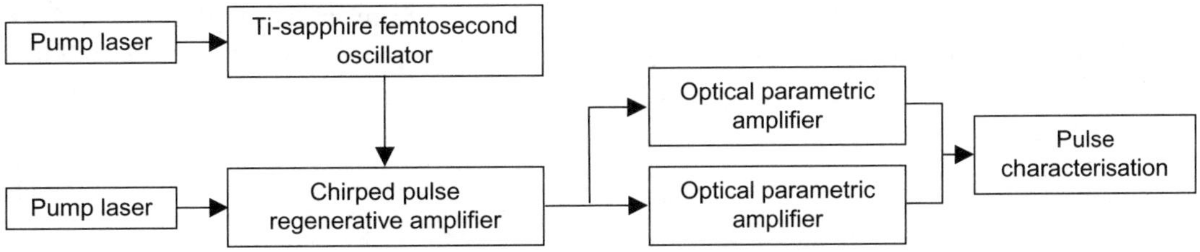

Figure 11.2 Block diagram for a typical femtosecond laser system. Femtosecond pulses at 800 nm with nJ energies are generated in the Ti:sapphire oscillator by Kerr-lens mode-locking (see Section 11.2.1). These are amplified by stretching, regenerative amplification, and then compression in the chirped pulse regenerative amplifier (see Section 11.2.2); wavelength conversion occurs in the optical parametric amplifier (OPA) to give pulses in the visible, UV or IR with pulse energies on the order of a few mJ (see Section 11.2.3); finally, pulses are characterized using autocorrelation or cross-correlation methods (see Section 11.2.4). Both the oscillator and amplifier are pumped by high power lasers.

*Some of these processes are considered in Chapter 8, and Study Box 8.1.

i.e. a situation in which an upper level is populated more than the ground state, and photons emitted by spontaneous emission then stimulate the emission of further photons. One of the cavity mirrors has a reflectivity of less than 100%, allowing a fraction of the trapped radiation to leave the cavity on each reflection. A cavity supports optical frequencies, ν, for which the length of the cavity is an integer or half integer number of wavelengths; these frequencies form standing waves within the cavity, with a node at each mirror, and comprise the longitudinal modes of the cavity. These modes have frequencies $\nu = nc/2L$, where L is the cavity length, c is the speed of light in a vacuum, and n is an integer. The number of longitudinal modes (values of n) that exist in the laser cavity is determined by the spectral bandwidth of the laser transition.

To obtain femtosecond laser pulses, the different longitudinal modes that propagate in the laser cavity must be mode-locked so that they oscillate in phase with one another (see Study Box 11.1 for more details about the phase of a short pulse). If many modes are allowed to coexist in the cavity and all their phases are aligned at one point, the result is a very small region of constructive interference that lasts for a few femtoseconds; the more modes there are, the shorter the resulting pulse, in accordance with the energy-time uncertainty principle, and as illustrated in Figure 11.4. This principle may be used to determine the precise limiting value for the product of the spectral width, $\Delta\nu$, and time duration, Δt, known as the time-bandwidth product (TBWP). For a pulse with a Gaussian distribution of modes and Gaussian temporal intensity distribution, the minimum time-bandwidth product is 0.441,

$$\Delta\nu\Delta t \geqslant \frac{4 \ln 2}{2\pi} = 0.441, \tag{11.1}$$

if $\Delta\nu$ and Δt, expressed in units of Hertz and seconds respectively, are both defined as full-widths at half maximum (FWHM). A Gaussian pulse is the most commonly modelled pulse shape produced by femtosecond oscillators. Ti:sapphire (titanium-doped sapphire, Ti^{3+}:Al_2O_3) lasers are the most common laboratory femtosecond lasers and can have a spectral bandwidth of around 150 nm (centred at 780 nm), that can support pulse durations down to $\geqslant$6 fs, which is the limit for a single cycle laser pulse in the near IR.

Mode-locking in a Ti:sapphire laser exploits the optical Kerr effect within the gain medium. The optical Kerr effect is the change in the refractive index of a medium induced by an intense incident electric field; it is a three-wave mixing $\chi^{(3)}$ effect (see Study Box 11.2). The refractive index of the medium is $n = n_0 + n_2 I$, where n_0 is the linear refractive index ($n_0 = 1.76$ at 800 nm in Ti:sapphire), n_2 is the non-linear refractive index ($n_2 = 3.1 \times 10^{-16}\,\mathrm{cm^2\,W^{-1}}$ at 800 nm in Ti:sapphire), and I is the laser intensity. The phase delay experienced by the propagating optical pulse is proportional to the refractive index,

$$\phi(\omega) = \frac{n(\omega)\omega L}{c}, \tag{11.2}$$

where ω is the angular frequency ($\omega = 2\pi\nu$), $n(\omega)$ is the frequency-dependent refractive index, and L is the path length through the material. The spatial dependence of the intensity of the laser pulse, $I(x)$, induces a spatial dependence to the refractive index, $n(x) = n_0 + n_2 I(x)$, which produces a position-dependent phase delay,

$$\Delta\phi(x) = \frac{n_2 I(x)\omega L}{c}. \tag{11.3}$$

Most lasers, femtosecond or otherwise, produce beams with a Gaussian spatial intensity profile. The Gaussian beam experiences a variation in refractive index over its profile that is equivalent to a

STUDY BOX 11.1: PHASE AND CHIRPED PULSES

As shown in Figure 11.3(a), the phase of a sine wave is simple to visualize: the peaks and troughs are separated by a phase of π. A femtosecond laser pulse is composed of many constituent frequencies of electric field oscillations. The relative phase of the constituent frequencies determines the interferences between the different frequencies and hence the temporal structure of the pulse (see Figure 11.3(b) and (c)). The superposition illustrated in **A** results in a bandwidth limited pulse, in which the oscillation frequency is constant over time and is determined by the 'energy-time uncertainty' relation given in Eq. (11.1) of the main text, whereas that in **B** results in a 'chirped pulse', *i.e.* a pulse for which the instantaneous frequency varies with time. Not only is chirp important in laser design, but it can also be exploited in pump-probe schemes, as illustrated in Section 11.5.

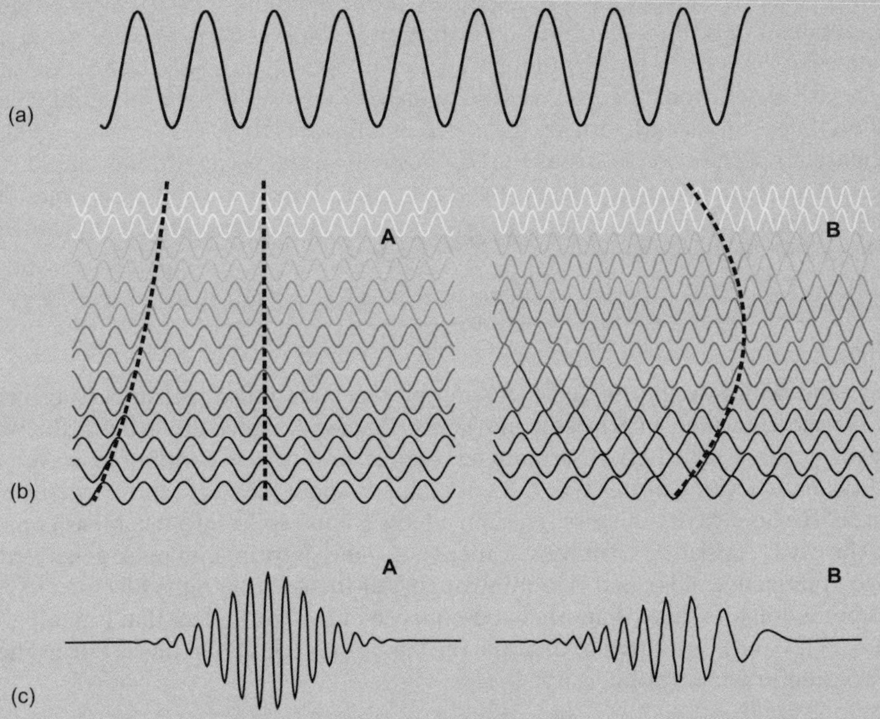

Figure 11.3 a) A simple sine wave. b) An illustration of the phase relationships between different frequency components in a bandwidth limited femtosecond pulse (**A**) and a chirped femtosecond pulse (**B**). c) The resulting time-dependent waves associated with the superposition of the frequency components shown in b). Note that in **A** the instantaneous frequency is constant over the duration of the pulse, whereas in **B** the instantaneous frequency varies linearly with time. The dashed lines connect points on the waves of constant phase.

The frequency dependent phase shift, $\phi(\omega)$, can be expanded as a Taylor series about the central frequency, ω_0,

$$\phi(\omega) = \phi(\omega_0) + \phi'(\omega - \omega_0) + \tfrac{1}{2}\phi''(\omega - \omega_0)^2 + \cdots. \qquad \text{(B11.1.1)}$$

The static absolute phase (the 1st term on the right hand side of Eq. (B11.1.1)) is the phase of the optical cycles relative to the pulse envelope profile. For femtosecond pulses of duration $\tau \sim 30$ fs, for which there are many optical cycles within the pulse envelope, this term is usually ignored. The linear phase or total group delay (the 2nd term on the right hand side of Eq. (B11.1.1)) causes a shift in the temporal envelope of the pulse relative to a reference time (see dotted lines on the left-hand panel of Figure 11.3(b)). A non-zero linear phase corresponds to a net delay in the pulse. The quadratic phase or 'linear chirp' (the 3rd term on the right hand side of Eq. (B11.1.1)) causes the instantaneous frequency to increase or decrease linearly with time, just as the sound of a bird's chirp has increasing vibrational frequency with time. ϕ'', the linear chirp, is also called the group delay dispersion (GDD). Linear chirp will be introduced to a pulse as it passes through any optical medium with a wavelength dependent refractive index. A linearly chirped pulse is necessarily of longer duration and lower peak intensity than the transform-limited pulse, and the frequency (the 'instantaneous frequency') of the pulse as a function of time will vary linearly. If the instantaneous frequency increases with time it is called a positive chirp and if it decreases with time it is called a negative chirp. A short pulse has a large bandwidth, so the temporal stretch (chirp) produced after it passes through an optical medium will be high. For example, a 30 fs pulse centered at 800 nm will be stretched to 51 fs when passing though 1 cm of an optical glass called BK7. The amount of chirp can be calculated from the group velocity dispersion (GVD) for a material. GVD is the amount of GDD per unit length, usually measured in units of $fs^2\,mm^{-1}$. For short <100 fs pulses, it is particularly important to control the GDD inherent in the oscillator and amplifier cavities. Finally, the fourth term in Eq. (B11.1.1) is the third order phase or second order chirp. This term and the higher order terms in the Taylor expansion become increasingly important with decreasing pulse duration and increasing bandwidth.

Helen H. Fielding and Abigail D. G. Nunn

positive lens, and which results in self-focussing of the laser pulse. Self-focussing provides the mechanism for mode-locking in femtosecond lasers. Placing an aperture around the waist of the focussed beam in the laser cavity discriminates against unfocussed continuous-wave (CW) laser light. Kerr-lens mode-locking of this type does not happen spontaneously, and is usually initiated by vibrating one of the laser cavity mirrors. This introduces a noise spike into the intensity profile of the radiation in the cavity, initiating effective self-focussing – the short intense pulse generated induces a larger change in refractive index and is more strongly self-focussed than the low intensity CW light.

In the temporal domain, there is another consequence of the Kerr effect that is analogous to self-focussing. Eq. (11.3) may re-written to describe the temporal variation in phase due to the normally Gaussian variation in temporal intensity, $I(t)$:

$$\Delta\phi(t) = \frac{n_2 I(t)\omega L}{c}. \tag{11.4}$$

The changing phase of the propagating pulse is associated with a change in the frequency of the light. For a pulse propagating in an optical medium of length L, the accumulated phase results in an instantaneous frequency (defined mathematically as the derivative of the phase with respect to time) shift,

$$\Delta\omega(t) = -n_2 \frac{\mathrm{d}I(t)}{\mathrm{d}t}\frac{\omega L}{c}. \tag{11.5}$$

The sign of the frequency shifts depends on the slope of the Gaussian intensity profile, resulting in new frequency components, redder and bluer at short and long times respectively. This broadens

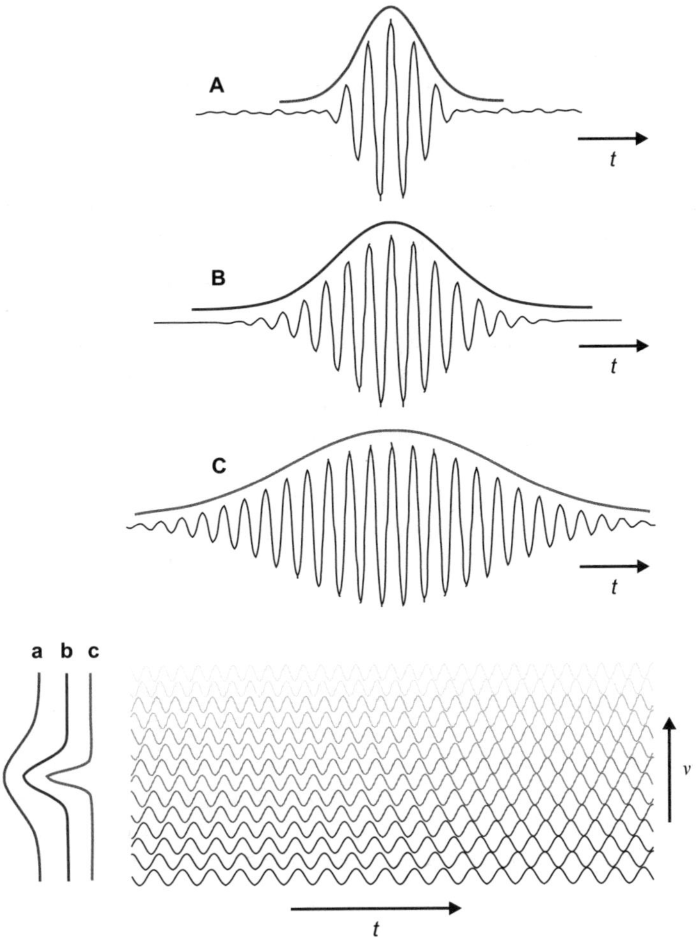

Figure 11.4 Illustration of the energy-time uncertainty principle. The electric fields of three transform-limited Gaussian pulses marked **A**, **B** and **C** are all generated by the interference of the CW waves shown at the bottom of the panel, with various spectral widths. The envelopes of the electric fields are marked in colour in each case. The lengths of the pulses increase in the order **A** < **B** < **C**. The spectrum for each of the pulses is illustrated with the waves in the lowest panel (spectrum **a** corresponds to pulse **A**, **b** to **B**, and **c** to **C**). The widths of the spectra increase in reverse order to the temporal lengths (**c** < **b** < **a**) in accordance with the energy-time uncertainty principle, $\Delta\nu\Delta t = 0.441$ (see Eq. (11.1)). **A** has 5 optical cycles; the shortest possible femtosecond pulse has a duration of 6 fs at 800 nm, which corresponds to a single optical cycle. In the limit of a delta function in the time-domain the corresponding spectrum is infinitely wide, whereas in the limit of infinite duration (CW) the spectrum corresponds to a single frequency.

the spectrum and therefore reduces the pulse duration, Δt. This phenomenon is called self-phase modulation (SPM) and is important in some femtosecond laser oscillators as a means of reducing the pulse duration.

The output of a typical mode-locked femtosecond Ti:sapphire oscillator is a train of 30–100 fs pulses, with central wavelength $\lambda = 750$–850 nm, at a repetition rate of 80–100 MHz and pulse energy of a few nJ. However, most processes of interest to the photochemist are initiated by absorption of photons in the ultraviolet wavelength range $\lambda = 200$–350 nm, and wavelengths further into the infrared region (μm) are important for vibrational spectroscopy experiments. The easiest way of generating a stable source of tuneable femtosecond pulses in the visible, ultraviolet or infrared is to exploit non-linear optical techniques in optical parametric amplifiers. Non-linear optical techniques are discussed further in Study Box 11.2.

11.2.2 Femtosecond Amplifiers

Before attempting frequency conversions to longer or shorter wavelengths, the output pulse energy of a Ti:sapphire oscillator must be amplified to improve the efficiency of the non-linear optical processes involved. Avoidance of, and compensation for, self-phase modulation is an important consideration in high power < 100 fs amplifiers. In order to amplify the short pulse without significant distortion of the phase, chirped pulse amplification is necessary, as explained in the following (see also Study Box 11.1 for a discussion of chirped pulses). Using chirped pulse and regenerative amplifier technology* it is possible to provide pulse energies of a few mJ at kHz repetition rates, corresponding to peak intensities in the TW cm^{-2} (1 TW = 10^{12} W) range.

When the modes of the laser are fully locked together, at the maximum intensity of the pulse, all of the individual frequencies have their phases aligned. Such a pulse is termed a transform-limited pulse because its duration is the shortest possible duration that can be given by Fourier-transforming the spectrum of the pulse, and its time and spectral FWHMs obey the limit given in Eq. (11.1). As discussed in detail in Study Box 11.1, more formally, in the Taylor series expansion of the phase as a function of the angular frequency,

$$\phi(\omega) = \phi(\omega_0) + \phi'(\omega - \omega_0) + \tfrac{1}{2}\phi''(\omega - \omega_0)^2 + \cdots, \quad (11.6)$$

we find that for the transform-limited pulse, only the first two terms are non-zero.

The central wavelength of a femtosecond laser pulse propagates at the phase velocity, while the modulating pulse envelope moves at the group velocity. In terms of the angular frequency, ω, and the wavenumber, $k = 2\pi/\lambda$, the phase velocity of the pulse is defined as

$$v_\mathrm{p} = \frac{\omega}{k}, \quad (11.7)$$

and the group velocity by

$$v_\mathrm{g} = \frac{\mathrm{d}\omega}{\mathrm{d}k}. \quad (11.8)$$

In vacuum, both the phase velocity and the group velocity are constant and equal to c. However, in a medium, which may be a solid, liquid or a gas, because the refractive index, n, is a function of ω, then $v_\mathrm{p} \neq v_\mathrm{g}$, and hence the temporal spreading of frequencies, dispersion, and non-zero group delay dispersion (GDD), occur. The term group velocity dispersion (GVD) is the amount of GDD per unit length, as mentioned in Study Box 11.1, but the term is often used as a general description for the stretching or compression of ultrafast pulses in dispersive media.

In a typical commercial, high-energy, sub-100 fs regenerative amplifier, the output of the femtosecond oscillator first enters a stretcher. A schematic of a regenerative amplifier is shown in Study Box 11.3, where they are discussed in more detail. The stretcher comprises a diffraction grating and gold-coated mirror, aligned so as to introduce a positive chirp to the laser pulses. This lengthens the pulses temporally, reducing the peak power. These chirped pulses are then injected into the regenerative amplifier cavity by an optical switch, usually either a Pockels cell, or a Pockels cell and

*Regenerative amplifiers are optical amplifiers with a resonator in which a light pulse can make many (possibly hundreds) of round trips so that it is amplified to a high level before it is coupled out of the resonator.

STUDY BOX 11.2: NON-LINEAR OPTICS

When light passes through a medium, its electric field distorts the electron density within the medium, and the sample becomes polarized. Under most circumstances, the induced polarization, $\boldsymbol{P}$, is proportional to the applied electric field $\boldsymbol{E}$,

$$\boldsymbol{P} = \varepsilon_0 \chi^{(1)} \boldsymbol{E}, \tag{B11.2.1}$$

where ε_0 is the permittivity of free space, and $\chi^{(1)} = n^2 - 1$ is the linear susceptibility of the medium. n is the (complex) refractive index of the sample. The real part of n determines the speed of propagation through the sample, $v = c/\mathrm{Re}(n)$, while the imaginary part describes absorption by the medium. In linear optics, the light may be deflected, delayed, or attenuated by passing through a medium, but its frequency remains unchanged.

When dealing with light of much higher intensities, such as that found in a laser beam, the induced polarization in the medium is often not a linear function of the electric field of the light, and we enter the regime of *non-linear optics*. The non-linear polarization is usually expanded in a power series,

$$\boldsymbol{P} = \varepsilon_0 \left(\chi^{(1)} \boldsymbol{E} + \chi^{(2)} \boldsymbol{E}^2 + \chi^{(3)} \boldsymbol{E}^3 + \cdots \right), \tag{B11.2.2}$$

where $\chi^{(n)}$ is the nth order susceptibility. For the series to converge, we must have $\chi^{(3)} \boldsymbol{E}^3 \ll \chi^{(2)} \boldsymbol{E}^2 \ll \chi^{(1)} \boldsymbol{E}$.

The electric field of the light in a laser beam of frequency ω takes the form $\boldsymbol{E} = \boldsymbol{E}_\omega \cos \omega t$. When we substitute this into Equation (B11.2.2) we find that when we expand out the terms, use a little trigonometrical trickery, and collect like terms, the result may be written

$$\boldsymbol{P} = \boldsymbol{P}_0 + \boldsymbol{P}_1 \cos \omega t + \boldsymbol{P}_2 \cos 2\omega t + \boldsymbol{P}_3 \cos 3\omega t + \cdots \tag{B11.2.3}$$

The first term in this series is the DC term, the second is the fundamental term, the third is the second harmonic term, and so on. We see that the non-linear response of the medium gives rise to components of light at frequencies ω, 2ω, 3ω, etc..

If we take a general electric field of the form $E = E_0 + E_\omega \cos \omega t$, the terms generated give rise to a whole slew of non-linear optical effects, summarized in Table 11.1. These phenomena are not all automatically observed when the intensity of light propagating through a medium becomes high enough, largely due to the additional constraints that both the energy and momentum of the photons involved must be conserved. To illustrate this, we will consider just one of these processes in more detail, that of second harmonic generation (SHG). SHG is widely used to generate laser light in spectral regions for which suitable laser media are not available. As an example, tuneable UV laser light at the wavelengths required for many LIF and REMPI detection schemes (see Study Box 5.2 on *LIF and REMPI detection schemes*) may be generated by frequency doubling the visible output of a tuneable dye laser. To treat the conditions for SHG rigorously requires a detailed analysis of the coupled wave equations for the light propagating at the fundamental and second harmonic frequencies. However, an appreciation of the basic principles may be obtained without resorting to such extensive mental gymnastics.

The second harmonic, 2ω, wave is initially generated with a well-defined phase relationship to the fundamental, ω, wave. However, if this relationship is not maintained as the two waves propagate, destructive interference between the two waves makes the SHG process extremely inefficient. Constructive interference requires that the phase relationship is maintained as the waves propagate through the crystal. Formally, this means that the two waves must propagate in the same direction with the same phase velocity *i.e.* $v(\omega) = v(2\omega)$. This requirement is called *phase matching*, and when it is satisfied, SHG becomes highly efficient. Requiring phase matching is equivalent to requiring conservation of photon energy and linear momentum. Energy is conserved since $\hbar(2\omega) = \hbar\omega + \hbar\omega$, and since linear momentum $\boldsymbol{p}$ is related to photon wavenumber $\boldsymbol{k}$ by $\boldsymbol{p} = \hbar\boldsymbol{k}$, conservation of momentum requires $k(2\omega) = 2k(\omega)$. This is achieved when the refractive index of the material is equal for the two frequencies.

Table 11.1 Non-linear optical effects on application of a field $E = E_0 + E_\omega \cos \omega t$ to a non-linear medium.

Term	*Frequency*	*Effect*
$\chi^{(1)}E_0$	ω	DC polarizability
$\chi^{(1)}E_\omega \cos \omega t$	ω	Optical polarizability
$\chi^{(2)}E_0^2$	DC	DC hyperpolarizability
$\chi^{(2)}E_0E_\omega \cos \omega t$	ω	Linear electro-optic effect (Pockels effect)
$\chi^{(2)}E_\omega^2$	DC	Optical rectification
$\chi^{(2)}E_\omega^2 \cos 2\omega t$	2ω	Second harmonic generation
$\chi^{(3)}E_0^2E_\omega \cos \omega t$	ω	DC Kerr effect
$\chi^{(3)}E_0E_\omega^2 \cos 2\omega t$	2ω	DC induced second harmonic generation
$\chi^{(3)}E_\omega^3 \cos 3\omega t$	3ω	Third harmonic generation
$\chi^{(3)}E_\omega^3 \cos \omega t$	ω	Optical frequency Kerr effect

There are several schemes for achieving phase matching, but the most commonly employed rely on the SHG crystal being anisotropic, such that its refractive index varies with propagation direction. Since the velocity v of light in a medium is directly related to the refractive index by $v = c/n$, such crystals are often said to have a 'fast' and a 'slow' axis for propagation of light. The refractive index is generally also wavelength dependent. Given these properties, if the fundamental and second harmonic are polarized at right angles to each other, it is generally fairly straightforward to find an angle of propagation through the crystal for which the refractive indices, and therefore the velocities, for the two beams, are identical *i.e.* $n_1(\omega) = n_2(2\omega)$. This scheme, in which two photons at frequency ω with identical polarizations mix to give a single photon at frequency 2ω with orthogonal polarization, is known as Type I phase matching. An alternative scheme is Type II phase matching, in which the two photons at frequency ω have orthogonal polarizations and the photon at frequency 2ω takes the polarization of one of the two input photons. The same type of mechanism can also be used to mix light waves of different frequencies in a sum frequency generation (SFG) process. Other second order non-linear effects include difference frequency generation (DFG), optical parametric generation (OPG), and optical parametric amplification (OPA).

Claire Vallance

quarter waveplate combination, which rotates the polarization of the light. The optical switch is on for about 1 ns, allowing just one of the femtosecond pulses to pass through it, to or from the cavity to the Brewster window at the crystal, which can selectively transmit the pulse depending on its polarization. The active medium in the regenerative amplifier is another Ti:sapphire crystal that is usually pumped by a solid-state, Q-switched, Nd:YLF laser. The chirped pulses make many round trips in the cavity before a second optical switch ejects the amplified pulse into a compressor. In the compressor another diffraction grating and mirror set are aligned to introduce a negative chirp to compensate for the positive chirp introduced by the stretcher, and sometimes to compensate for higher order phase errors introduced by the amplification process itself.

11.2.3 Wavelength Tuning of Femtosecond Pulses

The most straightforward and common techniques for extending the wavelength coverage of femtosecond Ti:sapphire lasers take advantage of the non-linear polarization induced in a transparent medium at high optical electric field intensities. As discussed in Study Box 11.2, the polarization is given by

$$\boldsymbol{P} = \varepsilon_0\left(\chi^{(1)}\boldsymbol{E} + \chi^{(2)}\boldsymbol{E}^2 + \chi^{(3)}\boldsymbol{E}^{(3)} + \cdots\right), \tag{11.9}$$

where ε_0 is the permittivity of free space, and $\chi^{(i)}$, with $i = 1, 2, 3, \ldots$, the electric susceptibilities. For low light intensities, $\boldsymbol{P}$ is linear in the electric field $\boldsymbol{E}$, whereas the higher order non-linear contributions become important with increasing intensity.

The second order non-linearity of non-centrosymmetric crystal materials is exploited in optical parametric amplifiers (OPAs). Signal beam photons, ω_s, and higher energy pump beam photons, ω_p, are combined in a nonlinear crystal where the pump photons are converted into an equal number of photons at the original signal frequency and at a new lower frequency called the idler frequency, ω_i. Thus energy is conserved and $\omega_p = \omega_s + \omega_i$. The signal and idler frequencies are determined by phase matching constraints $\boldsymbol{k}_p = \boldsymbol{k}_s + \boldsymbol{k}_i$, and can be tuned by rotating the optical axis of the crystal. These signal and idler frequencies then undergo further non-linear optical mixing with the intense pump beam, followed by second-harmonic generation (SHG), third-harmonic generation (THG), or fourth-harmonic generation (FHG), to generate tuneable femtosecond pulses across the visible and ultraviolet regions of the spectrum.

In the vacuum ultraviolet region all known crystalline materials absorb radiation, so it is necessary to use a gas as the non-linear medium to generate these shorter wavelengths. Focussing intense femtosecond laser radiation into a pulsed jet of gas, or into a gas-filled capillary tube,[753] generates high-order odd harmonics, *via* a non-linear optical process similar to that used in the REMPI and H-atom Rydberg tagging techniques, described in Study Box 5.3 and in Chapter 6.* This process is referred to as high-order harmonic generation (HHG) and is of particular relevance to the new generation of attosecond light sources (1 as $= 10^{-18}$ s) operating in the soft X-ray region at wavelengths around ~ 10 nm. Further details about the operation and optical layout of a typical femtosecond laser system are given in Study Box 11.3.

11.2.4 Pulse Characterization

The spectral profile of an ultrashort laser pulse can be measured using a spectrometer. If the laser system generating the pulse is ideal, the spectral bandwidth can then be related to the minimum pulse duration by the time-bandwidth product (Eq. (11.1)) to indicate the approximate duration of an ultrashort laser pulse. Usually, numerous optical elements are involved in the production of pulses at the desired wavelength, and this causes distortions in the phase of the pulse that need to be removed using feedback from a measurement of the true pulse duration. In practice, a measurement known as pulse autocorrelation, recorded using a second harmonic generation crystal, is the simplest method to obtain an indication of the true pulse duration. In an SHG autocorrelation measurement, the pulse is overlapped with a delayed replica of itself (produced by a beamsplitter and an optical delay line) in a non-linear crystal. A signal will be detected at twice the frequency of

*Note, however, that the examples referred to employ nanosecond rather than femtosecond lasers. One can only generate low order harmonics using the intensities associated with nanosecond lasers.

the input pulse (due to the second term in (11.9)). Because the efficiency of the frequency doubling depends on the total intensity of the input light, the measured signal is proportional to the amount of temporal overlap of the two pulses. By recording the SHG signal as a function of the optical delay between the two pulses, the temporal pulse shape may be determined. For a Gaussian pulse the autocorrelation intensity signal has a FWHM that is $\sqrt{2}$ times the input electric field pulse duration. A cross-correlation measurement may be obtained similarly by recording the second order non-linear signal produced as a function of delay between a reference pulse and an unknown pulse in a non-linear crystal. Alternatively, a useful method of measuring the cross-correlation of two pulses used in a two-colour pump-probe experiment is to characterize them in a molecular sample, similar to the one upon which the pump-probe experiments are being carried out. For example, a two-colour photoionization experiment may be adapted so that a molecule may be non-resonantly ionized using the two photons of the two-colour experiment; the cross-correlation is the electron or ion yield measured as a function of delay. In some cases, further characterization of the laser pulse is desirable. Complete characterization is the term given to the measurement of the intensity and phase of the electric field, as a function of either time or frequency. The most popular technique for complete characterization of femtosecond pulses is to retrieve the phase from a spectrogram produced by measuring the spectrum of the autocorrelation of a laser pulse. This technique is called FROG (frequency resolved optical gating), and produces the electric field from a phase retrieval algorithm applied to an experimental spectrogram trace of intensity *versus* delay and frequency of the SHG signal.[754] Alternative methods of complete characterization are based on spectral interferometry (SPIDER or spectral phase interferometry for direct electric field reconstruction)[755] and the MIIPS (multiphoton intrapulse interference phase scan) method, in which the diagnostic tool is SHG intensity measured as a function of reference phases added by a pulse shaper.[756]

11.3 PROBES FOR ULTRAFAST CHEMICAL DYNAMICS

Even the fastest electronic devices cannot measure femtosecond transients, so all femtochemistry experiments employ an optical pump-probe technique to follow the molecular dynamics. As we have already seen in Chapter 1, George Porter and Ronald Norrish were awarded part of the 1967 Nobel Prize for Chemistry for the development of the pump-probe method, and, in particular, for their investigations of 'extremely fast' chemical reactions, which at the time referred to reactions occurring on a microsecond timescale.[7] The concept is illustrated schematically in Figure 11.5. A pump pulse promotes the system of interest from its initial state into some excited state that subsequently evolves along a reaction or dynamic coordinate. This first pulse defines the zero-of-time. A probe pulse is then fired at a series of precisely timed delays, Δt, to transfer the excited state population into a new state whose population can be monitored. When delays on the femtosecond timescale are required, the delay is controlled using an optical delay line in which a moveable mirror is mounted on a computer-controlled translation stage that is capable of very precise movements ($\Delta t = 10$ fs corresponds to a 3 μm path difference between the pump and probe pulses).

In Chapters 4 to 8 of this book, some of the techniques used for studying the dynamics of gas-phase reactions are described. Many of these techniques – absorption spectroscopy, resonance-enhanced multi-photon ionization (REMPI) and laser-induced fluorescence (LIF) – may be used in combination with femtosecond lasers in order to deduce valuable time-resolved information. For gas-phase femtosecond experiments, an increasingly popular and versatile detection technique is time-resolved photoelectron spectroscopy (TRPES)[757,758] (see Section 11.4.3). The method selected to follow a particular molecular reaction often depends on the nature of the dynamics being probed, and hence we will discuss probe methods alongside our discussion of photoinduced processes.

STUDY BOX 11.3: FEMTOSECOND LASER OPTICAL LAYOUT

The three figures in this Study Box show a Ti:sapphire femtosecond oscillator (top), chirped pulse regenerative amplifier (middle) and optical parametric amplifier (OPA) with frequency conversion unit (bottom).

Femtosecond Oscillator

The pump beam (532 nm CW laser) enters the cavity *via* L1, passing through M4, to reach the Ti:Al_2O_3 crystal where the laser emission at 785 nm occurs. The laser cavity is defined by the mirror M7 and the partially reflecting mirror M0/M1. The prism pair BP1 and BP2 allow control of the chirp of the pulse. The starter introduces noise into the cavity to trigger the Kerr-lens mode-locking of the laser. The path defined by the dotted line is used for alignment purposes only, and is not part of the mode-locked beam path.

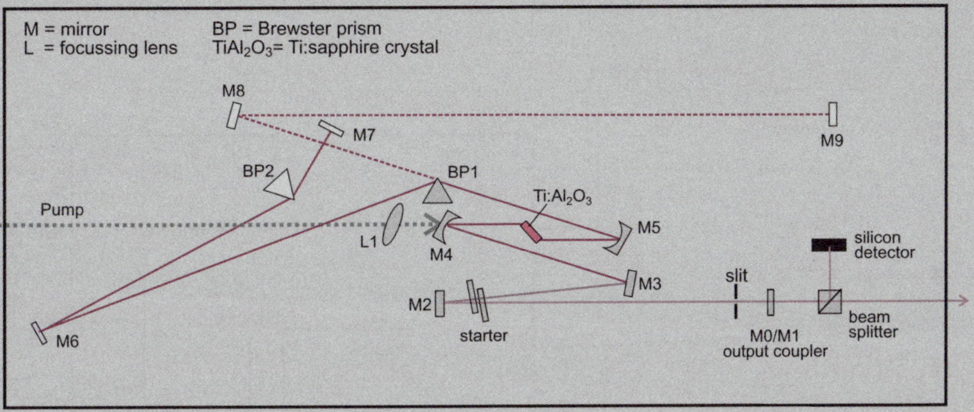

Regenerative Amplifier

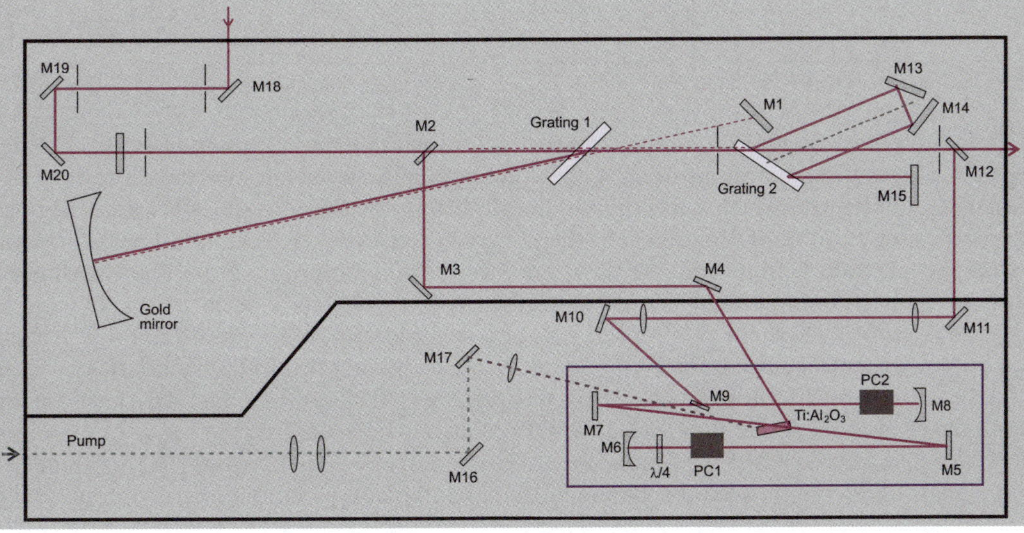

The output of the femtosecond oscillator first enters a stretcher *via* M18, M19 and M20. The stretcher comprises a diffraction grating (Grating 1) and gold-coated mirror (Gold mirror) that are aligned to introduce a positive chirp to the laser pulses. This lengthens the pulses temporally, reducing the peak power. These chirped pulses are then injected *via* M2 and M3 into the regenerative amplifier cavity (bounded by M6 and M8) by an optical switch, a Pockels cell (PC1) and quarter waveplate ($\lambda/4$) combination, to rotate the polarization of the light. The optical switch is on for about 1ns, allowing just one of the femtosecond pulses to pass into the cavity, to the Brewster window at the crystal. The Brewster window selectively transmits one polarization. The active medium in the regenerative amplifier is another Ti:sapphire crystal that is usually pumped by a solid-state, Q-switched, Nd:YLF laser that enters the cavity *via* M16 and M17. The chirped pulses make many round trips in the cavity before a second optical switch (PC2) ejects the pulse into a compressor *via* M9, M10, M11 and M12. In the compressor, another diffraction grating (Grating 2) and mirror set (M13, M14 and M15) are aligned to introduce a negative chirp to compensate for the positive chirp introduced by the stretcher.

OPA

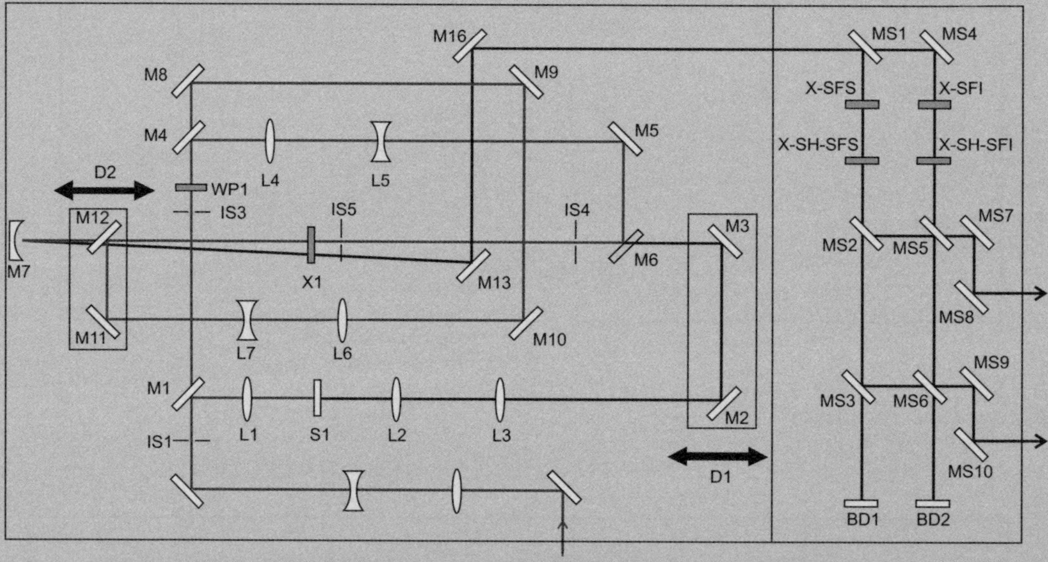

The beam enters the OPA and is collimated before white light is generated at S1. The delay stage D1 allows the temporal position of the white light pulse to be fine-tuned with respect to the fundamental pulse with which it recombines on M6 before the OPA crystal X1, where the optical parametric amplification (OPA) signal in the infrared is created. OPA is a non-linear three-wave mixing phenomenon, and in this case the input infrared signal photons from the white light are combined with the pump beam (795 nm, *via* transmission through M1, then *via* M4, M5 and M6) in the OPA crystal where the pump photons are converted into more signal photons, as well as idler photons. Signal wavelength flexibility is possible by mixing the fundamental in X1 with any of the broad bandwidth of white light. Selective mixing is achieved by phase matching *via* angle tuning the crystal (see Study Box 11.2). The OPA signal is reflected from M7 and then combined on M12 with another part of the fundamental beam and passed back through X1 to amplify the OPA signal. The delay between the amplifier fundamental and the OPA signal is optimized using the delay D2. If a different wavelength range is desired, the OPA idler may be optimized instead of the signal. The amplified OPA signal or idler leaves the OPA *via* M13 and M16 and enters the frequency conversion unit where wavelengths in the visible and UV are created by sum-frequency mixing with the fundamental (at X-SFS or X-SFI) and/or second harmonic generation (at X-SH-SFS or X-SH-SFI) of the resulting signal.

Helen H. Fielding and Abigail D. G. Nunn

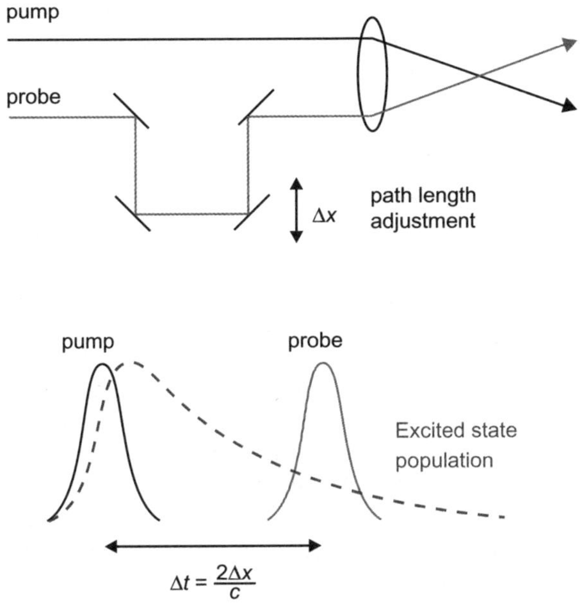

Figure 11.5 Two-colour pump-probe experiment. Two femtosecond laser beams are delayed with respect to one another using a delay line, an arrangement of four mirrors as shown in the figure, and the two beams are then focussed together into a molecular sample (top). The first beam, I_{pump} (black solid line), produces an excited state population in the sample (dashed line in bottom figure), that the second beam, I_{probe} (grey solid line), probes at different values of delay, Δt, by creating an observable signal.

11.4 ULTRAFAST PHOTOINDUCED PROCESSES

The large spectral width of a femtosecond light pulse results in a number of molecular eigenstates being excited coherently. This coherent superposition of wave functions then evolves in time. The production of coherent superpositions of states is a common feature of femtosecond experiments. This is an important distinguishing feature compared with the majority of the (nanosecond) experiments described so far in these Tutorials, which involve the preparation of single eigenstates of the system*.

*Although the majority of this book has focussed on experiments involving preparation of single eigenstates (or incoherent mixtures of states), aspects of coherent excitation of states have been described. The reader might find it helpful at this point to review the material in Sections 3.3.2.3, 7.4.2, and 8.4 and 8.5, as well as Problems 1 and 2 of Chapter 7. Coherence is also relevant to the description of angular momentum polarization discussed in Chapter 9.

11.4.1 Wavepackets

The most intuitive example of a coherent oscillation is one resulting from a coherent superposition of vibrational wave functions, $\psi_\upsilon(R)$. The resulting wavepacket can be written as

$$\Psi(R,t) = \sum_{\upsilon} a_\upsilon g_\upsilon \psi_\upsilon(R) e^{-i\omega_\upsilon t}, \tag{11.10}$$

where a_υ are the transition dipole matrix elements connecting the excited vibrational states to the initial state, and g_υ is the amplitude of the laser field at each excitation frequency within the spectral profile of the laser pulse. The wavepacket oscillates back and forth along the internuclear coordinate at the vibrational period (see Problem 3 of this chapter).

When projecting a molecular wavepacket onto a final state, it is necessary that the transition to the final state is fairly well understood in order for the dynamics of the excited state to be deconvoluted successfully from the final state excitation probabilities. The most versatile techniques are able to project the entire excited state wavepacket onto a measurable signal that can be dispersed in energy, to observe the beating in time between the different frequency components within the eigenstate composition of the wavepacket, or to observe the transfer of energy between different electronic energy channels. Thus, the choice of probe in a pump-probe experiment is critical in determining what aspect, or aspects, of the excited state dynamics may be observed.

A classic illustration of nuclear vibrational wavepackets is an elegant experiment on Na_2 performed by Gustav Gerber and his colleagues.[759] The excitation scheme is shown in Figure 11.6. From the Na_2 $X^1\Sigma_g^+$ ground electronic state, a femtosecond pump pulse creates a vibrational wavepacket in the $A^1\Sigma_u^+$ state that oscillates with a period of 306 fs, from an excitation that spans the vibrational levels $\upsilon' = 10-14$. The same femtosecond pump pulse also creates a wavepacket in the $2^1\Pi_g$ state ($\upsilon^* = 11-18$), which oscillates with a period of 363 fs. A delayed, but otherwise identical femtosecond probe pulse is used to probe the wavepacket motion in both states, and the detected signal is the number of Na_2^+ ions formed from both states. Two-photon ionization from the inner turning point of the A state creates a Na_2^+ ion in its $^2\Sigma_g^+$ state. Single-photon excitation from the outer turning point of the $2^1\Pi_g$ state wavepacket creates a doubly excited Rydberg state, which autoionizes to $Na_2^+(^2\Sigma_g^+)$, and dissociates to give Na^+ and Na(3s). A Fourier analysis of the pump-probe signal reveals the different vibrational spacings in the two wavepackets, and the phase of the envelope of the Na_2^+ ion signal structure relative to the zero-of-time reflects the phase-shift between the absorption of the probe photons by the two wavepackets at the inner and outer turning points of their origins.

11.4.2 Photodissociation

Photodissociation of a diatomic molecule is the simplest of all possible chemical reactions, and some examples of femtosecond studies of these types of systems have already been presented in Section 8.4. The photodissociation of NaI was a landmark experiment in the development of the field of femtochemistry.[760] The relevant potential energy (PE) curves for NaI photodissociation are presented in Figure 11.7. The lower energy adiabatic PE curve has ionic character for small internuclear separations *R*, and is covalent for large *R*. The opposite is true for the higher energy adiabatic PE curve, which is covalent for small *R* and ionic for large *R*. Ahmed Zewail's experiment used a femtosecond pump laser to promote the molecule from the ground state to a coherent superposition of excited vibrational states. The excited NaI molecules then oscillate within the excited PE well, but each time the Na-I bond length reaches about 0.25 nm, there is a 10% probability that the wavepacket crosses non-adiabatically to the lower PE curve to form neutral Na and I atoms (see Chapter 4). The remaining excited state population continues to oscillate within

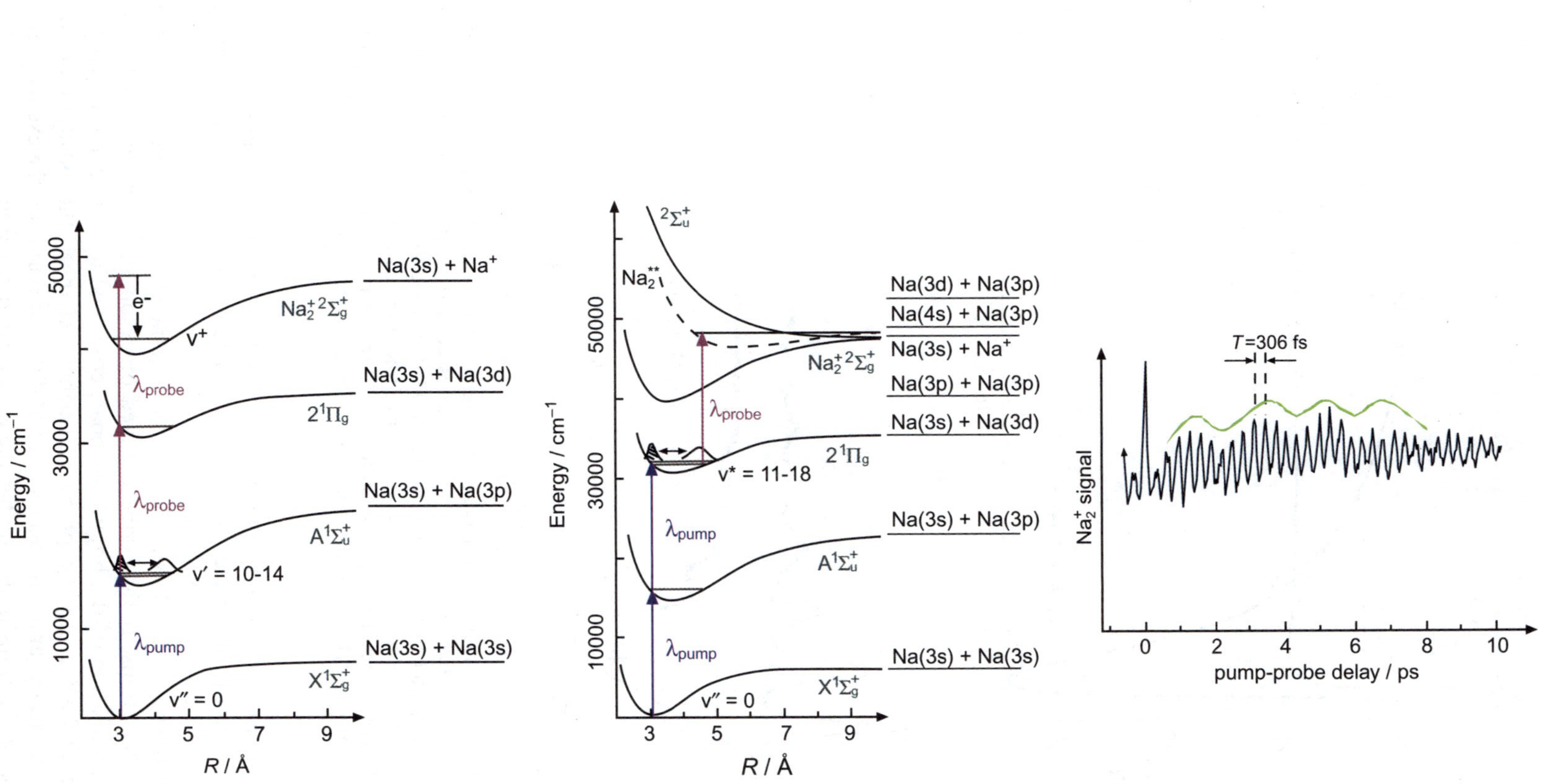

Figure 11.6 Na_2 potential energy curves (left and middle). A wavepacket is produced on each of the $A^1\Sigma_u^+$ (left) and $2^2\Pi_g$ (middle) states, and probed at the inner and outer turning points, respectively. The total population of $Na_2^+(^2\Sigma_g^+)$ is the probe signal (right), with the wavepacket oscillation period in the $A^1\Sigma_u^+$ state apparent at a maximum in the envelope structure (highlighted in the figure on the right). Adapted from Figures 1, 2, and 4 of T. Baumert *et al.*[759]

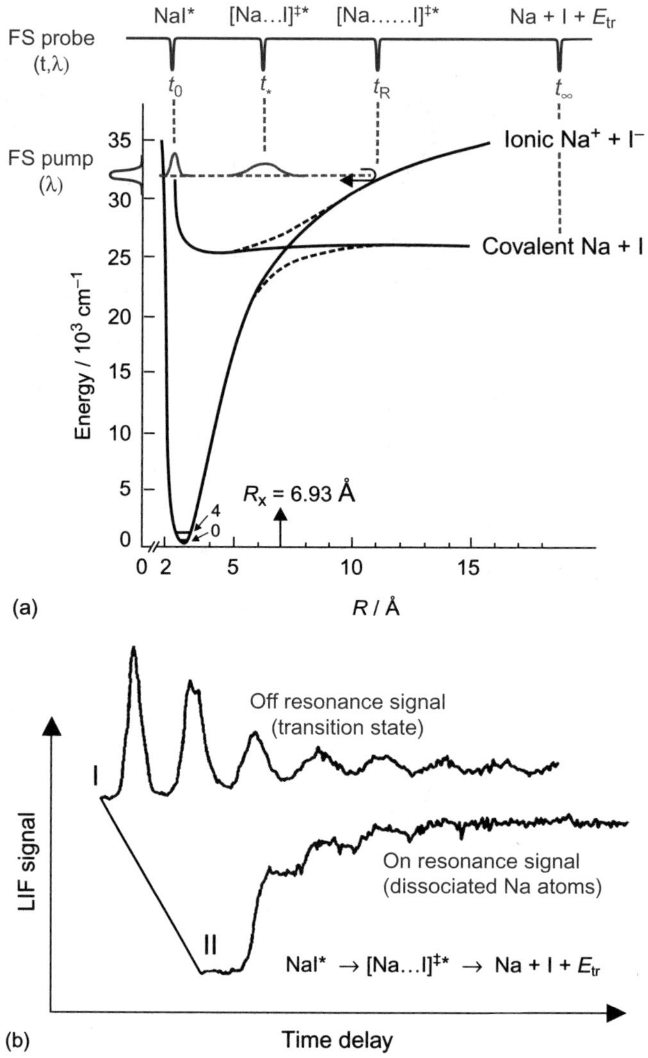

Figure 11.7 a) NaI potential energy curves and wavepacket dynamics schemes. The molecule is photo-excited to the upper potential energy curve by the femtosecond (fs) pump pulse and oscillates between the inner and outer turning points; at the avoided crossing there is a 10% probability the molecule will dissociate into neutral Na and I. b) Time-resolved LIF signal. The off resonance LIF wavelength probes the transition state (I), and the on resonance wavelength probes the dissociated Na atoms (II). Adapted from Figure 1 of T. S. Rose *et al.*[760]

the upper PE curve. The detection technique employed in the experiment is laser-induced fluorescence (see Study Box 5.2): when the probe excitation laser is tuned to 589 nm it is resonant with the Na D-line transitions of isolated atomic products. Since these transitions occur at very large (essentially infinite) Na-I bond lengths, the fluorescence signal obtained provides a measure of the total number of dissociated Na atoms. If the probe is tuned off-resonance from the Na D-line transitions, it monitors the population of the bound wavepacket oscillating within the upper PE curve, as shown in Figure 11.7. Varying the probe wavelength changes the position of the 'observation window', allowing the dynamics to be observed at different points on the excited state, and providing a direct probe of the transition-state for the dissociation reaction.

11.4.3 Femtochemistry of Larger Molecules

Laser-induced fluorescence proved a sensitive and extremely powerful tool in the experiment described above. Its generality, however, and that of the closely related technique of transient absorbance, is limited due to the lack of sensitivity for most large molecular systems. In larger molecules, where there is more than one bond to break, the different dissociation products can often be distinguished by femtosecond single-photon or multi-photon ionization followed by time-of-flight mass spectrometry.

Photoexcitation of larger molecules leads to the population of a high density of vibronic (electronic and vibrational) levels that are coupled to many other electronic and vibrational states. If the excited electronic state is well separated from all other electronic states, the coupling matrix elements between the electronic states are small, and there are no accessible conical intersections (CIs), then intramolecular vibrational redistribution (IVR) will be the only possible relaxation path available. If the potential energy surface of the excited electronic state crosses that of another electronic state, non-adiabatic coupling can result in a flow of energy between the electronic and vibrational degrees of freedom which is known as internal conversion when there is no change in spin multiplicity.*

Femtosecond time-resolved photoelectron spectroscopy (TRPES) has been exploited very successfully to observe ultrafast internal conversion (IC) in a number of polyatomic molecules. A classic example is S_2–S_1 IC in the linear polyene *all-trans* 2,4,6,8-decatetraene (DT).[761] The energy level scheme is presented in Figure 11.8. The first optically allowed transition is to the $S_2(1^1B_u)$ state, which is a singly excited configuration, correlating electronically with the $D_0(1^2B_g)$ ground electronic state of the cation. The $S_1(2^1A_g)$ state arises from a configuration interaction between singly and doubly excited configurations and correlates electronically with the $D_1(1^2A_u)$ first electronically excited state of the cation (see Section 2.4). This experiment uses a femtosecond pump laser to prepare the molecule in a vibrationless S_2 state. The latter then evolves into a

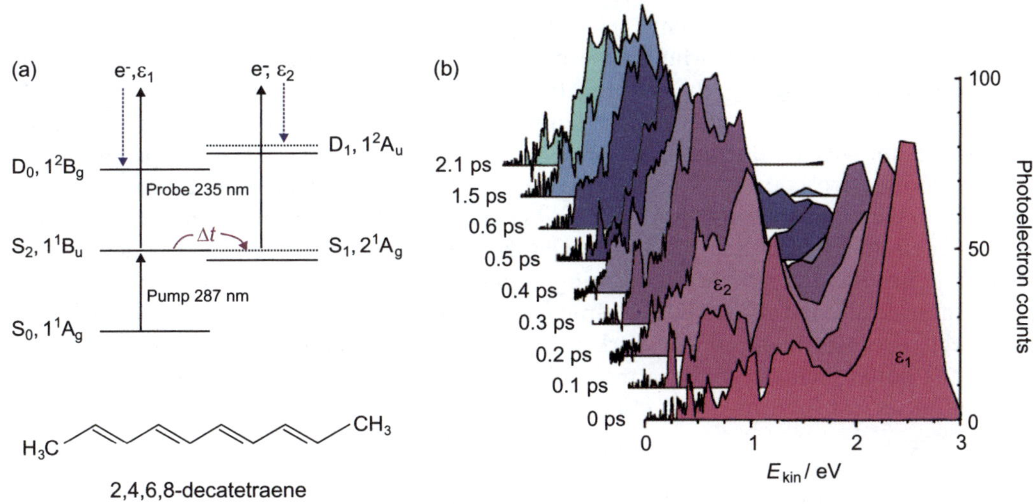

Figure 11.8 Pump-probe excitation scheme (a) and photoelectron kinetic energy as a function of delay (b) for *all-trans* 2,4,6,8-decatetraene. Internal conversion is observed through the change in kinetic energy of the photoelectrons from the ε_1 to ε_2 channel. Adapted from Figures 2 (left) and 3 (right) of V. Blanchet *et al.*[761]

*Internal conversion involves a breakdown in the Born-Oppenheimer approximation, as discussed in Chapters 4 and 8.

vibrationally hot S_1 state by IC. A delayed femtosecond probe laser ionizes the system and the photoelectron kinetic energy spectrum provides a measure of the S_1 and S_2 population.

TRPES has several key advantages over techniques such as LIF and time-resolved transient absorption, and the theory and applications of this method have been reviewed extensively elsewhere.[757,758] TRPES can be used to probe any excited state as long as the ionizing photon has sufficient energy. The changing photoelectron spectrum provides a picture of the excited state as it evolves. Imaging photoelectron spectrometers are now commonly used to record photoelectron angular distributions as well as photoelectron energies.* These have the advantage that they can also provide information about the evolution of the electronic angular momentum on the excited state. Indeed, femtosecond TRPEI (time-resolved photoelectron imaging) is now an established technique in its own right.[762] TRPES has also been combined with coincidence ion measurements to provide time-resolved molecular frame 3D dynamics of photodissociation reactions.[763] It has also been used to study intramolecular reactions such as tautomerization.[764]

Ultrafast electron diffraction is a powerful technique that allows the full 3D structure of a molecule, as it passes through a photoinduced reaction, to be reconstructed. The photoreaction is initiated by a laser pulse and then subsequent structures are measured using an ultrashort electron pulse as the probe. Recent developments in ultrafast electron diffraction, crystallography, and micrography have been reviewed by Ahmed Zewail.[765] One specific example is the ultrafast electron diffraction study of the non-concerted elimination reaction of $C_2F_4I_2$ in which the structure of the CF_4I intermediate was determined.[766]

11.4.4 The Condensed Phase

Studies of dynamics in the liquid phase are of considerable importance to a wide range of chemistry. The timescale of femtosecond experiments is well suited to investigations in the condensed phase, where the time between collisions is only a few hundred femtoseconds (see Study Box 7.4). A particular important area is the dynamics of biological processes, because femtochemistry techniques can be applied in environments closer to those found in nature than those used in single molecule gas-phase experiments. Many experimental techniques employed in the liquid phase are based on four-wave mixing, *i.e.* they exploit the third order polarizability (see Eq. (11.9)). Femtosecond 2D-IR spectroscopy enables the temporal coupling between different IR modes in a molecule to be unraveled from the spectral congestion associated with 1D IR spectroscopy. 2D vibrational echo spectroscopy[767,768] is an analogue of 2D NMR that has been extensively applied to biomolecules in the liquid phase, in both dynamics studies and for molecular identification. Like 2D NMR, this method can be used to study dynamics in thermally equilibrated condensed phase environments, but has the advantage of being applicable to the observation of dynamics on the ultrafast timescale. In the solid state, diffraction methods are important because electronic or vibronic spectroscopy is not as good a marker of the evolving geometry and dynamics of the system as it is for the liquid phase or a single molecule in an excited state. Pump-probe diffraction with short X-ray pulses has been reviewed elsewhere.[769,770]

11.5 FEMTOSECOND COHERENT CONTROL

In the broadest terms, coherent control may be described as the manipulation of molecular processes using the coherent properties of laser light. The first attempts to control molecular dynamics started soon after the invention of the laser, when the high intensity and narrow bandwidth of laser light relative to that produced by conventional light sources were exploited to populate vibrational

*The technique employed is usually a variant of velocity map ion imaging described in Study Box 5.3. It might also be helpful at this point to reread Section 8.4.

overtones, in an effort to dissociate the bond associated with a specific vibrational mode. Unfortunately, these attempts failed because intramolecular vibrational energy redistribution occurred before the excitation energy reached the dissociation limit for the target bond (see Chapter 7). Today, laser control of molecular dynamics and chemical reactivity is a major research field.[771] Scientists working in many branches of chemistry and physics continue to push the boundaries of technical and theoretical expertise in order to attempt to achieve the long-cherished goal of harnessing the power of laser light to manipulate and understand chemical reactivity and dynamics.

Over the last two decades, a number of coherent control schemes have been developed. Early methods were single-parameter schemes in which a single factor, such as the phase difference between two laser frequencies[772] or the time-difference between two pulses,[773] was used to alter the favoured pathway for a reaction. Time-domain and frequency-domain approaches to coherent control are in fact equivalent and in many ways overlap.[774] In the frequency domain, control over the population of a final quantum state is achieved by simultaneously accessing multiple independent excitation paths. The net excitation is a result of interference between the competing pathways, and manipulating the relative phases and amplitudes of these pathways changes the population transferred to the final state. This manipulation of multiple interfering pathways is analogous to the Young's double slit experiment. This methodology of coherent control was developed by Brumer and Shapiro.[775]

In the time-domain, the simplest control scheme employs pairs of optical pulses, as in a traditional pump-probe experiment, but instead of simply monitoring wavepacket evolution, in some systems controlling the timing of the probe pulse can select between two or more final states, as in Zewail's $I_2 + Xe$ experiment described in Section 11.1. This approach is a direct extension of the wavepacket dynamics experiments described in Section 11.4. The related 'pump-dump' approach was formalized by Tannor, Kosloff and Rice.[773] Their model (shown schematically in Figure 11.9) is based on a ground state potential energy surface of a molecule ABC featuring a minimum separated from two product channels, say AB + C and A + BC, by saddle points. Above the ground state surface is an excited state surface with a minimum that is displaced from that of the ground state, and has rotated normal coordinates. The femtosecond pump pulse excites vertically from the ground state, forming a vibrational wavepacket on the excited surface. The wavepacket evolves on the excited state surface and at some later time is pumped back down to the ground state surface by the second 'dump' femtosecond pulse. If the wavepacket on the excited surface takes the molecule to geometries corresponding to those in the product channels of the ground state, vertical transitions from these points will place the wavepacket in a product channel of the ground state. The second laser pulse can be timed so that it causes de-excitation into products.

The next step in complexity in a coherent control experiment involves changing the phase relationships between the eigenstates within a wavepacket or changing the wavepacket composition, and this can be achieved in a variety of ways.

11.5.1 Pulse Trains

Wavepacket interferometry experiments may be carried out using a train of identical pairs of pulses. Such pulse trains can be created using a Michelson interferometer. The electric field in the time domain, $\boldsymbol{E}(t)$, arising from two pulses, one delayed by a time t_0 relative to the other, can be written

$$\boldsymbol{E}(t) = |\boldsymbol{E}(t)|\exp[i(\omega t - \phi(t))] + |\boldsymbol{E}(t - t_0)|\exp[i(\omega(t - t_0) - \phi(t - t_0))], \quad (11.11)$$

where ω is the frequency of the radiation, and $\phi(t)$ is the time-dependent phase. In a wavepacket interferometry experiment, two or more excited state wavepackets are created and allowed to interfere

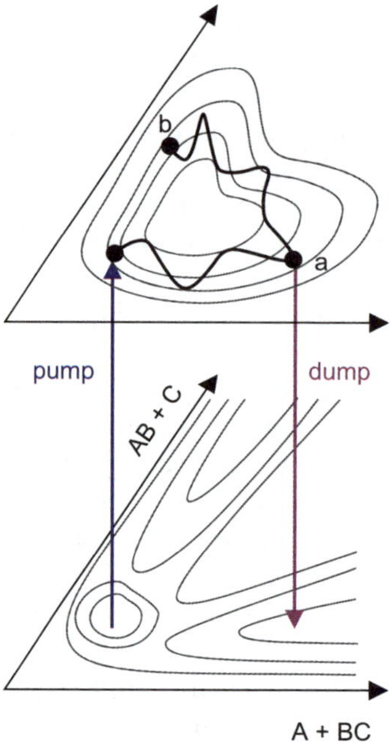

Figure 11.9 Schematic representation of the pump-dump experiments. The pump-dump delay can be set to catch the wavepacket at (a) or (b) to produce A + BC or AB + C, respectively.

with one another. A beautiful example of vibrational wavepacket interferometry is an experiment performed by Stuart Rice, Graham Fleming and coworkers involving pairs of phase-locked vibrational wavepackets in molecular iodine.[776] In these experiments, a first femtosecond pulse transfers population from the ground electronic state of the iodine molecule to the excited B-state, creating a vibrational wavepacket that evolves freely until the second femtosecond pulse transfers more population from the ground state of the iodine molecule to create a second, almost identical, vibrational wavepacket in the excited state. As illustrated in Figure 11.10, when the two wavepackets overlap spatially, they interfere and the interference is constructive or destructive, depending on the relative optical phase difference between the two femtosecond laser pulses, and on the time-dependent quantum mechanical phase difference between the two wavepackets. The resultant interference-dependent excited state population is then determined by measuring the total fluorescence signal from the excited state. This interference method has also been applied widely to electron wavepackets in highly excited Rydberg states of atoms and molecules.[777]

11.5.2 Chirped Pulses

As already discussed in Section 11.2 and Study Box 11.1, in a chirped laser pulse, the component frequencies are spread out in time, and the electric field can be written as,

$$\boldsymbol{E}(t) = |\boldsymbol{E}(t)| \exp\left[i\left(\omega t - \phi'' t^2\right)\right], \tag{11.12}$$

where ϕ'' is the linear chirp. A negatively chirped pulse (in which the instantaneous frequency decreases with time) can be used to step through the decreasing vibrational spacing in a diatomic molecule and thus enhance dissociation. Gustav Gerber and his coworkers[778] have used chirped pulses to control the multi-photon ionization of the sodium dimer, in a follow on from the experiments described in

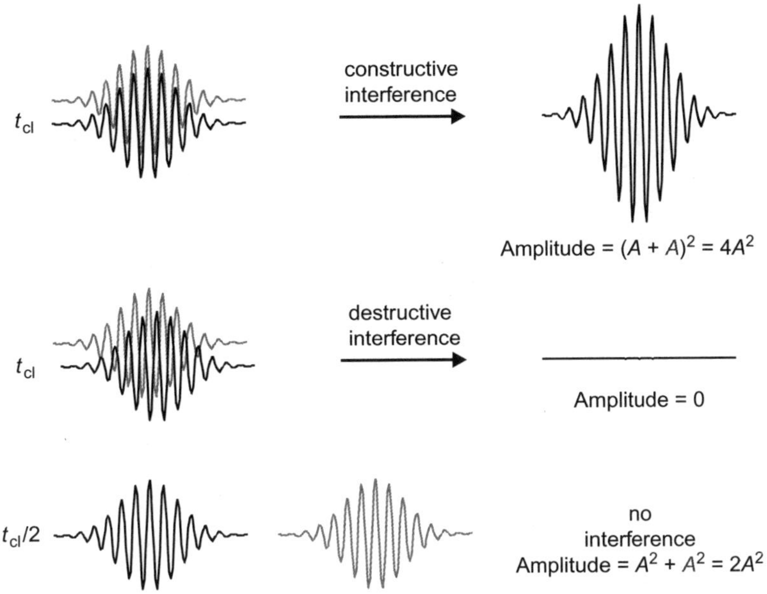

Figure 11.10 The 1st wavepacket (grey) is excited and allowed to evolve. At a later time, a 2nd identical wavepacket (black) is also excited. If the 2nd wavepacket is created when the 1st has returned to its original position, t_{cl}, the two wavepackets can interfere constructively (upper panel) or destructively (middle panel). The total excited state population oscillates between zero and a maximum of $4A^2$, where A is the amplitude of each individual wavepacket. Alternatively, if the 2nd wavepacket is created when the 1st wavepacket is far away from its original position $t_{cl}/2$, the two wavepackets do not overlap spatially and cannot interfere (lower panel). The total excited state population is simply the incoherent sum of the amplitudes of the individual wavepackets which is $2A^2$.

Section 11.4 and illustrated in Figure 11.11. Negatively chirped pulses ('down-chirped') were found to increase the population of the excited $2^1\Pi_g$ state (for reasons explained below), although the final $Na_2^+(^2\Sigma_g^+)$ ion population was enhanced by using a positively chirped ('up-chirped') pulse. The total excited $2^1\Pi_g$ state population is produced *via* a one-photon transition to the $A^1\Sigma_u^+$ state, upon which a wavepacket motion is initiated. The difference in the potentials for the $A^1\Sigma_u^+$ and $2^1\Pi_g$ states is such that the negatively chirped pulse is able to follow the outward motion of the $A^1\Sigma_u^+$ wavepacket with the correct, decreasing, excitation energy for the second photon to excite to the next $2^1\Pi_g$ state. The mechanism by which the ion population is maximized with the positively chirped pulse is different. The Frank-Condon maxima for both of the two-photon transitions leading to the excited $2^1\Pi_g$ state correspond to redder parts of the pulse, so absorption of the redder parts of the positively chirped pulse allows the $2^1\Pi_g$ population to reach a maximum early with respect to the pulse duration. This means that the final ionizing photon absorption is timed for the maximum in intensity of the same up-chirped pulse and a maximum is seen in the ion yield as a result.[778]

11.5.3 Programmable Pulse Shaping

Simple pulse shape manipulation in the form of pulse train and chirped pulse generation has proved valuable for coherent control experiments in small molecules. However, more complex pulse shapes are usually required to control the dynamics of larger molecules possessing more degrees of freedom. Commercially available pulse shaping instruments are now able to shape the phase and amplitude arbitrarily for each frequency present in a femtosecond laser pulse. Such a tailored pulse

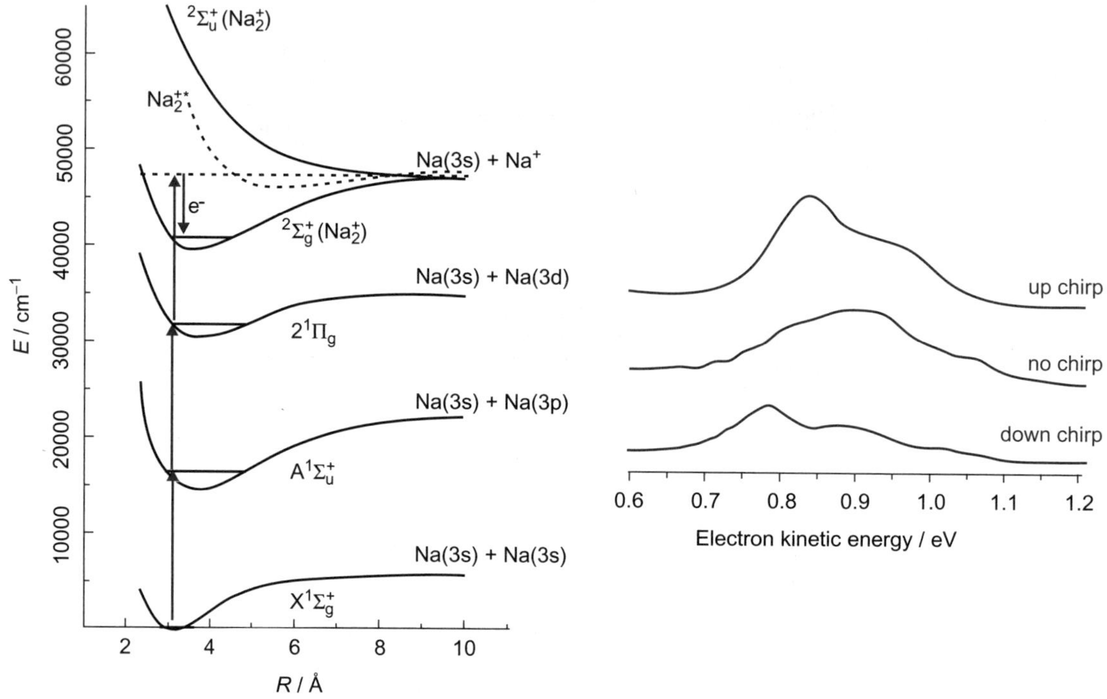

Figure 11.11 Left: Potential energy curves for the Na_2 and Na_2^+ dimer, showing one of the possible processes that occur when Na_2 is irradiated with 618 nm femtosecond radiation. Ionization to the $Na_2^+(^2\Sigma_g^+)$ state yields high energy photoelectrons (0.7–1.0 eV). Right: The measured photoelectron signal shows that the up-chirped pulse yields significantly more of the $Na_2^+(^2\Sigma_g^+)$ product, as the redder part of the pulse is energetically more optimal for the first two one-photon absorption processes, and thus the population in the $2^1\Pi_g$ state is maximized in time for the intensity maximum of the laser pulse, producing a larger ion yield after ionization by the third photon. Adapted from Figures 2 (left) and 3a (right) of A. Assion *et al.*[778]

may be described by the generalized function,

$$\boldsymbol{E}(\omega, t) = |\boldsymbol{E}(\omega, t)| \exp[i(\omega t - \phi(t))], \tag{11.13}$$

where $\phi(t)$ and $|\boldsymbol{E}(\omega,t)|$ may have almost any possible functional form. The most widely used commercial pulse shaping devices are spatial light modulators (SLMs). An SLM is comprised of an array of liquid crystals whose birefringences can be modified individually using electric fields, in order to control the phase retardance of the light passing through. Pulse shaping using SLMs requires the input pulse to be dispersed so that the different frequencies of the light can be addressed spatially by the SLM. The most common way of doing this is to place the SLM in the focal plane of a 4*f* zero-dispersion compressor (see Figure 11.12).

If a pair of SLM arrays is used in a pulse shaper, back-to-back with a pair of polarizers, the phase and amplitude of the output pulse are dependent on the sum and difference between the retardances of (*i.e.* the phase delays imposed by) each of the two arrays, and so the phase and amplitude of the pulse may be independently shaped.[779,780]* The output electric field, $\boldsymbol{E}_{out}$, is plane polarized, and written as,

$$\boldsymbol{E}_{out} = \begin{pmatrix} E_x \\ E_y \end{pmatrix} = \exp\left[i\frac{(\Delta\phi_1 + \Delta\phi_2)}{2}\right]\begin{pmatrix} \cos(\Delta\phi_1 - \Delta\phi_2)/2 \\ 0 \end{pmatrix}, \tag{11.14}$$

*Also see ref. [781], which is a helpful tutorial text that gives more detail on the origin of Eq. (11.14).

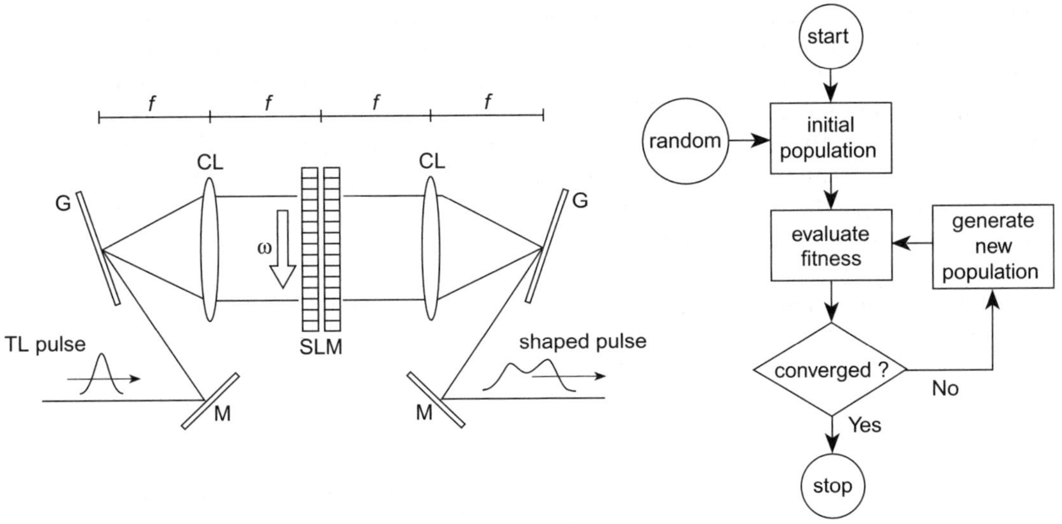

Figure 11.12 Left: 4*f* configuration for an SLM pulse shaper. The unshaped transform-limited (TL) pulse is diffracted by the grating, G, and then collimated at CL before passing through the SLM with its frequencies spatially dispersed. It is then focussed and recombined in a symmetrical arrangement. Right: The outline of a genetic algorithm. Populations represent a set of, initially random, pulse shapes.[779]

where E_x and E_y refer to the components of the electric field vector in the x and y directions respectively for light propagating in the z-direction, and $\Delta\phi_1$ and $\Delta\phi_2$ are the phase retardances of each of the two SLM arrays.[781] Currently available commercial dual array SLMs have 128 or 640 pixels, allowing for high resolution frequency domain pulse shaping. The 4*f* setup requires careful alignment to eliminate spatial aberrations, and when the pulses are very short, the lenses in a 4*f* setup (Figure 11.12) must be replaced by curved mirrors to minimize chirping of the pulse by the lenses. SLMs operate well over the wavelength range 488–1620 nm, so the output of a Ti:sapphire mode-locked laser can be shaped easily. Direct pulse shaping in the UV or far-IR is not possible using SLMs, but pulse shaping in the region of the electromagnetic spectrum necessary for the excitation of electronic states (UV) and molecular vibrations (IR) can be achieved indirectly. Various solutions are available and tend to be based on pulse shaping at around 800 nm followed by wavelength conversion (difference frequency generation (DFG), sum frequency generation (SFG) or second harmonic generation (SHG)) to the UV or IR.[782,783]

With the incredible flexibility in the achievable phase and intensity profiles, the question arises as to what is the best way to design a complex pulse to control the chemical products of a photochemical reaction. Optimal control theory, and other theoretical methods, have been developed to design *a priori* pulse shapes.[784] However, the need for theory was cleverly circumvented by Judson and Rabitz, who proposed using a learning algorithm (usually a genetic algorithm) in a closed loop within an experiment, using the SLM variables as the optimization parameters.[785] A basic genetic algorithm is outlined in Figure 11.12. Since the conception of this experimental design, optimal control has been realized in a huge number of experiments including energy transfer processes in light harvesting bacteria[786] and *cis-trans* isomerizations.[787,788] The first experimental demonstration of adaptive feedback control of photochemical dissociation, using automated pulse shaping and an evolutionary algorithm, was carried out using strong laser fields. Gustav Gerber and his coworkers employed intense shaped infrared laser pulses in a multiphoton ionization scheme to control the branching ratio of various photodissociation products of iron pentacarbonyl and dicarbonylchloro (η5-cyclopentadienyl) iron.[789]

In most gas-phase femtochemistry experiments, the laser beams are focussed into a molecular beam. For a 1 μJ, 100 fs pulse at 250 nm focussed to a diameter of 50 μm by a 50 cm focal length lens, the intensity is 10^{11} W cm^{-2} (the estimate of the beam waist at the focus is obtained using Gaussian optics, as discussed in ref. [752]). Typically, multiphoton excitation and ionization occur when the intensity is $>10^{13}$ W cm^{-2}, but just below this, in the perturbative regime, the electric field of the laser is weaker than the Coulomb electric field, but is strong enough to perturb the electronic energy levels through the Stark effect. This is another mechanism by which shaped pulses can control the outcome of chemical processes.

Unfortunately, it has generally proved difficult to unravel the photochemical control mechanisms from the rather complex pulse shapes obtained using these closed loop learning experiments. One of the exceptions is an experiment by Chantal Daniel and coworkers, illustrated in Figure 11.13, in which the photodissociation of the cyclopentadienyl (cp) manganese compound cpMn(CO)$_3$ was controlled.[790] The authors of this study showed that the ratio of the parent ion yield to the daughter ion yield was increased using a two-colour multiphoton ionization scheme analogous to their optimal control pulse: in the control pulse (centred at 800 nm) the first sub-pulse was bluer (~799 nm) and the second was redder (~800 nm); this suggested a two photon excitation followed by ionization. The details of the excited state dynamics were then uncovered by electronic structure and quantum dynamics calculations on the excited potential energy surfaces of the molecule.

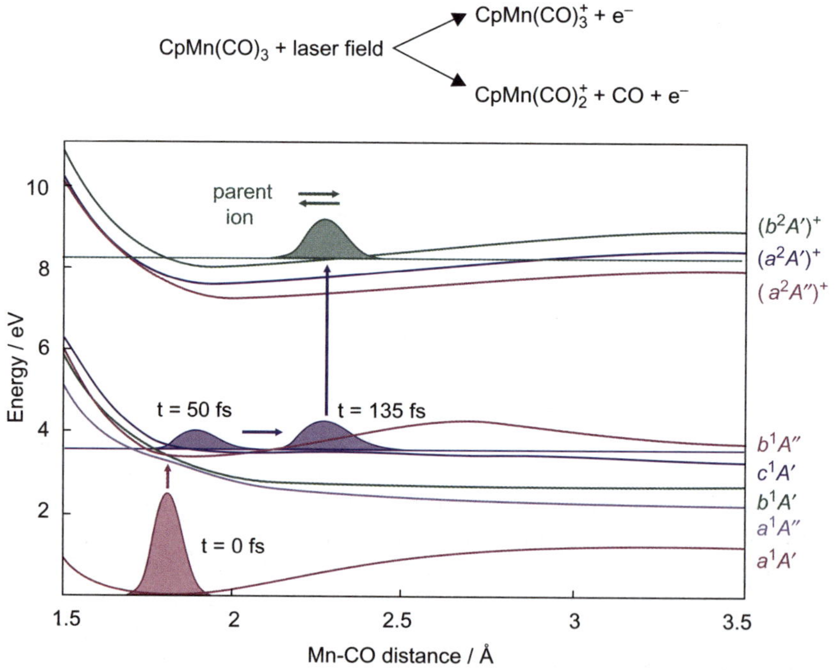

Figure 11.13 Exciting CpMn(CO)$_3$ with a femtosecond laser pulse centered at 800 nm results in two competing pathways: excitation to the c^{1}A$'$ state followed by ionization to the b^{2}A$'$ ionic state (CpMn(CO)$_3^+$ parent ion) and excitation to the b^{1}A$''$ state, non-adiabatic transfer to the a^{1}A$''$ dissociative state, and then ionization to the a^{2}A$''$ ionic state (CpMn(CO)$_2^+$ daughter ion). A learning algorithm was employed to optimize the yield of the parent ion. A detailed analysis of the resulting pulse shape together with quantum simulations revealed that the optimized pulse was composed of two dominant sub-pulses separated by 85 fs. The first sub-pulse, centered around 798.7 nm, excites the c^{1}A$'$ state preferentially. After a delay of 85 fs the population on the c^{1}A$'$ state is in the Franck-Condon window for ionization to the b^{2}A$'$ ionic state (parent ion) by a second sub-pulse centered around 800 nm. Adapted from ref. [790].

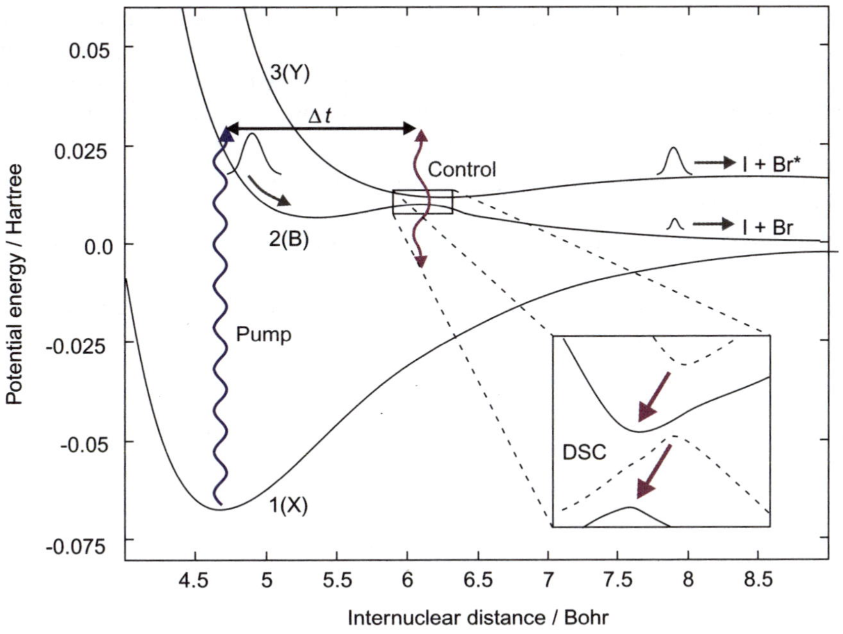

Figure 11.14 Illustration of the dynamic Stark control of the photodissociation of IBr. The molecule is pumped to the 2(B) state and as it dissociates, a strong femtosecond pulse in the IR (the "catalysis pulse") temporarily shifts the barrier formed by the crossing of the I + Br and I + Br* diabatic potentials. The barrier shifts because of the different Stark shifts of the diabatic states 2(B) and 3(Y). Product yields for the two channels are monitored by detection of the velocity of the I molecules in a REMPI scheme; the corresponding Br and Br* yields are given by conservation of momentum from the I product kinetic energies. Adapted from ref. [791].

An interesting example of a slightly different type of femtosecond laser control is the creation of spatially aligned molecules. This method can allow more precise dynamical information to be obtained from a subsequent femtosecond photochemical reaction such as a photodissociation.[792] Another example involves using a non-resonant intense laser field. The field is not strong enough to dissociate or ionize the molecule, but controls a photochemical reaction by modifying the shapes of the potential energy surfaces involved in the photoreaction through the dynamic Stark effect. The non-resonant laser field effectively acts as a catalyst for the photochemical reaction. Albert Stolow and his colleagues employed such an approach to control the ratio of spin-orbit states of Br atoms produced during the photodissociation of IBr. As illustrated in Figure 11.14, control was achieved by using the strong non-resonant field to lower the adiabatic barrier to the avoided crossing on the excited potential energy surface.[791] The controlling IR pulse is considered in this experiment to be a catalysis pulse because it temporarily changes the potential within which the reactive vibrational wavepacket evolves. The careful choice of wavelength for the control pulse in this experiment is crucial because it cannot be resonant with any (electronic, vibrational, rotational) transitions. After the wavepacket passes though a critical region on the potential energy surface, the control pulse intensity goes to zero and the surfaces relax back to their original forms (see Figure 11.14).

11.6 OUTLOOK

Femtosecond lasers have already had a huge impact on the field of molecular dynamics by providing a light source whose timescale is the same as that of molecular vibrations. Nevertheless, there have been some tremendous new advances in experimental techniques and computational capabilities during the last few years, which have placed us in the perfect position to investigate the dynamics of more complex molecular systems in unprecedented detail. New attosecond laser sources should also provide us with the exciting possibility of probing even faster timescales, including those of electron dynamics. For example, in a pioneering experiment, Krausz and coworkers have exploited attosecond technology to measure the relaxation dynamics of core-excited Kr with a lifetime of just 8 fs.[793]

The development of sophisticated femtosecond pulse shaping techniques has revolutionized the field of coherent control, but despite these technological advances, the field is still in its infancy and we still have a great deal to learn in terms of understanding why the optimized laser pulses control processes in the way that they do. One of the major current focusses of optimal coherent control experiments is to use the experiments to gain further insights into the mechanisms of photochemical processes, particularly in larger biological systems.[794,795]

There is absolutely no doubt that there will be many exciting new applications of femtosecond lasers in molecular dynamics and control throughout the 21st century.

11.7 PROBLEMS

1. Imagine a particle confined to an infinitely high walled box, between $0 \leqslant x \leqslant a$. Assume that it is prepared in a coherent superposition state described by the time-dependent wavefunction

$$\psi(t) = c_1(t)\phi_1 + c_2(t)\phi_2,$$

where the coefficients satisfy the equation $c_1(t)c_1^*(t) + c_2(t)c_2^*(t) = 1$, and the functions, ϕ_n, are the particle in a box eigenfunctions

$$\phi_n = \sqrt{\frac{2}{a}}\sin\left(\frac{n\pi x}{a}\right) \quad \text{with } n = 1, 2, \ldots.$$

(a) Use the time-dependent Schrödinger equation (see Eq. (3.11)) to show that

$$c_m(t) = c_m(0)\exp(-i\varepsilon_m t/\hbar),$$

where ε_m is the energy of eigenstate ϕ_m, with $m = 1$ or 2. c_m (0) refers to the value coefficient $c_m(t)$ at $t = 0$.

(b) Obtain an expression for $\psi(t)\psi^*(t)$ in terms of ϕ_1 and ϕ_2.

(c) Compare the temporal behaviour of $\psi(t)\psi^*(t)$ when $c_1(0) = 1$, $c_2(0) = 0$ and $c_1(0) = c_2(0) = 1/\sqrt{2}$. Comment on the answers that you obtain.

[Hint: Based on the expression you obtained in part (b), work out $\psi(t)\psi^*(t)$ in terms of ϕ_1 and ϕ_2 at $t = 0$ and $t = \pi/\omega$, where $\omega = (\varepsilon_2 - \varepsilon_1)/\hbar$.]

2. (a) For a pulse with a Gaussian distribution of modes (FWHM 100 nm, centered at 800 nm) and Gaussian temporal intensity distribution, what is the minimum FWHM pulse duration that can be achieved?

(b) For a pulse of intensity $I = 10^{14}$ W cm^{-2}, calculate the refractive index of Ti:sapphire at 800 nm. Determine the phase delay at 800 nm after travelling through:

i. 5 mm of Ti:sapphire ($n_0 = 1.76$, $n_2 = 3.1 \times 10^{-16}$ cm^2 W^{-1});

ii. 5 mm BK7 glass ($n_0 = 1.51$, $n_2 = 3.5 \times 10^{-16}$ cm^2 W^{-1}).

3. Show that the autocorrelation function is given by $\langle\Psi(t)|\Psi(0)\rangle = \sum_{\upsilon} a_{\upsilon}^2 g_{\upsilon}^2 e^{iw_{\upsilon}t}$. Plot the modulus of the autocorrelation function following excitation of a wavepacket created from vibrational states centered around $\bar{\upsilon}=21$, with a transform-limited Gaussian laser pulse of 40 fs duration, for
   (a) a harmonic oscillator with $\omega_e = 170\,\text{cm}^{-1}$;
   (b) an anharmonic oscillator with $\omega_e = 170\,\text{cm}^{-1}$ and $\omega_e x_e = 2\,\text{cm}^{-1}$.
   For a Gaussian pulse, $g_{\upsilon} = \exp[-2\ln 2(E_{\upsilon} - E_{\bar{\upsilon}})^2/\text{FWHM}^2]$. For simplicity, assume that a_{υ}^2 is approximately constant.

CHAPTER 12

Cold and Ultracold Collisions

GERRIT C. GROENENBOOM AND LIESBETH M. C. JANSSEN

Theoretical Chemistry, Institute for Molecules and Materials, Radboud University Nijmegen, The Netherlands

12.1 INTRODUCTION

In 1995 several groups succeeded in producing Bose-Einstein condensation in dilute gases of alkali atoms.[796–798] As will be discussed in detail in Section 12.5, Bose-Einstein condensation is a phenomenon that occurs when collections of bosons (particles of integer spin) macroscopically condense into their lowest quantum state. In 2003 molecular Bose-Einstein condensates of alkali dimers were reported.[799–801] The temperatures of these atomic and molecular condensates are on the order of nK-μK. Theoretical studies show that even at these ultracold temperatures, chemical reactions may occur, and may even be very fast.[802,803]

In this chapter we start our journey with Arrhenius' 19th century description of reactions at ambient temperatures and then work our way down to lower and lower temperatures. In the Arrhenius equation, the temperature (T) dependent reaction rate is given by

$$k(T) = Ae^{-E_a/k_B T}, \tag{12.1}$$

where k_B is the Boltzmann constant and A is a proportionality constant, usually referred to simply as the pre-exponential factor, or A-factor. The activation energy E_a is the energy required to pass the transition state. Eq. (12.1) can be derived using classical statistical mechanics. It predicts that the reaction rate drops to zero quickly when $k_B T \ll E_a$.

Some reactions, however, are barrierless ($E_a \sim 0$), and their rate may increase at lower temperatures. This is particularly true for ion-molecule reactions. As early as 1905, Langevin derived an expression for the reaction rate of ion-molecule reactions. As discussed in Section 12.2, this expression only depends on the long range part of the intermolecular potential, and the model is called a 'capture model'.[804] Later, it was found that neutral radical-radical reactions and even some radical-molecule reactions may also be fast at low temperatures. These barrierless reactions are very important in the lower parts of the stratosphere, where temperatures may be around 200 K.

Tutorials in Molecular Reaction Dynamics
Edited by Mark Brouard and Claire Vallance

Published by the Royal Society of Chemistry, www.rsc.org

The air around us contains on the order of 10^{19} molecules per cm^3. In interstellar space, areas where the density is on the order of $10^6\,cm^{-3}$ appear as clouds when observed through telescopes, against a background of even lower density in most of the interstellar medium. Interstellar clouds have temperatures in the range of 10–100 K. Still, chemical reactions occur and play a crucial role in, *e.g.*, the formation of stars.[805]

As discussed in previous chapters, the calculation of a reaction rate (or a cross section) requires knowledge of the potential energy surface. Depending on the chemical system under consideration, it may be sufficient to know only the potential around the transition state or only the long range part. Computation of the potential surface always requires quantum mechanics, since it involves the motion of the electrons in the system. However, to compute the nuclear dynamics on the surface, classical mechanics is generally a good starting point. At lower temperatures, one has to consider quantum effects, such as tunnelling, resonances, zero-point energy, quantization of the angular momenta of the reactants and products, and quantization of the angular momentum of the colliding complex as a whole.

Such quantum effects become dominant at around 1 K. The cosmic background radiation has a temperature of 2.76 K, and the coldest known place in the universe – outside the laboratory – is about 1 K. However, in laboratory experiments even lower temperatures can be reached. At temperatures around 1 K, molecules have a kinetic energy that is comparable to their interaction energy with electric and magnetic fields of the strengths achievable in laboratory-based experiments. This provides many opportunities to study and manipulate cold gases, and such studies have become an active area of research.[806,807]

The focus of this tutorial chapter is primarily theoretical. Collisions at low temperatures, or low collision energies, present a unique set of challenges and opportunities for theory, and we outline the main approaches that are currently available to study cold collisions from a theoretical point of view. Experimental aspects concerning the production of cold molecules and the study of cold collisions are covered in Study Boxes 12.1 and 12.5. The chapter starts with sections describing classical and quantum capture theory. Section 12.4 discusses the Wigner threshold laws, particular solutions of the Schrödinger equation that are valid at low energies. Finally, Section 12.5 outlines what happens to matter at ultra-low temperatures, and in particular describes the phenomenon of Bose-Einstein condensation.

12.2 CLASSICAL CAPTURE THEORY

The term 'capture model' reflects the fact that the reaction rate (or cross section) is approximated as the rate at which the reactants come together and reach the barrier or transition state. As already discussed, at low temperatures, or under low collision energy conditions, reactions that possess a potential energy barrier along the reaction coordinate tend to become very slow, and therefore only exothermic reactions taking place on attractive potential energy surfaces have significant reaction rates. For such processes, the only barrier to be traversed is the centrifugal barrier,[823] and once past this barrier, the reaction probability can often be assumed to be near unity. So called 'recrossing' of the barrier back to reactants is not usually accommodated in the capture models to be described. It is for this reason that capture theories tend to be most applicable to reactions on attractive potential energy surfaces, for which the capture process (crossing the barrier) is usually rate determining.

12.2.1 Classical Central Force Problem

To introduce the key concepts* of low energy scattering theory we review the problem of two point particles A and B interacting through a potential $V(r)$ that only depends on the distance r between

*At this point the reader may wish to review the material covered in Section 3.2.3.

STUDY BOX 12.1: MAKING COLD MOLECULES

Cooling of molecules into the 'cold' (10 K to 1mK) and 'ultracold' (sub 1 mK) regimes is currently an area of intense research interest,[807–810] and a number of techniques have been developed to achieve ultra-low temperatures in molecular gases. The aim of the following is to highlight a few of these techniques, rather than to provide an exhaustive review. For more detailed information the reader is referred to ref. [811].

Techniques for cooling molecules may be separated into 'first-stage' and 'second-stage' methods. First-stage methods are used to slow molecules down to temperatures of a few Kelvin or less, at which point they can be captured into an electromagnetic trap. Once the atoms or molecules have been trapped, second-stage techniques allow further cooling into the 'ultracold' regime.

First-stage Cooling Techniques

A number of first-stage cooling techniques for molecules are based on deceleration of molecules produced in a supersonic expansion. Supersonic jet expansions produce internally cold molecules with a narrow velocity distribution simply by expanding gas from a high pressure region to a low pressure region through a small orifice. Cooling of a molecule's internal degrees of freedom is achieved virtually 'for free', and beam temperatures in the region of 1–2 K are readily achievable. The key difference between molecules in a molecular beam and the cold molecules of interest is that in a molecular beam the narrow velocity distribution is centred around a velocity that is not zero. From the point of view of cold molecule production, the problem therefore lies in finding a suitable means to decelerate the beam.

There have been numerous approaches to this problem. The simplest method uses a rotating assembly that moves the beam source backwards as the gas expands in order to cancel out the beam velocity. The idea was in fact first used in reverse by Moon *et al.*[812] as a means of beam acceleration, but has more recently been adopted for cold molecule production by Gupta and Herschbach.[813,814] The technique has been demonstrated successfully for beams of O_2, SF_6 and CH_3F, with lowest speeds of 67, 55 and 91 m s^{-1}, respectively. However, while it is a completely general technique, there are considerable technical difficulties associated with the mechanical assembly, and it is also unlikely that the technique will prove capable of producing molecules in the 'ultracold' regime.

More sophisticated methods use electric or magnetic fields either to select the low-velocity tail of a molecular beam, or to slow a narrow 'slice' of the beam velocity distribution to much lower velocities. If the beam gas has a large enough dipole, velocity selection with a bent electrostatic quadrupole may be used to select slow molecules from the beam, as shown in Figure 12.1. A quadrupole may be used to select low-field-seeking states of strongly dipolar molecules in much the same way as a hexapole is used to perform a similar action on symmetric top molecules (see Section 9.3.2.2). When a bend is introduced into the quadrupole, only molecules moving slowly enough to interact strongly with the guiding field are transported around the bend, with the remainder of molecules being lost from the quadrupole. The fraction of the beam that is velocity-selected in this way depends on the potential applied to the quadrupole rods (with lower potentials yielding a lower-velocity fraction), the radius of the bend (with tighter bends yielding a lower-velocity fraction), and on the magnitude of the ratio between the Stark shift in the quadrupole field and the mass of the beam molecules (molecules with larger Stark shifts and smaller masses are more easily guided round the bend in the quadrupole).

An increasingly popular approach to molecular beam deceleration is the *Stark deceleration* technique pioneered by Gerard Meijer and coworkers.[421,808] A Stark decelerator uses time-varying electric fields to decelerate a 'velocity slice' from a molecular beam to zero velocity in the lab frame. Such a decelerator is shown schematically in Figure 12.2. Pairs of electrodes are placed along the beam axis, with every second electrode set to high voltage and the interleaving electrodes grounded. When travelling between a grounded electrode and one at high potential, a molecule in a low-field-seeking Stark state will be climbing a potential gradient and will be slowed down. If the potential applied to the rods were static, once the molecule had passed the high-field electrode it would then accelerate down the potential gradient on the other side and convert the potential energy it had gained back to kinetic energy. However, if as the molecule nears the high-potential electrode, the electrode potentials are switched (so that those at high potential are now grounded, and vice versa), the molecule again finds itself climbing a potential gradient, and is slowed further. Repeating this switching process as the molecule travels through the decelerator allows molecules to be slowed to velocities low enough that they may be captured into an electrostatic trap. Only a certain fraction of beam molecules are decelerated, whose initial velocity, position, and quantum state place them in phase with the switching frequency of the decelerator.

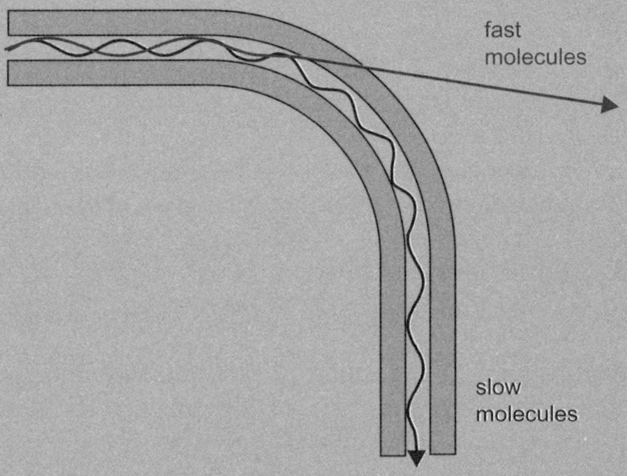

Figure 12.1 Selection of a low-velocity fraction from a molecular beam using a bent quadrupole.

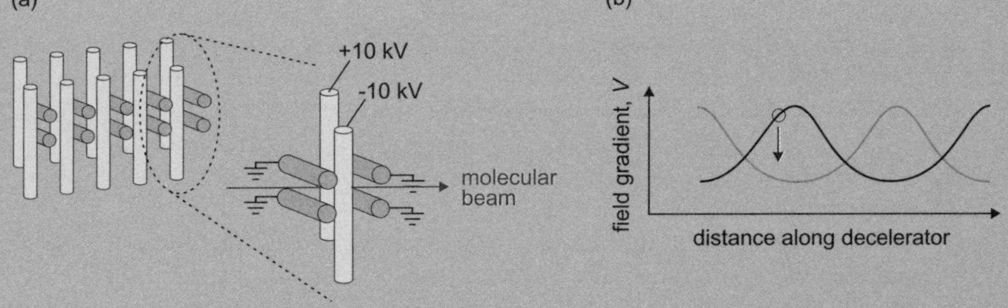

Figure 12.2 (a) A section of a Stark decelerator; (b) Inside the decelerator, low-field-seeking Stark states climb a potential gradient and are slowed. On switching the decelerator field, the molecules are returned to the 'bottom of the hill', and must climb the potential gradient again. This process is repeated along the length of the decelerator.

An alternative technique, known as Zeeman deceleration, uses a similar principle but employs a pulsed *magnetic field*.[815]

Another approach to first-stage deceleration relies on molecular beam scattering processes. Crossed molecular beam experiments or molecular photodissociation experiments often produce some fraction of products that are stationary in the laboratory frame. This occurs when the product recoil velocity exactly cancels the velocity of the centre of mass of the reaction system. As an example, Chandler and coworkers[423,424] have showed that in crossed beam inelastic scattering of NO with Argon, rovibrationally quantum-state selected NO molecules with velocities of less than $15\,m\,s^{-1}$ may be produced with densities of up to 10^8 molecules cm^{-3}.

Not all first-stage cooling techniques employ molecular beams. A popular approach to first-stage cooling is a method known as *buffer gas cooling,* in which the molecules of interest are cooled to temperatures of typically a few hundred milliKelvin through collisions with cryogenic helium gas stored in a cryostat. This is an entirely general and very promising technique, applicable to any molecule that can survive a large number of collisions with atomic helium. A disadvantage of the technique is the challenge of separating the cold molecules from the buffer gas after cooling has taken place.

Second Stage Cooling

Second-stage cooling techniques nearly all have their origins in laser-based methods for cooling and trapping of atoms. Steve Chu,[816] Claude Cohen-Tannoudji,[817] and William Phillips[818] were the first to produce ultracold atoms in this way, and won the 1997 Nobel Prize in Physics for their efforts.

Laser cooling is based on momentum transfer from the photons in a laser beam to the atoms to be cooled in order to slow the atoms down. Each time an atom absorbs a photon, a small amount of momentum is transferred to the atom, slightly changing its velocity. Many absorption-emission cycles are required to slow an atom down appreciably, and so an atomic transition is required that forms a *closed optical loop.* This means that after absorption, spontaneous emission returns the atom back to its original state. Atoms with suitable two-level closed systems include alkali metals, metastable rare gas atoms, or singly charged alkali-earth ions. Atoms that emit to other states can also be laser cooled, but an additional laser is required to excite a suitable transition to return the 'lost' atoms to the optical loop.

The most common laser cooling technique is called Doppler cooling, and usually involves an experimental setup consisting of six intersecting laser beams, as shown in Figure 12.3. This is usually achieved using three orthogonally propagating beams, each reflected back along its own beam path by a mirror. The frequency of each laser beam is tuned slightly to the red of an atomic transition. Due to the Doppler effect, atoms moving towards the light source absorb more photons than those moving away, and are slowed. The excited state will of course receive a second momentum 'kick' when it emits a photon to return to the ground state, but the direction of this second kick is random, so that the net force after many transitions is along the laser beam direction. The three-axis laser beam configuration means that atoms are slowed down no matter which direction they are moving in. Temperatures as low as 150 μK may be achieved for ^{85}Rb using this technique.

To prevent cooled atoms from falling out of the laser interaction region under the force of gravity, Doppler cooling is often combined with a magnetic trapping force to give a *magneto-optical trap* (MOT). Zeeman splitting of the energy levels in the magnetic field increases as atoms move away from the centre of the trap, shifting the atomic resonance closer to the laser frequency, and increasing the chance of the atom receiving a photon 'kick' towards the centre of the trap when the appropriate (circular) polarization of light is employed.

Other laser cooling methods are available to reduce the temperature of the atom cloud still further. Of these, 'Sysiphus cooling', also known as 'polarization gradient cooling',[819] also earned a share of the 1997 Nobel Prize.

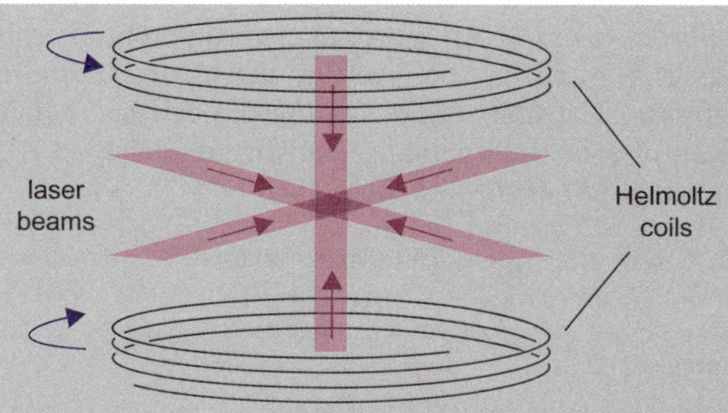

Figure 12.3 Laser beam geometry for Doppler cooling of an atomic gas.

While laser cooling has been highly successful for a number of atomic species, the absence of suitable closed optical cycle transitions within the much more complicated energy level structure of a molecule means that the technique cannot be extended into the molecular realm. A number of techniques have been developed to overcome this limitation.

Sympathetic cooling

Sympathetic cooling is an attractive approach in which an ensemble of laser-cooled and trapped ultracold atoms translationally cools an ensemble of co-trapped ions or molecules via elastic (thermalizing) collisions. One of the first molecular species to be cooled in this way was a 43-atom molecular ion of a dye called Alexa Fluor 350.[820] The method has since been applied to ions with masses of up to several tens of thousands of atomic mass units. The lowest achievable temperature for the molecular species is determined by the temperature of the trapped atoms.

Evaporative cooling

Evaporative cooling relies on elastic (thermalizing) collisions between trapped molecules, *without* the presence of a co-trapped laser-cooled atomic species. As the trap depth is reduced, the warmest molecules escape, and so long as thermal equilibrium is maintained through collisions within the trap, the Boltzmann distribution gradually shifts to lower and lower temperatures. In principle evaporative cooling could be an effective second-stage cooling technique for molecules, but in practice the number densities of trapped molecules required to maintain thermal equilibrium through collisions have not yet been achieved.

Other methods

Other methods of forming cold molecules using laser-cooled atoms include *photoassociation*[821] and *Feschbach resonance magnetic tuning*.[822] Methods of this type, which form cold molecules from cold atoms, are referred to as *indirect* cooling methods, to distinguish them from *direct* cooling of 'warm' molecules into the ultracold regime. Both of the indirect methods referred to above produce highly internally excited diatomic species (though it is possible to use shaped femtosecond laser pulses to coax the system into the ground state and reduce the vibrational excitation). These techniques have been used successfully to produce ultracold molecules, but they are limited to a few species, with little scope for generalization to larger systems.

Claire Vallance

the two particles (often referred to as a central force or central potential problem). Let the positions of the particles, with masses m_A and m_B, be given by the Cartesian coordinates $\boldsymbol{r}_A$ and $\boldsymbol{r}_B$ with respect to a space-fixed frame. The first step in finding the classical equations of motion of the particles is the introduction of Jacobi coordinates (shown previously in Figure 3.1), *i.e.* the coordinates of the centre-of-mass (CM) of the system,

$$\boldsymbol{R}_{CM} = \frac{m_A \boldsymbol{r}_A + m_B \boldsymbol{r}_B}{m_A + m_B}, \tag{12.2}$$

and the relative coordinates

$$\boldsymbol{r} = \boldsymbol{r}_B - \boldsymbol{r}_A. \tag{12.3}$$

The classical kinetic energy of the system is given by

$$T = \tfrac{1}{2} m_A \dot{\boldsymbol{r}}_A \cdot \dot{\boldsymbol{r}}_A + \tfrac{1}{2} m_B \dot{\boldsymbol{r}}_B \cdot \dot{\boldsymbol{r}}_B = \tfrac{1}{2} M \dot{\boldsymbol{R}}_{CM} \cdot \dot{\boldsymbol{R}}_{CM} + \tfrac{1}{2} \mu \dot{\boldsymbol{r}} \cdot \dot{\boldsymbol{r}}, \tag{12.4}$$

where $M = m_A + m_B$ is the total mass of the system, $\mu = (1/m_A + 1/m_B)^{-1}$ is the reduced mass, and the dot over a symbol indicates its time-derivative. In the absence of external forces, the CM of the system moves with a constant velocity throughout the collision. The relative motion is decoupled from that of the CM, and the equations of motion for $\boldsymbol{r}$ are equivalent to the equations of motion for a single particle with mass μ moving in a potential $V(r)$. The components of the conjugate momentum $\boldsymbol{p}$ are defined by

$$p_i = \frac{\partial T}{\partial \dot{r}_i} = \mu \dot{r}_i, \tag{12.5}$$

where $i = 1, 2, 3$ labels the Cartesian components of the vector. The classical Hamiltonian of the system can then be expressed as a function of coordinates and their conjugate momenta as (see Chapter 3)

$$H = \frac{p^2}{2\mu} + V(r), \tag{12.6}$$

where p is the magnitude of $\boldsymbol{p}$. The classical equations of motion in Jacobi coordinates, known as the Hamilton-Jacobi equations of motion, are given by

$$\dot{r}_i = \frac{\partial H}{\partial p_i} = \frac{p_i}{\mu} \tag{12.7}$$

$$\dot{p}_i = -\frac{\partial H}{\partial r_i} = -\frac{\partial V(r)}{\partial r_i} = -\frac{\partial r}{\partial r_i}\frac{\partial V(r)}{\partial r} = -\frac{r_i}{r}\frac{\partial V(r)}{\partial r}. \tag{12.8}$$

With $\boldsymbol{r} = r\hat{\boldsymbol{r}}$, where $\hat{\boldsymbol{r}}$ is a unit vector pointing along $\boldsymbol{r}$, these equations may be written in vector notation as

$$\dot{\boldsymbol{r}} = \mu^{-1} \boldsymbol{p} \tag{12.9}$$

$$\dot{\boldsymbol{p}} = -\hat{\boldsymbol{r}} \frac{\partial V(r)}{\partial r}. \tag{12.10}$$

For readers not familiar with the Hamilton-Jacobi equations, we note that from the last two equations one readily recovers Newton's equations of motion $\boldsymbol{F} = \mu \ddot{\boldsymbol{r}}$, where the force $\boldsymbol{F}$ is seen to be equal to $\dot{\boldsymbol{p}}$.

The angular momentum of the system,

$$\boldsymbol{\ell} = \boldsymbol{r} \times \boldsymbol{p}, \tag{12.11}$$

is conserved, *i.e.* independent of time, since

$$\dot{\boldsymbol{\ell}} = \dot{\boldsymbol{r}} \times \boldsymbol{p} + \boldsymbol{r} \times \dot{\boldsymbol{p}} = \mu^{-1}\boldsymbol{p} \times \boldsymbol{p} - r\frac{\partial V(r)}{\partial r}\hat{\boldsymbol{r}} \times \hat{\boldsymbol{r}} = \boldsymbol{0}. \tag{12.12}$$

Hence, the vectors $\boldsymbol{r}$, $\boldsymbol{p}$, and $\dot{\boldsymbol{r}}$ are always in a plane perpendicular to $\boldsymbol{\ell}$. The square of the length of $\boldsymbol{\ell}$ is given by

$$|\boldsymbol{\ell}|^2 = \boldsymbol{\ell} \cdot \boldsymbol{\ell} = (\boldsymbol{r} \times \boldsymbol{p}) \cdot (\boldsymbol{r} \times \boldsymbol{p}) = (\boldsymbol{r} \cdot \boldsymbol{r})(\boldsymbol{p} \cdot \boldsymbol{p}) - (\boldsymbol{r} \cdot \boldsymbol{p})(\boldsymbol{r} \cdot \boldsymbol{p}) = r^2p^2 - (\boldsymbol{r} \cdot \boldsymbol{p})^2. \tag{12.13}$$

Defining the momentum along the vector $\boldsymbol{r}$ as $p_r \equiv \hat{\boldsymbol{r}} \cdot \boldsymbol{p}$, we may rewrite the equation as

$$r^2p^2 = |\boldsymbol{\ell}|^2 + r^2p_r^2, \tag{12.14}$$

which we may use to write the Hamiltonian of the system as

$$H = \frac{p^2}{2\mu} + V(r) = \frac{|\boldsymbol{\ell}|^2}{2\mu r^2} + \frac{p_r^2}{2\mu} + V(r). \tag{12.15}$$

Hence, the problem of finding $r(t)$ is equivalent to solving a one-dimensional problem with an effective potential

$$V_{\text{eff}}(r) = \frac{|\boldsymbol{\ell}|^2}{2\mu r^2} + V(r), \tag{12.16}$$

where the first term is called the centrifugal term, and accounts for the rotational kinetic energy of the reactant pair (see Chapter 1, Section 1.4.3). The equation of motion for r is

$$\mu\ddot{r} = -\frac{\mathrm{d}V_{\text{eff}}(r)}{\mathrm{d}r}. \tag{12.17}$$

To find the complete solution $\boldsymbol{r}(t) = r(t)\,\hat{\boldsymbol{r}}(t)$ we expand $\hat{\boldsymbol{r}}$ as

$$\hat{\boldsymbol{r}} = \boldsymbol{e}_x \cos\varphi + \boldsymbol{e}_y \sin\varphi, \tag{12.18}$$

where $\boldsymbol{e}_x$ and $\boldsymbol{e}_y$ are two orthonormal vectors in the plane perpendicular to $\boldsymbol{\ell}$ and φ is a time-dependent polar angle. For the time derivative of the direction $\hat{\boldsymbol{r}}$ we have

$$\dot{\hat{\boldsymbol{r}}} = \dot{\varphi}(-\boldsymbol{e}_x \sin\varphi + \boldsymbol{e}_y \cos\varphi) = \dot{\varphi}\hat{\boldsymbol{r}}_\perp, \tag{12.19}$$

where $\hat{\boldsymbol{r}}_\perp$ is the unit vector perpendicular to both $\boldsymbol{\ell}$ and $\hat{\boldsymbol{r}}$. From Eqs. (12.11) and (12.9) we have

$$\boldsymbol{\ell} = \mu r^2\hat{\boldsymbol{r}} \times \dot{\hat{\boldsymbol{r}}} = \mu r^2\dot{\varphi}\hat{\boldsymbol{r}} \times \hat{\boldsymbol{r}}_\perp, \tag{12.20}$$

and so, since $|\hat{\boldsymbol{r}} \times \hat{\boldsymbol{r}}_\perp| = 1$, we find that

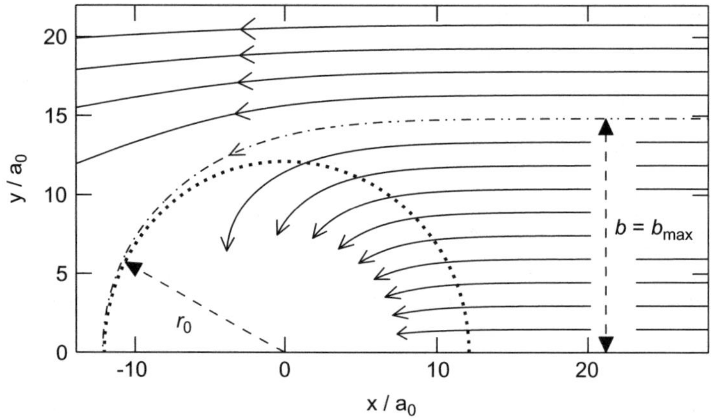

Figure 12.4 Classical capture theory for an isotropic potential: trajectories for a potential $V_6(r) = -50r^{-6}$, a total energy of $E = 10k_B$, and a reduced mass of $\mu = 7.5$ u (where u is the atomic mass unit), with impact parameters (b) less than $22a_0$. The dotted semi-circle indicates the distance r_0 for which the effective potential V_{eff} has a maximum, *i.e.* the "point of no return". All trajectories with $b < b_{max}$ contribute to the capture cross section and the trajectories with $b > b_{max}$ do not.

$$|\ell| = \mu r^2 \dot{\varphi}. \tag{12.21}$$

Since ℓ is a constant and $r(t)$ can be found from Eq. (12.17), the angle φ is given by the integral

$$\varphi(t) = \varphi(0) + \int_0^t \frac{|\ell|}{\mu r(t)^2} \mathrm{d}t. \tag{12.22}$$

As shown in Figure 12.4, our initial conditions ($t = 0$) assume that the particle is in the xy-plane with a large positive value of the x coordinate, moving in the negative $\boldsymbol{e}_x$ direction with initial relative speed,* v,

$$\dot{\boldsymbol{r}}(0) = -v\boldsymbol{e}_x, \tag{12.23}$$

and a position at ($t = 0$)

$$\boldsymbol{r}(0) = a\boldsymbol{e}_x + b\boldsymbol{e}_y. \tag{12.24}$$

The coefficient b, which is taken as positive, is the *impact parameter,* introduced in Chapter 1. If the potential $V(r)$ was zero then $\boldsymbol{r}(t)$ would move parallel to the x-axis and pass the origin at a distance b, *i.e.* the impact parameter would correspond to the nearest approach of the two particles in the absence of an interaction potential. By substituting Eqs. (12.23) and (12.24) into (12.11) we get

$$\ell = -\mu v(a\boldsymbol{e}_x + b\boldsymbol{e}_y) \times \boldsymbol{e}_x = \mu v b\boldsymbol{e}_z, \tag{12.25}$$

and we see that the impact parameter, the initial velocity, and the reduced mass determine the magnitude of the angular momentum of the system

$$|\ell| = \mu v b. \tag{12.26}$$

*Note that in this chapter we drop the 'rel' notation for relative velocity, v_{rel}.

When $|\ell|$ is known, the effective potential given in Eq. (12.16) is known, and Eq. (12.17) can be solved to find $r(t)$. The result may be substituted into Eq. (12.22) to determine the full trajectory.

12.2.2 Cross Sections

Collisions may be *elastic, inelastic,* or *reactive*. In an elastic collision, the direction of relative motion of the particles changes. In the CM-frame, the speeds of the particles are conserved. However, in the laboratory-fixed frame speeds may change as a result of collisions. It is through these elastic collisions that thermal equilibrium is reached after a hot gas is expanded into a cold gas. This principle is used in the buffer gas cooling technique, where, *e.g.*, laser-ablated CaH is cooled to 0.4 K by collisions with a cryogenically cooled helium buffer gas.[824] For this reason, in the context of cold chemistry, elastic collisions are sometimes called *good collisions*.

In an *inelastic* collision, the internal state of at least one of the colliding particles changes, *e.g.*, it is rotationally or vibrationally excited or relaxed. Experiments on molecules with non-zero spin, and thus with a non-zero magnetic moment, are often carried out by trapping them in a magnetic field. In a collision, the orientation of the spin may change into a state that is expelled from the trap, and therefore this kind of inelastic collision is sometimes called a *bad collision*.

In a reactive collision, the composition of the particles changes (*e.g.*, A + BC → AB + C). Whether or not a collision leads to a reaction depends on the impact parameter. As discussed in Chapters 1 and 3, and illustrated in Figure 12.4, if a reaction only occurs when the impact parameter b of the trajectory is less than some value b_{max}, and then occurs with unit probability, the cross section, σ, for that process is

$$\sigma = \pi b_{\max}^2. \tag{12.27}$$

The cross section has the dimensions of area, and in general depends on the kinetic energy of the particles, $E = \frac{1}{2}\mu v^2$. It is also possible that the reaction occurs with some probability that depends both on the impact parameter and on the collision energy, $0 \leqslant P(b,E) \leqslant 1$. In this case the cross section is given by

$$\sigma(E) = 2\pi \int_0^\infty P(b,E)b\,\mathrm{d}b. \tag{12.28}$$

As noted in Chapter 1, the function $P(b, E)$ is called the *opacity* function. Cross sections for elastic and inelastic processes are defined in a similar way.

12.2.3 Canonical Reaction Rates

For a binary process

$$\mathrm{A} + \mathrm{B} \rightarrow \mathrm{C}$$

the evolution in time of the concentrations [A], [B], and [C] (in units of molecules cm^{-3}) is given by

$$-\frac{\mathrm{d}[\mathrm{A}]}{\mathrm{d}t} = -\frac{\mathrm{d}[\mathrm{B}]}{\mathrm{d}t} = \frac{\mathrm{d}[\mathrm{C}]}{\mathrm{d}t} = k(T)[\mathrm{A}][\mathrm{B}], \tag{12.29}$$

where $k(T)$ is the temperature dependent reaction rate in units of $cm^3\ s^{-1}$ molecule^{-1}. It may be computed as the Boltzmann average of the cross section $\langle v\sigma(E)\rangle$,*

$$k(T) = \int_0^\infty v\sigma(E) f(v)\, dv, \tag{12.30}$$

where $f(v)$ is the Maxwell-Boltzmann speed distribution. At ultralow temperatures this expression completely breaks down, not only because the rate will depend on whether the particles are bosons or fermions, but also because Eq. (12.29) will no longer apply, as we shall see in Section 12.5.6. The Maxwell-Boltzmann speed distribution for particles with mass m is given by

$$f(v) = 4\pi\left(\frac{m}{2\pi k_B T}\right)^{3/2} v^2 e^{-mv^2/2k_B T}. \tag{12.31}$$

The distribution is normalized such that

$$\int_0^\infty f(v)\, dv = 1. \tag{12.32}$$

For the average kinetic energy, we have

$$\int_0^\infty \tfrac{1}{2} m v^2 f(v)\, dv = \tfrac{3}{2} k_B T, \tag{12.33}$$

and the average speed of the particles is

$$\bar{v} = \int_0^\infty v f(v)\, dv = \sqrt{\frac{8 k_B T}{\pi m}}. \tag{12.34}$$

It is possible to have a mixture of gases for which the speed distributions are characterized by different temperatures T_1 and T_2. If the masses of the different particles are m_1 and m_2, the relative speed distribution is found by replacing the temperature T by the mass-weighted average

$$\bar{T} = \frac{m_2 T_1 + m_1 T_2}{m_1 + m_2}, \tag{12.35}$$

and m by the reduced mass μ. If the two gases are in thermal equilibrium, we have $T_1 = T_2 = \bar{T} = T$. If, however, one gas is much colder than the other, *e.g.*, $T_2 \ll T_1$, then Eq. (12.35) simplifies to

$$\bar{T} \approx \frac{m_2}{m_1 + m_2} T_1 = \frac{\mu}{m_1} T_1. \tag{12.36}$$

*Apart from the change in notation for v_{rel}, this is the same as Eq. (1.12).

The rate constant in Eq. (12.30) may alternatively be written as an integral over the relative kinetic energy $E = \frac{1}{2}\mu v^2$, using $\mathrm{d}E = \mu v\, \mathrm{d}v$,

$$k(T) = \sqrt{\frac{8k_\mathrm{B}T}{\pi\mu}} \frac{1}{(k_\mathrm{B}T)^2} \int_0^\infty \sigma(E) e^{-E/k_\mathrm{B}T} E\, \mathrm{d}E. \tag{12.37}$$

Substituting $x = \frac{E}{k_\mathrm{B}T}$ and $\bar{v} = \sqrt{8k_\mathrm{B}T/\mu\pi}$ gives

$$k(T) = \bar{v} \int_0^\infty \sigma(xk_\mathrm{B}T) \mathrm{e}^{-x} x\, \mathrm{d}x = \bar{v}\bar{\sigma}. \tag{12.38}$$

This shows that if $\sigma(E)$ is constant, then $k(T) \propto \bar{v} \propto \sqrt{T}$. Eq. (12.30) shows that if $\sigma(E) \propto v^{-1} \propto E^{-\frac{1}{2}}$ then $k(T)$ is a constant. Below we will see that these two cases apply to elastic and inelastic collisions at low temperatures, respectively.

12.2.4 Isotropic Interactions

In capture theory we assume that cross sections are completely determined by long range attractive interactions between particles.* Collisions with zero impact parameter $b = 0$ are assumed to be reactive, with unit reaction probability. For non-zero impact parameters, $b > 0$, the system has non-zero angular momentum $|\ell| = \mu v b$ and the effective potential given by Eq. (12.16) contains a repulsive centrifugal term, which may give rise to a centrifugal barrier. It is assumed that trajectories contribute to the cross section if, and only if, they pass over this centrifugal barrier. In many cases the long range interaction is well described by the leading term of the potential when expanded in powers of $1/r$ (see Study Box 12.2). By assuming a long range interaction of the form

$$V_n(r) = -\frac{c_n}{r^n}, \tag{12.39}$$

where $c_n > 0$ is called the long range coefficient, we derive analytic formulas for cross sections and reaction rates in the capture model. We give the derivation only for $n > 2$. For $n = 2$ it is actually easier, and we leave this as an exercise for the reader. We do not consider $n = 1$, *i.e.* ion-ion collisions. First, we find the maximum in the effective potential by solving

$$\frac{\mathrm{d}}{\mathrm{d}r} V_\mathrm{eff}(r) = -\frac{|\ell|^2}{\mu r^3} + \frac{n c_n}{r^{n+1}} = 0. \tag{12.40}$$

The solution $r = r_0$ is

$$r_0 = \left(\frac{n\mu c_n}{|\ell|^2}\right)^{1/(n-2)}. \tag{12.41}$$

*It might be helpful to reread Section 1.4.3 at this point. See also ref. [823], particularly Chapter 4, page 175, for a recent general reference to much of the material covered in this section.

STUDY BOX 12.2: LONG RANGE INTERACTION POTENTIALS

As we will see, in order to apply capture models it is necessary to know the form of the long range interaction potential. Generally, one can write the radial dependence of the long range potential between two particles as a sum of dipole, quadrupole, octapole, etc. terms of the form

$$V(r) = -\sum_n \frac{c_n}{r^n}, \tag{B12.2.1}$$

where c_n is a constant depending on the charges q_1 and q_2, dipole moments μ_1 and μ_2, polarizabilities α_1 and α_2, and ionization potentials I_1 and I_2 of the two interacting species, and n depends on the type of interaction. The terms discussed here arise from first- and second-order perturbation theory, with a multipole expansion for the Coulomb interaction operator (and without inclusion of relativistic effects). Specifically for two multipoles, a 2^i-pole and a 2^j-pole, the electrostatic interaction varies with the power $n = i + j + 1$. In some cases, particular terms in the expansion will dominate over others at certain separations, and the expansion can be approximated by a single term.

Examples of interaction potentials between two particles are given in the following table, in which the polarizability volume α' is defined as $\alpha' = \alpha/4\pi\varepsilon_0$.

Interaction	n	c_n
ion-ion	1	$\frac{q_1 q_2}{4\pi\varepsilon_0}$
ion-dipole	2	$\frac{\mu_1 q_2}{4\pi\varepsilon_0}$
dipole-dipole	3	$\frac{\mu_1 \mu_2}{2\pi\varepsilon_0}$
ion-induced dipole	4	$\frac{\alpha_1 q_2^2}{32\pi^2\varepsilon_0^2}$
dipole-induced dipole	6	$\frac{\mu_2 \alpha_2'}{4\pi\varepsilon_0}$
induced dipole-induced dipole	6	$\frac{3\alpha_1'\alpha_2' I_1 I_2}{2(I_1+I_2)}$

Note that the dipole-dipole interaction is in general a function of the relative orientation of the two dipoles, as given in Section 12.2.5. If the dipoles are aligned parallel to each other the interaction reduces to the following

$$V(r) = -\frac{\mu_1\mu_2}{2\pi\varepsilon_0 r^3} P_2(\cos\theta), \tag{B12.2.2}$$

where θ defines the direction of the dipoles with respect to the vector $\boldsymbol{r}$ joining the two dipoles, and the second Legendre polynomial is defined as $P_2(\cos\theta) = \frac{1}{2}(3\cos^2\theta - 1)$. The dipole-dipole interaction given in the table is that for two dipoles fixed parallel to one another, in a linear configuration ($\theta = 0$), such that $P_2(\cos\theta) = 1$. The various induced-dipole interactions listed in the table represent averages over orientation angle. The induced dipole-induced dipole interaction is more commonly known as the dispersion or London interaction, and unlike the other interactions is purely quantum mechanical in origin, arising from the interactions between instantaneous or induced dipoles on the two particles. The dispersion interaction is only approximately given by the expressions given in the above table.

A realistic intermolecular potential has a long range attractive component, encompassing the above interactions, and a short range component arising from the overlap and subsequent bonding/antibonding interactions between partially or completely filled atomic and molecular orbitals. To obtain the precise form of this shorter range region of the potential requires detailed calculation using the techniques discussed in Chapter 2.

Mark Brouard and Claire Vallance

For the centrifugal barrier $V_{\text{eff}}(r_0)$ we find, after factorization,

$$V_{\text{eff}}(r_0) = \frac{|\ell|^2}{2\mu r_0^2} - \frac{c_n}{r_0^n} = \left(\frac{|\ell|^2}{\mu}\right)^{n/(n-2)} \frac{n-2}{2n}(nc_n)^{-2/(n-2)}. \tag{12.42}$$

Trajectories are reactive when $V_{\text{eff}}(r_0) \leqslant E$. This results in a maximum value for $|\ell|$:

$$|\ell_{\max}|^2 = \mu n (c_n)^{2/n}\left(\frac{2E}{n-2}\right)^{(n-2)/n}, \tag{12.43}$$

a corresponding maximum impact parameter

$$b_{\max} = \frac{|\ell_{\max}|}{\mu v}, \tag{12.44}$$

a cross section (assuming $P(b,E)=1$)

$$\sigma(E) = \pi b_{\max}^2 = \frac{\pi}{2} n \left(\frac{2}{n-2}\right)^{(n-2)/n}\left(\frac{c_n}{E}\right)^{2/n}, \tag{12.45}$$

and a rate

$$k(T) = \sqrt{\frac{2\pi}{\mu}} n \left(\frac{2}{n-2}\right)^{(n-2)/n} (c_n)^{2/n} (k_B T)^{(n-4)/2n} \Gamma\left(2 - \tfrac{2}{n}\right), \tag{12.46}$$

where the Gamma function is defined by

$$\Gamma(n) = \int_0^\infty x^{n-1} e^{-x} dx. \tag{12.47}$$

The Gamma function has the special value $\Gamma(1/2) = \sqrt{\pi}$, and it satisfies the recurrence relation $\Gamma(n+1) = n\Gamma(n)$, so we also have $\Gamma(3/2) = \frac{1}{2}\sqrt{\pi}$. For integer values $\Gamma(n+1) = n!$.

The position of the centrifugal barrier at collision energy E is found by substituting Eq. (12.43) into Eq. (12.41),

$$r_0 = \left(\frac{(n-2)c_n}{2E}\right)^{1/n}. \tag{12.48}$$

For the model to be valid, the potential for $r \geqslant r_0$ must be given to a good approximation by the leading long range term $V_n(r)$. The classical capture theory cross sections and rate constants for different values of n are given in Table 12.1.

Table 12.1 Classical capture theory energy dependent cross sections $\sigma_n(E)$ and temperature dependent reaction rates $k_n(T)$ for $-c_n/r^n$ long range potentials.

n	$\sigma_n(E)$	$k_n(T)$
2	$\pi c_2/E$	$2\sqrt{\frac{2\pi}{\mu}}c_2(k_BT)^{-1/2}$
3	$3\pi(c_3/2E)^{2/3}$	$4\sqrt{\frac{\pi}{3\mu}}\Gamma(1/3)(c_3)^{\frac{2}{3}}(k_BT)^{-1/6}$
4	$2\pi\sqrt{c_4/E}$	$2\pi\sqrt{\frac{2c_4}{\mu}}$
5	$\frac{5\pi}{2}\left(\frac{2}{3}\right)^{3/5}(c_5/E)^{2/5}$	$3\sqrt{\frac{2\pi}{\mu}}\Gamma(3/5)\left(\frac{2}{3}\right)^{3/5}(c_5)^{2/5}(k_BT)^{1/10}$
6	$3\pi\left(\frac{1}{2}\right)^{2/3}(c_6/E)^{1/3}$	$2^{\frac{11}{6}}\Gamma(2/3)\sqrt{\frac{\pi}{\mu}}(c_6)^{1/3}(k_BT)^{1/6}$

Capture theory was first developed in 1905 by Langevin, who used it to study reactions between ions and polarizable atoms.[825] In that case the interaction is proportional to r^{-4} and the long range coefficient is given by (see Study Box 12.2)

$$c_4 = \frac{1}{2}\frac{\alpha q^2}{(4\pi\varepsilon_0)^2}, \tag{12.49}$$

where q is the charge of the ion, and α is the polarizability of the atom. Substituting c_4 into the capture rate coefficient (see table 12.1) gives

$$k_{\text{Langevin}}(T) = \frac{2\pi q}{4\pi\varepsilon_0}\sqrt{\frac{\alpha}{\mu}}. \tag{12.50}$$

Note that this Langevin rate is independent of the temperature.

The expression for the rate for $n=6$ was first given by E. Gorin in 1939.[826] This is relevant to the case when there are no electrostatic interactions between the two atoms or molecules (*i.e.* when there are no permanent charges or dipoles on the particles involved). As shown in Study Box 12.2, the leading long range term is then proportional to r^{-6}.

12.2.5 Anisotropic Interactions

The first order electrostatic interaction between two neutral molecules with a non-zero dipole moment is proportional to r^{-3} (see Study Box 12.2). However, we cannot use the $k_3(T)$ capture rate formula directly in this case, because the interaction depends on the orientation of the molecules. For some orientations the interaction will be attractive, but for other orientations the interaction will be repulsive. To be precise, the interaction potential is given by

$$V(r,\theta_1,\phi_1,\theta_2,\phi_2) = -\frac{\mu_1\mu_2}{4\pi\varepsilon_0 r^3}[2\cos\theta_1\cos\theta_2 - \sin\theta_1\sin\theta_2\cos(\phi_1-\phi_2)], \tag{12.51}$$

where r is the distance between the centers of mass of the molecules, μ_1 and μ_2 are the magnitudes of the dipole moments of the molecules, and (θ_1, ϕ_1) and (θ_2, ϕ_2) are the spherical polar angles defining the orientations of the dipole vectors of the molecules in a dimer fixed frame, *i.e.* a frame in which the z-axis lies parallel to the vector pointing from the centre-of-mass of molecule 1 to the centre-of-mass of molecule 2.* In principle, the classical capture rate can be found by computing a large number of classical trajectories and by determining for each trajectory whether or not it crosses the centrifugal barrier. One strategy is to determine the opacity function as the fraction of "reactive trajectories" for a given impact parameter b, and use Eq. (12.28), where the integral over b is performed numerically.

An approximate analytical result can be found with the infinite order sudden (IOS) approximation. In this approximation the expression for $k_3(T)$ is found as the average over all orientations of both dipole vectors, setting the rate equal to zero whenever the interaction for a certain orientation is repulsive. When the interaction is attractive an orientation dependent c_3 coefficient is determined from Eq. (12.51) and the expression from Table 12.1 is used. Reorientation of the molecules during the collision is not taken into account. In atomic units (in which $4\pi\varepsilon_0 = 1$), the result of the procedure is

$$k_3^{\text{dip-dip}}(T) = 1.765\,(\mu_1\mu_2)^{\frac{2}{3}}\sqrt{\frac{\pi}{\mu}}\,(k_B T)^{-1/6}. \tag{12.52}$$

More cases can be found in ref. [804].

12.3 QUANTUM CAPTURE THEORY

So far, our treatment has been completely classical. Quantum mechanics requires several modifications of the model. First of all, in quantum mechanics angular momenta are quantized, and, in particular, the magnitude of the orbital angular momentum becomes $|\ell| = \hbar\sqrt{\ell(\ell+1)}$, where ℓ is a non-negative integer.† The wave functions corresponding to different values of the quantum number ℓ are referred to as *partial waves*. The classical expression $|\ell| = \mu v b$ [Eq. (12.26)] shows that for a fixed value of $|\ell|$, the impact parameter b goes to infinity when the velocity v goes to zero. This suggests that when the temperature approaches zero, only the $\ell = 0$ partial wave can contribute to cross sections, and that, in general, at lower temperatures fewer partial waves contribute than at higher temperatures.

When the interaction potential is anisotropic, which is usually the case in collisions involving molecules, we must also treat rotation of the colliding fragments. Since the rotational constants of the colliding molecules may be much larger than the rotational constant of the resulting collision complex, quantization of the rotation of the molecules may be important at higher temperatures where many partial waves still contribute to the cross sections. This is particularly true in molecular scattering experiments, in which the colliding molecules are cooled to the lowest rotational states, while the CM collision energy may still be high.

In the classical capture theory outlined above, trajectories are assumed reactive when the energy is above the centrifugal barrier, and nonreactive otherwise. In quantum mechanics, tunnelling may lead to reaction at energies below the barrier, while reflection may occur even if the energy is above the barrier.

To derive quantum capture theory, we start with the exact quantum mechanical expression for the energy dependent state-to-state differential cross section. This gives the most detailed information about a collision event.

---

*Note that for parallel dipoles, $\phi_1 - \phi_2 = 0$ and $\theta_1 = \theta_2$, and Eq. (12.51) reduces to Eq. (B12.2.2) of Study Box 12.2.
†Note that, as elsewhere in this text, we reserve the symbol ℓ for the orbital angular momentum quantum number.

12.3.1 Quantum Scattering Theory

If there is no interaction between the two particles, the wave function may be written as a plane wave.* We denote the quantum numbers describing the states of the particles collectively by $|n\rangle$. As an example, for a system consisting of two diatomic molecules in a certain rovibrational state we have $|n\rangle = |v_a j_a m_a, v_b j_b m_b\rangle$, where v_a and v_b are the vibrational quantum numbers, j_a and j_b are the diatomic angular momentum quantum numbers, and m_a and m_b are the projections of the angular momenta onto a space fixed axis. A flux normalized plane wave with wave vector $\boldsymbol{k}_n = k_n\hat{\boldsymbol{k}}$ is given by

$$\Psi_n^{\text{pw}} = |n\rangle v_n^{-\frac{1}{2}} e^{i\boldsymbol{k}_n\cdot\boldsymbol{r}} = |n\rangle v_n^{-\frac{1}{2}} \sum_\ell i^\ell (2\ell+1) j_\ell(k_n r) P_\ell(\hat{\boldsymbol{k}}\cdot\hat{\boldsymbol{r}}), \tag{12.53}$$

where P_ℓ is a Legendre polynomial and j_ℓ a spherical Bessel function of the first kind,[827] and the velocities v_n are given by $v_n = \hbar k_n/\mu$, with μ the reduced mass. Using the spherical harmonic addition theorem [239],

$$P_\ell(\hat{\boldsymbol{k}}\cdot\hat{\boldsymbol{r}}) = \frac{4\pi}{2\ell+1} \sum_{m_\ell=-\ell}^{\ell} Y_{\ell m_\ell}(\hat{\boldsymbol{r}})\, Y_{\ell m_\ell}(\hat{\boldsymbol{k}})^*, \tag{12.54}$$

and the asymptotic form of the spherical Bessel function

$$j_\ell(z) \approx \frac{\sin(z-\ell\pi/2)}{z} = \frac{e^{i(z-\ell\pi/2)} - e^{-i(z-\ell\pi/2)}}{2iz}, \tag{12.55}$$

the plane wave may be written, for large r, as

$$\Psi_n^{\text{pw}} \approx \frac{2\pi}{ik_n r} \sum_{\ell m_\ell} |n\rangle v_n^{-\frac{1}{2}} Y_{\ell m_\ell}(\hat{\boldsymbol{r}}) \left[e^{i(k_n r-\ell\pi/2)} - e^{-i(k_n r-\ell\pi/2)}\right] i^\ell\, Y_{\ell m_\ell}(\hat{\boldsymbol{k}})^*. \tag{12.56}$$

The effect of switching on the interaction is to modify the outgoing part of the wave function, so asymptotically the scattering wave function can be written as

$$\Psi_n^{\text{sc}} \approx \frac{2\pi}{ik_n r} \sum_{\ell m_\ell} \sum_{n'} \sum_{\ell' m'_\ell} |n'\rangle v_{n'}^{-\frac{1}{2}} Y_{\ell' m'_\ell}(\hat{\boldsymbol{r}}) \left[-e^{-i(k_n r-\ell\pi/2)} \delta_{n'n}\delta_{\ell'\ell}\delta_{m'_\ell m_\ell} + e^{i(k_n r-\ell\pi/2)} S_{n'\ell' m'_\ell;n\ell m_\ell}\right] i^\ell\, Y_{\ell m_\ell}(\hat{\boldsymbol{k}})^*, \tag{12.57}$$

where $S_{n'\ell' m'_\ell;n\ell m_\ell}$ are the scattering matrix (or S-matrix) elements, introduced in Chapter 3 and discussed further below. Solving the time-independent scattering problem amounts to finding solutions of the Schrödinger equation that satisfy the so-called S-matrix boundary conditions for large r,

$$\Psi_{n,\ell,m_\ell} = \frac{1}{r} \sum_{n'} \sum_{\ell' m'_\ell} |n'\rangle\, v_{n'}^{-\frac{1}{2}} Y_{\ell' m'_\ell}(\hat{\boldsymbol{r}}) \left[-e^{-i(k_n r-\ell\pi/2)} \delta_{n'n}\delta_{\ell'\ell}\delta_{m'_\ell m_\ell} + e^{i(k_n r-\ell\pi/2)} S_{n'\ell' m'_\ell;n\ell m_\ell}\right]. \tag{12.58}$$

*There is some overlap of material here with that covered in Chapter 3, and readers might wish to refresh their memory of Section 3.3.2.5 before embarking on this section.

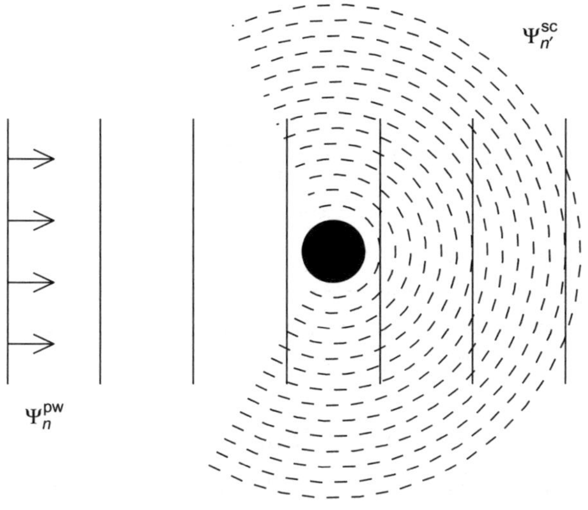

Figure 12.5 Quantum scattering: outside the scattering region (the solid circle in the centre) the wavefunction can be written as the sum of an unperturbed plane wave (Ψ_n^{pw}) for incoming channel $|n\rangle$ and a sum of outgoing spherical waves $\Psi_{n'}^{sc}$ for each product channel $|n'\rangle$.

As noted above, it is these individual solutions that are referred to as partial waves. The S-matrix is a complex symmetric unitary matrix. When the potential is zero, the S-matrix is a unit matrix, $S_{n'\ell' m'_\ell; n\ell m_\ell} = \delta_{n'n}\delta_{\ell'\ell}\delta_{m'_\ell m_\ell}$, and the partial waves add up to a plane wave again. The S-matrix is related to the T-matrix through $\boldsymbol{S} = \boldsymbol{I} - \boldsymbol{T}$, where $\boldsymbol{I}$ is a unit matrix.

As discussed in Chapter 3, the scattering wave function of Eq. (12.57) may be reorganized into an incoming plane wave plus an outgoing spherical wave, as illustrated in Figure 12.5,

$$\Psi_n^{sc} \approx |n\rangle v_n^{-\frac{1}{2}} e^{i\boldsymbol{k}_n \cdot \boldsymbol{r}} + \sum_{n'} |n'\rangle\, v_{n'}^{-\frac{1}{2}} \frac{e^{ik_{n'}r}}{r} f_{n'\leftarrow n}(\hat{\boldsymbol{r}}; \hat{\boldsymbol{k}}), \tag{12.59}$$

where the so called scattering amplitude, introduced in Chapter 3, is given by

$$f_{n'\leftarrow n}(\hat{\boldsymbol{r}}, \hat{\boldsymbol{k}}) = \frac{2\pi}{ik_n} \sum_{\ell m_\ell \ell' m'_\ell} i^{\ell-\ell'}\, Y_{\ell' m'_\ell}(\hat{\boldsymbol{r}})\, T_{n'\ell' m'_\ell; n\ell m_\ell}\, Y_{\ell m_\ell}(\hat{\boldsymbol{k}})^*. \tag{12.60}$$

The notation with the arrow is used because initial and final quantum numbers should not be interchanged, *i.e.* $f_{n'\leftarrow n} \neq f_{n\leftarrow n'}$. The state-to-state differential cross section for a particular incident direction $\hat{\boldsymbol{k}}$ is given by*

$$\sigma_{n'\leftarrow n}(\hat{\boldsymbol{r}}, \hat{\boldsymbol{k}}) = \left| f_{n'\leftarrow n}(\hat{\boldsymbol{r}}, \hat{\boldsymbol{k}}) \right|^2. \tag{12.61}$$

The state-to-state integral cross section for a particular incident direction is given by

$$\sigma_{n'\leftarrow n}(\hat{\boldsymbol{k}}) = \iint \sigma_{n'\leftarrow n}(\hat{\boldsymbol{r}}, \hat{\boldsymbol{k}}) \mathrm{d}\hat{\boldsymbol{r}}. \tag{12.62}$$

*Note that Eq. (12.61) is essentially the same as that given in Eq. (3.45), apart from notation.

Assuming that the incident directions are isotropically distributed, such as they would be in a bulk gaseous sample, the state-to-state integral cross section is obtained by taking an average over all incoming directions $\hat{\boldsymbol{k}}$

$$\sigma_{n'\leftarrow n} = \frac{1}{4\pi}\iint \sigma_{n'\leftarrow n}(\hat{\boldsymbol{k}})\,\mathrm{d}\hat{\boldsymbol{k}} = \frac{\pi}{k_n^2}\sum_{\ell m_\ell \ell' m'_\ell}\left|T_{n'\ell' m'_\ell;n\ell m_\ell}\right|^2 = \frac{\pi}{k_n^2}P_{n'n}, \tag{12.63}$$

where we have introduced the reaction probability matrix $\boldsymbol{P}$ in the last step. So far, the formalism applies to inelastic scattering. To extend it to reactive scattering we only have to include an *arrangement label* γ to the quantum numbers n that describe the molecules. This modification is sufficient as long as three-body breakup cannot occur.

12.3.2 Connection with Classical Capture Theory

The diagonal elements of the T-matrix determine the elastic scattering cross sections and the off-diagonal elements determine the inelastic and reactive cross sections. The off-diagonal elements of the T-matrix are equal to the off-diagonal elements of the S-matrix. The S-matrix is unitary, so the sum of the squares of the absolute values of all elements of a given column is equal to one. Hence, the sum over ℓ', m'_ℓ in Eq. (12.63), excluding the diagonal element, gives at most one for each column. This is still true if we sum over all possible reaction products n'. Thus, the maximum contribution of the partial wave with a given ℓ to the inelastic or reactive cross section for some initial state $|n\rangle$ is given by

$$\sigma_{n,\ell}^{\max} = \frac{\pi}{k_n^2}(2\ell+1), \tag{12.64}$$

where k_n is the wavenumber for channel n, and the factor $(2\ell+1)$ arises from the summation over m_ℓ. Assuming that all partial waves up to some maximum value $\ell_{\max}$ are fully reactive and higher partial waves are nonreactive, we obtain an initial state selected cross section

$$\sigma_n = \sum_{\ell=0}^{\ell_{\max}} \sigma_{n,\ell}^{\max} = \frac{\pi}{k_n^2}(\ell_{\max}+1)^2, \tag{12.65}$$

where we have used the fact that $\Sigma_{i=0}^{n} i = n(n+1)/2$. To compare this result to the classical expression $\mu b_{\max}^2$, given in Eq. (12.27), we associate the classical angular momentum squared $|\ell|^2$ with $\hbar^2\,\ell_{\max}\,(\ell_{\max}+1)$. Using Eq. (12.26) this gives

$$b_{\max}^2 = \frac{\hbar^2\,\ell_{\max}(\ell_{\max}+1)}{\mu^2 v^2}, \tag{12.66}$$

and with $\mu v = p = \hbar k_n$ we obtain

$$\sigma = \pi b_{\max}^2 = \frac{\pi}{k_n^2}\ell_{\max}(\ell_{\max}+1). \tag{12.67}$$

One expects the classical theory only to work when a sufficient number of partial waves contribute, in which case $\ell_{\max}(\ell_{\max}+1) \approx (\ell_{\max}+1)^2$.

12.3.3 Coupled Channels Capture Theory

In quantum capture theory it is assumed that the capture cross sections can be found by solving the Schrödinger equation in a restricted region of the potential energy surface that is located entirely in the reactant arrangement. The computation then becomes very similar to a coupled channels calculation for inelastic scattering, discussed in Chapter 3. The only difference lies in the boundary conditions at small internuclear separations r. The wave function is not assumed to be finite, but the flux is assumed to be inwards (towards small r) at some point $r = r_a$. Here we will not derive the coupled channels equation, but only summarize the main results and give the capture theory boundary conditions.

In the coupled channels approach (see also Chapter 3, Section 3.3.4.2), the Hamiltonian is written as the sum of the radial kinetic energy operator and the remainder ($\Delta\hat{H}$),

$$\hat{H} = -\frac{\hbar^2}{2\mu} r^{-1} \frac{\mathrm{d}^2}{\mathrm{d}r^2} r + \Delta\hat{H} \tag{12.68}$$

and the Schrödinger equation in the reactant arrangement is written as

$$\frac{\hbar^2}{2\mu} r^{-1} \frac{\mathrm{d}^2}{\mathrm{d}r^2} r\Psi = (\Delta\hat{H} - E)\Psi. \tag{12.69}$$

The wave function is expanded in channel functions $|n'\rangle$,

$$\Psi_n = r^{-1} \sum_{n'} |n'\rangle \, U_{n'n}(r), \tag{12.70}$$

where each column of the coefficient matrix $\boldsymbol{U}(r)$ defines a wave function. To simplify the notation, we assume that the partial wave quantum numbers ℓ and m_ℓ are included in n. By substituting the expansion into the Schrödinger equation and projecting onto the channel eigenfunctions, a set of coupled second order differential equations for the expansion coefficients is found*

$$\boldsymbol{U}''(r) = \boldsymbol{W}(r)\boldsymbol{U}(r), \tag{12.71}$$

where the primes denote derivatives with respect to r. The coupling matrix is given by

$$\boldsymbol{W}_{n'n}(r) = \frac{2\mu}{\hbar^2} \langle n'|\Delta\hat{H} - E|n\rangle. \tag{12.72}$$

In an inelastic scattering problem the condition that the wave function is finite gives the boundary condition that $\boldsymbol{U}(r=0) = 0$. As we have seen in Chapter 3, Section 3.3.4.3, this boundary condition, together with the coupled channels equation (12.71), defines a linear relation between the expansion coefficients and their derivatives with respect to r,

$$\boldsymbol{U}'(r) = \boldsymbol{Y}(r)\boldsymbol{U}(r), \tag{12.73}$$

where $\boldsymbol{Y}(r)$ is called the log-derivative matrix. In the capture problem, the boundary condition is that at some small value of r, inside the centrifugal barrier, the flux can only be inwards. For a one-dimensional single channel problem with $\Delta\hat{H} = V(r)$, this means that around some point $r = r_a$ the

*Note that apart from notation this equation is the same as Eq. (3.56) of Chapter 3.

wave function has the simple travelling wave form e^{-ikr}, where k is the wave number at $r=r_a$, *i.e.*

$$\frac{\hbar^2 k^2}{2\mu} = E - V(r_a), \tag{12.74}$$

and the boundary condition for the (1×1) log-derivative matrix is $\boldsymbol{Y}(r_a)=\boldsymbol{U}'(r_a)/\boldsymbol{U}(r_a)=-ik$. To define the boundary conditions in the multichannel case, the coupling matrix $\boldsymbol{W}(r_a)$ is diagonalized to obtain a set of uncoupled one dimensional problems,

$$\boldsymbol{W}(r_a)\,\boldsymbol{Q}(r_a) = \boldsymbol{Q}(r_a)\boldsymbol{\Lambda}(r_a), \tag{12.75}$$

where $\boldsymbol{\Lambda}(r_a)$ is a diagonal matrix of eigenvalues, and the columns of the matrix $\boldsymbol{Q}(r_a)$ are the eigenvectors of the matrix $\boldsymbol{W}(r_a)$. The negative eigenvalues correspond to open channel eigenfunctions, with $\Lambda_{oo}=-k_o^2$, and the positive eigenvalues correspond to closed channel eigenfunctions with $\Lambda_{cc}=k_c^2$. Transforming the coupled channels problem to the channel eigenfunction basis with

$$\tilde{\boldsymbol{U}}(r) = \boldsymbol{Q}^\dagger(r_a)\,\boldsymbol{U}(r), \tag{12.76}$$

gives

$$\tilde{\boldsymbol{U}}''(r) = \boldsymbol{Q}^\dagger(r_a)\boldsymbol{W}(r)\boldsymbol{Q}(r_a)\tilde{\boldsymbol{U}}(r) \approx \boldsymbol{\Lambda}(r_a)\tilde{\boldsymbol{U}}(r). \tag{12.77}$$

The approximation of assuming that the $\boldsymbol{W}(r)$ matrix is constant around r_a results in a set of one dimensional problems, and the matrix $\tilde{\boldsymbol{U}}(r)$ becomes diagonal. The inward flux boundary conditions for open channel eigenfunctions are now given by

$$\tilde{U}_{oo}(r) = e^{-ik_o r}, \tag{12.78}$$

and the boundary conditions for closed channels are

$$\tilde{U}_{cc}(r) = e^{k_c r}. \tag{12.79}$$

The log-derivative matrix in the channel eigenfunction basis is also diagonal at $r=r_a$, and the boundary conditions are given by

$$\tilde{Y}_{ii}(r_a) = \left[\boldsymbol{Q}^\dagger\boldsymbol{Y}(r_a)\boldsymbol{Q}\right]_{ii} = \begin{cases} -ik_i & \text{for open channels,} \\ k_i, & \text{for closed channels.} \end{cases} \tag{12.80}$$

Since the matrix $\boldsymbol{Q}$ with eigenvectors is unitary, the boundary conditions for the log-derivative matrix in the original basis are given by

$$\boldsymbol{Y}(r_a) = \boldsymbol{Q}(r_a)\tilde{\boldsymbol{Y}}(r_a)\boldsymbol{Q}^\dagger(r_a). \tag{12.81}$$

The boundary conditions of Eqs. (12.78) and (12.79) apply if the channel eigenvalues $\boldsymbol{\Lambda}(r)$ are approximately constant around $r=r_a$. Sometimes it is better to approximate the channel eigenvalues by a linear function of r. For that case the boundary conditions can be found in ref. [828].

The general technique for propagating the log-derivative matrix to some point $r=r_b$ sufficiently far outside the centrifugal barrier relies on dividing $[r_a, r_b]$ into a set of small sectors $[r_i, r_{i+1}]$.

In each sector one determines a so-called embedding type propagator defined by

$$\begin{bmatrix} \boldsymbol{U}_i' \\ \boldsymbol{U}_{i+1}' \end{bmatrix} = \begin{bmatrix} \mathcal{Y}_1^{(i)} & \mathcal{Y}_2^{(i)} \\ \mathcal{Y}_3^{(i)} & \mathcal{Y}_4^{(i)} \end{bmatrix} \begin{bmatrix} -\boldsymbol{U}_i \\ \boldsymbol{U}_{i+1} \end{bmatrix}, \tag{12.82}$$

where $\boldsymbol{U}_i = \boldsymbol{U}(r_i)$. The minus sign in the definition is not essential, but with this choice one can show that $\mathcal{Y}_2^{(i)} = \mathcal{Y}_3^{(i)}$. To find the propagator one can diagonalize the $\boldsymbol{W}$ matrix in the middle of the sector and assume it to be constant, which results in a set of uncoupled one-dimensional problems, as above. For the one-dimensional problems the propagator can be found analytically, and the result can be transformed back to the original basis. More accurate propagators have been developed, which, *e.g.*, assume the eigenvalues of the $\boldsymbol{W}(r)$ matrix to change linearly over the interval, and correct for non-zero coupling with a Green's function technique. Once the sector propagator is found it can be used to propagate the log-derivative matrix at $r = r_i$, defined by

$$\boldsymbol{U}_i' = \boldsymbol{Y}(r_i)\,\boldsymbol{U}_i, \tag{12.83}$$

to r_{i+1}:

$$\boldsymbol{Y}(r_{i+1}) = \mathcal{Y}_4^{(i)} - \mathcal{Y}_3^{(i)}\left[\boldsymbol{Y}(r_i) + \mathcal{Y}_1^{(i)}\right]^{-1}\mathcal{Y}_2^{(i)}. \tag{12.84}$$

In this way, the log-derivative matrix can be propagated to $r = r_\mathrm{b}$. For sufficiently large r the S-matrix boundary conditions for $\boldsymbol{U}(r)$ are given by [*cf.* Eq. (3.69)]

$$\boldsymbol{U}(r) = -\boldsymbol{I}(r) + \boldsymbol{O}(r)\boldsymbol{S}, \tag{12.85}$$

where $\boldsymbol{I}(r)$ is a diagonal matrix with flux normalized incoming waves

$$I_{n,n}(r) = v_n^{-\frac{1}{2}} \mathrm{e}^{-i(k_n r - \ell\pi/2)}, \tag{12.86}$$

and $\boldsymbol{O}(r) = \boldsymbol{I}(r)^*$ are the outgoing waves. By substituting the asymptotic form of the wave function into the defining relation of the log-derivative matrix [Eq. (12.73)] we can relate the $\boldsymbol{S}$ matrix to the log-derivative matrix

$$\boldsymbol{S}(E) = [\boldsymbol{Y}(r_\mathrm{b})\boldsymbol{O}(r_\mathrm{b}) - \boldsymbol{O}'(r_\mathrm{b})]^{-1}[\boldsymbol{Y}(r_\mathrm{b})\boldsymbol{I}(r_\mathrm{b}) - \boldsymbol{I}'(r_\mathrm{b})]. \tag{12.87}$$

Because of the complex boundary conditions at $r = r_\mathrm{a}$, the S-matrix is not unitary and the capture probability for a given incoming channel can be found by

$$P_{n\ell m_\ell}(E) = 1 - \sum_{n'\ell' m_\ell'} \left|S_{n'\ell' m_\ell';n\ell m_\ell}(E)\right|^2, \tag{12.88}$$

where we have written the partial wave quantum numbers explicitly again for clarity. The capture cross section for incoming channel n is found as

$$\sigma_n(E) = \frac{\pi}{k_n^2}\sum_{\ell m_\ell} P_{n\ell m_\ell}(E). \tag{12.89}$$

Note that this capture approximation provides initial state selected cross sections only and information about the product state distribution is lost. A model exists for complex-forming reactions in which capture theory ideas are used in reactant as well as product arrangements. Together with a statistical model to describe the complex, this provides partial information about the product state distribution.[828]

12.3.4 Quantum Adiabatic Capture Theory

At low temperatures the collision time is long compared with the characteristic vibrational and rotational timescales of the colliding molecules. This allows us to introduce an approximation analogous to the Born-Oppenheimer approximation, which exploits the difference in time scales of electronic and nuclear motion, and simplifies the dynamical calculation. Solving the Schrödinger equation for the fast motion amounts to diagonalizing the $\boldsymbol{W}(r)$ matrix on a grid of r points, as in Eq. (12.75), and treating the eigenvalues $\boldsymbol{\Lambda}_{mn}(r_a)$ as uncoupled one-dimensional potentials (multiplied by a factor of $2\mu/\hbar^2$). These potentials will asymptotically correlate with molecular states. For each molecular state asymptotically allowed at an energy E, the capture probability is computed by solving the one-dimensional quantum capture problem. This calculation is done exactly as the coupled channels equation, except that all matrices become scalars, and the propagators and log-derivative matrices in the channel eigenfunction basis are never transformed back to the original basis. The result is again a capture probability for each initial state n, and the capture cross section is obtained as in Eq. (12.89).

Often, the result of this approximation is in good agreement with full coupled channels capture theory.[804,829,830] When the coupling between different rotational states is strong, the initial state selected capture rates may not be very good. However, often one is only interested in the Maxwell-Boltzmann average of the capture rates over all possible initial states, in the case of thermal equilibrium. These thermally averaged rates may still be good, even for strong rotational coupling. In the next section we show that the thermal capture rate only depends on the cumulative capture rate at a given total energy. We derive an expression for the thermal rate coefficient in terms of the cumulative reaction probability, $N(E)$, introduced in Section 7.5.1.*

12.3.5 Thermal Capture Rates

For an initial state $|n\rangle$ with channel energy (*i.e.* internal energy) ε_n and kinetic energy $E_n^{(\mathrm{kin})}$ the total energy is

$$E = \varepsilon_n + E_n^{(\mathrm{kin})} = \varepsilon_n + \frac{\hbar^2 k_n^2}{2\mu}. \tag{12.90}$$

In time-independent scattering theory one computes the state-to-state cross section as a function of the total energy [Eq. (12.63)],

$$\sigma_{n'\leftarrow n}(E) = \frac{\pi}{k_n^2} P_{n'n}(E). \tag{12.91}$$

The state-to-state temperature dependent reaction rate is given by [Eq. (12.37)]

$$k_{n'\leftarrow n}(T) = \sqrt{\frac{8k_\mathrm{B}T}{\pi\mu}} \frac{1}{(k_\mathrm{B}T)^2} \int_0^\infty \sigma_{n'\leftarrow n}(E) \mathrm{e}^{-E_n^{(\mathrm{kin})}/k_\mathrm{B}T} E_n^{(\mathrm{kin})} \mathrm{d}E_n^{(\mathrm{kin})}. \tag{12.92}$$

*An alternative derivation of the expression given here for $k(T)$, and an introductory discussion about $N(E)$, can be found in ref. [10].

Sometimes, only the Boltzmann averaged reaction rate is required,

$$k(T) = Q_{\text{int}}^{-1}(T) \sum_{n'n} k_{n' \leftarrow n}(T) e^{-\varepsilon_n / k_B T}, \tag{12.93}$$

where the internal partition function is given by

$$Q_{\text{int}}(T) = \sum_{n} e^{-\varepsilon_n / k_B T}. \tag{12.94}$$

By substituting Eq. (12.92) into Eq. (12.93) and changing the order of integration and summation, one obtains a much simplified expression

$$k(T) = \frac{1}{2\pi\hbar Q_t Q_{\text{int}}} \int_{-\infty}^{\infty} N(E)\, e^{-E/k_B T} dE, \tag{12.95}$$

where the translational partition function per volume is given by

$$Q_t = \frac{1}{\hbar^3} \left(\frac{\mu k_B T}{2\pi} \right)^{3/2}, \tag{12.96}$$

and the cumulative reaction probability $N(E)$ is defined as the sum of the reaction probabilities over all open reactant and product states

$$N(E) = \sum_{n'n} P_{n'n}(E). \tag{12.97}$$

When there are no open channels at a given total energy, $N(E)$ is set to zero, so the range of integration can be taken from $-\infty$ to $+\infty$ in Eq. (12.95). Since $N(E)$ depends on the *unweighted* sum over initial states, one sees that an approximation that does not properly describe the mixing of initial states may still produce an accurate thermal rate. In particular, this explains why adiabatic capture theory may be much more accurate for the thermal capture rate than for initial state selected rates.

12.3.6 Total Angular Momentum Representation

When no external fields are present, both the collision complex and its Hamiltonian are invariant under overall rotation. This invariance gives rise to total angular momentum conservation. This symmetry can be exploited by using the total angular momentum representation for the wave function. In Section 12.3.1 we employed an uncoupled angular momentum basis for two colliding molecules. For an atom-diatom system the uncoupled rotational basis is $|jm_j\rangle|\ell m_\ell\rangle$, where $|jm_j\rangle$ is the rotational wave function of the molecule and $|\ell m_\ell\rangle$ is the orbital angular momentum of the collision partners. The coupled angular momentum basis is defined by

$$|(j\ell)JM_J\rangle = \sum_{m_j=-j}^{j} \sum_{m_\ell=-\ell}^{\ell} |jm_j\rangle|\ell m_\ell\rangle\langle jm_j \ell m_\ell|JM_J\rangle, \tag{12.98}$$

where $\langle jm_j\, \ell m_\ell|JM_J\rangle$ is a Clebsch-Gordan coefficient.[239] The coupled states are eigenfunctions of the square of the total angular momentum operator for the system ($\hat{J}^2$) and of its space-fixed

z-component $\hat{J}_z$. The total angular momentum quantum number J can take the values $|j-\ell|$, $|j-\ell|+1, \ldots, j+\ell$, and the projection quantum number M_J can take the values $-J, -J+1, \ldots, J$. For a given j and ℓ there are as many coupled basis functions as uncoupled ones

$$\sum_{J=|\ell-j|}^{\ell+j} (2J+1) = (2j+1)(2\ell+1), \tag{12.99}$$

and the coupled basis functions are orthonormal, *i.e.* the Clebsch-Gordan coefficients are elements of a unitary transformation. When a scattering wave function is written in the coupled basis, the S-matrix is related to the S-matrix in the uncoupled basis through

$$S_{n'j'\ell',nj\ell}^{J',M'_J;JM_J} = \sum_{m'_j m'_\ell m_j m_\ell} \left\langle J'M'_J \middle| j'm'_j\ell'm'_\ell \right\rangle S_{n'j'm'_j\ell'm'_\ell;njm_j\ell m_\ell} \left\langle jm_j\ell m_\ell \middle| JM_J \right\rangle. \tag{12.100}$$

When the system has cylindrical symmetry, *i.e.* when there is no external field, or when an external field is applied parallel to the space-fixed z-axis, the projection of the total angular momentum on the z-axis is conserved, *i.e.* $M'_J = m'_j + m'_\ell = m_j + m_\ell = M_J$, and the S-matrix elements are zero when $M'_J \neq M_J$. When the system has spherical symmetry, *i.e.* there are no external fields, then J is also a good quantum number, and the S-matrix is independent of M_J,

$$S_{n'j'\ell',nj\ell}^{J',M'_J;JM_J} = \delta_{J'J}\delta_{M'_J,M_J} S^J_{n'j'\ell';nj\ell}, \tag{12.101}$$

and Eq. (12.100) may be inverted to give

$$S_{n'j'm'_j\ell'm'_\ell;njm_j\ell m_\ell} = \sum_{JM_J} \left\langle j'm'_j\ell'm'_\ell \middle| JM_J \right\rangle S^J_{n'j'\ell';nj\ell} \left\langle JM_J \middle| jm_j\ell m_\ell \right\rangle. \tag{12.102}$$

The T-matrix satisfies an analogous relation, and hence, using the orthogonality relations of the Clebsch-Gordan coefficients, we may derive an expression for the sum of the reaction probabilities over all projection quantum numbers

$$\sum_{m'_j m'_\ell m_j m_\ell} P_{n'j'm'_j\ell'm'_\ell;njm_j\ell m_\ell} = \sum_J (2J+1) P^J_{n'j'\ell';nj\ell}, \tag{12.103}$$

where the factor $(2J+1)$ is the result of summing over M_J.

To construct a total angular momentum basis for two colliding molecules one must first couple the rotational wave functions of the molecules, $|j_a m_a\rangle$ and $|j_b m_b\rangle$ to give a coupled basis $|(j_a j_b j m_j)\rangle$, which in turn can be coupled with the end-over-end rotational function $|\ell m_\ell\rangle$ to obtain the total angular momentum basis functions $|\{(j_a j_b) j\ell\} JM_J\rangle$.

For cold collisions in electrostatic or magnetic traps it may be important to include the external field in the calculation. In such cases it is advantageous to use the uncoupled representation; otherwise the coupled representation is more efficient.

12.4 WIGNER THRESHOLD LAWS

Even in the early days of quantum mechanics, solutions of the Schrödinger equation in the limit of low collision energies were analyzed and expressions were found for the energy-dependence of the collision cross section. For example, it was found that for potentials of finite range, *i.e.* potentials that are vanishingly small when the distance between collision partners is larger than some value r_0, the elastic cross section at low energy is dominated by *s*-wave scattering, *i.e.* by the contribution of the $\ell=0$ partial wave, and the cross section is energy independent. The inelastic cross section is also dominated by the *s*-wave, but it increases as $E^{-\frac{1}{2}}$ when the collision energy $E \equiv E_c$ is sufficiently low. Here we will derive the general results obtained by Wigner in 1948.[831] These results are known as the Wigner threshold laws.

12.4.1 Bouncing Off a Cliff

To demonstrate the difference between quantum mechanics and classical mechanics in the limit of low energies we consider the one-dimensional problem shown in Figure 12.6. The potential takes the values $V(x)=0$ for $x<0$ and $V(x)=V_0<0$ for $x \geqslant 0$. A classical particle with mass μ moving in the positive *x*-direction will have zero probability of reflection at $x=0$ and unit probability of transmission.

In the quantum mechanical description, for $x<0$ the wave function contains contributions from a unit flux incoming part and a reflected part,

$$\Psi_L(x) = v^{-\frac{1}{2}}(\mathrm{e}^{ikx} - R\mathrm{e}^{-ikx}), \quad x<0 \tag{12.104}$$

and a contribution solely from a transmitted part for $x \geqslant 0$ from,

$$\Psi_R(x) = v_0^{-\frac{1}{2}} T \mathrm{e}^{ik_0 x}, \quad x \geqslant 0. \tag{12.105}$$

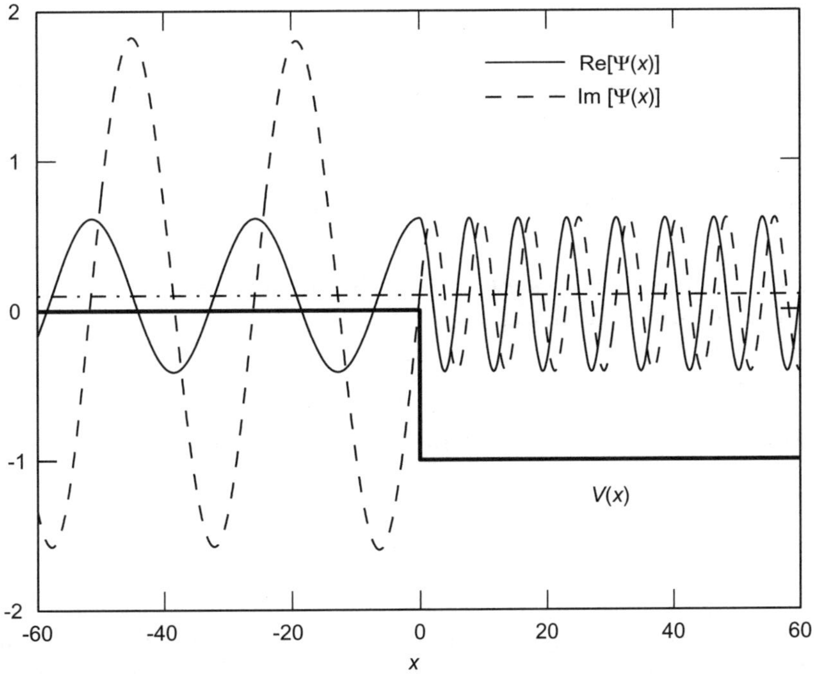

Figure 12.6 Real and imaginary part of the scattering wave function for a one-dimensional step-function potential $V(x)$.

The total energy is

$$E = \frac{\hbar^2 k^2}{2\mu} = V_0 + \frac{\hbar^2 k_0^2}{2\mu}, \tag{12.106}$$

with velocities to the left and right of the cliff $v = \hbar k/\mu$ and $v_0 = \hbar k_0/\mu$. Matching the wave functions $\Psi_L(x)$ and $\Psi_R(x)$ and their first derivatives at $x = 0$ gives

$$T = \frac{2\alpha}{\alpha^2 + 1} \tag{12.107}$$

$$R = \frac{1 - \alpha^2}{\alpha^2 + 1}, \tag{12.108}$$

where $\alpha = \sqrt{k/k_0}$ and $|T|^2 + |R|^2 = 1$. Hence, in this one-dimensional problem, if $k/k_0 \ll 1$ the probability of transmission is proportional to $|T|^2 \propto k \propto \sqrt{E}$ and the probability of reflection, $|R|^2$, approaches unity when the energy E approaches zero. This quantum result is very different from the classical result, in which the probability of reflection is always zero. The quantum effect is sometimes referred to as quantum *suppression*, because unlike for a classical particle the transmission probability is not unit. The classical result is recovered in the high energy limit ($E \gg V_0$, $k \approx k_0$, and $\alpha \approx 1$).

12.4.2 *s*-wave Elastic Scattering

Solutions of the quantum central force problem may be expanded in partial waves

$$\Psi_{\ell m_\ell}(\boldsymbol{r}) = r^{-1} u_\ell(\boldsymbol{r}) Y_{\ell m_\ell}(\hat{\boldsymbol{r}}), \tag{12.109}$$

where $u_\ell(r)$ describes the radial dependence of the wavefunction, and the spherical harmonic, $Y_{\ell m_\ell}(\hat{\boldsymbol{r}})$, its angular dependence. For $u_0(r)$, the so called *s*-wave, the Schrödinger equation for the radial motion is

$$\left[-\frac{\hbar^2}{2\mu} \frac{\mathrm{d}^2}{\mathrm{d}r^2} + V(r) - E \right] u_0(r) = 0, \tag{12.110}$$

and the boundary condition at $r = 0$ is $u_0(0) = 0$. As before, we assume that the potential is negligible for $r > r_0$. The log-derivative matrix at r_0, defined by

$$u_0'(r_0) = Y(r_0) u_0(r_0), \tag{12.111}$$

is found by propagation from $r = 0$ to $r = r_0$. In this region, the kinetic energy

$$E = \frac{\hbar^2 k^2}{2\mu}, \tag{12.112}$$

is assumed to be so small compared to the potential $V(r)$ that it may be neglected and $Y(r_0)$ becomes energy independent. To find the energy dependence of the T-matrix we first determine the radial wave function with K-matrix boundary conditions, which is convenient because it uses real functions,

$$u_0(r) = \sin(kr) - K\cos(kr), \quad \text{for } r > r_0. \tag{12.113}$$

Consider this equation in conjunction with Eq. (12.109). Note that the term $r^{-1}\sin(kr)$ is usually called the regular solution, since $r^{-1}\sin(kr)$ is finite for $r=0$, while $r^{-1}\cos(kr)$ is called the irregular solution. We now assume that $kr_0 \ll 1$ so that we can replace $\sin(kr)$ and $\cos(kr)$ by the leading term in their Taylor expansion for $r=r_0$,

$$u_0(r) \approx kr - K \tag{12.114}$$

and

$$u_0'(r) \approx k. \tag{12.115}$$

Since we assume the log-derivative matrix at $r=r_0$ to be energy-independent, we find that the K-matrix must be proportional to k,

$$K \propto k \propto \sqrt{E}. \tag{12.116}$$

The S-matrix is related to the K-matrix through (see Problem 2)

$$S = (1 - iK)(1 + iK)^{-1}. \tag{12.117}$$

For small K we have, to first order in K

$$(1 + iK)^{-1} \approx 1 - iK + \ldots, \tag{12.118}$$

so

$$T = 1 - S \approx 1 - (1 - iK)^2 \approx 2iK. \tag{12.119}$$

Hence, the T-matrix is also proportional to k, and the elastic s-wave scattering cross section,

$$\sigma(E) = \frac{\pi}{k^2}|T|^2, \tag{12.120}$$

is energy independent for small E.

12.4.3 Scattering Length

The boundary condition for scattering off a hard sphere with radius r_h is $u_0(r_h)=0$, *i.e.* the particle cannot penetrate the sphere so the wavefunction is zero at its boundary. The wave function $u_0(r)$ in Eq. (12.114) looks like the wave function corresponding to a hard-sphere problem, with radius $r_h = K/k$. This motivates the definition of the *scattering length* a as

$$a = \lim_{k\to 0} \frac{K}{k}. \tag{12.121}$$

The scattering length may also be defined in terms of a quantity known as the *phase shift*, δ, which is related to the S-matrix by $S=e^{2i\delta}$, and represents the shift in phase of the wavefunction on scattering. For small phase shifts we have $S\approx 1+2i\delta$ and $K\approx -\delta$. With Eq. (12.120) and $T\approx 2iK$, the elastic s-wave cross section in the limit of low energy is related to the scattering length through

$$\sigma = 4\pi a^2. \tag{12.122}$$

In the Born approximation, the total elastic cross section, in the case of an isotropic potential, is given by

$$\sigma = \frac{\mu^2}{4\pi\hbar^4}\left|\iiint V(\mathbf{r})\mathrm{d}\mathbf{r}\right|^2. \tag{12.123}$$

As we will see shortly, it is helpful to consider a δ-function potential of the form

$$V(\mathbf{r}) = \frac{4\pi a\hbar^2}{\mu}\delta(\mathbf{r}), \tag{12.124}$$

for which we have

$$\sigma = \frac{\mu^2}{4\pi\hbar^4}\left|\frac{4\pi a\hbar^2}{\mu}\right|^2 = 4\pi a^2. \tag{12.125}$$

The fact that this expression is the same as that given in Eq. (12.122) suggests that, to simplify the calculation of the properties of ultracold gases, one can often replace the actual potential by a δ-function potential that gives the same scattering length and hence the same elastic cross section. Notice that a positive scattering length corresponds to an effectively repulsive interaction, and a negative scattering length to an effectively attractive interaction.

12.4.4 Inelastic Scattering at Low Energy

We will now consider the energy dependence of the cross section for an inelastic or reactive process when the kinetic energy in the incoming channel approaches zero. As above, we assume that the potential is negligible for $r > r_0$. Furthermore, we will consider all partial waves, and not limit ourselves to *s*-wave scattering. This is important since we may be interested in a process that changes the angular momentum of the colliding molecules. In the absence of external fields, the total angular momentum is conserved, so the incoming and outgoing partial waves cannot both be *s*-waves if the angular momentum of one of the species changes.

As in the one-channel elastic case, we will use *K*-matrix boundary conditions, which give real wave functions. The expansion of the wave function,

$$\Psi_{n\ell} = r^{-1}\sum_{n'\ell'}|n'\ell'\rangle\, U_{n'\ell';n\ell}(r), \tag{12.126}$$

is similar to the expansion of the wave function with *S*-matrix boundary conditions, but instead of incoming and outgoing waves there are regular and irregular waves for large *r*,

$$U_{n'\ell';n\ell}(r) \approx f_{n\ell}(r)\delta_{n'n}\delta_{\ell'\ell} + g_{n'\ell'}(r)K_{n'\ell';n\ell}. \tag{12.127}$$

The regular waves are defined by

$$f_{n\ell}(r) = v_n^{-\frac{1}{2}}k_n r j_\ell(k_n r), \tag{12.128}$$

and the irregular waves by

$$g_{n\ell}(r) = v_n^{-\frac{1}{2}} k_n r y_\ell(k_n r), \tag{12.129}$$

where y_ℓ is a spherical Bessel function of the second kind.[827] For $\ell = 0$ we have the s-wave functions, $zj_0(z) = \sin z$ and $zy_0(z) = -\cos z$. The total energy is conserved,

$$E = \varepsilon_n + \frac{\hbar^2 k_n^2}{2\mu} = \varepsilon_{n'} + \frac{\hbar^2 k_{n'}^2}{2\mu}. \tag{12.130}$$

We will analyze the wave function for small kinetic energy in the incoming channel, $k_n \to 0$. This means that all inelastic processes must be exothermic, and the kinetic energy in the outgoing channel $|n'\ell'\rangle$, namely $\hbar^2 k_{n'}^2/2\mu$, is determined by $\varepsilon_n - \varepsilon_{n'} \gg \hbar^2 k_n^2/2\mu$. Since this implies that k_n' is approximately constant under these assumptions, we may assume that $f_{n'\ell'}(r)$ and $g_{n'\ell'}(r)$ are energy independent. As in the derivation for s-wave scattering, we assume that the kinetic energy of the incoming channel may also be neglected when propagating the log-derivative matrix $\boldsymbol{Y}(r_0)$ from $r = 0$ to $r = r_0$. When matching the wave function at $r = r_0$ to the K-matrix boundary conditions, we assume that for the incoming channel $k_n r_0 \ll 1$ and the regular and irregular functions are replaced by the leading term in the Taylor expansion. For the Bessel functions we have

$$j_\ell(k_n r) \approx (k_n r)^\ell, \tag{12.131}$$

$$y_\ell(k_n r) \approx (k_n r)^{-(\ell+1)}, \tag{12.132}$$

and hence for the regular and irregular waves around $r = r_0$,

$$f_{n\ell}(r) \propto k_n^{\ell+\frac{1}{2}}, \tag{12.133}$$

$$g_{n\ell}(r) \propto k_n^{-(\ell+\frac{1}{2})}. \tag{12.134}$$

The defining equation for the log-derivative matrix, Eq. (12.73), shows that $\boldsymbol{Y}(r_0)$ is energy independent only if each column of the matrix $\boldsymbol{U}(r_0)$ has the same dependence on k_n, since the log-derivative matrix is independent of scaling of the columns of $\boldsymbol{U}$. The k_n dependence of the matrix elements $K_{n'\ell';n\ell}$ therefore follows from Eq. (12.127) and from the k_n dependence of the regular and irregular waves for the elastic and inelastic channels. For the elastic K-matrix elements for the incoming channel with small kinetic energy we obtain

$$k_n^{\ell+\frac{1}{2}} \propto k_n^{-(\ell'+\frac{1}{2})} K_{n\ell';n\ell}, \tag{12.135}$$

or

$$K_{n\ell';n\ell} \propto k_n^{\ell+\ell'+1}. \tag{12.136}$$

Since we assumed the irregular waves for exothermic channels to be energy independent, we find for inelastic matrix elements

$$K_{n'\ell';n\ell} \propto k_n^{\ell+\frac{1}{2}}. \tag{12.137}$$

In the K-matrix there are also elements that relate two exothermic channels, and these must of course be independent of k_n. Hence, the K-matrix, which is real and symmetric, has a block structure

$$\boldsymbol{K} = \begin{pmatrix} \boldsymbol{K}_{n,n} & \boldsymbol{K}^T_{n',n} \\ \boldsymbol{K}_{n',n} & \boldsymbol{K}_{n',n'} \end{pmatrix}. \tag{12.138}$$

Because of the energy independent (n', n') block we cannot directly use the analogue of Eq. (12.118), but we use instead

$$\boldsymbol{T} = \boldsymbol{I} - \boldsymbol{S} = \boldsymbol{I} - \frac{\boldsymbol{I} - i\boldsymbol{K}}{\boldsymbol{I} + i\boldsymbol{K}} = 2i(\boldsymbol{I} + i\boldsymbol{K})^{-1}\boldsymbol{K}, \tag{12.139}$$

and, together with the general expression for the inverse of a block matrix

$$\begin{pmatrix} \boldsymbol{A} & \boldsymbol{B} \\ \boldsymbol{C} & \boldsymbol{D} \end{pmatrix}^{-1} = \begin{pmatrix} (\boldsymbol{A} - \boldsymbol{B}\boldsymbol{D}^{-1}\boldsymbol{C})^{-1} & -\boldsymbol{A}^{-1}\boldsymbol{B}(\boldsymbol{D} - \boldsymbol{C}\boldsymbol{A}^{-1}\boldsymbol{B})^{-1} \\ -\boldsymbol{D}^{-1}\boldsymbol{C}(\boldsymbol{A} - \boldsymbol{B}\boldsymbol{D}^{-1}\boldsymbol{C})^{-1} & (\boldsymbol{D} - \boldsymbol{C}\boldsymbol{A}^{-1}\boldsymbol{B})^{-1} \end{pmatrix}, \tag{12.140}$$

one may derive for the T-matrix elements for low energy elastic scattering

$$\left|T_{n\ell';n\ell}\right|^2 \propto k_n^{2\ell+2\ell'+2}, \tag{12.141}$$

and for low energy inelastic scattering

$$\left|T_{n'\ell';n\ell}\right|^2 \propto k_n^{2\ell+1}. \tag{12.142}$$

Hence, for elastic cross sections we find

$$\sigma_{n\ell'\leftarrow n\ell} \propto k_n^{2\ell+2\ell'}, \tag{12.143}$$

and for the inelastic cross sections

$$\sigma_{n'\ell'\leftarrow n\ell} \propto k_n^{2\ell-1}. \tag{12.144}$$

We find again that for low energy s-wave scattering, $\ell = \ell' = 0$, the elastic cross sections are energy-independent. The s-wave inelastic cross sections depend on the kinetic energy as $E_{\text{kin}}^{-\frac{1}{2}}$, which, as discussed in Section 12.2, results in a temperature independent rate constant. A temperature independent rate constant was also found for the classical Langevin ion-molecule capture rate, but that was the result of classical motion on the long range $1/r^4$ potential. The effect of the long range potential in the quantum regime is discussed in ref. [832]. It is concluded that the expression for low energy inelastic T-matrix elements, Eq. (12.142), is valid if the potential falls off in the long range more rapidly than $1/r^2$. The expression for single channel elastic scattering, Eq. (12.141), is valid if $\ell > (n\text{–}3)/2$ for a long range potential $-c_n/r^n$. Otherwise $T_{\ell,\ell} \propto k^{n-2}$.

For processes that result in a change of Δm in the angular momentum projection quantum number, but which do not change the internal energy, the threshold law for the cross section is[835]

$$\sigma_{\Delta m} \propto k_n^{2\Delta m} \quad \text{for } \Delta m \quad \text{even}, \tag{12.145}$$

and

$$\sigma_{\Delta m} \propto k_n^{2(\Delta m+1)} \quad \text{for} \quad \Delta m \quad \text{odd.} \tag{12.146}$$

By way of an illustration, we discuss the case of the scattering of OH by ^{3}He in a weak magnetic field. To start with, it is helpful to consider the energy level structure of OH($X^2\Pi_{3/2}$) in the presence of such a field, as illustrated in Figure 12.7. The figure shows the energies of the two lowest $j = 3/2$ states of OH($X^2\Pi_{3/2}$) with odd and even parity, labeled *e* and *f* respectively. These levels are $2j+1=4$ fold degenerate. In a magnetic field the degeneracy of the $m_j = 3/2, 1/2, -1/2, -3/2$ levels is lifted, as shown in the figure. The $m_j = 3/2$, *e* level is called a 'low-field seeking' state, and it can be trapped in a magnetic trap with minimum field strength in the centre. Such a trap can be created with two magnetic coils in the anti-Helmholtz configuration. The trap can be loaded with OH radicals using the buffer gas loading technique, in which the radicals are cooled by collisions with cryogenically cooled ^{3}He buffer gas of approximately 0.5 K. More details of buffer-gas cooling are given in Study Box 12.1. Collisions of OH with helium atoms can produce $m_j < 0$, 'high field seeking' states, which are expelled from the trap. Elastic collisions between helium and OH are necessary to cool the OH and bring it into thermal equilibrium with the helium buffer gas. Note that elastic collisions do not exchange energy in the CM-frame, where each collision is referenced to the velocity of the centre of mass of the collision partners, but in the LAB-frame they are responsible for producing thermal equilibrium (see Section 5.2.2).

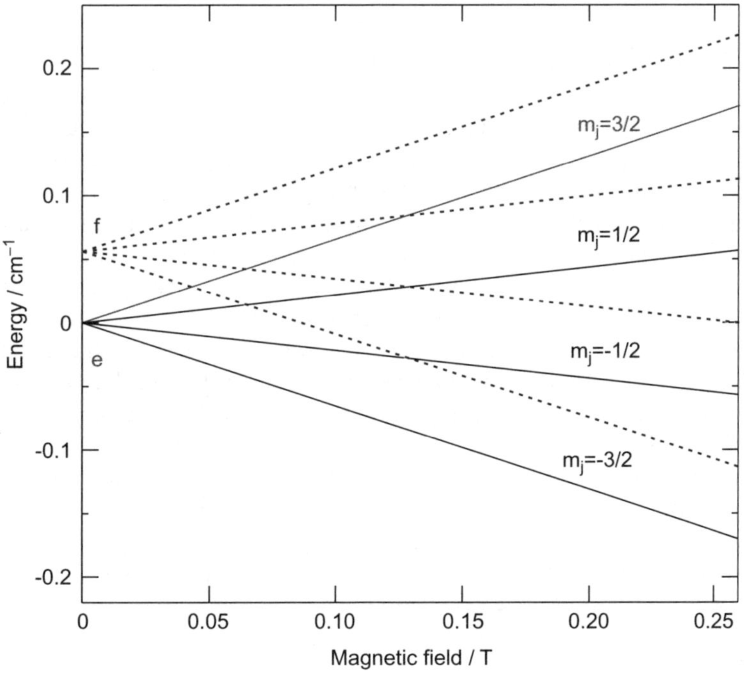

Figure 12.7 Illustration of the Zeeman effect for the lowest few rotational levels of the OH radical. The OH radical has a $^2\Pi_{3/2}$ ground state. The total electron spin *S* is 1/2, and the total angular momentum $j = 3/2$ has projections $\Omega = \pm 3/2$ on the diatomic axis. The two $|\Omega| = 3/2$ levels combine to give states of odd and even parity, labeled *e* and *f*, respectively, in the figure. (The spectroscopic parity of the states varies as $(-1)^{j-S}$, as is discussed further in Chapter 5.) See text for further details. Adapted from refs. [833,834].

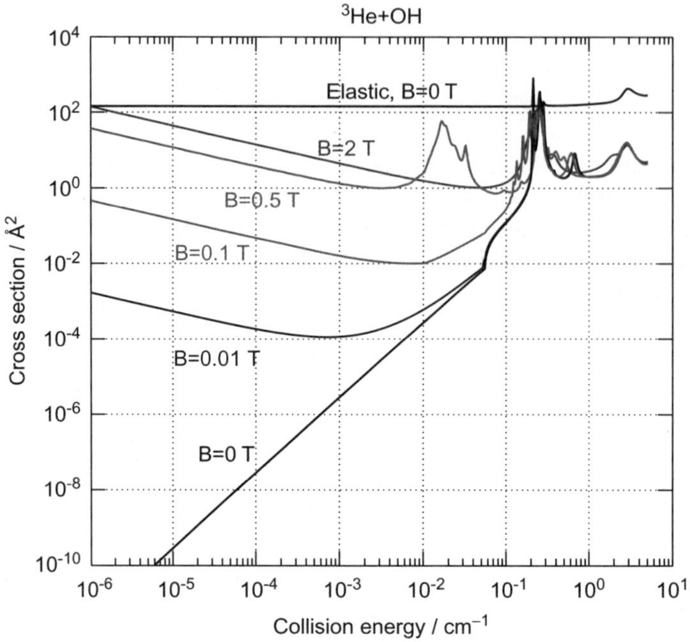

Figure 12.8 Cross sections for elastic and inelastic $^3He + OH(^2\Pi_{3/2}, m_j = 3/2, e)$ collisions, at several magnetic field strengths. See text for details. Adapted from refs. [833,834].

Cross sections for elastic and inelastic $^3He + OH(^2\Pi_{3/2}, m_j = 3/2, e)$ collisions, at several magnetic field strengths, are shown in Figure 12.8.[833,834] In the inelastic collisions the m_j quantum number can change. The parity of the OH radical can change if the collision energy and/or the magnetic field is sufficiently strong (see Figure 12.7). For collision energies above approximately 0.1 cm^{-1} many resonances are found in the inelastic cross sections. The ratio between elastic and inelastic cross sections approaches unity, which is very unfavorable for a buffer gas cooling experiment. At lower energies a resonance is seen at an energy of around 0.01–0.1 cm^{-1} for a magnetic field strength, B, of $B = 0.5$ T. For $B = 0$ and 0.01 T the onset of parity changing collisions is seen around 0.05 cm^{-1}. For lower collision energies the Wigner threshold laws are found: the elastic cross sections (upper black curve for $B = 0$ T) become constant; and for $B > 0$ the m_j changing collisions are inelastic and $\sigma \propto E_c^{-1/2}$. For $B = 0$ the m_j levels are degenerate and the "reorientation threshold laws" apply, as given in Eqs. (12.145) and (12.146). For $\Delta m = 1$, 2 these equations give $\sigma \propto E_c^2$, as can be seen in the figure.

Finally, for exothermic reactive processes, the threshold law is the same as for exothermic inelastic processes. This quantum result does not rely on a capture model, and so it also applies when there is a reaction barrier.

It should also be noted that here we have only considered processes in which there are at most two reactants or two products. More complicated processes including, *e.g.*, three-body breakup are discussed in ref. [836].

12.5 ULTRACOLD PHENOMENA

To explore the world of ultracold phenomena, we will consider ultracold gases that are confined in space, *e.g.*, in a three-dimensional box. Experimentally, gases have been confined in magnetic traps, which can be modelled as three-dimensional harmonic oscillators. In a $T = 0$ K ideal Bose gas, *i.e.* a gas of non-interacting bosons, the translational motion of each particle is described by the same ground state wave function of the trap. This is not true for fermions, since two fermions cannot be simultaneously in the exact same quantum state.

So far in this chapter we have assumed that the rate of a reaction is determined by the collision rate and the two-particle collision cross section. When considering reactions in a $T = 0$ K Bose gas, this is no longer appropriate; since the bosons occupy the same state, the "collisions" occur simultaneously throughout the trap, and a full quantum description of the system is required.

For a macroscopic system, *e.g.*, a cubic box with a volume of 1 cm^3, the excitation energy to the first excited quantum state is extremely small. For example, for a sodium atom in such a box it would be on the order of 10^{-15} cm^{-1} (1 cm^{-1} corresponds to 1.44 K). However, as a result of a quantum statistical effect described in detail in Section 12.5.2, the ground state of the trap will acquire a macroscopic population at much higher temperatures if the density of the gas is sufficiently high. This effect is called Bose-Einstein condensation. Bose-Einstein condensates, or BECs, were first predicted by Bose for photons in 1924,[837] and the ideas were rapidly developed by Einstein to yield a complete theory of condensate formation for particles.[838] However, it was not until a full 70 years later that a group of scientists led by Eric Cornell and Carl Weiman at JILA provided the first experimental proof,[796] forming a BEC in a vapour of rubidium-87 atoms at a temperature of less than 170 nK.

Figure 12.9 shows data from the first experiments on Bose-Einstein condensation in a dilute gas of Rb atoms.[796] The atoms were optically pre-cooled, loaded into a magnetic trap, and then further cooled by evaporation. Below the critical temperature, the onset of BEC formation is clearly observed in the experimentally measured velocity distributions as the growth of a peak at extremely low velocity, corresponding to atoms in the lowest energy state within the trap. The peak has a finite width due to the Heisenberg uncertainty principle, with the width of the velocity distribution being determined by the confinement of the atoms in space within the trap.

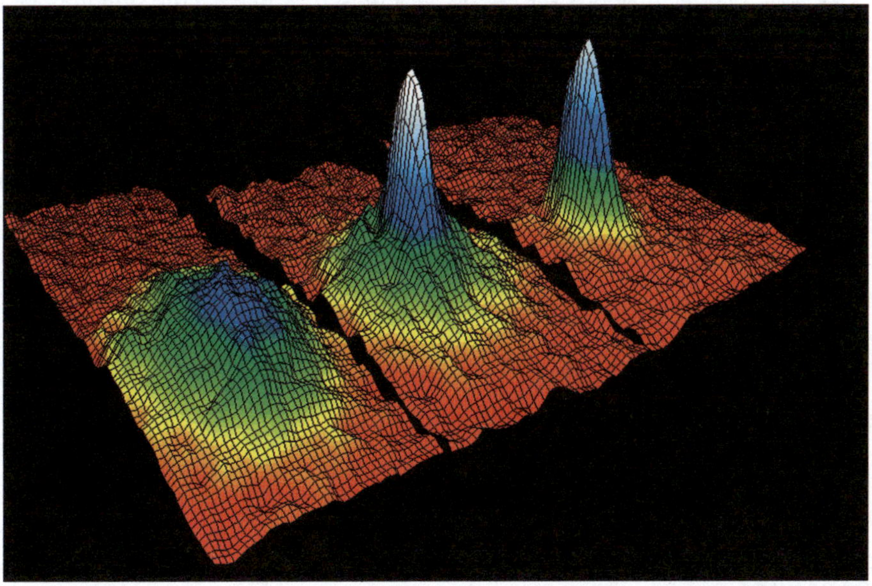

Figure 12.9 Velocity distribution data for a gas of Rb atoms, showing the onset of Bose-Einstein condensation below the critical temperature, T_c.[796] The three images show the velocity distribution just before the onset of BEC, just after the appearance of the condensate, and after further evaporation to leave behind a sample of nearly pure condensate. Reproduced from the NIST image gallery with permission from NIST/JILA/CU-Boulder.

Study Box 12.3 provides a brief introduction to this theory of Bose-Einstein condensation, which is discussed in more detail in the next few sections. In the next section we will derive the condensation phenomenon for an ideal Bose gas. Section 12.5.4, on the Gross-Pitaevskii equation, shows how the condensate wave function changes when interactions between the bosons are taken into account. In the last section we will show how the quantum statistics of bosons and fermions affects the rates of reactions.

12.5.1 Particle in a Box

The energy levels of a *one*-dimensional particle in a box with size a are given by

$$\varepsilon_n = \frac{\hbar^2 \pi^2}{2ma^2} n^2, \quad n = 1, 2, 3, \ldots, \tag{12.147}$$

where m is the mass of the particle. The corresponding wave functions are

$$\phi_n(x) = \sqrt{\frac{2}{a}} \sin\left(n \frac{\pi x}{a}\right). \tag{12.148}$$

The canonical partition function for N non-interacting *distinguishable particles* at a temperature T is

$$q_t(T) = \left(\sum_{n=1}^{\infty} \mathrm{e}^{-\beta \varepsilon_n} \right)^N, \tag{12.149}$$

where $\beta^{-1} = k_B T$. For a large cubic box the energy spacings are small, so the sum may be replaced by the integral

$$q_t(T) = \left(\int_0^{\infty} \mathrm{e}^{-\beta \frac{\hbar^2 \pi^2}{2ma^2} n^2} \, \mathrm{d}n \right)^N = \left(\frac{a}{\Lambda} \right)^N, \tag{12.150}$$

provided that the box is sufficiently large, or that the temperature is sufficiently high (and β small). In Eq. (12.150), Λ is the thermal de Broglie wavelength, defined as

$$\Lambda = \left(\frac{2\pi\hbar^2}{mk_B T} \right)^{1/2}. \tag{12.151}$$

For a three-dimensional box the energy levels are

$$\varepsilon_n = \varepsilon_{n_x} + \varepsilon_{n_y} + \varepsilon_{n_z}, \tag{12.152}$$

where ε_{n_i} are the energies associated with motion along the x, y, and z directions, as given in Eq. (12.147), and the wave functions are

$$\phi_n(\boldsymbol{r}) = \phi_{n_x}(x)\phi_{n_y}(y)\phi_{n_z}(z), \tag{12.153}$$

where ϕ_{n_i} are the one dimensional wavefunctions as given in Eq. (12.148).

STUDY BOX 12.3: AN INTRODUCTION TO THE THEORY OF BOSE-EINSTEIN CONDENSATION

Explaining the phase transition that leads to formation of a BEC in a gas of identical bosons below a critical temperature T_c requires a little statistical mechanics. You will already be familiar with the Boltzmann distribution, which describes the thermal distribution of atomic or molecular quantum states at a given temperature. The Boltzmann distribution becomes inaccurate at low temperatures, and we must use instead the Fermi-Dirac distribution (for fermions, with half integer spin), or the Bose-Einstein distribution (for bosons, with integer spin). According to Bose-Einstein statistics, the number of bosons n_i in a given state with energy ε_i is

$$n_i = \frac{1}{\exp\left(\frac{\varepsilon_i - \mu}{k_B T}\right) - 1}, \quad \text{(B12.3.1)}$$

where μ is the chemical potential. The total number of particles is $N = \Sigma_i n_i$. Since n_i cannot take negative values, it is clear from inspection of Eq. (B12.3.1) that this distribution only makes sense if μ is smaller than the lowest possible value of ε_i. Also, at sufficiently low temperature, this condition on μ is not compatible with conservation of the total number of particles, since as $T \to 0$, n_i becomes vanishingly small. The chemical potential reduces as the temperature is lowered (recall that μ is essentially the molar Gibbs free energy, that $\mathrm{d}G/\mathrm{d}T = -S$, and that S is always positive), and will eventually reach zero if $\varepsilon_0 = 0$. For N particles in a volume V, there is therefore a critical temperature T_c defined by the condition

$$N = \sum_i \frac{1}{\exp\left(\frac{\varepsilon_i - \mu}{k_B T_c}\right) - 1}. \quad \text{(B12.3.2)}$$

Because for a free particle the translational energy levels are so closely spaced as to form essentially a continuum, we can replace the sum in the above expression by an integral, and it then becomes possible to solve the resulting integral equation to find T_c. The solution is

$$T_c = \frac{\alpha \hbar^2}{m k_B} n^{2/3}, \quad \text{(B12.3.3)}$$

where m is the boson mass, $n = N/V$ is the number density, and the constant α is given by

$$\alpha = \frac{2\pi}{[\zeta(3/2)]^{2/3}} \approx 3.3, \quad \text{(B12.3.4)}$$

where $\zeta(x)$ is the Riemann zeta function.

Below the critical temperature, summing the state populations given by the Bose-Einstein distribution (Eq. (B12.3.1)) yields a number less than the total number of atoms N. Einstein postulated that the 'excess' atoms must be scattered into the ground state, corresponding to the lowest value of ε_i. At $T = T_c$, the fraction of atoms in this state is zero, rising to 1 at $T = 0$; quantitatively, we have for the fraction in the ground state $f = 1 - (T/T_c)^{3/2}$. This is the phenomenon known as Bose-Einstein condensation.

The condition for Bose-Einstein condensation may be reformulated in terms of the thermal de Broglie wavelength, $\Lambda = \sqrt{2\pi\hbar^2/(mk_\mathrm{B}T)}$, giving

$$\Lambda_\mathrm{c} = 1.377\left(\frac{V}{N}\right)^{1/3} \tag{B12.3.5}$$

BEC therefore occurs when the thermal wavelength associated with each particle becomes larger than the interparticle separation. The particles then 'overlap' in space and condense into a single quantum mechanical entity.

Claire Vallance

12.5.2 Bose-Einstein Condensation

To describe Bose-Einstein condensation we will use the grand canonical partition function

$$Z(V, T, \mu) = \sum_{N=0}^{\infty} \mathrm{e}^{\beta\mu N} Q(N, V, T), \tag{12.154}$$

where V is the volume of the container, μ is the chemical potential,* N is the number of bosons in the sample, and Q is the canonical partition function. For an ideal Bose gas, this expression becomes[839]

$$Z(V, T, \mu) = \prod_k \frac{1}{1 - \lambda \mathrm{e}^{-\beta\varepsilon_k}}, \tag{12.155}$$

where k runs over all single particle energy levels and λ is the fugacity

$$\lambda = \mathrm{e}^{\beta\mu} = \mathrm{e}^{\mu/k_\mathrm{B}T}. \tag{12.156}$$

The equation of state of the ensemble is

$$pV = k_\mathrm{B}T \ln Z, \tag{12.157}$$

where p is the pressure. The connection with thermodynamics is made through the relation

$$\mathrm{d}(pV) = S\,\mathrm{d}T + N\,\mathrm{d}\mu + p\,\mathrm{d}V, \tag{12.158}$$

where S is the entropy. For the total number of particles in the system N we find,

$$N = \left.\frac{\mathrm{d}(pV)}{\mathrm{d}\mu}\right|_{V,T} = k_\mathrm{B}T\left(\frac{\partial \ln Z}{\partial \mu}\right)_{V,T} = \sum_k \frac{\lambda \mathrm{e}^{-\beta\varepsilon_k}}{1 - \lambda \mathrm{e}^{-\beta\varepsilon_k}} = \sum_k \bar{n}_k. \tag{12.159}$$

*It should be clear from the context whether or not the symbol μ is referring to the chemical potential, as in this section, or to the reduced mass, as used in previous sections on collision theory, but take care to avoid confusion on this point.

Here $\bar{n}_k$ represents the average population of state k, with energy ε_k, and is defined as

$$\bar{n}_k = \frac{\lambda e^{-\beta\varepsilon_k}}{1 - \lambda e^{-\beta\varepsilon_k}}. \tag{12.160}$$

The fugacity λ and the related chemical potential μ are simply parameters that, together with the temperature and the energy levels ε_k, determine the populations $\bar{n}_k$, and thus the total number of particles N, through Eqs. (12.159) and (12.160). The possible values of λ and μ are determined by the requirement that populations cannot be negative. This condition is

$$0 \leq \lambda e^{-\beta\varepsilon_k} = e^{-\beta(\varepsilon_k - \mu)} < 1. \tag{12.161}$$

With ε_0 defined as the lowest energy, the condition is $\mu < \varepsilon_0$. In the expressions only $\varepsilon_k - \mu$ appears, so we can take the zero of energy as $\varepsilon_0 = 0$, and thus $\mu < 0$ and $0 \leqslant \lambda < 1$. When λ is small, the populations of the states are proportional to $e^{-\beta\varepsilon_k}$, which corresponds to the classical Maxwell-Boltzmann distribution. Notice that in this case the chemical potential $\mu \ll 0$, even though the interaction between the particles is zero for our ideal Bose gas. Next, consider the population of the ground state with $\varepsilon_0 = 0$,

$$\bar{n}_0 = \frac{\lambda}{1 - \lambda}. \tag{12.162}$$

When λ approaches 1 (μ approaches 0), the population of the ground state $\bar{n}_0$ can become arbitrarily large. The total number of particles N is the sum of $\bar{n}_0$ and the total population of the excited states (N–$\bar{n}_0$). We will now determine N–$\bar{n}_0$ as a function of λ in the statistical limit. This means that for a given temperature T, or a given β, we assume the box to be so large that the energy spacings between the levels ε_k are small compared to $k_B T$. Thus, for the lowest excited state, which is three fold degenerate, we assume that $\beta\varepsilon_1 \ll 1$, such that $e^{-\beta\varepsilon_1} \approx 1 - \beta\varepsilon_1$. We note that the population of the ground state can only be considerably larger than that of $k = 1$ and higher states if λ is very close to 1. If we write $\lambda = 1 - \delta_\lambda$, where $\delta_\lambda \approx 0$, we find that the ground state population is

$$\bar{n}_0 = \frac{\lambda}{1 - \lambda} \approx \frac{1}{\delta_\lambda}, \tag{12.163}$$

and the population of the first excited state is (from Eq. (12.160))

$$\bar{n}_1 \approx \frac{(1 - \delta_\lambda)(1 - \beta\varepsilon_1)}{1 - (1 - \delta_\lambda)(1 - \beta\varepsilon_1)} \approx \frac{1}{\delta_\lambda + \beta\varepsilon_1}. \tag{12.164}$$

Hence, $\bar{n}_0$ can only be considerably larger than $\bar{n}_1$ if δ is small compared to $\beta\varepsilon_1$. When λ increases, the population $\bar{n}_k$ of each level increases, since the numerator in Eq. (12.160) increases with λ and the denominator decreases with λ. We may therefore compute an upper limit to the population of all excited states by setting $\lambda = 1$,

$$N - \bar{n}_0 = \sum_{k>0} \frac{e^{-\beta\varepsilon_k}}{1 - e^{-\beta\varepsilon_k}}. \tag{12.165}$$

For bosons in a three-dimensional box, the energy levels ε_k are given by ε_n of Eq. (12.152) and the summation over k must be replaced by

$$\sum_k = \sum_{n_x=1}^{\infty} \sum_{n_y=1}^{\infty} \sum_{n_z=1}^{\infty}. \tag{12.166}$$

The energies ε_n can be written as bn^2, with $n^2 = n_x^2 + n_y^2 + n_z^2$ and

$$b = \frac{\hbar^2 \pi^2}{2ma^2}. \tag{12.167}$$

The summation can be approximated by an integral over one octant of a sphere,

$$\sum_{n_x>0} \sum_{n_y>0} \sum_{n_z>0} f(n) = \frac{\pi}{2} \int_0^\infty n^2 f(n) \mathrm{d}n, \tag{12.168}$$

or, with $\varepsilon = bn^2$ and $n^2 \, \mathrm{d}n = (1/2) b^{-3/2} \sqrt{\varepsilon} \, \mathrm{d}\varepsilon$,

$$N - \bar{n}_0 = \frac{\pi}{4} b^{-3/2} \int_0^\infty \frac{\sqrt{\varepsilon} \mathrm{e}^{-\beta\varepsilon}}{1 - \mathrm{e}^{-\beta\varepsilon}} \mathrm{d}\varepsilon. \tag{12.169}$$

To evaluate the integral we use the series expansion

$$(1 - x)^{-1} = \sum_{l=0}^{\infty} x^l, \quad \text{for } |x| < 1, \tag{12.170}$$

for $x = \mathrm{e}^{-\beta\varepsilon}$, which gives

$$N - \bar{n}_0 = \frac{\pi}{4} b^{-3/2} \sum_{l=1}^{\infty} \int_0^\infty \sqrt{\varepsilon} \mathrm{e}^{-\beta l \varepsilon} \mathrm{d}\varepsilon = \frac{\pi}{4} b^{-3/2} \sum_{l=1}^{\infty} \frac{\Gamma(3/2)}{(\beta l)^{3/2}}. \tag{12.171}$$

With $\Gamma(3/2) = \frac{1}{2}\sqrt{\pi}, a^3 = V$, the Riemann-zeta function defined by

$$\zeta(n) = \sum_{l=1}^{\infty} \frac{1}{l^n}, \tag{12.172}$$

and the de Broglie wavelength defined in Eq. (12.151), the contribution of the excited states to the density is at most

$$\rho_{\mathrm{ex}} = \frac{N - \bar{n}_0}{V} = \left(\frac{m k_{\mathrm{B}} T}{2\pi\hbar^2} \right)^{3/2} \zeta(3/2) = \Lambda^{-3} \zeta(3/2), \tag{12.173}$$

where $\zeta(3/2) \approx 2.612$. This equation shows that if the temperature is lowered, the excited states are less occupied. When ρ_{ex} drops significantly below the total density ρ, the ground state must accommodate the difference, *i.e.* $\rho_0 = \rho - \rho_{\mathrm{ex}}$, and a Bose-Einstein condensate is formed. This happens at the critical temperature T_{c}. The corresponding density is called the critical density ρ_{c}. Below the critical temperature the total density is given by

$$\rho = \rho_0 + \rho_{\mathrm{ex}} = \frac{1}{V} \frac{\lambda}{1 - \lambda} + \Lambda^{-3} \zeta(3/2). \tag{12.174}$$

The excited state density is related to the density ρ_{c} at the critical temperature through

$$\frac{\rho_{\mathrm{ex}}}{\rho_{\mathrm{c}}} = \left(\frac{\Lambda_{\mathrm{c}}}{\Lambda} \right)^3 = \left(\frac{T}{T_{\mathrm{c}}} \right)^{3/2}, \tag{12.175}$$

where Λ_c is the de Broglie wavelength at temperature T_c. Hence, the condensate fraction is

$$\frac{\rho_0}{\rho_c} = \frac{\rho_c - \rho_{ex}}{\rho_c} = 1 - \left(\frac{T}{T_c}\right)^{3/2}. \tag{12.176}$$

In this limit of a large volume there is a discontinuity in the derivative of the condensate fraction at $T = T_c$, which marks the phase transition. The process of Bose-Einstein condensation is explored in more detail in Study Box 12.4.

12.5.3 A Bose-Einstein Condensate in a Harmonic Trap

In contrast to the 'particle in a box' description of the previous sections, Bose-Einstein condensates of dilute gases created in experiments such as that shown in Figure 12.9 are not in practice confined by walls, but by magneto-optical traps. The confining field in such a trap may be modelled by a three-dimensional harmonic oscillator potential

$$V_{\text{trap}}(\boldsymbol{r}) = \tfrac{1}{2}\left(K_x x^2 + K_y y^2 + K_z z^2\right). \tag{12.177}$$

We will only consider isotropic potentials, for which the three force constants $K_x = K_y = K_z$ are equal to the same constant K. In this case, the one-particle energy levels are given by

$$\varepsilon_n = \left(n_x + n_y + n_z + \tfrac{3}{2}\right)\hbar\omega, \tag{12.178}$$

where $\omega = \sqrt{K/m}$ is 2π times the trap frequency, and the quantum numbers n_x, n_y, and n_z are non-negative integers. The ground state one-particle wave function is given by

$$\phi(\boldsymbol{r}) = \left(\frac{m\omega}{\pi\hbar}\right)^{3/4} \exp\left(-\frac{1}{2}\frac{m\omega}{\hbar} r^2\right). \tag{12.179}$$

The relation between the number of particles and the critical temperature is

$$k_B T_c = \hbar\omega\left(\frac{N}{\zeta(3)}\right)^{1/3} \approx 0.94\hbar\omega N^{1/3}, \tag{12.180}$$

and the formula for the condensate fraction is

$$\frac{N_0}{N} = 1 - \left(\frac{T}{T_c}\right)^3. \tag{12.181}$$

The derivation of this relation is similar to that of Eq. (12.176) in the previous section, and may be found in, *e.g.*, ref. [840].

12.5.4 The Gross-Pitaevskii Equation

In this section we develop the Schrödinger equation for the condensate, known as the Gross-Pitaevskii equation, the solution of which provides the condensate wave function. For N

STUDY BOX 12.4: A SIMULATION OF BOSE-EINSTEIN CONDENSATION

Here we describe a simple numerical simulation of Bose-Einstein condensation in a three-dimensional cubic box. Readers with some programming experience may like to try writing a code to carry out this simulation themselves. The results of the simulation are shown in Figure 12.10. In the calculations, the mass, m, is set to the mass of a sodium atom, and the temperature, T, is fixed at a value corresponding to a thermal de Broglie wavelength of $\Lambda = 100\ a_0$. The solid lines in the figure correspond to a large box of volume $V = a^3 = 1000\ \Lambda^3$ and the dashed lines correspond to a small box of volume $20\ \Lambda^3$. The dimensionless parameter χ is defined by

$$\chi = \frac{N^{1/3}\Lambda}{a} = \rho^{1/3}\Lambda, \tag{B12.4.1}$$

so at $\chi = 1$ there is one particle per Λ^3. In the simulation, all states with energies less than $15\ k_B T$ are included. To make the plot, a set of λ values is chosen and the populations $\bar{n}_k$ follow from Eq. (12.160), after which $N = \Sigma_k \bar{n}_k$ and χ can be computed. (For those interested in trying this for themselves it will be necessary to find a good grid of λ (or μ) values, which will require careful study of Section 12.5.2.)

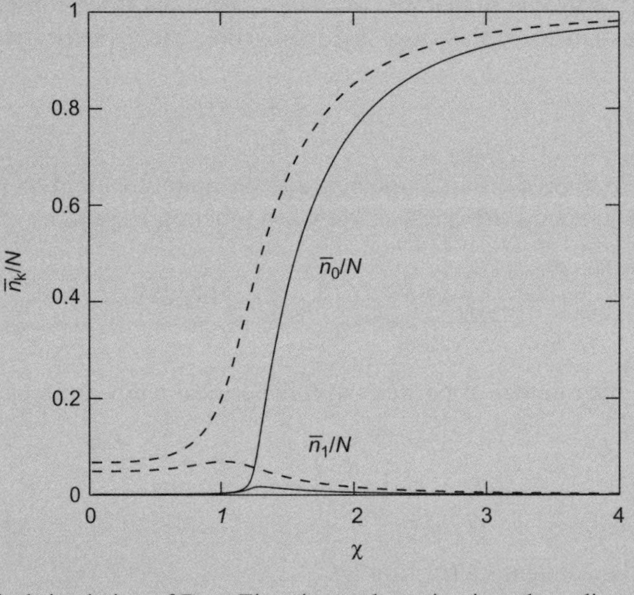

Figure 12.10 Numerical simulation of Bose-Einstein condensation in a three-dimensional cubic box. $\bar{n}_k/N$ represents the fractional population in level k, while χ is a dimensionless parameter which is inversely proportional to the size of the box. The solid lines correspond to a box of volume $V = a^3 = 1000\ \Lambda^3$, while the dashed lines correspond to a box of volume $20\ \Lambda^3$. See the text of this study box for details.

The plot in Figure 12.10 shows that at this fixed temperature T the fractional population of the ground state $\bar{n}_0/N$ approaches unity as the density (or χ) increases. The plot shows that the fractional population of the first excited state $\bar{n}_1/N$ remains small at all densities. Note that $\bar{n}_{k+1} < \bar{n}_k$ for all k. Also, for the larger box the maximum population of the first excited state is less than for the smaller box (dashed line). At infinite box size the fractional populations of all excited states are infinitesimal.

Gerrit Groenenboom

STUDY BOX 12.5: COLD AND ULTRACOLD CHEMISTRY

Chemists tend to be very familiar with chemistry occurring at or near room temperature or higher, a regime in which reaction rates are described by the Arrhenius equation, and kinetics and dynamics may be understood in terms of passage over energetic barriers along the reaction coordinate of the appropriate potential energy surface. When considering chemistry at very low temperatures, we enter an entirely different regime. Molecules simply do not have enough energy to surmount energetic barriers, and only barrierless reactions, such as ion-molecule or radical-radical reactions, are able to proceed through a classical mechanism. When a small barrier is present, reaction may still sometimes proceed, but only *via* a mechanism involving quantum-mechanical tunnelling through the barrier, leading to novel kinetic and dynamical behaviour. At still lower temperatures, the thermal wavelengths of the reacting molecules become larger than the molecules themselves, and the concept of reaction occurring as a result of a classical collision breaks down entirely. For example, at temperatures in the milliKelvin range, thermal wavelengths are often several nanometres or even tens of nanometres. Reactions can no longer be viewed in terms of a collision between two 'hard spheres', but must be considered in terms of interference between two waves.

The lowest known temperatures found in nature are in interstellar space. As noted in the Introduction to this chapter, the cosmic background radiation corresponds to a temperature of around 2.7 K, which sets an approximate lower limit to the temperatures a molecule might experience in space. Temperatures in interstellar gas clouds, in which most of the molecules known to exist in space have been observed, range from around 100 K near the edges of a cloud down to around 10 K in the 'dense' core. Temperatures down to a few Kelvin are the regime of 'cold chemistry', and the desire to understand the formation of molecular species in interstellar space has sparked a great deal of interest in chemistry at these temperatures.

Many reactions of interstellar significance have been studied using the CRESU technique.[829] The acronym stands for the French 'Cinétique de Réaction en Ecoulement Supersonique Uniforme', or (in English) Reaction Kinetics in Uniform Supersonic Flow. The technique is based on expanding a high-pressure gas through a specially shaped nozzle called a de Laval nozzle. The nozzle is designed to collimate the gas into a uniform, collision-free, supersonic beam at a characteristic temperature, and by adjusting the nozzle design, gas flows with temperatures from room temperature down to around 10 K may be achieved. The gas density in the expansion is relatively high (in the range from 10^{16} to $10^{18}\,\text{cm}^{-3}$), so that thermal equilibrium is maintained at all times, and the flow is uniform with respect to temperature, density, and velocity. This is in marked contrast to a pulsed molecular beam, in which no thermal equilibrium, and therefore no well-defined temperature, is established. Chemical kinetics measurements within the gas flow are generally performed using laser pump-probe techniques (see Chapter 1 and Study Box 5.2). The pump laser typically generates a reactive species through photolysis of a precursor molecule, and product detection in the probe step is often through laser-induced fluorescence (see Study Box 5.2). The CRESU technique is typically used to study reactions with essentially no activation barrier, in which the long-range part of the intermolecular potential provides the energetic driving force. These types of reactions are often well-described by the capture theories treated in detail in this chapter, and include radical reactions and ion-molecule reactions. A wide range of astrochemically relevant reactions have been studied using the CRESU technique.

Ion-molecule reactions have also been widely studied using the Selected-Ion Flow Tube (SIFT) technique,[845] in which mass-selected ions are flowed down a drift tube in an inert carrier gas and react with molecular species injected downstream of the ion inlet. Reaction products are then detected by a mass spectrometer.

Temperatures below 1 K take us into the regime of 'ultracold chemistry', and have only been achieved on earth in the laboratory. Study Box 12.1 outlined some of the ways in which molecular species may be cooled to such temperatures. In principle, any of these techniques could be used to create cold reactants for a scattering experiment. However, in practice studying the chemistry of these molecules presents huge challenges. Ultracold molecules can generally only be produced at extremely low number densities, orders of magnitude lower than those typically used in a crossed molecular beam experiment, so that exquisitely sensitive detection techniques are required in order to monitor the very small number of products formed through collisions. For this reason there have been very few successful experimental studies of ultracold chemistry. We will look at two examples.

Our first example involves a general technique for studying ultracold ion-molecule reactions in an electromagnetic trap. When ions are laser-cooled and trapped, they form a crystalline array in which the positions of the ions are approximately fixed (with some small degree of motion caused by the oscillating trap potentials). The relative positions of the ions are determined by the balance between the confining forces of the trap and the Coulomb repulsion of the ions, with typical spacings on the order of 10–20 μm. During the laser cooling process, the ions are continuously absorbing and emitting photons, and by detecting the emitted fluorescence, an image of the crystal may be obtained. Such an image is shown in Figure 12.11. Coulomb crystals may be stored for several hours in a suitable vacuum, and provide an ideal and general method for studying ultracold collisions between ions and other chemical species. When an individual ion reacts to form another chemical species, it is no longer resonant with the cooling laser, and its fluorescence is extinguished. For example, Softley *et al.*[846] have combined Coulomb crystals with a velocity-selected molecular beam to study the kinetics of the reaction $Ca^+ + CH_3F \rightarrow CaF^+ + CH_3$ at relative translational temperatures down to 1 K.

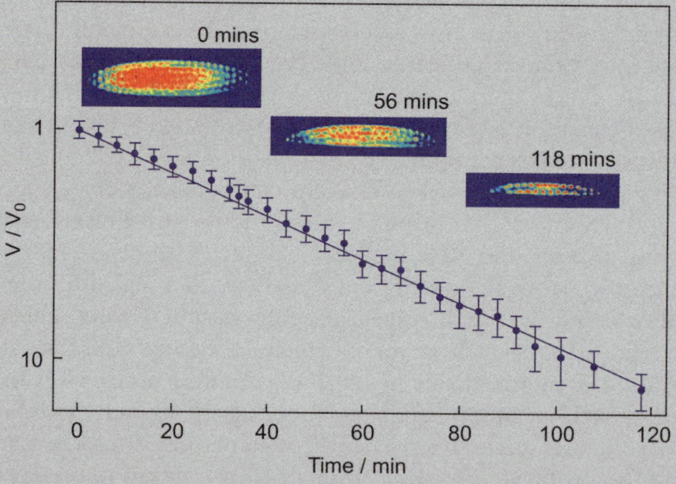

Figure 12.11 A logarithmic plot of the ratio of volumes $V(t)/V(0)$ occupied by Ca^+ ions of a Coulomb crystal containing 250 ions as a function of the reaction time. The straight line satisfies the pseudo-first-order rate law $V(t) = V(0)\exp - (N_{CH_3F}k_2t)$, where N_{CH_3F} is the number density of CH_3F, and k_2 is the bimolecular rate constant for the reaction between CH_3F and the Ca^+ ions trapped in the crystal. The inset pictures are images of the Coulomb crystal at a series of times, and illustrate the depletion of the Ca^+ ions due to reaction with CH_3F. Adapted from ref. [846].

Our second example demonstrates the dominant effects of quantum mechanical phenomena on chemical reactivity at ultralow temperatures. Ospelkaus *et al.*[847] have recently provided one of the first experimental demonstrations of chemistry in the nanoKelvin regime with their observation of the exchange reaction $KRb + KRb \rightarrow K_2 + Rb_2$ in an optical trap. In this temperature regime there is no energy available to overcome energetic barriers, including centrifugal barriers, which means that reaction can only occur at very small impact parameters, corresponding to orbital angular momentum quantum numbers of $\ell=0$ (s-wave scattering) or $\ell=1$ (p-wave scattering). The $^{40}K^{87}Rb$ reactant molecules are fermions, and according to the Pauli principle the total wavefunction for the KRb + KRb system must therefore be antisymmetric with respect to molecular exchange. When the molecules are prepared in the same internal state, it turns out that the p-wave ($\ell=1$) channel is the lowest symmetry allowed channel. However, the centrifugal barrier for this channel is more than an order of magnitude larger than $k_B T$, and reaction can only occur slowly *via* tunnelling through the centrifugal barrier. If the two KRb molecules are prepared in different internal states (by flipping the orientation of the nuclear spins), the symmetry constraint is lifted. s-wave scattering, for which there is no centrifugal barrier, becomes allowed, and the reaction rate increases by a factor of 10 to 100.

Claire Vallance

non-interacting bosons this wave function is the Hartree product,

$$\Psi(\boldsymbol{r}_1, \boldsymbol{r}_2, \ldots, \boldsymbol{r}_N) = \prod_{i=1}^{N} \phi(\boldsymbol{r}_i), \tag{12.182}$$

where the single particle wave function $\phi(\boldsymbol{r})$ is normalized,

$$\langle\phi|\phi\rangle = \int |\phi(\boldsymbol{r})|^2 \mathrm{d}\boldsymbol{r} = 1. \tag{12.183}$$

For particles in a cubic box, $\phi(\boldsymbol{r})$ is the ground state wave function $\phi_{0,0,0}$ from Eq. (12.153), and for an isotropic three-dimensional harmonic trap it is given by Eq. (12.179). To find an approximate wave function for the condensate in the presence of interactions between the particles, we will use the Hartree product (*i.e.* the non-interacting wave function) as an *ansatz* (starting function) for the wave function and variationally optimize the single particle wave function $\phi(\boldsymbol{r})$. This is a mean-field approach, and it is analogous to the Hartree-Fock approach for fermions. We consider a pairwise additive interaction between the particles, with the effective pair potential of Eq. (12.124)

$$\hat{U} = U_0 \sum_{i<j} \delta(\boldsymbol{r}_i - \boldsymbol{r}_j), \tag{12.184}$$

where U_0 is determined by the scattering length, $U_0 = 4\pi a\hbar^2/\mu$. The total Hamiltonian of the gas also contains the kinetic energy of the particles and the trap potential $V_{\mathrm{trap}}(\boldsymbol{r})$,

$$\hat{H} = \sum_{i=1}^{N} \left[\frac{\boldsymbol{p}_i^2}{2m} + V_{\mathrm{trap}}(\boldsymbol{r}_i)\right] + \hat{U}, \tag{12.185}$$

where $\boldsymbol{p} = \frac{\hbar}{i}\nabla$. For the trap potential the expectation value is given by

$$\langle\Psi|\sum_{i=1}^{N} V_{\mathrm{trap}}(\boldsymbol{r}_i)|\Psi\rangle = N\langle\phi|V_{\mathrm{trap}}(\boldsymbol{r})|\phi\rangle = \boldsymbol{N} \int \boldsymbol{V}_{\mathrm{trap}}(\boldsymbol{r})|\phi(\boldsymbol{r})|^2 \mathrm{d}\boldsymbol{r}, \tag{12.186}$$

and for the kinetic energy it is

$$\langle\Psi|\sum_{i=1}^{N}\frac{\boldsymbol{p}_i^2}{2m}|\Psi\rangle = \frac{N}{2m}\langle\phi|\boldsymbol{p}^2|\phi\rangle = \frac{N}{2m}\langle\boldsymbol{p}\phi|\boldsymbol{p}\phi\rangle = \frac{N\hbar^2}{2m}\int|\nabla\phi(\boldsymbol{r})|^2\mathrm{d}\boldsymbol{r}. \tag{12.187}$$

Finally, to evaluate the expectation value of the two-particle interaction operator for a single interaction between the particles i and j we use

$$\iint\phi(\boldsymbol{r}_i)^*\phi(\boldsymbol{r}_j)^*\delta(\boldsymbol{r}_i-\boldsymbol{r}_j)\phi(\boldsymbol{r}_i)\phi(\boldsymbol{r}_j)\mathrm{d}\boldsymbol{r}_i\mathrm{d}\boldsymbol{r}_j = \int|\phi(\boldsymbol{r}_i)|^4\mathrm{d}\boldsymbol{r}_i, \tag{12.188}$$

and note that the sum over all $i<j$ gives $N(N-1)/2$ identical contributions,

$$\langle\Psi|\hat{U}|\Psi\rangle = U_0\frac{N(N-1)}{2}\int|\phi(\boldsymbol{r})|^4\mathrm{d}\boldsymbol{r}. \tag{12.189}$$

We assume N to be large, so we set $N(N-1)/2\approx N^2/2$, and find for the total energy

$$E = N\int\left[\frac{\hbar^2}{2m}|\nabla\phi(\boldsymbol{r})|^2+V_{\text{trap}}(\boldsymbol{r})|\phi(\boldsymbol{r})|^2+\frac{1}{2}NU_0|\phi(\boldsymbol{r})|^4\right]\mathrm{d}\boldsymbol{r}. \tag{12.190}$$

Before we variationally minimize this energy expression, we consider the particles in a box problem, for which $V_{\text{trap}}(\boldsymbol{r})=0$. If we take the ground state particle in a box wave function [Eq. (12.153)] for $\phi(\boldsymbol{r})$ we find

$$E = \frac{3\pi^2\hbar^2}{2m}\rho a+\frac{27}{16}\rho^2VU_0, \tag{12.191}$$

where $V=a^3$ is the volume of the box, $\rho=N/V$ is the particle density, and we have used the integral

$$\int_0^a\sin^4\left(\frac{\pi x}{a}\right)\mathrm{d}x = \frac{3a}{8}. \tag{12.192}$$

We note that the second term in Eq. (12.191) scales with the volume of the box and the square of the density. Hence, if either the volume or the density is sufficiently high, this term will dominate. We may also compute the expectation value of the energy for the wave function for a homogeneous gas: $\phi(r)=V^{-\frac{1}{2}}$. Strictly, this wave function does not satisfy the particle in a box boundary conditions, but if we assume that the volume is large, we may neglect the effects at the boundary and we find

$$E = \frac{1}{2}\rho^2VU_0. \tag{12.193}$$

Thus, when the interaction is repulsive ($U_0>0$) and the volume and density are sufficiently high, we find that (invoking the variation principle) the homogeneous gas wave function describes the condensate better than the particle in a box wave function, since $\frac{1}{2}\rho^2VU_0<\frac{27}{16}\rho^2VU_0$. For attractive interactions ($U_0<0$) we observe that the nonlinear $|\phi(\boldsymbol{r})|^4$ term favours the least homogeneous solution, which in practice may result in collapse of the condensate.

We will now derive the Gross-Pitaevskii equation, which minimizes the energy of Eq. (12.190) variationally. It is convenient to introduce the function

$$\psi(\boldsymbol{r}) = N^{1/2}\phi(\boldsymbol{r}). \tag{12.194}$$

The normalization of ψ is such that the particle density is given by

$$\rho(\boldsymbol{r}) = |\psi(\boldsymbol{r})|^2. \tag{12.195}$$

The integral over the density is the total number of particles N. Thus, within the present *ansatz*, $\psi(\boldsymbol{r})$ contains all information about the condensate wave function and it is sometimes referred to as "the condensate wave function". The energy expression as a functional of ψ is

$$E[\psi] = \int \left[\frac{\hbar^2}{2m}|\nabla\psi(\boldsymbol{r})|^2 + V(\boldsymbol{r})|\psi(\boldsymbol{r})|^2 + \frac{1}{2}U_0|\psi(\boldsymbol{r})|^4\right]\mathrm{d}\boldsymbol{r}. \tag{12.196}$$

According to the variational principle, we must vary ψ to minimize the energy E, but we must satisfy the constraint that the number of particles

$$\int |\psi(\boldsymbol{r})|^2 \mathrm{d}\boldsymbol{r} = N[\psi] = N \tag{12.197}$$

is conserved. For this constraint minimization the Lagrange multiplier method is used. This amounts to introducing a new parameter, μ,* and performing the unconstrained minimization of $E[\psi]-\mu N[\psi]$, such that for first order variations $\delta\psi$,

$$E[\psi + \delta\psi] - \mu N[\psi + \delta\psi] = E[\psi] - \mu N[\psi]. \tag{12.198}$$

In principle the wave function may be complex, but instead of varying the real and imaginary part it is more convenient (and mathematically equivalent) to vary ψ and ψ^* separately. Therefore, we rewrite the energy expression as

$$E[\psi] = \int \left[-\frac{\hbar^2}{2m}\psi^*(\boldsymbol{r})\nabla^2\psi(\boldsymbol{r}) + V(\boldsymbol{r})\psi^*(\boldsymbol{r})\psi(\boldsymbol{r}) + \frac{1}{2}U_0\psi^*(\boldsymbol{r})^2\psi(\boldsymbol{r})^2\right]\mathrm{d}\boldsymbol{r}, \tag{12.199}$$

substitute $\psi^* \rightarrow \psi^* + \delta\psi^*$, and set terms linear in $\delta\psi^*$ equal to zero. The result

$$\left[-\frac{\hbar^2}{2m}\nabla^2 + V(\boldsymbol{r}) + U_0|\psi(\boldsymbol{r})|^2\right]\psi(\boldsymbol{r}) = \mu\psi(\boldsymbol{r}) \tag{12.200}$$

is known as the time-independent Gross-Pitaevskii equation. The variation $\psi \rightarrow \psi + \delta\psi$ can be used to show that μ must be real. If one finds a $\psi(\boldsymbol{r})$ that satisfies the Gross-Pitaevskii equation, one cannot normalize the result to obtain a condensate wave function for a given number of particles N, because the equation is nonlinear in ψ. Instead, one must choose a value for μ, find a solution $\psi(\boldsymbol{r})$, and determine the number of particles to which it corresponds.

*The choice of the symbol μ for this parameter will become obvious in the following discussion.

The parameter μ is the chemical potential, $\mu = \partial E/\partial N$. This follows from rewriting Eq. (12.198) as

$$\mu = \frac{E[\psi + \delta\psi] - E[\psi]}{N[\psi + \delta\psi] - N[\psi]} = \frac{\Delta E}{\Delta N}. \qquad (12.201)$$

12.5.5 Thomas-Fermi Approximation

We saw above that when the volume is large and the interaction U_0 is repulsive, the kinetic energy term in the Gross-Pitaevskii equation may be neglected. This is known as the Thomas-Fermi approximation. The solution is

$$|\psi(\boldsymbol{r})|^2 = \rho(\boldsymbol{r}) = \frac{\mu - V_{\text{trap}}(\boldsymbol{r})}{U_0} \qquad (12.202)$$

in regions where the right hand side is positive, and $\psi(\boldsymbol{r}) = 0$ otherwise. The discontinuity in the derivative of the wave function at the edge of the cloud defined by $V_{\text{trap}}(\boldsymbol{r}) = \mu$ is the result of neglecting the kinetic energy term. As before, we find that the density is constant when $V_{\text{trap}}(\boldsymbol{r}) = 0$. In that case we also find $\mu = NV^{-1}U_0$ and $E = N^2V^{-1}U_0/2$, which agrees with the definition of the chemical potential $\mu = \partial E/\partial N$.

The density profile of a Bose-Einstein condensate is very characteristic. The classical thermal Boltzmann distribution is proportional to $e^{-V(r)/k_B T}$. If the temperature is lowered and a condensate fraction is formed, it is easily recognized in the experiment as a separate phase with a localized density distribution near the minimum of the trap.

12.5.6 Bose-enhancement and Pauli-blocking

For a reaction occurring in a trap, the available product states are fully quantized. Not only the internal states of the atoms or molecules that are formed are quantized, but also their translational motion. When the products are fermions, they must satisfy Pauli's exclusion principle: the occupation of a given product state can be at most 1. This is called *Pauli-blocking*. It can only affect the rate of a process at ultralow temperatures, because at normal temperatures the number of available product states is enormous and the average population of product states will be much smaller than unity. For bosons, there is no restriction on the occupation of a given state. In fact, the rate of a process producing a product in an already occupied state is enhanced. This effect is called *Bose enhancement* or *Bose stimulation.*

To show the origin of this enhancement, we consider a model two level system of $N_1 + N_2 = N$ identical bosons. For instance, a Bose-Einstein condensate of N_2 ground state atoms in which N_1 atoms in some excited state are introduced. We also consider a perturbation that induces transitions between the two levels. For concreteness we take the component of the dipole operator that connects the two levels. Matrix elements of this operator determine the rate of spontaneous emission.

To define the wave functions we introduce the shorthand notation $a(i) = \phi_1(\boldsymbol{r}_i)$ and $b(i) = \phi_2(\boldsymbol{r}_i)$, where ϕ_1 and ϕ_2 are the one-particle wave functions corresponding to the two levels and i labels the particles. We will assume those functions to be normalized and orthogonal to each other. The wave function that describes a system with N_1 particles in level 1 and N_2 particles in level 2 may be written as

$$|N_1N_2\rangle = n(N_1, N_2)\hat{S}[a(1)a(2)\cdots a(N_1)b(N_1+1)b(N_1+2)\cdots b(N_1+N_2)], \qquad (12.203)$$

where $\hat{S}$ is the symmetrizer, which is the sum of all $N!$ permutations $\hat{P}_k$ of the $N = N_1 + N_2$ particle labels,

$$\hat{S} = \sum_{k=1}^{N!} \hat{P}_k \tag{12.204}$$

and $n(N_1, N_2)$ is the normalization constant. To compute it we note that

$$\hat{P}_l \hat{S} = \sum_k \hat{P}_l \hat{P}_k = \hat{S}, \tag{12.205}$$

because $\hat{P}_l \hat{P}_k$ runs over all permutations, although in a different order. We can use this result to show that

$$\hat{S}^2 = \left(\sum_l \hat{P}_l \right) \hat{S} = N! \hat{S}. \tag{12.206}$$

Since $\hat{S}$ is Hermitian, we may derive

$$\begin{aligned} \langle N_1 N_2 | N_1 N_2 \rangle &= |n(N_1, N_2)|^2 \langle \hat{S} a(1) \cdots b(N) | \hat{S} | a(1) \cdots b(N) \rangle \\ &= |n(N_1, N_2)|^2 \langle a(1) \cdots b(N) | \hat{S}^2 | a(1) \cdots b(N) \rangle \\ &= |n(N_1, N_2)|^2 N! \sum_k \langle a(1) \cdots b(N) | \hat{P}_k | a(1) \cdots b(N) \rangle. \end{aligned} \tag{12.207}$$

Out of the $N!$ terms, there will be $N_1! N_2!$ terms that yield 1, and all the others will be zero, so the normalization constant is

$$n(N_1, N_2) = (N! N_1! N_2!)^{-\frac{1}{2}}. \tag{12.208}$$

Let $\hat{d}$ be the component of the dipole operator that couples the levels a and b. It is written as the sum of one-particle operators

$$\hat{d} = \sum_{i=1}^{N} \hat{d}_i, \tag{12.209}$$

with matrix elements

$$\langle a(i) | \hat{d}_i | b(i) \rangle = d_{ab}. \tag{12.210}$$

The rate of spontaneous emission, a process in which the population of the upper level is decreased by one, and the population of the lower level is increased by one, is proportional to

$$\tilde{k} = \left| \langle N_1 N_2 | \hat{d} | N_1 - 1, N_2 + 1 \rangle \right|^2. \tag{12.211}$$

A derivation similar to the one used to derive the normalization of the wave function shows that

$$\tilde{k} = N_1 (N_2 + 1) |d_{ab}|^2. \tag{12.212}$$

What is remarkable about this result is that the rate not only depends on the number of particles N_1 in the upper level, but also on the number of particles in the lower level, N_2. This effect is called

Bose enhancement of the dipole matrix element. It should be noted that the above treatment of spontaneous emission is still lacking some essential ingredients. For example, photons carry momentum, and total momentum must be conserved. A more rigorous theoretical treatment[841] predicts that enhancement of spontaneous emission can occur in an interacting Bose-Einstein condensate.

The analogous relation for fermions contains a factor $N_1(1 - N_2)$, which shows that the process can only proceed if the initial state has an occupation of 1, and the final state is empty.

At ultralow temperatures there is no energy available for endothermic reactions, and one would expect that the energy released by an exothermic reaction would destroy the condensate. However, Bose enhancement may occur in resonance reactions, where the energy of the reactants is equal to the energy of the products. For example, two alkali atoms may form a diatomic molecule in its highest vibrational state. The molecular level can be tuned into resonance with the atomic level with a magnetic field, since the magnetic moment of the diatom will in general not be exactly two times the magnetic moment of the atom. Such a resonance is called a Feshbach resonance (see Chapter 6). This technique has been used to produce both ultracold molecules, and molecular Bose-Einstein condensates.[822] Other methods to achieve resonance are the use of Raman transitions[842] or photo-dissociation of Bose-Einstein condensates.[843]

Most theoretical papers on ultracold chemistry use the second quantization formalism to denote wave functions of the form Eq. (12.182) and the operators acting on them.[832] A short introduction to this formalism, with applications to Bose-Einstein condensates can be found in ref. [844].

12.6 OUTLOOK

Producing and manipulating cold and ultracold atoms and molecules is a very active and rapidly expanding field. In this chapter we only covered some of the basics of cold and ultracold collisions. Review papers[806,807,848–850] and books[811,851] describing the state of the art appear regularly, and major steps forward are reported frequently. To give just a few examples of recent progress, we mention the creation of a gas of ultracold polar molecules,[852] the creation of a Bose Einstein condensate of strontium atoms,[853] the observation of an ultracold atom-exchange reaction[847] (further details of this work may be found at the end of Study Box 12.5), and the trapping of molecules on a chip.[854]

12.7 PROBLEMS

1. (a) A simple classical one-dimensional model for the collision energy dependence of the cross section of a chemical reaction uses the following expression

$$E_t = \tfrac{1}{2}\mu\dot{R}^2 + V_{\text{eff}}(R),$$

where E_t is the initial collision energy, and $V_{\text{eff}}(R)$ is the radial dependence of the effective potential. Explain the origins of this expression, and present arguments to show that the effective potential may be written

$$V_{\text{eff}}(R) = V(R) + E_t\frac{b^2}{R^2}, \tag{12.213}$$

where $V(R)$ is the radial dependence of the potential energy, and b is the impact parameter for the collision. Recall that the kinetic energy associated with orbital motion can be written $E_{\text{cent}} = |\ell|^2/2\mu R^2$.

(b) For a reaction between an ion and a molecule, the long range ion-induced dipole interaction potential has the form

$$V(R) = -\frac{C_4}{R^4},$$

where the coefficient is defined $C_4 = \alpha' e^2/(8\pi\varepsilon_0)$, and the constant α' is the polarizability volume of the molecule. Use this expression, together with equation (12.213) above, to show that the location of the maximum in the centrifugal barrier on the effective potential is given by the expression

$$R_0 = \left(\frac{2C_4}{E_t b^2}\right)^{1/2}.$$

(c) Hence show that, for the radial kinetic energy at the barrier to be greater than zero, the impact parameter for the ion-molecule reaction is limited by the equation

$$b^2 \leqslant \left(\frac{4C_4}{E_t}\right)^{1/2}.$$

Use this expression to estimate the cross section for the reaction (known as the capture cross section), stating any additional assumptions made, and sketch the dependence of the cross section on collision energy (*i.e.* the excitation function).

(d) The polarizability volume, α', of O_2 is $1.4 \times 10^{-30}\,m^3$. Estimate the ion-induced dipole contribution to the capture cross section for the $Ca^+ + O_2$ reaction. Employ a mean collision energy corresponding to a temperature of 10 K.

2. The *K*-matrix boundary conditions for an *s*-wave are given in Eq. (12.113),

$$u_0^{(K)}(r) = \sin kr - K \cos kr. \qquad (12.214)$$

The *S*-matrix boundary conditions for this wave function are

$$u_0^{(S)}(r) = -e^{-ikr} + e^{ikr} S. \qquad (12.215)$$

The difference between these boundary conditions is only the normalization of the wave function, *i.e.*,

$$u_0^{(K)}(r) = A u_0^{(S)}(r), \qquad (12.216)$$

where A is a constant. Derive the relation between the *S*- and the *K*-matrix [Eq. (12.117)]. Hint: take the derivative with respect to r of the last equation:

$$\frac{d}{dr} u_0^{(K)}(r) = A \frac{d}{dr} u_0^{(S)}(r). \qquad (12.217)$$

Assume K to be known. For some fixed value of r Eqs. (12.216) and (12.217) are now two equations linear in A and S, so they can be solved to find S.

Further Reading

CHAPTER 1

1. P. W. Atkins and J. de Paula, *Physical Chemistry*, 8th Edition (Oxford University Press, Oxford, UK, 2006).
2. M. J. Pilling and P. W. Seakins, *Reaction Kinetics*, 2nd Edition (Oxford University Press, Oxford, UK, 1995).
3. G. Scoles, D. Bassi, U. Buck and D. C. Lainé, *Atomic and Molecular Beam Methods*, Volumes 1 and 2 (Oxford University Press, New York, USA, 1988).
4. R. D. Levine and R. B. Bernstein, *Molecular Reaction Dynamics and Chemical Reactivity* (Oxford University Press, New York, USA, 1987).
5. M. Brouard, *Reaction Dynamics*, Oxford University Chemistry Primers No. 61 (Oxford University Press, UK, 1998).
6. R. D. Levine, *Molecular Reaction Dynamics* (Cambridge University Press, Cambridge, UK, 2005).

CHAPTER 2

1. P. R. Bunker, *Molecular Symmetry and Spectroscopy* (Academic, New York, 1979).
2. H. Weyl, *The Classical Groups: Their Invariants and Representations*, 2nd Edition (Princeton, USA, 1946).
3. E. B. Wilson, J. C. Decius, and P. C. Cross, *Molecular Vibrations* (Dover, New York, USA, 1955).
4. A. Ben-Israel and T. N. Greville, *Generalised Inverses: Theory and Applications* (Wiley, New York, USA, 1974).
5. C. J. Cramer, *Essentials of Computational Chemistry: Theories and Models*, 2nd Edition (Wiley, New York, USA, 2004).
6. F. Jensen, *Introduction to Computational Chemistry*, 2nd Edition (Wiley, Chichester, UK, 2007).
7. A. Szabo and N. S. Ostlund, *Modern Quantum Chemistry: Introduction to Advanced Electronic Structure Theory* (MacMillan, New York, USA, 1989).

Tutorials in Molecular Reaction Dynamics
Edited by Mark Brouard and Claire Vallance

Published by the Royal Society of Chemistry, www.rsc.org

8. J. O. Hirschfelder, C. F. Curtiss, and R. B. Bird, *Molecular Theory of Gases and Liquids*, 2nd Edition, (Wiley, New York, USA, 1966).
9. A. J. Stone, *The Theory of Intermolecular Forces* (Clarendon, Oxford, UK, 1996).
10. J. I. Steinfeld, J. S. Francisco and W. L. Hase, *Chemical Kinetics and Dynamics* (Prentice-Hall, Englewood Cliffs, USA, 1989).
11. J. N. Murrell, S. Carter, S. C. Farantos, P. Huxley, and A. J. C. Varandas, *Molecular Potential Energy Functions* (Wiley, Chichester, UK, 1984).
12. W. H. Press, S. A. Teukolskly, W. T. Vetterling and B. P. Flannery, *Numerical Recipes in Fortran: the Art of Scientific Computing*, 2nd Edition (Cambridge University Press, Cambridge, UK, 1992).
13. D. Bonchev, D. H. Rouvray, M. A. Collins, and K. C. Thompson, *Chemical Group Theory: Techniques and Applications* (Gordon and Breach, Reading, 1995).

CHAPTER 3

1. H. Goldstein, *Classical Mechanics*, 2nd Edition (Addison Wesley, Reading MA, USA, 1981).
2. H. Goldstein, C. P. Poole, and J. L. Safko, *Classical Mechanics*, Pearson Education, 3rd Edition (2001).
3. C. Cohen-Tannoudji, B. Diu, and F. Laloë, *Quantum Mechanics*, (J. Wiley and Sons, New York USA, 1977).
4. A. Messiah, *Quantum Mechanics*, (North Holland, Amsterdam, NL, 1961).
5. J. J. Sakurai, *Modern Quantum Mechanics*, (Benjamin/Cummings, Menlo Park CA, USA, 1985).
6. J. Z. H. Zhang, *Theory and Application of Quantum Molecular Dynamics* (World Scientific, Singapore, 1999).
7. S. M. Blinder, *Foundations of Quantum Mechanics* (Academic Press, London, UK, 1974).
8. T. Bear and W. L. Hase, *Unimolecular Reaction Dynamics: Theory and Experiments*, (Oxford University Press, Oxford, UK, 1996).
9. R. B. Bernstein, *Atom-Molecule Collision Theory: A Guide to the Experimentalist* (Plenum Press, New York, USA, 1979).
10. H. Margenau and G. M. Murphy, *The Mathematics of Physics and Chemistry*, 2nd Edition (Van Nostrand Inc, Princeton, 1956).
11. P. W. Atkins and R. Friedman, *Molecular Quantum Mechanics*, 4th Edition (Oxford University Press, Oxford, UK, 2005).
12. A. Messiah, *Quantum Mechanics* (North Holland, Amsterdam, Netherlands, 1961).
13. R. E. Wyatt, J. Z. H. Zhang and W. H. Miller, *Dynamics of Molecules and Chemical Reactions* (Marcell Dekker, New York, USA, 1996).
14. R. Schinke, *Photodissociation Dynamics: Spectroscopy and Fragmentation of Small Polyatomic Molecules* (Cambridge University Press, Cambridge, UK, 1995).
15. W. C. Gardiner, *Combustion Chemistry* (Springer, Berlin, Germany, 1984).
16. R. N. Zare, *Angular Momentum*, (John Wiley and Sons, New York, USA, 1988).
17. M. E. Rose, *Elementary theory of Angular Momentum*, (J. Wiley and Sons, New York, USA, 1957).
18. A. R. Edmonds, *Angular Momentum in Quantum Mechanics*, (Princeton University Press, Princeton NJ, USA, 1960).
19. W. H. Press, S. A. Teukolsky, W. T. Vetterling, and B. P. Flannery, *Numerical Recipes in Fortran: the art of Scientific Computing*, 2nd Edition (Cambridge University Press, New York, USA, 1992).

CHAPTER 4

1. R. N. Zare, *Angular Momentum, Understanding Spatial Aspects in Chemistry and Physics* (Wiley, New York, USA, 1988).
2. P. F. Bernath, *Spectra of Atoms and Molecules*, 2nd Edition (Oxford University Press, Oxford, UK, 2005).
3. H. Lefebvre-Brion and R. W. Field, *The Spectra and Dynamics of Diatomic Molecules*, 2nd Edition (Elsevier, Amsterdam, The Netherlands, 2004).
4. G. Herzberg, *Molecular Spectra and Molecular Structure*, Volumes I–III, 2nd Edition (Van Nostrand, Inc., New York, USA, 1950).

CHAPTER 5

1. R. B. Bernstein, *Atom-Molecule Collision Theory: A Guide to the Experimentalist* (Plenum Press, New York, USA, 1979).
2. Y. T. Yardley, *Introduction to Molecular Energy Transfer* (Academic Press, New York, USA, 1980).
3. H. J. Metcalf and P. van der Straten, *Laser Cooling and Trapping*, (Springer Press, New York, USA, 1999).
4. J. M. Bowman, *Molecular Collision Dynamics* (Springer-Verlag, Berlin, Germany, 1983).
5. A. G. Suits and R. E. Continetti, *Imaging in Chemical Dynamics* (ACS Symposium Series 770, 2001).
6. R. Y. Rubinstein, *Simulation and the Monte Carlo Method* (John Wiley and Sons Inc., New York, USA, 1981).

CHAPTER 6

1. G. Scoles, D. Bassi, U. Buck and D. C. Lainé, *Atomic and Molecular Beam Methods*, Volumes 1 and 2 (Oxford University Press, New York, USA, 1988).
2. X. Yang and K. Liu, *Modern Trends in Chemical Reaction Dynamics: Experiment and Theory (Part I and II)*, Adv. Series in Phys. Chem. Vol. 14 (World Scientific, Singapore, 2004).

CHAPTER 7

1. K. A. Holbrook, M. J. Pilling, and S. Robertson, *Unimolecular Reactions*, 2nd Edition (John Wiley and Sons Ltd., Chichester, UK, 1996).
2. P. J. Robinson and K. A. Holbrook, *Unimolecular Reactions* (Wiley Interscience, New York, USA, 1972).
3. T. Baer and W. L. Hase, *Unimolecular Reaction Dynamics* (Oxford University Press, New York, USA, 1996).
4. R. G. Gilbert and S. C. Scott, *Theory of Unimolecular and Recombination Reactions* (Blackwell Scientific Publications, Oxford, UK, 1990).

CHAPTER 8

1. A. Gilbert and J. E. Baggott, *Essentials of Molecular Photochemistry* (Blackwell Scientific Publications, Oxford, UK, 1991).

2. N. J. Turro, V. Ramamurthy and J. C. Scaiano, *Principles of Molecular Photochemistry* (University Science Books, Sausalito, California, USA, 2009).
3. J. M. Brown, *Molecular Spectroscopy*, Oxford Chemistry Primers No. 55 (Oxford University Press, Oxford, UK, 1998).
4. E. E. Nikitin, *Theory of Elementary Atomic and Molecular Processes in Gases* (Clarendon Press, Oxford, UK, 1974).

CHAPTER 9

1. K. Blum, *Density Matrix Theory and Applications* (Plenum Press, New York, USA, 1996).
2. M. Auzinsh and R. Ferber, *Optical Polarization of Molecules* (Cambridge University Press, Cambridge, UK, 1995).
3. D. A. Varshalovich, A. N. Moskalev and V. K. Khersonskii, *Quantum Theory of Angular Momentum* (World Scientific, Singapore, 1988).
4. W. H. Miller *Dynamics of Molecular Collisions, Part B* (Plenum Press, New York, USA, 1976).
5. R. B. Bernstein, *Atom-Molecule Collision Theory: A Guide to the Experimentalist* (Plenum Press, New York, USA, 1979).
6. R. E. Wyatt and J. Z. H. Zhang, *Dynamics of Molecules and Chemical Reactions* (Marcel Dekker, New York, USA, 1996).
7. D. M. Hirst, *Potential Energy Surfaces: Molecular Structure and Reaction Dynamics* (Taylor and Francis, London, UK, 1985).
8. I. Bengsston and K. Życzkowski, *Geometry of Quantum States: an Introduction to Quantum Entanglement* (Cambridge University Press, Cambridge, UK, 2006).

CHAPTER 10

1. J. S. Blakemore, *Solid State Physics* (Cambridge University Press, Cambridge, UK, 1985).
2. A. Zangwill, *Physics at Surfaces* (Cambridge University Press, Cambridge, UK, 1988).

CHAPTER 11

1. H. Zewail, *Femtochemistry: Ultrafast Dynamics of the Chemical Bond*, Volumes I and II (World Scientific, New Jersey, Singapore, 1994).
2. W. Demtroder, *Laser Spectroscopy, Basic Concepts and Instrumentation* (Springer-Verlag, Berlin, Germany, 1996).
3. R. Trebino, *Frequency-Resolved Optical Gating: The Measurement of Ultrashort Laser Pulses*, 2nd Edition (Kluwer Academic Publishers, The Netherlands, 2000).
4. R. Trebino and J. Squier, *Ultrafast Optics Textbook* (available online at http://www.physics.gatech.edu/frog/ultratext.html, 2008).

CHAPTER 12

1. R. V. Krems, W. E. Stwalley and B. Friedrich, *Cold Molecules. Theory, Experiment, Applications* (CRC Press, Taylor and Francis Group, Florida, USA, 2009).
2. I. W. M. Smith, *Low Temperatures and Cold Molecules* (World Scientific, London, 2008).

3. M. Abramowitz and I. A. Stegun, *Handbook of Mathematical Functions* (National Bureau of Standards, Washington, D.C., 1964).
4. L. D. Landau and E. M. Lifshitz, *Quantum Mechanics, Non-Relativistic Theory* (Pergamon Press, London, 1959).
5. C. J. Pethick and H. Smith, *Bose-Einstein Condensation in Dilute Gases* (Cambridge University Press, Cambridge, UK, 2002).

Glossary of Acronyms

BEC	Bose-Einstein condensate
BO	Born-Oppenheimer
CASSCF	complete active space self-consistent field
CC	coupled channel
CC	coupled cluster
CCD	charge-coupled device
CCSD	coupled cluster with single and double excitations
CCSD(T)	coupled cluster with fully calculated single and double excitations and triple excitations calculated using perturbation theory
CCSDTQ	coupled cluster with single, double, triple and quadruple excitations
CDCS	(pair) correlated differential cross section
CG	Clebsch Gordan (coefficient)
CI	configuration interaction
CI	conical intersection
CID	configuration interaction with double excitations
CIS	configuration interaction with single excitations
CISD	configuration interaction with single and double excitations
CM	centre of mass
CMB	crossed molecular beams
CNP	complete nuclear permutation
CS	coupled states
CW	continuous wave
DAF	distributed approximating function
DC	direct current
DCS	differential cross section
DFFD	dispersion fitted finite difference
DFG	difference frequency generation
DFT	density functional theory
DI	direct inelastic (scattering)
DVR	discrete variable representation
EI	electron impact
FBR	finite basis representation
FD	finite difference

Tutorials in Molecular Reaction Dynamics
Edited by Mark Brouard and Claire Vallance

Published by the Royal Society of Chemistry, www.rsc.org

FFT	fast Fourier transform
FHG	fourth harmonic generation
FT	Fourier transform
FWHM	full width at half maximum
GDD	group delay dispersion
GP	geometric phase
GVD	group velocity dispersion
GWP	Gaussian wavepacket
HF	Hartree-Fock
HHG	high-order harmonic generation
HOMO	highest occupied molecular orbital
ICS	integral cross section
IMLS	interpolating moving least squares (interpolation method)
IOSA	infinite-order sudden approximation
IR	infra-red
IRP	intrinsic reaction path
ISC	intersystem crossing
IVR	intramolecular vibrational redistribution
KE	kinetic energy
KER	kinetic energy release
LAB	laboratory (frame)
LIF	laser-induced fluorescence
LUMO	lowest unoccupied molecular orbital
MCP	microchannel plate
MCSCF	multiconfigurational self-consistent field
MCTDH	multiconfigurational time-dependent Hartree
MEP	minimum energy path
MOT	magneto-optical trap
MP*n*	*n*th order Moller Plesset perturbation theory
MS	modified Shepard (interpolation method)
NGP	neglect of geometric phase
NMR	nuclear magnetic resonance
OPA	optical parametric amplifier
PC	personal computer
PDDCS	polarization-dependent differential cross section
PDF	probability density function
PE	potential energy
PES	potential energy surface
PHOTOLOC	Photoinitiated reaction analyzed by the law of cosines
PI	photoionization
PP	polarization parameter
PRB	laser beam probe frame
QBS	quantum bottleneck state
QCT	quasi-classical trajectory
QM	quantum mechanics or quantum mechanical
QQT	quasi quantum treatment
RAS	restricted active space
REMPI	resonance-enhanced multiphoton ionization
RKHS	reproducing kernel Hilbert space (interpolation method)
RP	reaction path

RRKM	Rice Rampsberger Kassel Marcus (theory)
SA	steric asymmetry
SCF	self-consistent field
SFG	sum frequency generation
SHG	second harmonic generation
SI	symplectic integrator
SLM	spatial light modulator
SOP	split operator
SSH	Schwartz Slawsky Herzfeld (theory)
TD	time-dependent
TD	trapping-desorbing
TDSE	time-dependent Schrödinger equation
TDWP	time-dependent wavepacket
THG	third harmonic generation
TI	time-independent
TISE	time-independent Schrödinger equation
TOF	time of flight
TRPEI	time-resolved photoelectron imaging
TRPES	time-resolved photoelectron spectroscopy
UMP2	spin unrestricted MP2 (see MP*n*)
UV	ultraviolet
VAP	vibrationally adiabatic potential
VDW	van der Waals (interaction)
VMI	velocity-mapped imaging
VUV	vacuum ultraviolet
XUV	extreme ultraviolet

Bibliography

1. P. W. Atkins and J. de Paula, *Physical Chemistry*, 8th Edition, (Oxford University Press, Oxford, UK), 2006.
2. M. J. Pilling and P. W. Seakins, *Reaction Kinetics*, 2nd Edition, (Oxford University Press, Oxford, UK), 1995.
3. G. Scoles, D. Bassi, U. Buck, and D. C. Lainé, Eds., *Atomic And Molecular Beam Methods: Volume 1*, (Oxford University Press, New York, USA), 1988.
4. G. Scoles, D. C. Lainé, and U. Valbusa, Eds., *Atomic And Molecular Beam Methods: Volume 2*, (Oxford University Press, New York, USA), 1992.
5. G. Hall, K. Liu, M. J. McAuliffe, C. F. Giese, and W. R. Gentry, *J. Chem. Phys.*, 1983, **78**, 5260.
6. R. G. Macdonald and K. Liu, *J. Chem. Phys.*, 1989, **91**, 821.
7. G. Porter, in *Nobel Lectures, Chemistry 1963–1970*, (Elsevier Publishing Company, Amsterdam, Netherlands), 1972.
8. A. A. Tsekouras, C. A. Leach, K. S. Kalogerakis, and R. N. Zare, *J. Chem. Phys.*, 1992, **97**, 7220.
9. R. D. Levine and R. B. Bernstein, *Molecular Reaction Dynamics and Chemical Reactivity*, (Oxford University Press, New York, USA), 1987.
10. M. Brouard, *Reaction Dynamics*, (Oxford Chemistry Primers 61, Oxford University Press, UK), 1998.
11. R. D. Levine, *Molecular Reaction Dynamics*, (Cambridge University Press, Cambridge, UK), 2005.
12. A. H. Zewail, *J. Chem. Phys.*, 1987, **87**, 2395.
13. R. B. Metz, T. Kitsopoulos, A. Weaver, and D. M. Neumark, *J. Chem. Phys.*, 1988, **88**, 1463.
14. D. E. Manolopoulos, K. Stark, H. J. Werner, D. W. Arnold, S. E. Bradforth, and D. M. Neumark, *Science*, 1993, **262**, 1852.
15. P. R. Bunker, *Molecular Symmetry and Spectroscopy*, (Academic, New York), 1979.
16. H. Weyl, *The Classical Groups: Their Invariants and Representations*, 2nd Ed., (Princeton, Princeton, USA), 1946.
17. M. A. Collins and D. F. Parsons, *J. Chem. Phys.*, 1993, **99**, 6756.
18. H. Le and D. Kendall, *Annals of Statistics*, 1993, **21**, 1225.
19. R. Littlejohn and M. Reinsch, *Rev. Mod. Phys.*, 1997, **69**, 213.
20. G. Simons, R. G. Parr, and J. M. Finlan, *J. Chem. Phys.*, 1973, **59**, 3229.
21. K. C. Thompson, M. J. T. Jordan, and M. A. Collins, *J. Chem. Phys.*, 1998, **108**, 8302.

Tutorials in Molecular Reaction Dynamics
Edited by Mark Brouard and Claire Vallance

Published by the Royal Society of Chemistry, www.rsc.org

22. E. B. Wilson, J. C. Decius, and P. C. Cross, *Molecular Vibrations*, (Dover, New York, USA), 1955.
23. A. Ben-Israel and T. N. Greville, *Generalised Inverses: Theory and Applications*, (Wiley, New York, USA), 1974.
24. C. J. Cramer, *Essentials of Computational Chemistry: Theories and Models*, 2nd Ed., (Wiley, New York, USA), 2004.
25. F. Jensen, *Introduction to Computational Chemistry*, 2nd Ed., (Wiley, Chichester, UK), 2007.
26. A. Szabo and N. S. Ostlund, *Modern Quantum Chemistry: Introduction to Advanced Electronic Structure Theory*, (MacMillan, New York, USA), 1989.
27. J. C. Polanyi, *Acc. Chem. Res.*, 1972, **5**, 161.
28. J. C. Polanyi, *Science*, 1987, **236**, 680.
29. J. C. Polanyi, *Angew. Chemie Int. Engl. Ed.*, 1987, **26**, 952.
30. J. O. Hirschfelder, C. F. Curtiss, and R. B. Bird, *Molecular theory of gases and liquids*, 2nd Ed., (Wiley, New York, USA), 1966.
31. A. J. Stone, *The Theory of Intermolecular Forces*, (Clarendon, Oxford, UK), 1996.
32. J. I. Steinfeld, J. S. Francisco, and W. L. Hase, *Chemical Kinetics and Dynamics*, (Prentice-Hall, Englewood Cliffs, USA), 1989.
33. K. Fukui, *Acc. Chem. Res.*, 1981, **14**, 363.
34. R. Elber and M. Karplus, *Chem. Phys. Lett.*, 1987, **139**, 375.
35. W. H. Miller, N. C. Handy, and J. E. Adams, *J. Chem. Phys.*, 1980, **72**, 99.
36. R. P. A. Bettens, T. A. Hansen, and M. A. Collins, *J. Chem. Phys.*, 1999, **111**, 6322.
37. J. C. Corchado, Y.-Y. Chuang, P. L. Fast, W.-P. Hu, Y. P. Liu, G. C. Lynch, K. A. Nguyen, C. F. Jackels, A. F. Ramos, B. A. Ellingson, B. J. Lynch, J. Zheng, V. S. Melissas, J. Villà, I. Rossi, E. L. Coitino, J. Pu, T. V. Albu, R. Steckler, B. C. Garrett, A. D. Isaacson, and D. G. Truhlar, *POLYRATE 9.7: Computer Program for the Calculation of Chemical Reaction Rates for Polyatomics*, http://comp.chem.umn.edu/polyrate/.
38. T. Helgaker, E. Uggerud, and H. J. A. Jensen, *Chem. Phys. Lett.*, 1990, **173**, 145.
39. W. Chen, W. L. Hase, and H. B. Schlegel, *Chem. Phys. Lett.*, 1994, **228**, 426.
40. W. L. Hase, K. Song, and M. Gordon, *Computing in Science and Engineering*, Vol. **5**, 2003.
41. H. B. Schlegel, J. M. Millam, S. S. Iyengarand, G. A. Voth, A. D. Daniels, and G. E. Scuseria, *J. Chem. Phys.*, 2001, **114**, 9758.
42. J. N. Murrell, S. Carter, S. C. Farantos, P. Huxley, and A. J. C. Varandas, *Molecular Potential Energy Functions*, (Wiley, Chichester, UK), 1984.
43. L. P. Viegas, A. Alijah, and A. J. C. Varandas, *J. Chem. Phys.*, 2007, **126**, 074309.
44. G. C. Schatz, *Rev. Mod. Phys.*, 1989, **61**, 669.
45. J. M. Bowman and G. C. Schatz, *Annu. Rev. Phys.*, 1995, **46**, 169.
46. R. J. Duchovic, Y. L. Volobuev, G. C. Lynch, A. W. Jasper, D. G. Truhlar, T. C. Allison, A. F. Wagner, B. C. Garrett, J. Espinosa-García, and J. C. Corchado, *POTLIB-online*, http://comp.chem.umn.edu/potlib/, 2004.
47. W. H. Press, S. A. Teukolsky, W. T. Vetterling, and B. P. Flannery, *Numerical Recipes in Fortran: the Art of Scientific Computing*, 2nd Ed., (Cambridge University Press, Cambridge, UK), 1992.
48. L. M. Raff, M. Malshe, M. Hagan, D. I. Doughan, M. G. Rockley, and R. Komanduri, *J. Chem. Phys.*, 2005, **122**, 084104.
49. S. Manzhosa and T. Carrington-Jr., *J. Chem. Phys.*, 2006, **125**, 084109.
50. A. Schmelzer and J. N. Murrell, *Int. J. Quant. Chem.*, 1985, **28**, 287.
51. A. J. C. Varandas and J. N. Murrell, *Chem. Phys. Lett.*, 1981, **84**, 440.
52. J. Ischtwan and M. A. Collins, *J. Chem. Phys.*, 1991, **94**, 7084.

53. D. Bonchev and D. H. Rouvray, Eds., M. A. Collins and K. C. Thompson, in *Chemical Group Theory: Techniques and Applications*, (Gordon and Breach, Reading), 1995.
54. A. Brown, B. J. Braams, K. Christoffel, Z. Jin, and J. M. Bowman, *J. Chem. Phys.*, 2003, **119**, 8790.
55. W. K. Park, J. Park, S. C. Park, B. J. Braams, C. Chen, and J. M. Bowman, *J. Chem. Phys.*, 2006, **125**, 081101.
56. X. Huang, B. J. Braams, J. M. Bowman, R. E. A. Kelly, J. Tennyson, G. C. Groenenboom, and A. van der Avoird, *J. Chem. Phys.*, 2008, **128**, 034312.
57. T.-S. Ho and H. Rabitz, *J. Chem. Phys.*, 1996, **104**, 2584.
58. T. Hollebeek, T.-S. Ho, and H. Rabitz, *J. Chem. Phys.*, 1997, **106**, 7223.
59. T. Hollebeek, T.-S. Ho, H. Rabitz, and L. B. Harding, *J. Chem. Phys.*, 2001, **114**, 3945.
60. G. C. Schatz, A. Papaioannou, L. A. Pederson, L. B. Harding, T. Hollebeek, T.-S. Ho, and H. Rabitz, *J. Chem. Phys.*, 1997, **107**, 2340.
61. J. Ischtwan and M. A. Collins, *J. Chem. Phys.*, 1994, **100**, 8080.
62. M. J. T. Jordan, K. C. Thompson, and M. A. Collins, *J. Chem. Phys.*, 1995, **102**, 5647.
63. R. P. A. Bettens and M. A. Collins, *J. Chem. Phys.*, 1999, **111**, 816.
64. M. A. Collins, *Theoretical Chemistry Accounts*, 2002, **108**, 313.
65. G. G. Maisuradze and D. L. Thompson, *J. Phys. Chem.*, 2003, **107**, 7118.
66. R. Dawes, D. L. Thompson, Y. Guo, A. F. Wagner, and M. Minkoff, *J. Chem. Phys.*, 2007, **126**, 184108.
67. O. Godsi, C. R. Evenhuis, and M. A. Collins, *J. Chem. Phys.*, 2006, **125**, 104105.
68. I. Antola, M. Eckert-Maksic, M. Barbatti, and H. Lischkaa, *J. Chem. Phys.*, 2007, **127**, 234303.
69. A. Toniolo and T. J. Martínez, *J. Chem. Phys.*, 2005, **123**, 234308.
70. P. W. Atkins and R. Friedman, *Molecular Quantum Mechanics*, 4th Edition, (Oxford University Press, Oxford, UK), 2005.
71. H. Goldstein, *Classical Mechanics*, (Addison Wesley, Reading MA, USA), 2nd Ed., 1981.
72. J. Z. H. Zhang, *Theory and Application of Quantum Molecular Dynamics*, (World Scientific, Singapore), 1999.
73. G. Nyman and H. G. Yu, *Reports on Progress in Physics*, 2000, **63**, 1001.
74. S. M. Blinder, *Foundations of Quantum Mechanics*, (Academic Press, London, UK), 1974.
75. M. Karplus, R. N. Porter, and R. D. Sharma, *J. Chem. Phys.*, 1965, **43**, 3259.
76. R. B. Bernstein, Ed., *Atom-Molecule Collision Theory: A Guide to the Experimentalist*, Plenum, 1979.
77. F. J. Aoiz, L. Bañares, and V. J. Herrero, *J. Chem. Soc. Faraday Trans.*, 1998, **94**, 2483.
78. G. J. Kroes and R. P. H. Rettschnick, *J. Chem. Phys.*, 1989, **91**, 1556.
79. A. J. H. M. Meijer, G. C. Groenenboom, and A. van der Avoird, *J. Chem. Phys.*, 1994, **101**, 7603.
80. S. E. Choi and R. B. Bernstein, *J. Chem. Phys.*, 1986, **85**, 150.
81. G. C. Groenenboom and A. J. H. M. Meijer, *J. Chem. Phys.*, 1994, **101**, 7592.
82. J. M. Millam, V. Bakken, W. Chen, W. L. Hase, and H. B. Schlegel, *J. Chem. Phys.*, 1999, **111**, 3800.
83. U. Lourderaj, K. Song, T. L. Windus, Y. Zhuang, and W. L. Hase, *J. Chem. Phys.*, 2007, **126**, 044105.
84. A. J. C. Varandas and D. M. C. Marques, *J. Chem. Phys.*, 1994, **100**, 1908.
85. M. Herman, *Annu. Rev. Phys. Chem.*, 1994, **45**, 83.
86. D. Sholl and J. C. Tully, *J. Chem. Phys.*, 1998, **109**, 7702.
87. Boris Podolsky, *Phys. Rev.*, 1928, **32**, 812.
88. H. Essén, *Am. J. Phys.*, 1978, **46**, 983.
89. H. Margenau and G. M. Murphy, *The Mathematics of Physics and Chemistry*, 2nd Ed., (van Nostrand Inc., Princeton, USA), 1956.

90. A. van der Avoird, P. E. S. Wormer, and R. Moszyński, *Chem. Rev.*, 1994, **94**, 1931.
91. F. Gatti, C. Iung, M. Menou, Y. Justum, A. Nauts, and X. Chapuisat, *J. Chem. Phys.*, 1998, **108**, 8804.
92. F. Gatti, C. Iung, M. Menou, and X. Chapuisat, *J. Chem. Phys.*, 1998, **108**, 8821.
93. M. Mladenović, *J. Chem. Phys.*, 2000, **112**, 1070.
94. R. G. Wooley, Ed., B. T. Sutcliffe in *The Dynamics of Molecules*, (NATO ASI ser., Plenum, New York), 1980.
95. X. Carbo, Ed., B. T. Sutcliffe, in *Proceedings of XII International congress of Latin-speaking Theoretical Chemists*, (North-Holland, Amsterdam, Netherlands), 1982.
96. A. Messiah, *Quantum Mechanics*, (North Holland, Amsterdam, Netherlands), 1961.
97. E. J. Heller, *J. Chem. Phys.*, 1975, **62**, 1544.
98. G. D. Billing, *Phys. Chem. Chem. Phys.*, 2002, **4**, 2865.
99. W. H. Miller, *Dynamics of Molecules and Chemical Reactions*, R. E. Wyatt and J. Z. H. Zhang, Eds., (Marcel Dekker, New York, NY, USA), 1996.
100. Y. Huang, W. Zhu, D. J. Kouri, and D. K. Hoffman, *Chem. Phys. Lett.*, 1993, **213**, 209.
101. V. A. Mandelshtam and H. S. Taylor, *J. Chem. Phys.*, 1995, **103**(8), 2903.
102. V. A. Mandelshtam and H. S. Taylor, *J. Chem. Phys.*, 1995, **102**(19), 7390.
103. R. Chen and H. Guo, *Chem. Phys. Lett.*, 1996, **261**, 605.
104. R. Chen and H. Guo, *J. Chem. Phys.*, 1996, **105**, 3569.
105. S. K. Gray and G. G. Balint-Kurti, *J. Chem. Phys.*, 1998, **108**, 950.
106. S. C. Althorpe, P. Soldán, and G. G. Balint-Kurti, Ed., S. C. Althorpe, in *Wavepacket calculations of differential cross sections for reactive scattering*, Collaborative Computational Project on Molecular Quantum Dynamics (CCP6), 2001.
107. S. C. Althorpe, *J. Chem. Phys.*, 2001, **114**, 1601.
108. G. G. Balint-Kurti, *Adv. Chem. Phys.*, 2003, **128**, 249.
109. E. B. Stechel, R. B. Walker, and J. C. Light, *J. Chem. Phys.*, 1978, **69**, 3518.
110. D. J. Zvijac and J. C. Light, *Chem. Phys.*, 1976, **12**, 237.
111. J. C. Light and R. B. Walker, *J. Chem. Phys.*, 1976, **65**, 4272.
112. D. E. Johnson, *J. Comp. Phys.*, 1973, **13**, 445.
113. D. E. Manolopoulos, *J. Chem. Phys.*, 1986, **85**, 6425.
114. D. E. Manolopoulos, M. D'Mello and R. E. Wyatt, *J. Chem. Phys.*, 1989, **91**, 6096.
115. R. T. Pack, *J. Chem. Phys.*, 1974, **60**, 633.
116. J. Tennyson and B. T. Sutcliffe, *J. Chem. Phys.*, 1982, **77**, 4061.
117. J. Tennyson and B. T. Sutcliffe, *J. Mol. Spectrosc.*, 1983, **101**, 71.
118. D. Kosloff and R. Kosloff, *J. Comp. Phys.*, 1983, **52**, 35.
119. R. Kosloff and D. Kosloff, *J. Chem. Phys.*, 1983, **79**, 1823.
120. D. Neuhauser, M. Bear, and D. Kouri, *J. Chem. Phys.*, 1990, **93**, 2499.
121. S. K. Gray and C. E. Wozny, *J. Chem. Phys.*, 1989, **91**, 7671.
122. A. S. Dickenson and P. R. Certain, *J. Chem. Phys.*, 1968, **49**, 4209.
123. J. C. Light, I. P. Hamilton, and J. V. Lill, *J. Chem. Phys.*, 1985, **82**, 1400.
124. J. V. Lill, G. A. Parker, and J. C. Light, *J. Chem. Phys.*, 1986, **85**, 900.
125. D. T. Colbert and W. H. Miller, *J. Chem. Phys.*, 1992, **96**, 1982.
126. C. Schwartz, *J. Math. Phys.*, 1985, **26**, 411.
127. J. Echave and D. C. Clary, *J. Chem. Phys.*, 1994, **100**, 402.
128. S. K. Gray and E. M. Goldfield, *J. Chem. Phys.*, 2001, **115**, 8331.
129. D. A. Mazziotti, *J. Chem. Phys.*, 2002, **117**, 2455.
130. D. A. Mazziotti, *Chem. Phys. Lett.*, 1999, **299**, 473.
131. D. K. Hoffman, M. Arnold, and D. J. Kouri, *J. Phys. Chem.*, 1992, **96**, 6539.
132. D. K. Hoffman and D. J. Kouri, *J. Phys. Chem.*, 1992, **96**(3), 1179.

133. C. Leforestier, R. H. Bisseling, C. Cerjan, M. D. Feit, R. Friesner, A. Guldberg, A. Hammerich, G. Jolicard, W. Karrlein, H. D. Meyer, N. Lipkin, O. Roncero, and R. Kosloff, *J. Comp. Phys.*, 1991, **94**, 59.
134. J. A. Fleck Jr, J. R. Morris, and M. D. Feit, *Appl. Phys.*, 1976, **10**, 129.
135. R. Kosloff, *J. Phys. Chem.*, 1988, **92**, 2087.
136. S. K. Gray and D. E. Manolopoulos, *J. Chem. Phys.*, 1996, **104**, 7099.
137. G. G. Balint-Kurti, R. N. Dixon, and C. C. Marston, *J. Chem. Soc. Faraday Trans.*, 1990, **86**, 1741.
138. A. J. H. M. Meijer, E. M. Goldfield, S. Gray, and G. G. Balint-Kurti, *Chem. Phys. Lett.*, 1998, **293**, 270.
139. D. H. Zhang and J. Z. H. Zhang, *J. Chem. Phys.*, 1994, **101**, 3671.
140. R. Schinke, *Photodissociation dynamics: spectroscopy and fragmentation of small polyatomic molecules*, (Cambridge University Press, Cambridge, UK), 1995.
141. A. J. H. M. Meijer and E. M. Goldfield, *J. Chem. Phys.*, 1998, **108**, 5404.
142. A. J. H. M. Meijer, *Comput. Phys. Commun.*, 2001, **141**, 330.
143. E. M. Goldfield and S. K. Gray, *Comput. Phys. Commun.*, 1996, **98**, 1.
144. S. Y. Lin, Z. G. Sun, H. Guo, D. H. Zhang, P. Honvault, D. Q. Xie, and S. Y. Leo, *J. Phys. Chem. A*, 2008, **112**, 602.
145. D. E. Skinner, T. C. German, and W. H. Miller, *J. Phys. Chem. A.*, 1998, **102**, 3828.
146. A. Baram, I. Last, and M. Baer, *Chem. Phys. Lett.*, 1993, **212**, 649 and references therein.
147. P. McGuire, *Chem. Phys. Lett.*, 1973, **23**, 575.
148. P. McGuire and D. J. Kouri, *J. Chem. Phys.*, 1974, **60**, 2488.
149. T. P. Tsien and R. T. Pack, *Chem. Phys. Lett.*, 1970, **6**, 54.
150. T. P. Tsien and R. T. Pack, *Chem. Phys. Lett.*, 1970, **6**, 600.
151. T. P. Tsien and R. T. Pack, *Chem. Phys. Lett.*, 1971, **8**, 59.
152. V. Khare, D. J. Kouri, and M. Baer, *J. Chem. Phys.*, 1979, **71**, 1179.
153. M. Baer, V. Khare, and D. J. Kouri, *Chem. Phys. Lett.*, 1979, **68**, 378.
154. D. C. Clary and A. J. H. M. Meijer, *J. Chem. Phys.*, 2002, **116**, 9829.
155. D. W. Zhang, M. L. Wang, and J. Z. H. Zhang, *J. Phys. Chem. A*, 2003, **107**, 7106.
156. T. F. Miller III, D. C. Clary, and A. J. H. M. Meijer, *J. Chem. Phys.*, 2005, **122**, 244323.
157. J. M. Bowman, *J. Phys. Chem.*, 1991, **95**, 4960.
158. M. H. Beck, A. Jackle, G. A. Worth, and H. D. Meyer, *Physics Reports-Review Section of Physics Letters*, 2000, **324**, 1.
159. H.-D. Meyer and G. A. Worth, *Theor. Chem. Acc.*, 2003, **109**, 251.
160. J. Z. H. Zhang, *J. Chem. Phys.*, 1999, **111**, 3929.
161. C. Zhu, S. Nangia, A. Jasper, and D. G. Truhlar, *J. Chem. Phys.*, 2004, **121**, 7658.
162. H. Wang, X. Sun, and W. H. Miller, *J. Chem. Phys.*, 1998, **108**, 9726.
163. C. Colletti and G. Billing, *J. Chem. Phys.*, 2000, **113**, 11101.
164. W. C. Gardiner, *Combustion Chemistry*, (Springer, Berlin FRG), 1984.
165. J. Troe and V. G. Ushakov, *J. Chem. Phys.*, 2001, **115**, 3621.
166. R. J. Duchovic and M. A. Parker, *J. Phys. Chem. A*, 2005, **109**, 5883.
167. S. M. Hwang, S.-O. Ryu, K. J. De Witt, and M. J. Rabinowitz, *Chem. Phys. Lett.*, 2005, **408**, 107.
168. C. Xu, D. Xie, D. H. Zhang, S. Y. Lin, and H. Guo, *J. Chem. Phys.*, 2005, **122**, 244305.
169. S. Y. Lin, E. J. Rackham, and H. Guo, *J. Phys. Chem. A*, 2006, **110**, 1534.
170. M. J. Bronikowski, R. Zhang, D. J. Rakestraw, and R. N. Zare, *Chem. Phys. Lett.*, 1989, **156**, 7.
171. R. T. Pack, E. A. Butcher, and G. A. Parker, *J. Chem. Phys.*, 1993, **99**, 9310.
172. R. T. Pack, E. A. Butcher, and G. A. Parker, *J. Chem. Phys.*, 1995, **102**, 5998.
173. B. Kendrick and R. T. Pack, *J. Chem. Phys.*, 1996, **104**, 7502.

174. J. Dai and J. Z. H. Zhang, *J. Phys. Chem.*, 1996, **100**, 6898.
175. R. Fei, X. S. Zheng, and G. E. Hall, *J. Phys. Chem. A*, 1997, **101**, 2541.
176. B. Kendrick and R. T. Pack, *J. Chem. Phys.*, 1997, **106**, 3519.
177. A. J. H. M. Meijer and E. M. Goldfield, *J. Chem. Phys.*, 1999, **110**, 870.
178. E. M. Goldfield and A. J. H. M. Meijer, *J. Chem. Phys.*, 2000, **113**, 11055.
179. A. J. H. M. Meijer and E. M. Goldfield, *Phys. Chem. Chem. Phys.*, 2001, **3**, 2811.
180. H. Zhang and S. C. Smith, *J. Chem. Phys.*, 2002, **117**, 5174.
181. H. Zhang and S. C. Smith, *J. Chem. Phys.*, 2002, **116**, 2354.
182. R. A. Sultanov and N. Balakrishnan, *J. Phys. Chem. A*, 2004, **108**, 8759.
183. T. González-Lezana, E. J. Rackham, and D. E. Manolopoulos, *J. Chem. Phys.*, 2004, **120**, 2247.
184. P. Bargueño, T. González-Lezana, P. Larrégaray, L. Bonnet, and J. C. Rayez, *Phys. Chem. Chem. Phys.*, 2007, **9**, 1127.
185. P. Bargueño, T. González-Lezana, P. Larrégaray, L. Bonnet, J.-C. Rayez, M. Hankel, S. C. Smith, and A. J. H. M. Meijer, *J. Chem. Phys.*, 2008, **128**, 244308.
186. M. Hankel, S. C. Smith, and A. J. H. M. Meijer, *J. Chem. Phys.*, 2007, **127**, 064316.
187. L. B. Harding, J. Troe, and V. G. Ushakov, *Phys. Chem. Chem. Phys.*, 2000, **2**, 631.
188. J. M. C. Marques and A. J. C. Varandas, *Phys. Chem. Chem. Phys.*, 2001, **3**, 505.
189. L. B. Harding, J. Troe, and V. G. Ushakov, *Phys. Chem. Chem. Phys.*, 2001, **3**, 2630.
190. J. M. C. Marques and A. J. C. Varandas, *Phys. Chem. Chem. Phys.*, 2001, **3**, 2632.
191. H. Zhang and S. C. Smith, *Phys. Chem. Chem. Phys.*, 2001, **3**, 2282.
192. H. Zhang and S. C. Smith, *Phys. Chem. Chem. Phys.*, 2004, **6**, 884.
193. H. Zhang and S. C. Smith, *J. Chem. Phys.*, 2004, **20**, 9583.
194. S. Y. Lin, D. Xie, and H. Guo, *J. Chem. Phys.*, 2006, **125**, 091103.
195. X. Wu and E. F. Hayes, *J. Chem. Phys.*, 1997, **107**, 2705.
196. K. Keßler and K. Kleinermanns, *J. Chem. Phys.*, 1992, **97**, 374.
197. K. Kleinermanns and R. Schinke, *J. Chem. Phys.*, 1984, **80**, 1440.
198. K. Kleinermanns and E. Linnebach, *J. Chem. Phys.*, 1985, **82**, 5012.
199. K. Kleinermanns, *Radiochimica Acta*, 1988, **43**, 118.
200. K. Kleinermanns, E. Linnebach, and M. Pohl, *J. Chem. Phys.*, 1989, **91**, 2181.
201. A. Jacobs, F. M. Schuler, H. R. Volpp, M. Wahl, and J. Wolfrum, *Ber. Bunsenges. Phys. Chem.*, 1990, **94**, 1390.
202. A. Jacobs, H. R. Volpp, and J. Wolfrum, *Chem. Phys. Lett.*, 1991, **177**, 200.
203. S. Sieger, V. Sick, H.-R. Volpp, and J. Wolfrum, *Isr. J. Chem.*, 1994, **34**, 5.
204. V. Ebert, C. Schulz, H.-R. Volpp, J. Wolfrum, and P. Monkhouse, *Isr. J. Chem.*, 1999, **39**, 1.
205. M. A. Bajeh, E. M. Goldfield, A. Hanf, C. Kappel, A. J. H. M. Meijer, H.-R. Volpp, and J. Wolfrum, *J. Phys. Chem. A*, 2001, **105**, 3359.
206. H. Yang, W. C. Gardiner, K. S. Shin, and N. Fujii, *Chem. Phys. Lett.*, 1994, **231**, 449.
207. C.-L. Yu, M. Frenklach, D. A. Masten, R. K. Hanson, and C. T. Bowman, *J. Phys. Chem.*, 1994, **98**, 4770.
208. H. Du and J. P. Hessler, *J. Chem. Phys.*, 1992, **96**, 1077.
209. S.-O. Ryu, S. M. Hwang, and M. J. Rabinowitz, *J. Phys. Chem.*, 1995, **99**, 13984.
210. K. Honma, *J. Chem. Phys.*, 1995, **102**, 7856.
211. H. L. Kim, M. A. Wickramaaratchi, X. Zheng, and G. E. Hall, *J. Chem. Phys.*, 1994, **101**, 2033.
212. H. Rubahn, W. J. van der Zande, R. Zhang, M. J. Bronikowski, and R. N. Zare, *Chem. Phys. Lett.*, 1991, **186**, 154.
213. J. A. Miller, R. J. Kee, and C. K. Westbrook, *Annu. Rev. Phys. Chem.*, 1990, **41**, 345.
214. C. Y. Yang and S. J. Klippenstein, *J. Chem. Phys.*, 1995, **103**, 7287.

215. M. R. Pastrana, L. A. M. Quintales, J. Brandão, and A. J. C. Varandas, *J. Phys. Chem.*, 1990, **94**, 8073.
216. B. K. Kendrick and R. T. Pack, *J. Chem. Phys.*, 1995, **102**, 1994.
217. C. F. Melius and R. J. Blint, *Chem. Phys. Lett.*, 1979, **64**, 183.
218. P. Honvault, S. Y. Lin, D. Xie, and H. Guo, *J. Phys. Chem. A*, 2007, **111**, 5349.
219. E. A. McCollough and R. E. Wyatt, *J. Chem. Phys.*, 1971, **54**, 3578.
220. J. C. Juanes-Marcos, S. C. Althorpe, and E. Wrede, *Science*, 2005, **309**, 1227.
221. S. C. Althorpe, F. Fernandez-Alonso, B. D. Bean, J. D. Ayers, A. E. Pomerantz, and E. Wrede R. N. Zare, *Nature*, 2002, **416**, 67.
222. D. M. Medvedev, S. K. Gray, E. M. Goldfield, M. J. Lakin, D. Troya, and G. C. Schatz, *J. Chem. Phys.*, 2004, **120**, 1231.
223. S. L. Mielke and D. G. Truhlar, *J. Phys. Chem. A*, 2009, **113**, 4817.
224. J. Cao and G. A. Voth, *J. Chem. Phys.*, 1994, **100**, 5106.
225. I. R. Craig and D. E. Manolopoulos, *J. Chem. Phys.*, 2004, **121**, 3368.
226. W. H. Miller, *J. Phys. Chem. A*, 2001, **105**, 2942.
227. A. Nakayama and N. Makri, *J. Chem. Phys.*, 2003, **119**, 8592.
228. Y. Wu and V. S. Batista, *J. Chem. Phys.*, 2004, **121**, 1676.
229. S. P. Webb, T. Iordanov, and S. Hammes-Schiffer, *J. Chem. Phys.*, 2002, **117**, 4106.
230. B. Lasorne, M. J. Bearpark, M. A. Robb, and G. A. Worth, *Chem. Phys. Lett.*, 2006, **432**, 604.
231. M. Born and J. R. Oppenheimer, *Ann. Physik*, 1927, **84**, 457.
232. D. R. Yarkony and H. Köppel, *Conical Intersections*, in Advanced Series in Physical Chemistry, Vol. **15**, W. Domcke, Ed., (World Scientific, Singapore), 2004.
233. G. A. Worth and L. S. Cederbaum, *Annu. Rev. Phys. Chem.*, 2004, **55**, 127.
234. D. R. Yarkony, *Rev. Mod. Phys.*, 1996, **68**, 985.
235. H. S. W. Massey, *Rep. Proc. Phys.*, 1948, **12**, 248.
236. C. Wittig, *J. Phys. Chem. B*, 2005, **109**, 8428.
237. L. J. Butler, *Annu. Rev. Phys. Chem.*, 1998, **49**, 125.
238. E. Wrede, S. Laubach, S. Schulenburg, A. Brown, E. R. Wouters, A. J. Orr-Ewing, and M. N. R. Ashfold, *J. Chem. Phys.*, 2001, **114**, 2629.
239. R. N. Zare, *Angular Momentum, Understanding Spatial Aspects in Chemistry and Physics*, (Wiley, New York, USA), 1988.
240. P. F. Bernath, *Spectra of Atoms and Molecules*, 2nd Edition (Oxford University Press, Oxford, UK), 2005.
241. M. H. Alexander, B. Pouilly, and T. Duhoo, *J. Chem. Phys.*, 1993, **99**, 1752.
242. P. M. Regan, D. Ascenzi, A. Brown, G. G. Balint-Kurti, and A. J. Orr-Ewing, *J. Chem. Phys.*, 2000, **112**, 10259.
243. R. N. Porter, R. M. Stevens, and M. Karplus, *J. Chem. Phys.*, 1968, **49**, 5163.
244. A. I. Boothroyd, W. J. Keogh, P. G. Martin, and M. R. Peterson, *J. Chem. Phys.*, 1996, **104**, 7139.
245. S. J. Greaves, D. Murdock, E. Wrede, and S. C. Althorpe, *J. Chem. Phys.*, 2008, **128**, 164306.
246. G. C. Schatz, L. A. Pederson, and P. J. Kuntz, *Faraday Discuss. Chem. Soc.*, 1997, **108**, 357.
247. F. J. Aoiz, L. Bañares, M. Brouard, J. F. Castillo, and V. J. Herrero, *J. Chem. Phys.*, 2000, **113**, 5339.
248. T. Takayanagi, *J. Chem. Phys.*, 2002, **116**, 2439.
249. M. Brouard, P. O'Keeffe, and C. Vallance, *J. Phys. Chem. A.*, 2002, **106**, 3629.
250. J. C. Tully and R. Preston, *J. Chem. Phys.*, 1971, **55**, 562.
251. J. C. Tully, *J. Chem. Phys.*, 1990, **93**, 1061.
252. G. Herzberg and H. C. Longuet-Higgins, *Discuss. Faraday Soc.*, 1963, **35**, 77.
253. M. S. Child, *Adv. Chem. Phys.*, 2002, **124**, 1.
254. M. V. Berry, *Proc. R. Soc. London Ser. A*, 1984, **392**, 45.

255. B. K. Kendrick, *J. Phys. Chem.*, 2003, **107**, 6739.
256. J. C. Juanes-Marcos and S. C. Althorpe, *J. Chem. Phys.*, 2005, **122**, 204324.
257. H. Lefebvre-Brion and R. W. Field, *The Spectra and Dynamics of Diatomic Molecules*, 2nd Edition, (Elsevier, Amsterdam, The Netherlands), 2004.
258. S. R. Langford, P. M. Regan, A. J. Orr-Ewing, and M. N. R. Ashfold, *Chem. Phys.*, 1998, **231**, 245.
259. H. M. Lambert, P. J. Dagdigian, and M. H. Alexander, *J. Chem. Phys.*, 1998, **108**, 4460.
260. P. M. Regan, S. R. Langford, A. J. Orr-Ewing, and M. N. R. Ashfold, *J. Chem. Phys.*, 1999, **110**, 281.
261. G. C. Schatz, P. McCabe, and J. N. L. Connor, *Faraday Discuss. Chem. Soc.*, 1998, **110**, 139.
262. G. Herzberg, *Molecular spectra and molecular structure*, Volumes **I–III**, 2nd Ed., (Van Nostrand, Inc., New York, USA), 1950.
263. G. M. Sweeney and K. G. McKendrick, *J. Chem. Phys.*, 1997, **106**, 9182.
264. P. C. Samartzis, B. L. G. Bakker, T. P. Rakitzis, D. H. Parker, and T. N. Kitsopoulos, *J. Chem. Phys.*, 1999, **110**, 5201.
265. S.-H. Lee and K. Liu, *J. Chem. Phys.*, 1999, **111**, 6253.
266. F. Dong, S.-H. Lee and K. Liu, *J. Chem. Phys.*, 2001, **115**, 1197.
267. L. Che, Z. Ren, X. Wang, W. Dong, D. Dai, X. Wong, D. H. Zhang, X. Yang, L. Sheng, G. Li, H.-J. Werner, F. Lique, and M. H. Alexander, *Science*, 2007, **317**, 1061.
268. F. J. Aoiz, L. Bañares, T. Bohm, A. Hanf, V. J. Herrero, K.-H. Jung, A. Läuter, K. W. Lee, M. Menendez, V. Sáez Rábanos, I. Tanarro, H.-R. Volpp, and J. Wolfrum, *J. Phys. Chem A*, 2000, **104**, 10452.
269. A. Hanf, A. Lauter, D. Suresh, H. R. Volpp, and J. Wolfrum, *Chem. Phys. Lett.*, 2001, **340**, 71.
270. B. Retail, J. K. Pearce, C. Murray, and A. J. Orr-Ewing, *J. Chem. Phys.*, 2005, **122**, 101101.
271. M. Brouard, I. Burak, S. D. Gatenby, and G. A. J. Markillie, *Chem. Phys. Lett.*, 1998, **287**, 682.
272. M. Brouard, I. Burak, S. Marinakis, L. Rubio Lago, P. Tampkins, and C. Vallance, *J. Chem. Phys.*, 2004, **121**, 10426.
273. G. M. Sweeney, A. Watson, and K. G. McKendrick, *J. Chem. Phys.*, 1997, **106**, 9172.
274. M. Ziemkiewicz, M. Vojcik, and D. J. Nesbitt, *J. Chem. Phys.*, 2005, **123**, 224307.
275. M. P. Deskevich, D. J. Nesbitt, and H.-J. Werner, *J. Chem. Phys.*, 2004, **120**, 7281.
276. F. Dong, S.-H. Lee, and K. Liu, *J. Chem. Phys.*, 2000, **113**, 3633.
277. S. A. Nizkorodov, W. W. Harper, and D. J. Nesbitt, *Faraday Discuss. Chem. Soc.*, 1999, **113**, 107.
278. W. W. Harper, S. A. Nizkorodov, and D. J. Nesbitt, *J. Chem. Phys.*, 2002, **116**, 5622.
279. F. Lique, M. H. Alexander, G. Li, H.-J. Werner, S. A. Nizkorodov, W. W. Harper, and D. J. Nesbitt, *J. Chem. Phys.*, 2008, **128**, 084313.
280. M. N. R. Ashfold, N. H. Nahler, A. J. Orr-Ewing, O. P. J. Vieuxmaire, R. L. Toomes, T. N. Kitsopoulos, I. A. Garcia, D. A. Chestakov, S. M. Wu, and D. H. Parker, *Phys. Chem. Chem. Phys.*, 2006, **8**, 26.
281. M. H. Alexander, G. Capecchi, and H.-J. Werner, *Science*, 2002, **296**, 715.
282. M. H. Alexander, G. Capecchi, and H.-J. Werner, *Faraday Discuss. Chem. Soc.*, 2004, **127**, 59.
283. X. Wang, W. Dong, C. Xiao, L. Che, Z. Ren, D. Dai, X. Wang, P. Casavecchia, X. Yang, B. Jiang, Z. Sun, S.-Y. Lee, D. H. Zhang, H. J. Werner, and M. H. Alexander, *Science*, 2008, **322**, 573.
284. E. Garand, J. Zhou, D. E. Manolopoulos, M. H. Alexander, and D. M. Neumark, *Science*, 2008, **319**, 72.
285. B. Retail, S. J. Greaves, J. K. Pearce, R. A. Rose, and A. J. Orr-Ewing, *Phys. Chem. Chem. Phys.*, 2007, **9**, 3261.

286. Y. Matsumi, K. Izumi, V. Skorokhodov, M. Kawasaki, and N. Tanaka, *J. Phys. Chem. A*, 1997, **101**, 1216.
287. B. Retail, J. K. Pearce, S. J. Greaves, R. A. Rose, and A. J. Orr-Ewing, *J. Chem. Phys.*, 2008, **128**, 184303.
288. N. Balucani, G. Capozza, F. Leonori, E. Segoloni, and P. Casavecchia, *Int. Rev. Phys. Chem.*, 2006, **25**, 109.
289. A. G. Suits and R. E. Continetti, Eds., E. R. Wouters, M. Ahmed, D. S. Peterka, A. S. Bracker, A. G. Suits, and O. S. Vasyutinskii, in *Imaging in Chemical Dynamics, ACS Symposium Series 770*, (American Chemical Society, Washington DC, USA), 2001.
290. M. G. D. Nix, A. L. Devine, B. Cronin, R. N. Dixon, and M. N. R. Ashfold, *J. Chem. Phys.*, 2006, **125**, 133318.
291. M. L. Hause, Y. H. Yoon, A. S. Case, and F. F. Crim, *J. Chem. Phys.*, 2008, **128**, 104307.
292. M. G. D. Nix, A. L. Devine, R. N. Dixon, and M. N. R. Ashfold, *Chem. Phys. Lett.*, 2008, **463**, 305.
293. P. Ehrenfest, *Koninkl. Ned. Akad. Wetenschap. Proc.*, 1914, **16**, 591.
294. G. E. Ewing and D. W. Chandler, *J. Chem. Phys.*, 1980, **73**, 4904.
295. L. D. Landau and E. Teller, *Phys. Z. Sowj. UN.*, 1936, **10**, 34.
296. M. S. Child, *Molecular Collision Theory*, (Dover publications, Mineola N.Y., USA), 1996.
297. J. D. Lambert, *International Series of Monographs on Chemistry: Vibrational and Rotational Relaxation in Gases*, (Clarendon Press, Oxford, UK), 1977.
298. E. E. Nikitin and J. Troe, *Phys. Chem. Chem. Phys.*, 2008, **10**, 1483.
299. R. N. Schwartz, Z. I. Slawsky, and K. F. Hertzfeld, *J. Chem. Phys.*, 1952, **20**, 1591.
300. R. B. Bernstein and K. H. Kramer, *J. Chem. Phys.*, 1966, **44**, 4473.
301. R. B. Bernstein, Ed., W. R. Gentry, in *Atom-Molecule Collision Theory; A Guide for the Experimentalist*, p. 391, (Plenum Press, New York, USA), 1979.
302. M. Faubel, *Adv. Atomic and Molec. Phys.*, 1983, **19**, 345.
303. U. Buck and V. Khare, *Chem. Phys.*, 1977, **26**, 215.
304. H. A. Rabitz and R. G. Gordon, *J. Chem. Phys.*, 1970, **53**, 1831.
305. V. Khare, D. J. Kouri, and R. T. Pack, *J. Chem. Phys.*, 1978, **69**, 4419.
306. D. Secrest, *J. Chem. Phys.*, 1975, **62**, 710.
307. J. T. Yardley, *Introduction to Molecule Energy Transfer*, (Academic Press, New York, USA), 1980.
308. A. Gijsbertsen, H. Linnartz, C. J. Taatjes, and S. Stolte, *J. Am. Chem. Soc.*, 2006, **128**, 8777.
309. A. Ballast, A. Gijsbertsen, H. Linnartz, and S. Stolte, *A Quasi-Quantum Treatment of Inelastic Molecular Collisions*, in *Proceedings of the 25th International Symposium on Rarefied Gas Dynamics*, M. S. Ivanov and A. K. Rebrov Eds., (Publishing House of the Siberian branch of the Russian Academy of Sciences, St. Petersburg, Russia), 2007, p. 1263.
310. A. Ballast, A. Gijsbertsen, H. Linnartz, and S. Stolte, *Mol. Phys.*, 2008, **85**, 5499.
311. C. A. Taatjes, A. Gijsbertsen, and M. J. L. de Lange, *J. Phys. Chem. A*, 2007, **111**, 7613.
312. J. P. Toennies, *Annu. Rev. Phys. Chem.*, 1976, **27**, 225.
313. R. B. Bernstein, *Comments At. Mol. Phys.*, 1973, **4**, 43.
314. M. Lemeshko and B. Friedrich, *J. Chem. Phys.*, 2008, **129**, 24301.
315. H. J. Metcalf and P. van der Straten, *Laser Cooling and Trapping*, (Springer Press, New York, USA), 1999.
316. P. McCabe, J. N. L. Connor, and D. Sokolovski, *J. Chem. Phys.*, 1998, **108**, 5695.
317. D. Beck, U. Ross, and W. Schepper, *Z. Physik A*, 1979, **293**, 107.
318. D. Beck, *Chem. Phys.*, 1988, **126**, 19.
319. S. D. Bosanac, *Phys. Rev. A*, 1980, **22**, 2617.
320. S. D. Bosanac and U. Buck, *Chem. Phys. Lett.*, 1981, **81**, 315.

321. D. W. Chandler and S. Stolte, *Inelastic Energy Transfer: The NO-rare Gas System*, in *Gas Phase Molecular Reaction and Photodissociation Dynamics*, K. C. Lin and P. D. Kleiber, Eds., (Transworld Research Network, Kerala, India), 2007, p. 1.
322. M. L. Costen, S. Marinakis, and K. G. McKendrick, *Chem. Soc. Rev.*, 2008, **37**, 732.
323. M. Faubel, *Status and Future Developments in the study of Transport Properties*, in: *NATO ASI Series C*, Vol. **361**, W. A. Wakeham, A. S. Dickinson, F. R. W. McCourt, and V. Vesovic, Eds., (Kluwer, Dordrecht, Netherlands), 1992, p. 73.
324. A. S. Dickinson and D. Richards, *Adv. At. Mol. Phys.*, 1982, **18**, 165.
325. R. Schinke and J. M. Bowman, in *Molecular Collision Dynamics, Chapt. 4 'Rotational Rainbows in Atom-Diatom Scattering*, J. M. Bowman, Ed., (Springer-Verlag, Berlin, Germany), 1983, p. 61.
326. H. J. Korsch and A. Ernesti, *J. Phys. B: At. Mol. Opt. Phys.*, 1992, **25**, 3565.
327. H.-J. Loesch, *Adv. Chem. Phys.*, 1980, **42**, 422.
328. H. J. Korsch and R. Schinke, *J. Chem. Phys.*, 1981, **75**, 3859.
329. A. G. Suits, L. S. Bontuyan, P. L. Houston, and B. J. Whitaker, *J. Chem. Phys.*, 1992, **96**, 8618.
330. L. S. Bontuyan, A. G. Suits, P. L. Houston, and B. J. Whitaker, *J. Phys. Chem.*, 1993, **97**, 6342.
331. V. Khare, D. J. Kouri, and D. K. Hoffman, *J. Chem. Phys.*, 1981, **74**, 2275.
332. U. Buck, F. Huisken, D. Otten, and R. Schinke, *Chem. Phys. Lett.*, 1983, **101**, 126.
333. U. Buck, D. Otten, R. Schinke, and D. Poppe, *J. Chem. Phys.*, 1985, **82**, 202.
334. S. D. Bosanac, *Chem. Phys. Lett.*, 1984, **103**, 484.
335. K. Bergmann, R. Engelhardt, U. Hefter, P. Hering, and J. Witt, *Phys. Rev. Lett.*, 1978, **40**, 1446.
336. K. Bergmann, U. Hefter, and J. Witt, *J. Chem. Phys.*, 1980, **72**, 4777.
337. J. A. Serri, A. Morales, W. Moskowitz, D. E. Pritchard, C. H. Becker, and J. L. Kinsey, *J. Chem. Phys.*, 1980, **72**, 6304.
338. J. A. Serri, C. H. Becker, M. B. Ebel, J. L. Kinsey, W. Moskowitz, and D. E. Pritchard, *J. Chem. Phys.*, 1981, **74**, 5116.
339. K. Bergmann, U. Hefter, A. Mattheus, J. Witt, and R. Schinke, *Phys. Rev. Lett.*, 1981, **46**, 915.
340. P. L. Jones, U. Hefter, A. Mattheus, J. Witt, K. Bergmann, W. Mller, W. Meyer, and R. Schinke, *Phys. Rev. A*, 1982, **26**, 1283.
341. P. L. Jones, E. Gottwald, U. Hefter, and K. Bergmann, *J. Chem. Phys.*, 1983, **78**, 3838.
342. A. Mattheus, A. Fischer, G. Ziegler, E. Gottwald, and K. Bergmann, *Phys. Rev. Lett.*, 1986, **56**, 712.
343. W. P. Moskowitz, B. Stewart, R. M. Bilotta, J. L. Kinsey, and D. E. Pritchard, *J. Chem. Phys.*, 1984, **80**, 5496.
344. W. R. Gentry and C. F. Giese, *Phys. Rev. Lett.*, 1977, **39**, 1259.
345. W. R. Gentry and C. F. Giese, *J. Chem. Phys.*, 1977, **67**, 5389.
346. W. R. Gentry and C. F. Giese, *Rev. Sci. Instr.*, 1978, **49**, 595.
347. U. Buck, F. Huisken, J. Schleusener, and H. Pauly, *Phys. Rev. Lett.*, 1977, **38**, 680.
348. U. Buck, *Faraday Discuss. Chem. Soc.*, 1982, **73**, 187.
349. M. Faubel, K. Kohl, and J. P. Toennies, *J. Chem. Phys.*, 1980, **73**, 2650.
350. M. Faubel and G. Kraft, *J. Chem. Phys.*, 1986, **85**, 2671.
351. L. J. Rawluk, Y. B. Fan, Y. Apelblat, and M. Keil, *J. Chem. Phys.*, 1991, **94**, 4205.
352. J. J. C. Barrett, H. R. Mayne, and M. Keil, *J. Chem. Phys.*, 1994, **100**, 304.
353. K. T. Lorenz, M. S. Westley, and D. W. Chandler, *Phys. Chem. Chem. Phys.*, 2000, **2**, 481.
354. J. M. Hutson, *J. Chem. Phys.*, 1988, **89**, 4550.
355. J. M. Hutson, *J. Phys. Chem.*, 1992, **96**, 4237.

356. J. M. Ceremia and H. Rabitz, *J. Chem. Phys.*, 2001, **115**, 8899.
357. D. W. Chandler and P. L. Houston, *J. Chem. Phys.*, 1987, **87**, 1445.
358. M. H. Alexander, *J. Chem. Phys.*, 1982, **76**, 5974.
359. R. de Vivie and S. D. Peyerimhoff, *J. Chem. Phys.*, 1989, **90**, 3660.
360. F. H. Geuzebroek, M. G. Tenner, A. W. Kleyn, H. Zacharias, and S. Stolte, *Chem. Phys. Lett.*, 1991, **187**, 520.
361. M. H. Alexander, *Chem. Phys.*, 1985, **92**, 337.
362. P. J. Dagdigian, M. H. Alexander, and K. Liu, *J. Chem. Phys.*, 1989, **91**, 839.
363. L. Bigio and E. R. Grant, *J. Chem. Phys.*, 1987, **87**, 5589.
364. M. H. Alexander, P. Andresen, R. Bacis, R. Bersohn, F. J. Comes, P. J. Dagdigian, R. N. Dixon, R. W. Field, G. W. Flynn, K.-H. Gericke, E. R. Grant, B. J. Howard, J. R. Huber, D. S. King, J. L. Kinsey, K. Kleinermanns, K. Kuchitsu, A. C. Luntz, A. J. McCaffery, B. Pouilly, H. Reisler, S. Rosenwaks, E. W. Rothe, M. Shapiro, J. P. Simons, R. Vasudev, J. R. Wiesenfeld, C. Wittig, and R. N. Zare, *J. Chem. Phys.*, 1988, **89**, 1749.
365. H. Joswig, P. Andresen, and R. Schinke, *J. Chem. Phys.*, 1986, **85**, 1904.
366. M. H. Alexander, *J. Chem. Phys.*, 1993, **99**, 7725.
367. Y. Sumiyoshi and Y. Endo, *J. Chem. Phys.*, 2007, **127**, 184309.
368. M. H. Alexander and S. Stolte, *J. Chem. Phys.*, 2000, **112**, 8017.
369. H. Kohguchi, T. Suzuki, and M. H. Alexander, *Science*, 2001, **294**, 832.
370. S. Stolte, J. Reuss, and H. L. Schwartz, *Physica*, 1973, **66**, 211.
371. H. H. W. Thuis, S. Stolte, and J. Reuss, *Chem. Phys.*, 1979, **43**, 351.
372. S. Stolte, J. Reuss, and H. L. Schwartz, *Physica*, 1972, **57**, 254.
373. H. H. W. Thuis, S. Stolte, J. Reuss, J. J. H. van den Biessen, and C. J. N. van den Meijdenberg, *Chem. Phys.*, 1980, **52**, 211.
374. N. J. Bridge and A. D. Buckingham, *Proc. Royal Soc. A*, 1966, **295**, 334.
375. H. L. Schwartz, S. Stolte, and J. Reuss, *Chem. Phys.*, 1973, **2**, 1.
376. P. Andresen, H. Joswig, H. Pauly, and R. Schinke, *J. Chem. Phys.*, 1982, **77**, 2204.
377. C. W. McCurdy and W. H. Miller, *J. Chem. Phys.*, 1977, **66**, 463.
378. M. H. Alexander, *J. Chem. Phys.*, 1999, **111**, 7426.
379. A. Lin, S. Antonova, A. P. Tsakotellis, and G. C. McBane, *J. Phys. Chem. A*, 1999, **103**, 1198.
380. M. Yang and M. H. Alexander, *J. Chem. Phys.*, 1995, **103**, 6973.
381. M. J. L. de Lange, J. J. van Leuken, M. M. J. E. Drabbles, J. Bulthuis, J. G. Snijders, and S. Stolte, *Chem. Phys. Lett.*, 1998, **294**, 332.
382. M. J. L. de Lange, M. Drabbels, P. T. Griffiths, J. Bulthuis, S. Stolte, and J. G. Snijders, *Chem. Phys. Lett.*, 1999, **313**, 491.
383. J. J. van Leuken, F. H. W. van Amerom, J. Bulthuis, J. G. Snijders, and S. Stolte, *J. Phys. Chem.*, 1995, **99**, 15573.
384. M. Brouard, S. P. Duxon, P. A. Enriquez, and J. P. Simons, *J. Chem. Phys.*, 1992, **97**, 7414.
385. F. J. Aoiz, M. Brouard, P. A. Enriquez, and R. Sayos, *J. Chem. Soc. Faraday Trans.*, 1993, **89**, 1427.
386. N. E. Shafer, A. J. Orr-Ewing, W. R. Simpson, H. Xu, and R. N. Zare, *Chem. Phys. Lett.*, 1993, **212**, 155.
387. W. R. Simpson, A. J. Orr-Ewing, and R. N. Zare, *Chem. Phys. Lett.*, 1993, **212**, 163.
388. T. N. Kitsopoulos, M. A. Buntine, D. P. Baldwin, R. N. Zare, and D. W. Chandler, *Science*, 1993, **260**, 1605.
389. N. E. Shafer-Ray, A. J. Orr-Ewing, and R. N. Zare, *J. Phys. Chem.*, 1995, **99**, 7591.
390. A. T. J. B. Eppink and D. H. Parker, *Rev. Sci. Instrum.*, 1997, **68**, 3477.
391. M. S. Westley, K. T. Lorenz, D. W. Chandler, and P. L. Houston, *J. Chem. Phys.*, 2001, **114**, 2669.
392. K. T. Lorenz, D. W. Chandler, and G. C. McBane, *J. Phys. Chem. A* 2002, **106**, 1144.

393. H. Kohguchi and T. Suzuki, *Annu. Rep. Prog. Chem., Sect. C*, 2002, **98**, 421.
394. N. Yonekura, C. Gebauer, H. Kohguchi, and T. Suzuki, *Rev. Sci. Instrum.*, 1990, **70**, 3265.
395. B.-Y. Chang, R. C. Hoetzlein, J. A. Mueller, J. D. Geiser, and P. L. Houston, *Rev. Sci. Instrum.*, 1998, **69**, 1665.
396. W. Li, S. D. Chambreau, S. A. Lahankar, and A. G. Suits, *Rev. Sci. Instrum.*, 2005, **76**, 063106.
397. B. J. Whitaker, *Image reconstruction: the Abel transform. In Imaging in chemical dynamics*, A. G. Suits & R. E. Continetti, Eds., (ACS Symposium Series 770, Ch. 5, p. 68), 2001.
398. A. S. Bracker, E. R. Wouters, A. G. Suits, and O. S. Vasyutinskii, *J. Chem. Phys.*, 1999, **110**, 6749.
399. M. J. Bass, M. Brouard, A. P. Clark, and C. Vallance, *J. Chem. Phys.*, 2002, **117**, 8723.
400. C. R. Gebhardt, T. P. Rakitzis, P. C. Samartzis, V. Ladopoulos, and T. N. Kitsopoulos, *Rev. Sci. Instrum.*, 2001, **72**, 3848.
401. D. Townsend, M. Minitti, and A. G. Suits, *Rev. Sci. Instrum.*, 2003, **74**, 2530.
402. J. J. Lin, J. Zhou, W. Shiu, and K. Liu, *Rev. Sci. Instrum.*, 2003, **74**, 2495.
403. L. Dinu, A. T. J. B. Eppink, F. Rosca-Pruna, H. L. Offerhaus, W. J. van der Zande, and M. J. J. Vrakking, *Rev. Sci. Instrum.*, 2002, **73**, 4206.
404. R. Y. Rubinstein, *Simulation and the Monte Carlo Method*, (John Wiley and Sons Inc., New York, USA), 1981.
405. S. D. Jons, J. E. Shirley, M. T. Vonk, C. F. Geise, and W. R. Gentry, *J. Chem. Phys.*, 1992, **97**, 7831.
406. P. A. Barrass, P. Sharkey, and I. W. M. Smith, *Phys. Chem. Chem. Phys.*, 2003, **5**, 1400.
407. A. Gijsbertsen, H. Linnartz, G. Rus, A. E. Wiskerke, S. Stolte, D. W. Chandler, and J. Kłos, *J. Chem. Phys.*, 2005, **123**, 224305.
408. A. Gijsbertsen, H. Linnartz, and S. Stolte, *J. Chem. Phys.*, 2006, **125**, 133112.
409. U. Fano and J. H. Macek, *Rev. Mod. Phys.*, 1973, **45**, 553.
410. A. J. Orr-Ewing and R. N. Zare, *Annu. Rev. Phys. Chem.*, 1994, **45**, 315.
411. V. K. Nestorov, R. D. Hinchliffe, R. Uberna, J. I. Cline, K. T. Lorenz, and D. W. Chandler, *J. Phys. Chem.*, 2001, **115**, 7881.
412. I. V. Hertel and W. Stoll, *Adv. At. Mol. Phys.*, 1978, **13**, 113.
413. F. J. Aoiz, J. E. Verdasco, V. J. Herrero, V. S. Rabanos, and M. A. Alexander, *J. Chem. Phys.*, 2003, **119**, 5860.
414. F. J. Aoiz, V. J. Herrero, V. S. Rabanos, and J. E. Verdasco, *Phys. Chem. Chem. Phys.*, 2004, **6**, 4407.
415. J. I. Cline, K. T. Lorenz, E. A. Wade, J. W. Barr, and D. W. Chandler, *J. Chem. Phys.*, 2001, **115**, 6277.
416. E. A. Wade, K. T. Thomas, D. W. Chandler, J. W. Barr, G. L. Bares, and J. I. Cline, *Chem. Phys.*, 2004, **301**, 261.
417. D. C. Jacobs and R. N. Zare, *J. Chem. Phys.*, 1986, **85**, 5499.
418. K. T. Lorenz, D. W. Chandler, J. W. Barr, W. W. Chen, G. L. Barnes, and J. I. Cline, *Science*, 2001, **293**, 2063.
419. J. P. Toennies, *Z. Physik.*, 1965, **182**, 257.
420. J. P. Toennies, *Z. Physik.*, 1966, **193**, 76.
421. H. L. Bethlem, G. Berden and G. Meijer, *Phys. Rev. Lett.*, 1999, **83**, 1558.
422. S. Y. T. van de Meerakker and G. Meijer, *Faraday Discuss.*, 2009, **142**, 113.
423. M. S. Elioff, J. J. Valentini and D. W. Chandler, *Science*, 2003, **302**, 1940.
424. M. S. Elioff, J. J. Valentini and D. W. Chandler, *Eur. Phys. J. D*, 2004, **31**, 385.
425. K. E. Strecker and D. W. Chandler, *Phys. Rev. A*, 2008, **78**, 063406.
426. J. J. Kay, S. Y. T. van de Meerakker, K. E. Strecker, and D. W. Chandler, *Faraday Discuss.*, 2009, **142**, 9.

427. K. E. Strecker and D. W. Chandler, in *Low Temperatures and cold molecules*, I. W. M. Smith, Ed., (World Scientific, London), 2008.
428. H. Meyer, *Chem. Phys. Lett.*, 1994, **230**, 519.
429. H. Meyer, *J. Chem. Phys.*, 1995, **102**, 3151.
430. Y. Kim and H. Meyer, *Chem. Phys.*, 2004, **301**, 237.
431. Y. T. Lee, in *Atomic and Molecular Beam Methods*, Vol. **1**, G. Scoles, Ed., (Oxford University Press, New York, USA), 1988.
432. P. Casavecchia, *Rep. Progr. Phys.*, 2000, **63**, 355.
433. D. R. Herschbach, *Angew. Chemie Int. Engl. Ed.*, 1987, **26**, 1223.
434. W. B. Miller, S. A. Safron, and D. R. Herschbach, *J. Chem. Phys.*, 1972, **56**, 3581.
435. L. Schnieder, K. Seekamp-Rahn, E. Wrede, and K. H. Welge, *J. Chem. Phys.*, 1997, **107**, 6175.
436. L. Schnieder, K. Seekamp-Rahn, J. Borkowski, E. Wrede, K. H. Welge, F. J. Aoiz, L. Bañares, M. J. Dmello, V. J. Herrero, V. S. Rabanos, and R. E. Wyatt, *Science*, 1995, **269**, 207.
437. X. Liu, J. J. Lin, S. A. Harich, G. C. Schatz, and X. Yang, *Science*, 2000, **285**, 1249.
438. C. Lin, M. F. Witinski, and H. F. Davis, *jcp*, 2003, **119**, 251.
439. M. F. Witinski, M. Ortiz-Suárez, and H. F. Davis, *J. Chem. Phys.*, 2006, **124**, 094307.
440. Y. T. Lee, J. D. McDonald, P. R. Le Breton, and D. R. Herschbach, *Rev. Sci. Instrum.*, 1969, **40**, 1402.
441. X. Yang and K. Liu, Eds., *Modern Trends in Chemical Reaction Dynamics: Experiment and Theory* (*Part I & II*), Adv. Series in Phys. Chem. Vol. **14**, (World Scientific, Singapore), 2004.
442. G. Scoles, Ed., W. R. Gentry, in *Atomic and Molecular Beam Methods*, Vol. **1**, p. 54, (Oxford University Press, New York, USA), 1987.
443. R. I. Kaiser, *Chem. Rev.*, 2002, **102**, 1309.
444. X. Yang and K. Liu, Eds., J. J. Schroden and H. F. Davis, in *Modern Trends in Chemical Reaction Dynamics: Experiment and Theory (Part II);* (Adv. Series in Phys. Chem. Vol. **14**), Chapter 5, (World Scientific, Singapore), 2004.
445. Y. T. Lee, *Science*, 1987, **236**, 793.
446. Y. T. Lee, *Angew. Chemie Int. Engl. Ed.*, 1987, **26**, 939.
447. D. M. Neumark, A. M. Wodtke, G. N. Robinson, C. C. Hayden, and Y. T. Lee, *Phys. Rev. Lett.*, 1984, **53**, 226.
448. D. M. Neumark, A. M. Wodtke, G. N. Robinson, C. C. Hayden, and Y. T. Lee, *J. Chem. Phys.*, 1985, **82**, 3045.
449. M. Baer, M. Faubel, B. Martinez-Haya, L. Y. Rusin, U. Tappe, and J. P. Toennies, *J. Chem. Phys.*, 1999, **110**, 10231.
450. J. H. Birely, R. R. Herm, K. R. Wilson, and D. R. Herschbach, *J. Chem. Phys.*, 1967, **47**, 993.
451. M. Alagia, N. Balucani, L. Cartechini, P. Casavecchia, E. H. van Kleef, G. G. Volpi, F. J. Aoiz, L. Bañares, D. W. Schwenke, T. C. Allison, S. L. Mielke, and D. G. Truhlar, *Science*, 1996, **273**, 1519.
452. W. B. Miller, S. A. Safron, and D. R. Herschbach, *Faraday Discuss. Chem. Soc.*, 1967, **44**, 108.
453. M. Alagia, N. Balucani, P. Casavecchia, D. Stranges, and G. G. Volpi, *J. Chem. Phys.*, 1993, **98**, 8341.
454. R. Grice, *Int. Rev. Phys. Chem.*, 1995, **14**, 315.
455. X. Yang, J. J. Lin, Y. T. Lee, D. A. Blank, A. G. Suits, and A. M. Wodtke, *Rev. Sci. Instrum.*, 1997, **68**, 3317.
456. C. C. Wang, J. Shu, J. J. Lin, Y. T. Lee, and X. Yang, *J. Chem. Phys.*, 2002, **117**, 153.
457. D. A. Blank, W. Sun, A. G. Suits, Y. T. Lee, S. W. North, and G. E. Hall, *J. Chem. Phys.*, 1998, **108**, 5414.

458. W. Sun, K. Yokoyama, J. C. Robinson, A. G. Suits, and D. M. Neumark, *J. Chem. Phys.*, 1999, **110**, 4363.
459. J. C. Robinson, S. A. Harris, W. Sun, N. E. Sveum, and D. M. Neumark, *J. Am. Chem. Soc.*, 2002, **124**, 10211.
460. J. J. Lin, Y. Chen, Y. Y. Lee, Y. T. Lee, and X. Yang, *Chem. Phys. Lett.*, 2002, **361**, 374.
461. S.-H. Lee, Y.-Y. Lee, Y. T. Lee, and X. Yang, *J. Chem. Phys.*, 2003, **119**, 827.
462. S.-H. Lee, W.-K. Chen, C. Chaudhuri, W.-J. Huang, and Y. T. Lee, *J. Chem. Phys.*, 2006, **125**, 144315.
463. L. R. McCunn, K.-C. Lau, M. J. Krisch, L. Butler, J.-W. Tsung, and J. J. Lin, *J. Phys. Chem. A*, 2006, **110**, 1625.
464. S.-H. Lee, *J. Chem. Phys.*, 2009, **131**, 174312.
465. D. A. Blank, N. Hemmi, A. G. Suits, and Y. T. Lee, *Chem. Phys.*, 1998, **231**, 261.
466. N. Hemmi and A. G. Suits, *J. Chem. Phys.*, 1998, **109**, 5338.
467. S.-H. Lee, W.-J. Huang, and W.-K. Chen, *Chem. Phys. Lett.*, 2007, **446**, 276.
468. S.-H. Lee, W.-K. Chen, and W.-J. Huang, *J. Chem. Phys.*, 2009, **130**, 054301.
469. C. Chaudhuri, I.-C. Lu, J. J. Lin, and S.-H. Lee, *Chem. Phys. Lett.*, 2007, **444**, 237.
470. P. A. Willis, H. U. Stauffer, R. Z. Hinrichs, and H. F. Davis, *Rev. Sci. Instrum.*, 1999, **70**, 2606.
471. X. Yang and K. Liu, Eds., P. Casavecchia, G. Capozza, and E. Segoloni, in *Modern Trends in Chemical Reaction Dynamics, Part II: Experiment and Theory*, (Adv. Ser. Phys. Chem., Vol. **14**), p. 329, (World Scientific, Singapore), 2004.
472. G. Capozza, E. Segoloni, F. Leonori, G. G. Volpi, and P. Casavecchia, *J. Chem. Phys.*, 2004, **120**, 4557.
473. P. Casavecchia, G. Capozza, E. Segoloni, F. Leonori, N. Balucani, and G. G. Volpi, *J. Phys. Chem. A*, 2005, **109**, 3527.
474. W. L. Fitch and A. D. Sauter, *Anal. Chem.*, 1983, **55**, 832.
475. H. Deutsch, K. Becker, and T. D. Märk, *Int. J. Mass Spectrom. & Ion Processes*, 1997, **167/168**, 503.
476. H. Deutsch, K. Becker, S. Matt, and Märk, *Int. J. Mass Spectrom. & Ion Processes*, 2000, **197**, 37.
477. T. L. Nguyen, L. Vereecken, H. J. Hou, M. T. Nguyen, and J. Peeters, *J. Phys. Chem. A*, 2005, **109**, 7489.
478. W. Hu, G. Lendvay, B. Maiti, and G. C. Schatz, *J. Phys. Chem. A*, 2008, **112**, 2093.
479. M. Costes, N. Daugey, C. Naulin, A. Bergeat, F. Leonori, E. Segoloni, R. Petrucci, N. Balucani, and P. Casavecchia, *Faraday Discuss. Chem. Soc.*, 2006, **133**, 157.
480. F. Leonori, R. Petrucci, E. Segoloni, A. Bergeat, K. M. Hickson, N. Balucani, and P. Casavecchia, *J. Phys. Chem. A*, 2008, **112**, 1363.
481. A. M. Mebel, V. V. Kislov, and M. Hayashi, *J. Chem. Phys.*, 2007, **126**, 204310.
482. Y. Guo, A. M. Mebel, X. Gu, and R. I. Kaiser, *J. Phys. Chem. A*, 2007, **111**, 2980.
483. R. N. Dixon, D. W. Huang, X. F. Yang, S. A. Harich, J. J. Lin, and X. Yang, *Science*, 1999, **285**, 1249.
484. X. Yang, *Int. Rev. Phys. Chem.*, 2005, **24**, 37.
485. X. Liu, J. J. Lin, S. A. Harich, G. C. Schatz, and X. Yang, *Science*, 2000, **289**, 1536.
486. S. A. Harich, D. Dai, C. C. Wang, X. Yang, S. D. Chao, and R. T. Skodje, *Nature*, 2002, **419**, 281.
487. D. Dai, C. C. Wang, S. A. Harich, X. Wang, X. Yang, S. D. Chao, and R. T. Skodje, *Science*, 2003, **300**, 1730.
488. J. Zhang, D. Dai, C. C. Wang, S. A. Harich, X. Wang, X. Yang, M. Gustaffson, and R. T. Skodje, *Phys. Rev. Lett.*, 2006, **96**, 093201.

489. M. Qiu, Z. Ren, L. Che, D. Dai, S. A. Harich, X. Wang, X. Yang, M. Gustafsson, R. T. Skodje, C. Xu, D. Xie, Z. Sun, and D. H. Zhang, *Science*, 2006, **311**, 1440.
490. M. Qiu, Z. Ren, L. Che, D. Dai, S. A. Harich, X. Wang, and X. Yang, *Chinese J. Chem. Phys.*, 2006, **19**, 93.
491. B. R. Strazisar, C. Lin, and H. F. Davis, *Science*, 2000, **290**, 958.
492. J. P. Marangos, N. Shen, H. Ma, M. H. R. Hutchinson, and J. P. Connerade, *J. Opt. Soc. Am. B*, 1990, **7**, 1254.
493. M. Brouard, S. P. Duxon, P. A. Enriquez, and J. P. Simons, *J. Chem. Soc. Faraday Trans.*, 1993, **89**, 1435.
494. M. J. Bass, M. Brouard, C. Vallance, T. N. Kitsopoulos, P. C. Samartzis, and R. L. Toomes, *J. Chem. Phys.*, 2003, **119**, 7168.
495. F. J. Aoiz, M. Brouard, and P. A. Enriquez, *J. Chem. Phys.*, 1996, **105**, 4964.
496. E. Wrede, *Ph. D. Thesis*, University of Bielefeld, Germany, 1998.
497. S. A. Harich, D. Dai, X. Yang, S. D. Chao, and R. T. Skodje, *J. Chem. Phys.*, 2002, **116**, 4769.
498. S. D. Chao, S. A. Harich, D. Dai, C. C. Wang, X. Yang, and R. T. Skodje, *J. Chem. Phys.*, 2002, **117**, 8341.
499. K. D. Rinnen, D. A. V. Kliner, M. A. Buntine, and R. N. Zare, *Chem. Phys. Lett.*, 1990, **169**, 365.
500. F. Fernandez-Alonso and R. N. Zare, *Annu. Rev. Phys. Chem.*, 2002, **53**, 67.
501. K. Liu, *Annu. Rev. Phys. Chem.*, 2001, **52**, 139.
502. F. Fernandez-Alonso, B. D. Bean, J. D. Ayers, A. E. Pomerantz, R. N. Zare, L. Bañares, and F. J. Aoiz, *Angew. Chem. Int. Edn. Engl.*, 2000, **39**, 2748.
503. R. T. Skodje and X. Yang, *Int. Rev. Phys. Chem.*, 2004, **23**, 253.
504. R. S. Friedman and D. G. Truhlar, *Chem. Phys. Lett.*, 1991, **183**, 539.
505. K. Liu, R. Skodje, and D. E. Manolopoulos, *Phys. Chem. Comm.*, 2002, **5**, 27.
506. F. J. Aoiz, L. Bañares, J. F. Castillo, V. J. Herrero, B. Martinez-Haya, P. Honvault, J. M. Launay, X. Liu, J. J. Lin, S. A. Harich, C. C. Wang, and X. Yang, *J. Chem. Phys.*, 2002, **116**, 10692.
507. S. K. Gray, G. G. Balint-Kurti, G. C. Schatz, J. J. Lin, X. Liu, S. A. Harich, and X. Yang, *J. Chem. Phys.*, 2000, **113**, 7330.
508. X. Liu, C. C. Wang, S. A. Harich, and X. Yang, *Phys. Rev. Lett.*, 2002, **89**, 133201.
509. K. Liu, Y.-T. Hsu, and J.-H. Wang, *J. Chem. Phys.*, 1997, **107**, 2351.
510. X. Liu, J. J. Lin, S. A. Harich, and X. Yang, *Phys. Rev. Lett.*, 2001, **86**, 408.
511. F. J. Aoiz, L. Banares, J. F. Castillo, M. Brouard, W. Denzer, C. Vallance, P. Honvault, J.-M. Launay, A. J. Dobbyn, and P. J. Knowles, *Phys. Rev. Lett.*, 2001, **86**, 1729.
512. A. J. Dobbyn and P. K. Knowles, *Mol. Phys.*, 1997, **91**, 1107.
513. A. J. Dobbyn and P. K. Knowles, *Faraday Discuss. Chem. Soc.*, 1998, **110**, 247.
514. F. J. Aoiz, L. Bañares, V. J. Herrero, V. Saez Rabanos, K. Stark, and H.-J. Werner, *Chem. Phys. Lett.*, 1994, **223**, 215.
515. J. F. Castillo, D. E. Manolopoulos, K. Stark, and H.-J. Werner, *J. Chem. Phys.*, 1996, **104**, 6531.
516. K. Stark and H.-J. Werner, *J. Chem. Phys.*, 1996, **104**, 6515.
517. R. T. Skodje, D. Skouteris, D. E. Manolopoulos, S. H. Lee, F. Dong, and K. Liu, *Phys. Rev. Lett.*, 2000, **85**, 1206.
518. X. Wang, W. Dong, M. Qiu, Z. Ren, L. Che, D. Dai, X. Wang, X. Yang, Z. Sun, B. Fu, S.-Y. Lee, X. Xu, and D. H. Zhang, *Proc. Nat. Acad. Sci.*, 2008, **105**, 6227.
519. W. Shiu, J. J. Lin, and K. Liu, *Phys. Rev. Lett.*, 2004, **92**, 103201.
520. A. Sinha, M. C. Hsiao, and F. F. Crim, *J. Chem. Phys.*, 1991, **94**, 4928.
521. M. J. Bronikowski, W. R. Simpson, and R. N. Zare, *J. Phys. Chem.*, 1993, **97**, 2194.

522. M. Alagia, N. Balucani, P. Casavecchia, S. Stranges, G. G. Volpi, D. C. Clary, A. Kliesch, and H.-J. Werner, *Chem. Phys.*, 1996, **207**, 389.
523. G. Ochoa de Aspuru, and D. C. Clary, *J. Phys. Chem.*, 1998, **102**, 9631.
524. X. Yang and K. Liu, Eds., S. H. Lee, and K. Liu, in *Modern Trends in Chemical Reaction Dynamics*, Advanced Series in Physical Chemistry, Vol. **14**, Part II, Chapt. 1, (World Scientific Pub., Singapore), 2004.
525. J. J. Lin, J. Zhou, W. Shiu, and K. Liu, *Science*, 2003, **300**, 966.
526. K. Liu, *Phys. Chem. Chem. Phys.*, 2007, **9**, 17.
527. J. Zhou, J. J. Lin, W. Shiu, and K. Liu, *Phys. Chem. Chem. Phys.*, 2006, **8**, 3000.
528. J. Zhou, J. J. Lin, W. Shiu, and K. Liu, *J. Chem. Phys.*, 2003, **119**, 4997.
529. K. Schulter and R. G. Gordon, *J. Chem. Phys.*, 1976, **64**, 2918.
530. J. Zhou, J. J. Lin, and K. Liu, *J. Chem. Phys.*, 2004, **121**, 813.
531. P. Casavecchia, F. Leonori, N. Balucani, R. Petrucci, G. Capozza, and E. Segoloni, *Phys. Chem. Chem. Phys.*, 2009, **11**, 46.
532. J. J. Schroden, H. F. Davis, and C. A. Bayse, *J. Phys. Chem. A*, 2007, **111**, 11421.
533. A. D. Estillore, C. Huang, W. Li, and A. G. Suits, *J. Chem. Phys.*, 2008, **129**, 074301.
534. D. R. Albert and H. F. Davis, *J. Phys. Chem. Lett.*, 2010, **1**, 1107.
535. B. Retail, R. A. Rose, J. K. Pearce, S. J. Greaves, and A. J. Orr-Ewing, *Phys. Chem. Chem. Phys.*, 2008, **10**, 1675.
536. T. M. Selby, G. Meloni, F. Goulay, S. R. Leone, A. Fahr, C. A. Taatjes, and D. L. Osborn, *J. Phys. Chem. A*, 2008, **112**, 9366.
537. C. A. Taatjes, N. Hansen, D. L. Osborn, K. Kohse-Hinghaus, T. A. Cool, and P. R. Westmoreland, *Phys. Chem. Chem. Phys.*, 2008, **10**, 20.
538. C. A. Taatjes, D. L. Osborn, T. M. Selby, G. Meloni, A. J. Trevitt, E. Epifanovsky, A. I. Krylov, B. Sirjean, E. Dames, and H. Wang, *J. Phys. Chem. A*, 2010, **114**, 3355.
539. B. L. FitzPatrick, K.-C. Lau, L. J. Butler, S.-H. Lee, and J. Jr.-M. Lin, *J. Chem. Phys.*, 2008, **129**, 084301.
540. S. Yan, Y. T. Wu, and K. P. Liu, *Proc. Natl. Acad. Sci.*, 2008, **105**, 12667.
541. G. Czako and J. M. Bowman, *J. Chem. Phys.*, 2009, **131**, 244302.
542. D. H. Zhang, M. Yang, M. A. Collins, and S.-Y. Lee, *Proc. Nat. Ac. Sci.*, 2002, **99**, 11579.
543. T. Odiorne, P. Brooks, and J. V. V. Kasper, *J. Chem. Phys.*, 1971, **55**, 1980.
544. J. G. Pruett, F. R. Grabiner, and P. R. Brooks, *J. Chem. Phys.*, 1975, **63**, 1173.
545. S. Yan, Y. T. Wu, B. L. Zhang, X. F. Yue, and K. P. Liu, *Science*, 2007, **316**, 1723.
546. F. F. Crim, *Proc. Nat. Acad. Sci.*, 2008, **105**, 12654.
547. S. Yoon, R. J. Holiday, E. L. Sibert, and F. F. Crim, *J. Chem. Phys.*, 2003, **119**, 9568.
548. M. S. Child, *Acc. Chem. Res.*, 1985, **18**, 45.
549. A. Sinha, J. D. Thoemke, and F. F. Crim, *J. Chem. Phys.*, 1992, **96**, 372.
550. G. P. Glass and B. K. Chaturvedi, *J. Chem. Phys.*, 1981, **75**, 2749.
551. R. Zellner and W. Steinert, *Chem. Phys. Lett.*, 1981, **81**, 568.
552. G. C. Light and J. H. Matsumoto, *Chem. Phys. Lett.*, 1978, **58**, 578.
553. I. W. M. Smith and F. F. Crim, *Phys. Chem. Phys. Chem.*, 2002, **4**, 3543.
554. R. J. Holiday, C. H. Kwon, C. J. Annesley, and F. F. Crim, *J. Chem. Phys.*, 2006, **125**, 133101.
555. R. B. Metz, J. D. Thoemke, J. M. Pfeiffer, and F. F. Crim, *J. Chem. Phys.*, 1993, **99**, 1744.
556. Z. H. Kim, H. A. Bechtel, and R. N. Zare, *J. Am. Chem. Soc.*, 2001, **123**, 12714.
557. H. A. Bechtel, Z. H. Kim, J. P. Camden, and R. N. Zare, *J. Chem. Phys.*, 2004, **120**, 791.
558. H. A. Bechtel, J. P. Camden, D. J. A. Brown, M. R. Martin, R. N. Zare, and K. Vodopyanov, *Angew. Chem. Int. Ed.*, 2005, **44**, 2382.
559. C. J. Annesley, A. E. Berke, and F. F. Crim, *J. Phys. Chem. A*, 2008.
560. P. Maroni, D. C. Papageorgopoulos, M. Sacchi, T. T. Dang, R. D. Beck, and T. R. Rizzo, *Phys. Rev. Lett.*, 2005, **94**, 246104.

561. D. R. Killelea, V. L. Campbell, N. S. Shuman, and A. L. Utz, *Science*, 2008, **319**, 790.
562. F. F. Crim, *Accounts Chem. Res.*, 1999, **32**, 877.
563. D. J. Nesbitt and R. W. Field, *J. Phys. Chem.*, 1996, **100**, 12735.
564. J. D. Rynbrandt and B. S. Rabinovitch, *J. Chem. Phys.*, 1971, **54**, 2275.
565. J. D. Rynbrandt and B. S. Rabinovitch, *J. Phys. Chem.*, 1971, **75**, 2164.
566. P. R. Stannard and W. M. Gelbart, *J. Phys. Chem.*, 1981, **85**, 3592.
567. J. F. Kauffman, M. J. Cote, P. G. Smith, and J. D. McDonald, *J. Chem. Phys.*, 1989, **90**, 2874.
568. L. R. Kundkar and A. H. Zewail, *Ann. Rev. Phys. Chem.*, 1990, **41**, 15.
569. M. J. DeWitt, H. S. Yoo, and B. H. Pate, *J. Phys. Chem. A*, 2004, **108**, 1348.
570. Y. Yamada, J. Okano, N. Mikami, and T. Ebata, *J. Chem. Phys.*, 2005, **123**, 124316.
571. K. A. Holbrook, M. J. Pilling, and S. Robertson, *Unimolecular Reactions*, 2nd Edition, (John Wiley and Sons Ltd., Chichester, UK), 1996.
572. G. M. Stewart and J. D. McDonald, *J. Chem. Phys.*, 1983, **78**, 3907.
573. M. Gruebele, *Adv. Chem. Phys.*, 2001, **114**, 193.
574. M. Gruebele and P. G. Wolynes, *Accounts Chem. Res.*, 2004, **37**, 261.
575. W. H. Miller, *J. Phys. Chem. A*, 1998, **102**, 793.
576. P. J. Robinson and K. A. Holbrook, *Unimolecular Reactions*, (Wiley-Interscience, New York, USA), 1972.
577. T. Baer and W. L. Hase, *Unimolecular Reaction Dynamics*, (Oxford University Press, New York, USA), 1996.
578. R. G. Gilbert and S. C. Scott, *Theory of Unimolecular and Recombination Reactions*, (Blackwell Scientific Publications, Oxford, UK), 1990.
579. B. H. Mahan, *J. Chem. Ed.*, 1974, **51**, 709.
580. A. Callegari and T. R. Rizzo, *Chem. Soc. Rev.*, 2001, **30**, 214.
581. I. C. Chen, W. H. Green, and C. B. Moore, *J. Chem. Phys.*, 1988, **89**, 314.
582. S. K. Kim, E. R. Lovejoy, and C. B. Moore, *Science*, 1992, **256**, 1541.
583. R. J. Barnes, G. Dutton, and A. Sinha, *J. Phys. Chem. A*, 1997, **101**, 8374.
584. R. J. Barnes, G. Dutton, and A. Sinha, *J. Chem. Phys.*, 1997, **107**, 3730.
585. A. Callegari, J. Rebstein, J. S. Muenter, R. Jost, and T. R. Rizzo, *J. Chem. Phys.*, 1999, **111**, 123.
586. G. Dutton, R. J. Barnes, and A. Sinha, *J. Chem. Phys.*, 1999, **111**, 4976.
587. J. K. Agbo, D. M. Leitner, D. A. Evans, and D. J. Wales, *J. Chem. Phys.*, 2005, **123**, 124304.
588. D. M. Leitner and P. G. Wolynes, *Chem. Phys.*, 2006, **329**, 163.
589. J. T. Hynes, *Annu. Rev. Phys. Chem.*, 1985, **36**, 573.
590. J. C. Owrutsky, D. Raftery, and R. M. Hochstrasser, *Annu. Rev. Phys. Chem.*, 1994, **45**, 519.
591. E. T. J. Nibbering, H. Fidder, and E. Pines, *Annu. Rev. Phys. Chem.*, 2005, **56**, 337.
592. C. G. Elles and F. F. Crim, *Annu. Rev. Phys. Chem.*, 2006, **57**, 273.
593. Y.-C. Cheng and G. R. Fleming, *Annu. Rev. Phys. Chem.*, 2009, **60**, 241.
594. J. P. Toennies and A. F. Vilesov, *Annu. Rev. Phys. Chem.*, 1998, **49**, 1.
595. F. Stienkemeier and K. K. Lehmann, *J. Phys. B: At. Mol. Opt. Phys.*, 2006, **39**, R127.
596. A. McIlroy and D. J. Nesbitt, *J. Chem. Phys.*, 1990, **92**, 2229.
597. A. Gilbert and J. E. Baggott, *Essentials of Molecular Photochemistry*, (Blackwell Scientific Publications, Oxford, UK), 1991.
598. N. J. Turro, V. Ramamurthy, and J. C. Scaiano, *Principles of Molecular Photochemistry*, (University Science Books, Sausalito, CA, USA), 2009.
599. C. Vallance, *New J. Chem.*, 2005, **29**, 876.
600. E. F. van Dishoeck and A. Dalgarno, *J. Chem. Phys.*, 1983, **79**, 873.
601. A. P. Clark, R. Cireasa, M. Brouard, F. Quadrini, and C. Vallance, in *Gas Phase Molecular Reaction and Photodissociation Dynamics*, Eds. K. C. Lin and P. D. Kleiber, (Transworld Research Network, Kerala, India), 2007.

602. Y. Asano and S. Yabushita, *Chem. Phys. Lett.*, 2003, **372**, 348.
603. E. P. Wigner and E. E. Witmer, *Z. Physik.*, 1928, **51**, 859.
604. R. S. Mulliken, *Phys. Rev.*, 1930, **36**, 1440.
605. Ph. Wernet, M. Odelius, K. Godehusen, J. Gaudin, O. Schwarzkopf, and W. Eberhardt, *Phys. Rev. Lett.*, 2009, **103**, 013001.
606. D. Strasser, F. Goulay, and S. R. Leone, *J. Chem. Phys.*, 2007, **127**, 184305.
607. A. H. Zewail, *J. Phys. Chem. A*, 2000, **104**, 5660.
608. A. V. Baklanov, L. M. C. Janssen, D. H. Parker, L. Poisson, B. Soep, J. M. Mestagh, and O. Gobert, *J. Chem. Phys.*, 2008, **129**, 214306.
609. R. N. Dixon, *J. Chem. Phys.*, 2005, **122**, 194302.
610. J. M. Brown, *Molecular Spectroscopy*, Oxford Chemistry Primers, No. 55, (Oxford University Press, Oxford, UK), 1998.
611. P. Kruit and F. H. Read, *J. Phys. E*, 1983, **16**, 313.
612. R. N. Zare and D. R. Herschbach, *Proc. IEEE*, 1963, **51**, 173.
613. T. P. Rakitzis, S. A. Kandel, A. J. Alexander, Z. H. Kim, and R. N. Zare, *Science*, 1998, **281**, 1346.
614. L. D. A. Siebbeles, M. Glass-Maujean, O. S. Vasyutinskii, J. A. Beswick, and O. J. Roncero, *J. Chem. Phys.*, 1994, **100**, 3610.
615. S. Mukamel and J. Jortner, *J. Chem. Phys.*, 1974, **61**, 5348.
616. M. D. Morse, Y. B. Band, and K. F. Freed, *J. Chem. Phys.*, 1983, **78**, 6066.
617. Y.-J. Jee, M. S. Park, Y. S. Kim, Y.-J. Jung, and K.-H. Jung, *Chem. Phys. Lett.*, 1998, **287**, 701.
618. T. P. Rakitzis and T. N. Kitsopoulos, *J. Chem. Phys.*, 2002, **116**, 9228.
619. R. J. Van Brunt and R. N. Zare, *J. Chem. Phys.*, 1968, **48**, 4304.
620. A. P. Clark, M. Brouard, F. Quadrini, and C. Vallance, *Phys. Chem. Chem. Phys.*, 2006, **8**, 5591.
621. A. G. Suits and O. S. Vasyutinskii, *Chem. Phys.*, 2008, **108**, 3706.
622. D. A. Chestakov, D. H. Parker, K. V. Vidma, and T. P. Rakitzis, *J. Chem. Phys.*, 2006, **124**, 024315.
623. Y. Asano and S. Yabushita, *Bull. Korean Chem. Soc.*, 2003, **24**, 703.
624. T. P. Rakitzis and R. N. Zare, *J. Chem. Phys.*, 1999, **110**, 3341.
625. R. N. Dixon, *J. Chem. Phys.*, 1986, **85**, 1866.
626. B. V. Picheyev, A. G. Smolin, and O. S. Vasyutinskii, *J. Phys. Chem. A*, 1997, **101**, 7614.
627. A. J. Alexander, *J. Chem. Phys.*, 2003, **118**, 6234.
628. E. E. Nikitin, *Theory of elmentary atomic and molecular processes in gases*, (Clarendon Press, Oxford, UK), 1974.
629. G. G. Balint-Kurti, A. J. Orr-Ewing, J. A. Beswick, and A. Brown, *J. Chem. Phys.*, 2002, **116**, 10760.
630. D. Townsend, S. A. Lahankar, S. K. Lee, S. D. Chambreau, A. G. Suits, X. Zhang, J. Rheinecker, L. B. Harding, and J. M. Bowman, *Science*, 2004, **306**, 1158.
631. T. P. Rakitzis, P. C. Samartzis, R. L. Toomes, T. N. Kitsopoulos, Alex Brown, G. G. Balint-Kurti, O. S. Vasyutinskii, and J. A. Beswick, *Science*, 2003, **300**, 1936.
632. D. J. Leahy, D. L. Osborn, D. R. Cyr, and D. M. Neumark, *J. Chem. Phys.*, 1995, **103**, 2495.
633. A. J. Alexander, *Phys. Chem. Chem. Phys.*, 2005, **7**, 3693.
634. D. C. Radenovic, A. J. A. van Roij, S.-M. Wu, J. J. Ter Meulen, D. H. Parker, M. P. J. van der Loo, L. M. C. Janssen, and G. C. Groenenboom, *Mol. Phys.*, 2008, **106**, 557.
635. D. C. Radenovic, A. J. A. van Roij, S.-M. Wu, J. J. Ter Meulen, D. H. Parker, M. P. J. van der Loo, and G. C. Groenenboom, *Phys. Chem. Chem. Phys.*, 2009, **11**, 4754.
636. H. Kim, K. S. Dooley, S. W. North, G. E. Hall, and P. L. Houston, *J. Chem. Phys.*, 2006, **125**, 133316.

637. V. V. Kuznetsov and O. S. Vasyutinskii, *J. Chem. Phys.*, 2007, **127**, 044308.
638. G. Parlant and D. R. Yarkony, *J. Chem. Phys.*, 1999, **110**, 363.
639. A. Brown, G. G. Balint-Kurti, and O. S. Vasyutinskii, *J. Phys. Chem. A*, 2004, **108**, 7790.
640. A. G. Smolin, O. S. Vasyutinskii, G. G. Balint-Kurti, and A. Brown, *J. Phys. Chem. A*, 2006, **110**, 5371.
641. A. Brown, *Int. J. Quantum Chem.*, 2007, **107**, 2665.
642. J. Aldegunde, M. P. de Miranda, J. M. Haigh, B. K. Kendrick, V. Sáez-Rábanos, and F. J. Aoiz, *J. Phys. Chem. A*, 2005, **109**, 6200.
643. K. Blum, *Density Matrix Theory and Applications*, (Plenum, New York, USA), 1996.
644. M. Auzinsh and R. Ferber, *Optical Polarization of Molecules*, (Cambridge University Press, Cambridge, UK), 1995.
645. M. Jacob and G. C. Wick, *Ann. Phys.*, 1959, **7**, 404.
646. J. Aldegunde, F. J. Aoiz, V. Sáez-Rábanos, B. K. Kendrick, and M. P. de Miranda, *Phys. Chem. Chem. Phys.*, 2007, **9**, 5794.
647. D. A. Varshalovich, A. N. Moskalev, and V. K. Khersonskii, *Quantum Theory of Angular Momentum*, (World Scientific, Singapore), 1988.
648. M. P. de Miranda, F. J. Aoiz, V. Sáez-Rábanos, and M. Brouard, *J. Chem. Phys.*, 2004, **121**, 9830.
649. J. Aldegunde, F. J. Aoiz, and M. P. de Miranda, *Phys. Chem. Chem. Phys.*, 2008, **10**, 1139.
650. M. P. de Miranda, F. J. Aoiz, and V. Sáez-Rábanos, *J. Chem. Phys.*, 1999, **111**, 5368.
651. W. H. Miller, Ed., R. N. Porter, and L. M. Raff, in *Dynamics of Molecular Collisions*, Part B, p. 1, (Plenum Press, New York, USA), 1976.
652. W. H. Miller, Ed., W. L. Hase, in *Dynamics of Molecular Collisions*, Part B, p. 121, (Plenum Press, New York, USA), 1976.
653. R. B. Bernstein, Eds., D. G. Truhlar, and J. T. Muckerman, in *Atom–Molecule Collision Theory: A guide for experimentalists*, p. 505, (Plenum Press, New York, USA), 1979.
654. M. Baer, Ed., L. M. Raff, and D. L. Thompson, in *The Theory of Chemical Reaction Dynamics*, vol. 111, p. 1, (CRC Press, Boca Raton, USA), 1985.
655. R. E. Wyatt, and J. Z. H. Zhang, Ed., H. R. Mayne, in *Dynamics of Molecules and Chemical Reactions*, p. 589, (Marcel Dekker, New York, USA), 1996.
656. D. M. Hirst, *Potential Energy Surfaces: Molecular Structure and Reaction Dynamics*, (Taylor & Francis, London, UK), 1985.
657. R. N. Zare, *Ber. Bunsenges. Phys. Chem.*, 1982, **86**, 422.
658. R. Altkorn, R. N. Zare, and C. H. Greene, *Mol. Phys.*, 1985, **65**, 1.
659. H.-J. Loesch and F. Stienkemeier, *J. Chem. Phys.*, 1993, **98**, 9570.
660. O. Höbel and H.-J. Loesch, *Faraday Discuss. Chem. Soc.*, 1999, **113**, 337.
661. F. J. Aoiz, M. T. Martínez, and V. Sáez-Rábanos, *J. Chem. Phys.*, 2001, **114**, 8880.
662. S. A. Kandel, A. J. Alexander, Z. H. Kim, R. N. Zare, F. J. Aoiz, L. Bañares, J. F. Castillo, and V. Sáez-Rábanos, *J. Chem. Phys.*, 2000, **112**, 670.
663. H. Kramer and R. B. Bernstein, *J. Phys. Chem.*, 1965, **42**, 767.
664. P. R. Brooks and E. M. Jones, *J. Chem. Phys.*, 1966, **45**, 3449.
665. R. B. Bernstein, R. J. Beuhler, and K. H. Kramer, *J. Am. Chem. Soc.*, 1966, **88**, 5331.
666. P. R. Brooks, *Science*, 1976, **193**, 11.
667. S. Stolte, *Ber. Bunsenges. Phys. Chem.*, 1982, **86**, 413.
668. D. H. Parker and R. B. Bernstein, *Ann. Rev. Phys. Chem.*, 1989, **40**, 561.
669. D. H. Parker, H. Jalink, and S. Stolte, *J. Phys. Chem.*, 1987, **91**, 5427.
670. M. C. van Beek, J. J. ter Meulen, and M. H. Alexander, *J. Chem. Phys.*, 2000, **113**, 637.
671. C. G. Aitken, D. A. Blunt, and P. W. Harland, *Int. J. Mass. Spec.*, 1995, **149**, 279.
672. R. S. MacKay, T. J. Curtiss, and R. B. Bernstein, *J. Chem. Phys.*, 1990, **92**, 801.
673. M. Okada, S. Goto, and T. Kasai, *J. Am. Chem. Soc.*, 2007, **129**, 10052.

674. T. P. Rakitzis, M. H. M. Janssen, and A. J. van den Brom, *Science*, 2004, **303**, 1852.
675. P.-Y. Tsai, D.-C. Che, M. Nakamura, K.-C. Lin, and T. Kasai, *Phys. Chem. Chem. Phys.*, 2010, **12**, 2532.
676. B. Friedrich and D. R. Herschbach, *Z. Phys. D*, 1991, **18**, 153.
677. B. Friedrich, D. P. Pullman, and D. R. Herschbach, *J. Phys. Chem.*, 1991, **95**, 8118.
678. H.-J. Loesch and A. Remscheid, *J. Chem. Phys.*, 1990, **93**, 4779.
679. H.-J. Loesch and A. Remscheid, *J. Phys. Chem.*, 1991, **95**, 8194.
680. H.-J. Loesch and J. Möller, *J. Phys. Chem.*, 1993, **97**, 2158.
681. H.-J. Loesch and J. Möller, *Faraday Discuss. Chem. Soc.*, 1999, **113**, 241.
682. K. Härtelt and B. Friedrich, *J. Chem. Phys.*, 2008, **128**, 224313.
683. K. F. Lee, D. M. Villeneuve, P. B. Corkum, A. Stolow, and J. G. Underwood, *Phys. Rev. Lett.*, 2006, **97**, 173001.
684. F. Rosca-Pruna and M. J. J. Vrakking, *J. Chem. Phys.*, 2002, **116**, 6567.
685. A. P. Clark, M. Brouard, F. Quadrini, and C. Vallance, *Phys. Chem. Chem. Phys.*, 2006, **8**, 5591.
686. G. G. Balint-Kurti, A. Brown, and O. S. Vasyutinskii, *Phys. Scr.*, 2006, **73**.
687. D. Sofikitis, L. Rubio-Lago, M. R. Martin, D. J. A. Brown, N. C.-M. Bartlett, A. J. Alexander, R. N. Zare, and P. T. Rakitzis, *J. Chem. Phys.*, 2007, **127**, 144307.
688. D. Sofikitis, L. Rubio-Lago, M. R. Martin, D. J. A. Brown, N. C.-M. Bartlett, A. J. Alexander, R. N. Zare, and P. T. Rakitzis, *Phys. Rev. A*, 2007, **76**, 012503.
689. P. D. D. Monks, C. Xiahou, and J. N. L. Connor, *J. Chem. Phys.*, 2006, **125**, 133504.
690. M. P. de Miranda and R. Gargano, *Chem. Phys. Lett.*, 1999, **309**, 257.
691. D. Skouteris, S. Crocchianti, and A. Laganà, *Chem. Phys. Lett.*, 2007, **440**, 1.
692. V. Aquilanti, S. Cavalli, G. Grossi, and R. W. Anderson, *J. Phys. Chem.*, 1991, **95**, 8184.
693. V. Aquilanti, S. Cavalli, G. Grossi, and R. W. Anderson, *J. Phys. Chem.*, 1993, **97**, 2443.
694. J. C. Juanes-Marcos, S. C. Althorpe, and E. Wrede, *J. Chem. Phys.*, 2007, **126**, 044317.
695. I. Bengtsson and K. Życzkowski, *Geometry of Quantum States: an Introduction to Quantum Entanglement*, (Cambridge University Press, Cambridge, UK), 2006.
696. J. Aldegunde, J.-M. Alvariño, M. P. de Miranda, V. Sáez-Rábanos, and F. J. Aoiz, *J. Chem. Phys.*, 2006, **125**, 133104.
697. D. P. Woodruff, Ed., *The Chemical Physics of Solid Surfaces*, Vol. **11**, Surface Dynamics, (Elsevier, Amsterdam, The Netherlands), 2003.
698. H. A. Michelsen, C. T. Rettner, and D. J. Auerbach, *The adsorption of hydrogen at copper surfaces: A model system for the study of activated adsorption*, in *Surface Reactions*, R. J. Madix, Ed., (Springer, Berlin, Germany), 1993, p. 123.
699. G. R. Darling and S. Holloway, *Rep. Prog. Phys.*, 1995, **58**, 1595.
700. A. Gross, *Surf. Sci. Rep.*, 1998, **32**, 291.
701. A. Hodgson, *Prog. Surf. Sci.*, 2000, **63**, 1.
702. G. J. Kroes, A. Gross, E. J. Baerends, M. Scheffler, and D. A. McCormack, *Acc. Chem. Res.*, 2002, **35**, 193.
703. G. O. Sitz, *Rep. Prog. Phys.*, 2002, **65**, 1165.
704. A. W. Kleyn, *Chem. Soc. Rev.*, 2003, **32**, 87.
705. J. Libuda and H. J. Freund, *Surf. Sci. Rep.*, 2005, **57**, 157.
706. K. Watanabe, D. Menzel, N. Nilius, and H. J. Freund, *Chem. Rev.*, 2006, **106**, 4301.
707. M. Dürr and U. Höfer, *Surf. Sci. Rep.*, 2006, **61**, 465.
708. A. M. Wodtke, D. Matsiev, and D. J. Auerbach, *Prog. Surf. Sci.*, 2008, **83**, 167.
709. J. C. Tully, *Faraday Disc. Chem. Soc.*, 2004, **127**, 463.
710. A. D. Kinnersley, G. R. Darling, S. Holloway, and B. Hammer, *Surf. Sci.*, 1996, **364**, 219.
711. J. S. Blakemore, *Solid State Physics*, (Cambridge University Press, Cambridge, UK), 1985.

712. K. Schönhammer and O. Gunnarsson, *Many-Body Phenomena at Surfaces*, in *Many-Body Phenomena at Surfaces*, D. Langreth and H. Suhl, Eds., (Academic Press, New York, USA), 1984, p. 421.
713. S. Andersson, L. Wilzén, and M. Persson, *Phys. Rev. B*, 1988, **38**, 2967.
714. A. Zangwill, *Physics at surfaces*, (Cambridge University Press, Cambridge, UK), 1988.
715. M. Head-Gordon, J. C. Tully, C. T. Rettner, C. B. Mullins, and D. J. Auerbach, *J. Chem. Phys.*, 1991, **94**, 1516.
716. R. J. W. E. Lahaye, S. Stolte, S. Holloway, and A. W. Kleyn, *J. Chem. Phys.*, 1996, **104**, 8301.
717. D. P. Woodruff, Ed., A. W. Kleyn, in *Molecular beam scattering at surfaces*, in The Chemical Physics of Solid Surfaces, Vol. **11**, Theory of Low Temperature Gas-Phase Reactions, p. 79, (Elsevier, Amsterdam, The Netherlands), 2003.
718. J. D. White, J. Chen, D. Matsiev, D. J. Auerbach, and A. M. Wodtke, *Nature*, 2005, **433**, 503.
719. G. R. Darling, M. Kay, and S. Holloway, *Surf. Sci.*, 1998, **400**, 314.
720. C. T. Rettner, E. K. Schweizer, and H. Stein, *J. Chem. Phys.*, 1990, **93**, 1442.
721. C. R. Arumainayagam and R. J. Madix, *Prog. Surf. Sci.*, 1991, **38**, 1.
722. A. Gross and M. Scheffler, *Chem. Phys. Lett.*, 1995, **256**, 417.
723. D. A. McCormack, G. J. Kroes, R. A. Olsen, J. A. Groeneveld, J. N. P. van Stralen, E. J. Barends, and R. C. Mowrey, *Chem. Phys. Lett.*, 2000, **328**, 317.
724. G. J. Kroes and M. F. Somers, *J. Theo. Comp. Chem.*, 20025, **4**, 493.
725. G. R. Darling and S. Holloway, *Faraday Discuss.*, 1998, **110**, 153.
726. R. Bisson, M. Sacchi, T. T. Dang, B. Yoder, P. Maroni, and R. D. Beck, *J. Phys. Chem. A*, 2007, **111**, 12679.
727. A. C. Luntz and D. S. Bethune, *J. Chem. Phys.*, 1989, **90**, 1274.
728. A. Bukoski, H. L. Abbott, and I. Harrison, *J. Chem. Phys.*, 2005, **123**, 094707.
729. E. Hasselbrink, *Cur. Opin. Solid State Mat. Sci.*, 2006, **10**, 192.
730. A. Böttcher, R. Imbeck, A. Morgante, and G. Ertl, *Phys. Rev. Lett.*, 1990, **65**, 2035.
731. M. Binetti, O. Weisse, E. Hasselbrink, G. Katz, R. Kosloff, and Y. Zeiri, *Chem. Phys. Lett.*, 2003, **373**, 366.
732. J. Behler, B. Delley, S. Lorenz, K. Reuter, and M. Scheffler, *Phys. Rev. Lett.*, 2005, **94**, 036104.
733. M. J. Cardillo, M. Balooch, and R. E. Stickney, *Surf. Sci.*, 1975, **50**, 263.
734. G. P. Brivio and T. B. Grimley, *Surf. Sci. Rep.*, 1993, **17**, 1.
735. W. van Willigen, *Phys. Letts.*, 1968, **28A**, 80.
736. M. J. Murphy and A. Hodgson, *Phys. Rev. Lett.*, 1997, **78**, 4458.
737. A. Namiki, *Prog. Surf. Sci.*, 2006, **81**, 337.
738. H. A. Michelsen, C. T. Rettner, and D. J. Auerbach, *Phys. Rev. Lett.*, 1992, **69**, 2678.
739. H. Hou, S. J. Gulding, C. T. Rettner, A. M. Wodtke, and D. J. Auerbach, *Science*, 1997, **277**, 80.
740. S. Wright, J. Skelly, and A. Hodgson, *Faraday Discuss Chem. Soc.*, 2000, **117**, 113.
741. T. Matsushima, *Progress in Surface Science*, 2007, **82**, 435.
742. R. N. Carter, M. J. Murphy, and A. Hodgson, *Surface Science*, 1997, **387**, 102.
743. M. J. Murphy, J. F. Skelly, and A. Hodgson, *J. Chem. Phys.*, 1998, **109**, 3619.
744. E. Fridell, A. Rosen, and B. Kasemo, *Langmuir*, 1994, **10**, 699.
745. T. Yamanaka and T. Matsushima, *Phys. Rev. Lett.*, 2008, **100**, 026104.
746. Y. S. Ma, A. Kokalj, and T. Matsushima, *Phys. Chem. Chem. Phys.*, 2005, **7**, 3716.
747. E. H. G. Backus, A. Eichler, A. W. Kleyn, and M. Bonn, *Science*, 2005, **310**, 1790.
748. E. D. Potter, J. L. Herek, S. Pederson, Q. Liu, and A. H. Zewail, *Nature*, 1993, **355**, 66.
749. I. Grenthe, Ed., A. H. Zewail, in *Nobel Lectures, Chemistry 1996–2000*, (World Scientific Co., Singapore), 2003.
750. A. H. Zewail, *Femtochemistry-Ultrafast Dynamics of the Chemical Bond*, Vols. **I** and **II**, (World Scientific, Singapore), 1994.

751. P. M. W. French, *Reports on Progress in Phys.*, 1995, **58**, 169.
752. W. Demtroder, *Laser spectroscopy, basic concepts and instrumentation*, (Springer-Verlag, Berlin, Germany), 1996.
753. A. Paul, E. A. Gibson, X. S. Zhang, A. Lytle, T. Popmintchev, X. B. Zhou, M. M. Murnane, I. P. Christov, and H. C. Kapteyn, *IEEE J. Quant. Electron.*, 2006, **42**, 14.
754. R. Trebino, *Frequency-resolved Optical Gating: The Measurement of Ultrashort Laser Pulses*, 2nd Ed., (Kluwer Academic Publishers, The Netherlands), 2000.
755. C. Iaconis and I. A. Walmsley, *IEEE J. Quantum Electron.*, 1999, **35**, 501.
756. V. V. Lozovoy, I. Pastirk, and M. Dantus, *Optics Letters*, 2004, **29**, 775.
757. A. Stolow, *Ann. Rev. Phys. Chem.*, 2003, **54**, 89.
758. A. Stolow, *Int. Rev. Phys. Chem.*, 2003, **22**, 377.
759. T. Baumert, M. Grosser, R. Thalweiser, and G. Gerber, *Phys. Rev. Lett.*, 1991, **67**, 3753.
760. T. S. Rose, M. J. Rosker, and A. H. Zewail, *J. Chem. Phys.*, 1988, **88**, 6672.
761. V. Blanchet, M. Z. Zgierski, T. Seideman, and A. Stolow, *Nature*, 1999, **401**, 52.
762. T. Suzuki, *Ann. Rev. Phys. Chem.*, 2006, **57**, 555.
763. O. Gessner, A. M. D. Lee, J. P. Shaffer, H. Reisler, S. V. Levchenko, A. I. Krylov, J. G. Underwood, H. Shi, A. L. L. East, D. M. Wardlaw, E. T. Chrysostom, C. C. Hayden, and A. Stolow, *Science*, 2006, **311**, 219.
764. S. Lochbrunner, T. Schultz, M. Schmitt, J. P. Shaffer, M. Z. Zgierski, and A. Stolow, *J. Chem. Phys.*, 2001, **114**, 2519.
765. A. H. Zewail, *Ann. Rev. Phys. Chem.*, 2006, **57**, 65.
766. H. Ihee, V. A. Lobastov, U. M. Gomez, B. M. Goodson, R. Srinivasan, C. Y. Ruan, and A. H. Zewail, *Science*, 2001, **291**, 458.
767. I. J. Finkelstein, J. R. Zheng, H. Ishikawa, S. Kim, K. Kwak, and M. D. Fayer, *Phys. Chem. Chem. Phys.*, 2007, **9**, 1533.
768. S. Park, K. Kwak, and M. D. Fayer, *Laser Phys. Lett.*, 2007, **4**, 704.
769. M. Bargheer, N. Zhavoronkov, M. Woerner, and T. Elsaesser, *ChemPhysChem*, 2006, **7**, 783.
770. K. J. Gaffney and H. N. Chapman, *Science*, 2007, **316**, 1444.
771. M. Dantus and V. V. Lozovoy, *Chem. Rev.*, 2004, **104**, 1813.
772. L. C. Zhu, V. Kleiman, X. N. Li, S. P. Lu, K. Trentelman, and R. J. Gordon, *Science*, 1995, **270**, 77.
773. D. J. Tannor, R. Kosloff, and S. A. Rice, *J. Chem. Phys.*, 1986, **85**, 5805.
774. M. Shapiro and P. Brumer, *J. Chem. Phys.*, 1986, **84**, 540.
775. M. Shapiro and P. Brumer, *J. Chem. Phys.*, 1986, **84**, 4103.
776. N. F. Scherer, R. J. Carlson, A. Matro, M. Du, A. J. Ruggiero, V. Romerorochin, J. A. Cina, G. R. Fleming, and S. A. Rice, *J. Chem. Phys.*, 1991, **95**, 1487.
777. H. H. Fielding, *Annu. Rev. Phys. Chem.*, 2005, **56**, 91.
778. A. Assion, T. Baumert, J. Helbing, V. Seyfried, and G. Gerber, *Chem. Phys. Lett.*, 1996, **259**, 488.
779. A. M. Weiner, *Rev. Sci. Instrum.*, 2000, **71**, 1929.
780. A. M. Weiner, *Prog. Quant. Electr.*, 1995, **19**, 161.
781. J. Vaughan, *Ultrafast Pulse Shaping*, in Ultrafast Optics Textbook, R. Trebino and J. Squier, Eds., available online at http://www.physics.gatech.edu/frog/ultratext.html, 2008.
782. P. Nuernberger, G. Vogt, R. Selle, S. Fechner, T. Brixner, and G. Gerber, *Appl. Phys. B*, 2007, **88**, 519.
783. D. S. N. Parker, A. D. G. Nunn, R. S. Minns, and H. H. Fielding, *Appl. Phys. B – Lasers and Optics*, 2009, **94**, 181.
784. R. Chakrabarti and H. Rabitz, *Int. Rev. Phys. Chem.*, 2007, **26**, 671.
785. R. S. Judson and H. Rabitz, *Phys. Rev. Lett.*, 1992, **68**, 1500.
786. J. L. Herek, W. Wohlleben, R. J. Cogdell, D. Zeidler, and M. Motzkus, *Nature*, 2002, **417**, 533.

787. G. Vogt, G. Krampert, P. Niklaus, P. Nuernberger, and G. Gerber, *Phys. Rev. Lett.*, 2005, **94**, 068305.
788. V. I. Prokhorenko, A. M. Nagy, S. A. Waschuk, L. S. Brown, R. R. Birge, and R. J. D. Miller, *Science*, 2006, **313 (5791)**, 1257.
789. A. Assion, T. Baumert, M. Bergt, T. Brixner, B. Kiefer, V. Seyfried, M. Strehle, and G. Gerber, *Science*, 1998, **282**, 919.
790. C. Daniel, J. Full, L. Gonzalez, C. Lupulescu, J. Manz, A. Merli, S. Vajda, and L. Woste, *Science*, 2003, **299**, 536.
791. B. J. Sussman, D. Townsend, M. Y. Ivanov, and A. Stolow, *Science*, 2006, **314**.
792. J. G. Underwood, M. Spanner, M. Y. Ivanov, J. Mottershead, B. J. Sussman, and A. Stolow, *Phys. Rev. Lett.*, 2003, **90**, 223001.
793. M. Drescher, M. Hentschel, R. Kienberger, M. Uiberacker, V. Yakovlev, A. Scrinzi, Th. Westerwalbesloh, U. Kleineberg, U. Heinzmann, and F. Krausz, *Nature*, 2002, **419**, 803.
794. A. Nagy, V. Prokhorenko, and R. J. D. Miller, *Current Opinion in Structural Biology*, 2006, **16**, 654.
795. W. Wohlleben, T. Buckup, J. L. Herek, and M. Motzkus, *ChemPhysChem*, 2005, **6**, 850.
796. M. H. Anderson, J. R. Ensher, M. R. Matthews, C. E. Wieman, and E. A. Cornell, *Science*, 1995, **269**, 198.
797. K. B. Davis, M.-O. Mewes, M. R. Andrews, N. J. van Druten, D. S. Durfee, D. M. Kurn, and W. Ketterle, *Phys. Rev. Lett.*, 1995, **75**, 3969.
798. C. C. Bradley, C. A. Sackett, J. J. Tollett, and R. G. Hulet, *Phys. Rev. Lett.*, 1995, **75**, 1687.
799. S. Jochim, M. Bartenstein, A. Altmeyer, G. Hendl, S. Riedl, C. Chin, J. Hecker Denschlag, and R. Grimm, *Science*, 2003, **302**, 2101.
800. M. Greiner, C. A. Regal, and D. S. Jin, *Nature*, 2003, **426**, 537.
801. M. W. Zwierlein, C. A. Stan, C. H. Schunck, S. M. F. Raupach, S. Gupta, Z. Hadzibabic, and W. Ketterle, *Phys. Rev. Lett.*, 2003, **91**, 250401.
802. P. F. Weck and N. Balakrishnan, *Eur. Phys. J. D*, 2004, **31**, 417.
803. E. Bodo, F. A. Gianturco, N. Balakrishnan, and A. Dalgarno, *J. Phys. B: At. Mol. Opt. Phys.*, 2004, **37**, 3641.
804. D. C. Clary, *Annu. Rev. Phys. Chem.*, 1990, **41**, 61.
805. I. W. M. Smith, *Angew. Chem. Int. Ed.*, 2006, **45**, 2842.
806. R. V. Krems, *Phys. Chem. Chem. Phys.*, 2008, **10**, 4079.
807. M. T. Bell and T. P. Softley, *Mol. Phys.*, 2009, **107**, 99.
808. H. L. Bethlem and G. Meijer, *Int. Rev. Phys. Chem.*, 2003, **22**, 73.
809. S. Y. T. van de Meerakker, H. L. Bethlem, and G. Meijer, *Nat. Phys.*, 2008, **4**, 595.
810. S. Willitsch, M. T. Bell, A. D. Gingell, and T. P. Softley, *Phys. Chem. Chem. Phys.*, 2008, **10**, 7200.
811. R. V. Krems, W. C. Stwalley, and B. Friedrich, Eds., *Cold molecules. Theory, Experiment, Applications*, (CRC Press, Taylor & Francis Group, Florida, USA), 2009.
812. P. B. Moon, C. T. Rettner, and J. P. Simons, *J. Chem. Soc. Faraday Trans.*, 1978, **74**, 630.
813. M. Gupta and D. Herschbach, *J. Phys. Chem. A*, 1999, **103**, 10670.
814. M. Gupta and D. Herschbach, *J. Phys. Chem. A*, 2001, **105**, 1626.
815. N. Vanhaecke, U. Meier, M. Andrist, B. H. Meier, and F. Merkt, *Phys. Rev. A*, 2007, **75**, 031402.
816. S. Chu, *Rev. Mod. Phys.*, 1998, **70**, 685.
817. C. N. Cohen-Tannoudji, *Rev. Mod. Phys.*, 1998, **70**, 707.
818. W. D. Phillips, *Rev. Mod. Phys.*, 1998, **70**, 721.
819. J. Dalibard and C. Cohen-Tannoudji, *J. Opt. Soc. Am. B*, 1985, **2**, 1707.
820. A. Ostendorf, C. B. Zhang, M. A. Wilson, D. Offenberg, B. Roth, and S. Schiller, *Phys. Rev. Lett.*, 2006, **97**, 243005.

821. K. M. Jones, E. Tiesinga, P. D. Lett, and P. S. Julienne, *Rev. Mod. Phys.*, 2006, **78**, 483.
822. T. Koëhler, K. Góral, and P. S. Julienne, *Rev. Mod. Phys.*, 2006, **78**, 1311.
823. I. W. M. Smith, Ed., *Low Temperatures and cold molecules*, (World Scientific, London, UK), 2008.
824. J. D. Weinstein, R. deCarvalho, T. Guillet, B. Friedrich, and J. M. Doyle, *Nature*, 1998, **395**, 148.
825. P. M. Langevin, *Ann. Chim. Phys.*, 1905, **5**, 245.
826. E. Gorin, *J. Chem. Phys.*, 1939, **7**, 256.
827. M. Abramowitz and I. A. Stegun, *Handbook of Mathematical Functions*, (National Bureau of Standards, Washington, D.C., USA), 1964.
828. M. H. Alexander, E. J. Rackham, and D. E. Manolopoulos, *J. Chem. Phys.*, 2004, **121**, 5221.
829. I. R. Sims and I. W. M. Smith, *Annu. Rev. Phys. Chem.*, 1995, **46**, 109.
830. S. J. Klippenstein and Y. Georgievskii, *Theory of Low Temperature Gas-Phase Reactions*, (World Scientific, London, UK), 2008 in ref. [823].
831. E. P. Wigner, *Phys. Rev.*, 1948, **73**, 1002.
832. L. D. Landau and E. M. Lifshitz, *Quantum mechanics, non-relativistic theory*, (Pergamon Press, London, UK), 1959.
833. Z. Pavlovic, T. V. Tscherbul, H. R. Sadeghpour, G. C. Groenenboom, and A. Dalgarno, *J. Phys. Chem. A*, 2009, **113**, 14670.
834. T. V. Tscherbul, G. C. Groenenboom, R. V. Krems, and A. Dalgarno, *Faraday Discuss.*, 2009, **142**, 127.
835. R. V. Krems and A. Dalgarno, *Phys. Rev. A*, 2003, **67**, 050704.
836. H. R. Sadeghpour, J. L. Bohn, M. J. Cavagnero, B. D. Esry, I. I. Fabrikant, J. H. Macek, and A. R. P. Rau, *J. Phys. B: At. Mol. Opt. Phys.*, 2000, **33**, 93.
837. S. N. Bose, *Zeit. Phys.*, 1924, **26**, 178.
838. A. Einstein, *Sitzungsberichte der Preussischen Akademie der Wissenschaften*, 1925, **1**, 3.
839. D. A. McQuarrie and H. B. Levine, *Physica*, 1965, **31**, 749.
840. C. J. Pethick and H. Smith, *Bose-Einstein Condensation in Dilute Gases*, (Cambridge University Press, Cambridge, UK), 2002.
841. A. Görlitz, A. P. Chikkatur, and W. Ketterle, *Phys. Rev. A*, 2001, **63**, 041601(R).
842. D. J. Heinzen, R. Wynar, P. D. Drummond, and K. V. Kheruntsyan, *Phys. Rev. Lett.*, 2000, **84**, 5029.
843. M. G. Moore and A. Vardi, *Phys. Rev. Lett.*, 2002, **88**, 160402.
844. P. Meystre, *Adv. At. Mol. Opt. Phys.*, 2001, **47**, 1.
845. D. Smith and P. Spanel, *Mass Spectrom. Rev.*, 1995, **14**, 255.
846. S. Willitsch, M. T. Bell, A. D. Gingell, S. R. Procter, and T. P. Softley, *Phys. Rev. Lett.*, 2008, **100**, 043203.
847. S. Ospelkaus, K.-K. Ni, D. Wang, M. H. G. de Miranda, B. Neyenhuis, G. Quéméner. P. S. Julienne, J. L. Bohn, D. S. Jin, and J. Ye, *Science*, 2010, **327**, 853.
848. M. Schnell and G. Meijer, *Angew. Chem. Int. Ed.*, 2009, **48**, 6010.
849. O. Dulieu and C. Gabbanini, *Rep. Prog. Phys.*, 2009, **72**, 086401.
850. D. W. Chandler, *J. Chem. Phys.*, 2010, **132**, 110901.
851. M. Weidemüller and C. Zimmermann, Eds., *Cold atoms and molecules*, (Wiley, Weinheim, Germany), 2009.
852. K.-K. Ni, S. Ospelkaus, M. H. G. de Miranda, A. Pe'er, B. Neyenhuis, J. J. Zirbel, S. Kotochigova, P. S. Julienne, D. S. Jin, and J. Ye, *Science*, 2008, **322**, 231.
853. S. Stellmer, M. K. Tey, B. Huang, R. Grimm, and F. Schreck, *Phys. Rev. Lett.*, 2009, **103**, 200401.
854. S. A. Meek, H. Conrad, and G. Meijer, *Science*, 2009, **324**, 1699.

Subject Index

References to figures are given in italic type; references to tables are given in bold type.

ab initio modelling 41–42
Abel inversion 154
acetylene 185–186
adiabatic models 18–19
  capture theory 414
  photodissociation 265–267
  see also Born-Oppenheimer approximation
adsorption 343–344, 355–358
adsorption-desorption 343-358, 355–358, 359
alignment *see* polarization
alignment moment 159
angular momentum
  centrifugal barriers 21–23
  conservation 19–20, 50, 174, 175–176
  coupling 108
  measurement 158–163, 303–304
  see also rotational energy
angular momentum catastrophe 82
angular scattering distribution 7–8, 174–179
  photodissociation 252–258
  see also scattering angle
anisotropic interactions 406–407
argon 140–144, 160–161
Arrhenius equation 392
atomic polarization 258–263
  adiabatic model 266–267

basis sets 62–63, 74
Berry phase 103–104
bolometry 138
Boltzmann distribution 8, 427–428
bond-selective chemistry 223–224
Born-Oppenheimer approximation 18–19, 28–31, 88–89, 414
  breakdown 89–90
  see also potential energy surfaces, coupled
  see also adiabatic models
Bose enhancement 438–440
Bose-Einstein condensation 392–393
  canonical description 429–431
  Gross-Pitaevskii equation 431–437
  harmonic trap 431
  phase transition 427–428
  simulation 432
  Thomas-Fermi approximation 438
bosons 429–430
bromine 174–175
  photodissociation 244–248, 249–251, 254–257
    atomic polarization 260–263
brute force orientation 308–310

capture theory 393–401
  classical 393–407
  coupled channels 411–414
  quantum 408–410
    adiabatic 414
    potential step 417–418
    thermal capture rates 414–415
centre-of-mass (CM) frame 11–12, 172–174
  differential cross-section measurement 203–205
centrifugal barriers 21–22
Chebyshev polynomials 79–80
chirped laser pulses 367–368
chlorine
  hydrogen reaction 111–112
  methane reaction 219
chloromethane 1–2

classical mechanics 49–50, 120–121
capture theory 393–407
Hamiltonian formulation 51
Lagrangian formulation 50–51
scattering in central potential 51–53
close coupling 69–70
closed optical loops 396
CM frame *see* centre-of-mass frame
CNP group 44
coherent control 382–383
cold regime 394, 433–434
anisotropic interactions 406–407
inelastic scattering 420–424
isotropic interactions 403–406
reaction cross-section 401
s-wave scattering 419–420
collision energy 13
commutators 58
Complete Nuclear Permutation (CNP) group 44
condensed matter 236–237, 382
see also Bose-Einstein condensation; liquids; solid surfaces
configuration interaction (CI) 34
conical intersections 99–105
conservation rules 13, 19–21
classical 49–50
cooling 129, 394–397
coordinate systems 11–13, 31–33, 37–38, 52, 172–174
molecular beam experiments 172–174, 204
polarization moments 293–294
transformation 14–15
variable changes 14–15
copper 339
Coriolis coupling 82, 106–107
correlated differential cross-section (CDCS) 208–210
Coulomb crystals 434
coupled channels capture theory 411–414
coupling (potential energy surfaces) *see* potential energy surfaces, coupled
CRESU 433
cross-section 5–6, 401
cold regime 401, 417–424
control by polarization 326–328
integral 7–8, 8
measurement 144–148
partial differential 123
see also differential cross-section
crossed molecular beam experiments *see* molecular beam experiments
crystal structures 340
cumulative reaction probability 230–231, 233–235
cyclopentadienyl manganese 388

de Laval nozzle 433
Debye-Waller effect 334–335
all-trans-2,4,6,8-decatraene 381–382
delta function 62–63
density functional theory (DFT) 338–339
density matrix 289–290
multipolar expansion 290
density-to-flux transformation 172–173
desorption 355–358
detailed balance principle 356
deuterium 188–193, 334–335, 359
diabatic processes 90–109, 111
coupling 93–94
photodissociation 266–271
diatomic molecules 35–36, 93, 133–135
differential cross-section (DCS) 7–8
correlated 208–210
impact parameter and 127
measurement 135–139, 150–153, 171
in CM frame 203–205
nitric oxide-argon system 144–145
polarization dependent 292–293
product-pair correlated 208–210
quantum state selective 155–157, 171
state-averaged 180–185
state-to-state 8–9, 66, 150–151, 167–169, 218–220
diffractive scattering 125
diffusion-limited reactions 236
dipole-dipole interactions 404
direct dissociation 351–355
direct reactions 174–175
discrete variable representation (DVR) 75–76
dispersion fitted finite difference representation 77–78
dissociation at surfaces 348–350
classification 348–349
direct dissociation 351–355
photodissociation *see* photodissociation
distinguished coordinates 41
DMBE IV surface 84
Doppler cooling 396

Doppler-resolved laser-induced fluorescence 136–137
Doppler-selected time-of-flight mass spectrometric detection 203
dye lasers 18
dynamical resonance 195

effective potential 21
Ehrenfest theorem 60, 117–118
Einstein coefficients 16–17
elastic scattering 127–128
  at surfaces 342–344
  cold regime 418–419
electron correlation 34
electron impact ionization 168–169, 179–180
electronic energies 18–19, 29
  see also Born-Oppenheimer approximation
electronic wavefunction 29, 90–91, 100
  see also Born-Oppenheimer approximation
ellipse 130–135
energetic corrugation 353
energy conservation 13, 19–21
energy transfer 116–117
  alignment and orientation 161–163
  experimental studies 135–139
  inelastic collisions 120–121
  measurements, vector 148–163
  nitric oxide-argon system 139–148
  principles 117–118
  rotational angular momentum 158–163
  scattering angle and 126–129
  steric asymmetry 149–150
  translational to rotational 130–135
ethane 181–185
evaporative cooling 397
excimer lasers 18
excitation *see* vibrational states
excitation function 23

Fast Fourier Transform (FFT) 75
femtochemistry 25, 377–382
  coherent control 382–384
  photodissociation 249–251
  spatially aligned molecules 389
  see also pump-probe spectroscopy
femtosecond lasers
  amplifiers 370–373, 375
  oscillators 365–370, 375
  programmable pulse shaping 385–389
  wavelength tuning 373
Fermi blocking 439–440
Fermi golden rule 106, 245, 247
Feshbach resonance 199, *201*, 397–398
finite basis representation 76
flash photolysis 13–14
fluorescence depletion 228–229
fluorine 1–2, 25–26, *229*
  hydrogen reaction 198–199
  methane reaction 199–200, 205
Fourier transform 61, 75
frames of reference 52–53
  energy transfer 123–125
  molecular beam experiments 172–174, 204
  see also coordinate systems
free motion 60–64
frequency resolved optical gating (FROG) 374
fugacity 429
functional form fitting 43–45

Gaussian wavepacket 61
geometric corrugation 353
geometric phase 103–104
glory scattering 126–128
greenhouse gases 116–117
group delay dispersion (GDD) 370

Hamilton-Jacobi equations 398
Hamiltonian mechanics 51
hard sphere collisions 121–130
harmonic oscillator 55–56
harpoon mechanism 88–89
Hartree-Fock approximation 34–35
Heisenberg uncertainty principle 54
helicity-conserving approximation 82
helium-3 423–424
Herzberg type I dissociation 241–243
hexapole state selection 302–307
highest occupied molecular orbital (HOMO) 247–248
Hund coupling 108
hydrogen 25–26, 188–194
  chlorine reaction 111–112
  fluorine reaction 198–199
  molecular hydrogen reaction 103–104
  oxygen reaction 193–197
  oxygen reaction 83–85
  surface dissociation 354
  surface recombination 359–361

hydrogen chloride 110, 112, 219
  photodissociation 95–99
hydrogen fluoride 198–199
hydrogen halides 95–99, 107, 274
hydroxyl radical 84, 200–201, 222, 241
  helium-3 scattering 423–424
  photodissociation 267–271
  solid surface scattering 361–362

impact parameter 4, 52, 127, 400
incident flux density 125
inelastic scattering 120–121, 401
  at solid surfaces 344–347
  low energy 420–424
insulators 333
integral cross-section 7–8, 8
interaction potential 65, 117
  long-range 404–405
interfacial scattering *see* surface scattering
interference 128
interpolated moving least squares (IMLS) 46–47
interstellar clouds 393, 433
intramolecular vibrational redistribution (IVR) 23, 228–230, 381
intrinsic reaction path (IRP) 40
iodine 260–263, 363, *364*
ion imaging 152, 153–155, 160–163, *270*
ionization techniques 179–180
isolated collisions 9–10
IVR *see* intramolecular vibrational redistribution

Jacobian determinants 14–15, 398
Jahn-Teller effect 101–102

Kerr effect 366, 368
ketene 233–234
kinematic effects 20
kinetic energy
  quantum operator 55–56
  see also translational energy; vibrational energy

laboratory (LAB) frame 11–13, 123–124, 172–174
  see also coordinate systems
Lagrangian mechanics 50–51
Landau-Zener model 95
Langmuir-Hinshelwood reactions 356
large molecules 86
laser spectroscopy 211
laser vibrational excitation 226–228
laser-induced fluorescence (LIF) 110, 135, 136–138, 380
lasers 16–18
  amplifiers 370–373
  chirped pulses 367–368, 384–385
  femtosecond 363–365, 375
  oscillators 365–370
  pulse characterization 373–374
  pulse shaping 386–389
  pulse trains 383–384
  wavelength tuning 373
  see also non-linear optics; pump-probe experiments
lattice vectors 340
Legendre polynomials 252
Lennard-Jones potential 117
Lippmann-Schwinger equation 65–66
liquids 236–237, 382
local mode description 220–221
long range interaction potential 404–405
low-energy collisions *see* cold collisions
lowest occupied molecular orbital (LUMO) 247–248

M quantum number 259
Madelung energy 340
magnetic moment 108
magneto optical traps 129–130, 396
mass spectrometry 135, 169–170
  Dopper-selected time-of-flight 203–204
  soft ionization 179–180
  time-of-flight 135, 203–204
mass-weighted coordinates 40, 216
Maxwell-Boltzmann distribution 9
methane 199–200, 205–208, 219
methyl chloride 1–2
methyl radical 110, 112, 205
methylene 233–234
Michelson interferometer 383–384
microcanonical transition state theory 232–233
minimum energy path 37, 41
modified Sheperd (MS) interpolation 45–46
molecular beam experiments 10–11, 168–169
  argon-oxygen 140–144, 160–161
  cooling 394–397

molecular beam experiments (*continued*)
detectors 169–171
differential cross-section determination 168–169
mass spectrometric detection 169–171
measurement 163–165
measurement of energy transfer 151
pulsed 171
reaction mechanisms 174–179
with REMPI detection 201–210
terminal velocity 10
molecular wavefunction 28–29
molecular polarization *see* polarization
Moller-Plesset perturbation theory 34
momentum (linear) conservation 13, 19–21
monogenic systems 50
Monte Carlo simulations 156
multi-configurational time-dependent Hartree method 83
multipoles 285

neural networks 44
Newtonian mechanics 49–50
nickel 354
nitric oxide 140–144, 160–161, 306–307, 345–346
nitrogen 361
noble gases 140–144
non-adiabatic processes 88–89
experimental probes 109–113
surface scattering 355
see also potential energy surfaces, coupled
non-crossing rule 93–94
non-linear optics (NLO) 17–18, 371–372, 373, 387
non-resonant multiphoton ionization 137–138
nuclear wavefunction 29–30, 89–90
see also Born-Oppenheimer approximation

Ω quantum number 247–248
opacity function 5, 53
operators 58
optical parametric amplification 376–377
optical parametric oscillators 17–18
orbital angular momentum 4
orientation 280, 353–354
see also polarization
orthonormality 62
oxygen 181–185
hydrogen reaction 83–85, 193–197
photodissociation 251–254

parallel transition 255
partial differential cross-section 123–124
partial waves 407
pendular states 309–310
perpendicular transition 255
perturbation methods 34–35
phase boundaries 333–334
see also surface scattering
phase (laser pulses) 367
phase shift (capture theory) 419–420
phase space 233
photoassociation 397
photodissociation 82, 240–244
adiabatic model 265–267
atomic product polarization 258–263
bromine 244–248, 249–251
theoretical treatment 263–265
electric dipole transitions 245–247
femtosecond probes 378–379
fragment angular distributions 254–255
hydrochloric acid 95–99
hydrogen halides 107
hydroxyl radical 267–274
interference effects 271–274
oxygen 251–254
polyatomic molecules 242
potential energy surfaces 240–244
sudden model 267
translational anisotropy 252
PHOTOLOC 152, 191–192
platinum 354
Poisson brackets 58
Polanyi's rules 38–39, 215, 216, 351–355
polarization 279–280
atomic 258–263, 266–267
optical 303–304, 371
vector correlations 280–281
see also orientation; polarized molecules
polarization moment 285–288
analysis 314–316
calculation 292–299, 312–314
conversion to real values 288–289
density matrices 290–291
physical interpretation 287–288
polarization parameters (PP) 262–263, 324
polarization-dependent differential cross-sections (PDDCS) 292–293

polarized molecules
  production 299–310, 389
    brute force 308–310
  reaction cross-section control 326–328
  stereodynamic portraits 316–321
polyatomic molecules, photodissociation 242
population inversion 17
potassium 174–175, 218–219
potential barriers 21–22
potential energy surface (PES) 2–3
  adiabatic, non-adiabatic coupling 90–92
  Born-Oppenheimer approximation 28–31
  bromine photodissociation 247–248
  construction 42–47
  coordinate systems 31–33
  coupled 90–92
    conical intersections 99–105
    diabatic representation 92–99
    non-crossing 106–109
  diatomic molecules 35–36
  energy calculation 33–35
  functional form fitting 43–45
  hydrogen-oxygen reaction 84–86
  interpolation 45–47
  minimum energy path 37–38, 41
  molecule-solid surface interactions 336–342
  non-adiabatic 95–99
  photodissociation and 240–244
  Polanyi's rules 38–39
  polyatomic molecules 37–42
  reaction pathways and 220–223
  state-resolved 214–218
  triatomic molecules 36–37
POTLIB 43
probability density functions 54, 283–289
  polarization moment 285–289
  see also density matrix
product angular distribution 283–285
product pair correlations 202–208
pseudoquantization 295
pulsed molecular beams 171
pump-probe spectroscopy 10, 13–15, 25, 249–250, 374–377
  oxygen-hydrogen reaction 197
  see also femtochemistry

quantum beats 227–228
quantum bottleneck states 194–195
quantum capture theory 408–410
quantum mechanics 53–54
  approximation methods 82–83
  comparison with classical mechanics 59–60, **60**
  dynamics 298–299
  free motion 60–64
  see also Schrödinger equation
quantum suppression 418
quasi quantum treatment 127
quasi-classical trajectory (QCT) 53, 293–298

R-matrix 70–73
Rabinovitch's bicycle 225–226
rainbow scattering 126–128, 138
rare gases 140–144
rate constants 3–4, *6*, 9
  cold regime 401–403
  cumulative reaction probability 230–233
reaction cross-section *see* cross-section
reaction pathways
  bond selection 223–224
  see also potential energy surface
reaction rates 3–4, 392
  see also cross-section; rate constants
reciprocal space 340
recombinative desorption 355–356
Renner-Teller interaction 85, 102
reproducing kernel Hilbert space (RKHS) 45
resonance-enhanced multiphoton ionisation (REMPI) 110, 136–138, 202–210
rotational energy 130–135
rotational portrait 317
RRKM theory 232
rubidium 425–426
Rydberg tagging 189–190

S-matrix 64–67, 69, 81, 408–410
s-wave scattering 419–420
scattering angle *6*, 7–8, 126–129
  energy transfer and 126–128
  reaction mechanism and 174–179
  see also angular scattering distribution
scattering length 419–420
scattering matrix *see* S-matrix
scattering resonances 188–199
  hydrogen-fluorine 198
  hydrogen-hydrogen 188–193
Schrödinger equation 30–31, 54–55
  time-dependent *see* time-dependent Schrödinger equation

Schrödinger equation (*continued*)
  time-independent *see* time-independent Schrödinger equation
second harmonic generation 371–372
selected-ion flow tube (SIFT) 433–434
Sheperd interpolation 45–46
single-collision conditions 9–10
skew angle 216–217
sodium 378
sodium iodide 378–380
soft ionization 179–180
solvents 236
spatial light modulators (SLM) 386–389
spectral phase interferometry for direct electric field reconstruction (SPIDER) 374
spin-orbit coupling 107–109
split operator method (SOP) 79
Stark deceleration 394–395
Stark effect 304–305
state-selective chemistry 223–224
state-to-state cross-sections 8–9, 66, 150–151, 167–169, 218–220
statistical reactions 23–24, 230
  cumulative reaction probability 230–231
  microcanonical transition state theory 232–233
stereodynamics
  analysis 316–321
  density matrices 289–290
  incomplete information 321–324
  polarization moments 285–289
  probability density functions 283–285
  quantum 298–299
  see also polarization
steric asymmetry 149–150
stimulated emission 16
stripping 174–175
sudden model 267
supercells *339*
surface scattering
  direct dissociation 351–355
  elastic 342–344
  inelastic 344–347
  molecular rotation 353–354
  non-activated dissociation 348–350
  non-adiabatic reactivity 355
  potential energy surfaces 335–342
  reaction channels 334–336
  recombinitive desorption 355–358
sympathetic cooling 397
symplectic propagators 80–81
synchrotron radiation 180
Sysiphus cooling 396

T-matrix 66, 409–410
temperature
  energy transfer and 118
  surface scattering and 356
  see also cold regime; ultracold regime
Thomas-Fermi approximation 438
three-body systems 74
time-bandwidth product (TBWP) 366
time-dependent Schrödinger equation (TDSE) 30, 56–64, 73–74
  approximations 82–83
  final state analysis 81–82
  kinetic energy operator 55–56
  nucleus 30–31
  physical interpretation 56–60
  propagation 78–81
  representation 74–78
  see also S-matrix
time-independent Schrödinger equation (TISE) 64–67, 68–73
  asymptotic wave function 69
  close coupling 69–70
  R-matrix theory 70–73
time-of-flight (TOF) mass spectrometry 135, 203–204
time-resolved photoelectron spectroscopy (TRPES) 381
transition matrix 66, 409–410
transition states 24–26
translational moderation 124
trapping-mediated dissociation 348–349
two-atom molecules 35–36, 93, 133–135

ultracold regime 394, 424–440
uncertainty principle 54
unpolarized molecules 322–324

variable changes 14–15
velocity-map imaging 153–155, 155–158, 160–163, *270*
vibrational energy
  relaxation 118–120
  solid surface scattering 346–347, 351–352
  transfer to translational energy 117–120
  wavepackets 378
  see also excitation modes

vibrational states
  chemical activation 225–226
  effect on reaction pathway 218–220
  energy flow 224
  intramolecular coupling 228–230
  laser excitation 226–227
  relaxation 118–120
  water 220–222
  zero-order 226–227
vibrationally adiabatic potential (VAP) curves 198, *199*
vibronic coupling 106–109

water, excitation modes 220–222
wavepacket interferometry 383–384
wavepackets 61, 61–64, 64, 378
Wigner threshold laws 416–424
Wilson-B matrix 33

xenon iodide 363, *364*
XXZLG surface 84

Zeeman deceleration 395
Zeeman effect 108
zero-order states 226–227